Protein Misfolding, Aggregation, and Conformational Diseases

PROTEIN REVIEWS

Recent Volumes in this Series

VIRAL MEMBRANE PROTEINS: STRUCTURE, FUNCTION, AND DRUG DESIGN
Edited by Wolfgang B. Fischer

THE p53 TUMOR SUPPRESSOR PATHWAY AND CANCER
Edited by Gerard P. Zambetti

PROTEOMICS AND PROTEIN-PROTEIN INTERACTIONS: BIOLOGY, CHEMISTRY, BIOINFORMATICS, AND DRUG DESIGN
Edited by Gabriel Waksman

PROTEIN MISFOLDING, AGGREGATION AND CONFORMATIONAL DISEASES PART A: PROTEIN AGGREGATION AND CONFORMATIONAL DISEASES
Edited by Vladimir N. Uversky and Anthony L. Fink

PROTEIN INTERACTIONS: BIOPHYSICAL APPROACHES FOR THE STUDY OF COMPLEX REVERSIBLE SYSTEMS
Edited by Peter Schuck

PROTEIN MISFOLDING, AGGREGATION, AND CONFORMATIONAL DISEASES PART B: MOLECULAR MECHANISMS OF CONFORMATIONAL DISEASES
Edited by Vladimir N. Uversky and Anthony L. Fink

Protein Misfolding, Aggregation, and Conformational Diseases

Part B: Molecular Mechanisms of Conformational Diseases

Edited by

VLADIMIR N. UVERSKY

Department of Biochemistry and Molecular Biology, and the Center for Computational Biology and Bioinformatics, Indiana University School of Medicine, Indianapolis, Indiana; Institute for Biological Instrumentation, Russian Academy of Sciences, Pushchino, Moscow Region, Russia

ANTHONY L. FINK

Department of Chemistry and Biochemistry, University of California, Santa Cruz, California

Springer

Vladimir N. Uversky
Department of Biochemistry and Molecular Biology
Center for Computational Biology and Bioinformatics
Indiana University School of Medicine
Indianapolis, IN 46202

and

Institute for Biological Instrumentation
Russian Academy of Sciences
Pushchino, Moscow Region 142290
Russia

Anthony L. Fink
Department of Chemistry and Biochemistry
University of California
Santa Cruz, CA 95064
USA

Library of Congress Control Number: 2005926771

ISBN-10: 0-387-36529-X
ISBN-13: 978-0-387-36529-9
e-ISBN-10: 0-387-36534-6
e-ISBN-13: 978-0-387-36534-3

Printed on acid-free paper

9 8 7 6 5 4 3 2 1

springer.com

Contents

Prion Protein and Prion Diseases

Section IV: Changes in Supramolecular Structure

Contributors

Zafer Ali-Han
Department of Microbiology and Immunology
McGill University
Montreal H2B 3A4
Quebec
Canada

Liana G. Apostolova
Department of Neurology
David Geffen School of Medicine
University of California at Los Angeles
Los Angeles, CA 90095-1769
USA

Elizabeth M. Baden
Department of Biochemistry and Molecular Biology
Mayo Clinic College of Medicine
Rochester, MN 55905
USA

Kevin J. Barnham
Department of Pathology
The University of Melbourne
The Mental Health Research Institute of Victoria
Victoria 3010
Australia

Ilia V. Baskakov
Medical Biotechnology Center
University of Maryland Biotechnology Institute
Department of Biochemistry and Molecular Biology
University of Maryland School of Medicine
Baltimore, MD 21201
USA

Vittorio Bellotti
Biotechnology Laboratories IRCCS Policlinico San Matteo
Department of Biochemistry
University of Pavia
27100 Pavia
Italy

Ashley I. Bush
Laboratory for Oxidation Biology
Genetics and Aging Research Unit and Department of Psychiatry
Harvard Medical School
Massachusetts General Hospital East
Charlestown, MA 02129
USA

Joel N. Buxbaum
Division of Research Rheumatology
Department of Molecular and Experimental Medicine
The Scripps Research Institute
La Jolla, CA 92037
USA

Robin W. Carrell
Departments of Medicine and Haematology
Cambridge Institute for Medical Research
University of Cambridge
Cambridge, CB2 2XY
UK

Erica S. Chevalier-Larsen
Department of Biochemistry and Molecular Biology
Thomas Jefferson University
Philadelphia, PA 19107
USA

Anne Clark
Diabetes Research Laboratories
Oxford Centre for Diabetes
Endocrinology and Metabolism
Churchill Hospital
Oxford, OX3 7LJ
UK

Fred E. Cohen
Institute for Neurodegenerative Diseases
Department of Cellular and Molecular Pharmacology
Department of Biochemistry and Biophysics
University of California at San Francisco
San Francisco, CA 94143
USA

Jeffrey L. Cummings
Departments of Neurology and Psychiatry and Biobehavioral Sciences
University of California at Los Angeles
Los Angeles, CA 90095-1769
USA

Cyril C. Curtain
School of Physics and Materials Engineering
Monash University
Victoria 3168
Australia

Stephen J. DeArmond
Institute for Neurodegenerative Diseases
Department of Pathology
University of California at San Francisco
San Francisco, CA 94143
USA

Janelle K. De Stigter
Department of Biochemistry and Molecular Biology
Mayo Clinic College of Medicine
Rochester, MN 55905
USA

Christopher M. Dobson
Department of Chemistry
University of Cambridge
Cambridge, CB2 1EW
UK

Mireille Dumoulin
Department of Chemistry
University of Cambridge
Cambridge, CB2 1EW
UK

Alan R. Fersht
MRC Centre for Protein Engineering
Cambridge, CB2 2QH
UK

Assaf Friedler
MRC Centre for Protein Engineering
Cambridge, CB2 2QH
UK

Ahmad Galaleldeen
Department of Biochemistry
X-ray Crystallography Core Laboratory
University of Texas Health Science Center at San Antonio
San Antonio, TX 78229
USA

Jean-Marie Gillis
Department of Physiology
Catholic University of Louvain
1200 Bruxelles
Belgium

Hans H. Goebel
Department of Neuropathology
Johannes Gutenberg University Medical Center
55101 Mainz
Germany

John J. Harding
Center for Geriatric Neuroscience Research
Institute of Biomedical Science and Technology
Department of Biomedical Science and Technology
Konkuk University
Seoul 143-701, Korea

P. John Hart
Department of Biochemistry
X-ray Crystallography Core Laboratory
University of Texas Health Science Center at San Antonio
San Antonio, TX 78229
USA

Eric W. Hewitt
Astbury Centre for Structural Molecular Biology
School of Biochemistry and Microbiology
University of Leeds
Leeds, LS2 9JT
UK

Andreas C. Joerger
MRC Centre for Protein Engineering
Cambridge, CB2 2QH
UK

Russell J.K. Johnson
Department of Chemistry
University of Cambridge
Cambridge, CB2 1EW
UK

J. William Langston
The Parkinson's Institute
Sunnyvale, CA 94089
USA

Yoon Suk Kim
The Parkinson's Institute
Sunnyvale, CA 94089
USA

Joanna Krzewska
Laboratoire d'Enzymologie et Biochimie Structurales
CNRS
91198 Gif-sur-Yvette Cedex
France

Giuseppe Legname
SISSA - Neurobiology Sector
Ed. Q1 AREA SCIENCE PARK
S.S. 14 Km 163, 5 Basovizza
34012 Trieste
Italy

Seung-Jae Lee
The Parkinson's Institute
Sunnyvale, CA 94089
USA

Paul T. Martin
Columbus Children's Research Institute
Department of Pediatrics and Neurology
Ohio State University
Columbus, OH 43205
USA

Richard W. McLaughlin
Department of Biochemistry and Molecular Biology
Mayo Clinic College of Medicine
Rochester, MN 55905
USA

Ronald Melki
Laboratoire d'Enzymologie et Biochimie Structurales
CNRS
91198 Gif-sur-Yvette Cedex
France

Diane E. Merry
Department of Biochemistry and Molecular Biology
Thomas Jefferson University
Philadelphia, PA 19107
USA

Victor M. Miller
Department of Neurology
University of Iowa College of Medicine
Iowa City, IA 52242
USA

Jenni Moffitt
Diabetes Research Laboratories
Oxford Centre for Diabetes
Endocrinology and Metabolism
Churchill Hospital
Oxford, OX3 7LJ
UK

Isobel J. Morten
Astbury Centre for Structural Molecular Biology
School of Biochemistry and Microbiology
University of Leeds
Leeds, LS2 9JT
UK

Hitoshi Okazawa
Department of Neuropathology
Medical Research Institute
Tokyo Medical and Dental University
PRESTO
Japan Science and Technology Agency (JST)
Tokyo, 113-8510
Japan

Henry L. Paulson
Department of Neurology
University of Iowa College of Medicine
Iowa City, IA 52242
USA

Stanley B. Prusiner
Institute for Neurodegenerative Diseases
Department of Neurology
Department of Biochemistry and Biophysics
University of California at San Francisco
San Francisco, CA 94143
USA

Sheena E. Radford
Astbury Centre for Structural Molecular Biology
School of Biochemistry and Microbiology
University of Leeds
Leeds, LS2 9JT
UK

Marina Ramirez-Alvarado
Department of Biochemistry and Molecular Biology
Mayo Clinic College of Medicine
Rochester, MN 55905
USA

Laura A. Sikkink
Department of Biochemistry and Molecular Biology
Mayo Clinic College of Medicine
Rochester, MN 55905
USA

Anya L. Taboas
Department of Biochemistry and Molecular Biology
Mayo Clinic College of Medicine
Rochester, MN 55905
USA

Roger John Willis Truscott
Australian Cataract Research Foundation
University of Wollongong
Save Sight Institute
Sydney, NSW 2001
Australia

Vladimir N. Uversky
Department of Biochemistry and Molecular Biology
Center for Computational Biology and Bioinformatics
Indiana University School of Medicine
Indianapolis, IN 46202
USA

Mariz Vainzof
Human Genome Research Center
Department of Biology
University of Sao Paulo
05508-900 São Paulo, SP
Brazil

Alexandra Vrabie
Department of Neuropathology
Johannes Gutenberg University Medical Center
Langenbeckstrasse 1
55101 Mainz
Germany

Mayana Zatz
Human Genome Research Center
Department of Biology
University of Sao Paulo
05508-900 São Paulo, SP
Brazil

I

Altered Protein Structure and Enhanced Aggregation/Deposition

1

The Pathogenesis of Alzheimer's Disease: General Overview

Liana G. Apostolova and Jeffrey L. Cummings

Abstract

Alzheimer's disease (AD) is the most common neurodegenerative disease and the leading cause for cognitive decline in the elderly. The disease results in relentlessly progressive decline in intellectual function and gradual loss of activities of daily living. Ninety percent of AD patients have late onset "sporadic" AD. Only 10% become symptomatic before age 65. Two percent of AD is familial and is caused by autosomal dominant genetic mutations. AD is caused by the aberrant folding and aggregation of two proteins: β-amyloid and tau. The disease is also characterized by synaptic dysfunction and neuronal loss. Protein metabolic disturbances lead to a variety of secondary cell injury and death mechanisms including oxidation, inflammation, excitotoxicity, and apoptosis. Emerging therapies are largely directed at amyloid beta protein production or accumulation or the cascade of secondary consequences. The current U.S. FDA–approved therapy includes the anticholinesterase inhibitors and the *N*-methyl-D-aspartate blocker memantine. New drug trials with disease-modifying agents are currently on their way.

1.1. Introduction

The large family of neurodegenerative disorders is composed of many acquired neurologic diseases with distinct phenotypic and pathologic expressions. Until recently, a link between Huntington's disease, amyotrophic lateral sclerosis, and Alzheimer's disease (AD) was not feasible to imagine, but the fascinating advances in molecular biology, immunopathology, and genetics indicate that these nosologic entities share common pathophysiologic mechanisms. Each results from derangement in the processing and/or folding of specific proteins that eventually accumulate within the cell.

1.2. Classification of the Proteinopathies

The most popular classification of the proteinopathies is based on the causative protein. However, they can also be divided by the location of the aggregates: intracellular versus extracellular; neuronal versus glial, or by clinical presentation. Table 1.1 summarizes the major characteristics of the neuroproteinopathies. Three proteins—β-amyloid, α-synuclein, and tau—cause 70% of the dementias in the elderly and more than 90% of all dementias.

Table 1.1. Classification of the proteinopathies

Misfolded Protein	Disease	Location of inclusions	Location	Common phenotype
Aβ	Alzheimer's disease	Extracellular	Cerebral cortex	Memory loss followed by deterioration of all other cognitive domains and functional impairment
	Mixed dementia with Lewy bodies	Extracellular	Cerebral cortex	Features from both AD and DLB
α-Synuclein	Parkinson's disease	Intraneuronal	SN, BS nuclei	Parkinsonism ± late dementia
	Dementia with Lewy bodies	Intraneuronal	SN, BS nuclei, cerebral cortex	Dementia, parkinsonism, fluctuations, hallucinations, delusions, RBD
	Multiple system atrophy	Intraglial	Cerebral cortex, BG, BS nuclei, cerebellar WM	Parkinsonism, autonomic failure, cerebellar symptoms, RBD, rare dementia
	Neurodegeneration with brain iron accumulation	Intraneuronal		Childhood/adolescence presentation with chorea, dystonia, spasticity, dysarthria
Tau	Frontotemporal dementia	Intraneuronal intraglial	Cerebral and cerebellar cortex	Apathy, disinhibition, impulsivity, obsessive-compulsive behavior, poor planning and judgment
	Progressive supranuclear palsy	Intraneuronal, intraglial	BG, BS nuclei, cerebral cortex, dentate nucleus	Supranuclear ocular dysmotility, parkinsonism, axial rigidity, postural instability, dementia
	Corticobasal degeneration	Intraneuronal, intraglial	Cerebral cortex, BG, BS nuclei	Dementia, unilateral parkinsonism and apraxia, dystonia, myoclonus, alien limb
	Parkinsonism-dementia-ALS complex of Guam	Intraneuronal, intra-axonal, intradendritic, extracellular	Cerebral cortex, forebrain nuclei	Dementia, gait disorder, spasticity, ALS, parkinsonism, neuropathy
	Alzheimer's disease	Intracellular	Hippocampus	Cortex
Prion protein	Creutzfeldt-Jakob disease	Extraneuronal	Cerebral and cerebellar cortex	Dementia, myoclonus, ataxia, pyramidal and extrapyramidal signs
	Gerstman-Straussler-Schenker disease	Extraneuronal	Cerebellum	Ataxia, spastic paraparesis, ocular dysmotility, ± dementia
	Fatal familial insomnia	Rare extracellular	Thalamus, temporal lobes	Insomnia, behavioral changes, dysautonomia
	Kuru	Extraneuronal	Cerebellum	Ataxia, dysarthria, tremor, behavioral changes, ± dementia
Huntingtin	Huntington's disease	Intraneuronal-intranuclear	Neostriatum, cerebral cortex, hippocampus	Chorea, behavioral changes, dementia
Superoxide dismutase	Amyotrophic lateral sclerosis		Cerebral cortex, spinal cord	Spasticity, hyperreflexia, muscle atrophy, fasciculations
Atrophin	Dentatorubral-pallidoluysian atrophy	Intraneuronal-intranuclear	Cerebral cortex	Parkinsonism, epilepsy, ataxia, dementia, choreoathetosis, myoclonus, dystonia
Ataxin	Spinocerebellar ataxias	Intraneuronal-intranuclear	Cerebellum, BS	Cerebellar ataxia, ± parkinsonism, ± dysarthria, ± ocular dysmotility, ± dysphagia, ± dementia, ± neuropathy

BG, basal ganglia; BS, brain stem; SN, substantia nigra; RBD, REM behavior disorder; ALS, amyotrophic lateral sclerosis.

1.3. Alzheimer's Disease

AD is the most common cause of cognitive decline among people age 65 and older. It currently affects 4.5 million Americans and is projected to afflict 13.2 million by the year 2050 in the United States alone. This colossal rise parallels the declining mortality among the elderly combined with the well-established age-related exponential increase in AD prevalence (Hebert *et al.*, 2003). AD ranks third in total health care cost after heart disease and cancer. The national direct and indirect annual cost of AD approaches $100 billion dollars per year (Schumock, 1998).

Sporadic AD is a disease of the elderly; most patients are diagnosed after 65 years of age. About 10% of AD cases present under age 65 and have been referred to as having early onset AD (EOAD). Three causative autosomal dominant mutations have been described: the APP gene mutation on chromosome 21, the presenilin 1 gene mutation on chromosome 14, and the presenilin 2 gene mutation on chromosome 1. These autosomal dominant forms comprise only about 2% of all AD (Campion *et al.*, 1999). Additionally having an extra copy of the APP gene, as do Down's patients (trisomy 21), also leads to early pathologic and clinical changes of AD.

1.3.1. Mild Cognitive Impairment as a Prodromal Stage of AD

Mild cognitive impairment (MCI) is a heterogeneous cognitive state between normal aging and dementia (Petersen *et al.*, 1999; Petersen, 2000; Petersen *et al.*, 2001a; Petersen *et al.*, 2001b; Wahlund *et al.*, 2003; Winblad *et al.*, 2004). MCI patients suffer from cognitive impairment while still enjoying functional lifestyles (Petersen *et al.*, 1999; Petersen, 2000; Petersen *et al.*, 2001a; Petersen *et al.*, 2001b; Winblad *et al.*, 2004). They frequently express concern over their cognitive performance and score 1.5 SD below age- and education-adjusted norms on one or more formal neuropsychological measures of cognition. MCI has recently been stratified into four major subtypes. Single domain amnestic MCI presents with isolated memory impairment (Petersen *et al.*, 2001b; Winblad *et al.*, 2004). Cognitive testing reveals poor learning and retention of verbal and/or nonverbal information over time (Petersen *et al.*, 1999; Rivas-Vazquez *et al.*, 2004). It is most commonly due to impending AD or rarely to hippocampal sclerosis. Multidomain amnestic and nonamnestic MCI involve other cognitive domains such as language, attention, executive function, praxis, and/or visuospatial abilities. These types of MCI convert to AD, vascular dementia, dementia with Lewy bodies, or frontotemporal dementia (Petersen *et al.*, 2001b; Winblad *et al.*, 2004). Single domain nonamnestic MCI indicates impairment of one domain other than memory and predates the onset of frontotemporal or vascular dementia, dementia with Lewy bodies, Parkinson's disease dementia, progressive supranuclear palsy, or very rarely AD (Petersen *et al.*, 2001b; Winblad *et al.*, 2004). Because about 80% of all senile dementias are due to AD, it is not surprising that most patients with MCI have the pathologic hallmarks of AD—neocortical senile plaques, neurofibrillary tangles, atrophy and neuronal loss in layer II of the entorhinal cortex (Gomez-Isla *et al.*, 1996; Troncoso *et al.*, 1996; Haroutunian *et al.*, 1998; Price and Morris, 1999; Kordower *et al.*, 2001). The annual conversion of MCI to AD is 12–15% (Petersen *et al.*, 1999; Petersen, 2000).

1.3.2. First Stage (Early AD)

The diagnosis of dementia according to the *Diagnostic and Statistical Manual of Mental Disorders*, 4th edition (1994), requires deficits in one cognitive domain in addition to memory and some functional impairment. Classically, AD patients have declarative short-term and episodic memory deficits. Their long-term memory is generally well preserved in the early stages. While clueing and multiple choices initially help the retrieval process, these compensatory

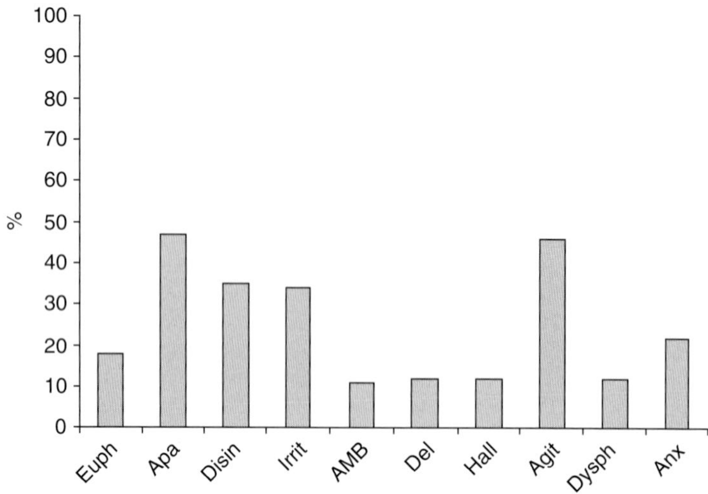

Figure 1.1. Frequency of neuropsychiatric symptoms in mild AD.

mechanisms gradually fail. Even during the mild stage, formal cognitive testing reveals additional disturbances such as inefficient set shifting, concrete thinking, deficient judgment, word finding deficits, and/or mild impairment of visuospatial skills such as copying three-dimensional figures or drawing a clock (Pasquier, 1999).

In addition to the cognitive symptoms of mild AD, patients display a variety of neuropsychiatric symptoms. Perhaps the most pervasive one, frequently apparent during the MCI stage (Hwang *et al.*, 2004), is apathy. It manifests with loss of interest, motivation, volition, enjoyment, spontaneity, and emotional behavior and occurs in 40% of the patients with mild, 80% with moderate, and 90% with advanced AD (Mega *et al.*, 1996). Depressed mood (dysthymia) is likewise common in the early stages, as are irritability, disinhibition, and anxiety (Cummings, 2003) (Fig. 1.1).

1.3.3. Intermediate Stage (Moderate AD)

In this stage of the disease, the deficits in nonmemory cognitive domains are often pronounced. There is increasing difficulty operating appliances. Fluent anomic aphasia occurs early on and evolves to transcortical sensory aphasia where comprehension is diminished, while reading and repetition are spared. There is relative preservation of syntactic and phonologic abilities until later in the disease course (Pasquier, 1999). Visuospatial deficits become more pronounced. Judgment and insight decline rapidly. Sleep-wake cycle disruption may emerge during this stage of the disease with nighttime awakenings and sundowning. Appetite changes (most commonly loss of appetite) become more pronounced, and weight loss is common (Cummings, 2003).

The repertoire of neuropsychiatric symptoms may include delusions, most commonly paranoid, of theft or infidelity, hallucinations, agitation, and hoarding behaviors, in addition to worsening apathy, dysthymia, irritability, disinhibition, and anxiety (Fig. 1.2). A deficit of emotional processing is also common in the moderate stages of AD. It manifests with difficulties interpreting facial emotions and lack of emotional prosody in spontaneous speech (Cummings, 2003).

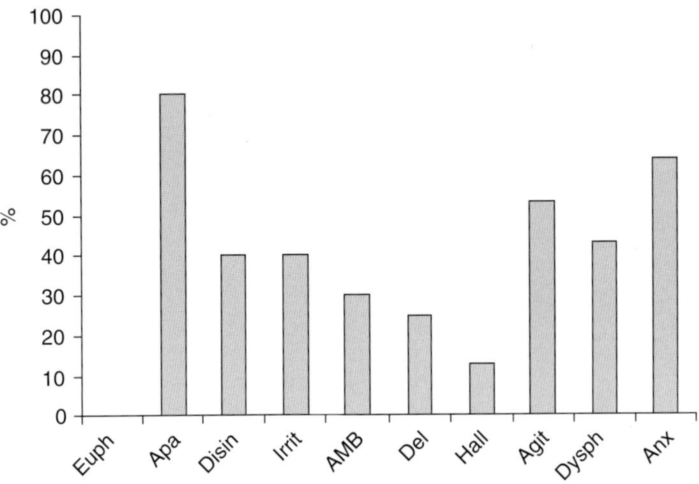

Figure 1.2. Frequency of neuropsychiatric symptoms in moderate AD.

1.3.4. Late Stage (Severe AD)

As patients continue to decline, they gradually lose their ability to communicate and even to recognize familiar faces. They fail in their basic activities of daily living such as eating, walking, and continence and may reach a bedridden state. They most commonly succumb to pneumonia, urosepsis, or other complication associated with immobility (Kukull *et al.*, 1994; Beard *et al.*, 1996; Hoyert and Rosenberg, 1999; Kalia, 2003). The neurobehavioral features shift toward positive behaviors such as agitation, belligerence, irritability, violence, and away from negative behaviors such as apathy and dysthymia (Fig. 1.3).

1.3.5. Variants of AD

This general clinical overview provides an overview of how AD presents and evolves. Many patients defy this classic progression and produce the many cognitive/behavioral faces of AD (Cummings, 2000). A few relatively common variants are discussed here.

In the frontal variant of AD, patients present with prominent behavioral and/or personality changes in addition to short-term memory loss. These patients often are impatient, irritable, impulsive, and/or disinhibited. They show impairments on tests of frontal executive performance such as categorical verbal fluency and Trail Making A (Chen *et al.*, 1998; Johnson *et al.*, 1999), as well as on response inhibition and set shifting (Chen *et al.*, 1998). They were shown to have higher neurofibrillary tangle (NFT) burden in the frontal neocortex (Johnson *et al.*, 1999).

Posterior cortical atrophy is a relatively rare presentation of AD. The patients present with prominent visuospatial dysfunction such as partial or full Balint syndrome (simultanagnosia, ocular apraxia, and ocular ataxia) and/or partial or full Gerstmann syndrome (acalculia, agraphia, right/left disorientation, finger agnosia). In addition, they may have visual field deficits or constructional, dressing, and ideomotor apraxia. They typically have relatively preserved memory and insight especially early in the disease course (Mendez *et al.*, 2002; Renner *et al.*, 2004;

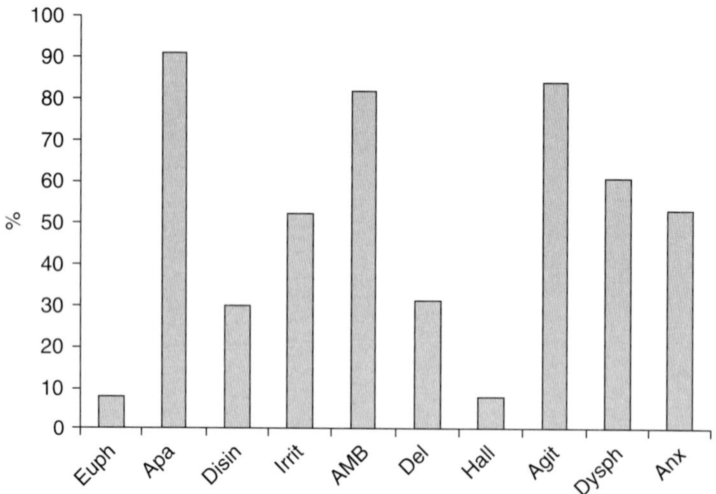

Figure 1.3. Frequency of neuropsychiatric symptoms in severe AD.

Tang-Wai *et al.*, 2004). In addition to severe parieto-occipital atrophy, these patients have higher NFT burden in primary visual and secondary visual association cortices and relative hippocampal sparing (Renner *et al.*, 2004; Tang-Wai *et al.*, 2004).

Finally, the occasional AD patient may also present with early progressive language involvement (Kramer and Miller, 2000) or significant parkinsonian signs and symptoms (Cummings, 2000; Kurlan *et al.*, 2000).

1.4. Pathogenesis of AD

The major proteinaceous deposit in AD is β-amyloid (also called Aβ). *Amyloid* is a descriptive term for proteins that stain with Congo red and thioflavin S and demonstrate birefringence in polarized light. Amyloid deposits are found in most dementias. The amyloid fibrils in all neurodegenerative disorders are approximately 10-nm-wide and up to 10-μm-long filaments with characteristic tertiary structure—the cross β-structure, where the individual protein strands run perpendicularly to and the stabilizing hydrogen bonds are parallel to the long axis of the fiber (Ross and Poirier, 2004).

1.4.1. Protein Folding and Misfolding

All proteins implicated in neurodegeneration are found in many mammalian species, and each of them has its unique physiologic role. For instance, the amyloid precursor protein (APP), the parent molecule of Aβ, is thought to play a role in synaptic stabilization and plasticity, neuritic outgrowth, regulation of neuronal survival, and cell adhesion (Van Gassen *et al.*, 2000; Mattson, 2004). APP's secreted form, called nexin-2, has been shown to inhibit coagulation factor XIa (Van Nostrand *et al.*, 1992; Schmaier *et al.*, 1993). APP's C-terminal fragment post γ-secretase cleavage has been reported to mediate nuclear signaling and modulate gene expression (Gao and

Pimplikar, 2001; Leissring *et al.*, 2002; Cao and Sudhof, 2004; von Rotz *et al.*, 2004). Tau is a vital component of the cytoskeleton and serves as microtubule promoter and stabilizer (Friedhoff *et al.*, 1998; Ghetti *et al.*, 2003; Giasson *et al.*, 2003; Revesz and Holton, 2003). α-Synuclein is thought to act presynaptically by modulation of synaptic transmission and synaptic vesicle transport and to play a role in neuronal plasticity (Giasson *et al.*, 2003; Jellinger, 2003; Spillantini, 2003).

The initial faulty steps in the pathogenesis involves the ribosome. Here the newly formed polypeptide chains start folding co-translationally as they emerge from the ribosomal complex (Sitia and Braakman, 2003). The precision of the conformational process is critical for protein function. The secondary, tertiary, and quaternary protein structures depend on formation of hydrogen bonds, electrostatic and hydrophobic interactions and are governed by the unique amino acid sequence (e.g., by the primary peptide structure). In its unfolded and partially folded state, the protein undergoes dynamic conformational changes to reach the most functional and ergonomic state. Specific intermediate folding states substantially promote and facilitate efficient folding (Uversky, 2003). Nevertheless, misfolding occurs. It may result from a DNA mutation in the protein gene, faulty transcription of the genetic code of the protein into messenger RNA, faulty translation into polypeptides at the ribosome, or from the presence of local environmental changes that impede correct folding. Indeed, most mutations that lead to fibrillation and protein deposition do so by means of destabilization and partial unfolding of the natural protein (Uversky, 2003). Misfolding occurs constantly and is reversed by the actions of an elaborate group of chaperone proteins dedicated to facilitating efficient folding and reversing misfolding. They do so by binding to the aggregation-promoting regions such as hydrophobic residues, unpaired cysteines, or immature glycans normally destined for the core of the protein. By shielding these regions, they confine the peptide to an intermediate conformation that promotes correct folding (Sherman and Goldberg, 2001; Hartl and Hayer-Hartl, 2002; Sitia and Braakman, 2003). Decreased capacity for chaperone induction in the endoplasmic reticulum is seen in normal aging (Lansbury, 1999) and to a greater extent in presenilin 1 mutation carriers (Katayama *et al.*, 1999). Under normal circumstances, once multiple attempts for correct folding fail, the misfolded peptides are conjugated with ubiquitin and targeted for proteolysis in the 26S proteosome (Goldberg, 2003). If the misfolded proteins are not efficiently degraded by the ubiquitin-proteasome system due to proteasome inhibition, overwhelming misfolded protein production, or other factors, protein accumulation occurs. Once the critical concentration is reached, aggregates form and exert further negative feedback upon the ubiquitin-proteasome system (Bence *et al.*, 2001; Sitia and Braakman, 2003).

1.4.2. Fibrillogenesis

Fibrillogenesis includes three major steps: structural transformation of the protein with production of cross-β-sheets, nucleation, and fibril growth (Uversky, 2003). The common theme among neurodegenerative proteins is their propensity to form a specific tertiary structure, the cross-β-sheet, where the polypeptide β-strands are perpendicularly oriented to the long axis of the fibril (Uversky, 2003). Increasing evidence sugests several factors that may initiate fibrillation. Many disease-causing mutations influence the peptide's affinity to form cross-β-sheets by changing the primary structure of the protein. Two possible mechanisms are the reduction of the net charge of the peptide with resultant decline in electrostatic repulsion and/or by interruption of long contiguous hydrophobic residue stretches. A highly significant negative correlation between the net charge and the aggregation rate has been described (Chiti *et al.*, 2002). Another potential contributing mechanism is a change in stretch of amino acids critical for the proper formation of secondary and tertiary structures (Chiti *et al.*, 2002). Environmental factors such as low pH, organic solvents, metal ions, and high temperature promote aggregation and fibril growth *in vitro* by shifting the equilibrium toward the more unstable peptide species (Kelly, 1998;

Harper *et al.*, 1999; Dobson, 2003; Uversky, 2003). Alternatively unfolded proteins appear to stabilize after binding to their specific targets and ligands *in vivo*. Slight modifications in the environment and/or the availability of binding ligands impedes normal interaction and increases the concentration of unbound peptides and shifts the equilibrium toward fibril formation (Uversky, 2003). It seems that once a threshold concentration of partially unfolded species is reached, the aggregation cascade is initiated. The aggregates then provide a nucleus that promotes further aggregation and accelerated fibril growth (Harper *et al.*, 1999; Rochet and Lansbury, 2000; Dobson, 2003). The latter occurs through both association of oligomers (protofibrils) and simple end addition of monomers (as opposed to disassembly that tends to occur only via monomer release) (Harper *et al.*, 1999; Nybo *et al.*, 1999; Rochet and Lansbury, 2000).

1.5. Pathology of AD

AD is the only neurodegenerative disease where three misfolded proteins contribute to the spectrum of pathologic changes.

1.5.1. β-Amyloid

The Aβ fragment of the APP protein is a by-product of APP processing. The normally prevailing α-secretase–mediated APP processing splits the large APP molecule in the middle of the Aβ sequence and does not liberate pathogenic Aβ species. However, alternative splicing by two enzymes (the β- and γ-secretases) results in generation of the pathogenic Aβ fragment. Depending on the exact site of action of γ-secretase, several Aβ peptides with 39–43 amino acids are produced (Mesulam, 2000). The longer moieties are more amyloidogenic (Kayed *et al.*, 2003). Although β- and γ-secretase are active throughout the life span, plaques rarely form in young individuals, but after the age of 60 nearly all elderly develop some Aβ deposits (Berg *et al.*, 1998; Price and Morris, 1999).

The amyloid fibrils tend to clump together and form inclusions (Table 1.1). β-Amyloid forms several different extracellular deposits: focal plaques, diffuse plaques, fleecy lake-like deposits, punctate and stelate deposits, and cotton-wool plaques.

Focal plaques are the hallmark pathology of AD (Figs. 1.4A–1.4D). Most commonly they are in the form of neuritic plaques (also called senile plaques) with a dense central core rich in Aβ40 and peripheral halo of diffuse amyloid rich in Aβ42. Activated astrocytes are frequently seen just beyond the outer edge of the halo. They send multiple cytoplasmic processes around and infiltrate the plaque. The astrocytic perinuclear cytoplasm may contain prominent granules composed of Aβ42 probably as a result of endocytosis and phagocytosis of dying neuronal synaptic structures and dendrites that have accumulated Aβ42 (Nagele *et al.*, 2004). The neuritic plaque contains many dystrophic neurites (Figs. 1.4A and 1.4B). The second type of focal plaque, the compact or burned-out plaque, lacks the peripheral layer rich in Aβ42 and has only the dense central core of Aβ40. Such plaques are commonly seen in the cerebellum and globus pallidus (Duyckaerts and Dickson, 2003). Plaques of similar appearance were noted after amyloid vaccination (Ferrer *et al.*, 2004). Both types of focal plaques have other cellular elements such as activated microglia embedded peripherally as evidence of an ongoing inflammatory response. Several inflammatory markers have been identified (e.g., early members of the complement cascade, cytokines, and α_1-antichymotrypsin). Cholesterol and Apo E are present in the plaques. The latter is seen more commonly in the compact variety (Duyckaerts and Dickson, 2003). Focal aggregates are necessary for the definite pathologic diagnosis of AD. Nevertheless, it is the soluble species and not the plaques that correlate best with cognitive decline (McLean *et al.*, 1999).

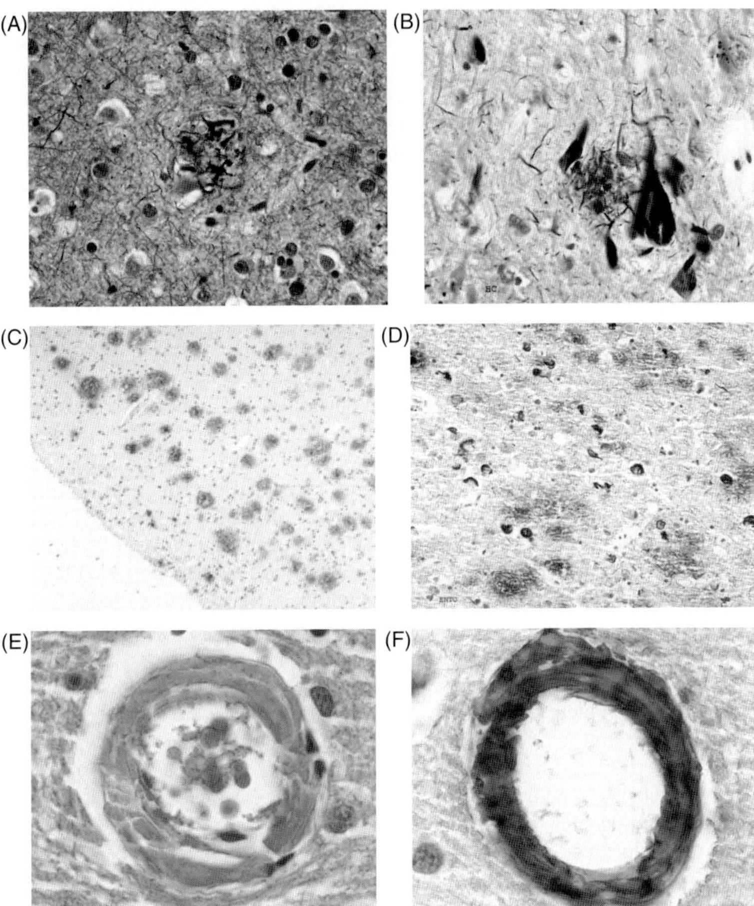

Figure 1.4. Amyloid and tau pathology in AD. (A) Neuritic plaque and neuropil threads (Bielchowski stain); (B) neuritic plaque, neurofibrillary tangles, and neuropil threads (Bielchowski stain); (C) amyloid plaques in cerebral cortex (Abeta 42 immunostaining); (D) diffuse and neuritic plaques in cerebral cortex (Abeta 42 immunostaining); (E) amyloid angiopathy (hematoxylin-eosin stain); (F) amyloid angiopathy (Abeta 40 immunostaining). Images courtesy of Ivan Klement, M.D.

The diffuse deposits consist of noncompact amyloid and vary in size from a few to more than 100 μm (Fig. 1.4D). They are commonly seen in cognitively normal elderly sometimes in very high concentrations. Some consider them a sign of "pathological aging" and others especially when present in the cortex as indicative of AD. They consist of a different type of Aβ: the amino-truncated Aβ (as opposed to the carboxy-terminal modified Aβ). These deposits typically do not stain with Congo red and hematoxylin-eosin (Duyckaerts and Dickson, 2003).

Fleecy and lake-like deposits are very large diffuse Aβ deposits that can be seen in the superficial layers of the presubiculum and the internal entorhinal layers. They consist of the amino-terminal truncated Aβ and thus share the staining characteristics of the diffuse deposits.

In very advanced AD patients, subpial and juxta-arterial fleecy deposits can be seen. These deposits stain for Aβ42 but only weekly with Congo red (Duyckaerts and Dickson, 2003).

Punctate and stelate deposits are less than 5 μm in size and are frequently found in the vicinity of focal or diffuse plaques (Duyckaerts and Dickson, 2003).

Cotton-wool plaques were first described in presenilin 1 patients with deletion of exon 9 (Crook *et al.*, 1998; Houlden *et al.*, 2000; Steiner *et al.*, 2001). Recently, they have been observed in sporadic AD in association with typical pathology (Le *et al.*, 2001). They consist of Aβ42 and stain with hematoxylin-eosin but not with Congo red. As their name implies, they look like cotton balls. They are frequently seen in association with moderate to advanced cerebral amyloidosis (Duyckaerts and Dickson, 2003).

In addition to plaques, Aβ deposits are found in the small arteries, veins, and capillaries of cerebral and cerebellar cortices and the leptomeninges. Up to 80% of AD patients have vascular amyloidosis. The ratio of Aβ40 to Aβ42 in cerebral amyloidosis is in favor of Aβ40 (Duyckaerts and Dickson, 2003).

The first amyloid plaques form in the temporo-occipital association regions (Mesulam, 2000; Duyckaerts and Dickson, 2003) followed by the perirhinal/entorhinal area (Braak and Braak, 1997), the parietal and later the frontal neocortex (Braak and Braak, 1997). The highest plaque densities are seen in temporal and occipital association cortices, intermediate in parietal and lowest in limbic and frontal association cortices (Duyckaerts and Dickson, 2003). There is predilection of amyloid deposition in cortical layers III and Va, followed by layers IV and Vb, while layers II and VI are frequently spared (Braak and Braak, 1997). The basal ganglia are differentially affected with those receiving cortical projections being more affected then those projecting to the cortex (Mesulam, 2000). The amygdala and especially its mediobasal nucleus is also involved (Hof, 1997). The hippocampus can be relatively spared (Berg *et al.*, 1998; Duyckaerts and Dickson, 2003). Neuritic plaques are occasionally found in the limbic regions of nondemented elderly over 75 years of age. They increase in density in AD with disease progression (Price and Morris, 1999). Diffuse plaques are widely distributed throughout the cerebrum. They are frequently found in cerebellum and in basal ganglia (Duyckaerts and Dickson, 2003). They are equally prominent in all stages of AD and are frequently found in nondemented elderly (Berg *et al.*, 1998; Price and Morris, 1999). The other major type of amyloid deposit, the cerebral amyloid angiopathy, affects predominately the occipital neocortex. Neocortical amyloid deposits ordinarily precede neocortical NFTs (Duyckaerts and Dickson, 2003).

1.5.2. Tau

Tau is a vital structural element of the microtubular transport system in the nervous system. Its main function is microtubule stabilization (Friedhoff *et al.*, 1998; Ghetti *et al.*, 2003; Giasson *et al.*, 2003; Revesz and Holton, 2003). The integrity of microtubules is critical to normal neuronal function (Duyckaerts and Dickson, 2003).

The tau gene located on chromosome 17 has four 31- to 32-amino-acid tandem repeats close to its carboxyl end. As a result of alternative RNA splicing, tau has six isoforms: three are three-repeat and three are four-repeat tau. In normal cortex, the three- and four-repeat forms are equally expressed. In tauopathies, the ratio is changed. AD is the only dementia with both three- and four-repeat tau (Goedert, 2003).

Tau inclusions while staining with hematoxylin-eosin and amyloid stains are much easier to visualize after silver impregnation. The most sensitive and specific contemporary method is tau immunohistochemistry. This technique also is capable of demonstrating pretangle intraneuronal pathology (Duyckaerts and Dickson, 2003). There are three types of tau deposits in AD: NFTs, neuropil threads, and dystrophic neurites.

NFTs are composed of 22-nm paired helical filaments (PHF), and each PHF is composed of 8–14 tau monomers (Friedhoff *et al.*, 1998). They commonly affect the pyramidal cortical neurons and assume a flame-like shape (Fig. 1.4B). In subcortical locations such as the forebrain and brain-stem nuclei, where the neuronal bodies have more oval form, the NFTs are globose. The tau in NFTs undergoes several major structural transformations such as truncation, glycation, and phosphorylation. Tau's affinity for microtubules is controlled by phosphorylation/dephosphorylation. Diseased tau is resistant to dephosphorylation and thus its main function is severely compromised. Extracellular NFTs are rare and are referred to as ghost tangles. They are presumed to be the remnants of dead neurons and are most commonly seen in the hippocampus. When surrounded by dystrophic neuritis, they are called tangle-associated neuritic clusters (Duyckaerts and Dickson, 2003).

Neuropil threads are most commonly seen in AD and only rarely identified in other tauopathies such as corticobasal dgeneration (Feany and Dickson, 1996). They are short tortuous neuronal dendrites filled with abnormal tau (Duyckaerts and Dickson, 2003) (Figs. 1.4A and 1.4B).

Dystrophic neurites are tau-containing dendritic structures that are seen in the periphery of the senile plaques. Some are also immunoreactive for APP and ubiquitin (Duyckaerts and Dickson, 2003) (Figs. 1.4A and 1.4B).

NFTs initially accumulate in the transentorhinal area, layer II of the lateral entorhinal cortex, and the perirhinal cortex (Hof, 1997; Braak *et al.*, 2000; Van Hoesen *et al.*, 2000; Duyckaerts and Dickson, 2003). As AD progresses, the NFTs affect layer II of the medial entorhinal cortex and the rest of the parahippocampal gyrus. In the later stages, NFTs are seen in layer IV of the entorhinal cortex (Van Hoesen *et al.*, 2000). Spreading to the hippocampus, they first affect the subiculum and the CA1 area and later CA2, CA3, and CA4 areas. The tangles then "spill" into the temporal neocortex first on the medial and inferior and later on the lateral surface. There is subsequent involvement of the parietal and frontal association cortices (Mesulam, 2000; Duyckaerts and Dickson, 2003). In the cortex, tangles affect mostly the large pyramidal neurons of layers III and V (Mesulam, 2000). The general topographic pattern of NFT spread is from the limbic system to the neocortex (Hof, 1997). Subcortical tangles are seen very early in the basal forebrain nuclei and the corticomedial nuclear groups of the amygdala (Duyckaerts and Dickson, 2003). These nuclei also receive projections from the entorhinal cortex, CA1 and subiculum of the hippocampus (Hof, 1997). In the basal ganglia, tangles are generally seen in the nuclei projecting to the cortex but not all are equally involved (e.g., the limbic thalamus is heavily involved while most of the other thalamic nuclei are spared). In the brain stem, tangles can be found in the medial substantia nigra, locus coeruleus, raphe nuclei, medial and lateral parabrachial, subpeduncular, and reticular formation nuclei of the medulla (Duyckaerts and Dickson, 2003). The temporal progression of tau pathology is the framework of the widely used Braak and Braak staging criteria for AD (Braak and Braak, 1991).

NFTs correlate better with dementia severity than amyloid plaques (Berg *et al.*, 1998). However, they can be absent in the neocortex in 10% of patients with AD and in as many as 50% of mild AD cases (Tiraboschi *et al.*, 2004). They also were found in the entorhinal and perirhinal cortices and the anterior olfactory nucleus of nondemented elderly under the age of 75, and in the CA1 of the hippocampus in those over the age of 75 (Price and Morris, 1999; Knopman *et al.*, 2003).

1.5.3. α-Synuclein

α-Synuclein is a natively unfolded 140-amino-acid polypeptide that becomes more structured upon binding to the cell membrane. It localizes predominately at the presynaptic terminals.

It is the major constituent of Lewy bodies (LBs) (Klucken *et al.*, 2003; Spillantini, 2003). In addition to α-synuclein, LBs contain ubiquitin, proteasome subunits, heat-shock proteins, lipids, redox-active iron, Aβ, and hyperphosphorylated neurofilaments (Jellinger, 2002; Klucken *et al.*, 2003). Classic LBs are round cytoplasmic inclusions with an eosinophilic core and a pale halo (Figs. 1.5C and 1.5D). They can be found in the brain-stem nuclei of Parkinson's disease (PD), dementia with Lewy bodies (DLB), AD, and occasionally normal elderly. Cortical LBs have a more variable shape and no halo (Figs. 1.5A and 1.5B). They are seen predominately in layers V and VI of the temporal, insular, and cingulate cortices of patients with AD, DLB, and PD with dementia (Jellinger, 2003). Coexistence of LBs, AD plaques, and tangles in AD was first described by Kosaka in 1979 (Kosaka and Mehraein, 1979) and has been reported in 7–71% of sporadic AD. The presence of LBs in AD correlates with more severe neuronal loss and depigmentation in the brain-stem nuclei, lower cortical choline acetyltransferase levels, and less neurofibrillary pathology compared with pure AD. The amygdala is a common location for LBs in AD. Even in the absence of cortical LBs, amygdalar LBs are found in 15.8% of the patients (Jellinger, 2003).

Recently an interaction between Aβ, tau and α-synuclein was described. Giasson *et al.* demonstrated that α-synuclein can induce tau fibrillization and that both tau and α-synuclein promote mutual polymerization (Giasson *et al.*, 2003). Support for this pathophysiologic interaction comes from the colocalization of tau and α-synuclein in the same inclusion (Schmidt *et al.*, 1996; Marui *et al.*, 2000; Ishizawa *et al.*, 2003) (Fig. 1.6). Aβ can likewise promote LB (Masliah *et al.*, 2001) and NFT pathology in transgenic animals (Gotz *et al.*, 2001; Lewis *et al.*, 2001; Masliah *et al.*, 2001). Soluble Aβ levels correlate very strongly with cortical NFT density (McLean *et al.*, 1999).

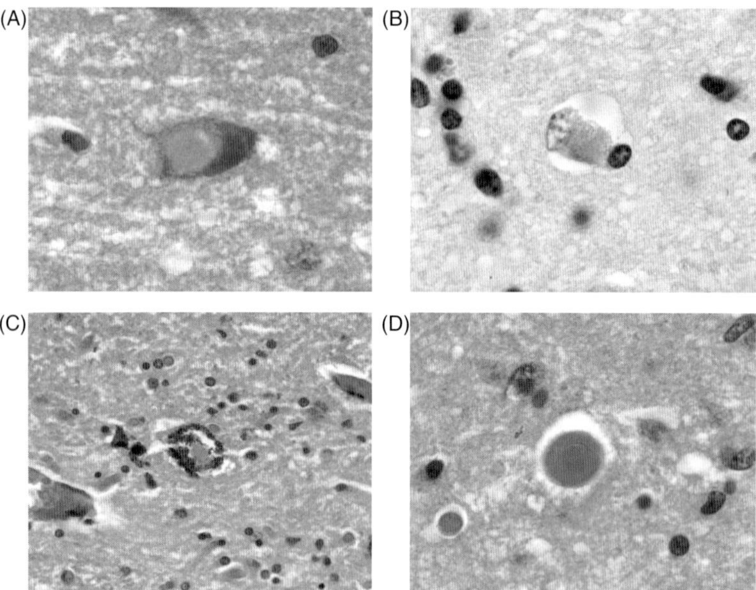

Figure 1.5. α-Synuclein pathology. (A) Cortical LB (hematoxylin-eosin stain); (B) cortical LB (α-synuclein immunostaining); (C) LB in a pigmented neuron of substantia nigra (hematoxylin-eosin stain); (D) classic LB in locus coeruleus (hematoxylin-eosin stain). Images courtesy of Ivan Klement, M.D.

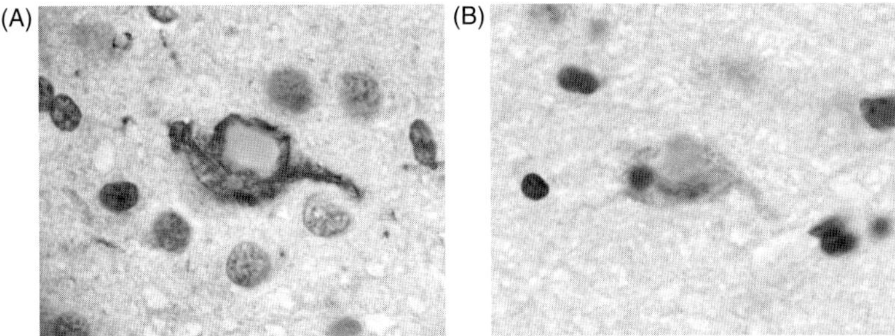

Figure 1.6. Colocalization of NFT and LB within the same neuron. (A) Tau immunostaining; (B) α-synuclein immunostaining. Images courtesy of Ivan Klement, M.D.

1.5.4. Synaptic Toxicity and Dysfunction

Synaptic dysfunction is a very early event (Masliah, 2000) preceding the decline of choline acetyltransferase activity (Tiraboschi *et al.*, 2000) and plaque and tangle formation (Braak *et al.*, 1994; Hsia *et al.*, 1999). The proposed mechanisms are disrupted transmitter synthesis, transport, release and/or reuptake, disturbed vesicle trafficking, alterations of presynaptic and/or postsynaptic receptors and ion channels, and disruption of second messenger chains (Selkoe, 2002, 2003; Coleman *et al.*, 2004). Synaptic loss follows. A 25–35% decrease in synaptic density can be found in the temporal and frontal neocortex 2–4 years after AD is diagnosed (Masliah, 2000), and loss of synapses is ubiquitous throughout the hippocampus and neocortex at autopsy in advanced cases (Honer, 2003). However, synaptic loss is not merely consequence of neuronal death: 50% decrease in synapse-to-neuron ratio in AD suggests that many surviving neurons have lost their synaptic apparatus prior to cell death (Bertoni-Freddari *et al.*, 1996; Bertoni-Freddari *et al.*, 2003; Coleman *et al.*, 2004).

The causes of synaptic loss are not completely understood. In AD, synaptic density correlates inversely with soluble Aβ levels but not plaque burden (Mucke *et al.*, 2000). In young APP transgenic mice, loss of presynaptic terminals precedes plaque formation (Hsia *et al.*, 1999). The pre-fibril oligomeric Aβ forms are synaptotoxic and interfere with long-term potentiation (Walsh *et al.*, 2002). Aβ oligomer-specific antibodies can abolish Aβ-induced neurotoxicity in cell culture (Kayed *et al.*, 2003). The soluble forms of Aβ including the oligomers correlate better than amyloid plaque density with cognitive function in MCI and AD and age of death in AD (Lue *et al.*, 1999; McLean *et al.*, 1999; Naslund *et al.*, 2000). Subjects who meet pathologic criteria for AD but had no symptoms of dementia premortem have synaptophysin levels similar to those of normal elderly without AD-type pathology and soluble Aβ levels much lower than those of AD subjects (Lue *et al.*, 1999). Other factors that may contribute to synaptic dysfunction are disrupted transport of essential synaptic components from the perikaryon to the synapse by NFT accumulation or deafferentation (e.g., diaschisis) (Falke *et al.*, 2003).

1.5.5. Neuronal Loss

The earliest decline in neuronal counts is observed in layer II of the entorhinal cortex. It is later evident in selected hippocampal regions such as the subiculum, CA1, and the hilus (West *et al.*, 1994; Price *et al.*, 2001; Klucken *et al.*, 2003; West *et al.*, 2004). Hippocampal neuronal loss correlates well with gross volume change (Bobinski *et al.*, 2000).

Neuronal loss is ubiquitous in AD. Generally, the sites of most severe involvement where more than 50% of the neuronal population is affected are (in decreasing order) entorhinal cortex, anterior olfactory nucleus, hippocampus, amygdala, and temporal neocortex. Neuronal loss is variable in nucleus basalis of Meynert. In the brain stem, significant neuronal loss is observed in locus coeruleus and raphe nuclei (Duyckaerts and Dickson, 2003).

1.5.6. Phenotype-Proteotype Relationship

All neurodegenerative disorders result from protein misfolding and aggregation into amyloid deposits with cross-β structure, yet they have completely different phenotypic presentations (Table 1.1).

Aβ has predilection for the associative neocortex and the limbic system and usually spares the primary motor and sensory cortices. The phenotypical expression of Aβ pathology is dementia of the Alzheimer's type with relentless memory loss and subsequent involvement of other cognitive spheres. Neuritic plaque density in the prosubiculum correlates with psychosis (Zubenko *et al.*, 1991). The relatively rare plaque-predominant variant of AD is characterized by a milder and slower progression as compared with classic AD with both plaque and tangle pathology (Berg *et al.*, 1998; Jellinger, 2003). The atypical variants of AD result from selective involvement of specific brain areas as in posterior cortical atrophy.

Besides the occurrence of NFT in AD, tau inclusions are found in frontotemporal dementia, corticobasal degeneration, progressive supranuclear palsy, pallidopontonigral degeneration, and parkinsonism-dementia complex of Guam. What these diseases have in common is disruption of the fronto-subcortical circuits resulting in impaired executive functions, poor judgment, and personality changes. In AD, tau pathology correlates with both behavioral and cognitive changes. Neocortical NFT counts correlate with dementia severity (Berg *et al.*, 1998). Disproportionate NFT involvement of the frontal lobes underlies the frontal AD variant with significant agitation, irritability, and disinhibition (Johnson *et al.*, 1999). Agitation also may correlate with orbitofrontal and anterior cingulate NFT burden (Tekin *et al.*, 2001). Apathy correlates with anterior cingulate NFTs (Tekin *et al.*, 2001) and psychosis with NFTs in the middle frontal cortex (Zubenko *et al.*, 1991). NFT involvement of the substantia nigra can trigger extrapyramidal features (Liu *et al.*, 1997). The rare tangle-predominant variant of AD can present with cognitive decline of lesser severity and slower progression compared with classical AD (Jellinger, 2003).

α-Synuclein aggregates (e.g., LBs), are the third pathologic inclusion that can be found in AD. Cognitively, α-synuclein pathology contributes to greater impairment of attention and visuoperception and lower verbal fluency scores (Hansen *et al.*, 1990). Neurologically, these patients typically have prominent extrapyramidal features and behaviorally more complex psychopathology (Del Ser *et al.*, 2001). They are more likely to have acute-subacute onset and cognitive fluctuations relative to patients with pure AD (Del Ser *et al.*, 2001). Rapid eye movement (REM) sleep behavior disorder is a distinctive clinical feature of the synucleinopathies (Schenck *et al.*, 1997; Boeve *et al.*, 1998; Ferman *et al.*, 1999; Boeve *et al.*, 2001, 2003).

1.6. Theories Behind Selective Vulnerability in AD

AD has a unique pattern of progression and striking predilection for specific cortical and subcortical regions and selected cell types. This selective vulnerability cannot be explained by simple "spread" to neighboring cortical areas or layers. In an attempt to clarify this perplexing selectivity, several theories have emerged.

The plasticity theory posits that the areas most capable of neuroplasticity are the first to be affected. According to this theory, in AD the high demands placed by synaptic and neuronal loss and/or dysfunction put the cellular networks at chronically high and eventually unsustainable demands for neuroplasticity. The hightened level of neuroplasticity upregulates tau production and its phosphorylation and APP turnover. Because the limbic regions have the greatest potential for neuroplasticity, they naturally host the first NFTs while the areas least capable of neuroplasticity—the primary motor and sensory cortices—host the least. Once limbic dendritic and synaptic loss occurs, upregulation of neuroplasticity in the association cortices to which limbic areas connect follows (Mesulam, 2000; Arendt, 2004).

The connectivity model for AD proposes that the selective neuronal vulnerability is determined by afferent and efferent connections with other cortical areas. The connectionist's argument is based on the fact that the long corticocortical projection neurons are subjected to alzheimerization, whereas the short corticocortical projection neurons remain unaffected. They also argue that the AD stages represent a stepwise retrograde hierarchical "travel" of the pathology along intrahemispheric corticocortical tracts from the anatomically simpler allo- and periallocortical areas to the very complex neocortex. They base this theory on the pathologic evidence that NFTs are located in the cell bodies of neurons whose axons form the long corticocortical connections while the neuritic plaques concentrate at the specific cortical layers where their axons terminate. Furthermore, the subcortical nuclei that are affected by AD are the ones that provide significant corticopetal input (De Lacoste and White, 1993; Giannakopoulos *et al.*, 1998).

The myelination theory proposes that regional vulnerability is heavily influenced by neocortical myelination patterns (Braak *et al.*, 2000; Bartzokis *et al.*, 2003; Bartzokis, 2004; Bartzokis *et al.*, 2004; Braak and Del Tredici, 2004). Neocortical myelination is a prolonged process that continues well into adulthood. It starts in the primary fields and progresses through the first-order unimodal association areas to the high-order multimodal association areas (Braak *et al.*, 2000; Bartzokis *et al.*, 2003; Bartzokis, 2004; Bartzokis *et al.*, 2004; Braak and Del Tredici, 2004; Jernigan and Fennema-Notestine, 2004). Cortical and white matter myelin changes are not unidirectional because myelin remodeling and even breakdown occur during normal aging. The direction of myelin breakdown is opposite that of myelination with the high-order association cortices affected first (Braak and Braak, 1997; Bartzokis, 2004; Bartzokis *et al.*, 2004; Jernigan and Fennema-Notestine, 2004). Successful myelination ensures accelerated conduction velocity, reduces the metabolic demands and energy expenditure by as much as 5000-fold, and effectively suppresses axonal spouting (Braak and Del Tredici, 2004). Thus, thinly myelinated or unmyelinated neurons are subjected to higher levels of metabolism and therefore higher oxidative stress, which makes them prone to neurodegeneration. An important role in this theory is attributed to the highly metabolically active late myelinating oligodendrocytes (Bartzokis *et al.*, 2004; Benes, 2004; Connor, 2004; Whitman and Cotman, 2004). These glial cells are the major source of cholesterol production in the brain, and their function is crucial for axonal and dendritic sprouting and thus to plasticity and learning (Benes, 2004). Compared with normal aging, AD has more extensive myelin breakdown, and it is feasible that dysmyelination after oxidative damage to oligodendrocytes may play a role in AD pathogenesis and selective vulnerability (Bartzokis *et al.*, 2004; Connor, 2004; Noble, 2004; Whitman and Cotman, 2004).

Finally, morphologically identical neighboring neurons can vary in their biochemical composition, receptor expression, delicate cytoskeletal properties, neurotransmitter responsiveness and/or synthesis, metabolic activity, and so forth, and thus have different protein-specific vulnerability. The phenotype of the proteinopathy may depend on the cell-specific response to injury and result in cell-specific protein accumulation. This susceptibility and the neuronal response to injury are then further modified by individual genetic and epigenetic factors (Morrison, 1993; Cummings, 2000, 2003). Some examples of influential epigenetic factors in AD are age, gender, education, past history of head trauma, diet, and physical and mental activity. They are known to have a beneficial or detrimental impact on disease onset and/or progression.

1.7. Genotype-Phenotype Relationship

As described above, three genes have been reported to cause autosomal dominant EOAD: the APP gene mutation on chromosome 21, the presenilin 1 gene mutation on chromosome 14, and the presenilin 2 gene mutation on chromosome 1. Their clinical presentation can be markedly different from that of sporadic late onset AD. EOAD may have remarkably different clinicopathologic behavior from the late onset variety, namely atypical clinical features and shorter survival. Patients with presenilin 1 and 2 or APP gene mutations frequently have significant language impairment commonly described as halting speech or anomia that can progress to mutism (Lampe *et al.*, 1994; Kennedy *et al.*, 1995; Lopera *et al.*, 1997; Bertoli Avella *et al.*, 2002; Binetti *et al.*, 2003; Rippon *et al.*, 2003). Dysathria and/or dysphagia have been reported in several presenilin 1 kindreds and one APP family (Bird *et al.*, 1989; Nagano *et al.*, 1992; Yasuda *et al.*, 1999; Miklossy *et al.*, 2003; Moretti *et al.*, 2004). Autosomal dominant EOAD can manifest with early myoclonus and tonic-clonic or complex partial seizures; both are occasional late features of the sporadic late onset disease.(Bird *et al.*, 1989; Kennedy *et al.*, 1993; Lampe *et al.*, 1994; Newman *et al.*, 1994; Kennedy *et al.*, 1995; Campion *et al.*, 1996; Fox *et al.*, 1997; Lopera *et al.*, 1997; Bertoli Avella *et al.*, 2002; Binetti *et al.*, 2003; Miklossy *et al.*, 2003; Rippon *et al.*, 2003). Spastic paraplegia or quadriplegia is a frequently seen in EOAD (Farlow *et al.*, 1994; Verkkoniemi *et al.*, 2001; Miklossy *et al.*, 2003; Rippon *et al.*, 2003; Moretti *et al.*, 2004). Dystonia was reported in two presenilin 1-positive families (Rippon *et al.*, 2003; Moretti *et al.*, 2004). Neuropsychiatric symptoms are abundant and include visual and/or auditory hallucinations (Harvey *et al.*, 1998; Rippon *et al.*, 2003), delusions (Rippon *et al.*, 2003), paranoia (Farlow *et al.*, 1994; Queralt *et al.*, 2002; Miklossy *et al.*, 2003; Rippon *et al.*, 2003), emotional lability (Nagano *et al.*, 1992; Farlow *et al.*, 1994; Rippon *et al.*, 2003), depression (Kennedy *et al.*, 1993; Lopera *et al.*, 1997; Bertoli Avella *et al.*, 2002; Queralt *et al.*, 2002; Binetti *et al.*, 2003; Rippon *et al.*, 2003), disinhibition (Binetti *et al.*, 2003; Rippon *et al.*, 2003), aggression (Farlow *et al.*, 1994; Lopera *et al.*, 1997; Bertoli Avella *et al.*, 2002; Miklossy *et al.*, 2003; Rippon *et al.*, 2003), obsessive-compulsive behavior (Queralt *et al.*, 2002; Miklossy *et al.*, 2003; Rippon *et al.*, 2003), and apathy (Queralt *et al.*, 2002; Miklossy *et al.*, 2003; Rippon *et al.*, 2003). Presenilin patients rarely have been reported to present with frontotemporal dementia where behavioral symptoms compose the phenotypic foreground. (Raux *et al.*, 2000; Tang-Wai *et al.*, 2002). One presenilin 1 case even demonstrated Pick bodies and gliosis and was completely devoid of amyloid pathology (Dermaut *et al.*, 2004). Another extreme presenilin 1 case manifested the Kluver Bucy syndrome with hyperoral, hyperphagic, and hypersexual behavior (Tang-Wai *et al.*, 2002), and hypersexual behavior with exhibitionism and masochistic tendencies was described in another patient with presenilin 1 mutation (Queralt *et al.*, 2002). Cases with parkinsonism and LB pathology in addition to amyloid plaques and NFTs have been described further supporting the hypothesis of an interaction between the three proteins (Lantos *et al.*, 1994; Revesz *et al.*, 1997; Yokota *et al.*, 2002; Lleo *et al.*, 2004). Cerebellar ataxia was noted in one presenilin 1 patient (Miklossy *et al.*, 2003), while marked fluctuations were documented in another with APP mutation (Kennedy *et al.*, 1993). Presenilin 2 can cause severe amyloid angiopathy with rare amyloid plaques and NFTs (Nochlin *et al.*, 1998).

Age of onset and clinical manifestation can vary greatly within a single pedigree (Gomez-Isla *et al.*, 1999). A very notable example is the florid dementia with prominent cerebral atrophy affecting one presenilin 1 twin while the other was asymptomatic and showed no brain atrophy

(Binetti *et al.*, 2003). Thus modifying genetic and epigenetic factors also play a role. Variable age at onset between presenilin 1 families has been attributed to the locus of the mutation, where transmembrane domains 2 and 4 present early and transmembrane domains 6 or 8 present at a later age. Later onset is more typical of presenilin 2 mutations (Kennedy *et al.*, 2001; Lleo *et al.*, 2004).

Finally, the apolipoprotein E gene polymorphism on chromosome 19 modifies the risk of getting AD and the age of onset. The gene has three major alleles e2, e3, and e4. Its product is a 299-amino-acid glycoprotein that functions as cholesterol transporter. The Apo e4 allele has been implicated in both early and late onset AD. It has been demonstrated to promote Aβ aggregation as compared with Apo e3 (Esler *et al.*, 2002) and to have a negative effect on neural plasticity (Teter *et al.*, 1999; Nathan *et al.*, 2002). Unlike the three autosomal dominant mutations described above, it does not trigger atypical clinical features. It accelerates disease onset in a dose-dependent fashion (Corder *et al.*, 1993; Yoshizawa *et al.*, 1994; Blacker *et al.*, 1997; Payami *et al.*, 1997; Hsiung *et al.*, 2004; Khachaturian *et al.*, 2004). The effect seems to be stronger in the Caucasian as opposed to the African-American ethnic group (Evans *et al.*, 2003) and before age 70 (Blacker *et al.*, 1997). Behaviorally, Apo e4 genotype correlates with aggression in advanced AD (Craig *et al.*, 2004) and pathologically with more extensive amyloid angiopathy (Duyckaerts and Dickson, 2003).

1.8. Disease-Modifying Therapy

The evidence for the role of Aβ in AD is overwhelming, and antiamyloid therapies are actively under investigation for potential disease-modifying effects. In 1996, Solomon *et al.* were first to report that anti-Aβ monoclonal antibodies can prevent Aβ aggregation *in vitro* and later reported that the antibodies are capable of promoting dissolution of Aβ deposits and inhibition of Aβ cytotoxic effects (Solomon *et al.*, 1996, 1997). Schenk *et al.* then advanced these ideas to active immunization with Aβ and showed no deposition of amyloid plaques in young immunized APP transgenic mice and marked reduction of the amyloid deposition in older mice (Schenk *et al.*, 1999). These results led to the great excitement and high expectations for the first active immunization trials in humans. The phase I study included 70 AD patients and showed adequate antibody response (Schenk, 2002). The phase II trial included 375 patients with 300 randomized to active vaccination. Unfortunately, 6% of the patients developed meningoencephalitis, and the trial was aborted (Orgogozo *et al.*, 2003). The autopsies of two of these patients showed greatly diminished amyloid plaque burden relative to their stage of AD without effect on NFT or vessel amyloid load. There was evidence of ongoing T cell–mediated autoimmune meningoencephalitis (Nicoll *et al.*, 2003; Ferrer *et al.*, 2004). A subgroup of patients who responded with antibody synthesis did cognitively better than those who did not in post-trial follow-up (Mattson, 2004). Currently, researchers are actively investigating various paths to modifying the antigen such as using Aβ fragments or plasmid DNA coding Aβ (Manea *et al.*, 2004; Schiltz *et al.*, 2004).

Passive immunization with mid-domain or amino-terminal Aβ antibodies results in decreased plaque burden in APP mice (Bard *et al.*, 2000, 2003; Cribbs *et al.*, 2003; Bussiere *et al.*, 2004). Passive immunization may be safer because it is unlikely to ilicit a T-cell autoimmune reaction and because the antibodies are short-lived and can be removed via plasmapheresis. However, until tested in humans, no one can predict whether passive immunization will be effective and safe.

Another therapeutic target is Aβ production. Many scientists have focused on developing and testing β- and γ-secretase inhibitors. Likewise, various peptide or nonpeptide inhibitors of Aβ polymerization are being explored (Wolfe, 2002; Cohen and Kelly, 2003).

Aβ aggregation can be approached by chelating important cofactors such as copper and iron. Clioquinol, originally an antimalarial compound, was shown to be safe in a pilot phase II clinical trial and is being tested further (Ritchie *et al.*, 2004).

Inflamation may play a role in AD. Activated microglia are frequent residents of amyloid plaques. So far, the trials of prednisone (Aisen, 2000; Aisen *et al.*, 2000), diclofenac (Scharf *et al.*, 1999), rofecoxib and naproxen (Aisen *et al.*, 2003) have all been negative despite good evidence that they are antiamyloidogenic *in vitro* (Weggen *et al.*, 2003).

Antioxidant trials have produced mixed results. Currently, only vitamin E in a 2000 IU dose is commonly used as therapy in AD. This approach is based on a large trial that showed vitamin E monotherapy to slow AD progression (Sano *et al.*, 1997). Vitamin E, however, failed to delay conversion of MCI to AD in a recent MCI trial (Petersen, 2003). Vitamin C is empirically recommended by some based on epidemiologic data that showed both vitamin E and C to lower the risk of dementia (Zandi *et al.*, 2004). *Ginkgo biloba* was tested in two human trials with conflicting results (Le Bars *et al.*, 2000; van Dongen *et al.*, 2003). Curcumin, a curry compound, has been shown to decrease Aβ deposition *in vitro* and in mice (Lim *et al.*, 2001; Ono *et al.*, 2004). Curcumin therapeutic trials are currently on their way.

The role of cholesterol in Aβ metabolism has been emphasized (Puglielli *et al.*, 2003), and statins reduce Aβ accumulation (Petanceska *et al.*, 2002). Trials of statins for AD are under way.

Tau therapies have drawn relatively less attention. Nevertheless, several compounds that can effectively inhibit various enzymes playing a role in tau's hyperphosphorylation are currently being explored (Iqbal *et al.*, 2002). Most attractive so far appear to be lithium (Bhat *et al.*, 2004), valproate (Loy and Tariot, 2002), and the *N*-methyl-D-aspartate (NMDA) receptor blocker memantine (Li *et al.*, 2004).

1.9. Symptomatic Therapy

The cholinesterase inhibitors donepezil (Doody, 2003), galantamine (Raskind, 2003), and rivastigmine (Farlow, 2003) are the current U.S. FDA–approved agents for mild to moderate AD. They work by increasing the concentration of acetylcholine in the synapse. Rivastigmine is known to inhibit butyrylcholin esterase, while galantamine is an allosteric modulator of the nicotinic presynaptic receptors (Geerts *et al.*, 2002). All three agents have been extensively tested and produce modest cognitive, functional, and behavioral effects. Their side-effect profile is well established—they are well tolerated when slowly titrated to avoid gastrointestinal upset.

Memantine is a newly approved drug for moderate to severe AD. It acts by weakly blocking the NMDA receptors, preventing the effects of detrimental continuous low levels of glutamate while permitting the physiologically advantageous large glutamate surge that normally results in postsynaptic action potential. Additionally, memantine may stimulate long-term potentiation and block tau hyperphosphorylation (Li *et al.*, 2004; Voisin *et al.*, 2004).

1.10. Future Directions

The advancing understanding of the intricate neurodegenerative phenotype-proteotype relationship allows for more precise diagnosis. In the near future, it may help identify disease-specific phenotype-guided therapeutic agents that will target those crucial pathogenic steps. With several disease-modifying approaches being tested, disease-modifying therapies are plausible.

Acknowledgments

The authors would like to thank Ivan Klement, M.D., for providing the pathology images, as well as John Ringman, M.D., George Bartzokis, M.D., and Magdalena Ivanova, Ph.D., for their advice. This project was supported by a grant from the National Institute on Aging for the UCLA Alzheimer's Disease Research Center (P50 16570), the Sidell-Kagan Foundation, and the Kassel Parkinson's Disease Foundation.

References

American Psychiatric Association (1994). Diagnostic and Statistical Manual of Mental Disorders, 4th edition. Washington, DC, American Psychiatric Association Press.

Aisen, P. S. (2000). Anti-inflammatory therapy for Alzheimer's disease: implications of the prednisone trial. *Acta Neurol Scand Suppl* 176: 85–89.

Aisen, P. S., Davis, K. L., Berg, J. D., Schafer, K., Campbell, K., Thomas, R. G., Weiner, M. F., Farlow, M. R., Sano, M., Grundman, M. and Thal, L. J. (2000). A randomized controlled trial of prednisone in Alzheimer's disease. Alzheimer's Disease Cooperative Study. *Neurology* 54: 588–593.

Aisen, P. S., Schafer, K. A., Grundman, M., Pfeiffer, E., Sano, M., Davis, K. L., Farlow, M. R., Jin, S., Thomas, R. G. and Thal, L. J. (2003). Effects of rofecoxib or naproxen vs placebo on Alzheimer disease progression: a randomized controlled trial. *JAMA* 289: 2819–2826.

Arendt, T. (2004). Neurodegeneration and plasticity. *Int J Dev Neurosci* 22: 507–514.

Bard, F., Cannon, C., Barbour, R., Burke, R. L., Games, D., Grajeda, H., Guido, T., Hu, K., Huang, J., Johnson-Wood, K., Khan, K., Kholodenko, D., Lee, M., Lieberburg, I., Motter, R., Nguyen, M., Soriano, F., Vasquez, N., Weiss, K., Welch, B., Seubert, P., Schenk, D. and Yednock, T. (2000). Peripherally administered antibodies against amyloid beta-peptide enter the central nervous system and reduce pathology in a mouse model of Alzheimer disease. *Nat Med* 6: 916–919.

Bard, F., Barbour, R., Cannon, C., Carretto, R., Fox, M., Games, D., Guido, T., Hoenow, K., Hu, K., Johnson-Wood, K., Khan, K., Kholodenko, D., Lee, C., Lee, M., Motter, R., Nguyen, M., Reed, A., Schenk, D., Tang, P., Vasquez, N., Seubert, P. and Yednock, T. (2003). Epitope and isotype specificities of antibodies to beta-amyloid peptide for protection against Alzheimer's disease-like neuropathology. *Proc Natl Acad Sci USA* 100: 2023–2028.

Bartzokis, G. (2004). Age-related myelin breakdown: a developmental model of cognitive decline and Alzheimer's disease. *Neurobiol Aging*. 25: 5–18; author reply 49–62.

Bartzokis, G., Cummings, J. L., Sultzer, D., Henderson, V. W., Nuechterlein, K. H. and Mintz, J. (2003). White matter structural integrity in healthy aging adults and patients with Alzheimer disease: a magnetic resonance imaging study. *Arch Neurol* 60: 393–398.

Bartzokis, G., Sultzer, D., Lu, P. H., Nuechterlein, K. H., Mintz, J. and Cummings, J. L. (2004). Heterogeneous age-related breakdown of white matter structural integrity: implications for cortical "disconnection" in aging and Alzheimer's disease. *Neurobiol Aging* 25: 843–851.

Beard, C. M., Kokmen, E., Sigler, C., Smith, G. E., Petterson, T. and O'Brien, P. C. (1996). Cause of death in Alzheimer's disease. *Ann Epidemiol* 6: 195–200.

Bence, N. F., Sampat, R. M. and Kopito, R. R. (2001). Impairment of the ubiquitin-proteasome system by protein aggregation. *Science* 292: 1552–1555.

Benes, F. M. (2004). A disturbance of late myelination as a trigger for Alzheimer's disease. *Neurobiol Aging* 25: 41–43.

Berg, L., McKeel, D. W., Jr., Miller, J. P., Storandt, M., Rubin, E. H., Morris, J. C., Baty, J., Coats, M., Norton, J., Goate, A. M., Price, J. L., Gearing, M., Mirra, S. S. and Saunders, A. M. (1998). Clinicopathologic studies in cognitively healthy aging and Alzheimer's disease: relation of histologic markers to dementia severity, age, sex, and apolipoprotein E genotype. *Arch Neurol* 55: 326–335.

Bertoli Avella, A. M., Marcheco Teruel, B., Llibre Rodriguez, J. J., Gomez Viera, N., Borrajero Martinez, I., Severijnen, E. A., Joosse, M., van Duijn, C. M., Heredero Baute, L. and Heutink, P. (2002). A novel presenilin 1 mutation (L174 M) in a large Cuban family with early ons *et al* zheimer disease. *Neurogenetics* 4: 97–104.

Bertoni-Freddari, C., Fattoretti, P., Casoli, T., Caselli, U. and Meier-Ruge, W. (1996). Deterioration threshold of synaptic morphology in aging and senile dementia of Alzheimer's type. *Anal Quant Cytol Histol* 18: 209–213.

Bertoni-Freddari, C., Fattoretti, P., Solazzi, M., Giorgetti, B., Di Stefano, G., Casoli, T. and Meier-Ruge, W. (2003). Neuronal death versus synaptic pathology in Alzheimer's disease. *Ann N Y Acad Sci* 1010: 635–638.

Bhat, R. V., Budd Haeberlein, S. L. and Avila, J. (2004). Glycogen synthase kinase 3: a drug target for CNS therapies. *J Neurochem* 89: 1313–1317.

Binetti, G., Signorini, S., Squitti, R., Alberici, A., Benussi, L., Cassetta, E., Frisoni, G. B., Barbiero, L., Feudatari, E., Nicosia, F., Testa, C., Zanetti, O., Gennarelli, M., Perani, D., Anchisi, D., Ghidoni, R. and Rossini, P. M. (2003). Atypical dementia associated with a novel presenilin-2 mutation. *Ann Neurol* 54: 832–836.

Bird, T. D., Sumi, S. M., Nemens, E. J., Nochlin, D., Schellenberg, G., Lampe, T. H., Sadovnick, A., Chui, H., Miner, G. W. and Tinklenberg, J. (1989). Phenotypic heterogeneity in familial Alzheimer's disease: a study of 24 kindreds. *Ann Neurol* 25: 12–25.

Blacker, D., Haines, J. L., Rodes, L., Terwedow, H., Go, R. C., Harrell, L. E., Perry, R. T., Bassett, S. S., Chase, G., Meyers, D., Albert, M. S. and Tanzi, R. (1997). ApoE-4 and age at onset of Alzheimer's disease: the NIMH genetics initiative. *Neurology* 48: 139–147.

Bobinski, M., de Leon, M. J., Wegiel, J., Desanti, S., Convit, A., Saint Louis, L. A., Rusinek, H. and Wisniewski, H. M. (2000). The histological validation of post mortem magnetic resonance imaging-determined hippocampal volume in Alzheimer's disease. *Neuroscience* 95: 721–725.

Boeve, B. F., Silber, M. H., Ferman, T. J., Kokmen, E., Smith, G. E., Ivnik, R. J., Parisi, J. E., Olson, E. J. and Petersen, R. C. (1998). REM sleep behavior disorder and degenerative dementia: an association likely reflecting Lewy body disease. *Neurology* 51: 363–370.

Boeve, B. F., Silber, M. H., Ferman, T. J., Lucas, J. A. and Parisi, J. E. (2001). Association of REM sleep behavior disorder and neurodegenerative disease may reflect an underlying synucleinopathy. *Mov Disord* 16: 622–630.

Boeve, B. F., Silber, M. H., Parisi, J. E., Dickson, D. W., Ferman, T. J., Benarroch, E. E., Schmeichel, A. M., Smith, G. E., Petersen, R. C., Ahlskog, J. E., Matsumoto, J. Y., Knopman, D. S., Schenck, C. H. and Mahowald, M. W. (2003). Synucleinopathy pathology and REM sleep behavior disorder plus dementia or parkinsonism. *Neurology* 61: 40–45.

Braak, E., Braak, H. and Mandelkow, E. M. (1994). A sequence of cytoskeleton changes related to the formation of neurofibrillary tangles and neuropil threads. *Acta Neuropathol (Berl)* 87: 554–567.

Braak, H. and Braak, E. (1991). Neuropathological stageing of Alzheimer-related changes. *Acta Neuropathol (Berl)* 82: 239–259.

Braak, H. and Braak, E. (1997). Frequency of stages of Alzheimer-related lesions in different age categories. *Neurobiol Aging* 18: 351–357.

Braak, H. and Del Tredici, K. (2004). Poor and protracted myelination as a contributory factor to neurodegenerative disorders. *Neurobiol Aging* 25: 19–23.

Braak, H., Del Tredici, K., Schultz, C. and Braak, E. (2000). Vulnerability of select neuronal types to Alzheimer's disease. *Ann N Y Acad Sci* 924: 53–61.

Bussiere, T., Bard, F., Barbour, R., Grajeda, H., Guido, T., Khan, K., Schenk, D., Games, D., Seubert, P., Buttini, M., Cannon, C., Carretto, R., Fox, M., Hoenow, K., Hu, K., Johnson-Wood, K., Kholodenko, D., Lee, C., Lee, M., Motter, R., Nguyen, M., Reed, A., Tang, P., Vasquez, N., Yednock, T., Burke, R. L., Huang, J., Lieberburg, I., Soriano, F., Weiss, K. and Welch, B. (2004). Morphological characterization of thioflavin-S-positive amyloid plaques in transgenic Alzheimer mice and effect of passive Abeta immunotherapy on their clearance. *Am J Pathol* 165: 987–995.

Campion, D., Brice, A., Dumanchin, C., Puel, M., Baulac, M., De La Sayette, V., Hannequin, D., Duyckaerts, C., Michon, A., Martin, C., Moreau, V., Penet, C., Martinez, M., Clerget-Darpoux, F., Agid, Y. and Frebourg, T. (1996). A novel presenilin 1 mutation resulting in familial Alzheimer's disease with an onset age of 29 years. *Neuroreport* 7: 1582–1584.

Campion, D., Dumanchin, C., Hannequin, D., Dubois, B., Belliard, S., Puel, M., Thomas-Anterion, C., Michon, A., Martin, C., Charbonnier, F., Raux, G., Camuzat, A., Penet, C., Mesnage, V., Martinez, M., Clerget-Darpoux, F., Brice, A. and Frebourg, T. (1999). Early-onset autosomal dominant Alzheimer disease: prevalence, genetic heterogeneity, and mutation spectrum. *Am J Hum Genet* 65: 664–670.

Cao, X. and Sudhof, T. C. (2004). Dissection of amyloid-beta precursor protein-dependent transcriptional transactivation. *J Biol Chem*. 279: 24601–24611.

Chen, S. T., Sultzer, D. L., Hinkin, C. H., Mahler, M. E. and Cummings, J. L. (1998). Executive dysfunction in Alzheimer's disease: association with neuropsychiatric symptoms and functional impairment. *J Neuropsychiatry Clin Neurosci* 10: 426–432.

Chiti, F., Calamai, M., Taddei, N., Stefani, M., Ramponi, G. and Dobson, C. M. (2002). Studies of the aggregation of mutant proteins *in vivo* provide insights into the genetics of amyloid diseases. *Proc Natl Acad Sci USA*. 99 Suppl 4: 16419–16426.

Cohen, F. E. and Kelly, J. W. (2003). Therapeutic approaches to protein-misfolding diseases. *Nature* 426: 905–909.

Coleman, P., Federoff, H. and Kurlan, R. (2004). A focus on the synapse for neuroprotection in Alzheimer disease and other dementias. *Neurology* 63: 1155–1162.

Connor, J. R. (2004). Myelin breakdown in Alzheimer's disease: a commentary. *Neurobiol Aging* 25: 45–47.

Corder, E. H., Saunders, A. M., Strittmatter, W. J., Schmechel, D. E., Gaskell, P. C., Small, G. W., Roses, A. D., Haines, J. L. and Pericak-Vance, M. A. (1993). Gene dose of apolipoprotein E type 4 allele and the risk of Alzheimer's disease in late onset families. *Science* 261: 921–923.

Craig, D., Hart, D. J., McCool, K., McIlroy, S. P. and Passmore, A. P. (2004). Apolipoprotein E e4 allele influences aggressive behaviour in Alzheimer's disease. *J Neurol Neurosurg Psychiatry* 75: 1327–1330.

Cribbs, D. H., Ghochikyan, A., Vasilevko, V., Tran, M., Petrushina, I., Sadzikava, N., Babikyan, D., Kesslak, P., Kieber-Emmons, T., Cotman, C. W. and Agadjanyan, M. G. (2003). Adjuvant-dependent modulation of Th1 and Th2 responses to immunization with beta-amyloid. *Int Immunol* 15: 505–514.

Crook, R., Verkkoniemi, A., Perez-Tur, J., Mehta, N., Baker, M., Houlden, H., Farrer, M., Hutton, M., Lincoln, S., Hardy, J., Gwinn, K., Somer, M., Paetau, A., Kalimo, H., Ylikoski, R., Poyhonen, M., Kucera, S. and Haltia, M. (1998). A variant of Alzheimer's disease with spastic paraparesis and unusual plaques due to deletion of exon 9 of presenilin 1. *Nat Med* 4: 452–455.

Cummings, J. L. (2000). Cognitive and behavioral heterogeneity in Alzheimer's disease: seeking the neurobiological basis. *Neurobiol Aging* 21: 845–861.

Cummings, J. L. (2003). Alzheimer's disease. In: J. L. Cummings (ed.), The Neuropsychiatry of Alzheimer's Disease and Related Dementias. Martin Dunitz Ltd, London, pp. 57–116.

Cummings, J. L. (2003). Toward a molecular neuropsychiatry of neurodegenerative diseases. *Ann Neurol* 54: 147–154.

De Lacoste, M. C. and White, C. L., 3rd. (1993). The role of cortical connectivity in Alzheimer's disease pathogenesis: a review and model system. *Neurobiol Aging* 14: 1–16.

Del Ser, T., Hachinski, V., Merskey, H. and Munoz, D. G. (2001). Clinical and pathologic features of two groups of patients with dementia with Lewy bodies: effect of coexisting Alzheimer-type lesion load. *Alzheimer Dis Assoc Disord* 15: 31–44.

Dermaut, B., Kumar-Singh, S., Engelborghs, S., Theuns, J., Rademakers, R., Saerens, J., Pickut, B. A., Peeters, K., van den Broeck, M., Vennekens, K., Claes, S., Cruts, M., Cras, P., Martin, J. J., Van Broeckhoven, C. and De Deyn, P. P. (2004). A novel presenilin 1 mutation associated with Pick's disease but not beta-amyloid plaques. *Ann Neurol* 55: 617–626.

Dobson, C. M. (2003). Protein folding and misfolding. *Nature* 426: 884–890.

Doody, R. S. (2003). Update on Alzheimer drugs: Donepezil. *Neurologist* 9: 225–229.

Duyckaerts, C. and Dickson, D. W. (2003). Neuropathology of Alzheimer's disease. In: D. W. Dickson (ed.), Neurodegeneration: The Molecular Pathology of Dementia and Movement Disorders. ISN Neuropath Press, Basel, pp. 47–65.

Esler, W. P., Marshall, J. R., Stimson, E. R., Ghilardi, J. R., Vinters, H. V., Mantyh, P. W. and Maggio, J. E. (2002). Apolipoprotein E affects amyloid formation but not amyloid growth *in vivo*: mechanistic implications for apoE4 enhanced amyloid burden and risk for Alzheimer's disease. *Amyloid* 9: 1–12.

Evans, D. A., Bennett, D. A., Wilson, R. S., Bienias, J. L., Morris, M. C., Scherr, P. A., Hebert, L. E., Aggarwal, N., Beckett, L. A., Joglekar, R., Berry-Kravis, E. and Schneider, J. (2003). Incidence of Alzheimer disease in a biracial urban community: relation to apolipoprotein E allele status. *Arch Neurol* 60: 185–189.

Falke, E., Nissanov, J., Mitchell, T. W., Bennett, D. A., Trojanowski, J. Q. and Arnold, S. E. (2003). Subicular dendritic arborization in Alzheimer's disease correlates with neurofibrillary tangle density. *Am J Pathol* 163: 1615–1621.

Farlow, M., Murrell, J., Ghetti, B., Unverzagt, F., Zeldenrust, S. and Benson, M. (1994). Clinical characteristics in a kindred with early-onset Alzheimer's disease and their linkage to a G→T change at position 2149 of the amyloid precursor protein gene. *Neurology* 44: 105–111.

Farlow, M. (2003). Update on rivastigmine. *Neurologist* 9: 230–234.

Feany, M. B. and Dickson, D. W. (1996). Neurodegenerative disorders with extensive tau pathology: a comparative study and review. *Ann Neurol* 40: 139–148.

Ferman, T. J., Boeve, B. F., Smith, G. E., Silber, M. H., Kokmen, E., Petersen, R. C. and Ivnik, R. J. (1999). REM sleep behavior disorder and dementia: cognitive differences when compared with AD. *Neurology* 52: 951–957.

Ferrer, I., Boada Rovira, M., Sanchez Guerra, M. L., Rey, M. J. and Costa-Jussa, F. (2004). Neuropathology and pathogenesis of encephalitis following amyloid-beta immunization in Alzheimer's disease. *Brain Pathol* 14: 11–20.

Fox, N. C., Kennedy, A. M., Harvey, R. J., Lantos, P. L., Roques, P. K., Collinge, J., Hardy, J., Hutton, M., Stevens, J. M., Warrington, E. K. and Rossor, M. N. (1997). Clinicopathological features of familial Alzheimer's disease associated with the M139V mutation in the presenilin 1 gene. Pedigree but not mutation specific age at onset provides evidence for a further genetic factor. *Brain* 120 (Pt 3): 491–501.

Friedhoff, P., von Bergen, M., Mandelkow, E. M., Davies, P. and Mandelkow, E. (1998). A nucleated assembly mechanism of Alzheimer paired helical filaments. *Proc Natl Acad Sci USA* 95: 15712–15717.

Gao, Y. and Pimplikar, S. W. (2001). The gamma -secretase-cleaved C-terminal fragment of amyloid precursor protein mediates signaling to the nucleus. *Proc Natl Acad Sci USA* 98: 14979–14984.

Geerts, H., Finkel, L., Carr, R. and Spiros, A. (2002). Nicotinic receptor modulation: advantages for successful Alzheimer's disease therapy. *J Neural Transm Suppl* 203–216.

Ghetti, B., Hutton, M. L. and Wszolek, Z. K. (2003). Frontotemporal dementia and parkinsonism linked to Chromosome 17 associated with tau gene mutations. In: D. W. Dickson (ed.), Neurodegeneration: The Molecular Pathology of Dementia and Movement Disorders. ISN Neuropath Press, Basel, pp. 86–102.

Giannakopoulos, P., Hof, P. R. and Bouras, C. (1998). Selective vulnerability of neocortical association areas in Alzheimer's disease. *Microsc Res Tech* 43: 16–23.

Giasson, B. I., Lee, V. M. and Trojanowski, J. Q. (2003). Interactions of amyloidogenic proteins. *Neuromolecular Med* 4: 49–58.

Giasson, B. I., Forman, M. S., Higuchi, M., Golbe, L. I., Graves, C. L., Kotzbauer, P. T., Trojanowski, J. Q. and Lee, V. M. (2003). Initiation and synergistic fibrillization of tau and alpha-synuclein. *Science* 300: 636–640.

Goedert, M. (2003). Introduction to the Tauopathies. In: D. W. Dickson (ed.), Neurodegeneration: The Molecular Pathology of Dementia and Movement Disorders. ISN Neuropath Press, Basel, pp. 82–85.

Goldberg, A. L. (2003). Protein degradation and protection against misfolded or damaged proteins. *Nature* 426: 895–899.

Gomez-Isla, T., Price, J. L., McKeel, D. W., Jr., Morris, J. C., Growdon, J. H. and Hyman, B. T. (1996). Profound loss of layer II entorhinal cortex neurons occurs in very mild Alzheimer's disease. *J Neurosci* 16: 4491–4500.

Gomez-Isla, T., Growdon, W. B., McNamara, M. J., Nochlin, D., Bird, T. D., Arango, J. C., Lopera, F., Kosik, K. S., Lantos, P. L., Cairns, N. J. and Hyman, B. T. (1999). The impact of different presenilin 1 andpresenilin 2 mutations on amyloid deposition, neurofibrillary changes and neuronal loss in the familial Alzheimer's disease brain: evidence for other phenotype-modifying factors. *Brain* 122 (Pt 9): 1709–1719.

Gotz, J., Chen, F., van Dorpe, J. and Nitsch, R. M. (2001). Formation of neurofibrillary tangles in P301l tau transgenic mice induced by Abeta 42 fibrils. *Science* 293: 1491–1495.

Hansen, L., Salmon, D., Galasko, D., Masliah, E., Katzman, R., DeTeresa, R., Thal, L., Pay, M. M., Hofstetter, R., Klauber, M. *et al.* (1990). The Lewy body variant of Alzheimer's disease: a clinical and pathologic entity. *Neurology* 40: 1–8.

Haroutunian, V., Perl, D. P., Purohit, D. P., Marin, D., Khan, K., Lantz, M., Davis, K. L. and Mohs, R. C. (1998). Regional distribution of neuritic plaques in the nondemented elderly and subjects with very mild Alzheimer disease.[see comment]. *Arch. Neurol* 55: 1185–1191.

Harper, J. D., Wong, S. S., Lieber, C. M. and Lansbury, P. T., Jr. (1999). Assembly of A beta amyloid protofibrils: an *in vivo* model for a possible early event in Alzheimer's disease. *Biochemistry* 38: 8972–8980.

Hartl, F. U. and Hayer-Hartl, M. (2002). Molecular chaperones in the cytosol: from nascent chain to folded protein. *Science* 295: 1852–1858.

Harvey, R. J., Ellison, D., Hardy, J., Hutton, M., Roques, P. K., Collinge, J., Fox, N. C. and Rossor, M. N. (1998). Chromosome 14 familial Alzheimer's disease: the clinical and neuropathological characteristics of a family with a leucine→serine (L250S) substitution at codon 250 of the presenilin 1 gene. *J Neurol Neurosurg Psychiatry* 64: 44–49.

Hebert, L. E., Scherr, P. A., Bienias, J. L., Bennett, D. A. and Evans, D. A. (2003). Alzheimer disease in the US population: prevalence estimates using the 2000 census. *Arch Neurol* 60: 1119–1122.

Hof, P. R. (1997). Morphology and neurochemical characteristics of the vulnerable neurons in brain aging and Alzheimer's disease. *Eur Neurol* 37: 71–81.

Honer, W. G. (2003). Pathology of presynaptic proteins in Alzheimer's disease: more than simple loss of terminals. *Neurobiol Aging* 24: 1047–1062.

Houlden, H., Baker, M., McGowan, E., Lewis, P., Hutton, M., Crook, R., Wood, N. W., Kumar-Singh, S., Geddes, J., Swash, M., Scaravilli, F., Holton, J. L., Lashley, T., Tomita, T., Hashimoto, T., Verkkoniemi, A., Kalimo, H., Somer, M., Paetau, A., Martin, J. J., Van Broeckhoven, C., Golde, T., Hardy, J., Haltia, M. and Revesz, T. (2000). Variant Alzheimer's disease with spastic paraparesis and cotton wool plaques is caused by PS-1 mutations that lead to exceptionally high amyloid-beta concentrations. *Ann Neurol* 48: 806–808.

Hoyert, D. L. and Rosenberg, H. M. (1999). Mortality from Alzheimer's disease: an update. *Natl Vital Stat Rep.* 47: 1–8.

Hsia, A. Y., Masliah, E., McConlogue, L., Yu, G. Q., Tatsuno, G., Hu, K., Kholodenko, D., Malenka, R. C., Nicoll, R. A. and Mucke, L. (1999). Plaque-independent disruption of neural circuits in Alzheimer's disease mouse models. *Proc Natl Acad Sci USA* 96: 3228–3233.

Hsiung, G. Y., Sadovnick, A. D. and Feldman, H. (2004). Apolipoprotein E epsilon4 genotype as a risk factor for cognitive decline and dementia: data from the Canadian Study of Health and Aging *CMAJ.* 171: 863–867.

Hwang, T. J., Masterman, D. L., Ortiz, F., Fairbanks, L. A. and Cummings, J. L. (2004). Mild cognitive impairment is associated with characteristic neuropsychiatric symptoms. *Alzheimer Dis Assoc Disord* 18: 17–21.

Iqbal, K., Alonso Adel, C., El-Akkad, E., Gong, C. X., Haque, N., Khatoon, S., Tsujio, I. and Grundke-Iqbal, I. (2002). Pharmacological targets to inhibit Alzheimer neurofibrillary degeneration. *J Neural Transm Suppl.* 309–319.

Ishizawa, T., Mattila, P., Davies, P., Wang, D. and Dickson, D. W. (2003). Colocalization of tau and alpha-synuclein epitopes in Lewy bodies. *J Neuropathol Exp Neurol* 62: 389–397.

Jellinger, K. A. (2002). Recent developments in the pathology of Parkinson's disease. *J Neural Transm Suppl* 347–376.

Jellinger, K. A. (2003). Neuropathological spectrum of synucleinopathies. *Mov Disord.* 18 Suppl 6: S2–12.

Jellinger, K. A. (2003). Plaque-predominant and tangle-predominant variants of Alzheimer's disease. In: D. W. Dickson (ed.), Neurodegeneration: The Molecular Pathology of Dementia and Movement Disorders. ISN Neuropath Press, Basel, pp. 66–68.

Jernigan, T. L. and Fennema-Notestine, C. (2004). White matter mapping is needed. *Neurobiol Aging* 25: 37–39.

Johnson, J. K., Head, E., Kim, R., Starr, A. and Cotman, C. W. (1999). Clinical and pathological evidence for a frontal variant of Alzheimer disease. *Arch Neurol* 56: 1233–1239.

Kalia, M. (2003). Dysphagia and aspiration pneumonia in patients with Alzheimer's disease. *Metabolism* 52: 36–38.

Katayama, T., Imaizumi, K., Sato, N., Miyoshi, K., Kudo, T., Hitomi, J., Morihara, T., Yoneda, T., Gomi, F., Mori, Y., Nakano, Y., Takeda, J., Tsuda, T., Itoyama, Y., Murayama, O., Takashima, A., St George-Hyslop, P., Takeda, M. and Tohyama, M. (1999). Presenilin-1 mutations downregulate the signalling pathway of the unfolded-protein response. *Nat Cell Biol* 1: 479–485.

Kayed, R., Head, E., Thompson, J. L., McIntire, T. M., Milton, S. C., Cotman, C. W. and Glabe, C. G. (2003). Common structure of soluble amyloid oligomers implies common mechanism of pathogenesis. *Science* 300: 486–489.

Kelly, J. W. (1998). The alternative conformations of amyloidogenic proteins and their multi-step assembly pathways. *Curr Opin Struct Biol* 8: 101–106.

Kennedy, A. M., Newman, S., McCaddon, A., Ball, J., Roques, P., Mullan, M., Hardy, J., Chartier-Harlin, M. C., Frackowiak, R. S., Warrington, E. K. *et al.* (1993). Familial Alzheimer's disease. A pedigree with a mis-sense mutation in the amyloid precursor protein gene (amyloid precursor protein 717 valine→glycine). *Brain* 116 (Pt 2): 309–324.

Kennedy, A. M., Newman, S. K., Frackowiak, R. S., Cunningham, V. J., Roques, P., Stevens, J., Neary, D., Bruton, C. J., Warrington, E. K. and Rossor, M. N. (1995). Chromosome 14 linked familial Alzheimer's disease. A clinico-pathological study of a single pedigree. *Brain* 118 (Pt 1): 185–205.

Kennedy, A. M., Rossor, M. N. and Hodges, J. R. (2001). Familial and sporadic Alzheimer's disease. In: J. H. Hodges (ed.), Early-Onset Dementia. Oxford University Press, New York, pp. 263–283.

Khachaturian, A. S., Corcoran, C. D., Mayer, L. S., Zandi, P. P. and Breitner, J. C. (2004). Apolipoprotein E epsilon4 count affects age at onset of Alzheimer disease, but not lifetime susceptibility: The Cache County Study. *Arch Gen Psychiatry* 61: 518–524.

Klucken, J., McLean, P. J., Gomez-Tortosa, E., Ingelsson, M. and Hyman, B. T. (2003). Neuritic alterations and neural system dysfunction in Alzheimer's disease and dementia with Lewy bodies. *Neurochem Res.* 28: 1683–1691.

Knopman, D. S., Parisi, J. E., Salviati, A., Floriach-Robert, M., Boeve, B. F., Ivnik, R. J., Smith, G. E., Dickson, D. W., Johnson, K. A., Petersen, L. E., McDonald, W. C., Braak, H. and Petersen, R. C. (2003). Neuropathology of cognitively normal elderly. *J Neuropathol Exp Neurol* 62: 1087–1095.

Kordower, J. H., Chu, Y., Stebbins, G. T., DeKosky, S. T., Cochran, E. J., Bennett, D. and Mufson, E. J. (2001). Loss and atrophy of layer II entorhinal cortex neurons in elderly people with mild cognitive impairment. *Ann Neurol.* 49: 202–213.

Kosaka, K. and Mehraein, P. (1979). Dementia-Parkinsonism syndrome with numerous Lewy bodies and senile plaques in cerebral cortex. *Arch Psychiatr Nervenkr.* 226: 241–250.

Kramer, J. H. and Miller, B. L. (2000). Alzheimer's disease and its focal variants. *Semin Neurol* 20: 447–454.

Kukull, W. A., Brenner, D. E., Speck, C. E., Nochlin, D., Bowen, J., McCormick, W., Teri, L., Pfanschmidt, M. L. and Larson, E. B. (1994). Causes of death associated with Alzheimer disease: variation by level of cognitive impairment before death. *J Am Geriatr Soc.* 42: 723–726.

Kurlan, R., Richard, I. H., Papka, M. and Marshall, F. (2000). Movement disorders in Alzheimer's disease: more rigidity of definitions is needed. *Mov Disord* 15: 24–29.

Lampe, T. H., Bird, T. D., Nochlin, D., Nemens, E., Risse, S. C., Sumi, S. M., Koerker, R., Leaird, B., Wier, M. and Raskind, M. A. (1994). Phenotype of chromosome 14-linked familial Alzheimer's disease in a large kindred. *Ann Neurol* 36: 368–378.

Lansbury, P. T., Jr. (1999). Evolution of amyloid: what normal protein folding may tell us about fibrillogenesis and disease. *Proc Natl Acad Sci USA* 96: 3342–3344.

Lantos, P. L., Ovenstone, I. M., Johnson, J., Clelland, C. A., Roques, P. and Rossor, M. N. (1994). Lewy bodies in the brain of two members of a family with the 717 (Val to Ile) mutation of the amyloid precursor protein gene. *Neurosci Lett.* 172: 77–79.

Le Bars, P. L., Kieser, M. and Itil, K. Z. (2000). A 26-week analysis of a double-blind, placebo-controlled trial of the ginkgo biloba extract EGb 761 in dementia. *Dement Geriatr Cogn Disord* 11: 230–237.

Le, T. V., Crook, R., Hardy, J. and Dickson, D. W. (2001). Cotton wool plaques in non-familial late-onset alzheimer disease. *J Neuropathol Exp Neurol* 60: 1051–1061.

Leissring, M. A., Murphy, M. P., Mead, T. R., Akbari, Y., Sugarman, M. C., Jannatipour, M., Anliker, B., Muller, U., Saftig, P., De Strooper, B., Wolfe, M. S., Golde, T. E. and LaFerla, F. M. (2002). A physiologic signaling role for the gamma-secretase-derived intracellular fragment of APP. *Proc Natl Acad Sci USA* 99: 4697–4702.

Lewis, J., Dickson, D. W., Lin, W. L., Chisholm, L., Corral, A., Jones, G., Yen, S. H., Sahara, N., Skipper, L., Yager, D., Eckman, C., Hardy, J., Hutton, M. and McGowan, E. (2001). Enhanced neurofibrillary degeneration in transgenic mice expressing mutant tau and APP. *Science* 293: 1487–1491.

Li, L., Sengupta, A., Haque, N., Grundke-Iqbal, I. and Iqbal, K. (2004). Memantine inhibits and reverses the Alzheimer type abnormal hyperphosphorylation of tau and associated neurodegeneration. *FEBS Lett.* 566: 261–269.

Lim, G. P., Chu, T., Yang, F., Beech, W., Frautschy, S. A. and Cole, G. M. (2001). The curry spice curcumin reduces oxidative damage and amyloid pathology in an Alzheimer transgenic mouse. *J Neurosci* 21: 8370–8377.

Liu, Y., Stern, Y., Chun, M. R., Jacobs, D. M., Yau, P. and Goldman, J. E. (1997). Pathological correlates of extrapyramidal signs in Alzheimer's disease. *Ann Neurol* 41: 368–374.

Lleo, A., Berezovska, O., Growdon, J. H. and Hyman, B. T. (2004). Clinical, pathological, and biochemical spectrum of Alzheimer disease associated with PS-1 mutations. *Am J Geriatr Psychiatry* 12: 146–156.

Lopera, F., Ardilla, A., Martinez, A., Madrigal, L., Arango-Viana, J. C., Lemere, C. A., Arango-Lasprilla, J. C., Hincapie, L., Arcos-Burgos, M., Ossa, J. E., Behrens, I. M., Norton, J., Lendon, C., Goate, A. M., Ruiz-Linares, A., Rosselli, M. and Kosik, K. S. (1997). Clinical features of early-onset alzheimer disease in a large kindred with an E280A presenilin-1 mutation. *JAMA* 277: 793–799.

Loy, R. and Tariot, P. N. (2002). Neuroprotective properties of valproate: potential benefit for AD and tauopathies. *J Mol Neurosci* 19: 303–307.

Lue, L. F., Kuo, Y. M., Roher, A. E., Brachova, L., Shen, Y., Sue, L., Beach, T., Kurth, J. H., Rydel, R. E. and Rogers, J. (1999). Soluble amyloid beta peptide concentration as a predictor of synaptic change in Alzheimer's disease. *Am J Pathol* 155: 853–862.

Manea, M., Mezo, G., Hudecz, F. and Przybylski, M. (2004). Polypeptide conjugates comprising a beta-amyloid plaque-specific epitope as new vaccine structures against Alzheimer's disease. *Biopolymers* 76: 503–511.

Marui, W., Iseki, E., Ueda, K. and Kosaka, K. (2000). Occurrence of human alpha-synuclein immunoreactive neurons with neurofibrillary tangle formation in the limbic areas of patients with Alzheimer's disease. *J Neurol Sci.* 174: 81–84.

Masliah, E. (2000). The role of synaptic proteins in Alzheimer's disease. *Ann N Y Acad Sci.* 924: 68–75.

Masliah, E., Rockenstein, E., Veinbergs, I., Sagara, Y., Mallory, M., Hashimoto, M. and Mucke, L. (2001). beta-amyloid peptides enhance alpha-synuclein accumulation and neuronal deficits in a transgenic mouse model linking Alzheimer's disease and Parkinson's disease. *Proc Natl Acad Sci USA* 98: 12245–12250.

Mattson, M. P. (2004). Pathways towards and away from Alzheimer's disease. *Nature* 430: 631–639.

McLean, C. A., Cherny, R. A., Fraser, F. W., Fuller, S. J., Smith, M. J., Beyreuther, K., Bush, A. I. and Masters, C. L. (1999). Soluble pool of Abeta amyloid as a determinant of severity of neurodegeneration in Alzheimer's disease. *Ann Neurol* 46: 860–866.

Mega, M. S., Cummings, J. L., Fiorello, T. and Gornbein, J. (1996). The spectrum of behavioral changes in Alzheimer's disease. *Neurology* 46: 130–135.

Mendez, M. F., Ghajarania, M. and Perryman, K. M. (2002). Posterior cortical atrophy: clinical characteristics and differences compared to Alzheimer's disease. *Dement Geriatr Cogn Disord* 14: 33–40.

Mesulam, M. M. (2000). Aging, Alzheimer's disease and dementia. In: M. M. Mesulam (ed.), Principles of Behavioral and Cognitive Neurology. Oxford University Press, Oxfrod, pp. 439–510.

Miklossy, J., Taddei, K., Suva, D., Verdile, G., Fonte, J., Fisher, C., Gnjec, A., Ghika, J., Suard, F., Mehta, P. D., McLean, C. A., Masters, C. L., Brooks, W. S. and Martins, R. N. (2003). Two novel presenilin-1 mutations (Y256S and Q222H) are associated with early-onset alzheimer's disease. *Neurobiol Aging* 24: 655–662.

Moretti, P., Lieberman, A. P., Wilde, E. A., Giordani, B. I., Kluin, K. J., Koeppe, R. A., Minoshima, S., Kuhl, D. E., Seltzer, W. K. and Foster, N. L. (2004). Novel insertional presenilin 1 mutation causing Alzheimer disease with spastic paraparesis. *Neurology* 62: 1865–1868.

Morrison, J. H. (1993). Differential vulnerability, connectivity, and cell typology. *Neurobiol Aging.* 14: 51–54; discussion 55–56.

Mucke, L., Masliah, E., Yu, G. Q., Mallory, M., Rockenstein, E. M., Tatsuno, G., Hu, K., Kholodenko, D., Johnson-Wood, K. and McConlogue, L. (2000). High-level neuronal expression of abeta 1–42 in wild-type human amyloid protein precursor transgenic mice: synaptotoxicity without plaque formation. *J Neurosci* 20: 4050–4058.

Nagano, K., Miki, T., Yoshioka, K., Katsumi, D., Katsuya, T., Takeda, M., Ikeda, M., Tanabe, H., Nishimura, T., Sakai, Y. *et al.* (1992). [Two kindreds with familial Alzheimer's disease—analysis of the APP717 mutation and the mutated genes for the prion protein]. *Nippon Ronen Igakkai Zasshi* 29: 509–514.

Nagele, R. G., Wegiel, J., Venkataraman, V., Imaki, H. and Wang, K. C. (2004). Contribution of glial cells to the development of amyloid plaques in Alzheimer's disease. *Neurobiol Aging* 25: 663–674.

Naslund, J., Haroutunian, V., Mohs, R., Davis, K. L., Davies, P., Greengard, P. and Buxbaum, J. D. (2000). Correlation between elevated levels of amyloid beta-peptide in the brain and cognitive decline. *JAMA* 283: 1571–1577.

Nathan, B. P., Jiang, Y., Wong, G. K., Shen, F., Brewer, G. J. and Struble, R. G. (2002). Apolipoprotein E4 inhibits, and apolipoprotein E3 promotes neurite outgrowth in cultured adult mouse cortical neurons through the low-density lipoprotein receptor-related protein. *Brain Res.* 928: 96–105.

Newman, S. K., Warrington, E. K., Kennedy, A. M. and Rossor, M. N. (1994). The earliest cognitive change in a person with familial Alzheimer's disease: presymptomatic neuropsychological features in a pedigree with familial Alzheimer's disease confirmed at necropsy. *J Neurol Neurosurg Psychiatry* 57: 967–972.

Nicoll, J. A., Wilkinson, D., Holmes, C., Steart, P., Markham, H. and Weller, R. O. (2003). Neuropathology of human Alzheimer disease after immunization with amyloid-beta peptide: a case report. *Nat Med.* 9: 448–452.

Noble, M. (2004). The possible role of myelin destruction as a precipitating event in Alzheimer's disease. *Neurobiol Aging* 25: 25–31.

Nochlin, D., Bird, T. D., Nemens, E. J., Ball, M. J. and Sumi, S. M. (1998). Amyloid angiopathy in a Volga German family with Alzheimer's disease and a presenilin-2 mutation (N141I). *Ann Neurol* 43: 131–135.

Nybo, M., Svehag, S. E. and Holm Nielsen, E. (1999). An ultrastructural study of amyloid intermediates in A beta1–42 fibrillogenesis. *Scand J Immunol* 49: 219–223.

Ono, K., Hasegawa, K., Naiki, H. and Yamada, M. (2004). Curcumin has potent anti-amyloidogenic effects for Alzheimer's beta-amyloid fibrils *in vivo*. *J Neurosci Res.* 75: 742–750.

Orgogozo, J. M., Gilman, S., Dartigues, J. F., Laurent, B., Puel, M., Kirby, L. C., Jouanny, P., Dubois, B., Eisner, L., Flitman, S., Michel, B. F., Boada, M., Frank, A. and Hock, C. (2003). Subacute meningoencephalitis in a subset of patients with AD after Abeta42 immunization. *Neurology* 61: 46–54.

Pasquier, F. (1999). Early diagnosis of dementia: neuropsychology. *J Neurol* 246: 6–15.

Payami, H., Grimslid, H., Oken, B., Camicioli, R., Sexton, G., Dame, A., Howieson, D. and Kaye, J. (1997). A prospective study of cognitive health in the elderly (Oregon Brain Aging Study): effects of family history and apolipoprotein E genotype. *Am J Hum Genet.* 60: 948–956.

Petanceska, S. S., DeRosa, S., Olm, V., Diaz, N., Sharma, A., Thomas-Bryant, T., Duff, K., Pappolla, M. and Refolo, L. M. (2002). Statin therapy for Alzheimer's disease: will it work? *J Mol Neurosci* 19: 155–161.

Petersen, R. C., Smith, G. E., Waring, S. C., Ivnik, R. J., Tangalos, E. G. and Kokmen, E. (1999). Mild cognitive impairment: clinical characterization and outcome.[erratum appears in *Arch Neurol* 1999; 56(6):760]. *Arch Neurol* 56: 303–308.

Petersen, R. C. (2000). Aging, mild cognitive impairment, and Alzheimer's disease. *Neurol Clin.* 18: 789–806.

Petersen, R. C., Stevens, J. C., Ganguli, M., Tangalos, E. G., Cummings, J. L. and DeKosky, S. T. (2001a). Practice parameter: early detection of dementia: mild cognitive impairment (an evidence-based review). Report of the Quality Standards Subcommittee of the American Academy of Neurology. *Neurology* 56: 1133–1142.

Petersen, R. C., Doody, R., Kurz, A., Mohs, R. C., Morris, J. C., Rabins, P. V., Ritchie, K., Rossor, M., Thal, L. and Winblad, B. (2001b). Current concepts in mild cognitive impairment. *Arch Neurol* 58: 1985–1992.

Petersen, R. C. (2003). Donepezil and vitamin E as treatments for mild cognitive impairment. 9th International Conference on Alzheimer's Disease and Related Disorders, Philadelphia, Pennsylvania.

Price, J. L. and Morris, J. C. (1999). Tangles and plaques in nondemented aging and "preclinical" Alzheimer's disease. *Ann Neurol* 45: 358–368.

Price, J. L., Ko, A. I., Wade, M. J., Tsou, S. K., McKeel, D. W. and Morris, J. C. (2001). Neuron number in the entorhinal cortex and CA1 in preclinical Alzheimer disease. *Arch Neurol* 58: 1395–1402.

Puglielli, L., Tanzi, R. E. and Kovacs, D. M. (2003). Alzheimer's disease: the cholesterol connection. *Nat Neurosci* 6: 345–351.

Queralt, R., Ezquerra, M., Lleo, A., Castellvi, M., Gelpi, J., Ferrer, I., Acarin, N., Pasarin, L., Blesa, R. and Oliva, R. (2002). A novel mutation (V89L) in the presenilin 1 gene in a family with early onset alzheimer's disease and marked behavioural disturbances. *J Neurol Neurosurg Psychiatry* 72: 266–269.

Raskind, M. A. (2003). Update on Alzheimer drugs: Galantamine. *Neurologist* 9: 225–229.

Raux, G., Gantier, R., Thomas-Anterion, C., Boulliat, J., Verpillat, P., Hannequin, D., Brice, A., Frebourg, T. and Campion, D. (2000). Dementia with prominent frontotemporal features associated with L113P presenilin 1 mutation. *Neurology* 55: 1577–1578.

Renner, J. A., Burns, J. M., Hou, C. E., McKeel, D. W., Jr., Storandt, M. and Morris, J. C. (2004). Progressive posterior cortical dysfunction: a clinicopathologic series. *Neurology* 63: 1175–1180.

Revesz, T. and Holton, J. L. (2003). Anatamopathological spectrum of tauopathies. *Mov Disord.* 18 Suppl. 6: S13–20.

Revesz, T., McLaughlin, J. L., Rossor, M. N. and Lantos, P. L. (1997). Pathology of familial Alzheimer's disease with Lewy bodies. *J Neural Transm Suppl.* 51: 121–135.

Rippon, G. A., Crook, R., Baker, M., Halvorsen, E., Chin, S., Hutton, M., Houlden, H., Hardy, J. and Lynch, T. (2003). Presenilin 1 mutation in an african american family presenting with atypical Alzheimer dementia. *Arch Neurol.* 60: 884–888.

Ritchie, C. W., Bush, A. I. and Masters, C. L. (2004). Metal-protein attenuating compounds and Alzheimer's disease. *Expert Opin Invest Drugs* 13: 1585–1592.

Rivas-Vazquez, R. A., Mendez, C., Rey, G. J. and Carrazana, E. J. (2004). Mild cognitive impairment: new neuropsychological and pharmacological target. *Arch Clin Neuropsychol* 19: 11–27.

Rochet, J. C. and Lansbury, P. T., Jr. (2000). Amyloid fibrillogenesis: themes and variations. *Curr Opin Struct Biol.* 10: 60–68.

Ross, C. A. and Poirier, M. A. (2004). Protein aggregation and neurodegenerative disease. *Nat Med.* 10 Suppl: S10–17.

Sano, M., Ernesto, C., Thomas, R. G., Klauber, M. R., Schafer, K., Grundman, M., Woodbury, P., Growdon, J., Cotman, C. W., Pfeiffer, E., Schneider, L. S. and Thal, L. J. (1997). A controlled trial of selegiline, alpha-tocopherol, or both as treatment for Alzheimer's disease. The Alzheimer's Disease Cooperative Study. *N Engl J Med.* 336: 1216–1222.

Scharf, S., Mander, A., Ugoni, A., Vajda, F. and Christophidis, N. (1999). A double-blind, placebo-controlled trial of diclofenac/misoprostol in Alzheimer's disease. *Neurology* 53: 197–201.

Schenck, C. H., Mahowald, M. W., Anderson, M. L., Silber, M. H., Boeve, B. F. and Parisi, J. E. (1997). Lewy body variant of Alzheimer's disease (AD) identified by postmortem ubiquitin staining in a previously reported case of AD associated with REM sleep behavior disorder. *Biol Psychiatry* 42: 527–528.

Schenk, D., Barbour, R., Dunn, W., Gordon, G., Grajeda, H., Guido, T., Hu, K., Huang, J., Johnson-Wood, K., Khan, K., Kholodenko, D., Lee, M., Liao, Z., Lieberburg, I., Motter, R., Mutter, L., Soriano, F., Shopp, G., Vasquez, N., Vandevert, C., Walker, S., Wogulis, M., Yednock, T., Games, D. and Seubert, P. (1999). Immunization with amyloid-beta attenuates Alzheimer-disease-like pathology in the PDAPP mouse. *Nature* 400: 173–177.

Schenk, D. (2002). Amyloid-beta immunotherapy for Alzheimer's disease: the end of the beginning. *Nat Rev Neurosci* 3: 824–828.

Schiltz, J. G., Salzer, U., Mohajeri, M. H., Franke, D., Heinrich, J., Pavlovic, J., Wollmer, M. A., Nitsch, R. M. and Moelling, K. (2004). Antibodies from a DNA peptide vaccination decrease the brain amyliod burden in a mouse model of Alzheimer's disease. *J Mol Med.* 82: 706–714.

Schmaier, A. H., Dahl, L. D., Rozemuller, A. J., Roos, R. A., Wagner, S. L., Chung, R. and Van Nostrand, W. E. (1993). Protease nexin-2/amyloid beta protein precursor. A tight-binding inhibitor of coagulation factor IXa. *J Clin Invest.* 92: 2540–2545.

Schmidt, M. L., Martin, J. A., Lee, V. M. and Trojanowski, J. Q. (1996). Convergence of Lewy bodies and neurofibrillary tangles in amygdala neurons of Alzheimer's disease and Lewy body disorders. *Acta Neuropathol (Berl).* 91: 475–481.

Schumock, G. T. (1998). Economic considerations in the treatment and management of Alzheimer's disease. *Am J Health-System Pharmacy* 55: S17–21.

Selkoe, D. J. (2002). Alzheimer's disease is a synaptic failure. *Science* 298: 789–791.

Selkoe, D. J. (2003). Folding proteins in fatal ways. *Nature* 426: 900–904.

Sherman, M. Y. and Goldberg, A. L. (2001). Cellular defenses against unfolded proteins: a cell biologist thinks about neurodegenerative diseases. *Neuron* 29: 15–32.

Sitia, R. and Braakman, I. (2003). Quality control in the endoplasmic reticulum protein factory. *Nature.* 426: 891–894.

Solomon, B., Koppel, R., Hanan, E. and Katzav, T. (1996). Monoclonal antibodies inhibit *in vivo* fibrillar aggregation of the Alzheimer beta-amyloid peptide. *Proc Natl Acad Sci USA* 93: 452–455.

Solomon, B., Koppel, R., Frankel, D. and Hanan-Aharon, E. (1997). Disaggregation of Alzheimer beta-amyloid by site-directed mAb. *Proc Natl Acad Sci USA* 94: 4109–4112.

Spillantini, M. G. (2003). Introduction to synucleinopathies. In: D. W. Dickson (ed.), Neurodegeneration: The Molecular Pathology of Dementia And Movement Disorders. ISN Neuropath Press, Basel, pp. 156–158.

Steiner, H., Revesz, T., Neumann, M., Romig, H., Grim, M. G., Pesold, B., Kretzschmar, H. A., Hardy, J., Holton, J. L., Baumeister, R., Houlden, H. and Haass, C. (2001). A pathogenic presenilin-1 deletion causes abberrant Abeta 42 production in the absence of congophilic amyloid plaques. *J Biol Chem.* 276: 7233–7239.

Tang-Wai, D., Lewis, P., Boeve, B., Hutton, M., Golde, T., Baker, M., Hardy, J., Michels, V., Ivnik, R., Jack, C. and Petersen, R. (2002). Familial frontotemporal dementia associated with a novel presenilin-1 mutation. *Dement Geriatr Cogn Disord* 14: 13–21.

Tang-Wai, D. F., Graff-Radford, N. R., Boeve, B. F., Dickson, D. W., Parisi, J. E., Crook, R., Caselli, R. J., Knopman, D. S. and Petersen, R. C. (2004). Clinical, genetic, and neuropathologic characteristics of posterior cortical atrophy. *Neurology* 63: 1168–1174.

Tekin, S., Mega, M. S., Masterman, D. M., Chow, T., Garakian, J., Vinters, H. V. and Cummings, J. L. (2001). Orbitofrontal and anterior cingulate cortex neurofibrillary tangle burden is associated with agitation in Alzheimer disease. *Ann. Neurol* 49: 355–361.

Teter, B., Xu, P. T., Gilbert, J. R., Roses, A. D., Galasko, D. and Cole, G. M. (1999). Human apolipoprotein E isoform-specific differences in neuronal sprouting in organotypic hippocampal culture. *J Neurochem* 73: 2613–2616.

Tiraboschi, P., Hansen, L. A., Alford, M., Masliah, E., Thal, L. J. and Corey-Bloom, J. (2000). The decline in synapses and cholinergic activity is asynchronous in Alzheimer's disease. *Neurology* 55: 1278–1283.

Tiraboschi, P., Hansen, L. A., Thal, L. J. and Corey-Bloom, J. (2004). The importance of neuritic plaques and tangles to the development and evolution of AD. *Neuro.* 62: 1984–1989.

Troncoso, J. C., Martin, L. J., Dal Forno, G. and Kawas, C. H. (1996). Neuropathology in controls and demented subjects from the Baltimore Longitudinal Study of Aging. *Neurobiol Aging* 17: 365–371.

Uversky, V. N. (2003). Protein folding revisited. A polypeptide chain at the folding-misfolding-nonfolding cross-roads: which way to go? *Cell Mol Life Sci.* 60: 1852–1871.

van Dongen, M., van Rossum, E., Kessels, A., Sielhorst, H. and Knipschild, P. (2003). Ginkgo for elderly people with dementia and age-associated memory impairment: a randomized clinical trial. *J Clin Epidemiol* 56: 367–376.

Van Gassen, G., Annaert, W. and Van Broeckhoven, C. (2000). Binding partners of Alzheimer's disease proteins: are they physiologically relevant? *Neurobiol Dis.* 7: 135–151.

Van Hoesen, G. W., Augustinack, J. C., Dierking, J., Redman, S. J. and Thangavel, R. (2000). The parahippocampal gyrus in Alzheimer's disease. Clinical and preclinical neuroanatomical correlates. *Ann N Y Acad Sci.* 911: 254–274.

Van Nostrand, W. E., Schmaier, A. H. and Wagner, S. L. (1992). Potential role of protease nexin-2/amyloid beta-protein precursor as a cerebral anticoagulant. *Ann N Y Acad Sci.* 674: 243–252.

Verkkoniemi, A., Kalimo, H., Paetau, A., Somer, M., Iwatsubo, T., Hardy, J. and Haltia, M. (2001). Variant Alzheimer disease with spastic paraparesis: neuropathological phenotype. *J Neuropathol Exp Neurol* 60: 483–492.

Voisin, T., Reynish, E., Portet, F., Feldman, H. and Vellas, B. (2004). What are the treatment options for patients with severe Alzheimer's disease? *CNS Drugs* 18: 575–583.

von Rotz, R. C., Kohli, B. M., Bosset, J., Meier, M., Suzuki, T., Nitsch, R. M. and Konietzko, U. (2004). The APP intracellular domain forms nuclear multiprotein complexes and regulates the transcription of its own precursor. *J Cell Sci.* 117: 4435–4448.

Wahlund, L. O., Pihlstrand, E. and Jonhagen, M. E. (2003). Mild cognitive impairment: experience from a memory clinic. *Acta Neurol Scand Suppl.* 179: 21–24.

Walsh, D. M., Klyubin, I., Fadeeva, J. V., Cullen, W. K., Anwyl, R., Wolfe, M. S., Rowan, M. J. and Selkoe, D. J. (2002). Naturally secreted oligomers of amyloid beta protein potently inhibit hippocampal long-term potentiation *in vivo*. *Nature* 416: 535–539.

Weggen, S., Eriksen, J. L., Sagi, S. A., Pietrzik, C. U., Ozols, V., Fauq, A., Golde, T. E. and Koo, E. H. (2003). Evidence that nonsteroidal anti-inflammatory drugs decrease amyloid beta 42 production by direct modulation of gamma-secretase activity. *J Biol Chem.* 278: 31831–31837.

West, M. J., Coleman, P. D., Flood, D. G. and Troncoso, J. C. (1994). Differences in the pattern of hippocampal neuronal loss in normal ageing and Alzheimer's disease. *Lancet* 344: 769–772.

West, M. J., Kawas, C. H., Stewart, W. F., Rudow, G. L. and Troncoso, J. C. (2004). Hippocampal neurons in pre-clinical Alzheimer's disease. *Neurobiol Aging* 25: 1205–1212.

Whitman, G. T. and Cotman, C. W. (2004). Oligodendrocyte degeneration in AD. *Neurobiol Aging* 25: 33–36.

Winblad, B., Palmer, K., Kivipelto, M., Jelic, V., Fratiglioni, L., Wahlund, L. O., Nordberg, A., Backman, L., Albert, M., Almkvist, O., Arai, H., Basun, H., Blennow, K., de Leon, M., DeCarli, C., Erkinjuntti, T., Giacobini, E., Graff, C., Hardy, J., Jack, C., Jorm, A., Ritchie, K., van Duijn, C., Visser, P. and Petersen, R. C. (2004). Mild cognitive impairmentbeyond controversies, towards a consensus: report of the International Working Group on Mild Cognitive Impairment. *J Intern Med.* 256: 240–246.

Wolfe, M. S. (2002). Therapeutic strategies for Alzheimer's disease. *Nat Rev Drug Discov.* 1: 859–866.

Yasuda, M., Maeda, K., Hashimoto, M., Yamashita, H., Ikejiri, Y., Bird, T. D., Tanaka, C. and Schellenberg, G. D. (1999). A pedigree with a novel presenilin 1 mutation at a residue that is not conserved in presenilin 2. *Arch Neurol.* 56: 65–69.

Yokota, O., Terada, S., Ishizu, H., Ujike, H., Ishihara, T., Nakashima, H., Yasuda, M., Kitamura, Y., Ueda, K., Checler, F. and Kuroda, S. (2002). NACP/alpha-synuclein, NAC, and beta-amyloid pathology of familial Alzheimer's disease with the E184D presenilin-1 mutation: a clinicopathological study of two autopsy cases. *Acta Neuropathol (Berl)* 104: 637–648.

Yoshizawa, T., Yamakawa-Kobayashi, K., Komatsuzaki, Y., Arinami, T., Oguni, E., Mizusawa, H., Shoji, S. and Hamaguchi, H. (1994). Dose-dependent association of apolipoprotein E allele epsilon 4 with late-onset, sporadic Alzheimer's disease. *Ann Neurol* 36: 656–659.

Zandi, P. P., Anthony, J. C., Khachaturian, A. S., Stone, S. V., Gustafson, D., Tschanz, J. T., Norton, M. C., Welsh-Bohmer, K. A. and Breitner, J. C. (2004). Reduced risk of Alzheimer disease in users of antioxidant vitamin supplements: the Cache County Study. *Arch Neurol* 61: 82–88.

Zubenko, G. S., Moossy, J., Martinez, A. J., Rao, G., Claassen, D., Rosen, J. and Kopp, U. (1991). Neuropathologic and neurochemical correlates of psychosis in primary dementia. *Arch Neurol* 48: 619–624.

2

Free Radicals, Metal Ions, and Aβ Aggregation and Neurotoxicity

Kevin J. Barnham, Cyril C. Curtain, and Ashley I. Bush

Abstract

Two of the characteristic pathologic features present in the brains of Alzheimer's disease (AD) patients are the deposition of aggregated amyloid-β peptide (Aβ) and high levels of oxidative stress. Both these phenomena can be explained by Aβ's interactions with metal ions. When Aβ coordinates Zn, Cu, and Fe, the peptide aggregates. If the metals are redox active such as Cu and Fe, then reactive oxygen species (ROS) are generated. The generation of ROS has been implicated in the toxic mechanism of Aβ. A class of metal-protein attenuating compounds (MPACs) that are capable of inhibiting Aβ-metal interactions have been developed. The prototypic MPAC, clioquinol, has shown efficacy in cell and animal models of AD and promising results in a small-scale phase IIa clinical trial.

2.1. Introduction

Metal ions are essential for life, and approximately 30% of all proteins are metalloproteins. The unique chemical properties of the various metal ions allow the coordinating proteins to carry out vital cellular processes. Consequently, cells have developed highly elaborate means of regulating metal ion interactions; because the biochemical use of metal ions is not without hazard. The same properties that cells harness for beneficial means can become destructive when not properly regulated. This is particularly so for the most abundant redox-active metals copper and iron; the ability of these metal ions to occupy multiple valence states and activate molecular oxygen has been utilized by a variety of enzymes including those involved in cellular respiration. However, unregulated, redox-active metals will inappropriately react with oxygen to generate toxic ROS.

The brain is an organ that concentrates metal ions, and recent evidence suggests that a breakdown in metal homeostasis plays a critical role in a variety of age-related neurodegenerative diseases (Bush, 2003; Barnham *et al.*, 2004b). Common features of these diseases include the deposition of misfolded protein (specific to the disease) and substantial cellular damage as a result of oxidative stress. As a general principle, the chemical origin of this oxidative damage comes from reactions of molecular oxygen with the redox-active metals copper and iron that are resident in the tissues (Halliwell and Gutteridge, 1999). We have noted that abnormal metal interaction with the target protein in several age-dependent degenerative diseases could contribute to the etiology (Bush, 2003; Barnham *et al.*, 2004b). Hence, copper has been implicated in amyotrophic lateral sclerosis (the aggregating protein is superoxide dismutase 1) and

Creutzfeldt-Jakob disease (prion) while iron may play a significant role in Parkinson's disease (α-synuclein). Evidence is emerging that implicates copper, iron, and redox-silent zinc in the pathogenesis of AD, and modulating these interactions may have therapeutic potential.

Genetic evidence from cases of familial AD indicates that Aβ metabolism is linked to the disease (Hardy, 1997; Price *et al.*, 1998). AD is characterized by the deposition of amyloid plaques; the major constituent of AD plaques is Aβ, which is cleaved from the membrane-bound amyloid precursor protein (APP) (Glenner and Wong, 1984; Masters *et al.*, 1985; Kang *et al.*, 1987).

The importance of Zn^{2+} in plaque formation has been highlighted by the finding that age and female sex-related plaque formation in Tg2576 transgenic mice was greatly reduced upon the genetic ablation of the zinc transporter 3 protein, which is required for zinc transport into synaptic vesicles (Lee *et al.*, 2002). The amyloid plaques could be described as metallic sinks as remarkably high concentrations of Cu (400 μM), Zn (1 mM), and Fe (1 mM) have been found in amyloid deposits in AD-affected brains (Smith *et al.*, 1997; Lovell *et al.*, 1998).

In vitro studies have shown that low micromolar levels of Zn^{2+} are sufficient to induce protease-resistant aggregation and precipitation of Aβ (Huang *et al.*, 1997). Cu^{2+} and Fe^{3+} also induce peptide aggregation that is exaggerated at acidic pH (Atwood *et al.*, 1998). These metals are found in relatively high concentrations in the region of the brain most susceptible to AD neurodegeneration, where they play important roles in normal physiology. During neurotransmission, high concentrations of Zn (300 μM) and Cu (30 μM) are released, which may explain why Aβ precipitation into amyloid commences in the synapse (Terry, 1996). It has recently been shown that the *N*-methyl-D-aspartate (NMDA) receptor mediates copper homeostasis in neurons (Schlief *et al.*, 2005) and conversely that copper has an inhibitory effect on NMDA receptors (Trombley and Shepherd, 1996).

Another factor that has been suggested to contribute to AD is a breakdown in the degradation/clearance mechanisms of Aβ with age. One of the genetic risk factors associated with AD is apolipoprotein E (Apo E) allotype ($\varepsilon4 > \varepsilon3 > \varepsilon2$) (Strittmatter *et al.*, 1993). Apo E is the principal cholesterol carrier in the brain and may be involved in the degradation/clearance of Aβ (Puglielli *et al.*, 2003). It is not yet clear how the various Apo E isoforms act to impact upon AD pathophysiology. However, Apo E is also a metal chelator (Miyata and Smith, 1996). This property may contribute to the ability of Apo E to protect neuronal cell cultures against both H_2O_2 and Aβ-mediated toxicity, with the isoform ε2 giving greater protection than ε3, with ε4 giving the least protection (Miyata and Smith, 1996). The level of protection parallels the ability of the isoforms to coordinate metal ions and to inhibit Zn- or Cu-induced aggregation of Aβ (Moir *et al.*, 1999).

The cause of the neuronal cell loss in AD may be related to increased oxidative stress from excessive free-radical generation (Martins *et al.*, 1986; Smith *et al.*, 1997; Sayre *et al.*, 2000). The major source of free-radical production in the brain is from the reactivity of transition metals, Cu and Fe. These metals do not exist in a free ionic form but exert biochemical redox activity in the bound state. Despite being vital, if the redox activity of Cu and Fe is not strictly regulated, toxic ROS such as the hydroxyl radical (•OH) can be generated (Halliwell and Gutteridge, 1999). High oxygen consumption, relatively low antioxidant levels, and limited regenerative capacity results in brain tissue being particularly susceptible to oxidative damage from ROS. There is a large body of evidence indicating that the homeostasis of Zn, Cu, and Fe and their respective binding proteins are significantly altered in the AD brain (Atwood *et al.*, 1999). Microparticle-induced x-ray emission analysis of the cortical and accessory basal nuclei of the amygdala indicated these metals accumulate in the neuropil of the AD brain in concentrations that are three- to fivefold increased compared with age-matched controls (Lovell *et al.*, 1998). Importantly, these metal ions are normally concentrated in those regions of the brain most affected by AD pathology. Evidence for abnormal Cu homeostasis in AD includes a 2.2-fold increase in the concentration of Cu in cerebrospinal fluid (Basun *et al.*, 1991), and an accompanying increase of plasma Cu in AD (Squitti *et al.*, 2002). There is an extensive literature describing abnormal levels of Fe and Fe-binding proteins in AD (Bishop *et al.*, 2002). Importantly, the Fe that is found within the amyloid deposits of human brain and in amyloid-bearing APP transgenic mice is redox active (Smith *et al.*, 1997, 1998). Raman spectroscopy studies have demonstrated that Zn^{2+} and Cu^{2+} are coordinated to the histidine residues of the deposited Aβ in the senile plaque cores from diseased brain tissue and that the sulfur atom of Met35 of Aβ is oxidized, indicative of a pro-oxidant environment (Dong *et al.*, 2003).

2.2. APP/Aβ and Metal Homeostasis

To prevent transition metal-mediated oxidative stress, cells have evolved complex metal transport systems that deliver Cu and Fe to metalloenzymes and proteins. These include mammalian Cu chaperones that are involved in intracellular Cu trafficking to Cu/Zn superoxide dismutase and the Wilson's disease Cu ATPase (Waggoner *et al.*, 1999). The chaperones direct the Cu atoms to specific intracellular proteins, which results in unbound Cu being essentially absent in the intracellular environment (Rae *et al.*, 1999). Therefore, cuproproteins have an important role in maintaining cellular Cu metabolism (Andrews, 2001). APP is a ubiquitously expressed high-turnover protein, and data suggests that APP and Aβ may have some role in metal homeostasis. Using transgenic mouse models, it has been shown that overexpression of the carboxyl-terminal fragment of APP, containing Aβ, results in significantly reduced copper and iron levels in transgenic mouse brain; overexpression of APP in Tg2576 transgenic mice results in significantly reduced copper, but not iron (Maynard *et al.*, 2002), while APP knockout mice have increased copper levels in the brain and liver (White *et al.*, 1999). Consistent with this observation is the rise of intracellular copper in primary mouse cortical neurons and embryonic fibroblasts as a result of gene knockout of APP and APLP2 (Bellingham *et al.*, 2004). Recently, gene expression profiling has been used to show that APP expression is increased in response to chronic copper overload (Armendariz *et al.*, 2004), whereas copper deficiency leads to a downregulation of the APP gene (Bellingham *et al.*, 2004b). Copper levels can modulate APP processing, such that higher copper levels result in a reduction in Aβ production and a consequential increase in the non-amyloidogenic p3 form of the peptide and an increase in the secretion of the APP ectodomain (Borchardt *et al.*, 1999). Consistent with these cell-based findings are studies with APP transgenic mice that had their copper levels raised either by dietry supplement (Bayer *et al.*, 2003) or genetic manipulation (Phinney *et al.*, 2003), which showed lower levels of amyloid burden.

Independent copper binding sites have been identified on both Aβ and APP, which we hypothesize subserve a physiologic relationship with Cu. The molecular structure of the APP copper-binding domain has been solved and found to contain a novel copper-binding site that favors Cu(I) coordination (Barnham *et al.*, 2003a). The solvent accessibility of this site, structural homology to copper chaperones, and the role of APP in neuronal copper homeostasis are consistent with APP acting as a neuronal metallotransporter. Additionally, a zinc-binding site has also been observed within this domain encompassing residues 181–198 (Bush *et al.*, 1993).

It has recently been shown that soluble APP is able to stimulate NMDA receptor activation (Xiong *et al.*, 2004); given the role that Cu plays in modulating NMDA receptor activity (Schlief *et al.*, 2005) and APP plays in regulating Cu levels, it is possible that this APP activation of the NMDA is mediated through modulating Cu concentrations at the synapse.

2.3. Cu^{2+}- and Zn^{2+}-Induced Aggregation of Aβ

In vitro studies have shown that Aβ will coordinate Cu^{2+}, Zn^{2+}, and Fe^{3+} with high affinity (Bush *et al.*, 1994b), explaining the presence of these metals in amyloid plaques. This study also showed stabilization by Cu^{2+} of an apparent Aβ1-40 dimer on gel chromatography. It has been shown that $^{65}Zn^{2+}$ is displaced from Aβ when co-incubated with excess Cu^{2+} (Clements *et al.*, 1996), and that Cu^{2+} and Zn^{2+} share a common binding site (Yang *et al.*, 2000). Atwood *et al.*, 1998) found that Cu^{2+} was bound to soluble Aβ via histidine residues, and the precipitation of soluble Aβ by Cu^{2+} was reversibly modulated by pH with mildly acid conditions (pH 6.6) promoting Cu^{2+}-mediated precipitation, whereas raising the pH dissolved precipitated Aβ:Cu^{2+} complexes. Zn^{2+}-induced aggregation of soluble Aβ at pH 7.4 *in vitro* was totally reversible with chelation (Cherny *et al.*, 1999). It has also been reported (Cherny *et al.*, 1999) that marked Cu^{2+}-induced aggregation of Aβ1-40 occurred as the solution pH was lowered from 7.4 to 6.8 and that the reaction was completely reversible with either chelation or raising the pH. Aβ1-40 was reported to bind three to four Cu^{2+} ions when precipitated at pH 7.0. Rapid, pH-sensitive aggregation occurred at low nanomolar concentrations of both Aβ1-40 and Aβ1-42 with submicromolar concentrations of Cu^{2+}. Unlike Aβ1-40, Aβ1-42 was precipitated by submicromolar Cu^{2+} concentrations at pH 7.4. Rat Aβ1-40 and histidine-modified human Aβ1-40 were not aggregated by Zn^{2+}, Cu^{2+}, or Fe^{3+}, indicating that histidine residues are essential for metal-mediated Aβ assembly. Subsequently, it was shown that Cu^{2+}- and Zn^{2+}-selective chelators enhanced the dissolution of amyloid deposits in postmortem brain specimens from AD subjects (Cherny *et al.*, 1999) and from amyloid precursor protein overexpressing transgenic mice (Cherny *et al.*, 2001), confirming the part played by these metal ions in cerebral amyloid assembly. In particular, Zn^{2+} efficiently induces aggregation of synthetic Aβ under conditions similar to the physiologic ones in the normal brain, that is, at nanomolar and submicromolar concentrations of Aβ and Zn^{2+}, respectively (Bush *et al.*, 1994b). Recently, it has been demonstrated that Aβ does not precipitate when trace metal ions are rigorously excluded (Huang *et al.*, 2004). The effect of Cu^{2+} on the aggregation of Aβ is ambiguous compared with Zn^{2+}· Cu^{2+} has been shown to be a strong inducer of Aβ aggregation under certain conditions (Garzon-Rodriguez *et al.*, 1997). In contrast with the Zn^{2+}-induced Aβ aggregation that occurs over a wide pH range (5.5–7.5), the Cu^{2+}-induced aggregation occurs primarily at mildly acidic pH (Atwood *et al.*, 1998).

2.4. The Metal Binding Site(s) of Aβ

In aqueous solution, Aβ undergoes rapid conformational exchanges making it difficult to determine the precise nature of the metal binding sites. The question has been approached using various spectroscopic techniques, including Raman, CD, and magnetic resonance. Raman spectroscopy has been used to study the binding modes of Zn^{2+} and Cu^{2+} to Aβ in solution and insoluble aggregates (Miura *et al.*, 2000). Two different modes of metal-Aβ binding were reported, one characterized by metal binding to the imidazole N_τ atom of histidine, producing insoluble aggregates, the other involving metal binding to the N_π, but not the N_τ atom of histidine as well as to main-chain amide nitrogens, giving soluble complexes. Zn^{2+} binds to Aβ only via the N_τ regardless of pH, while the Cu^{2+} binding mode is pH dependent. At mildly acidic pH, Cu^{2+} binds to Aβ in the former mode, whereas the latter mode is predominant at neutral pH. From this data, it was proposed that the transition from one binding mode to the other explained the strong pH dependence of Cu^{2+}-induced Aβ aggregation (Miura *et al.*, 2000). Raman microscopy was also employed to study the metal binding sites in amyloid plaque cores, using the spectra-structure correlations for Aβ-transition metal binding (Dong *et al.*, 2003). Again it was observed that Zn^{2+} was coordinated to the histidine N_τ and the Cu^{2+} to the N_π, confirming that the metal binding mode was the same in both the synthetic peptide and its aggregates and the naturally occurring plaques.

Huang *et al.* (1999a) used multifrequency electron paramagnetic resonance (EPR; S-, X-, and Q-band) to show that copper coordinates tightly to Aβ1-40 and that an approximately equimolar mixture of peptide and $CuCl_2$ produced a single Cu^{2+}-peptide complex. Computer simulation of the S band spectrum with an axially symmetrical spin Hamiltonian and the *g* and *A* matrices ($g_\|$, 2.295; $g_\perp$, 2.073; $A_\|$, 163.60; $A_\perp$ 10.0×10^{-4} cm^{-1}) suggested a tetragonally distorted geometry, which is commonly found in type 2 copper proteins. Expansion of the $M_I = -1/2$ resonance revealed nitrogen ligand hyperfine coupling. Computer simulation of these resonances indicated the presence of at least three nitrogen atoms. This and the magnitude of the $g_\|$ and $A_\|$ values, together with Peisach and Blumberg plots (Peisach and Blumberg, 1974), are consistent with a fourth equatorial ligand binding to copper via an oxygen. Therefore, the coordination sphere for the copper-peptide complex was considered to be 3N1O.

These authors also used EPR spectroscopy to measure residual Cu^{2+} remaining after incubating stoichiometric ratios of $CuCl_2$ with Aβ1-40. There was a 76% loss of the Cu^{2+} signal, compatible with peptide-mediated reduction of Cu^{2+} to diamagnetic Cu^+, which is undetectable by EPR, in agreement with the corresponding concentration of Cu^+ measured by bioassay. There was no evidence of free, uncoordinated Cu^{2+} remaining after addition of the peptide.

Using a combination of NMR and EPR spectroscopy, we (Curtain *et al.*, 2001) proposed a structure for the high-affinity site and drew some conclusions about the interaction of the peptide with lipids and its modification by Cu^{2+}, Zn^{2+}, and pH. NMR studies on Aβ1-28 and Aβ1-40/2 indicated that both peptides were undergoing significant conformational exchange in aqueous solution. NMR and EPR spectra were also recorded for Aβ1-28 where the Nε2 nitrogens of the imidazole ring of the His residues 6, 13, and 14 were methylated (Me-Aβ1-28). The NMR spectra of Me-Aβ1-28 were virtually identical to Aβ1-28, the only significant differences being three strong singlets in the ^1H spectrum at 3.80, 3.82, and 3.83 ppm from the methyl groups attached to the His imidazole rings. A precipitate formed when Zn^{2+} was added to the solutions of Aβ1-28 or Aβ in PBS. NMR spectra of the supernatant of Aβ1-28 treated with Zn^{2+} showed that peaks assigned to C2H and C4H of His6, His13, and His14 of Aβ1-28 had broadened significantly. However, there was little or no change in the rest of the spectrum compared with Aβ1-28 prior to the addition of Zn^{2+}. This broadening of the NMR peaks due to the histidine residues is the result of the interaction of these residues with Zn^{2+}. The histidyl side chain is a well established ligand of zinc in proteins and peptides, and this result suggested that three of the ligands bound to Zn^{2+} were most likely the imidazole rings of the histidine residues. The broadening of these peaks is the result of chemical exchange between free and metal-bound states or among different metal-bound states. The broadening of peaks indicated intermediate exchange, which on the NMR timescale suggests that the metal binding affinity is in the micromolar range, in agreement with the low-affinity site described previously (Bush *et al.*, 1994b). When Cu^{2+} or Fe^{3+} was titrated into an aqueous solution of Aβ1-28, similar changes were observed in the ^1H spectrum, with the peaks assigned to the C2H and C4H of His6, His13, and His14 disappearing from the spectrum. A slight broadening of all peaks in the spectrum (associated with the paramagnetism of Cu^{2+} and Fe^{3+}) was also observed, but there were no other major changes after the addition of the metal ions. Metal-induced precipitation blocked attempts to saturate the metal-binding site. The precipitate made the collection of NMR spectra difficult, and few definitive conclusions could be drawn from spectra of peptide remaining in solution. When Zn^{2+} was added to an aqueous solution of Me-Aβ1-28, the changes observed in the spectrum were identical to those observed for Zn^{2+} added to Aβ1-28, but there was no visible precipitate.

In aqueous solution and lipid environments, coordination of metal ions to Aβ is the same, with His6, His13, and His14 all involved. The X-band EPR spectrum of Cu^{2+} bound to the peptides had the unsplit intense $g_{\perp}$ resonance characteristic of an axially symmetric square planar 3N1O or 4N coordination, $g_{\parallel} = 2.28$ and $g_{\perp} = 2.03$, $A_{\parallel} = 173.8$ Gauss. A notable finding was that increasing the Cu^{2+} concentration above ~0.3 mol/mol of peptide induced line broadening in the Cu^{2+} EPR spectra, over a pH range of 5.5 to 7.5, suggesting the presence of dipolar or exchange effects. These would be observed if two or more Cu ions were within approximately 6 Å of each other. These effects could be explained if at Cu^{2+}/peptide molar ratios >0.3, Aβ coordinated a second Cu^{2+} atom cooperatively. The effects were abolished if the histidine residues were methylated at either Nδ1 or Nε2, suggesting that bridging histidine residues were being formed (Curtain *et al.*, 2001; Tickler *et al.*, 2005). One consequence of coordination by a metal ion to the Nδ1 of a histidine residue is a reduction in the pK_a of Nε2 NH, making this nitrogen more suitable for metal binding (Sundberg and Martin, 1974), resulting in a histidine residue that can bridge metal ions; a good example being His63 at the active site of superoxide dismutase (Parge *et al.*, 1992). Similar bridging histidine residues have been proposed in the octarepeat region of the prion protein (Viles *et al.*, 1999), which has been shown to possess significant superoxide dismutase (SOD) activity in the presence of Cu^{2+} (Brown *et al.*, 1999). The line-broadening effects observed in the EPR spectra at Cu^{2+}/Aβ molar fractions up to 1.0 by us were not observed by Syme *et al.* (2004), or Antzutkin (2004). However, it is relevant that NaCl has a marked effect on metal-induced aggregation of Aβ (Huang *et al.*, 1997; Narayanan and Reif, 2005). We (Curtain *et al.*, 2001) obtained spectra from samples in phosphate-buffered saline at pH 7.4, Antzutkin (2004) adjusted the pH of the sample to pH 7.4 and dialyzed against distilled water, while Syme *et al.* (2004) used the nonphysiologic ethyl morpholine buffers. Recently, the existence of bridging Cu^{2+} dimers in Aβ1-28 in pH 7.4 PBS at Cu^{2+}/peptide ratios > 0.6 has been argued from the occurrence of *g* ~4 transitions in the EPR spectra (Curtain, private communication) with (Smith *et al.*, 2006). Neither *g* ~2 line broadening nor the *g* ~4 transitions occurred in pH 7.4 ethyl morpholine buffer, further emphasizing the importance of using physiologic buffers in studying Aβ-metal ion interactions. Similar line-broadening phenomena to that observed by us have been observed in the EPR spectra of imidazole bridged-copper complexes designed as SOD mimetics (Ohtsu *et al.*, 2000).

The bridging histidine may be responsible for the reversible metal-induced aggregation of Aβ. The bridging histidine residues may also explain the multiple metal-binding sites observed for each peptide and the high degree of cooperativity evident for subsequent metal binding (Curtain *et al.*, 2001). With three histidines bound to the metal center, a large scope exists for metal-mediated cross-linking of the peptides leading to aggregation, which will be reversible when the metal is removed by chelation. It is quite possible that metal-induced precipitation of Aβ is quite different from that induced by prolonged incubation of monomeric peptide in the putative absence of metal. For example, Miura *et al.* (2000) strongly suggested that the metal-induced aggregation of Aβ was promoted by cross-linking of the peptides through metal-His$\{N_\tau\}$ bonds, most likely through His$\{N_\tau\}$-metal-His$\{N_\tau\}$ bridges at three histidine residues.

Observations that rat Aβ by which differs from human Aβ by three substitutions, with Arg5, Tyr10, and His13 of human Aβ becoming Gly5, Phe10, and Arg13 (Shivers *et al.*, 1988), does not reduce Cu^{2+} and Fe^{3+}, is not readily precipitated by Zn^{2+} or Cu^{2+}, does not produce ROS as strongly as the human sequence, and does not produce plaques highlight the importance of the three histidine residues (Atwood *et al.*, 1998; Huang *et al.*, 1999b). Rat Aβ forms a metal complex via two histidine residues and two oxygen ligands rather than three histidine residues and one oxygen ligand, compared with human Aβ where the side-chain of His13 of human Aβ is ligated to the metal ion. This was borne out by the EPR spectrum, which was typical of a square planar 2N2O Cu^{2+} coordination (Curtain *et al.*, 2001).

Further advances in understanding the N coordination of Cu^{2+} will require more sophisticated EPR techniques than have been used so far, supported by input from other methods such as extended X-ray absorption fine structure (EXAFS). Equally, there remains uncertainty as to the nature of the potential O ligand. Raman data (Miura *et al.*, 2000) suggest that the ligand was the O of the tyrosine hydroxyl, which plays an important role in Cu^{2+}-induced aggregation. They were able to assign the 1504 cm^{-1} band in the Raman spectra of insoluble Cu^{2+}-Aβ1-16 aggregates to Cu^{2+}-bound tyrosinate, and the high intensity of the 1604 cm^{-1} band was attributed to a contribution from the Y8a band of tyrosinate. Unlike Zn^{2+}, Cu^{2+} binds to tyrosine in the insoluble aggregates of Aβ1-16. However, NMR (Syme *et al.*, 2004) and EPR data combined with mutagenesis (Karr *et al.*, 2005) has suggested that Tyr10 does not coordinate Cu^{2+}. Again, a variety of different solution conditions were used for these experiments, Miura *et al.* (2000) used phosphate-buffered saline, which might have had the effect of encouraging peptide association. In considering the issue of monomeric versus dimeric Cu^{2+}, it is important to remember that Aβ may form oligomers and multimers in a variety of ways, some more relevant to its neurotoxicity than others (Roher *et al.*, 1996; Walsh *et al.*, 2002; Barnham *et al.*, 2004a; Cleary *et al.*, 2005).

2.5. Aβ Redox Activity

Oxidative stress markers characterize the neuropathology both of Alzheimer's disease and of amyloid-bearing transgenic mice. The neurotoxicity of Aβ has been linked to hydrogen peroxide generation in cell cultures by a mechanism that is still being fully described but is likely to be dependent on Aβ coordinating redox active metal ions. Huang *et al.* (1999a,b) showed that human Aβ directly produces H_2O_2 by a mechanism that involves the reduction of metal ions, Fe^{3+} or Cu^{2+}. Spectrophotometry was used to show that the Aβ peptide reduced Fe^{3+} and Cu^{2+} to Fe^{2+} and Cu^{+} and that molecular oxygen was then trapped by Aβ and reduced to H_2O_2 in a reaction that is driven by substoichiometric amounts of Fe^{2+} or Cu^{+}. In the presence of Cu^{2+} or Fe^{3+}, Aβ produced a positive thiobarbituric-reactive substance, compatible with the generation of the hydroxyl radical (•OH). Tabner *et al.* (2002) used spin-trapping to identify the radical produced by Aβ in the presence of Fe^{2+}, concluding that it was •OH. However, they also found •OH was produced in the presence of Fe^{2+} by Aβ25-35, which does not contain a strong metal binding site.

Incubation with Cu^{2+} causes SDS-resistant oligomerization of Aβ (Atwood *et al.*, 2004), which is also found in the neurotoxic soluble Aβ extracted from the AD brain. Atwood *et al.* (2004) found that Cu^{2+} induced SDS-resistant oligomers of Aβ gave a fluorescence signal characteristic of the cross-linking of the peptide's tyrosine 10. This finding was confirmed by directly identifying the dityrosine by electrospray ionization mass spectrometry and by the use of a specific dityrosine antibody. The addition of H_2O_2 strongly promoted Cu^{2+}-induced dityrosine cross-linking of Aβ1-28, Aβ1-40, and Aβ1-42, and Atwood *et al.* (2004) suggested that the oxidative coupling was initiated by interaction of H_2O_2 with a Cu^{2+} tyrosinate. The dityrosine modification is significant because it is highly resistant to proteolysis and would be important in increasing the structural strength of the plaques.

We have (Barnham *et al.*, 2004a) used density functional theory calculations to elucidate the chemical mechanisms underlying the catalytic production of H_2O_2 by Aβ/Cu and the production of dityrosine. Here, tyrosine 10 (Y10) was identified as the critical residue. This finding accords with the growing awareness that the O_2 activation ability of many cupro-enzymes is also coupled to the redox properties of tyrosine and the relative stability of tyrosyl radicals. The latter play important catalytic roles in photosystem II, ribonucleotide reductase, COX-2, DNA photolyase, galactose oxidase, and cytochrome C oxidase (Whittaker, 2003).

With ascorbate as the electron donor, the first step in the catalytic production of H_2O_2 is the reduction of Cu^{2+} to Cu^+, and we proposed that the transfer could take place via a proton-coupled electron transfer (PCET) mechanism (Barnham *et al.*, 2004a). Reactions involving PCET are being increasingly implicated in a range of biological systems, including charge transport in DNA and enzymatic oxygen production (Whittaker, 2003). In this system, the electron transfer involves both p- and d-orbitals on the ascorbate, Y10, and the copper ion, while proton transfer involves p-orbitals on the O_2-atom of ascorbate and the side-chain oxygen of Y10. The significant change in electron spin on the copper ion going from the ground state to the transition state suggests that the proton and the electron are transferred within different molecular orbitals, as is predicted to be necessary for PCET to occur (Cukier and Nocera, 1998).

The Cu/tyrosinate hypothesis was tested using an Aβ1-42 peptide with tyrosine 10 substituted with alanine (Y10A). Both peptides gave rise to similar ^{65}Cu EPR spectra with the strong single $g_\perp$ resonance characteristic of an axially symmetric square planar complex, although there was a significant increase in the $g_\parallel$ value of Y10A. The increase was probably due to some distortion of the coordination sphere because the oxygen ligand, which was possibly from Y10, was now derived from another oxygen donor (e.g., H_2O, phosphate, or carboxylate from the peptide). While wild-type Aβ1-42 rapidly reduces Cu^{2+} to Cu^+ in aqueous solution, with near-complete reduction taking 80 min, the mutation of Y10 to alanine markedly decreased the ability of Aβ to reduce Cu^{2+}. Further, spin trapping studies also confirmed the DFT observation that Y10 acts as a gate that facilitates the electron transfer needed to reduce Cu^{2+} to Cu^+. When the spin trap 2-methyl-2-nitrosopropane was added to the reaction mixture *w.t.* Aβ1-42/Cu^{2+}/ascorbate, a broad line triplet characteristic of a trapped carbon-centered radical bound to a peptide appeared in the EPR spectra. However, if Y10A peptide were substituted for the *w.t.*, formation of this triplet was inhibited.

2.6. The Effect of Metal Binding on the Interaction of Aβ with Membranes

Numerous reports have described the effects of Aβ on membranes and lipid systems and their possible roles in its neurotoxicity. Structural studies in different membrane-mimetic systems have demonstrated considerable variation in peptide conformation. There is much experimental evidence from far-ultraviolet circular dichroism spectroscopy and FT-IR spectroscopy that the Aβ peptides can be membrane associated in the β-configuration (Choo-Smith *et al.*, 1997), although there are reports of membrane-associated α-helices being found in the presence of gangliosides (McLaurin *et al.*, 1998), cholesterol (Ji *et al.*, 2002), and Cu^{2+} or Zn^{2+} (Curtain *et al.*, 2001, 2003).

This variability under different conditions can be understood because most of the amyloidogenic peptides have been identified as being exceptionally pleiomorphic in structure. As the cell membrane is a lipid mosaic, it is possible that the peptides will exhibit different structures with different properties in different parts of the mosaic. The pleiomorphism is highly relevant to the cytoxicity of the peptide, because factors influencing it could act as switches to determine whether the peptide is a β-sheet with the potential to form amyloid or be membrane surface seeking or a membrane-penetrant α-helix.

We have used a combination of EPR and CD spectroscopy to study the effect of metal ions, pH, and cholesterol on the interaction of Aβ with bilayer membranes (Curtain *et al.*, 2001; Curtain *et al.*, 2003). EPR spectroscopy, using spin-labeled lipid chains or protein segments, has been used extensively to study translational and rotational dynamics in biological membranes. Lipids at the hydrophobic interface between lipid and transmembrane protein segments and peptides in their monomeric and oligomeric states have their rotational motion restricted (Curtain *et al.*, 2003). This population of lipids can be resolved in the EPR spectrum as a motionally restricted component distinct from the fluid bilayer lipids, which can be quantified to give both the stoichiometry and selectivity of the first shell of lipids interacting directly with membrane-penetrant peptides. The stoichiometric data can give an estimate of the number of subunits in a membrane-penetrant oligomeric structure. Using this approach, it was shown that Aβ1-40 and Aβ1-42 bound to Cu^{2+} or Zn^{2+} penetrated bilayers of negatively charged, but not zwitterionic lipid, giving rise to such a partly immobilized component in the spectrum. When the peptide:lipid was increased, the relationship between the mole fraction of peptide and proportion of slow component was linear. Even at a fraction of 15%, all of the peptide was associated with the lipid, suggesting that the structure penetrating the lipid membrane was well defined. The lipid:peptide ratio is approximately 4:1. This stoichiometry can be satisfied by six helices arranged in a pore surrounded by 24 boundary lipids. In the presence of Zn^{2+}, Aβ1-40 and Aβ1-45 both inserted into the bilayer over the pH range 5.5–7.5, as did Aβ1-42 in the presence of Cu^{2+}. However, only Aβ40 penetrated the lipid bilayer in the presence of Cu^{2+} at pH 5.5–6.5; at higher pH, there was a change in the Cu^{2+} coordination sphere that inhibited membrane insertion. The addition of cholesterol upto 0.2 mole fraction of the total lipid inhibited insertion of both peptides under all conditions investigated. CD spectroscopy revealed that the Aβ peptides had a high α-helix content when membrane penetrant, but were predominately β-strand when not. It is also possible to gain an estimate of the secondary structure of the peptide from the degree of immobilization of the lipid in the shell; the more immobilized, the more likely the peptide is present as a β structure. Simulation of the spectra and calculation of the on-off rates suggested that the peptide was most likely penetrating as an α-helix (Curtain *et al.*, 2003). In membrane-mimetic environments, coordination of the metal ion is the same as in aqueous solution, with the three-histidine residues, at sequence positions 6, 13, and 14, all involved in the coordination, along with an oxygen ligand. As had been observed at Cu^{2+}/peptide molar ratios > 0.3 in aqueous solution, line broadening was detectable in the EPR spectra, indicating that the peptide was coordinating a second Cu^{2+} atom in a highly cooperative manner at a site less than 6 Å from the initial binding site. So, there appear to be two switches, metal ions (Zn^{2+} and Cu^{2+}) and negatively charged lipids, needed to change the conformation of the peptide from β-strand nonpenetrant to α-helix penetrant.

2.7. Toxic Mechanism of Aβ

Synthetic Aβ is toxic to cells in the presence of Cu^{2+}, but this toxicity is inhibited by catalase implicating H_2O_2 in the toxic pathway (Behl *et al.*, 1994; Opazo *et al.*, 2002). A simple way of generating H_2O_2 and other ROS is the interaction between redox-active metal ions such as Cu and Fe with O_2, which is why cells have developed elaborate defense mechanisms for dealing with these metals. When Cu^{2+} or Fe^{3+} coordinate Aβ, extensive redox chemical reactions take place, reducing the oxidation state of both metals and producing H_2O_2 from O_2 in a catalytic manner (Huang *et al.*, 1999a,b; Cuajungco *et al.*, 2000; Opazo *et al.*, 2002; Barnham *et al.*, 2004a). The generation of H_2O_2 in the presence of the reduced form of the metal ion sets up conditions for Fenton chemistry where the generation of the highly toxic OH• radical can occur (Huang *et al.*, 1999a).

The various forms of Aβ in the AD brain are usually oxidatively modified (Head *et al.*, 2001), and reaction with Cu^{2+} can induce covalent cross-linking of Aβ yielding soluble oligomers and adducts on the side chains (Atwood *et al.*, 2000a,b; Barnham *et al.*, 2003; Atwood *et al.*, 2004; Barnham *et al.*, 2004a). We have hypothesized that such oxidative modification may contribute to the release of abnormal soluble forms of Aβ from the membrane, which correlates with dementia (McLean *et al.*, 1999). In the normal brain, most of the Aβ is found associated with membranes (Cherny *et al.*, 1999). We hypothesize that when Aβ coordinates redox-active metal ions, subsequent oxidation releases soluble oxidized forms of the peptide that resist clearance. This may explain how zinc originating from the synapse becomes so enriched in amyloid in AD. Our model suggests that in health, soluble Aβ is not present in the cortical synapse. In AD, soluble oxidized Aβ accumulates within the synapse where the high Zn^{2+} concentrations precipitate the copper/iron coordinated Aβ, creating a reservoir of potentially toxic Aβ that is in dissociable equilibrium with the soluble pool. The Zn^{2+} in the amyloid mass partially quenches H_2O_2 production, which explains why plaque amyloid burden correlates poorly with clinical dementia (Cuajungco *et al.*, 2000), whereas soluble Aβ levels correlate well with clinical severity (McLean *et al.*, 1999).

One of the consequences of oxidative modification to Aβ is the formation of a sulfoxide on Met35, and Met(O)Aβ peptide has been isolated from AD amyloid brain deposits (Naslund *et al.*, 1994; Kuo *et al.*, 2001; Dong *et al.*, 2003). A Raman spectroscopic study of senile plaque cores isolated from diseased brains has shown that much of the Aβ in these deposits contained methionine sulfoxide with copper and zinc coordinated to the histidine residues (Dong *et al.*, 2003).

In vitro, the methionine at position 35 can act as an electron donor for the reduction of the metal ions bound to Aβ (Curtain *et al.*, 2001). Although there are several potential electron donors such as GSH and ascorbic acid, *in vivo* it is likely that M35 occupies a privileged position being part of the Aβ sequence. Indeed, it has been shown using photoaffinity cross-linking experiments that in β-strand fibrils, M35 is cross-linked to the N-terminal region of Aβ (Egnaczyk *et al.*, 2001), that is, within close proximity to the redox-active metal binding site. When it is missing as in Aβ(1-28), the addition of exogenous methionine permits redox reactions to proceed, but with slower kinetics. When M35 is sequestered within a lipid environment, there is also no metal reduction. M35 oxidation also alters the physical properties of the peptide. Met(O)Aβ is more soluble in aqueous solution, and there is a disruption of the local helical structure when the peptide is dissolved in SDS micelles (Watson *et al.*, 1998). The formation of trimers and tetramers by Met(O)Aβ is significantly attenuated (Palmblad *et al.*, 2002), and fibril formation is inhibited (Hou *et al.*, 2002; Palmblad *et al.*, 2002). We (Barnham *et al.*, 2003a) showed by solid-state NMR that when Aβ coordinates and reduces Cu^{2+} to Cu^{+}, the M35 is oxidized. Although the Cu^{2+} coordination of the oxidized peptide is identical to non-oxidized Aβ and it will produce H_2O_2, it cannot penetrate lipid bilayers either in the presence or absence of Cu^{2+} or Zn^{2+}. On the other hand, Met(O)Aβ is toxic to neuronal cell cultures, a toxicity that is rescued by catalase and the metal chelator clioquinol. These results suggest that fibril formation and membrane penetration by Aβ could be epiphenomena and that the main requirement for cytoxicity is redox competence. In this connection, it is important to note that the oxidized M35 has the potential for further reduction to the sulfone and could thus still act as a Cu^{2+} reductant, acting *in vivo* in concert with agents such as ascorbic acid and GSH.

2.8. Aβ Membrane Association and Cytotoxicity

We (Ciccotosto *et al.*, 2004) further probed the role of M35 by preparing Aβ1-42 in which it was replaced with valine (AβM35V). The neurotoxic activity on primary mouse neuronal cortical cells of this peptide was enhanced, and this diminished cell viability occurred at a much faster rate compared with wild-type Aβ1-42. When cortical cells were treated with the peptides for only a short 1 h duration to minimize the incidence of cell death, and the amount of peptide bound to cortical cell extracts was quantitated by Western blotting, it was found that twice as much AβM35V compared with wild-type Aβ peptide bound to the cells after a 1-h cell exposure. It was suggested that the increased toxicity was related to the increased binding.

AβM35V bound Cu^{2+} with the same coordination sphere as *w.t.* Aβ and produced similar amounts of H_2O_2 as Aβ1-42 *in vitro*. The neurotoxic activity was rescued by catalase. The redox activity of the mutated peptide was followed by measuring the decline in time of the strength of the Cu^{2+}-AβM35V EPR signal, which showed that the reduction of Cu^{2+} to the EPR silent Cu^{+} was much slower compared with *w.t.* Aβ1-42, confirming that the M35 residue in Aβ42 plays an important part in the redox behavior of this peptide in solution. Like *w.t.* Cu^{2+}-Aβ1-42, Cu^{2+}-AβM35V inserted into a spin-labeled lipid bilayer gave a partially immobilized component in the EPR spectrum. This component had a narrower linewidth than that found for the similar component obtained with *w.t.* Cu^{2+}-Aβ1-42, suggesting that the valine substitution made the mutant peptide less rigid in the bilayer region and possibly easier to insert, thus explaining the increased cell membrane binding. The on- and off-rate constants estimated from the simulation experiments showed that AβM35V had a higher affinity for the lipid bilayer compared with Aβ42. CD analysis showed that AβM35V had a higher proportion of β-sheet structure and random coil than Aβ1-42, which would also suggest a more flexible structure in the bilayer. In summary, these and the results described above tell us that the wild-type Aβ, its oxidized form, Met(O)Aβ, and the mutant peptide, AβM35V, induce cell death via similar pathways that are metal-dependent and can generate H_2O_2 in the absence of a methionine residue. Fibril formation as a toxic species is not responsible for cell death. Membrane association *per se* may play a part in localizing the peptide; perhaps in domains particularly susceptible to oxidative damage.

The important role that membrane association plays in the toxicity was emphasized by the observation that Aβ peptides where the histidine residues were methylated (Me-His Aβ) with the aim of inhibiting metal mediated histidine bridged oligomerization resulted in nontoxic forms of Aβ. Even though the modified peptides were four times more efficient than wild-type Aβ in generating H_2O_2, unlike wild-type Aβ they did not bind to the cell membrane (Tickler *et al.*, 2005), indicating that the redox-associated toxicity of Aβ is a site-specific phenomena.

2.9. Therapeutic Potential of Inhibiting Aβ-Metal Interactions

Current U.S. FDA–approved drugs for AD provide modest functional improvement (e.g., acetylcholinesterase inhibitors) but do not retard the progression of the underlying disease. Hence, there is an urgent need to identify drugs that confront the disease in the central pathway of its pathogenesis.

Although we have identified metal-Aβ interactions as likely being central to AD pathogenesis, we are only beginning to understand the way by which the peptide coordinates up to 3.5 metal ions. Nevertheless, the efficacy of clioquinol (CQ; 5-chloro-7-iodo-8-hydroxyquinoline) *in vitro* and *in vivo* (see below) has encouraged us to proceed by empirical observation as we gather information about the qualities of the compounds that will attenuate metal-Aβ interactions. Broadly speaking, there are two potential classes of molecule capable of preventing such pathologic metal interactions. The first involves inhibiting Aβ-metal interactions by selectively occupying the metal-binding site on Aβ, thus preventing Cu/Zn/Fe coordination. The initial metal binding site coordinates His6, His13, and His14. However, these residues are natively unstructured in the absence of metal ions, and as a result the design of classic "lock and key" inhibitors is problematic. Interestingly, the epitopes of the vaccines that have been reported to inhibit Aβ cerebral accumulation in transgenic mice involve residues in this region of the peptide (1-15) (Spooner *et al.*, 2002).

A second pharmacological approach is to identify a class of molecules that will effectively compete with the peptide for the metal ions, that is, a metal-protein attenuating compound (MPAC). Cherny *et al.* (1999) have shown using a range of commonly available metal chelators [tetrakis-(2-pyridylmethyl)-ethylenediamine (TPEN), ethylene glycol bis(2-aminoethyl ether)-*N,N,N'N'*-tetraacetic acid (EGTA), and bathcuporine] that such compounds with a reasonable affinity for Cu, Zn, and Fe can solubilize deposited Aβ from the postmortem brain tissue of AD patients. Conversely, chelation of Ca and Mg inhibited solubilization of Aβ. However, traditional hydrophilic chelators do not pass well through the blood-brain barrier (BBB).

Our requirement for a metal-attenuating compound as a possible therapeutic for AD should not be confused with the concept of "chelation therapy" for the treatment of AD. "Chelation therapy" is a term associated with the removal of bulk metals such as in Wilson's disease (Cu) and β-thallesemia (Fe). The breakdown in metal homeostasis in these diseases leads to tissue saturation of metal. The mechanism of action of such compounds is that metal is sequestered peripherally by the chelators and cleared by excretion. In AD, the metals become repartitioned into the amyloid mass. This is probably why treatment with CQ (see below) induced an increase ($\approx$15%) in brain Cu and Zn levels in transgenic mice (Cherny *et al.*, 2001) and also induced a similar increase in plasma Zn levels in AD subjects (+30%) (Ritchie *et al.*, 2003). The intention of the metal-protein attenuating compound (MPAC) is to disrupt an abnormal metal-protein interaction to achieve a subtle repartitioning of metals and a subsequent normalization of metal distribution. Once the toxic cycle is inhibited, endogenous clearance processes can cope more effectively with the accumulated Aβ.

CQ was given via oral gavage over a 9-week period in a blind study to Tg2576 transgenic mice (Cherny *et al.*, 2001). The results showed a 49% decrease in brain Aβ burden compared with nontreated controls, there was no evidence of any toxicity, and the general health and body weight parameters were more stable in the treated animals. Treatment with CQ did not lead to a systematic decrease in metal levels most likely due to its moderate binding affinities. The metal ions removed from Aβ are redistributed rather than excreted; as a result, Cu and Zn levels rose ~15% in the treated transgenic mice, possibly because the metals were no longer prevented from normal cellular uptake by being incorporated into the amyloid mass. This suggests that CQ is able to some extent to redress the balance in metal homeostasis that breaks down with age. Interestingly, in the same study, triethylenetetramine (TETA), a hydrophilic high-affinity metal chelator that is incapable of crossing the BBB and has been used to treat Wilson's disease, did not inhibit Aβ deposition, indicating that systemic depletion of metal ions ("chelation therapy") is not likely to be an effective therapeutic strategy for the treatment of AD.

The success of CQ in the transgenic mouse trials has encouraged the use of this molecule in clinical trials. The effects of oral CQ treatment in a randomized, double-blind, placebo-controlled pilot phase 2 clinical trial of moderately severe AD patients were evaluated (Ritchie *et al.*, 2003). Thirty-six subjects were randomized (18 placebo and 18 CQ, with 32 completions) and stratified into more severely or less severely affected groups. The effect of treatment was statistically significant in preventing cognitive deterioration over 36 weeks in the more severely affected patients (baseline ADAS-cog $\geq$ 25). The performance of the less severely affected group (ADAS-cog $<$ 25) deteriorated negligibly over this interval, so cognitive changes could not be discriminated in this stratum. Plasma Aβ42 declined in the CQ group but increased in the placebo group ($p < 0.001$). Plasma Zn levels rose significantly (~30 %) in the CQ group. These results reproduce the lowering of plasma Aβ and the paradoxical rise in brain zinc observed in APP transgenic mice treated with the drug. The drug was generally well tolerated by participants.

References

Andrews, N. C. (2001). Mining copper transport genes. *Proc Natl Acad Sci USA* 98: 6543–6545.

Antzutkin, O. N. (2004). Amyloidosis of Alzheimer's Abeta peptides: solid-state nuclear magnetic resonance, electron paramagnetic resonance, transmission electron microscopy, scanning transmission electron microscopy and atomic force microscopy studies. *Magn Reson Chem* 42: 231–246.

Armendariz, A. D., Gonzalez, M., Loguinov, A. V. and Vulpe, C. D. (2004). Gene expression profiling in chronic copper overload reveals upregulation of Prnp and App. *Physiol Genomics* 20: 45–54.

Atwood, C. S., Moir, R. D., Huang, X., Scarpa, R. C., Bacarra, N. M., Romano, D. M., Hartshorn, M. A., Tanzi, R. E. and Bush, A. I. (1998). Dramatic aggregation of Alzheimer abeta by Cu(II) is induced by conditions representing physiological acidosis. *J Biol Chem* 273: 12817–12826.

Atwood, C. S., Huang, X., Moir, R. D., Tanzi, R. E. and Bush, A. I. (1999). Role of free radicals and metal ions in the pathogenesis of Alzheimer's disease. *Met Ions Biol Syst* 36: 309–364.

Atwood, C. S., Huang, X., Khatri, A., Scarpa, R. C., Kim, Y. S., Moir, R. D., Tanzi, R. E., Roher, A. E. and Bush, A. I. (2000a). Copper catalyzed oxidation of Alzheimer Abeta. *Cell Mol Biol (Noisy-Le-Grand)* 46: 777–783.

Atwood, C. S., Scarpa, R. C., Huang, X., Moir, R. D., Jones, W. D., Fairlie, D. P., Tanzi, R. E. and Bush, A. I. (2000b). Characterization of copper interactions with alzheimer amyloid beta peptides: identification of an attomolar-affinity copper binding site on amyloid beta1–42. *J Neurochem* 75: 1219–1233.

Atwood, C. S., Perry, G., Zeng, H., Kato, Y., Jones, W. D., Ling, K. Q., Huang, X., Moir, R. D., Wang, D., Sayre, L. M., Smith, M. A., Chen, S. G. and Bush, A. I. (2004). Copper mediates dityrosine cross-linking of Alzheimer's amyloid-beta. *Biochemistry* 43: 560–568.

Barnham, K. J., Ciccotosto, G. D., Tickler, A. K., Ali, F. E., Smith, D. G., Williamson, N. A., Lam, Y. H., Carrington, D., Tew, D., Kocak, G., Volitakis, I., Separovic, F., Barrow, C. J., Wade, J. D., Masters, C. L., Cherny, R. A., Curtain, C. C., Bush, A. I. and Cappai, R. (2003a). Neurotoxic, redox-competent Alzheimer's {beta}-amyloid is released from lipid membrane by methionine oxidation. *J Biol Chem* 278: 42959–42965.

Barnham, K. J., McKinstry, W. J., Multhaup, G., Galatis, D., Morton, C. J., Curtain, C. C., Williamson, N. A., White, A. R., Hinds, M. G., Norton, R. S., Beyreuther, K., Masters, C. L., Parker, M. W. and Cappai, R. (2003b). Structure of the Alzheimer's disease amyloid precursor protein copper binding domain. A regulator of neuronal copper homeostasis. *J Biol Chem* 278: 17401–17407.

Barnham, K. J., Haeffner, F., Ciccotosto, G. D., Curtain, C. C., Tew, D., Mavros, C., Beyreuther, K., Carrington, D., Masters, C. L., Cherny, R. A., Cappai, R. and Bush, A. I. (2004a). Tyrosine gated electron transfer is key to the toxic mechanism of Alzheimer's disease beta-amyloid. *FASEB J* 18: 1427–1429.

Barnham, K. J., Masters, C. L. and Bush, A. I. (2004b). Oxidative Stress in Neurodegenerative diseases. *Nat Rev Drug Disc* 3: 205–214.

Basun, H., Forssell, L. G., Wetterberg, L. and Winblad, B. (1991). Metals and trace elements in plasma and cerebrospinal fluid in normal aging and Alzheimer's disease. *J Neural Transm Park Dis Dement Sect* 3: 231–258.

Bayer, T. A., Schafer, S., Simons, A., Kemmling, A., Kamer, T., Tepests, R., Eckert, A., Schussel, K., Eikenberg, O., Sturchler-Pierrat, C., Abramowski, D., Staufenbiel, M. and Multhaup, G. (2003). Dietary Cu stabilizes brain superoxide dismutase 1 activity and reduces amyloid A{beta} production in APP23 transgenic mice. *Proc Natl Acad Sci USA* 100: 14187–14192.

Behl, C., Davis, J. B., Lesley, R. and Schubert, D. (1994). Hydrogen peroxide mediates amyloid beta protein toxicity. *Cell* 77: 817–827.

Bellingham, S. A., Ciccotosto, G. D., Needham, B. E., Fodero, L. R., White, A. R., Masters, C. L., Cappai, R. and Camakaris, J. (2004a). Gene knockout of amyloid precursor protein and amyloid precursor-like protein-2 increases cellular copper levels in primary mouse cortical neurons and embryonic fibroblasts. *J Neurochem* 91: 423–428.

Bellingham, S. A., Lahiri, D. K., Maloney, B., La Fontaine, S., Multhaup, G. and Camakaris, J. (2004b). Copper depletion down-regulates expression of the Alzheimer's disease amyloid-beta precursor protein gene. *J Biol Chem* 279: 20378–20386.

Bishop, G. M., Robinson, S. R., Liu, Q., Perry, G., Atwood, C. S. and Smith, M. A. (2002). Iron: a pathological mediator of Alzheimer disease? *Dev Neurosci* 24: 184–187.

Borchardt, T., Camakaris, J., Cappai, R., Masters, C. L., Beyreuther, K. and Multhaup, G. (1999). Copper inhibits beta-amyloid production and stimulates the non-amyloidogenic pathway of amyloid-precursor-protein secretion. *J Biol Chem* 344: 461–467.

Brown, D. R., Wong, B. S., Hafiz, F., Clive, C., Haswell, S. J. and Jones, I. M. (1999). Normal prion protein has an activity like that of superoxide dismutase. *J Biol Chem* 344: 1–5.

Bush, A. I. (2003). The metallobiology of Alzheimer's disease. *Trends Neurosci* 26: 207–214.

Bush, A. I., Multhaup, G., Moir, R. D., Williamson, T. G., Small, D. H., Rumble, B., Pollwein, P., Beyreuther, K. and Masters, C. L. (1993). A novel zinc(II) binding site modulates the function of the beta A4 amyloid protein precursor of Alzheimer's disease. *J Biol Chem* 268: 16109–16112.

Bush, A. I., Pettingell, W. H., Jr., Paradis, M. D. and Tanzi, R. E. (1994a). Modulation of Abeta adhesiveness and secretase site cleavage by zinc. *J Biol Chem* 269: 12152–12158.

Bush, A. I., Pettingell, W. H., Multhaup, G., d Paradis, M., Vonsattel, J. P., Gusella, J. F., Beyreuther, K., Masters, C. L. and Tanzi, R. E. (1994b). Rapid induction of Alzheimer Abeta amyloid formation by zinc. *Science* 265: 1464–1467.

Cherny, R. A., Legg, J. T., McLean, C. A., Fairlie, D. P., Huang, X., Atwood, C. S., Beyreuther, K., Tanzi, R. E., Masters, C. L. and Bush, A. I. (1999). Aqueous dissolution of Alzheimer's disease Abeta amyloid deposits by biometal depletion. *J Biol Chem* 274: 23223–23228.

Cherny, R. A., Atwood, C. S., Xilinas, M. E., Gray, D. N., Jones, W. D., McLean, C. A., Barnham, K. J., Volitakis, I., Fraser, F. W., Kim, Y., Huang, X., Goldstein, L. E., Moir, R. D., Lim, J. T., Beyreuther, K., Zheng, H., Tanzi, R. E., Masters, C. L. and Bush, A. I. (2001). Treatment with a copper-zinc chelator markedly and rapidly inhibits beta-amyloid accumulation in Alzheimer's disease transgenic mice. *Neuron* 30: 665–676.

Choo-Smith, L. P., Garzon-Rodriguez, W., Glabe, C. G. and Surewicz, W. K. (1997). Acceleration of amyloid fibril formation by specific binding of Abeta-(1-40) peptide to ganglioside-containing membrane vesicles. *Biochemistry* 36: 12862–12868.

Ciccotosto, G. D., Tew, D., Curtain, C. C., Smith, D., Carrington, D., Masters, C. L., Bush, A. I., Cherny, R. A., Cappai, R. and Barnham, K. J. (2004). Enhanced toxicity and cellular binding of a modified amyloid-beta peptide with a methionine to valine substitution. *J Biol Chem* 279: 42528–42534.

Cleary, J. P., Walsh, D. M., Hofmeister, J. J., Shankar, G. M., Kuskowski, M. A., Selkoe, D. J. and Ashe, K. H. (2005). Natural oligomers of the amyloid-beta protein specifically disrupt cognitive function. *Nat Neurosci* 8: 79–84.

Clements, A., Allsop, D., Walsh, D. M. and Williams, C. H. (1996). Aggregation and metal-binding properties of mutant forms of the amyloid A beta peptide of Alzheimer's disease. *J Neurochem* 66: 740–747.

Cuajungco, M. P., Goldstein, L. E., Nunomura, A., Smith, M. A., Lim, J. T., Atwood, C. S., Huang, X., Farrag, Y. W., Perry, G. and Bush, A. I. (2000). Evidence that the beta-amyloid plaques of Alzheimer's disease represent the redox-silencing and entombment of Abeta by zinc. *J Biol Chem* 275: 19439–19442.

Cukier, R. I. and Nocera, D. G. (1998). Proton-coupled electron transfer. *Annu Rev Phys Chem* 49: 337–369.

Curtain, C. C., Ali, F. E., Smith, D. G., Bush, A. I., Masters, C. L. and Barnham, K. J. (2003). Metal ions, pH, and cholesterol regulate the interactions of Alzheimer's disease amyloid-beta peptide with membrane lipid. *J Biol Chem* 278: 2977–2982.

Curtain, C. C., Ali, F., Volitakis, I., Cherny, R. A., Norton, R. S., Beyreuther, K., Barrow, C. J., Masters, C. L., Bush, A. I. and Barnham, K. J. (2001). Alzheimer's disease amyloid-beta binds copper and zinc to generate an allosterically ordered membrane-penetrating structure containing superoxide dismutase-like subunits. *J Biol Chem* 276: 20466–20473.

Dong, J., Atwood, C. S., Anderson, V. E., Siedlak, S. L., Smith, M. A., Perry, G. and Carey, P. R. (2003). Metal binding and oxidation of amyloid-beta within isolated senile plaque cores: Raman Microscopic Evidence. *Biochemistry* 42: 2768–2773.

Egnaczyk, G. F., Greis, K. D., Stimson, E. R. and Maggio, J. E. (2001). Photoaffinity cross-linking of Alzheimer's disease amyloid fibrils reveals interstrand contact regions between assembled beta-amyloid peptide subunits. *Biochemistry* 40: 11706–11714.

Garzon-Rodriguez, W., Sepulveda-Becerra, M., Milton, S. and Glabe, C. G. (1997). Soluble amyloid Abeta-(1-40) exists as a stable dimer at low concentrations. *J Biol Chem* 272: 21037–21044.

Glenner, G. G. and Wong, C. W. (1984). Alzheimer's disease: initial report of the purification and characterization of a novel cerebrovascular amyloid protein. *Biochem Biophys Res Commun* 120: 885–890.

Halliwell, B. and Gutteridge, J. (1999). *Free Radicals in Biology and Medicine*. Oxford, Oxford University Press.

Hardy, J. (1997). Amyloid, the presenilins and Alzheimer's disease. *Trends Neurosci* 20: 154–159.

Head, E., Garzon-Rodriguez, W., Johnson, J. K., Lott, I. T., Cotman, C. W. and Glabe, C. (2001). Oxidation of Abeta and plaque biogenesis in Alzheimer's disease and Down syndrome. *Neurobiol Dis* 8: 792–806.

Hou, L., Kang, I., Marchant, R. E. and Zagorski, M. G. (2002). Methionine 35 oxidation reduces fibril assembly of the amyloid Abeta-(1-42) Peptide of Alzheimer's disease. *J Biol Chem* 277: 40173–40176.

Huang, X., Atwood, C. S., Moir, R. D., Hartshorn, M. A., Vonsattel, J. P., Tanzi, R. E. and Bush, A. I. (1997). Zinc-induced Alzheimer's Abeta 1-40 aggregation is mediated by conformational factors. *J Biol Chem* 272: 26464–26470.

Huang, X., Cuajungco, M. P., Atwood, C. S., Hartshorn, M. A., Tyndall, J. D., Hanson, G. R., Stokes, K. C., Leopold, M., Multhaup, G., Goldstein, L. E., Scarpa, R. C., Saunders, A. J., Lim, J., Moir, R. D., Glabe, C., Bowden, E. F., Masters, C. L., Fairlie, D. P., Tanzi, R. E. and Bush, A. I. (1999a). Cu(II) potentiation of Alzheimer Abeta neurotoxicity. Correlation with cell-free hydrogen peroxide production and metal reduction. *J Biol Chem* 274: 37111–37116.

Huang, X., Atwood, C. S., Hartshorn, M. A., Multhaup, G., Goldstein, L. E., Scarpa, R. C., Cuajungco, M. P., Gray, D. N., Lim, J., Moir, R. D., Tanzi, R. E. and Bush, A. I. (1999b). The Abeta peptide of Alzheimer's disease directly produces hydrogen peroxide through metal ion reduction. *Biochemistry* 38: 7609–7616.

Huang, X., Atwood, C. S., Moir, R. D., Hartshorn, M. A., Tanzi, R. E. and Bush, A. I. (2004). Trace metal contamination initiates the apparent auto-aggregation, amyloidosis, and oligomerization of Alzheimer's Abeta peptides. *J Biol Inorg Chem* 9: 954–960.

Ji, S. R., Wu, Y. and Sui, S. F. (2002). Cholesterol is an important factor affecting the membrane insertion of beta-amyloid peptide (Abeta 1-40), which may potentially inhibit the fibril formation. *J Biol Chem* 277: 6273–6279.

Kang, J., Lemaire, H. G., Unterbeck, A., Salbaum, J. M., Masters, C. L., Grzeschik, K. H., Multhaup, G., Beyreuther, K. and Muller-Hill, B. (1987). The precursor of Alzheimer's disease amyloid A4 protein resembles a cell-surface receptor. *Nature* 325: 733–736.

Karr, J. W., Akintoye, H., Kaupp, L. J. and Szalai, V. A. (2005). N-Terminal deletions modify the Cu(2+) binding site in amyloid-beta. *Biochemistry* 44: 5478–5487.

Kuo, Y. M., Kokjohn, T. A., Beach, T. G., Sue, L. I., Brune, D., Lopez, J. C., Kalback, W. M., Abramowski, D., Sturchler-Pierrat, C., Staufenbiel, M. and Roher, A. E. (2001). Comparative analysis of amyloid-beta chemical structure and amyloid plaque morphology of transgenic mouse and Alzheimer's disease brains. *J Biol Chem* 276: 12991–12998.

Lee, J. Y., Cole, T. B., Palmiter, R. D., Suh, S. W. and Koh, J. Y. (2002). Contribution by synaptic zinc to the gender-disparate plaque formation in human swedish mutant APP transgenic mice. *Proc Natl Acad Sci USA* 99: 7705–7710.

Lovell, M. A., Robertson, J. D., Teesdale, W. J., Campbell, J. L. and Markesbery, W. R. (1998). Copper, iron and zinc in Alzheimer's disease senile plaques. *J Neurol Sci* 158: 47–52.

Martins, R. N., Harper, C. G., Stokes, G. B. and Masters, C. L. (1986). Increased cerebral glucose-6-phosphate dehydrogenase activity in Alzheimer's disease may reflect oxidative stress. *J Neurochem* 46: 1042–1045.

Masters, C. L., Simms, G., Weinman, N. A., Multhaup, G., McDonald, B. L. and Beyreuther, K. (1985). Amyloid plaque core protein in Alzheimer disease and Down syndrome. *Proc Natl Acad Sci USA* 82: 4245–4249.

Maynard, C. J., Cappai, R., Volitakis, I., Cherny, R. A., White, A. R., Beyreuther, K., Masters, C. L., Bush, A. I. and Li, Q. X. (2002). Overexpression of Alzheimer's disease amyloid-beta opposes the age-dependent elevations of brain copper and iron. *J Biol Chem* 277: 44670–44676.

McLaurin, J., Franklin, T., Fraser, P. E. and Chakrabartty, A. (1998). Structural transitions associated with the interaction of Alzheimer beta-amyloid peptides with gangliosides. *J Biol Chem* 273: 4506–4515.

McLean, C. A., Cherny, R. A., Fraser, F. W., Fuller, S. J., Smith, M. J., Beyreuther, K., Bush, A. I. and Masters, C. L. (1999). Soluble pool of Abeta amyloid as a determinant of severity of neurodegeneration in Alzheimer's disease. *Ann Neurol* 46: 860–866.

Miura, T., Suzuki, K., Kohata, N. and Takeuchi, H. (2000). Metal binding modes of Alzheimer's amyloid beta-peptide in insoluble aggregates and soluble complexes. *Biochemistry* 39: 7024–7031.

Miyata, M. and Smith, J. D. (1996). Apolipoprotein E allele-specific antioxidant activity and effects on cytotoxicity by oxidative insults and beta-amyloid peptides. *Nat Genet* 14: 55–61.

Moir, R. D., Atwood, C. S., Romano, D. M., Laurans, M. H., Huang, X., Bush, A. I., Smith, J. D. and Tanzi, R. E. (1999). Differential effects of Apolipoprotein E isoforms on metal-induced aggregation of Abeta using physiological concentrations. *Biochemistry* 38: 4595–4603.

Narayanan, S. and Reif, B. (2005). Characterization of chemical exchange between soluble and aggregated states of beta-amyloid by solution-state NMR upon variation of salt conditions. *Biochemistry* 44: 1444–1452.

Naslund, J., Schierhorn, A., Hellman, U., Lannfelt, L., Roses, A. D., Tjernberg, L. O., Silberring, J., Gandy, S. E., Winblad, B., Greengard, P. and et al. (1994). Relative abundance of Alzheimer Abeta amyloid peptide variants in Alzheimer disease and normal aging. *Proc Natl Acad Sci USA* 91: 8378–8382.

Ohtsu, H., Shimazaki, Y., Odani, A., Yamauchi, O., Mori, W., Itoh, S. and Fukuzumi, S. (2000). Synthesis and characterization of imidazolate-bridged dinuclear complexes as active site models of Cu,Zn-SOD. *J Am Chem Soc* 122: 5733–5741.

Opazo, C., Huang, X., Cherny, R. A., Moir, R. D., Roher, A. E., White, A. R., Cappai, R., Masters, C. L., Tanzi, R. E., Inestrosa, N. C. and Bush, A. I. (2002). Metalloenzyme-like activity of Alzheimer's disease beta-amyloid. Cu-dependent catalytic conversion of dopamine, cholesterol, and biological reducing agents to neurotoxic H(2)O(2). *J Biol Chem* 277: 40302–40308.

Palmblad, M., Westlind-Danielsson, A. and Bergquist, J. (2002). Oxidation of methionine 35 attenuates formation of amyloid betapeptide 1-40 oligomers. *J Biol Chem* 277: 19506–19510.

Parge, H. E., Hallewell, R. A. and Tainer, J. A. (1992). Atomic structures of wild-type and thermostable mutant recombinant human Cu,Zn Superoxide Dismutase. *Proc Natl Acad Sci USA* 89: 6109–6113.

Peisach, J. and Blumberg, W. E. (1974). Structural implications derived from the analysis of electron paramagnetic resonance spectra of natural and artificial copper proteins. *Arch Biochem Biophys* 165: 691–708.

Phinney, A. L., Drisaldi, B., Schmidt, S. D., Lugowski, S., Coronado, V., Liang, Y., Horne, P., Yang, J., Sekoulidis, J., Coomaraswamy, J., Chishti, M. A., Cox, D. W., Mathews, P. M., Nixon, R. A., Carlson, G. A., St George-Hyslop, P. and Westaway, D. (2003). In vivo reduction of amyloid-beta by a mutant copper transporter. *Proc Natl Acad Sci USA* 100: 14193–14198.

Price, D. L., Tanzi, R. E., Borchelt, D. R. and Sisodia, S. S. (1998). Alzheimer's disease: genetic studies and transgenic models. *Annu Rev Genet* 32: 461–493.

Puglielli, L., Tanzi, R. E. and Kovacs, D. M. (2003). Alzheimer's disease: the cholesterol connection. *Nat Neurosci* 6: 345–351.

Rae, T. D., Schmidt, P. J., Pufahl, R. A., Culotta, V. C. and O'Halloran, T. V. (1999). Undetectable intracellular free copper: the requirement of a copper chaperone for superoxide dismutase. *Science* 284: 805–808.

Ritchie, C. W., Bush, A. I., Mackinnon, A., Macfarlane, S., Mastwyk, M., MacGregor, L., Kiers, L., Cherny, R. A., Li, Q.-X., Tammer, A., Carrington, D., Mavros, C., Volitakis, I., Xilinas, M., Ames, D., Davis, S., Beyreuther, K., Tanzi, R. E. and Masters, C. L. (2003). Metal-protein attenuation with iodochlorhydroxyquin (clioquinol) targeting Abeta amyloid deposition and toxicity in Alzheimer's disease: biochemical and clinical responses in a pilot phase 2 clinical trial. *Arch Neurol* 60: 1685–1691.

Roher, A. E., Chaney, M. O., Kuo, Y. M., Webster, S. D., Stine, W. B., Haverkamp, L. J., Woods, A. S., Cotter, R. J., Tuohy, J. M., Krafft, G. A., Bonnell, B. S. and Emmerling, M. R. (1996). Morphology and toxicity of Abeta-(1-42) dimer derived from neuritic and vascular amyloid deposits of Alzheimer's disease. *J Biol Chem* 271: 20631–20635.

Sayre, L. M., Perry, G., Harris, P. L., Liu, Y., Schubert, K. A. and Smith, M. A. (2000). *In situ* oxidative catalysis by neurofibrillary tangles and senile plaques in Alzheimer's disease: a central role for bound transition metals. *J Neurochem* 74: 270–279.

Schlief, M. L., Craig, A. M. and Gitlin, J. D. (2005). NMDA receptor activation mediates copper homeostasis in hippocampal neurons. *J Neurosci* 25: 239–246.

Shivers, B. D., Hilbich, C., Multhaup, G., Salbaum, M., Beyreuther, K. and Seeburg, P. H. (1988). Alzheimer's disease amyloidogenic glycoprotein: expression pattern in rat brain suggests a role in cell contact. *EMBO J* 7: 1365–1370.

Smith, M. A., Harris, P. L., Sayre, L. M. and Perry, G. (1997). Iron accumulation in Alzheimer disease is a source of redox-generated free radicals. *Proc Natl Acad Sci USA* 94: 9866–9868.

Smith, M. A., Sayre, L. M., Anderson, V. E., Harris, P. L., Beal, M. F., Kowall, N. and Perry, G. (1998). Cytochemical demonstration of oxidative damage in Alzheimer disease by immunochemical enhancement of the carbonyl reaction with 2,4-dinitrophenylhydrazine. *J Histochem Cytochem* 46: 731–735.

Smith, D. P., Smith, D. G., Curtain, C. C., Boas, J. F., Pilbrow, J. R., Ciccotosto, G. D., Lau, T. L., Tew, D. J., Perez, K., Wade, J. D., Bush, A. I., Drew, S. C., Separovic, F., Masters, C. L., Cappai, R. and Barnham, K. J. (2006). Copper-mediated amyloid-beta toxicity is associated with an intermolecular histidine bridge. *J. Biol Chem* 281(22): 15145–15154.

Spooner, E. T., Desai, R. V., Mori, C., Leverone, J. F. and Lemere, C. A. (2002). The generation and characterization of potentially therapeutic Abeta antibodies in mice: differences according to strain and immunization protocol. *Vaccine* 21: 290–297.

Squitti, R., Lupoi, D., Pasqualetti, P., Dal Forno, G., Vernieri, F., Chiovenda, P., Rossi, L., Cortesi, M., Cassetta, E. and Rossini, P. M. (2002). Elevation of serum copper levels in Alzheimer's disease. *Neurology* 59: 1153–1161.

Strittmatter, W. J., Saunders, A. M., Schmechel, D., Pericak-Vance, M., Enghild, J., Salvesen, G. S. and Roses, A. D. (1993). Apolipoprotein E: high-avidity binding to beta-amyloid and increased frequency of type 4 allele in late-onset familial Alzheimer disease. *Proc Natl Acad Sci USA* 90: 1977–1981.

Sundberg, R. J. and Martin, R. B. (1974). Interactions of histidine and other imidazole derivatives with transition metal ions in chemical and biological systems. *Chem Rev* 74: 471–517.

Syme, C. D., Nadal, R. C., Rigby, S. E. and Viles, J. H. (2004). Copper binding to the amyloid-beta (Abeta) peptide associated with Alzheimer's disease: folding, coordination geometry, pH dependence, stoichiometry, and affinity of Abeta-(1-28): insights from a range of complementary spectroscopic techniques. *J Biol Chem* 279: 18169–18177.

Tabner, B. J., Turnbull, S., El-Agnaf, O. M. and Allsop, D. (2002). Formation of hydrogen peroxide and hydroxyl radicals from (Abeta)and alpha-synuclein as a possible mechanism of cell death in Alzheimer's disease and Parkinson's disease. *Free Radic Biol Med* 32: 1076–1083.

Terry, R. D. (1996). The pathogenesis of Alzheimer disease: an alternative to the amyloid hypothesis. *J Neuropathol Exp Neurol* 55: 1023–1025.

Tickler, A. K., Smith, D. G., Ciccotosto, G. D., Tew, D. J., Curtain, C. C., Carrington, D., Masters, C. L., Bush, A. I., Cherny, R. A., Cappai, R., Wade, J. D. and Barnham, K. J. (2005). Methylation of the imidazole side chains of the Alzheimer disease amyloid-beta peptide results in abolition of superoxide dismutase-like structures and inhibition of neurotoxicity. *J Biol Chem* 280: 13355–13363.

Trombley, P. Q. and Shepherd, G. M. (1996). Differential modulation by zinc and copper of amino acid receptors from rat olfactory bulb neurons. *J Neurophysiol* 76: 2536–2546.

Viles, J. H., Cohen, F. E., Prusiner, S. B., Goodin, D. B., Wright, P. E. and Dyson, H. J. (1999). Copper binding to the prion protein: structural implications of four identical cooperative binding sites. *Proc Natl Acad Sci USA* 96: 2042–2047.

Waggoner, D. J., Bartnikas, T. B. and Gitlin, J. D. (1999). The role of copper in neurodegenerative disease. *Neurobiol Dis* 6: 221–230.

Walsh, D. W., Klyubin, I., Fadeva, J. V., Cullen, W. K., Anwyl, R., Wolfe, M. S., Rowan, M. J. and Selkoe, D. J. (2002). Naturally secreted oliogmers of amyloid beta protein potently inhibit hippocampal long-term potentiation *in vivo*. *Nature* 416: 535–539.

Watson, A. A., Fairlie, D. P. and Craik, D. J. (1998). Solution structure of methionine-oxidized amyloid-beta peptide (1-40). Does oxidation affect conformational switching? *Biochemistry* 37: 12700–12706.

White, A. R., Reyes, R., Mercer, J. F., Camakaris, J., Zheng, H., Bush, A. I., Multhaup, G., Beyreuther, K., Masters, C. L. and Cappai, R. (1999). Copper levels are increased in the cerebral cortex and liver of APP and APLP2 knockout mice. *Brain Res* 842: 439–444.

Whittaker, J. W. (2003). Free radical catalysis by galactose oxidase. *Chem. Rev.* 103: 2347–2363.

Xiong, H., McCabe, L., Costello, J., Anderson, E., Weber, G. and Ikezu, T. (2004). Activation of NR1A/NR2B receptors by soluble factors from APP-stimulated monocyte-derived macrophages: implications for the pathogenesis of Alzheimer's disease. *Neurobiol Aging* 25: 905–911.

Yang, D. S., McLaurin, J., Qin, K., Westaway, D. and Fraser, P. E. (2000). Examining the zinc binding site of the amyloid-beta peptide. *Eur J Biochem* 267: 6692–6698.

3

Progress in Understanding the Mechanisms of Neuronal Dysfunction and Degeneration in Parkinson's Disease

J. William Langston

Abstract

During the past 30 years, a number of different hypotheses on cause of neuronal degeneration and dysfunction in Parkinson's disease have been intensively investigated. Roles have been postulated for oxidative stress, excitotoxcity, nitric oxide, mitochondrial dysfunction, and inflammation. The possibility of an environmental cause was highlighted by the discovery of a simple molecule known as 1-methyl-4-phenyl-1,2,3,6-tetrahydropyridine (MPTP), a compound that is selectively toxic to the same cells in the brain that die in Parkinson's disease. Yet the most recent hypothesis to come under scrutiny did not emerge from any of this previous work, but rather from a rare form of genetic parkinsonism known as PARK 1. This form of autosomal dominant parkinsonism is caused by mutations in the gene that encodes for the protein α-synuclein. Unexpectedly, this protein has been found to accumulate in nerve cells and their processes in all patients with Parkinson's disease, raising the possibility that abnormal protein folding and aggregation represent a fundamental feature of the disease. These observations have already ushered in a new era of research on the disease and stimulated novel strategies directed toward disease modification by providing new therapeutic targets for drug development. Only time will tell if this most recent chapter in the search for the molecular basis of neurodegeneration in Parkinson's disease will be one that holds the key to understanding this complex disorder and the highly characteristic pattern of selective vulnerability exhibited by the neuronal populations it affects. But an exciting body of accumulating scientific evidence is pointing in that direction.

3.1. Introduction

Of all of the neurodegenerative diseases, Parkinson's disease has one of the richest histories when it comes to theories on its cause and underlying molecular mechanisms. Why this is the case is not clear. Brissaud may have come the closest to an explanation when he stated: "Parkinson's disease remains so utterly inexplicable . . . that we are constantly drawn to it by the lure of the mysterious" (Brissaud, 1895). The disease is the second most common neurodegenerative disease of aging, exceeded only by Alzheimer's disease. Clinically, it is traditionally characterized by the triad of rest tremor, muscular rigidity, and slowness of movement or bradykinesia. The disease is slowly progressive and currently incurable. Neuropathologically, the disease is typically described as characterized by a loss of dopaminergic neurons in the substantia nigra and the presence of eosinophilic inclusion bodies in certain areas of the brain, including the substantia nigra, and known as Lewy bodies (Forno, 1986). As will become clear at the end of this chapter, this clinicopathologic concept may be changing, perhaps radically, as the result of new clinical and pathologic observations (Langston, 2006), but more on that later.

3.2. Hypotheses on the Cause of Parkinson's Disease

Interestingly, the first to speculate on cause was none other than James Parkinson himself, who postulated that the disease might be caused by stress or fright (Parkinson, 1817). The first major debate regarding the cause began later in the 19th century and centered around whether the disease was inherited or due to something in the environment. The great French neurologist Charcot favored the view that the disease did not have a "familial basis" (Charcot, 1878), whereas the distinguished English neurologist Gowers had noted that 15% of his patients had a strong family history, and for this reason he believed that inheritance did play a role in the disease (Gowers, 1888). The pendulum of opinion regarding genes versus environment has swung repeatedly between these two extremes at approximately 40-year intervals since that time. Remarkably, this debate continues to be as spirited as ever, and if anything, has even sharpened during the past several decades as the result of several recent discoveries, as we will see later in this chapter.

At the molecular level, the earliest of theory to gain traction, which continues to thrive right up to the present time, is the so-called free-radical hypothesis of Parkinson's disease, which was first put forth by Doyle Graham (Graham, 1978). This hypothesis posits that the oxidation of dopamine plays a central role in the degeneration of dopaminergic neurons through its enzymatic oxidation and the subsequent production of potentially damaging oxy-radicals, as well as its nonenzymatic oxidation to potentially toxic quinines. This is an attractive hypothesis because oxidation via monoamine oxidase generates hydrogen peroxide, which in the presence of iron, could lead to the formation of potentially damaging hydroxyl radicals (Youdim *et al.*, 1989). A corollary hypothesis that continues to fascinate neuroscientists interested in Parkinson's disease relates to an enigmatic substance known as neuromelanin, which at least in part is a by-product of the breakdown of catecholamines. Neuromelanin increases in catecholaminergic neuronal populations throughout life and may be deleterious to the cells in which it accumulates (Graham, 1978; Mann and Yates, 1983; Hirsch *et al.*, 1988; Prota and d'Ischia, 1993; Zecca *et al.*, 2003).

In addition to these early (and enduring) hypotheses, the past several decades have witnessed the advent of a number of other new and attractive concepts that could explain the mechanisms of cell death in Parkinson's disease, including excitotoxcity (Beal, 1998), a perturbation of energy production (Schulz and Beal, 1994), a role for nitrous oxide (Chabrier *et al.*, 1999), and the possibility that inflammation plays a contributory or even primary role in the disease (Hirsch *et al.*, 2003; McGeer and McGeer, 2004; Wersinger and Sidhu, 2006). Not surprisingly, these hypotheses have frequently dovetailed to make an even more compelling biological story about the molecular mechanisms that could underlie neuronal death and dysfunction in Parkinson's disease. For example, the effects of the mitochrondrial Complex I inhibitor rotenone, which causes selective nigrostriatal damage after chronic infusion in rodent, has been shown to result from the oxidative stress that occurs in the setting of failing mitochondrial energy production (Testa *et al.*, 2005).

3.3. The Discovery of the Biological Effects of MPTP

A number of these hypotheses have also been closely intertwined with the discovery of the parkinsongenic neurotoxin know as 1-methyl-4-phenyl-1,2,3,6-tetrahydropyridine (MPTP).

The biological effects of this compound were discovered after a group of young heroin addicts mysteriously developed full-blown and unalloyed parkinsonism after self-administering a new "synthetic heroin"(Langston *et al.*, 1983). Through some medical detective work, it was discovered that the substance they injected was tainted with MPTP, which proved to be the offending agent. We now know that MPTP is selectively toxic to the dopaminergic neurons in the substantia nigra, the same cells that degenerate in Parkinson's disease. It was quickly found that MPTP is rapidly biotransformed into 1-methyl-4-phenyl pyridinium ion (MPP+), its toxic metabolite, via the enzyme monoamine oxidase (MAO) B (Chiba *et al.*, 1984). MAO B inhibitors block MPTP toxicity (Cohen *et al.*, 1984; Heikkila *et al.*, 1984; Langston *et al.*, 1984).

These observations were important because they raised the possibility that MAO B inhibitors might also block progress of Parkinson's disease itself, either by inhibiting biotransformation of xenobiotics to potentially toxic metabolites or by blocking oxidative stress resulting from the oxidation of dopamine by MAO. And in fact these studies inspired clinical trials to test this hypothesis, both of which appeared to confirm that this approach did in fact slow disease progression (Tetrud and Langston, 1989; Parkinson's Study Group, 1989). However, selegiline, the selective MAO B inhibitor that was used for these studies, was also found to have a mild symptomatic effect, which unfortunately confounded interpretation of the trial results, as it was argued that its day to day symptomatic effects could not be separated from a true slowing of disease progress.

Another clinical trial that was inspired by MPTP was based on the original observations that MPP+ is taken up by mitochondria, where it proved to be a potent inhibitor of complex I of the mitochondrial respiratory chain (Ramsay *et al.*, 1986). It was not long before complex I deficits were discovered in the brain and platelets of patients with Parkinson's disease (Mizuno *et al.*, 1989; Parker *et al.*, 1989; Schapira *et al.*, 1989). Subsequently, Coenzyme Q_{10}, which enhances Complex I function, was found to be at least partially protective in animal models of Parkinson's disease (Beal *et al.*, 1998). This in turn lead to a clinical trial of this nutritional supplement to see if it could be used to slow progression of Parkinson's disease (Shults *et al.*, 2002). The results of this trial were recently reported (Shults *et al.*, 2002) and provided enough evidence for such an effect to warrant a much larger multicenter trial.

3.4. Environment and Genetic Contributions

Another major impact of the discovery of MPTP was its effect on the gene versus environment debate. Several observations served as a flashpoints tilting interest toward the environmental hypothesis in the final decades of the 20th century. The first was the proof of principle that a relatively simple compound such as MPTP could, after systemic exposure, induce virtually all of the motor features of the disease. The second was the striking similarity between its toxic metabolite, MPP+, and the well-known pesticide paraquat; this similarity stimulated tremendous interest in the possibility that pesticides and herbicides might play a role in causing or triggering the disease. Indeed, there is now a voluminous body of research on environmental risk factors and Parkinson's disease (Tanner and Ben-Shlomo, 1999). Further highlighting the potential importance of the environment are several large twin studies, both of which failed to provide evidence that heredity plays a major role in Parkinson's disease (Tanner *et al.*, 1999; Wirdefeldt *et al.*, 2004). However, just when we might have thought the pendulum would not swing back in the direction of genetics, precisely that has happened, but this has not occurred in quite the way that this might have been expected.

Rather, it began somewhat unexpectedly with the identification of large Italian family known as the Contursi kindred. In this family, a condition highly similar to Parkinson's disease appeared to be inherited in an autosomal dominant fashion. Although the family had been followed for some time (Golbe *et al.*, 1990, 1996), it was not until 1996 that the gene was mapped to chromosome 4q21, q23 (Polymeropoulos *et al.*, 1996). The following year, the causative mutation was identified, which proved to be an A53T mutation in the gene encoding for a protein known as α-synuclein (Polymeropoulos *et al.*, 1997). Shortly thereafter, a seminal paper was published in which Spillantini and colleagues (Spillantini *et al.*, 1997) reported that α-synuclein is a major component of the Lewy bodies, which, as noted earlier, is a cardinal neuropathologic feature of Parkinson's disease. There is little doubt that this observation began a new epoch in the history of Parkinson's disease (Langston *et al.*, 2002), one that focuses on the disease as a protein-folding disorder. Since that time, there has been increasing acceptance of the hypothesis that pathologic aggregation and/or misfolding of α-synuclein is a critical step in the process of neurodegeneration in the disease.

Although mutations in the gene encoding for α-synuclein cause a rare genetic form of parkinsonism, and this protein represents a major component of Lewy bodies, it has been unclear what role this protein played in typical sporadic disease, as most patients do not have mutations in α-synuclein. A critical observation that gave credence to the possibility that normal synuclein could be playing a "toxic role" in sporadic cases came as a result of solving the genetic basis of the Spellman-Muenter kindred (more recently known as the Iowa kindred). This kindred had been followed since the 1930s (Muenter *et al.*, 1998). Affected members exhibited young-onset parkinsonism, as well as dementia and marked autonomic involvement in some cases (Muenter *et al.*, 1998). Although a point mutation in the α-synuclein gene had been ruled out in this remarkable kindred, in 2003 it was discovered that affected patients had a triplication on one allele of the gene (Singleton *et al.*, 2003) and were therefore producing twice as much synuclein as would normally be present. We subsequently reported a second kindred (Farrer *et al.*, 2004) with a very similar parkinsonian phenotype.

Why are these cases so important? There are two reasons. First, after the description of the Contursi kindred with the A53T mutation in the α-synuclein [as well as two additional mutations that also appear to cause a PARK 1 parkinsonism (Kruger *et al.*, 1998; Zarranz *et al.*, 2004)], it was still far from clear how this related to typical Parkinson's disease, even though α-synuclein was found in Lewy bodies in all patients with Parkinson's disease (and other Lewy bodies diseases as well for that matter (Spillantini *et al.*, 1998)). The multiplication cases added an entirely new dimension to this question by demonstrating for the first time that native (or wild-type) synuclein could itself cause neuronal degeneration and Lewy body formation. Second, these cases suggest that α-synuclein may provide an answer to one of the great mysteries of Parkinson's disease (and in fact almost all neurologic diseases), and that is what explains the selective vulnerability of certain neuron populations and not others in the disease. Why, for example, are substantia nigra and locus ceruleus so vulnerable, while other areas, such as the cerebellum, are completely spared.

3.5. Selective Vulnerability and Parkinson's Disease

In regard to Parkinson's disease, one of the earliest theories attempting to explain selective vulnerability was that it could be attributed to the presence of neuromelanin. Because catecholaminergic neurons seem to be particularly susceptible in Parkinson's disease, neuromelanin has been repeatedly implicated in causing cell death (as described earlier in this chapter). But a number of investigators have also suggested that neuromelanin might also explain the pattern of selective vulnerability in Parkinson's disease. However, in the 1980s, David Marsden, a leading proponent of neuromelanin as a cause of neuronal degeneration in Parkinson's disease, was forced to abandon his theory because it did not explain the true selective vulnerability in the disease, primarily because Lewy bodies, a incontrovertible marker of the disease process, are clearly found in noncatecholaminergic, nonmelanized areas of the brain (e.g., the nucleus basalis of Mynert, a cholinergic nucleus) (Marsden, 1983). After discovery of MPTP, it was hoped that the distribution of monoamine oxidase might explain the distribution of disease in the brain, but it eventually became clear that there was not a correlation between the two.

How do the multiplication cases shed light on the question of selective vulnerability in Parkinson's disease? There are now a number of pathologically examined cases (Muenter *et al.*, 1998), including a patient who was followed at our institute (Farrer *et al.*, 2004), clearly showing that these cases, which overexpress native α-synuclein, exhibit the same selective vulnerability that is seen in idiopathic Parkinson's disease. Thus normal α-synuclein, at least when there is too much of it, seems capable of triggering changes in virtually all of the areas in the CNS that have been reported to be affected in the idiopathic disease, indicating that this protein may be providing some fundamental insights into the question of selective vulnerability in Parkinson's disease and therefore could be a key to understanding the disease process itself. If there is a fly in the ointment when it comes to the synuclein hypothesis based on these triplication cases, it is that the disease is more aggressive, both clinically and pathologically, than that seen in either Parkinson's disease or even dementia with Lewy bodies (DLB), a Lewy body disorder in which cognition and psychosis are typically more prominent than parkinsonism (McKeith *et al.*, 2005). For example, dementia and autonomic dysfunction are more prominent in these cases, and at the pathologic level, there is more cortical cell loss, Lewy neuritic pathology, and gliosis than is typically seen in DLB (D. Dickson, personal communication, 2006). It may be that what we are seeing is simply a quantitative phenomenon, as the duplication cases are said to have a more benign course with an older age of onset, more typical of sporadic Parkinson's disease (Chartier-Harlin *et al.*, 2004; Nishioka *et al.*, 2006). However, as there are no pathologically studied duplication cases to date, the final verdict on whether or not they will more perfectly mirror the changes seen in Parkinson's disease remains out, at least for the time being.

3.6. Redefining the Parkinson's Complex

Another, and perhaps unexpected impact of the identification of α-synuclein relates to its use as a new immunohistochemical tool. Before the discovery of α-synuclein, neuropathologists traditionally used hematoxylin-eosin (H&E) staining to identify Lewy bodies. However, this technique was only moderately sensitive and very poor at identifying abnormal protein depositions in neuronal processes, which are referred to as Lewy neurites. Synuclein immunohistochemistry has greatly facilitated our ability to visualize the distribution of abnormal protein deposits throughout the brain and the peripheral nervous system, particularly Lewy neurites (Spillantini *et al.*, 1998). Braak and colleagues (Braak *et al.*, 2002) have taken full advantage of this technique to develop a staging system for the disease. Interestingly, this staging system begins with lower brain stem and the olfactory bulb involvement, and only later (in the midstages of the disease) does the dopaminergic nigrostriatal system become involved (Braak *et al.*, 2002). While their hypothesis on the anatomic origin and order of disease progression continues to be debated, what is becoming increasing clear is that way of the areas of involvement which are becoming more clearly defined by synuclein immunohistochemistry (which as far a field as the neuronal plexuses of the gastrointestinal tract and the bladder) also causes clinical signs and symptoms (Pfeiffer and Bodis-Wollner, 2005; Langston, 2006). Indeed, the way these clinical and pathologic observations are coming together is having a dramatic impact on how we view the disease (Langston, 2006). It seems reasonable to consider using the term Parkinson's complex because of increasing diversity of features that this disease seems capable of causing, such as sleep disorders and constipation, which extend well beyond the traditional concept of parkinsonism (tremor, bradykinesia, and rigidity). As part of this effort to broaden the way of thinking about the disease, it might be worthwhile to consider expanding the name of the disease to Parkinson-Lewy disease, which would both honor Lewy's contribution to the neuropathology of the disorder and clearly denote the widespread nature of the disease beyond the dopaminergic nigrostriatal system (Langston, 2006).

Better defining the nature and anatomy of the disease, particularly the site of origin, could also provide critical insights into the cause of the disease. For example, as Braak and colleagues (Braak *et al.*, 2003) have suggested, if the disease originates in the olfactory bulb and enteric nervous system, it would point to mucous membranes as a portal of entry of an offending environmental agent. While this hypothesis is far from proven, it provides an excellent example of hypotheses and research leads that can be derived from better understanding the true nature of where the disease originates and its subsequent course of evolution.

3.7. Implications for Future Research and Identifying Disease-Modifying Therapies

Given the mounting evidence that abnormal protein handling may be a fundamental feature of Parkinson's disease, and the increasing likelihood that α-synuclein is a key player in this process, one can begin to map out a comprehensive research strategy for the next chapter in our quest to unravel the underlying mechanisms of cell death in Parkinson's disease. As a first step, there is a critical need to fully elucidate the protein chemistry of α-synuclein and its molecular and cellular effects in both health and disease. Both of these aspects are dealt with in great detail in the next two chapters of this book (Uversky, this volume; Lee and Kim, this volume). Clearly, it is essential to understand the normal functions of this protein, how it is synthesized, handled, and degraded if we are to eventually understand the role of this or other proteins in causing or propagating the neurodegenerative process in Parkinson's disease.

In the past, attempts to modify the course of Parkinson's disease have been driven by the theories on the mechanisms of cell death that were current at the time. Consider, for example, the clinical trials with MAO B inhibitors and Coenzyme Q_{10} outlined at the beginning of this chapter. The evolving protein misfolding/aggregation hypothesis represents no exception to this rule. The need for a successful strategy to slow or halt progression of Parkinson's disease needs little in the way of justification. With time, virtually all currently known symptomatic therapies lead to side effects and/or lose effectiveness, so alternatives are desperately needed. Importantly, compared with some other neurologic diseases, such as stroke, where much of the damage is done by the time patients receive medical attention, Parkinson's disease is very slowly progressive, particularly at onset, so much so, that the average time from first symptom to diagnosis is 2 years. Because the disease is typically so mild and nondisabling when first diagnosed, the ability to stop disease progression could achieve something that is not too different from a cure, as it would bestow near-normal functional and a meaningful quality of life for most if not all recently diagnosed patients.

For these reasons, many groups, including our own, are intensively exploring ways to inhibit or even reverse the pathologic misfolding and presumably toxic aggregation of α-synuclein. A wide variety of approaches should be considered, including finding small molecules that might achieve this to discovering ways to downregulate the production of α-synuclein. In this regard, it is worth noting that Li and colleagues (Li *et al.*, 2004) have found that the antituberculosis agent, rifampicin, is capable of doing just this in an *in vitro* setting. Other techniques to consider are ways to enhance the degradation of α-synuclein or even to block the inflammatory process that its accumulation appears to evoke in some models. It is by no means premature to predict that the coming decade will see a large body of research directed toward one or more of these approaches.

Importantly, we now have an increasing number of models in which to test new agents or strategies. For example, there are transgenic models in yeast (Griffioen *et al.*, 2006), *Drosophila* (Auluck and Bonini, 2002), and *C. elegans* (Cao *et al.*, 2005; Lakso *et al.*, 2003). In the latter two, overexpression of α-synuclein actually leads to dopaminergic neuronal degeneration. Each of these models is already being used to screen for small molecules that might be used to modify disease progress. Furthermore, viral vectors have now been successfully used to induce α-synuclein overexpression and nigral cell degeneration in rodents (Lo Bianco *et al.*, 2002; Yamada *et al.*, 2004; Maingay *et al.*, 2005), and our own group is in the process of applying this approach to primate models as well. Thus it appears that from both technical and practical standpoints, we may be nearing a time when new compounds and strategies for disease modification based on abnormal synuclein aggregation can be tested at multiple levels of biological complexity, with a goal of taking successful agents to clinical trials as expeditiously as possible.

3.8. PARK 8: The Latest Form of Genetic Parkinsonism

A review of the evolution of Parkinson's disease research would not be complete without at least briefly discussing the most recently discovered form of genetic parkinsonism, known as PARK 8. The first reports on this mutation appeared in 2004, in which several mutations were identified in the leucine-rich repeat kinase 2 (LRRK 2) gene in families with autosomal dominant parkinsonism (Paisan-Ruiz *et al.*, 2004; Zimprich *et al.*, 2004). However, the clinical phenotype seemed to be quite variable, with parkinsonism being the most common clinical manifestation, but dementia and amyotrophy were also seen in some cases. Perhaps the most astonishing example of phenotypic variation, however, was observed in one family in which pathology was available from four different affected members. Different pathologic features were found in each case, including typical brain-stem Lewy body parkinsonism, pure nigral cell degeneration, DLB type pathology, and a progressive supranuclear parkinsonism-like pathology. Even more remarkable, however, is that one of the known PARK 8 mutations, the G2019S variation, is now known to occur in approximately 1.5% of cases diagnosed as having typical late-onset, apparently sporadic Parkinson's disease in the United States and Europe (Gilks *et al.*, 2005; Kay *et al.*, 2005). In Ashkenazi Jew (Ozelius *et al.*, 2006) and in North African Arab (Lesage *et al.*, 2006) populations, this figure climbs to 18% to 41%, respectively. The range of age of onset is highly variable and can be in the nineties. It is now clear that the majority of these cases have typical Lewy body pathology, with Lewy bodies and Lewy neurite formation in typical areas of the brain, and the peripheral autonomic nervous system highly reminiscent if not identical to Parkinson's disease (Ross *et al.*, 2006). For all of these reasons, this new form of parkinsonism is likely to teach us a great deal about idiopathic Parkinson's disease and the role of α-synclein. The discovery of causative mutations in the LRRK 2 gene may have therapeutic implications, as the gene contains a kinase domain, and kinases are already a well-established drug target. Ultimately, determining how abnormalities in the LRRK 2 gene orchestrate a pattern of protein misfolding/aggregation that is so typical of Parkinson's disease seems destined to provide critical clues as to how this happens in the many patients who do not harbor such mutations.

3.9. Conclusion

As emphasized at the beginning of this chapter, research on the cause of Parkinson's disease and the mechanisms of cell death that underlie it has a long and rich history that has many chapters. Interestingly, at least two of these, the discovery of MPTP and the identification of the relationship of α-synuclein and Parkinson's disease, occurred as the result of clinical observation, and both have had a huge impact on the course of research in Parkinson's disease. Indeed, it would not be an overstatement to say that the PARK 1 story has ushered in an entirely new era of research on the mechanisms that underlie cellular death and dysfunction in Parkinson's disease (Langston *et al.*, 2002). This new realm of research might be best categorized as the "protein misfolding/aggregation" hypothesis and seems destined to extend well into future. For the many reasons outlined in this chapter, it seems likely that this new research direction will lead to new insights into the cause of the disease. Of course, the ultimate goal is to find new research strategies that will lead to treatments for Parkinson's disease that are truly disease modifying. As has been pointed out in these pages, this is not the first time the Parkinson's field has witnessed the advent of new theories that beckoned with the promise of disease modification. Indeed, the history of research on Parkinson's disease is replete with such efforts and similar levels of enthusiasm. The real question that lies before us is whether this will be the final chapter in what has been a remarkable history of hypothesis generation, exuberant research, and not infrequent disappointments. There are many reasons to believe that this new research direction may succeed where others have failed, but only time will tell. For now, all of the evidence suggests that we should proceed with a full-court press in pursuit of this exciting new research direction, with hope that the many patients who suffer with this terribly disabling disease will reap the benefits.

References

Auluck, P. K., and Bonini, N. M. (2002). Pharmacological prevention of Parkinson disease in Drosophila. *Nat Med* 8: 1185–1186.

Beal, M. F. (1998). Excitotoxicity and nitric oxide in Parkinson's disease pathogenesis. *Ann Neurol* 44: S110–S114.

Beal, M. F., Matthews, R. T., Tieleman, A., and Shults, C. W. (1998). Coenzyme Q_{10} attenuates the 1-methyl-4-phenyl-1,2,3,tetrahydropyridine (MPTP) induced loss of striatal dopamine and dopaminergic axons in aged mice. *Brain Res* 783: 109–114.

Braak, H., Del Tredici, K., Bratzke, H., Hamm-Clement, J., Sandmann-Keil, D., and Rub, U. (2002). Staging of the intracerebral inclusion body pathology associated with idiopathic Parkinson's disease (preclinical and clinical stages). *J Neurol* 249 Suppl 3: III/1–5.

Braak, H., Rub, U., Gai, W. P., and Del Tredici, K. (2003). Idiopathic Parkinson's disease: possible routes by which vulnerable neuronal types may be subject to neuroinvasion by an unknown pathogen. *J Neural Transm* 110: 517–536.

Brissaud, E. (1895) *Lecons sur les maladies nerveuses. Salpetriere, 1893–1894*, ed. H Miege. Paris: G. Masson.

Cao, S., Gelwix, C. C., Caldwell, K. A., and Caldwell, G. A. (2005). Torsin-mediated protection from cellular stress in the dopaminergic neurons of Caenorhabditis elegans. *J Neurosci* 25: 3801–3812.

Chabrier, P. E., Demerle-Pallardy, C., and Auguet, M. (1999). Nitric oxide synthases: targets for therapeutic strategies in neurological diseases. *Cell Mol Life Sci* 55: 1029–1035.

Charcot, J. M. (1878) *Lectures on the diseases of the nervous system*. London: The New Sydenham Society.

Chartier-Harlin, M. C., Kachergus, J., Roumier, C., Mouroux, V., Douay, X., Lincoln, S., Levecque, C., Larvor, L., Andrieux, J., Hulihan, M., Waucquier, N., Defebvre, L., Amouyel, P., Farrer, M., and Destee, A. (2004). Alpha-synuclein locus duplication as a cause of familial Parkinson's disease. *Lancet* 364: 1167–1169.

Chiba, K., Trevor, A., and Castagnoli, N., Jr. (1984). Metabolism of the neurotoxic tertiary amine, MPTP, by brain monoamine oxidase. *Biochem Biophys Res Commun* 120: 574–578.

Cohen, G., Pasik, P., Cohen, B., Leist, A., Mytilineou, C., and Yahr, M. D. (1984). Pargyline and deprenyl prevent the neurotoxicity of 1-methyl-4-phenyl-1,2,3,6-tetrahydropyridine (MPTP) in monkeys. *Eur J Pharmacol* 106: 209–210.

Farrer, M., Kachergus, J., Forno, L., Lincoln, S., Wang, D. S., Hulihan, M., Maraganore, D., Gwinn-Hardy, K., Wszolek, Z., Dickson, D., and Langston, J. W. (2004). Comparison of kindreds with parkinsonism and alpha-synuclein genomic multiplications. *Ann Neurol* 55: 174–179.

Forno, L. S. (1986). Lewy bodies. *N Engl J Med* 314: 122.

Gilks, W. P., Abou-Sleiman, P. M., Gandhi, S., Jain, S., Singleton, A., Lees, A. J., Shaw, K., Bhatia, K. P., Bonifati, V., Quinn, N. P., Lynch, J., Healy, D. G., Holton, J. L., Revesz, T., and Wood, N. W. (2005) A common LRRK2 mutation in idiopathic Parkinson's disease. *Lancet* 365: 415–416.

Golbe, L. I., Di Iorio, G., Bonavita, V., Miller, D. C., and Duvoisin, R. C. (1990). A large kindred with autosomal dominant Parkinson's disease. *Ann Neurol* 27: 276–282.

Golbe, L. I., Di Iorio, G., Sanges, G., Lazzarini, A. M., La Sala, S., Bonavita, V., and Duvoisin, R. C. (1996). Clinical genetic analysis of Parkinson's disease in the Contursi kindred. *Ann Neurol* 40: 767–775.

Gowers, W. R. (1888). *Diseases of the nervous system*. Philadelphia: P. Blakiston, Son and Company.

Graham, D. G. (1978). Oxidative pathways for catecholamines in the genesis of neuromelanin and cytotoxic quinones. *Mol Pharmacol* 14: 633–643.

Griffioen, G., Duhamel, H., Van Damme, N., Pellens, K., Zabrocki, P., Pannecouque, C., van Leuven, F., Winderickx, J., and Wera, S. (2006). A yeast-based model of alpha-synucleinopathy identifies compounds with therapeutic potential. *Biochim Biophys Acta* 1762: 312–318.

Heikkila, R. E., Manzino, L., Cabbat, F. S., and Duvoisin, R. C. (1984). Protection against the dopaminergic neurotoxicity of 1-methyl-4-phenyl-1,2,5,6-tetrahydropyridine by monoamine oxidase inhibitors. *Nature* 311: 467–469.

Hirsch, E., Graybiel, A. M., and Agid, Y. A. (1988). Melanized dopaminergic neurons are differentially susceptible to degeneration in Parkinson's disease. *Nature* 334: 345–348.

Hirsch, E. C., Breidert, T., Rousselet, E., Hunot, S., Hartmann, A., and Michel, P. P. (2003). The role of glial reaction and inflammation in Parkinson's disease. *Ann N Y Acad Sci* 991: 214–228.

Kay, D. M., Zabetian, C. P., Factor, S. A., Nutt, J. G., Samii, A., Griffith, A., Bird, T. D., Kramer, P., Higgins, D. S., and Payami, H. (2005). Parkinson's disease and LRRK2: Frequency of a common mutation in U.S. movement disorder clinics. *Mov Disord* 21: 519–523.

Kruger, R., Kuhn, W., Muller, T., Woitalla, D., Graeber, M., Kosel, S., Przuntek, H., Epplen, J. T., Schols, L., and Riess, O. (1998). Ala30Pro mutation in the gene encoding α-synuclein in Parkinson's disease. *Nat Genet* 18: 106–108.

Lakso, M., Vartiainen, S., Moilanen, A. M., Sirvio, J., Thomas, J. H., Nass, R., Blakely, R. D., and Wong, G. (2003). Dopaminergic neuronal loss and motor deficits in Caenorhabditis elegans overexpressing human alpha-synuclein. *J Neurochem* 86: 165–172.

Langston, J. W. (2006). The parkinson's complex: Parkinsonism is just the tip of the iceberg. *Ann Neurol* 59: 591–596.

Langston, J. W., Ballard, P., Tetrud, J. W., and Irwin, I. (1983). Chronic Parkinsonism in humans due to a product of meperidine-analog synthesis. *Science* 219: 979–980.

Langston, J. W., Irwin, I., Langston, E. B., and Forno, L. S. (1984). Pargyline prevents MPTP-induced parkinsonism in primates. *Science* 225: 1480–1482.

Langston, J. W., Golbe, L. I, and Lee, S. J. (2002). PARK 1 and α-Synuclein: a new era in Parkinson's research. In *Genetics of Movement Disorder*, ed. S Pulst, pp. 277–304. San Diego: Academic Press.

Lesage, S., Durr, A., Tazir, M., Lohmann, E., Leutenegger, A. L., Janin, S., Pollak, P., and Brice, A.; French Parkinson's Disease Genetics Study Group (2006).. LRRK2 G2019S as a cause of Parkinson's disease in North African Arabs. *N Engl J Med* 354: 422–423.

Li, J., Zhu, M., Rajamani, S., Uversky, V.N., and Fink, A. L. (2004). Rifampicin inhibits alpha-synuclein fibrillation and disaggregates fibrils. *Chem Biol* 11: 1513–1521.

Lo Bianco, C., Ridet, J. L., Schneider, B. L., Deglon, N., and Aebischer, P. (2002). alpha -Synucleinopathy and selective dopaminergic neuron loss in a rat lentiviral-based model of Parkinson's disease. *Proc Natl Acad Sci USA* 99: 10813–10818.

Maingay, M., Romero-Ramos, M., and Kirik, D. (2005). Viral vector mediated overexpression of human alpha-synuclein in the nigrostriatal dopaminergic neurons: a new model for Parkinson's disease. *CNS Spectr* 10: 235–244.

Mann, D. M., and Yates, P. O. (1983). Possible role of neuromelanin in the pathogenesis of Parkinson's disease. *Mech Ageing Dev* 21: 193–203.

Marsden, C. D. (1983). Neuromelanin and Parkinson's disease. *J Neural Transm Suppl* 19: 121–141.

McGeer, P. L., and McGeer, E. G. (2004). Inflammation and neurodegeneration in Parkinson's disease. *Parkinsonism Relat Disord* 10 Suppl 1: S3–S7.

McKeith, I. G., Dickson, D. W., Lowe, J., Emre, M., O'Brien, J. T., H, Cummings, J., Duda, J. E., Lippa, C., Perry, E. K., Aarsland, D., Arai, H., Ballard, C. G., Boeve, B., Burn, D. J., Costa, D., Del Ser, T., Dubois, B., Galasko, D., Gauthier, S., Goetz, C. G., Gomez-Tortosa, E., Halliday, G., Hansen, L. A., Hardy, J., Iwatsubo, T., Kalaria, R. N., Kaufer, D., Kenny, R. A., Korczyn, A., Kosaka, K., Lee, V. M., Lees, A., Litvan, I., Londos, E., Lopez, O. L., Minoshima, S., Mizuno, Y., Molina, J. A., Mukaetova-Ladinska, E. B., Pasquier, F., Perry, R. H., Schulz, J. B., Trojanowski, J. Q., and Yamada, M.; Consortium on DLB (2005). Diagnosis and management of dementia with Lewy bodies: third report of the DLB Consortium. *Neurology* 65: 1863–1872.

Mizuno, Y., Ohta, S., Tanaka, M., Takamiya, S., Suzuki, K., Sato, T., Oya, H., Ozawa, T., and Kagawa, Y. (1989). Deficiencies in complex I subunits of the respiratory chain in Parkinson's disease. *Biochem Biophys Res Commun* 163: 1450–1455.

Muenter, M. D., Forno, L. S., Hornykiewicz, O., Kish, S. J., Maraganore, D. M., Caselli, R. J., Okazaki, H., Howard, F. M. Jr., Snow, B. J., and Calne, D. B. (1998) Hereditary form of parkinsonism-dementia. *Ann Neurol* 43: 768–781.

Nishioka, K., Hayashi, S., Farrer, M. J., Singleton, A. B., Yoshino, H., Imai, H., Kitami, T., Sato. K., Kuroda, R., Tomiyama, H., Mizoguchi, K., Murata, M., Toda, T., Imoto, I., Inazawa, J., Mizuno, Y., and Hattori, N. (2006). Clinical heterogeneity of alpha-synuclein gene duplication in Parkinson's disease. *Ann Neurol* 59: 298–309.

Ozelius, L. J., Senthil, G., Saunders-Pullman, R., Ohmann, E., Deligtisch, A., Tagliati, M., Hunt, A. L., Klein, C., Henick, B., Hailpern, S. M., Lipton, R. B., Soto-Valencia, J., Risch, N., and Bressman, S. B. (2006). LRRK 2 G2019S as a cause of Parkinson's disease in Ashkenazi Jews. *N Engl J Med* 354: 424–425.

Paisan-Ruiz, C., Jain, S., Evans, E. W., Gilks, W. P., Simon, J., van der Brug, M., Lopez de Munain, A., Aparicio, S., Gil, A. M., Khan, N., Johnson, J., Martinez, J. R., Nicholl, D., Carrera, I. M., Pena, A. S., de Silva, R., Lees, A., Marti-Masso, J. F., Perez-Tur, J., Wood, N. W, and Singleton, A. B. (2004). Cloning of the gene containing mutations that cause PARK 8-linked Parkinson's disease. *Neuron* 44: 595–600

Parker, W. D., Jr., Boyson, S. J., and Parks, J. K. (1989). Abnormalities of the electron transport chain in idiopathic Parkinson's disease. *Ann Neurol* 26: 719–723.

Parkinson, J. (1817). *An Essay on the Shaking Palsy*. London: Whittingham and Rowland.

Parkinson's Study Group. (1989). Effect of deprenyl on the progression of disability in early Parkinson's disease. *N Engl J Med* 321: 1364–137.

Pfeiffer, R. F., and Bodis-Wollner, I. (2005). *Parkinson's Disease and Nonmotor Dysfunction*. Totowa, NJ: Humana Press.

Polymeropoulos, M. H., Higgins, J. J., Golbe, L. I., Johnson, W. G., Ide, S. E., Di Iorio, G., Sanges, G., Stenroos, E. S., Pho, L. T., Schaffer, A. A., Lazzarini, A. M., Nussbaum, R. L., and Duvoisin, R. C. (1996). Mapping of a gene for Parkinson's disease to chromosome 4q21–q23. *Science* 274: 1197–1199.

Polymeropoulos, M. H., Lavedan, C., Leroy, E., Ide, S. E., Dehejia, A., Dutra, A., Pike, B., Root, H., Rubenstein, J., Boyer, R., Stenroos, E. S., Chandrasekharappa, S., Athanassiadou, A., Papapetropoulos, T., Johnson, W. G., Lazzarini, A. M., Duvoisin, R. C., Di Iorio, G., Golbe, L. I., and Nussbaum, R. L. (1997). Mutation in the alpha-synuclein gene identified in families with Parkinson's disease. *Science* 276: 2045–2047.

Prota, G., and d'Ischia, M. (1993). Neuromelanin: a key to Parkinson's disease. *Pigment Cell Res* 6: 333–335.

Ramsay, R. R., Salach, J. I., and Singer, T. P. (1986). Uptake of the neurotoxin 1-methyl-4-phenylpyridine (MPP+) by mitochondria and its relation to the inhibition of the mitochondrial oxidation of NAD+-linked substrates by MPP+. *Biochem Biophys Res Commun* 134: 743–748.

Ross, O. A., Toft, M., Whittle, A. J., Johnson, J. L., Papapetropoulos, S., Mash, D. C., Litvan, I., Gordon, M. F., Wszolek, Z. K., Farrer, M. J., and Dickson, D. W. (2006). LRRK2 and Lewy body disease. *Ann Neurol* 59: 388–393.

Schapira, A. H., Cooper, J. M., Dexter, D., Jenner, P., Clark, J. B., and Marsden, C. D. (1989). Mitochondrial complex I deficiency in Parkinson's disease. *Lancet* 1: 1269.

Schulz, J. B., and Beal, M. F. (1994). Mitochondrial dysfunction in movement disorders. *Curr Opin Neurol* 7: 333–339.

Shults, C. W., Oakes, D., Kieburtz, K., Beal, M. F., Haas, R., Plumb, S., Juncos, J. L., Nutt, J., Shoulson, I., Carter, J., Kompoliti, K., Perlmutter, J. S., Reich, S., Stern, M., Watts, R. L., Kurlan, R., Molho, E., Harrison, M., and Lew, M.; Parkinson Study Group (2002). Effects of coenzyme Q_{10} in early Parkinson disease: evidence of slowing of the functional decline. *Arch Neurol* 59: 1541–1550.

Singleton, A. B., Farrer, M., Johnson, J., Singleton, A., Hague, S., Kachergus, J., Hulihan, M., Peuralinna, T., Dutra, A., Nussbaum, R., Lincoln, S., Crawley, A., Hanson, M., Maraganore, D., Adler, C., Cookson, M. R., Muenter, M., Baptista, M., Miller, D., Blancato, J., Hardy, J., and Gwinn-Hardy, K. (2003). alpha-Synuclein locus triplication causes Parkinson's disease. *Science* 302: 841.

Spillantini, M. G., Crowther, R. A., Jakes, R., Hasegawa, M., and Goedert, M. (1998). alpha-Synuclein in filamentous inclusions of Lewy bodies from Parkinson's disease and dementia with lewy bodies. *Proc Natl Acad Sci USA* 95: 6469–6473.

Spillantini, M. G., Schmidt, M. L., Lee, V. M., Trojanowski, J. Q., Jakes, R., and Goedert, M. (1997). Alpha-synuclein in Lewy bodies. *Nature* 388: 839–840.

Tanner, C. M., and Ben-Shlomo, Y. (1999). Epidemiology of Parkinson's disease. *Adv Neurol* 80: 153–159

Tanner, C. M., Ottman, R., Goldman, S. M., Ellenberg, J., Chan, P., Mayeux, R., and Langston, J. W. (1999). Parkinson disease in twins: an etiologic study. *JAMA* 281: 341–346.

Testa, C. M., Sherer, T. B., and Greenamyre, J. T. (2005). Rotenone induces oxidative stress and dopaminergic neuron damage in organotypic substantia nigra cultures. *Brain Res Mol Brain Res* 134: 109–118.

Tetrud, J. W., and Langston, J. W. (1989). The effect of deprenyl (selegiline) on the natural history of Parkinson's disease. *Science* 245: 519–522.

Wersinger, C., and Sidhu, A. (2006). An inflammatory pathomechanism for Parkinson's disease? *Curr Med Chem* 13: 591–602.

Wirdefeldt, K., Gatz, M., Schalling, M., and Pedersen, N. L. (2004). No evidence for heritability of Parkinson disease in Swedish twins. *Neurology* 63: 305–311.

Yamada, M., Iwatsubo, T., Mizuno, Y., and Mochizuki, H. (2004). Overexpression of alpha-synuclein in rat substantia nigra results in loss of dopaminergic neurons, phosphorylation of alpha-synuclein and activation of caspase-9: resemblance to pathogenetic changes in Parkinson's disease. *J Neurochem* 91: 451–461.

Youdim, M. B., Ben-Shachar, D., and Riederer, P. (1989). Is Parkinson's disease a progressive siderosis of substantia nigra resulting in iron and melanin induced neurodegeneration? *Acta Neurol Scand Suppl* 126: 47–54.

Zarranz, J. J., Alegre, J., Gomez-Esteban, J. C., Lezcano, E., Ros, R., Ampuero, I., Vidal, L., Hoenicka, J., Rodriguez, O., Atares, B., Llorens, V., Gomez Tortosa, E., del Ser, T., Munoz, D. G., and de Yebenes, J. G. (2004). The new mutation, E46K, of alpha-synuclein causes Parkinson and Lewy body dementia. *Ann Neurol* 55: 164–173.

Zecca, L., Zucca, F.A., Wilms, H., and Sulzer, D. (2003). Neuromelanin of the substantia nigra: a neuronal black hole with protective and toxic characteristics. *Trends Neurosci* 26: 578–580.

Zimprich, A., Biskup, S., Leitner, P., Lichtner, P., Farrer, M., Lincoln, S., Kachergus, J., Hulihan, M., Uitti, R. J., Calne, D. B., Stoessl, A. J., Pfeiffer, R. F., Patenge, N., Carbajal, I. C., Vieregge, P., Asmus, F., Muller-Myhsok, B., Dickson, D. W., Meitinger, T., Strom, T. M., Wszolek, Z. K., and Gasser, T. (2004) Mutations in LRRK 2 cause autosomal-dominant parkinsonism with pleomorphic pathology. *Neuron* 44: 601–607.

4

α-Synuclein Aggregation and Parkinson's Disease

Vladimir N. Uversky

Abstract

Parkinson's disease (PD) is a multifactorial movement disorder in which both genetic and especially environmental factors play important roles. Substantial evidence implicates the aggregation of α-synuclein, an abundant and conservative presynaptic brain protein with unknown function, as a critical factor in PD. Rare familial cases of PD are associated with the mutations A30P (Ala to Pro substitution at position 30), E46K (Glu to Lys substitution at position 46), and A53T (Ala to Thr substitution at position 53) in α-synuclein. The primary structure of α-synuclein is characterized by several unusual motifs, and this protein was shown to have two closely related homologues, β-synuclein and γ-synuclein. Under the physiologic conditions *in vitro*, α-synuclein is characterized by the lack of rigid well-defined 3-D structure (i.e., it belongs to the class of natively unfolded proteins). Intriguingly, α-synuclein is known to possess remarkable conformational plasticity. The structure of this protein depends dramatically on the environment, and a number of absolutely unrelated conformations have been observed, including a partially folded intermediate that is a key intermediate in aggregation and fibrillation, several oligomeric species, and fibrillar and amorphous aggregates. A number of factors that either accelerate or inhibit the rate of fibrillation *in vitro* have been described. Accelerators include environmental factors such as certain pesticides and metals, molecular crowding, and various natural and synthetic charged polymers. Inhibitors include high concentrations of stabilizers such as trimethylamine *N*-oxide (TMAO), certain catechols, rifampicin, baicalein, acidic lipid vesicles, and protein homologues (β- and γ-synucleins). Oxidation of the four methionine residues in α-synuclein leads to the abolishment of fibrillation, as does the nitration of tyrosine residues. There is a strong correlation between the conformation of α-synuclein (induced by various factors) and its rate of fibrillation. The aggregation process appears to be branched, with one pathway leading to fibrils and another to oligomeric intermediates that may ultimately form amorphous deposits. The molecular basis of Parkinson's disease appears to be tightly coupled to the aggregation of α-synuclein and the factors that affect its conformation.

4.1. Introduction

Although clinical symptoms of Parkinson's disease (PD) were first described less than 200 years ago (Parkinson, 1817), there are reports of possible parkinsonian syndromes dating back thousands of years (Duvoisin, 1992a; Gourie-Devi *et al.*, 1991). Today, PD is the second most common neurodegenerative disorder after Alzheimer's disease. It is estimated that ~1.5 million

Americans are affected by PD. Because only a small percentage of patients are diagnosed before age 50, PD is generally considered an aging-related disease, and approximately 1 of every 100 persons over the age of 60 in the United States suffers from this disorder. PD is a slowly progressive disease that affects neurons of the substantia nigra, a small area of cells in the midbrain. Gradual degeneration of the dopaminergic neurons causes a reduction in the dopamine content. This, in turn, can produce one or more of the classic signs of Parkinson's disease: resting tremor on one (or both) side(s) of the body; generalized slowness of movement (bradykinesia); stiffness of limbs (rigidity); and gait or balance problems (postural dysfunction). The precise mechanisms of neuronal death are unknown as of yet. Some surviving nigral dopaminergic neurons contain cytosolic filamentous inclusions known as Lewy bodies (LBs) when found in the neuronal cell body or Lewy neurites (LNs) when found in axons (Forno, 1996; Lewy, 1912). Abundant LBs and LNs in the cerebral cortex are also neuropathologic hallmarks of dementia with Lewy bodies (DLB), a common late-life dementia that is clinically similar to Alzheimer's disease, the Lewy body variant of Alzheimer's disease, the diffuse Lewy body disease, the multiple system atrophy, and the neurodegeneration with brain iron accumulation type I (Arawaka *et al.*, 1998; Gai *et al.*, 1998; Lucking and Brice, 2000; Okazaki *et al.*, 1961; Spillantini *et al.*, 1998a; Spillantini *et al.*, 1997; Takeda *et al.*, 1998; Trojanowski *et al.*, 1998; Wakabayashi *et al.*, 1997; Wakabayashi *et al.*, 1998). The pathognomonic cellular lesions found in oligodendrocytes in multiple system atrophy are known as glial cytoplasmic inclusions, or GCIs (Forno, 1996; Papp *et al.*, 1989). It has been pointed out that all aforementioned disorders are brain amyloidoses unified by pathologic intracellular inclusions of aggregates having the α-synuclein protein as a key component (Lundvig *et al.*, 2005; Spillantini *et al.*, 1998a; Wakabayashi *et al.*, 1997).

Several observations implicate α-synuclein in the pathogenesis of PD and several other neurodegenerative disorders known as synucleinopathies (Trojanowski and Lee, 2003). Furthermore, it has been emphasized that because of the accumulation of the evidence connecting α-synuclein to mechanisms underlying PD and related neurodegenerative brain amyloidoses, the year 1997 became a turning point in the reassessment of the molecular basis of Parkinson's disease (Trojanowski, 2003). For example, a direct role for α-synuclein in the neurodegenerative processes in PD and Lewy body dementia is demonstrated by genetic evidence. Autosomal dominant early-onset Parkinson's disease and Lewy body dementia was shown to be induced in a small number of kindreds as a result of three different missense mutations in the α-synuclein gene, corresponding with Ala to Pro substitution at position 30 (A30P), Glu to Lys substitution at position 46 (E46K), and Ala to Thr substitution at position 53 (A53T) in α-synuclein (Kruger *et al.*, 1998; Polymeropoulos *et al.*, 1997; Zarranz *et al.*, 2004) or as a result of the hyperexpression of the wild-type α-synuclein protein due to gene triplication (Farrer *et al.*, 2004; Singleton *et al.*, 2004; Singleton *et al.*, 2003). Antibodies to α-synuclein detect this protein in LBs and LNs, the hallmark lesions of PD. Therefore, a substantial portion of fibrillar material in these specific inclusions was shown to be comprised of α-synuclein, and insoluble α-synuclein filaments can be recovered from purified LBs (Spillantini *et al.*, 1998b; Spillantini *et al.*, 1997). The production of wild-type (WT) α-synuclein in transgenic mice (Masliah *et al.*, 2000) or of WT, A30P, and A53T in transgenic flies (Feany and Bender, 2000), leads to motor deficits and neuronal inclusions reminiscent of PD. Interestingly, it has been established that the peptide derived from the central hydrophobic region of α-synuclein represents a second major intrinsic constituent of the Alzheimer's plaques. Under the particular conditions, cells transfected with α-synuclein might develop LB-like inclusions (see Chapter 5). Other important observations correlating a-synuclein and PD pathogenesis, being reviewed in more detail elsewhere (Dev *et al.*, 2003; Dickson, 2001; Goedert, 2001a; Goedert, 2001b; Trojanowski and Lee, 2003; Uversky and Fink, 2002b), are briefly outlined below. By numerous studies from different laboratories, it has been established that the recombinant α-synuclein easily assembles into amyloid-like fibrils *in vitro*, and this process is modulated by familial point mutations. LBs and LNs detected in familial and sporadic PD, dementia with LBs, LB variant of Alzheimer's disease (AD), familial and sporadic AD, and elderly Down's syndrome were shown to contain α-synuclein. Similarly, this protein is a building block of GCIs in neurodegeneration with brain iron accumulation type I and multiple system atrophy. α-Synuclein is abnormally phosphorylated, ubiquitinated, and nitrated in pathology-related inclusions (LBs, LNs, GCIs, and related lesions). Coexpression of chaperones or β-synuclein with α-synuclein in transgenic animals was shown to suppress the neurodegeneration. α-Synuclein-positive proteinaceous deposits were shown to accumulate in several animal models where parkinsonism was induced by exposure to different neurotoxicants.

All these findings indicate that α-synuclein is a key player in the pathogenesis of several neurodegenerative disorders. Based on these observations, it has been concluded that PD and related synucleinopathies are brain amyloidoses that may share similar mechanisms and targets for drug discovery (Trojanowski and Lee, 2003). This review is focused on structural properties and the aggregational behavior of human α-synuclein. It also considers the effect of PD-related cellular and environmental factors on this protein and its aggregation, aiming to shed more light on the molecular basis of Parkinson's disease, which appears to be tightly related to the α-synuclein aggregation.

4.2. Structural Characteristics of Human α-Synuclein

4.2.1. Peculiarities of α-Synuclein Amino Acid Sequence

α-Synuclein was first described as a neuron-specific protein localized to the nucleus and presynaptic nerve terminals (Maroteaux *et al.*, 1988). α-Synuclein from different organisms possesses a high degree of sequence conservation. For example, mouse and rat α-synucleins are identical throughout the first 93 residues, whereas human and canary proteins differ from them by only two residues (Clayton and George, 1998). Human α-synuclein is composed of 140 amino acid residues and can be divided into three regions (Fig. 4.1A):

(a) Residues 1–60 form the N-terminal region. It includes the sites of both familial PD mutations and contains four 11-amino-acid imperfect repeats with a highly conservative hexameric motif (KTKEGV). The N-terminal region is predicted to form amphipathic α-helices, typical of the lipid-binding domain of apolipoproteins (Clayton and George, 1998; George *et al.*, 1995).
(b) The central region, residues 61–95, comprises the highly amyloidogenic NAC sequence (NAC stands for *non-A*β component of Alzheimer's disease amyloid) (Han *et al.*, 1995; Ueda *et al.*, 1993). NAC contains two additional KTKEGV repeats and represents a second major intrinsic constituent of Alzheimer's plaques, amounting to about 10% of these inclusions (Ueda *et al.*, 1993). An 11-amino-acid segment within the central part of the NAC domain (corresponding with residues 73 to 83 of α-synuclein) is missing in β-synuclein. An important characteristic of the α-synuclein primary structure is six imperfect repeats within the first 95 residues, which result in a variation in hydrophobicity (Clayton and George, 1998; Davidson *et al.*, 1998; George *et al.*, 1995; Han *et al.*, 1995; Ueda *et al.*, 1993; Uversky, 2003b; Uversky and Fink, 2002a; Uversky and Fink, 2002b) with a strictly conserved periodicity of 11 (George *et al.*, 1995). Such a periodicity is characteristic of the amphipathic lipid-binding α-helical domains of apolipoproteins (Clayton and George, 1998; George *et al.*, 1995), which have been extensively studied and assigned to subclasses according to their unique structural and functional properties (Segrest *et al.*, 1990; Segrest *et al.*, 1992). α-Synuclein shares the defining properties of the class A2 lipid-binding helix, distinguished

by its clustered basic residues at the polar-apolar interface, positioned ±100 degrees from the center of apolar face, predominance of lysines relative to arginines among these basic residues, and several glutamate residues at the polar surface (Segrest *et al.*, 1990; Segrest *et al.*, 1992; Perrin *et al.*, 2000). In agreement with these structural features, α-synuclein was shown to bind specifically to synthetic vesicles containing acidic phospholipids (Davidson *et al.*, 1998; Perrin *et al.*, 2000). This binding was shown to be accompanied by a dramatic increase in α-helix content (Davidson *et al.*, 1998; Perrin *et al.*, 2000).

(c) The highly charged C-terminal region is constituted by residues 96–149. This part of α-synuclein is highly enriched in acidic residues and prolines, suggesting that it adopts a disordered conformation. Three highly conserved tyrosine residues, which are considered as a family signature of α- and β-synucleins, are located in this region.

4.2.2. α-Synuclein is a Natively Unfolded Protein

α-Synuclein is an intrinsically unstructured, or natively unfolded, or intrinsically disordered protein, possessing little or no ordered structure under the "physiologic" conditions (i.e., conditions of neutral pH and low to moderate ionic strength) (Weinreb *et al.*, 1996; Uversky *et al.*, 2001b). The recently discovered class of natively unfolded proteins is gaining considerable attention from researchers due to unique capability of such proteins to perform numerous biological functions despite the lack of unique structure (Wright and Dyson, 1999; Uversky *et al.*, 2000; Dunker *et al.*, 2001; Dunker and Obradovic, 2001; Dunker *et al.*, 2002a; Dunker *et al.*, 2002b; Tompa, 2002; Uversky, 2002a; Uversky, 2002b; Uversky, 2003b; Daughdrill *et al.*, 2005; Dyson and Wright, 2005; Fink, 2005). The sequence of the typical intrinsically disordered proteins is characterized by an amino acid compositional bias and the existence of highly predictable flexibility (Dunker *et al.*, 1998). Furthermore, the majority of the intrinsically disordered proteins, being substantially depleted in such amino acid residues as Ile, Leu, Val, Trp, Phe, Tyr, Cys, and Asn, are enriched in Glu, Lys, Arg, Gly, Gln, Ser, Pro, and Ala (Dunker *et al.*, 2001). These features may account for the low hydrophobicity and high net charge of the intrinsically unstructured proteins, a property that has been pointed out for a few individual natively unfolded proteins (Hemmings *et al.*, 1984; Weinreb *et al.*, 1996; Gast *et al.*, 1995) and that recently has been used to develop a predictor showing whether a given amino acid sequence encodes a globular (folded) or an intrinsically unstructured protein (Oldfield *et al.*, 2005; Uversky *et al.*, 2000). The differences in amino acid compositions between ordered and intrinsically disordered proteins described above constituted a ground for the development of numerous algorithms aiming for the prediction of disordered proteins/regions [reviewed in Bracken *et al.* (2004); Romero *et al.* (2004)]. Figure 1B represents the results of disorder prediction on human α-synuclein sequence using one of these predictors, PONDR VL3 (Li *et al.*, 1999; Romero *et al.*, 2001). It can be seen that α-synuclein is predicted to be completely disordered (as A PONDR score ≥ 0.5 corresponds with a prediction of disorder), emphasizing that its sequence is typical of the intrinsically disordered proteins.

In agreement with this prediction, α-synuclein was shown to possess little ordered structure under physiologic conditions (Weinreb *et al.*, 1996; Uversky *et al.*, 2001b; Eliezer *et al.*, 2001; Uversky, 2003b). For example, at neutral pH it is characterized by far-UV CD and FTIR spectra typical of a substantially unfolded polypeptide chain with a low content of ordered secondary structure (Uversky *et al.*, 2001b). The hydrodynamic properties of α-synuclein are in a good agreement with the results of far-UV circular dichroism (CD) and Fourier transform infrared (FTIR) studies. In fact, it has been established that α-synuclein, being essentially expanded, does not have a tightly packed globular structure but is slightly more compact than expected for a random coil (Uversky *et al.*, 2001b; Uversky *et al.*, 2002c; Uversky, 2003b). It has been shown that α-synuclein sedimented more slowly than globular proteins of similar molecular

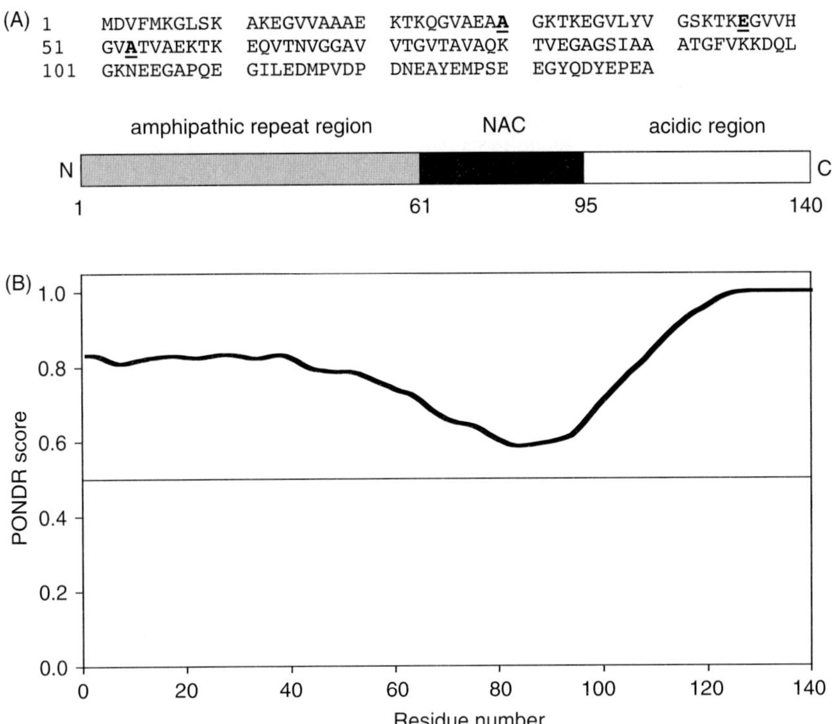

Figure 4.1. (A) Peculiarities of human α-synuclein amino acid sequence. The thee sites of early-onset PD-linked mutations (positions 30, 46, and 53) are highlighted. (B) Intrinsic disorder prediction on human α-synuclein sequence using one of these predictors, PONDR VL3.

weight, indicating that this protein is not compact (Weinreb *et al.*, 1996). Furthermore, based on the results of pulsed-field gradient NMR (which allows an estimation of the hydrodynamic radii), it has been concluded that α-synuclein is slightly collapsed (Morar *et al.*, 2001). Thus, at neutral pH, α-synuclein was shown to be essentially disordered but slightly more compact than a random coil. In agreement with this conclusion, a high-resolution NMR analysis of the protein under these conditions revealed that α-synuclein is largely unfolded in a solution but exhibits a region between residues 6 and 37 with a preference for helical conformation (Eliezer *et al.*, 2001). Interestingly, the results of recent studies on Raman optical activity spectra were consistent with the conclusion that α-synuclein may contain some poly-(L-proline) II (PPII) helical conformation (Syme *et al.*, 2002).

4.2.3. Structural Consequences of Amino Acid Substitutions

4.2.3.1. Familial Parkinson's Disease Point Mutations in α-Synuclein

There is considerable interest in the potential role of genetic factors in the etiology of PD (Golbe, 1990), as a small fraction of PD patients have a familial form of parkinsonism with an autosomal-dominant pattern of inheritance. Large pedigrees have been identified where members in different generations suffer from PD, with incidence of PD in family members being greater than in age-matched controls (Wood, 1998; Olanow and Tatton, 1999). Subsequently, a twin study revealed no difference in concordance between monozygotic and dizygotic twins of PD patients aged 60 years or older, but a significantly increased incidence was observed in monozygotic twins who developed PD at less than 50 years of age (Tanner *et al.*, 1999). This suggests that genetic factors are important in young-onset patients but are not likely to play a role in patients with sporadic PD (Olanow and Tatton, 1999). In agreement with this conclusion, PD was recently linked to the q21–23 region of chromosome 4 in a large Italian-American family, known as the Contursi kindred, members of which had a relatively early age of PD onset (Polymeropoulos *et al.*, 1997). A mutation in the gene encoding for α-synuclein was detected in this family, as well as in several apparently unrelated Greek families (Polymeropoulos *et al.*, 1997). Sequence analysis demonstrated that the mutation consisted of a single base change from G to A at position 209 (G209A), resulting in an Ala to Thr substitution at position 53 (A53T) in the α-synuclein. In the affected families, 85% of patients who expressed the mutant gene had clinical features of PD (Polymeropoulos *et al.*, 1997). An unusual aspect of the mutation is that the amino acid is already a threonine in rodents and other species (Hamilton, 2004).

A second mutation in the α-synuclein (A30P) has been described in a German family (Kruger *et al.*, 1998). Recently, one more α-synuclein mutation E46K was described in a Spanish family with autosomal dominant parkinsonism, dementia, and visual hallucinations of variable severity (Zarranz *et al.*, 2004). Furthermore, another genetic aberration, the triplication of the wild-type gene, has been reported in a large family from Iowa (Singleton *et al.*, 2003; Farrer *et al.*, 2004; Singleton *et al.*, 2004). Pathology from three of these kindreds is available, and the postmortem examination showed α-synuclein-positive Lewy bodies in the brain stem as well as nigral cell loss. However, α-synuclein pathology is not limited to the substantia nigra in many of these cases. In fact, the clinical descriptions of many of the patients with α-synuclein mutations reflect a disease with prominent dementia, presumably a reflection of the widespread cortical Lewy bodies in these cases. The Spanish E46K mutation was reported as Lewy body dementia (Zarranz *et al.*, 2004). In the Iowan kindred, glial cell inclusions are found (Singleton *et al.*, 2003; Farrer *et al.*, 2004; Singleton *et al.*, 2004), which would otherwise be typical of multiple system atrophy. Based on these observations, it has been emphasized that mutations in α-synuclein produce a fulminant disease that includes parkinsonism but is much more widespread and may resemble dementia with LB disease (Cookson, 2004). Furthermore, the disorder was shown also to be more progressive, tending to have an earlier onset than sporadic PD (Cookson, 2004). These findings strongly indicate that a single mutation in the human α-synuclein gene is sufficient to account for the PD phenotype.

Importantly, all three Parkinson's disease-related point mutations, A30P, E46K, and A53T, were shown to accelerate the α-synuclein aggregation (but not necessarily fibrillation) *in vitro* (El-Agnaf *et al.*, 1998c; Conway *et al.*, 1998; Narhi *et al.*, 1999; Giasson *et al.*, 1999; Conway *et al.*, 2000a; Conway *et al.*, 2000b; Conway *et al.*, 2000c; Li *et al.*, 2001; Li *et al.*, 2002; Greenbaum *et al.*, 2005). This raises the fundamental question on how a single point mutation in a disordered ("natively unfolded") conformation can affect the aggregation properties of the protein (see below).

It has been pointed out that the discovery of a genetic link between PD and α-synuclein shows that α-synuclein might play an important role in the pathogenesis of Parkinson's disease. Nevertheless, the majority of PD cases occur in individuals who carry the wild-type α-synuclein gene. It seems likely, then, that the familial early-onset mutations emphasize or enhance some property that is already inherent in wild-type properties. Thus, it has been anticipated that the A30P, E46K, and A53T mutations would change the structural properties of α-synuclein, affecting their predisposition to aggregate. To check this assumption, the structural properties of WT,

A30P, E49K, and A53T α-synucleins have been compared using a set of low-resolution techniques, such as circular dichroism, infrared spectroscopy, fluorescence, and several hydrodynamic approaches (El-Agnaf *et al.*, 1998c; Conway *et al.*, 1998; Giasson *et al.*, 1999; Narhi *et al.*, 1999; Conway *et al.*, 2000a; Conway *et al.*, 2000b; Conway *et al.*, 2000c; Greenbaum *et al.*, 2005; Li *et al.*, 2001; Li *et al.*, 2002;). Interestingly, detailed analysis revealed that the PD-related point mutations do not affect the overall structure of human α-synuclein, which remains natively unfolded (Li *et al.*, 2001; Li *et al.*, 2002; Greenbaum *et al.*, 2005). However, when high-resolution solution NMR spectroscopy was used to analyze the residual structure in human α-synuclein and its A30P and A53T mutants, it has been shown that the A30P mutation strongly attenuates the helical propensity found in the N-terminal region of in the wild-type α-synuclein, whereas the A53T mutation leaves this region unperturbed and exerts a more modest and local influence on structural propensity, resulting in a slight enhancement of preference for extended conformations in a small region around the site of mutation (Bussell and Eliezer, 2001). This was an important finding that provided some clues promoting better understanding of the molecular mechanism of α-synuclein aggregation and the effect PD-related mutations. In fact, the discovered residual helical structure does not represent an exclusive property of natively unfolded α-synuclein, as it is observed in unfolded states of several other proteins. Such residual structure has been proposed to play a potential role in early intramolecular protein folding events. Based on these observations, it has been concluded that this region of transient helical structure could play a role in α-synuclein aggregation, which is essentially an intermolecular folding process (Bussell and Eliezer, 2001). Obviously, the formation of a transient helical structure does not provide a clear path to the final product of the intermolecular folding/aggregation process, which is a β-sheet-rich structure (Li *et al.*, 2001; Li *et al.*, 2002). On the other hand, the observation that the A30P mutation eliminates this transient helicity, combined with the fact that this mutation accelerates early events in the oligomerization process, suggests that the residual helical structure may in fact retard or interfere with the initial intermolecular interactions of α-synuclein (Bussell and Eliezer, 2001).

4.2.3.2. Structural Properties of β- and γ-Synucleins

Synucleins belong to a family of closely related presynaptic proteins that arise from three distinct genes, described only in vertebrates (Clayton and George, 1998; Lavedan, 1998), which includes α-synuclein, which is also known as the non-amyloid component precursor protein, NACP, or synelfin (Ueda *et al.*, 1993; Jakes *et al.*, 1994; George *et al.*, 1995; Maroteaux *et al.*, 1988); β-synuclein, also referred to as phosphoneuro-protein 14, or PNP14 (Tobe *et al.*, 1992; Nakajo *et al.*, 1993; Jakes *et al.*, 1994; Lavedan *et al.*, 1998b); and γ-synuclein, also known as breast cancer-specific gene 1, BCSG1, and persyn (Ji *et al.*, 1997; Buchman *et al.*, 1998a; Buchman *et al.*, 1998b; Lavedan *et al.*, 1998a; Ninkina *et al.*, 1998).

Human β-synuclein is a 134-amino-acid neuronal protein which is 78% identical to α-synuclein (see Fig. 4.2). The α- and β-synucleins share a conserved C-terminus with three identically placed tyrosine residues. However, β-synuclein is missing 11 residues within the specific NAC region (Clayton and George, 1998; Lucking and Brice, 2000). The activity of β-synuclein may be regulated by phosphorylation (Nakajo *et al.*, 1993). Similar to α-synuclein, β-synuclein is expressed predominately in the brain. In contrast with α-synuclein, it is distributed more uniformly throughout the brain (Shibayama-Imazu *et al.*, 1993; Nakajo *et al.*, 1994).

Finally, γ-synuclein is the third member of the human synuclein family, which has 127 amino acids and shares 60% similarity with α-synuclein (see Fig. 4.2) (Clayton and George, 1998; Lucking and Brice, 2000). The specific feature separating this protein from α- and β-synucleins is a lack of the tyrosine-rich C-terminal signature (Clayton and George, 1998). γ-Synuclein is abundant in the spinal cord and sensory ganglia (Buchman *et al.*, 1998a; Buchman *et al.*,

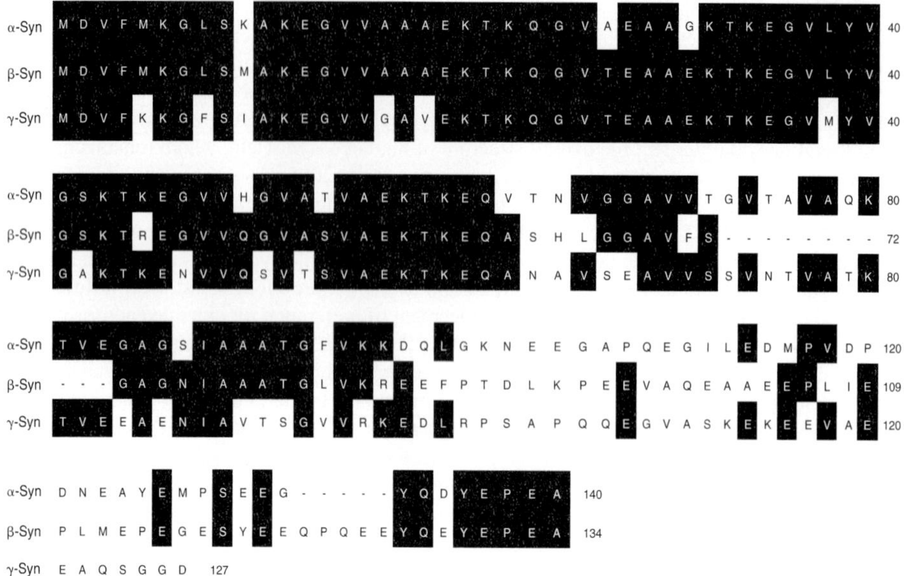

Figure 4.2. Optimally aligned sequences of α-, β-, and γ-synucleins. The total number of residues in each protein is 140, 134, and 127, respectively. Gaps in the β- and γ-sequences relative to α-synuclein are indicated with dashes. Residues in the β- and γ-synucleins similar to those in the α-sequence are highlighted.

1998b). Interestingly, this protein is more widely distributed within the neuronal cytoplasm than α- and β-synucleins, being present throughout the cell body and axons (Buchman *et al.*, 1998b).

The analysis of tissue distribution of the members of the synuclein family revealed that besides the central nervous system, these proteins can be found in several other tissues: α-synuclein in platelets (Hashimoto *et al.*, 1997), β-synuclein in Sertoli cells of the testis (Nakajo *et al.*, 1996; Ninkina *et al.*, 1999), and γ-synuclein in metastatic breast cancer tissue (Ji *et al.*, 1997) and epidermis (Ninkina *et al.*, 1999). Importantly, members of synuclein family are involved in the development of several pathologic conditions, including neurodegenerative diseases and breast cancer.

Structural properties of the members of the synuclein family have been compared using several physicochemical methods (Uversky *et al.*, 2002c). It has been established that all three proteins showed far-UV CD spectra typical of an unfolded polypeptide chain. Interestingly, α- and γ-synucleins possessed almost indistinguishable spectra, whereas the far UV-CD spectrum of β-synuclein showed a slightly increased degree of disorder. The suggestion on the increased unfoldedness of β-synuclein was further confirmed by hydrodynamic studies performed by size-exclusion chromatography and small angle x-ray scattering (SAXS). In fact, size-exclusion chromatographic analysis showed that β-synuclein was slightly more extended than α- and γ-synucleins: the R_S of β-synuclein was typical of a completely unfolded polypeptide chain, while α- and γ-synucleins were more compact than expected for a random coil (Uversky *et al.*, 2002c). This followed from comparison of values of the measured Stokes radius, R_S, with those calculated for a completely unfolded polypeptide chain of the appropriate molecular mass (Uversky, 1993; Uversky, 2002a; Uversky, 2002b; Uversky, 2003a). In the case of β-synuclein, the experimentally determined value (33.9 ± 0.4 Å) perfectly matched the calculated one (34.1 Å), but the Stokes radii measured for α- (31.8 ± 0.4 Å) and γ-synucleins (30.4 ± 0.4 Å) were notably lower than the corresponding calculated values (34.3 and 32.8 Å, respectively) (Uversky *et al.*, 2002c). These data emphasized the importance of the NAC region to maintain the residual partially collapsed structure in α- and γ-synucleins.

SAXS data further confirmed this conclusion. Guinier analysis of the scattering data shows that the synucleins are characterized by rather different R_g values at neutral pH. The observed R_g value for α-synuclein at neutral pH (40 ± 1 Å) is smaller than that estimated for a random coil conformation for a protein of this size (52 Å), indicating that the natively unfolded conformation of this protein is more compact than that of a random coil. On the other hand, the observed R_g value for β-synuclein (49 ± 1 Å) matches that expected for a completely unfolded polypeptide chain of this length (51 Å), which indicates the random coil conformation for this protein. γ-Synuclein had a very large R_g (61 ± 1 Å) under the conditions studied. This may be due to the very significant asymmetry of this protein or because of its self-association. Analysis of the SAXS forward-scattering intensity values, *I(0)* (which is proportional to the molecular weight of the molecule), confirmed that the large R_g is due to association. In fact, the *I(0)* value for γ-synuclein is more than twice that of for α- and β-synucleins (Uversky *et al.*, 2002c). Finally, the Kratky plots showed that the α- and β-synucleins do not have a well-developed globular structure, whereas γ-synuclein showed a characteristic maximum at low angles, indicating the presence of some globular structure in the oligomer (Uversky *et al.*, 2002c).

4.3. Conformational Behavior of Synucleins

4.3.1. Wild-Type α-Synuclein

Prologists have often wondered: what forces or factors will cause an intrinsically unstructured protein to fold. Because a natively unfolded protein possesses a low overall hydrophobicity and a large net charge, it is reasonable to assume that any changes in its environment leading to an increase in protein hydrophobicity and/or decrease in its net charge should be accompanied by at least a partial folding. The excess negative charge of α-synuclein at neutral pH (pI = 4.7) would be neutralized at lower pH values, and the overall hydrophobicity of a protein will increase with increasing temperature. Thus, a partial folding of α-synuclein under conditions of high temperature and/or low pH is expected. In agreement with this suggestion, it has been shown that α-synuclein adopts a partially folded conformation at acidic pH or at high temperatures (Uversky *et al.*, 2001b; Uversky and Fink, 2002b; Uversky, 2003b). In fact, both far-UV CD and FTIR spectra revealed the existence of pH- and temperature-induced formation of some ordered secondary structure, which was accompanied by a substantial decrease in hydrodynamic dimensions and development of the beginnings of a tightly packed core (Uversky, 2003b; Uversky and Fink, 2002b; Uversky *et al.*, 2001b). Thus, α-synuclein is unstructured under conditions of neutral pH. Either a decrease in pH or an increase in temperature transformed α-synuclein into a partially folded conformation. This illustrates "turned out" response to the changes in the environment typical for the natively unstructured proteins, which, unlike "normal" globular proteins, gain rather than lose ordered structure at extreme pH and high temperatures (Uversky, 2003b). The structure-forming effects of low pH were attributed to the minimization of the large net negative charge present at neutral pH, thereby decreasing intramolecular charge-charge repulsion and permitting hydrophobic-driven collapse to the partially folded conformation. The effect of elevated temperatures was attributed to the increased strength of the hydrophobic interaction at higher temperatures, leading to a stronger hydrophobic driving force for folding (Uversky *et al.*, 2001b).

4.3.2. PD-related α-Synuclein Mutants

In order to gain better understanding of the effects of PD-related mutations in α-synuclein on the conformational behavior of this protein, the structural properties of A30P and A53T mutants were compared with those of wild-type protein under the various conditions. Interestingly, the conformational transitions induced in these three proteins by a decrease in pH or an increase in temperature or methanol concentration were shown to be indistinguishable (Li *et al.*, 2001; Li *et al.*, 2002). Likewise, wild-type α-synuclein and the A30P and A53T mutants may be transformed into the partially folded intermediate state by decreasing the pH or increasing the temperature (Li *et al.*, 2001; Li *et al.*, 2002). Importantly, the structure of this intermediate state was shown to be independent of the mutations. Thus, the monomeric forms of WT, A30P, and A53T α-synucleins exhibit identical structural properties and conformational behavior (Li *et al.*, 2001; Li *et al.*, 2002).

4.3.3. Studies on β- and γ-Synucleins

The comparison of conformational behavior of different members of the synuclein family revealed that these proteins possess a comparable response to the changes in their environment. It has been established that although far-UV CD spectra of α-, β-, and γ-synucleins were slightly different at neutral pH, as the pH was decreased, changes were observed in the spectral shape for all three proteins, where the decrease in the minimum at 196 nm was accompanied by an increase in negative intensity around 222 nm, reflecting the pH-induced secondary structure formation (Uversky *et al.*, 2002c). All three proteins possessed almost identical far-UV CD spectra at acidic pH. These results were confirmed by the results of gel-filtration analysis, which showed that although β-synuclein was slightly more extended than α- and γ-synucleins at neutral pH, all three proteins possessed the same degree of compaction in acidic solutions (Uversky *et al.*, 2002c). Importantly, the pH-induced changes in the far-UV CD spectra of the synucleins were completely reversible. It has been suggested that, much like α-synuclein, the pH-induced increase in the structure of β- and γ-synuclein represents an intramolecular process involving the formation of a partially folded intermediate, and not self-association (Uversky *et al.*, 2002c).

4.4. Aggregation of α-Synuclein and Neurodegenerative Diseases

The presence of aggregated α-synuclein in a form of intraneuronal inclusions is a shared property of sporadic and familial autosomal dominant early-onset Parkinson's disease, Lewy body dementia, multiple system atrophy, neurodegeneration with brain iron accumulation type I, and several other neurodegenerative disorders. Thus, the strong correlation between α-synuclein-induced diseases and the aggregation of this protein justifies the study of putative pathogenic effects of α-synuclein aggregates. Several important aspects related to the peculiarities of α-synuclein aggregation are presented below. First, the molecular mechanism of α-synuclein fibrillation *in vitro* is described. Then the effects of PD-related point mutations on aggregation of this protein are considered, followed by the overview of cytotoxicity of different aggregated forms of α-synuclein. PD is believed to be an environmental disorder, as exposure to different environmental neurotoxicants was shown to contribute to the pathogenesis of this disorder. We are showing below that the modulation of α-synuclein aggregation induced by different neurotoxicants could be used to link the effect of environmental factors to the PD development. Finally, numerous cellular factors and processes were shown to contribute to α-synuclein aggregation and thus could be related to the pathogenesis of synucleinopathies. To illustrate this point, the effects of oxidative modifications, phosphorylation, C-terminal truncations, interaction with membranes and phospholipids, as well as binding to different proteins and charged macromolecules on α-synuclein aggregation behavior are considered. Furthermore, the brief analysis of molecular crowding and the structure of water as factors dramatically affecting α-synuclein aggregation is present.

4.4.1. Aggregation of Wild-Type Protein *In Vitro*

Although α-synuclein is a natively unfolded protein, it forms fibrils of highly organized secondary structure. For example, the FTIR spectrum of α-synuclein fibrils shows the major contribution from β-sheet (Uversky *et al.*, 2001b). Furthermore, x-ray diffraction analysis of α-synuclein fibrils showed the characteristic pattern of a cross-β-sheet structure in which the β-strands lie perpendicular to the long fiber axis, typical of all amyloid fibrils (Serpell *et al.*, 2000). Electron microscopy analysis indicates that human α-synuclein filaments are typically 6–9 nm in width and several micrometers long (Serpell *et al.*, 2000). Atomic force microscopy images of α-synuclein fibrillation reveal three different fibrillar species, corresponding with protofilaments, protofibrils, and mature fibrils (Khurana *et al.*, 2003). Clearly, α-synuclein cannot be present as an extended linear polymer within the fibril, because in a fully extended conformation the maximal linear dimension of a polypeptide with n residues is $n \times 3.63$ Å (Creighton, 1993). This gives ~51 nm for α-synuclein, which is at least 5 times greater than the diameter of the filament. Thus, α-synuclein must be folded within a fibril. It has been shown that the decrease in pH or an increase in temperature induces partial folding of α-synuclein. In contrast with an unfolded protein, a partially folded intermediate is anticipated to have contiguous hydrophobic patches on its surface, which are likely to foster self-association and hence potentially fibrillation (Uversky *et al.*, 2001b).

The histologic dye thioflavine T (TFT) is widely used for the detection of amyloid fibrils (Levine, 1995; Naiki *et al.*, 1989; Naiki *et al.*, 1990). In the presence of fibrils, TFT gives rise to a new excitation maximum at 450 nm and enhanced emission at 482 nm, whereas unbound TFT is essentially nonfluorescent at these wavelengths. Thus, the binding of TFT to α-synuclein fibrils is a very convenient method for studying the kinetics of fibril formation. The analysis of time-dependent changes in the TFT fluorescence during the process of α-synuclein fibril formation as a function of pH or as a function of temperature revealed that this process can be described by characteristic sigmoidal curves, which have an initial lag phase, a subsequent exponential growth phase, and a final equilibrium phase. Such curves are consistent with a nucleation-dependent polymerization model, in which the lag corresponds with the nucleation phase and the exponential part to fibril growth (elongation) (Jarrett and Lansbury, 1992; Jarrett and Lansbury, 1993; Lomakin *et al.*, 1997; Wood *et al.*, 1999). It has been established that decreasing the pH or increasing the temperature resulted in a very substantial acceleration of the kinetics of α-synuclein fibrillation (Uversky *et al.*, 2001b). pH has similar effects on both the lag time and the elongation rate (Uversky *et al.*, 2001b). Furthermore, the pH- and temperature-induced changes in the α-synuclein fibrillation kinetics were shown to be coincident with the pH- and temperature-driven structural transformations (Uversky *et al.*, 2001b). In other words, an excellent correlation between intramolecular conformational change and fibril formation has been established. This was a very important observation, and it was consistent with the conclusion that the process of α-synuclein fibrillation is dramatically accelerated by the partial folding of the natively unfolded protein, suggesting that this intermediate is a key species on the fibril-forming pathway (Uversky *et al.*, 2001b). The results were consistent with the following minimum scheme for α-synuclein fibrillation (Scheme 4.1) (Uversky *et al.*, 2001b):

$$U_N \rightleftarrows I \rightleftarrows Nucleus \longrightarrow F \qquad \text{(Scheme 4.1)}$$

where U_N, I, $Nucleus$, and F correspond with natively unfolded conformation, partially folded intermediate, fibril nucleus, and fibril, respectively. From this model, the two key kinetic steps are the structural transformation leading to the intermediate, and the formation of the nucleus (Uversky *et al.*, 2001b). Thus, any environmental or genetic factors that shift the equilibrium in favor of the intermediate will facilitate fibril formation, as observed. It has been also assumed that fibrillation of α-synuclein, leading to Lewy body formation and Parkinson's and related Lewy body diseases, as well as the development of other synucleinopathies, may arise from various factors that would significantly populate or increase the concentration of the aggregation-competent intermediate (Uversky *et al.*, 2001b). Possibilities include point mutations, nonpolar molecules, such as some pesticides, which might preferentially bind to the partially folded intermediate, and cations, which might mimic the effect of low pH (high proton concentration), as well as factors that result in an increase in the concentration of α-synuclein itself or cause modification of the protein (e.g., via oxidative damage, etc.).

Due to the conformational breathing of the α-synuclein molecule, even under the physiologic conditions *in vitro* there is equilibrium between natively unfolded and partially folded conformations. Obviously, high protein concentration might be predicted to increase the rate of fibrillation due to the increased concentration of the partially folded amyloidogenic intermediate. In agreement with this assumption, it has been shown that an increase in α-synuclein concentration resulted in a decrease in the lag time and an increase in the rate constant for fibril growth (Uversky *et al.*, 2001c). Furthermore, it has been established that there is an inverse linear correlation between the logarithm of α-synuclein concentration and the duration of lag time (Uversky *et al.*, 2001c). This reflects that the formation of the aggregation-prone partially folded conformation represents one of the rate-limiting steps in α-synuclein nucleation, and this process might be the first-order one. This would mean that the key partially folded conformation, once formed, oligomerizes very rapidly to form insoluble fibrils (Uversky *et al.*, 2001c). Similarly, a linear dependence of the first-order rate constant for fibril growth (elongation) on the protein concentration was observed (Uversky *et al.*, 2001c). The kinetics of elongation have been shown to follow first-order kinetics for several other proteins, including amyloid β-protein and insulin. The simplest explanation for this fact is that increasing protein concentration leads to increasing numbers of fibrils, due to increasing concentration of nuclei. Strong dependence of the fibrillation kinetics on α-synuclein concentration is potentially very important for the pathogenesis of PD and other synucleinopathies. In fact, it has been recently shown that a large family develops autosomal dominant PD (average age of onset, 34 years), ranging clinically from DLB to typical PD, due to the triplication of α-synuclein locus, thus, due to potentially enhanced level of α-synuclein production (Farrer *et al.*, 2004; Singleton *et al.*, 2004; Singleton *et al.*, 2003).

4.4.2. Effect of PD-Related Mutations on α-Synuclein Aggregation

Detailed analysis of the effect of point mutations on aggregation behavior revealed that wild-type and mutant α-synucleins can form soluble oligomers, amorphous aggregates, and fibrils (Li *et al.*, 2001; Narhi *et al.*, 1999). A53T showed the fastest rate of fibril formation, whereas A30P formed fibrils more slowly than the WT protein (Conway *et al.*, 1998; Li *et al.*, 2001; Li *et al.*, 2002; Narhi *et al.*, 1999). On the other hand, both mutant proteins were shown to be more prone to aggregate than wild-type protein, and the total aggregation (fibrillar and nonfibrillar) rates decreased in the following order A53T > A30P > WT (Li *et al.*, 2001; Narhi *et al.*, 1999). The absence of detectable structural differences between partially folded conformations of α-synucleins raises the question of how the A30P and A53T mutations affect the aggregation propensity of the protein. To explain this contradiction, the amino acid sequences of WT, A30P, and A53T have been analyzed on the basis of hydrophobicity and propensities to form β-sheet or α-helix (Li *et al.*, 2001). Surprisingly, both mutations have been shown to reduce hydrophobicity in the vicinity of the substitution. This was rather unexpected, because the hydrophobic interactions are assumed to be important in aggregation (Li *et al.*, 2001). At the same time, the propensity to form α-helical structure was shown to be somewhat diminished in the N-terminal region of both mutants. Because this region is believed to be involved in binding to lipids, this may also play a role in the physiologic consequences of the mutations (Bussell and Eliezer, 2004). Importantly, as it has been discussed above, high-resolution solution NMR spectroscopic analysis revealed that A30P mutation disrupts a region of residual helical structure that exists in the wild-type protein, whereas the A53T mutation results in a slight enhancement of a small region around the site of mutation with a preference for extended conformation (Bussell and Eliezer, 2001). Finally, the analysis of amino acid sequences predicted that both A30P and A53T mutants might form β-structure more readily than WT α-synuclein (Li *et al.*, 2001). The aggregated species of many proteins are enriched in β-structure. Furthermore, it has been established that the transformation of α-helical (or disordered) structure to β-sheets (including intermolecular ones) is a hallmark of aggregation and fibrillation processes. Taking these observations into account, it has been hypothesized that the increased internal susceptibility of A30P and A53T to form β-sheets may not be strong enough to alter the structure of the monomeric proteins but may affect the aggregation behavior of the α-synuclein mutants through specific stabilization of an intermolecular β-structure. This model for the effect of the familial Parkinson's disease mutations was illustrated by the comparison of the aggregation models for the WT (Scheme 4.2) and the mutants (Scheme 4.3) (Li *et al.*, 2001):

$$U_N \underset{}{\overset{\text{I}}{\rightleftarrows}} I \overset{\text{II}}{\rightleftarrows} \textit{Nucleus} \longrightarrow F, \qquad I \xrightarrow{\text{III}} A \qquad \text{(Scheme 4.2)}$$

$$U_N \underset{}{\overset{\text{I}}{\rightleftarrows}} I \overset{\text{II}}{\rightleftarrows} \textit{Nucleus} \longrightarrow F, \qquad I \xrightarrow{\text{III}} A \qquad \text{(Scheme 4.3)}$$

In this model, ***F*** and ***A*** represent fibrils and amorphous aggregates, respectively, U_N is the natively unfolded state, and ***I*** represents the partially folded intermediate. The Roman numerals indicate the major stages of the aggregation process. It has been proposed that the structural properties of U_N and ***I***, as well as the rate of their interconversion (stage I), are unaffected by the A30P and A53T mutations. However, the rates of stages II and III are facilitated by the familial Parkinson's disease point mutations (thicker arrows). This would result from the mutation-enhanced probability of forming intermolecular β-structure (Li *et al.*, 2001). As a result, the mutants show a faster rate of fibrillation (A53T) or amorphous aggregation (A30P).

It has been suggested that the acceleration of the initial oligomerization of A30P α-synuclein may be a result of the destabilization of a transient helical propensity that interferes with the requisite intermolecular interactions (Bussell and Eliezer, 2004). On the other hand, it is also likely that this mutation can also hinder the formation of β-sheet structure in the same region. In fact, proline is known to be a very poor β-sheet-forming residue (Chou and Fasman, 1978). All this suggests that A30P mutation might retard the formation of the β-sheet-rich mature fibril structure. Consequently, the initial oligomerization steps probably do not involve β-sheet structure in the neighborhood of the residue 30. However the subsequent stages of the mature fibril formation process may require this region of α-synuclein to participate in β-sheet structure, and these steps would likely be hindered by the presence of the proline mutation (Bussell and Eliezer, 2004).

Interestingly, the sequence of mouse α-synuclein differs from the human sequence only at seven positions, with one of these differences being at position 53, where the mouse protein contains a threonine residue (Clayton and George, 1998; Rochet *et al.*, 2000). Therefore, the A53T mutation in human protein is in fact a "reversion" to a residue found in a lower organism. This raised the question of why neither parkinsonian symptoms nor Lewy bodies have been observed in aged mice. These observations resulted in detailed characterization of murine protein. It has been shown that mouse α-synuclein also adopts a natively unfolded conformation. At elevated concentrations, mouse protein fibrillated faster than the three human synucleins. The fibrillation of mouse protein is inhibited by WT and A53T α-synuclein, leading to an accumulation of nonfibrillar oligomers (Rochet *et al.*, 2000).

Recently, the effect of E46K mutation on aggregation of α-synuclein has been analyzed (Greenbaum *et al.*, 2005). It has been established that the E46K mutation is also able to increase the propensity of α-synuclein to fibrillate, but this effect was less pronounced than that of the A53T mutation (Greenbaum *et al.*, 2005). Interestingly, the E46K mutation is located in the fourth KTKEGV-type repeat in the amino-terminal region of α-synuclein. It has been emphasized that a Glu residue similar to E46 is present in five of the six degenerative repeats in α-synuclein, and the only repeat that does not have such a residue (repeat 2) has Glu residues adjacent to each side of the repeat (Greenbaum *et al.*, 2005). Based on these observations, it has been suggested that the N-terminal region of α-synuclein and, more specifically, Glu residues in the repeats may be important in regulating the ability of α-synuclein to polymerize into amyloid fibrils (Greenbaum *et al.*, 2005).

In vitro findings of the effects of the A53T and E46K mutations on α-synuclein polymerization were shown to be consistent with pathogenesis in human subjects. In fact, patients with the A53T mutation have an earlier age of disease onset (average age of onset, 45 years) than those carrying the E46K mutation (average age of onset, 60 years) (Greenbaum *et al.*, 2005). These observations, together with the described above results showing that the A53T, E46K, and A30P mutations in α-synuclein lead to increased propensity to aggregate compared with the wild-type protein, provide strong support for a direct and critical role of α-synuclein aggregation in the etiology of Parkinson's disease.

4.4.3. Cytotoxicity of α-Synuclein Aggregates

The next important question is which species present during the fibrillation of α-synuclein could be responsible for the neuronal death in PD. Is it the natively unfolded monomer, the fibrillogenic monomeric intermediate, some oligomeric species (protofibrils), or mature fibril? Several lines of evidence indicated that this killing conformation is an oligomer. For example, it has been recently shown that oligomeric α-synuclein is deposited in detergent insoluble fractions from the brains of patients with the triplication mutation (Miller *et al.*, 2004). In fact, the puzzling property of the A30P mutant to dramatically accelerate the initial oligomerization of α-synuclein and to significantly retard the formation of mature fibrils (Conway *et al.*, 2000c; Li *et al.*, 2001; Li *et al.*, 2002) represents a key factor leading to the theory that oligomeric intermediates of α-synuclein, rather than mature fibrils, may in fact be the disease-associated species of the protein (Lansbury, 1999) and that oligomers, not fibrils, are toxic (Lashuel *et al.*, 2002a; Lashuel *et al.*, 2002b). It has been pointed out that oligomers, also known as protofibrils (Lashuel *et al.*, 2002b), can form annular (doughnut-like) structures that may have pore-like properties and might damage membranes (Volles and Lansbury, 2002). In agreement with this theory, annular α-synuclein oligomers have been recently isolated from human brain samples (Pountney *et al.*, 2004).

Additional evidence to support the idea of oligomer toxicity has been uncovered (Cookson, 2004):

(a) In cell models, toxicity is usually seen without heavily aggregated α-synuclein leading to the suggestion that soluble species mediate toxicity (Xu *et al.*, 2002).
(b) Detectable aggregation of α-synuclein and deposition of this protein into insoluble fractions occurs later than cell death *in vitro* (Gosavi *et al.*, 2002).
(c) Transgenic mice, which express human wild-type α-synuclein at a level comparable with the level of endogenous mouse protein, developed nonfibrillar intraneuronal inclusions in several brain areas, including the substantia nigra pars compacta (Masliah *et al.*, 2000).
(d) Transgenic mice expressing A53T and WT exhibited neurodegeneration outside the substantia nigra without fibrillar inclusions (van der Putten *et al.*, 2000).
(e) Lentiviral-based expression of human α-synuclein in rat substantia nigra resulted in selective dopaminergic toxicity with nonfibrillar inclusions (Lo Bianco *et al.*, 2002).
(f) The α-synuclein-containing inclusions in some animal models do not contain fibrils, and the fibril-containing inclusions of the fly can occur in the absence of neurodegeneration (Auluck and Bonini, 2002; Auluck *et al.*, 2002).
(g) Although involvement of α-synuclein in PD is often associated with the fact that this protein represents a major constituent of LBs, α-synuclein aggregation has been identified in neurons in forms other than Lewy bodies. Engorged neuronal processes containing ovoid, fusiform, club-shaped, and spherical α-synuclein inclusions have been found in substantia nigra pars compacta and other structures in both PD and LB disease brains (Galvin *et al.*, 1999; Irizarry *et al.*, 1998). Furthermore, thread-like aggregates of α-synuclein, which can be either coarse or fine, have also been detected in both dendrites and axons (Arima *et al.*, 1998; Braak *et al.*, 1999; Crowther *et al.*, 2000).

Based on these and several other observations, it has been suggested that the Lewy body formation does not cause but protects against PD (Goldberg and Lansbury, 2000). This strongly suggests that the death of neurons can be provoked by the formation of protofibrilar species. It has been pointed out that the process of α-synuclein fibrillation *in vitro* is frequently accompanied by the formation of protofibrils, which are usually observed by AFM and EM as a heterogeneous mixture of morphologies, including spherical, annular, pore-like, tube-like, and chain-like structures (Volles and Lansbury, 2003). Recently, it has been demonstrated that soluble 30- to 50-nm-sized annular α-synuclein oligomers could be released by a mild detergent treatment from GCIs purified from multiple system atrophy brain tissue (Pountney *et al.*, 2004), suggesting that pathologic synucleinopathy aggregates can be a source of annular α-synuclein species. Interestingly, the vast majority of recombinant α-synuclein was shown to yield spherical oligomers after detergent treatment. However, *in vitro* binding of Ca^{2+} to monomeric α-synuclein was shown to induce the rapid formation of annular oligomers (Lowe *et al.*, 2004). Based on these observations, it has been suggested that an intracytoplasmic Ca^{2+} concentration might affect the speciation of α-synuclein and that soluble α-synuclein annular oligomers may be cytotoxic species, either by interacting with cell membranes or components of the ubiquitin proteasome system (Pountney *et al.*, 2005).

4.4.4. Linking the Effect of Environmental Factors, α-Synuclein Aggregation, and PD

Although the exact cause of PD is unknown as of yet, considerable evidence suggests a multifactorial etiology involving genetic and environmental factors. The clinical and pathologic syndromes associated with both the familial PD cases and those caused by environmental agents are remarkably similar. This supports the hypothesis that PD is a heterogenous disease likely to be caused by more than one specific etiologic factor (Orth and Tabrizi, 2003). As it has been already pointed out, a significant genetic element in the etiology of PD was suggested by results from twin studies (Tanner *et al.*, 1999), case-control analysis (Gasser, 1998a; Gasser, 1998b; Gasser, 2001; Sveinbjornsdottir *et al.*, 2000), and the identification of mutations in genes of α-synuclein (Kruger *et al.*, 1998; Polymeropoulos *et al.*, 1997; Singleton *et al.*, 2003; Zarranz *et al.*, 2004), parkin (Kitada *et al.*, 1998), and DJ1 (Bonifati *et al.*, 2003), associated with autosomal inherited forms of PD. The search for environmental factors that influence the pathogenesis of PD has been significantly slower, and clinical and experimental studies in this field have only just begun. However, PD is now considered likely to be an "environmental" disease, as several lines of evidence point to environmental exposures being the contributing factors in the pathogenesis of this disorder (Di Monte, 2001; Di Monte, 2003; Di Monte *et al.*, 2002; McCormack *et al.*, 2002; Tanner, 1989; Tanner *et al.*, 1989), and a positive correlation between the prevalence of PD and industrialization has been recognized (Tanner, 1989).

It has been shown that certain neurotoxins could promote and accelerate the development of PD, whereas other environmental agents may be neuroprotective. For example, several epidemiologic studies revealed that caffeine consumption and cigarette smoking may decrease the risk of PD (Ragonese *et al.*, 2003; Ross *et al.*, 2000; Ross and Petrovitch, 2001). On the other hand, significant data are currently accumulated convicting a number of environmental agents to be the neurotoxins causing nigrostriatal damage and therefore contributing to the PD development. Among these PD-promoting environmental toxins are metals (Altschuler, 1999; Good *et al.*, 1992; Gorell *et al.*, 1999a; Gorell *et al.*, 1999b; Hirsch *et al.*, 1991; Tanner, 1989; Yasui *et al.*, 1992); solvents (Davis and Adair, 1999; Hageman *et al.*, 1999; Pezzoli *et al.*, 1996; Seidler *et al.*, 1996; Uitti *et al.*, 1994); carbon monoxide (Klawans *et al.*, 1982); and 1-methyl-4-phenyl-1,2,3,6 tetrahydropyridine (MPTP) (Langston and Ballard, 1983). Finally, one of the most consistent findings of epidemiologic studies is the association between increased PD risk and such environmental factors as rural residence (Liou *et al.*, 1997; Marder *et al.*, 1998; Morano *et al.*, 1994); farming (Fall *et al.*, 1999; Gorell *et al.*, 1998; Liou *et al.*, 1997; Semchuk *et al.*, 1993); well-water drinking (Liou *et al.*, 1997; Marder *et al.*, 1998; Morano *et al.*, 1994); exposure to agricultural chemicals, pesticides and/or herbicides (Fall *et al.*, 1999; Gorell *et al.*, 1998; Liou *et al.*, 1997; Semchuk *et al.*, 1992; Semchuk *et al.*, 1993; Vanacore *et al.*, 2002). The effect of several environmental factors as well as the influence of some cellular factors on structure and aggregation of human α-synuclein is considered below. This analysis provides basis for the understanding of the molecular mechanisms of PD pathogenesis, as it is very possible that α-synuclein is involved in the environmental hypothesis of PD neurodegeneration, and the direct interaction between this protein and neurotoxicants might contribute to the development of PD (Di Monte, 2001).

4.4.4.1. Effect of Pesticides and Herbicides

About a billion pounds of pesticides are used annually in the United States. Both epidemiologic and clinical observations reveal pesticides and herbicides as an important environmental PD risk factor (Fall *et al.*, 1999; Gorell *et al.*, 1998; Liou *et al.*, 1997; Semchuk *et al.*, 1992; Semchuk *et al.*, 1993; Vanacore *et al.*, 2002), with paraquat being the most often implicated as a potential neurotoxicant in PD (Fall *et al.*, 1999; Gorell *et al.*, 1998; Hertzman *et al.*, 1994; Liou *et al.*, 1997; Semchuk *et al.*, 1992; Semchuk *et al.*, 1993; Vanacore *et al.*, 2002). Thus, the analysis of α-synuclein interaction with these environmental agents could provide a critical link between toxicant exposures and the pathogenesis of α-synuclein-containing inclusions in idiopathic PD.

In fact, at least four lines of experimental evidences support a relationship between α-synuclein and toxicant exposure. First, injections of mice with MPTP have been reported to enhance α-synuclein mRNA and protein levels in midbrain extracts and to increase the number of α-synuclein-positive neurons in the substantia nigra (Vila *et al.*, 2000). Second, same toxicant-induced upregulation in α-synuclein production has been observed when mice were administered paraquat. Furthermore, in this case, paraquat exposure led to the formation of protein aggregates, which being specifically accumulated within neurons of the substantia nigra pars compacta, are thioflavine S–positive (i.e., possess properties of amyloid-like fibrils), and dual labeling and confocal imaging confirmed that these aggregates contained α-synuclein (Manning-Bog *et al.*, 2002). Third, α-synuclein-positive inclusions have been described as a feature of the neurodegenerative process triggered by rotenone infusion into rats (Betarbet *et al.*, 2000; Sherer *et al.*, 2003). Finally, experiments with the purified protein *in vitro* revealed that fibrillation α-synuclein is dramatically accelerated by several common pesticides (including rotenone, paraquat, and maneb), and this fibril-promoting effect was clearly dose-dependent (Manning-Bog *et al.*, 2002; Uversky *et al.*, 2002b; Uversky *et al.*, 2001d). On the other hand, 1-methyl-4-phenylpyridinium (MPP^+) did not affect the α-synuclein fibrillation (Uversky *et al.*, 2002b). These facts might give a simple mechanical explanation for the presence of proteinaceous lesions in rotenone and paraquat + maneb models and the preferential lack of such deposits in MPTP model—cellular deposits develop only when toxins are directly involved in the interaction with α-synuclein and promote its aggregation (this can be easily tested in simple *in vitro* experiments). This also provides a rationale for using both *in vitro* and *in vivo* approaches for the screening of putative neurotoxicants, which, by affecting α-synuclein conformation and aggregation, may play a role in the pathogenesis of synucleinopathies (Manning-Bog *et al.*, 2002).

In vitro analysis revealed that certain pesticides, representing different chemical classes, significantly stimulate the rate of formation of fibrils when incubated with α-synuclein at pH 7.5, 37°C, and the efficiency to accelerate fibrillation was shown to be correlated directly with the potency to induce a partial folding in α-synuclein (Li *et al.*, 2002; Manning-Bog *et al.*, 2002; Uversky *et al.*, 2002b; Uversky *et al.*, 2001d). Thus, the acceleration of α-synuclein aggregation by environmental PD risk factors is directly correlated with an increase in the concentration of critical partially folded conformation.

The results of these *in vitro* studies were confirmed by the analysis of the effect of paraquat in a mouse model. It has been shown that exposure of mice to paraquat led to the significant increase in brain levels of α-synuclein and was accompanied by the accumulation of α-synuclein-containing lesions within neurons of the substantia nigra pars compacta (Manning-Bog *et al.*, 2002). Furthermore, it has been pointed out that the pattern of paraquat-induced α-synuclein upregulation (with α-synuclein levels being enhanced at 2 days after each of three weekly toxin administrations and being returned to the basal values within 7 days post-treatment) suggests that an increased expression of this protein is a part of a neuronal response to toxic insults (Manning-Bog *et al.*, 2002). The question then arose concerning the significance of this toxin-modulated α-synuclein expression. In fact, if one assumes a harmful role of α-synuclein, then its toxin-induced upregulation is expected to contribute to neuronal injury (Feany and Bender, 2000; Forloni *et al.*, 2000; Maries *et al.*, 2003; Masliah *et al.*, 2000; Uversky, 2004; Uversky and Fink, 2002b). On the other hand, studies of cell cultures showed that an overexpression of α-synuclein may delay cell death caused by toxic agents (Lee *et al.*, 2001) and protect against apoptosis (da Costa *et al.*, 2000), assuming that increased levels of the protein, triggered by paraquat, MPTP, or other neurotoxicants, may minimize toxin damage. Alternatively, overexpression of α-synuclein might have both protective and harmful roles. In this scenario, the initial neuronal response may represent an attempt of neurons to counteract and to minimize injury. However, in the case of a severe or persistent insult, or if the toxin is able to interact with α-synuclein, protein upregulation may result in the development of pathology (Manning-Bog *et al.*, 2002). Therefore, it is most likely that α-synuclein deposition in the toxicant-induced animal models is a consequence of both the protein upregulation and the interaction of α-synuclein with neurotoxins. It is also quite possible that oxidative processes triggered by the toxin administration may contribute to the development of α-synuclein pathology (Manning-Bog *et al.*, 2002).

To clarify the relationship between α-synuclein expression and neuronal injury, paraquat neurotoxicity was compared in control animals to mice with transgenic expression of human α-synuclein driven by the tyrosine hydroxylase promoter (Manning-Bog *et al.*, 2002). It has been shown that paraquat caused both the formation of α-synuclein-containing intraneuronal deposits and the degeneration of nigrostriatal neurons in control mice. On the other hand, mice overexpressing α-synuclein displayed paraquat-induced protein aggregates but were completely protected against neurodegeneration, supporting a protective role of α-synuclein against toxic insults (Manning-Bog *et al.*, 2002). Based on these results, it has been concluded that toxicant-induced α-synuclein deposition is not necessarily associated with neurodegeneration, as previously proposed. Furthermore, the involvement of this protein in human neurodegenerative processes may arise not only from aggregation and gain of toxic function but also from a loss of defensive properties (Manning-Bog *et al.*, 2002).

It is necessary to emphasize that the data described above are not limited to paraquat. In fact, it has been shown that α-synuclein-positive inclusions represent a specific feature of the neurodegenerative process triggered by rotenone infusion into rats (Betarbet *et al.*, 2000; Sherer *et al.*, 2003). Importantly, *in vitro* analysis revealed that fibrillation α-synuclein is dramatically accelerated by several common pesticides, with rotenone being one of the most effective enhancers of the fibril formation (Uversky *et al.*, 2002b; Uversky *et al.*, 2001d).

4.4.4.2. Effect of Metal Cations

The possible involvement of heavy metals in the etiology of PD follows primarily from the results of epidemiologic studies (Altschuler, 1999; Gorell *et al.*, 1999a; Gorell *et al.*, 1999b; Rybicki *et al.*, 1993; Zayed *et al.*, 1990). Postmortem analysis of the brain tissues from PD patients gives further confirmation for the potential involvement of heavy metals in this disorder (Riederer *et al.*, 1989; Dexter *et al.*, 1991; Hirsch *et al.*, 1991); for example, it has been shown that LBs might contain high levels of iron and the presence of aluminum (Hirsch *et al.*, 1991). Several possible mechanisms for metal-stimulated fibrillation of α-synuclein can be envisioned. The simplest would involve direct interactions between the protein and the metal, leading to structural changes in α-synuclein, and resulting in the enhanced propensity to aggregate. In fact, it has been reported that α-synuclein aggregation is facilitated in the presence of Cu^{2+} (Paik *et al.*, 1999) and that Al^{3+} may induce structural perturbations in this protein (Paik *et al.*, 1997). Furthermore, a number of mono-, di-, and trivalent metal ions were shown to accelerate significantly the process of α-synuclein fibril formation (Uversky *et al.*, 2001c). The effectiveness of metal cations to induce fibrillation was shown to be correlated with the increasing ion charge density (Uversky *et al.*, 2001c). Importantly, a direct correlation between the accelerated fibrillation and the ability of cations to induce formation of an amyloidogenic partially folded species has been established (Uversky *et al.*, 2001c).

Thus, both pesticides/herbicides and metals are able to promote α-synuclein fibrillation via the specific stabilization of the amyloidogenic partially folded conformation. The mechanisms of this amyloidogenic conformation formation are different for metal cations and pesticides. The C-terminal region of α-synuclein contains a large amount of acidic residues. This brings about considerable repulsive interactions leading to the natively unfolded conformation of α-synuclein. The dominant effect of the metal ions on α-synuclein conformational change and fibrillation has been assumed to be due to the masking of the Coulombic charge-charge repulsion (Uversky *et al.*, 2002b; Uversky *et al.*, 2001c). For polyvalent cations, an additional important factor has been proposed, namely the potential for cross-linking, or bridging, between two or more carboxylates (Uversky *et al.*, 2001c). Structure-forming and fibrillation-accelerating effects of hydrophobic pesticides were assumed to be consistent with the idea that the pesticide binds to the partially folded conformation, which is in equilibrium with the intrinsically unstructured state, thereby increasing the population of the former. The driving force for this is the existence of contiguous patches of hydrophobic surface in the partially folded species, which are absent in the intrinsically unstructured conformation (Uversky *et al.*, 2002b; Uversky *et al.*, 2001d). Consequently, there are at least two different pathways, charge neutralization and the hydrophobic effects, by which environmental factors may stimulate the formation of α-synuclein fibrils at a fundamental molecular level. The synergy between cations and pesticides may arise from the different natures of their interactions with dimers or higher oligomers of the partially folded conformation (Uversky *et al.*, 2002b).

4.4.4.3. Effect of Organic Solvents

It has been pointed out that the exposure to solvents might represent another PD risk factor (Davis and Adair, 1999; Hageman *et al.*, 1999; Pezzoli *et al.*, 1996; Seidler *et al.*, 1996; Uitti *et al.*, 1994). To understand the potential link between exposure to organic solvents and PD development, the structural properties and aggregation/fibrillation propensities of α-synuclein in water organic solvent mixtures have been analyzed (Munishkina *et al.*, 2003). It has been shown that the addition of alcohols increased the content of ordered secondary structure (Munishkina *et al.*, 2003). Interestingly, low concentrations of all the solvents studied were shown to induce in α-synuclein an amyloidogenic partially folded conformation favoring a very rapid formation of fibrils. However, the structural transformations induced by higher solvent concentrations were dependent on the type of alcohol, with simple alcohols inducing a β-sheet-enriched conformation, whereas fluorinated alcohols promoted α-helix-rich species. Interestingly, both α-helical and β-structural species were shown to be initially monomeric but underwent association over longer times, and β-sheet-rich conformations were strongly prone to form amorphous aggregates (Munishkina *et al.*, 2003).

4.4.5. Effect of Oxidative Modification

4.4.5.1. Methionine Oxidation

It is well-known that the oxidation-modified proteins accumulate during normal aging (Leeuwenburgh *et al.*, 1998; Smith *et al.*, 1991; Stadtman *et al.*, 1993). Furthermore, oxidative injury has been implicated in the pathogenesis of several disorders including AD (Markesbery and Carney, 1999), PD (Duda *et al.*, 2000), dementia with LB (Lyras *et al.*, 1998), amyotrophic lateral sclerosis (Cookson and Shaw, 1999), Huntington's disease (Browne *et al.*, 1999), arthrosclerosis, inflammatory diseases (Witko-Sarsat *et al.*, 1998b), chronic renal falure (Witko-Sarsat *et al.*, 1998a), cataractogenesis (Fu *et al.*, 1998), and brain ischemia and carcinogenesis (Floyd, 1990). Importantly, all amino acids are susceptible to oxidation, although their levels of susceptibility vary greatly (Stadtman, 1993), but methionine is one of the most readily oxidized amino acid constituents of proteins. It is easily oxidized to methionine sulfoxide, MetO, by H_2O_2, hydroxyl radicals, hypochlorite, chloramines, and peroxynitrite; all these oxidants are produced in biological systems (Vogt, 1995). However, this modification can be repaired by methionine sulfoxide reductase, which catalyzes the thioredoxin-dependent reduction of MetO back to methionine (Moskovitz *et al.*, 1996; Sun *et al.*, 1999). Based on these observations, it was assumed that the oxidation of surface-exposed methionines may serve to protect other functionally essential residues from oxidative damage (Levine *et al.*, 1996).

It has been shown that under mild oxidative conditions (1–2% H_2O_2), all four methionines of α-synuclein, Met1, Met5, Met116, and Met127, located outside the repeat-containing region, are successfully oxidized to methionine sulfoxides (Uversky *et al.*, 2002d). The oxidized form of α-synuclein was shown to be more unfolded than a non-oxidized protein as manifested by the larger contribution of unordered structure to both FTIR and far-UV CD spectra (Uversky *et al.*, 2002d) and a detectable decrease in the α-synuclein-MetO compactness (Glaser *et al.*, 2005; Uversky *et al.*, 2002d; Yamin *et al.*, 2003a). This was attributed to the decreased hydrophobicity of oxidized methionine leading to a decrease in the overall hydrophobicity of the protein. Given the decrease in hydrophobicity, it was not a big surprise that the oxidized protein is less prone to oligomerize and aggregate, being substantially non-amyloidogenic and even able to inhibit the fibrillation of nonmodified α-synuclein (Uversky *et al.*, 2002d). In subsequent study, it has been shown that the degree of inhibition of α-synuclein fibrillation by methionine oxidation is proportional to the number of oxidized methionines (Hokenson *et al.*, 2004). This was accomplished by selectively converting Met residues into Leu, prior to Met oxidation. The results showed that with one oxidized Met, the kinetics of fibrillation were comparable with those for the control (non-oxidized), and with increasing numbers of methionine sulfoxides the kinetics of fibrillation became progressively slower (Hokenson *et al.*, 2004). Importantly, it has been shown that although the fibrillation of α-synuclein at neutral pH was completely inhibited by methionine oxidation (see below), the presence of certain metals (Ti^{3+}, Zn^{2+}, Al^{3+}, and Pb^{2+}) overcame this inhibition (Yamin *et al.*, 2003a). These observations brought about interesting speculations related to the biological function of α-synuclein as a natural scavenger of reactive oxygen species.

Many environment factors can potentially affect the rate of α-synuclein fibrillation, suggesting that in dopaminergic neurons there is a balance between factors that can accelerate fibrillation and those that inhibit or prevent it. For example, chaperones or chaperone-like species might be important in minimizing α-synuclein aggregation under normal conditions (Auluck *et al.*, 2002). Based on the fact that the addition of methionine-oxidized α-synuclein to the non-oxidized protein inhibited fibrillation of the latter (Glaser *et al.*, 2005; Uversky *et al.*, 2002d), it has been suggested that the methionine residues in α-synuclein may be used by the cells as a natural scavenger of reactive oxygen species. This assumption was based on the following facts: methionine can react with essentially all of the known oxidants found in normal and pathologic tissues; the concentration of α-synuclein may be as high as 1% of the total protein in soluble cytosolic brain fractions (Iwai *et al.*, 1995a); the concentration of α-synuclein could be increased significantly as a result of the neuronal response to toxic insult (Manning-Bog *et al.*, 2002); and methionine sulfoxide residues in proteins can be cycled back to their native methionines by methionine sulfoxide reductase, a process that might protect other functionally essential residues from oxidative damage (Glaser *et al.*, 2005; Uversky *et al.*, 2002d). The balance between the protective antioxidant role of the methionine residues that is enhanced by this recycling and the protective antifibrillation effect of oxidized methionine residues in α-synuclein may fail under conditions of industrial pollution due to exposure of a person to lead, aluminum, zinc, titanium, and other metals. Thus, in the presence of the enhanced concentrations of such industrial pollutants, toxic insult-induced upregulation of α-synuclein no longer plays a protective role; rather, it may represent a risk factor, leading to the effective metal-triggered fibrillation of the methionine-oxidized protein (Yamin *et al.*, 2003a).

4.4.5.2. Nitration of Tyrosines

Proteins are targets for numerous oxidative modifications induced by their interaction with reactive oxygen species (ROS) or reactive nitrogen species (RNS), produced as a result of different physiologic and nonphysiologic processes. The spectrum of oxidative modifications in proteins is very wide. For example, interaction of a protein with nitric oxide can lead to the S-nitrosylation of reduced cysteine residues, the nitration of tyrosine and tryptophan residues, and the formation of nitric oxide–iron heme adducts (Ischiropoulos, 2003b). One of the most frequent oxidative modifications of tyrosines is their nitration, which is a covalent protein modification resulting from the addition of a nitro ($-NO_2$) group onto one of the two equivalent *ortho* carbons of the aromatic ring of tyrosine residues (Ischiropoulos, 1998; Ischiropoulos, 2003a). Interestingly, not all proteins are uniformly susceptible to oxidative damage and only a few of them undergo nitration. Recently, it has been shown that there were only about 40 nitrated proteins out of 1000 during an inflammatory challenge. These included a large number of mitochondrial proteins, which regulate cellular energy metabolism (Aulak *et al.*, 2001). In many cases, nitration has been shown to affect protein function. For example, peroxynitrite-mediated nitration of lymphocyte-specific tyrosine kinase is known to inhibit the ability of this protein to phosphorylate tyrosine residues (Kong *et al.*, 1996). Similarly, manganese superoxide dismutase has been shown to be inactivated by selective nitration (MacMillan-Crow and Thompson, 1999), and creatine kinase, another key intracellular enzyme regulating energy metabolism, is nitrated and inactivated by peroxynitrite (Stachowiak *et al.*, 1998).

Oxidative injury has been implicated in the pathogenesis of PD (Duda *et al.*, 2000; Giasson *et al.*, 2000; Ischiropoulos, 2003c; Tieu *et al.*, 2003). The existence of extensive and widespread accumulation of nitrated α-synuclein (i.e., protein containing the product of the tyrosine oxidation, 3-nitrotyrosine) in Lewy bodies has been demonstrated, using antibodies to specific nitrated tyrosine residues in α-synuclein (Duda *et al.*, 2000; Giasson *et al.*, 2000). It was proposed that the selective and specific nitration of α-synuclein in different neurodegenerative synucleinopathies may directly link oxidative and nitrative damage to the onset and progression of these disorders (Giasson *et al.*, 2000).

To understand the potential structure-modifying and aggregation-modulating role of α-synuclein nitration, the protein was oxidatively modified *in vitro*. In α-synuclein, there are four tyrosines, Tyr39, Tyr125, Tyr132, and Tyr135, which are easily nitrated under the appropriate conditions (Uversky *et al.*, 2005; Yamin *et al.*, 2003b). The properties of nitrated α-synuclein were investigated using a variety of biophysical and biochemical techniques, which revealed that nitration led to formation of a partially folded conformation with increased secondary structure relative to the intrinsically disordered structure of the monomer and to oligomerization at neutral pH. The degree of self-association was concentration dependent, but at 1 mg/mL, nitrated α-synuclein was predominately an octamer. At low pH, small-angle x-ray scattering data indicated that the nitrated protein was monomeric (Uversky *et al.*, 2005; Yamin *et al.*, 2003b). Interestingly, the comparison of the accessibility of nonmodified and nitrated α-synuclein to proteolysis by trypsin at neutral pH revealed that the nonmodified α-synuclein degraded relatively fast under the conditions studied (10 min), whereas nitration increased the resistance of this protein to trypsinolysis and nitrated α-synuclein was digested significantly slower under the same conditions. Thus, oxidative modification increases the conformational stability and decreases the structural flexibility of α-synuclein, most likely due to the formation of stable soluble oligomers (Uversky *et al.*, 2005).

α-Synuclein fibrillation at neutral pH was completely inhibited by nitrotyrosination and was attributed to the formation of stable soluble oligomers. The presence of heparin or metals did not overcome the inhibition; however, the inhibitory effect was eliminated at low pH (Uversky *et al.*, 2005; Yamin *et al.*, 2003b). Furthermore, the addition of nitrated α-synuclein was shown to lead to a substantial inhibition of fibrillation of nonmodified α-synuclein. The extent of this inhibitory effect was shown to depend on the relative content of the modified protein, with larger concentrations showing stronger inhibition effects. In fact, no fibrillation was observed (at least within the timescale studied) when the ratio of modified over the nonmodified α-synuclein was

2:1 and higher (Uversky *et al.*, 2005; Yamin *et al.*, 2003b). Interestingly, the fibrillation of Tyr-free mutant of α-synuclein was shown to be relatively unaffected by nitrative oxidation, demonstrating that Tyr and Tyr nitration are not required for fibrillation (Norris *et al.*, 2003).

Overall, these *in vitro* studies showed that the nitration of α-synuclein has a significant effect on its structural properties and propensity for aggregation. Nitration leads to formation of a partially folded conformation, which is stabilized by the formation of specific oligomers, most likely octamers. Importantly, these octamers are stable and possess decreased conformational flexibility, as seen in the increase in stability toward limited proteolysis. Because this nitration-induced oligomerization inhibits fibrillation of human recombinant α-synuclein *in vitro*, the

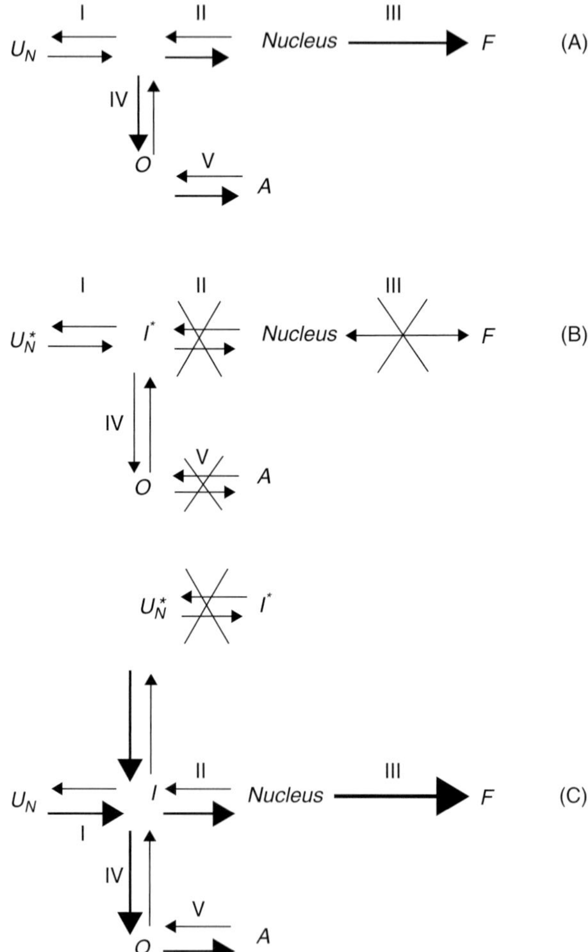

Figure 4.3. Models of nonmodified and nitrated α-synuclein aggregation under the different experimental conditions (see the text for explanations). (A) Aggregation of the nonmodified protein at neutral pH. (B) Aggregation of the nitrated protein at neutral pH. (C) Aggregation of nonmodified and nitrated proteins at acidic pH.

stable oligomers must be located off the fibrillation pathway. The fact that the nitrated protein inhibits fibrillation of nonmodified α-synuclein at substoichiometric concentrations suggests that the interaction between nitrated and nonmodified forms of human α-synuclein leads to the formation of stable off-pathway hetero-oligomers (Uversky *et al.*, 2005).

Models for the aggregation of nonmodified and nitrated α-synuclein at neutral pH are shown in Figure 4.3. The model for aggregation of both forms of the protein at acidic pH is also shown. In the model ***O***, ***F***, and ***A*** represent soluble oligomers, fibrils, and amorphous aggregates, respectively; $\boldsymbol{U_N}$ and $\boldsymbol{U_N}$* are the natively unfolded states of nonmodified and nitrated α-synuclein, respectively; whereas ***I*** and ***I***** represent the productive (leading to fibrils) and nonproductive partially folded conformations. The Roman numerals indicate the major stages of the aggregation process. Taking into account the observations presented in the current study, we propose that nitration of tyrosines leads to partial folding of α-synuclein; that is, leads to a shift in the equilibrium in stage I for the nitrated protein, $\boldsymbol{U_N}^* \leftrightarrow \boldsymbol{I}^*$, in favor of the partially folded conformation $\boldsymbol{I}^*$. This facilitates the formation of soluble oligomers, stage IV, whereas stages II, III, and V are arrested (or at least strongly inhibited) at neutral pH (see Fig. 4.3). It has been assumed that the partially folded conformation $\boldsymbol{I}^*$ is different from that of the nonmodified protein, ***I***, as the formation of ***I*** is known to accelerate rather than inhibit the fibrillation of α-synuclein (Uversky *et al.*, 2001b).

Both nonmodified and nitrated forms of α-synuclein show similar aggregation behavior at acidic pH (see Fig. 4.3). It is reasonable to suggest that the nonmodified and nitrated α-synucleins are more prone to form the productive partially folded conformation ***I*** (thicker arrow at stage I) and, as a consequence, show a faster rate of fibrillation (stages II–V). We assume that the pathways of nitrated α-synuclein to $\boldsymbol{I}^*$ and to stable oligomers are considerably diminished at low pH due to the pH-induced destabilization of these complexes and stabilization of productive partially folded conformation ***I*** (Fig. 4.3). A model for the inhibitory effect of nitrated α-synuclein on the non-oxidized protein fibrillation is shown in Figure 4.4, in which the conformational equilibrium is shifted toward the formation of soluble hetero-oligomers.

How can these findings be implicated with the etiology of PD? The extensive and widespread presence of nitrated α-synuclein has been shown in the signature inclusions of such synucleinopathies as Parkinson's disease, dementia with Lewy bodies, the Lewy body variant of Alzheimer's disease, and multiple system atrophy. Interestingly, nitrated α-synuclein was detected in the major filamentous building blocks of these inclusions, as well as in the insoluble

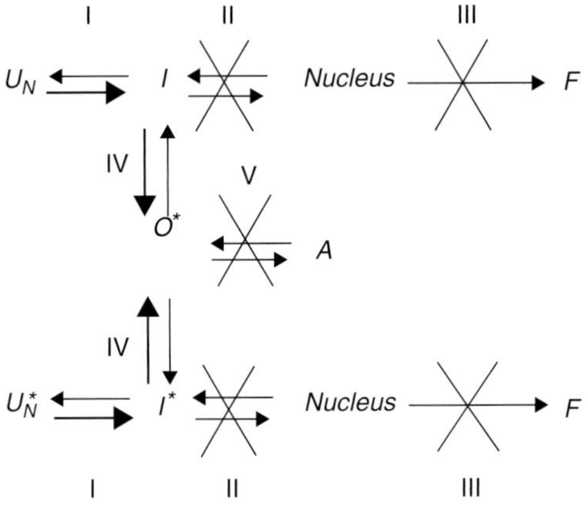

Figure 4.4. Model of the inhibition of nonmodified α-synuclein fibrillation in the presence of nitrated protein (see the text for explanations).

fractions of the affected brain regions of the different synucleinopathies (Giasson *et al.*, 2000). This led to the conclusion that the nitration of α-synuclein reflects a direct link between oxidative and nitrative damage and the onset and progression of these neurodegenerative disorders (Giasson *et al.*, 2000). However, it has been established that nitration effectively inhibits fibrillation of α-synuclein *in vitro*. Furthermore, the ability of the nonmodified α-synuclein to form fibrils is effectively inhibited by the addition of the substoichiometric concentrations of nitrated α-synuclein. These findings suggest that the nitration detected in the proteinaceous deposits accumulated in the synucleinopathy-affected brain most likely occurred after, rather than before, fibril formation (Uversky *et al.*, 2005).

4.4.6. Effect of Membranes

Although the normal function of α-synuclein remains poorly understood, it is believed that this protein is involved in interaction with synaptic vesicles, which is likely to be transient and/or reversible (Clayton and George, 1998). Such reversible lipid membrane binding is suggested by the sequence of the protein, which includes a number of imperfect 11-mer tandem repeats that are reminiscent in character to those observed in apolipoprotein sequences (George *et al.*, 1995; Weinreb *et al.*, 1996). In agreement with this suggestion, α-synuclein was shown to interact specifically with brain vesicles (Jensen *et al.*, 1998), synthetic vesicles containing acidic phospholipids (Davidson *et al.*, 1998; Jo *et al.*, 2000; Perrin *et al.*, 2000), and sodium dodecyl sulfate (SDS) micelles (Eliezer *et al.*, 2001). In all cases studied, this binding was accompanied by a dramatic increase in α-helix content. NMR analysis revealed that as a result of specific interaction with synthetic lipid vesicles and detergent micelles, the N-terminal region of α-synuclein adopts a highly helical conformation, consistent with predictions based on sequence analysis. The C-terminal part of the protein does not associate with either vesicles or micelles, remaining free and unfolded (Eliezer *et al.*, 2001). Interestingly, membrane-bound α-synuclein isolated from rat brain, which represents only ~15% of total α-synuclein, was shown to have a high aggregation propensity and the ability to seed aggregation of the cytosolic form of the protein (Lee *et al.*, 2002). Similarly, studies with synthetic vesicles revealed that interactions of α-synuclein with long-chain polyunsaturated fatty acids triggered rapid multimerization of the protein (Perrin *et al.*, 2000).

An analysis of α-synuclein–membrane interactions revealed additional interesting consequences of familial PD mutations. The two point mutations associated with Parkinson's disease were shown to have little (A30P) or no (A53T) effect on lipid binding or α-helicity (Perrin *et al.*, 2000). Similar to wild-type protein, these mutants showed tendencies to oligomerize in the presence of long-chain polyunsaturated fatty acids (Perrin *et al.*, 2000). In the rat optic system, it has been demonstrated that a portion of α-synuclein is carried by the vesicle-moving fast component of axonal transport and that it binds to brain vesicles throughout its amino-terminal repeat region. However, the A30P mutant was shown to be devoid of vesicle-binding activity. Based on these observations, it was proposed that mutant α-synuclein may accumulate, leading to assembly into Lewy body filaments (Jensen *et al.*, 1998).

4.4.7. Effect of Molecular Crowding

The concentration of macromolecules, including proteins, nucleic acids, carbohydrates, and small solutes, within a living cell can be as high as 400 g/L (Zimmerman and Trach, 1991), with the intracellular solutes taking up about a half of the total cellular volume (Ellis, 2001; Fulton, 1982; Yancey *et al.*, 1982; Zimmerman and Trach, 1991). Obviously, the volume occupied by solutes is unavailable to other molecules, a phenomenon known as "excluded volume effects" (Zimmerman and Trach, 1991; Minton, 2001), which may have large effects on the stability of biological macromolecules (Minton, 2000a; Eggers and Valentine, 2001a; Eggers and Valentine, 2001b; Bismuto *et al.*, 2002) and on macromolecular equilibria, such as protein-protein interactions (Minton, 2000b; Morar *et al.*, 2001). The decreased water activity in the crowded solutions was assumed to be the additional factor contributing to the stability and solubility of macromolecules. It has been suggested that volume exclusion in physiologic media could modulate the rate and extent of amyloid formation *in vivo* (Minton, 2000b). The validity of this hypothesis has been confirmed recently for the *in vitro* fibrillation of human α-synuclein (Shtilerman *et al.*, 2002; Uversky *et al.*, 2002a). Various types of polymers, from neutral polyethylene glycols and polysaccharides (Ficolls, dextrans) to inert proteins (lysozyme, BSA), were shown to accelerate α-synuclein fibrillation *in vitro*. The stimulation of fibrillation increases with a increasing length of polymer, as well as an increasing polymer concentration. At lower polymer concentrations (typically up to approximately 100 mg/mL) the major effect was ascribed to excluded volume, whereas at higher polymer concentrations evidence of opposing viscosity effects became apparent. Pesticides and metals, which are linked to increased risk of PD by epidemiologic studies, were shown to further accelerate α-synuclein fibrillation under conditions of molecular crowding (Munishkina *et al.*, 2004a).

4.4.8. Effect of Anions and Structure of Water

It has been shown recently that anions induce partial folding of α-synuclein at neutral pH, forming a critical amyloidogenic intermediate, which leads to significant acceleration of the rate of fibrillation. The magnitude of the fibrillation-accelerating effect did not depend on the efficiency of the amyloidogenic intermediate stabilization, as all anions were shown to induce comparable degree of folding in α-synuclein (Munishkina *et al.*, 2004b). On the contrary, at moderate salt concentrations, the fibril-promoting capability generally followed the position of the anions in the Hofmeister series, indicating that the major role of anions in fibrillation is their modulation of protein-water interaction. From this perspective, fibrillation could be considered a particular case of the more general phenomenon of protein salting-out. Nevertheless, the degree to which anions interact with water, stabilizing the hydrophobic interactions within the α-synuclein molecule or between protein molecules, is not the only factor determining the stability and solubility of this protein, and electrostatic effects have to be taken into account as well (Munishkina *et al.*, 2004b). Consequently, the enhanced fibrillation of α-synuclein in the presence of anions is the result of the loss of the uncompensated charge, which is a factor promoting the soluble unfolded conformation, and an increase in the preferential hydration, which promotes partial folding and aggregation by strengthening hydrophobic interactions. Both key steps of fibrillation, nucleation and fibril growth, are affected by a combination of these two effects. The addition of the first small quantities of salts eliminates the strong repulsion due to the net electrical charges, giving rise to partial folding of α-synuclein, whereas the continued addition of the salt brings about dehydration that results ultimately in fibrillation of the protein (Munishkina *et al.*, 2004b).

4.4.9. Effect of Phosphorylation

α-Synuclein, being predominately nonphosphorylated *in vivo* under normal conditions (Fujiwara *et al.*, 2002), is extensively phosphorylated at Ser129 in α-synucleinopathic lesions (Fujiwara *et al.*, 2002; Nishie *et al.*, 2004) and aging human brain (Saito *et al.*, 2003), as it has been demonstrated by mass-spectrometric and immunohistochemical analyses. For example, it has been shown that only ~4% of total brain α-synuclein freshly prepared from normal adult rat brains was phosphorylated at Ser129, being readily dephosphorylated postmortem. On the other hand, ~90% of α-synuclein isolated from dementia with LB brain was shown to be phosphorylated at Ser129 (Fujiwara *et al.*, 2002). Importantly, a similar extesive accumulation of Ser129 phosphorylated α-synuclein has been demonstrated in neurons of transgenic α-synuclein mouse and fly models (Kahle *et al.*, 2002; Neumann *et al.*, 2002; Takahashi *et al.*, 2003).

To examine whether phosphorylation of α-synuclein at Ser129 can affect fibrillation of this protein, recombinant human α-synuclein was incubated with or without prior phosphorylation by casein kinase 2, which has been shown to selectively phosphorylate Ser129 (Okochi *et al.*, 2000; Pronin *et al.*, 2000), and compared with the kinetics of fibril formation *in vitro* (Fujiwara *et al.*, 2002). It was not a big surprise to find that the phosphorylation accelerated oligomerization *in vitro* and promoted the formation of more fibrils as compared with non-phosphorylated α-synuclein (Fujiwara *et al.*, 2002). Importantly, these *in vitro* studies suggested that the phosphorylation of Ser129 promotes aggregation of α-synuclein in a form of fibrils and oligomers. The latter are thought to be pathogenic and act as nuclei for fibril propagation (Conway *et al.*, 2000c; Goldberg and Lansbury, 2000). Furthermore, these observations strongly suggest that excessive phosphorylation of α-synuclein in synucleinopathy brains (including those in transgenic animals), as well as the underlying alterations in the activities of various kinases and phosphatases, and in the conformation of α-synuclein, may contribute to abnormalities of α-synuclein. In conjunction with other post-translational modifications of α-synuclein, these changes would cause aggregation, eventually leading to neuronal cell death (Fujiwara *et al.*, 2002).

4.4.10. Effect of Sequence Truncations

It has been shown that in addition to the full-length protein, Lewy bodies and Lewy neuritis accumulated in sporadic PD and Lewy body dementia (Baba *et al.*, 1998), as well as glial cytoplasmic inclusions of multiple systems atrophy (Gai *et al.*, 1999), usually contain C-terminally truncated α-synuclein. The analysis of the aggregation behavior of the recombinant C-terminally truncated α-synuclein peptides *in vitro* revealed that the truncated constructs assembled into fibrils much more readily than the full-length protein (Crowther *et al.*, 1998; Murray *et al.*, 2003; Nielsen *et al.*, 2001; Serpell *et al.*, 2000). This demonstrates that the C-terminal region apparently exerts an inhibitory effect on the aggregation propensity of α-synuclein, suggesting that the C-terminus may play the role of intramolecular chaperone preventing α-synuclein from fibrillation. Based on these observations, it has been hypothesized that proteolytic degradation of α-synuclein may be an important factor in the assembly of α-synuclein in the various neurodegenerative diseases (Crowther *et al.*, 1998). Interestingly, it now becomes evident that the acidic C-terminal domain of α-synuclein has several biological functions, alterations of which (induced by the proteolytic truncation) might promote development of several neurodegenerative disorders. For example, it has been recently shown that human dopaminergic neuroblastoma SH-SY5Y cells expressing C-terminally truncated α-synuclein, particularly the 1–120 residue protein, were significantly more vulnerable to oxidative stress than the control cells (Kanda *et al.*, 2000). Furthermore, it has been shown that the C-terminally localized acidic 32-amino-acid domain of human α-synuclein contains the Ca^{2+}-binding site that binds Ca^{2+} with an IC_{50} of about 2–300 μM and that this reaction can be uninhibited by a 50-fold excess of Mg^{2+} (Nielsen *et al.*, 2001). Importantly, Ca^{2+} binding was shown to affect the aggregation and functionality of α-synuclein, as Ca^{2+} ions increased the efficiency of α-synuclein binding to microtubule-associated protein 1A and promoted oligomer formation. This suggests that Ca^{2+} ions may both participate in normal α-synuclein functions in the nerve terminal and exercise pathologic effects involved in the formation of Lewy bodies (Nielsen *et al.*, 2001).

Although the involvement of α-synuclein in the PD development represents a major focus of numerous current studies, this protein was first implicated in the pathogenesis of Alzheimer's disease (Ueda *et al.*, 1993). In fact, it has been shown that, in addition to the "normal" amyloid-β peptide (Aβ), the amyloid plaques isolated from Alzheimer's disease brains contain a novel peptide named non-Aβ component of Alzheimer's disease amyloid (NAC). Sequencing of this peptide, which represents about 10% of the non-SDS-soluble material, revealed that NAC comprised at least 35 amino acids derived from the central region of α-synuclein (residues 61–95) (Han *et al.*, 1995; Ueda *et al.*, 1993). *In vitro* analysis of NAC aggregation behavior revealed that this peptide readily forms amyloid-like fibrils, which was accompanied by a transition from totally random structure to predominately β-sheet (Han *et al.*, 1995; Iwai *et al.*, 1995b; El-Agnaf *et al.*, 1998a; El-Agnaf *et al.*, 1998b). Importantly, NAC was shown to be responsible for the α-synuclein interaction with Aβ peptides (Jensen *et al.*, 1997), which dramatically accelerated the aggregation of Aβ *in vitro* (Yoshimoto *et al.*, 1995). This raised the possibility that α-synuclein and its NAC fragment may play a role in the development of AD amyloid.

It has been shown that aggregates of NAC-related peptides are toxic to cells, as indicated by their inhibition of cellular 3-(4,5-dimethylthiazol-2-yl)-2,5-diphenyltetrazolium bromide (MTT) reduction to MTT formazan, an assay that is widely used for measuring cell viability (El-Agnaf *et al.*, 1998b; Bodles *et al.*, 2000; Bodles *et al.*, 2001). Furthermore, morphologic examination of cell nuclei, stained with the DNA-binding fluorochrome Hoechst 33258, showed that several cells exposed to NAC presented typical apoptotic morphology, including condensation of chromatin and nuclear fragmentation (El-Agnaf *et al.*, 1998b).

Altogether, the data presented above suggest that accumulation of ordered NAC and C-terminally truncated α-synuclein aggregates in the synapse may be responsible for the neurodegeneration observed in AD, PD, and other synucleinopathies. Alternatively, neurodegeneration may be caused by the loss of α-synuclein function induced by sequence truncation (Han *et al.*, 1995).

4.4.11. Effect of Charged Polymers

4.4.11.1. Polyanions

Different glycosaminoglycans (GAGs) are involved in the formation of the amyloid deposits found in a variety of human diseases. Proteoglycans (PGs) containing heparan sulfate, chondroitin sulfate, keratan sulfate, and/or dermatan sulfate have been found in all extracellular amyloid deposits examined (Kisilevsky, 2000; McLaurin *et al.*, 2000). GAGs and PGs stimulate *in vitro* formation of amyloid fibrils from the Alzheimer Aβ protein (Miller *et al.*, 1997; Castillo *et al.*, 1997; McLaurin *et al.*, 1999; Cotman *et al.*, 2000). Heparan sulfate was found associated with the intracellular neurofibrillary tangles (NFTs), which are found in neurons of Alzheimer patients and which are made from the microtubule-associated protein tau (Goedert *et al.*, 1996; Goedert, 1999). Recently, it has been established that different GAGs (heparin, heparan sulfate) and other highly sulfated polymers (dextran sulfate) are able bind to α-synuclein and stimulate its fibrillation *in vitro* (Cohlberg *et al.*, 2002). Heparin binding sites on a variety of proteins are known to comprise some clusters of basic amino acid residues capable of binding to the negatively charged heparin polymer (Cardin and Weintraub, 1989). The N-terminal region of α-synuclein contains multiple repeats of the consensus sequence KTKEGV (see above), and pairs of lysine residues spaced two residues apart occur at positions 10/12, 21/23, 33/35, 42/44, and 58/60. It is likely that this region of α-synuclein constitutes the GAG binding region and that it is the transition of the protein from an unstructured state to a partially folded conformation that creates a site for GAG binding (Cohlberg *et al.*, 2002).

Agrin is an extracellular matrix and transmembrane heparan sulfate proteoglycan in the central neuronal system (Tsen *et al.*, 1995; Cohen *et al.*, 1997; Halfter *et al.*, 1997). Recent studies have begun to investigate the role of agrin in brain and suggest that its function likely extends beyond that of a synaptogenic protein. In fact, it has been shown that agrin localizes to all lesion types in AD (Donahue *et al.*, 1999; Verbeek *et al.*, 1999; Cotman *et al.*, 2000) and potentiates Aβ fibril formation (Cotman *et al.*, 2000). In light of evidence that heparan sulfate may also contribute to the pathophysiology of PD (Cohlberg *et al.*, 2002), the role of agrin in regulating the aggregation state of α-synuclein has been examined (Cole *et al.*, 2005). To this end, the interaction of agrin with α-synuclein was studied *in vitro*. It has been demonstrated that agrin is able to bind to and promote α-synuclein insolubility, accelerating the formation of α-synuclein protofibrils and fibrils (Cole *et al.*, 2005). Furthermore, it has been demonstrated that agrin and α-synuclein co-localize to Lewy bodies and neurites in the substantia nigra of PD patients, indicating that agrin may contribute to the etiology of PD by modulating the aggregation state of α-synuclein in dopaminergic neurons (Cole *et al.*, 2005). Agrin's ability to modulate both α-synuclein and Aβ fibrillation would raise the possibility that these two neurodegenerative diseases might share similar molecular mechanisms that contribute to their respective protein aggregations (Cole *et al.*, 2005).

4.4.11.2. Polycations

Comparable fibrillation-promoting effects have been described for several unstructured polycations including polyethyleneimine, spermine, spermidine, polyLys, polyArg (Goers *et al.*, 2003b), and histones (Goers *et al.*, 2003a). This interaction leads to the oligomerization of α-synuclein and brings about minor conformational changes consistent with the formation of amyloidogenic partially folded conformation (Goers *et al.*, 2003a; Goers *et al.*, 2003b). The rate of α-synuclein fibrillation increases in the presence of polycations, with the magnitude of the acceleration effect depending on the chemical nature of the polycation, its length and concentration. It has been assumed that polycations accelerate α-synuclein fibrillation by suppressing electrostatic repulsion between the negatively charged residues within the α-synuclein polypeptide chain and due to the increase in local concentration of this protein (Goers *et al.*, 2003a; Goers *et al.*, 2003b).

The fact that α-synuclein is able to physically interact with histones *in vitro* brought about an interesting hypothesis that this interaction also can take place in a cell. In fact, α-synuclein was initially described as a neuron-specific protein localized to the presynaptic nerve terminals, *syn*apses, and the *nucl*eus (Maroteaux *et al.*, 1988). However, most subsequent studies have shown α-synuclein localized only within nerve terminals in the central nervous system (Clayton and George, 1998; Clayton and George, 1999). Contrary to these latter conclusions, it has been recently shown that paraquat administration triggers an upregulation of α-synuclein in the mouse brain, which has been interpreted as part of the response of neurons to toxic insults (Manning-Bog *et al.*, 2002), with this paraquat-induced injury being accompanied by the evidence of nuclear localization of α-synuclein (Goers *et al.*, 2003a). Although the nuclei of nigral neurons were shown to lack α-synuclein immunoreactivity in normal mice (i.e., animals injected with saline), they were stained with an α-synuclein antibody in mice injected with paraquat. Furthermore, co-localization of α-synuclein with neuron-specific nuclear protein (NeuN) and histone H3, two specific nuclear markers, was also consistent with the possibility that, after paraquat-induced neuronal insult, α-synuclein is translocated into the nuclei where it may interact with histones (Goers *et al.*, 2003a). Nuclear translocation of α-synuclein and formation of histone-α-synuclein complexes was suggested to provide a mechanism by which an α-synuclein-related neuronal response may be activated or sustained. In this model, α-synuclein-histone complexes were suggested to have a regulatory role by decreasing the pool of free histones available for DNA binding. The subsequent destabilization of nucleosome and enhanced manifestation of the DNA matrix activity could lead to increased transcription and ultimately to the production of proteins in response to a variety of stimuli, including toxic insults (Goers *et al.*, 2003a).

Finally, it has been pointed out that the pathophysiologic implications of histone-induced α-synuclein aggregation are not very obvious, as Lewy bodies in PD are usually found in the cytosol and never in the nucleus. This might suggest that histone-α-synuclein interactions may not underlie the development of these typical inclusions. However, α-synuclein-containing aggregates have been observed in the nuclei of transgenic mice that overexpress α-synuclein (Masliah *et al.*, 2000). This strongly suggests that the ability of histones to promote aggregation *in vitro* may contribute to pathologic changes *in vivo*, at least in mice (Goers *et al.*, 2003a).

4.4.12. Effect of Protein-Protein Interactions

4.4.12.1. Spontaneous Dimerization

The temperature-induced formation of partially folded conformation in human α-synuclein was completely reversible when the heat treatment was transient (Uversky *et al.*, 2001b). However, the incubation of this protein (both in the purified form and in the crude cytosolic preparation from COS-7 cells) at different temperatures for several days resulted in a temperature-dependent, progressive aggregation, with dimers being formed first (Uversky *et al.*, 2001a). It is important to note that the Western blot profile of small oligomers formed in cytosolic preparation appears identical to that formed from the purified protein. Conformational analysis revealed that a small amount of ordered secondary structure was stabilized by the self-assembly of partially folded α-synuclein as a result of prolonged incubation at elevated temperatures (Uversky *et al.*, 2001a). The efficiency of α-synuclein oligomerization was shown to increase proportionately with the temperature increase, both in purified form and in crude cytosolic preparation. Interestingly, this trapped conformation was structurally similar to the amyloidogenic partially folded monomeric confomer induced by low pH or high temperature. In fact, it has been shown that dimerization coincided with a small but reproducible change in the circular dichroism spectrum and an increase in the 1-anilinonaphthalene-8-sulfonic acid binding (Uversky *et al.*, 2001a). The hydrodynamic dimensions of the dimer measured by size-exclusion chromatography suggest a premolten, globule-like structure. Therefore, it has been concluded that the partially folded premolten, globule-like conformation of α-synuclein, which is unstable in the monomeric form, can be stabilized as the protein undergoes a highly selective self-assembly process during prolonged incubation at elevated temperatures, and that these oligomers may evolve into the fibril nucleus (Uversky *et al.*, 2001a).

The fact that the heat treatment induced oligomerization not only in the purified α-synuclein but also in the crude cytosolic preparation from COS-7 cells expressing α-synuclein is really important and suggests that the oligomerization was initiated by selective self-association between the partially folded α-synuclein molecules rather than by a heteromeric association of α-synuclein with other proteins in a complex cytosolic milieu (Uversky *et al.*, 2001a). This model is supported by the biochemical analyses of brain extracts from PD, DLB, or multiple system atrophy. Western blotting showed that brains from the individuals suffering from these diseases contained the 36-kDa α-synuclein immunoreactive protein (which corresponds with the dimeric form of the protein), but age-matched controls did not (Langston *et al.*, 1998; Campbell *et al.*, 2000; Campbell *et al.*, 2001). The fact that the above-described *in vitro* oligomerization produced the same oligomeric species found in human LB preparations suggests that these experimental observations may in fact reflect the *in vivo* mechanism of α-synuclein aggregation in the early stage of pathogenesis (Uversky *et al.*, 2001a).

4.4.12.2. Oxidative Dimers

The formation of oxidative dimers under the conditions of oxidative stress may represent an important prerequisite for the fibrillogenesis of α-synuclein (Krishnan *et al.*, 2003). Impirtantly, dimerization was shown to be accelerated for the pathogenic A30P and A53T mutants of α-synuclein because of their greater propensity to self-interact. As α-synuclein does not contain cysteine residues, the observed covalent aggregate formation caused by oxidation is probably restricted to dityrosine cross-links. In fact, it has been shown that the exposure of this protein to oxidation or nitration stresses *in vitro* induces formation of dimers and higher order oligomers with dityrosine cross-links (Souza *et al.*, 2000).

These results emphasized that the formation of oxidative dimer could be a critical rate-limiting step for the α-synuclein fibrillogenesis and that the inherent propensity of the A30P and A53T mutant proteins to form dimers more readily could account for their more rapid aggregation (Krishnan *et al.*, 2003). It has been also pointed out that these data could be used to link oxidation and synuclein aggregation to the same pathogenic pathway (Krishnan *et al.*, 2003), providing a strong support for the hypothesis that impairments of cellular antioxidative mechanisms and/or overproduction of reactive oxidative species may provoke the initiation and progression of neurodegenerative synucleinopathies (Beal, 1995; Duda *et al.*, 2000; Giasson *et al.*, 2000; Ko *et al.*, 2000; Zhou *et al.*, 2000; Butterfield and Kanski, 2001; Paxinou *et al.*, 2001).

4.4.12.3. Proteins Inhibiting α-Synuclein Aggregation

As it has been already pointed out, conformational analysis revealed that all three members of the synuclein family, α-, β-, and γ-synucleins, are natively unfolded under physiologic conditions *in vitro*, with β-synuclein possessing a slightly increased degree of disorder and γ-synuclein possessing larger probability to form stable dimers (Uversky *et al.*, 2002c). Furthermore, all three proteins formed comparable partially folded conformations at acidic pH or at high temperature (Uversky *et al.*, 2002c). It has been shown that α- and γ-synucleins are able to form fibrils, whereas β-synuclein did not fibrillate, being incubated under the same conditions (Uversky *et al.*, 2002c). However, more detailed analysis revealed that non-amyloidogenic β-synuclein can be forced to fibrillate under a variety of conditions (Yamin *et al.*, 2005). Particularly, some metals (Zn^{2+}, Pb^{2+}, and Cu^{2+}) were shown to induce a novel partially folded character of β-synuclein that triggers its rapid fibrillation. Importantly, in the presence of these metals, the mixtures of α- and β-synucleins exhibited rapid fibrillation. Furthermore, the metal-induced fibrillation of β-synuclein was further accelerated by the addition of GAGs or high concentrations of crowding agents. β-Synuclein was shown to undergo fast oligomerization and fibrillation in the presence of pesticides, whereas the addition of low concentrations of organic solvents induced the formation of amorphous aggregates. This new revelation demonstrated the effect of environmental metal and pesticide pollutants on their ability to develop an amyloidogenic, and potentially neurotoxic, conformation in an otherwise benign protein (Yamin *et al.*, 2005).

Intriguingly, the addition of either β- or γ-synuclein in a 1:1 molar ratio to α-synuclein solution substantially reduced the elongation rate and increased the duration of the lag-time of α-synuclein fibrillation (Uversky *et al.*, 2002c). A further increase in the relative concentration of β- or γ-synuclein (2:1 molar ratio) led to additional inhibition of α-synuclein fibrillation, and finally, at a 4:1 molar excess of β- over α-synuclein, the latter protein did not form fibrils within 2 to 3 weeks (Uversky *et al.*, 2002c). In agreement with these observations, it has been shown that β-synuclein inhibited α-synuclein aggregation in an animal model (doubly transgenic mice expressing human α- and β-synucleins) (Hashimoto *et al.*, 2001). These observations indicate that β- and γ-synucleins may act as regulators of α-synuclein fibrillation *in vivo*, in effect acting as chaperones to minimize the aggregation of α-synuclein. In particular, excess stoichiometries of β-synuclein would be expected to limit the amount of fibrillation. Consequently, one possible factor in the etiology of Parkinson's disease would be a decrease in the levels of β- or γ-synuclein (Uversky *et al.*, 2002c).

However, a set of conditions effectively inducing fibrillation and aggregation of "non-amyloidogenic" β-synuclein has been recently found (Yamin *et al.*, 2005). This included some metals (Zn^{2+}, Pb^{2+}, and Cu^{2+}) alone or in combination with crowding agents and glycosaminoglycans, as well as pesticides (rotenone, paraquat, and dieldrin) and low concentrations of organic solvents. Interestingly, the molecular mechanism of β-synuclein fibrillation and aggregation was shown to be very different from that of α-synuclein. It has been shown that β-synuclein did not fibrillate in crowded environments or in the presence of glycosaminoglycans (i.e., under the conditions known to induce very fast fibrillation of α-synuclein) when metals were not present in media (Yamin *et al.*, 2005).

The fact of establishing of the conditions favoring fibrillation of a protein is not an unexpected one, as there is an increasing belief that the ability to fibrillate is a generic property of a polypeptide chain, and all proteins are potentially able to form amyloid fibrils under appropriate conditions (Dobson, 1999; Dobson, 2001a; Dobson, 2001b; Fandrich *et al.*, 2001; Pertinhez *et al.*, 2001; Uversky and Fink, 2004). What was absolutely unique about these findings is that the established fibrillation-promoting conditions included the environmental pollutants (metals and pesticides) assumed to be involved in the pathogenesis of PD, and that they induced fibrillation and aggregation of β-synuclein, which was assumed to be a benign counterpart of amyloidogenic α-synuclein. Importantly, such factors as macromolecular crowding and glycosaminoglycans, being applied alone, did not promote fibrillation of β-synuclein. However, this protein was shown to form fibrils in a very efficient way when these same factors were applied in a combination with metals (Yamin *et al.*, 2005). Furthermore, it has been established that in the presence of some metals, β-synuclein does not inhibit the formation of α-synuclein fibrils, and mixtures of α- and β-synucleins exhibited rapid fibrillation, almost as fast as that of α-synuclein incubated with these metals. This demonstrates the crucial role of environmental pollutants, such as metals and pesticides/herbicides, in the induction of an amyloidogenic, and potentially neurotoxic, conformation in an otherwise benign protein (Yamin *et al.*, 2005). Furthermore, based on the fact that metals can eliminate the β-synuclein-induced fibrillation of α-synuclein, it has been concluded that the protective antifibrillation role of β-synuclein may fail under conditions of industrial pollution due to the exposure to certain metals or pesticides. It has been hypothesized that in the presence of the enhanced concentrations of these industrial pollutants, β-synuclein no longer plays a protective role; rather, it may represent a risk factor, being able to fibrillate effectively in metal- and pesticide-dependent manner (Yamin *et al.*, 2005).

4.4.12.4. Proteins Promoting α-Synuclein Aggregation

α-Synuclein is known to interact with several proteins [for reviews, see Dev *et al.* (2003); Lindersson and Jensen (2004); Lundvig *et al.* (2005)]. Importantly, some of these proteins have been shown to stimulate α-synuclein aggregation *in vitro* at substoichiometric concentrations including tau (Giasson *et al.*, 2002), histones (Goers *et al.*, 2003a), brain-specific protein p25α. (Lindersson *et al.*, 2005), tubulin (Alim *et al.*, 2002), and agrin (Cole *et al.*, 2005). Furthermore, it has been shown recently that the transcriptional cofactor high mobility group protein 1 (HMGB-1) is able to bind preferentially to aggregated α-synuclein and is present in α-synuclein filament-containing Lewy bodies isolated from brain tissue affected with dementia with Lewy bodies or Parkinson's disease (Lindersson *et al.*, 2004). Except for the histones and agrin, these proteins have all been identified as components of Lewy bodies and/or glial cytoplasmic inclusions (Alim *et al.*, 2002; Giasson *et al.*, 2002; Lindersson *et al.*, 2004; Lindersson *et al.*, 2005). As discussed above, α-synuclein was shown to co-localize with histones in murine nigral neurons upon toxic insults (Goers *et al.*, 2003a). The mechanism by which these proteins interact with α-synuclein is unknown, but they all contain basic motifs, suggesting that the interaction with α-synuclein may be mediated through ionic interactions. It has been emphasized that the identification of proteins regulating α-synuclein aggregation *in vitro* indicates that endogenous protein factors may regulate α-synuclein aggregation *in vivo* (Lundvig *et al.*, 2005).

4.4.13. Small Molecules Inhibiting α-Synuclein Fibrillation

4.4.13.1. Dopamine, L-DOPA, and Other Catecholamines

PD exists as both an idiopathic and familial disorder. Although the exact mechanisms underlying PD pathogenesis are unknown, the common pathway is the damage and subsequent loss of dopamine neurons (Maguire-Zeiss and Federoff, 2003). Importantly, the decline in the number of dopaminergic neurons below a critical threshold produces early symptomatic PD (Duvoisin, 1992b) as a result of the systematic reduction in the dopamine (DA) content. At the onset of symptoms, putamenal DA is depleted ~80%, and ~60% of dopaminergic neurons have already been lost in the the substantia nigra pars compacta. Besides the loss of the nigrostriatal dopaminergic neurons, the pathologic hallmark of PD is the presence of intraneuronal proteinacious cytoplasmic inclusions, LBs and LNs, enriched in aggregated α-synuclein. These observations obviously bring together α-synuclein, its aggregation, and dopamine.

One of the key neuromuscular neurotransmitters, dopamine, is relatively unstable and decomposes to quinones and dopaminochrome, as well as reactive oxygen species (Asanuma *et al.*, 2003; Davies, 1987; Davies *et al.*, 1987; LaVoie and Hastings, 1999; Spencer *et al.*, 1996). The key steps in the dopamine metabolism include the initial hydroxylation of tyrosine (by tyrosine hydroxylase) to form L-3,4-dihydroxyphenylalanine (L-DOPA, or levodopa), the immediate precursor of dopamine. DA is enzymatically decarboxylated to yield 3,4-dihydroxyphenylacetic acid (DOPAC), which is converted into 3-methoxy-4-hydroxy-phenylacetic acid (homovanillic acid; HVA). Epinephrine and norepinephrine are structurally related catecholamines, which are also found in the brain.

It has been shown recently that L-DOPA, dopamine, and other catecholamines dissolve fibrils of α-synuclein and Aβ peptide generated *in vitro* (Li *et al.*, 2004a). The catecholamines also inhibited the fibrillation of these proteins. In addition, intraneuronal α-synuclein deposits formed in a mouse model were dissolved by the incubation of tissue slices with L-DOPA. Importantly, because these catecholamines are known to be susceptible to oxidative breakdown, the inhibitory effects of different oxidation products were also studied, revealing that the products of oxidation were more effective than the parent compounds in inhibition of α-synuclein fibrillation (Li *et al.*, 2004a). In similar study, the inhibitory effect of a library of 169 drug-like molecules on α-synuclein fibrillation has been analyzed (Conway *et al.*, 2001). It has been shown that all but one of 15 fibril inhibitors were catecholamines related to dopamine. Furthermore, it has been established that the inhibitory activity of dopamine might depend on its oxidative ligation to α-synuclein. This modification was selective for the protofibril-to-fibril conversion, causing an accumulation of the α-synuclein protofibril. Adduct formation was assumed to provide an explanation for the dopaminergic selectivity of α-synuclein-associated neurotoxicity in PD and has implications for current and future PD therapeutic and diagnostic strategies (Conway *et al.*, 2001).

Thus, the presence of micromolar concentrations of dopamine or L-DOPA was shown to be sufficient to significantly inhibit fibril formation or disaggregate existing fibrils of Aβ or α-synuclein *in vitro*. It has been suggested that the molecular mechanisms involved in the inhibition and disaggregation effects may be different (Li *et al.*, 2004a). A potential mechanism of fibrillation inhibition might involve a formation of a covalent adduct between the orthoquinone derivative of dopamine and 5–10% of the α-synuclein. This covalently modified α-synuclein would then inhibit fibrillation, leading to the accumulation of protofibrils (Conway *et al.*, 2001). The dissolution of fibrils was suggested to involve preferential noncovalent interactions of the catechol with either soluble oligomers in equilibrium with fibrils or with the fibrils themselves, leading to the formation of stable, soluble oligomers. Alternately, the covalent modification by reactive oxidation products could lead to the weakening of the intermolecular forces in the fibrils, resulting in disaggregation and formation of soluble monomers or oligomers (Li *et al.*, 2004a). Interestingly, if fibrillar α-synuclein contributes to neurodegeneration, then the disassembly of fibrils by catechols may reverse or slow down disease progression. However, if the intermediates of fibrillation and/or the products of fibril disaggregation are neurotoxic, then the accumulation of stable fibrillar deposits, such as Lewy bodies in Parkinson's disease, could be a marker of neurons that are relatively resistant to the neurodegenerative process (Li *et al.*, 2004a).

4.4.13.2. Baicalein

Flavonoids represent a group of polyphenolic compounds that are important components in the human diet and medicinal plants. Flavonoids have broad pharmacological activities, due to the inhibition of certain enzymes and/or antioxidant activity. The normal intake of flavonoids is in the range of 50 to 800 mg/day, depending on the types of vegetables, fruits, and beverages consumed. Some of the dietary flavonoids were shown to possess neuroprotective properties (Joseph *et al.*, 1999; Schroeter *et al.*, 2000; Datla *et al.*, 2001). For example, the consumption of flavonoid-rich blueberries or strawberries can reverse cognitive and motor behavior deficits in rats (Joseph *et al.*, 1999), whereas intake of antioxidant flavonoids is associated with a lower incidence of dementia (Commenges *et al.*, 2000).

Baicalein is a typical flavonoid and is the main component of the traditional Chinese herbal medicine *Scutellaria* baicalensis. It is known that baicalein has multiple biological activities including antiallergic, anticarcinogenic, and anti-HIV properties (Li *et al.*, 2000; Shieh *et al.*, 2000; Gao *et al.*, 2001; Ikezoe *et al.*, 2001; Wu *et al.*, 2001), and natural medicines containing these compounds have been reported to have beneficial effects in treating memory loss and dementia (Lebeau *et al.*, 2001; Perry *et al.*, 1999). Furthermore, it has been reported that baicalein might protect rat cortical neurons from amyloid β-induced toxicity by its inhibition of lipoxygenase (Lebeau *et al.*, 2001).

Recently, it has been shown that micromolar concentrations of baicalein, or its oxidized forms, inhibit the formation of α-synuclein fibrils and disaggregate the preformed fibrils *in vitro* (Zhu *et al.*, 2004). It has also been established that the product of the inhibition reaction is predominately a soluble oligomeric α-synuclein, some lysine side chains of which were shown to be covalently modified by baicalein quinone to form a Schiff base. In a disaggregation reaction, baicalein was shown to cause fragmentation throughout the length of the fibril (Zhu *et al.*, 2004). These observations suggested that baicalein and similar compounds may have potential as therapeutic leads in PD treatment and that diets rich in flavonoids may be effective in preventing the disorder.

4.4.13.3. Rifampicin

Rifampicin is a semisynthetic derivative of the rifamycins, a class of antibiotics that are fermentation products of *Nocardia mediterranei* (Furesz, 1970). Epidemiologic studies revealed that leprosy patients might have significantly lower probability of senile dementia development if they had been under antileprosy treatment with dapsone, or rifampicin, and closely related drugs for the preceding several years (Chui *et al.*, 1994; Namba *et al.*, 1992). In addition to treatment for leprosy, rifampicin is widely used in some countries for the treatment of tuberculosis (Hartmann *et al.*, 1967). It has been shown that the overall prevalence of senile dementia was

2.9% in 1410 patients who were continuously on antileprosy treatment, compared with 6.25% in 1761 untreated patients (Mcgeer *et al.*, 1992). Based on this observation, it has been suggested that some drugs being used for leprosy might prevent Aβ aggregation, thus resulting in the absence of amyloid deposition (Tomiyama *et al.*, 1996). This hypothesis has been tested with two antileprosy drugs, depasone and rifampicin, and it has been found that rifampicin and its analogues, *p*-benzoquinone and hydroquinone, were able to inhibit Aβ1–40 or Aβ1–42 aggregtion and neurotoxicity *in vitro* (Tomiyama *et al.*, 1994; Tomiyama *et al.*, 1997; Tomiyama *et al.*, 1996). Similar rifampicin effects have been recently reported for human islet amyloid polypeptide, amylin (Tomiyama *et al.*, 1997). The ability of rifampicin to inhibit the development of senile dementia suggests that this drug can pass the blood-brain barrier.

These observations initiated the investigation of the effect of rifampicin on α-synuclein fibrillation *in vitro* (Li *et al.*, 2004b). It has been established that the antibiotic rifampicin is able to inhibit α-synuclein fibrillation and to disaggregate preformed fibrils in a concentration-dependent manner. Using size-exclusion chromatography, it has been indicated that rifampicin stabilized α-synuclein as both a monomer and soluble oligomers comprised of partially folded α-synuclein (Li *et al.*, 2004b). Furthermore, the inhibitory and disaggregating effects were significantly increased when aged samples of rifampicin were used. This indicates that the most active species in inhibiting fibrillation and disaggregating fibrils is an oxidation product of rifampicin. This conclusion was confirmed by performing similar experiments under anaerobic conditions. Based on these studies, it has been concluded that rifampicin-mediated inhibition of α-synuclein fibrillation and disaggregation of fibrils involves preferential stabilization of monomeric and soluble oligomeric forms and that rifampicin may potentially have some therapeutic application for PD (Li *et al.*, 2004b).

4.5. Conclusion

Data accumulated so far and briefly considered in this review leave no doubts that α-synuclein, and especially its aggregation, might play an important role in the pathology of several neurodegenerative disorders. Importantly, α-synuclein is a natively unfolded protein, which being considerably unstructured does not represent a real random coil, having instead some residual structure. Furthermore, human α-synuclein was shown to possess amazing structural plasticity and adopt several absolutely different conformations. The choice between these conformations is determined by the peculiarities of the protein environment that force α-synuclein to fold on a template-dependent manner. This exceptional environmental adaptability is illustrated in Figure 4.5, which represents α-synuclein as a chameleon (Uversky, 2003b), which is able to adopt different monomeric, oligomeric, and insoluble conformations (drawn at sides).

A general model of α-synuclein aggregation is depicted in Figure 4.6. Prior to aggregation, α-synuclein exists as a mixture of the natively unfolded and partially folded conformations. Equilibrium is essentially shifted toward the unfolded conformation under the normal physiologic conditions. Different factors, such as a decrease in pH, increase in temperature, the presence of amphipathic molecules (e.g., herbicides or pesticides), or presence of metal ions and other charged molecules, interaction with charged biopolymers, other proteins, or membrane, or the effect of macromolecular crowding, and numerous other environmental factors, stabilize structural rearrangements within the natively unfolded α-synuclein, thus shifting equilibrium and populating the partially folded conformation. It is important to remember that although the model represents a unique partially folded intermediate, this could be an oversimplification, and different factors may stabilize slightly different partially folded conformations. The process of self-association is facilitated due to the appearance of solvent-exposed hydrophobic clusters on

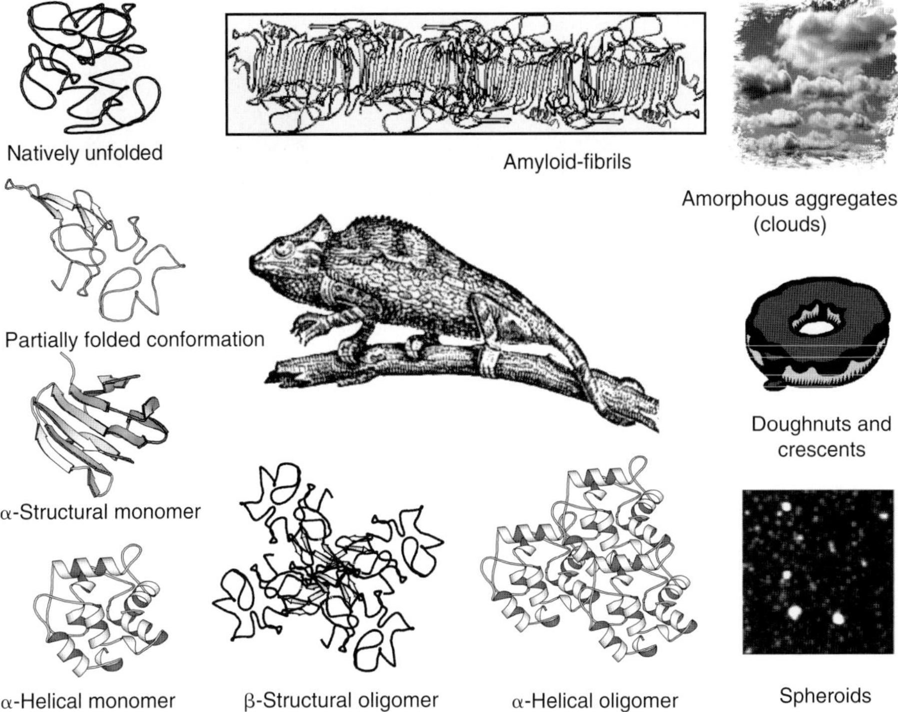

Figure 4.5. Being a protein-chameleon, α-synuclein is able to adopt absolutely different conformations in a template-dependent manner. Modified from Uversky (2003b).

the surface of a partially folded protein. The type of end-product depends on the particular experimental conditions, and association may result in the formation of amorphous aggregates, soluble oligomers, or amyloid-like fibrils. Any of these three species could be neurotoxic, leading to the specific loss of dopaminergic neurons in substantia nigra and, thus, giving rise to the Parkinson's disease.

It has been pointed out that synuclein toxicity may be exerted by specific populations of α-synuclein aggregates and/or mediated via various routes through proteins involved in different cellular processes (Lundvig *et al.*, 2005). The mechanisms proposed to describe the neurotoxicity of α-synuclein and its aggregates could be grouped into three major classes: mechanical distortion of cellular compartments/processes, toxic gain of function, and toxic loss of function. An analysis of the peculiarities of α-synuclein interaction with lipids revealed that some lipids can induce α-synuclein oligomerization upon binding, which in particular circumstances can disrupt membrane bilayers. Furthermore, some protofibrillar forms of α-synuclein were shown to penetrate membranes forming pore-like channels. These structures were proposed to be the neurotoxic species, which kill neurons via the abnormal increase in membrane permeability [see, e.g., Volles and Lansbury (2003)]. Alternatively, impairment of α-synuclein degradation has been proposed as a mechanism for neurotoxicity. In fact, it is known that α-synuclein is primarily degraded by proteasome. Importantly, the degradation of mutant α-synucleins is less efficient

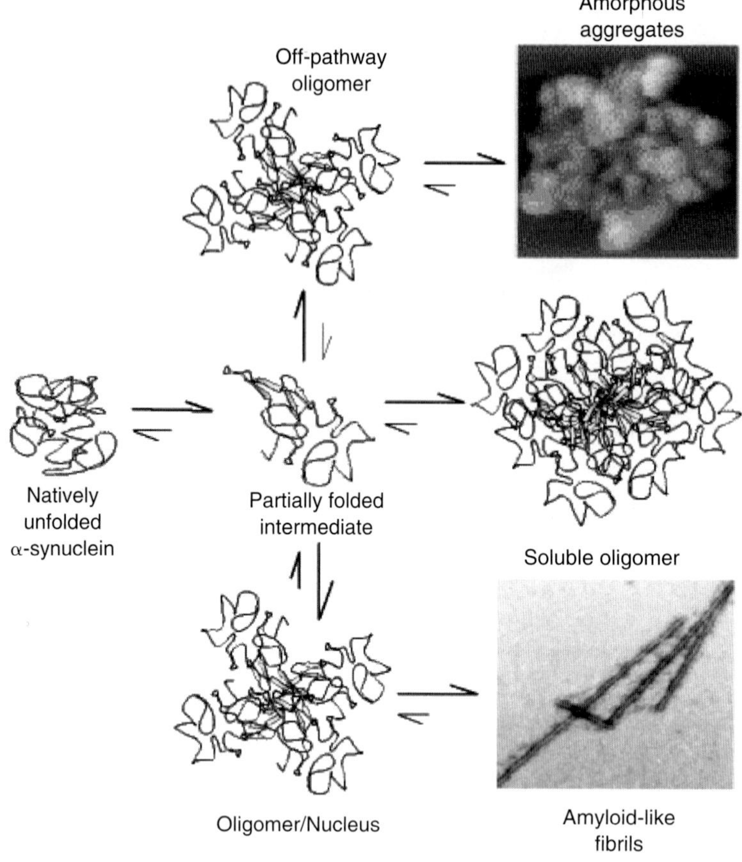

Figure 4.6. Multiple pathways of human α-synuclein aggregation.

than that of wild-type protein. Furthermore, other modifications of α-synuclein may impair proteasome degradation as well. This could lead to the elevation of cellular concentrations of α-synuclein, aggregate formation, and neurotoxicity (Bennett, 2005).

Numerous mechanisms of toxicity based on the toxic gain of function by α-synuclein have been proposed (Bennett, 2005). In fact, cytotoxicity could be exerted via the various forms of aggregated α-synuclein. This hypothesis is based on the evidence that α-synuclein, the known mutations, and the NAC fragment all can function as nucleation sites and promote aggregation under certain conditions. Cytotoxicity could also originate from the modifications of α-synuclein and transformation of this protein into neurotoxic species. This could be a result of α-synuclein exposure to metals, pesticides, dopamine metabolites, oxidative stress, exogenous toxins, and mitochondrial insufficiency. The covalently cross-linked oligomers (homopolymers or heteropolymers of α-synuclein), which might be responsible for the insolubility of pathogenic inclusions such as Lewy bodies, could also be cytotoxic. On the other hand, several studies emphasize that α-synuclein-related neurotoxicity might arise from a loss of function [summarized in Bennett

(2005)]. Obviously, all the factors mentioned above are not necessarily mutually exclusive but instead may be synergistic (Bennett, 2005).

Finally, a few words on the potential implications for therapeutics should be added. Current PD medications treat symptoms; none halt or retard dopaminergic neuron degeneration. The main obstacle to developing neuroprotective therapies is a limited understanding of the key molecular events that provoke neurodegeneration (Dauer and Przedborski, 2003). It has been also pointed out that the current difficulty of finding means for the successful inhibition of synucleinopathies is due to the fact that these diseases, being multifactorial, are known to be initiated by numerous factors (Bennett, 2005). However, many of them, if not all, are somehow related to the abnormal α-synuclein processing, functioning, or aggregation. This suggests that some effective therapeutic protocols could be developed based on the detailed analysis of α-synuclein function, misfunction, and aggregation. The identification of molecules that inhibit protein deposition or reverse fibril formation may be a critical step toward a better understanding of the pathophysiology of inclusions and deposits in human diseases. It may allow us to evaluate the role of fibrils in neurodegenerative processes and disease progression, and, should these molecules exert neuroprotective effects, their discovery could open new avenues for therapeutic intervention (Li *et al.*, 2004a). Importantly, recent studies showing the inhibitory and disaggregating effects of dopamine and related catecholamines (Li *et al.*, 2004a), baicalein (Zhu *et al.*, 2004), and rifampicin (Li *et al.*, 2004b) on α-synuclein fibrillation might represent the first steps in these new avenues.

Acknowledgments

The author expresses his deepest gratitude to Alexey Uversky for careful reading and editing of the manuscript. This work was supported in part by a grant from INTAS (2001–2347).

References

Alim, M. A., Hossain, M. S., Arima, K., Takeda, K., Izumiyama, Y., Nakamura, M., Kaji, H., Shinoda, T., Hisanaga, S., and Ueda, K. (2002). Tubulin seeds alpha-synuclein fibril formation. *J Biol Chem* 277: 2112–2117.

Altschuler, E. (1999). Aluminum-containing antacids as a cause of idiopathic Parkinson's disease. *Med Hypotheses* 53: 22–23.

Arawaka, S., Saito, Y., Murayama, S., and Mori, H. (1998). Lewy body in neurodegeneration with brain iron accumulation type 1 is immunoreactive for alpha-synuclein. *Neurology* 51: 887–889.

Arima, K., Ueda, K., Sunohara, N., Hirai, S., Izumiyama, Y., Tonozuka-Uehara, H., and Kawai, M. (1998) Immunoelectron-microscopic demonstration of NACP/alpha-synuclein-epitopes on the filamentous component of Lewy bodies in Parkinson's disease and in dementia with Lewy bodies. *Brain Res* 808: 93–100.

Asanuma, M., Miyazaki, I., and Ogawa, N. (2003). Dopamine- or L-DOPA-induced neurotoxicity: the role of dopamine quinone formation and tyrosinase in a model of Parkinson's disease. *Neurotox Res* 5: 165–176.

Aulak, K. S., Miyagi, M., Yan, L., West, K. A., Massillon, D., Crabb, J. W., and Stuehr, D. J. (2001). Proteomic method identifies proteins nitrated *in vivo* during inflammatory challenge. *Proc Natl Acad Sci USA* 98: 12056–12061.

Auluck, P. K., and Bonini, N. M. (2002). Pharmacological prevention of Parkinson disease in Drosophila. *Nat Med* 8: 1185–1186.

Auluck, P. K., Chan, H. Y., Trojanowski, J. Q., Lee, V. M., and Bonini, N. M. (2002). Chaperone suppression of alpha-synuclein toxicity in a Drosophila model for Parkinson's disease. *Science* 295: 865–868.

Baba, M., Nakajo, S., Tu, P. H., Tomita, T., Nakaya, K., Lee, V. M., Trojanowski, J. Q., and Iwatsubo, T. (1998). Aggregation of alpha-synuclein in Lewy bodies of sporadic Parkinson's disease and dementia with Lewy bodies. *Am J Pathol* 152: 879–884.

Beal, M. F. (1995). Aging, energy, and oxidative stress in neurodegenerative diseases. *Ann Neurol* 38: 357–366.

Bennett, M. C. (2005). The role of alpha-synuclein in neurodegenerative diseases. *Pharmacol Ther* 105: 311–331.

Betarbet, R., Sherer, T. B., MacKenzie, G., Garcia-Osuna, M., Panov, A. V., and Greenamyre, J. T. (2000). Chronic systemic pesticide exposure reproduces features of Parkinson's disease. *Nat Neurosci* 3: 1301–1306.

Bismuto, E., Martelli, P. L., De Maio, A., Mita, D. G., Irace, G., and Casadio, R. (2002). Effect of molecular confinement on internal enzyme dynamics: frequency domain fluorometry and molecular dynamics simulation studies. *Biopolymers* 67: 85–95.

Bodles, A. M., Guthrie, D. J., Harriott, P., Campbell, P., and Irvine, G. B. (2000). Toxicity of non-abeta component of Alzheimer's disease amyloid, and N-terminal fragments thereof, correlates to formation of beta-sheet structure and fibrils. *Eur J Biochem* 267: 2186–2194.

Bodles, A. M., Guthrie, D. J., Greer, B., and Irvine, G. B. (2001). Identification of the region of non-Abeta component (NAC) of Alzheimer's disease amyloid responsible for its aggregation and toxicity. *J Neurochem* 78: 384–395.

Bonifati, V., Rizzu, P., van Baren, M. J., Schaap, O., Breedveld, G. J., Krieger, E., Dekker, M. C., Squitieri, F., Ibanez, P., Joosse, M., van Dongen, J. W., Vanacore, N., van Swieten, J. C., Brice, A., Meco, G., van Duijn, C. M., Oostra, B. A., and Heutink, P. (2003). Mutations in the DJ-1 gene associated with autosomal recessive early-onset parkinsonism. *Science* 299: 256–259.

Braak, H., Sandmann-Keil, D., Gai, W., and Braak, E. (1999). Extensive axonal Lewy neurites in Parkinson's disease: a novel pathological feature revealed by alpha-synuclein immunocytochemistry. *Neurosci Lett* 265: 67–69.

Bracken, C., Iakoucheva, L. M., Romero, P. R., and Dunker, A. K. (2004). Combining prediction, computation and experiment for the characterization of protein disorder. *Curr Opin Struct Biol* 14: 570–576.

Browne, S. E., Ferrante, R. J., and Beal, M. F. (1999). Oxidative stress in Huntington's disease. *Brain Pathol* 9: 147–163.

Buchman, V. L., Adu, J., Pinon, L. G., Ninkina, N. N., and Davies, A. M. (1998a). Persyn, a member of the synuclein family, influences neurofilament network integrity. *Nat Neurosci* 1: 101–103.

Buchman, V. L., Hunter, H. J., Pinon, L. G., Thompson, J., Privalova, E. M., Ninkina, N. N., and Davies, A. M. (1998b). Persyn, a member of the synuclein family, has a distinct pattern of expression in the developing nervous system. *J Neurosci* 18: 9335–9341.

Bussell, R., Jr., and Eliezer, D. (2001). Residual structure and dynamics in Parkinson's disease-associated mutants of alpha-synuclein. *J Biol Chem* 276: 45996–46003.

Bussell, R., Jr., and Eliezer, D. (2004). Effects of Parkinson's disease-linked mutations on the structure of lipid-associated alpha-synuclein. *Biochemistry* 43: 4810–4818.

Butterfield, D. A., and Kanski, J. (2001). Brain protein oxidation in age-related neurodegenerative disorders that are associated with aggregated proteins. *Mech Ageing Dev* 122: 945–962.

Campbell, B. C., Li, Q. X., Culvenor, J. G., Jakala, P., Cappai, R., Beyreuther, K., Masters, C. L., and McLean, C. A. (2000). Accumulation of insoluble alpha-synuclein in dementia with Lewy bodies. *Neurobiol Dis* 7: 192–200.

Campbell, B. C., McLean, C. A., Culvenor, J. G., Gai, W. P., Blumbergs, P. C., Jakala, P., Beyreuther, K., Masters, C. L., and Li, Q. X. (2001). The solubility of alpha-synuclein in multiple system atrophy differs from that of dementia with Lewy bodies and Parkinson's disease. *J Neurochem* 76: 87–96.

Cardin, A. D., and Weintraub, H. J. (1989). Molecular modeling of protein-glycosaminoglycan interactions. *Arteriosclerosis* 9: 21–32.

Castillo, G. M., Ngo, C., Cummings, J., Wight, T. N., and Snow, A. D. (1997). Perlecan binds to the beta-amyloid proteins (A beta) of Alzheimer's disease, accelerates A beta fibril formation, and maintains A beta fibril stability. *J Neurochem* 69: 2452–2465.

Chou, P. Y., and Fasman, G. D. (1978). Empirical predictions of protein conformation. *Annu Rev Biochem* 47: 251–276.

Chui, D. H., Tabira, T., Izumi, S., Koya, G., and Ogata, J. (1994). Decreased beta-amyloid and increased abnormal Tau deposition in the brain of aged patients with leprosy. *Am J Pathol* 145: 771–775.

Clayton, D. F., and George, J. M. (1998). The synucleins: a family of proteins involved in synaptic function, plasticity, neurodegeneration and disease. *Trends Neurosci* 21: 249–254.

Clayton, D. F., and George, J. M. (1999). Synucleins in synaptic plasticity and neurodegenerative disorders. *J Neurosci Res* 58: 120–129.

Cohen, N. A., Kaufmann, W. E., Worley, P. F., and Rupp, F. (1997). Expression of agrin in the developing and adult rat brain. *Neuroscience* 76: 581–596.

Cohlberg, J. A., Li, J., Uversky, V. N., and Fink, A. L. (2002). Heparin and other glycosaminoglycans stimulate the formation of amyloid fibrils from alpha-synuclein *in vitro*. *Biochemistry* 41: 1502–1511.

Cole, G. J., Liu, I. H., Uversky, V. N., Munishkina, L. A., Fink, A. L. and Halfter, W., (2005). Agrin binds alpha-synuclein and modulates alpha-synuclein fibrillation. *Glycobiology* 15: 1320–1331.

Commenges, D., Scotet, V., Renaud, S., Jacqmin-Gadda, H., Barberger-Gateau, P., and Dartigues, J. F. (2000). Intake of flavonoids and risk of dementia. *Eur J Epidemiol* 16: 357–363.

Conway, K. A., Harper, J. D., and Lansbury, P. T. (1998). Accelerated *in vitro* fibril formation by a mutant alpha-synuclein linked to early-onset Parkinson disease. *Nat Med* 4: 1318–1320.

Conway, K. A., Harper, J. D., and Lansbury, P. T., Jr. (2000a). Fibrils formed *in vitro* from alpha-synuclein and two mutant forms linked to Parkinson's disease are typical amyloid. *Biochemistry* 39: 2552–2563.

Conway, K. A., Lee, S. J., Rochet, J. C., Ding, T. T., Harper, J. D., Williamson, R. E., and Lansbury, P. T., Jr. (2000b). Accelerated oligomerization by Parkinson's disease linked alpha-synuclein mutants. *Ann N Y Acad Sci* 920: 42–45.

Conway, K. A., Lee, S. J., Rochet, J. C., Ding, T. T., Williamson, R. E., and Lansbury, P. T., Jr. (2000c). Acceleration of oligomerization, not fibrillization, is a shared property of both alpha-synuclein mutations linked to early-onset Parkinson's disease: implications for pathogenesis and therapy. *Proc Natl Acad Sci USA* 97: 571–576.

Conway, K. A., Rochet, J. C., Bieganski, R. M., and Lansbury, P. T., Jr. (2001). Kinetic stabilization of the alpha-synuclein protofibril by a dopamine-alpha-synuclein adduct. *Science* 294: 1346–1349.

Cookson, M. R. (2004) The Biochemistry of Parkinson's Disease. *Annu Rev Biochem*: In press.

Cookson, M. R., and Shaw, P. J. (1999). Oxidative stress and motor neurone disease. *Brain Pathol* 9: 165–186.

Cotman, S. L., Halfter, W., and Cole, G. J. (2000). Agrin binds to beta-amyloid (Abeta), accelerates abeta fibril formation, and is localized to Abeta deposits in Alzheimer's disease brain. *Mol Cell Neurosci* 15: 183–198.

Creighton, T. E. (1993). *Proteins. Structures and Molecular Properties*. New York: W. H. Freeman and Company.

Crowther, R. A., Jakes, R., Spillantini, M. G., and Goedert, M. (1998). Synthetic filaments assembled from C-terminally truncated alpha-synuclein. *FEBS Lett* 436: 309–312.

Crowther, R. A., Daniel, S. E., and Goedert, M. (2000). Characterisation of isolated alpha-synuclein filaments from substantia nigra of Parkinson's disease brain. *Neurosci Lett* 292: 128–130.

da Costa, C. A., Ancolio, K., and Checler, F. (2000). Wild-type but not Parkinson's disease-related ala-53 → Thr mutant alpha-synuclein protects neuronal cells from apoptotic stimuli. *J Biol Chem* 275: 24065–24069.

Datla, K. P., Christidou, M., Widmer, W. W., Rooprai, H. K., and Dexter, D. T. (2001). Tissue distribution and neuroprotective effects of citrus flavonoid tangeretin in a rat model of Parkinson's disease. *Neuroreport* 12: 3871–3875.

Dauer, W., and Przedborski, S. (2003). Parkinson's disease: mechanisms and models. *Neuron* 39: 889–909.

Daughdrill, G. W., Pielak, G. J., Uversky, V. N., Cortese, M. S., and Dunker, A. K. (2005). Natively disordered proteins. In *Handbook of Protein Folding*, ed. J. Buchner, and T. Kiefhaber, pp. 271–353. Weinheim, Germany: Wiley-VCH, Verlag GmbH & Co. KGaA.

Davidson, W. S., Jonas, A., Clayton, D. F., and George, J. M. (1998). Stabilization of alpha-synuclein secondary structure upon binding to synthetic membranes. *J Biol Chem* 273: 9443–9449.

Davies, K. J. (1987). Protein damage and degradation by oxygen radicals. I. general aspects. *J Biol Chem* 262: 9895–9901.

Davies, K. J., Lin, S. W., and Pacifici, R. E. (1987) Protein damage and degradation by oxygen radicals. IV. Degradation of denatured protein. *J Biol Chem* 262: 9914–9920.

Davis, L. E., and Adair, J. C. (1999). Parkinsonism from methanol poisoning: benefit from treatment with anti-Parkinson drugs. *Mov Disord* 14: 520–522.

Dev, K. K., Hofele, K., Barbieri, S., Buchman, V. L., and van der Putten, H. (2003). Part II: alpha-synuclein and its molecular pathophysiological role in neurodegenerative disease. *Neuropharmacology* 45: 14–44.

Dexter, D. T., Carayon, A., Javoy-Agid, F., Agid, Y., Wells, F. R., Daniel, S. E., Lees, A. J., Jenner, P., and Marsden, C. D. (1991). Alterations in the levels of iron, ferritin and other trace metals in Parkinson's disease and other neurodegenerative diseases affecting the basal ganglia. *Brain* 114 (Pt 4): 1953–1975.

Di Monte, D. A. (2001). The role of environmental agents in Parkinson's disease. *Clin Neurosci Res* 1: 419–426.

Di Monte, D. A. (2003). The environment and Parkinson's disease: is the nigrostriatal system preferentially targeted by neurotoxins? *Lancet Neurol* 2: 531–538.

Di Monte, D. A., Lavasani, M., and Manning-Bog, A. B. (2002). Environmental factors in Parkinson's disease. *Neurotoxicology* 23: 487–502.

Dickson, D. W. (2001). Alpha-synuclein and the Lewy body disorders. *Curr Opin Neurol* 14: 423–432.

Dobson, C. M. (1999). Protein misfolding, evolution and disease. *Trends Biochem Sci* 24: 329–332.

Dobson, C. M. (2001a). Protein folding and its links with human disease. *Biochem Soc Symp*: 1–26.

Dobson, C. M. (2001b). The structural basis of protein folding and its links with human disease. *Philos Trans R Soc Lond B Biol Sci* 356: 133–145.

Donahue, J. E., Berzin, T. M., Rafii, M. S., Glass, D. J., Yancopoulos, G. D., Fallon, J. R., and Stopa, E. G. (1999). Agrin in Alzheimer's disease: altered solubility and abnormal distribution within microvasculature and brain parenchyma. *Proc Natl Acad Sci USA* 96: 6468–6472.

Duda, J. E., Giasson, B. I., Chen, Q., Gur, T. L., Hurtig, H. I., Stern, M. B., Gollomp, S. M., Ischiropoulos, H., Lee, V. M., and Trojanowski, J. Q. (2000). Widespread nitration of pathological inclusions in neurodegenerative synucleinopathies. *Am J Pathol* 157: 1439–1445.

Dunker, A. K., Brown, C. J., Lawson, J. D., Iakoucheva, L. M., and Obradovic, Z. (2002a). Intrinsic disorder and protein function. *Biochemistry* 41: 6573–6582.

Dunker, A. K., Brown, C. J., and Obradovic, Z. (2002b). Identification and functions of usefully disordered proteins. *Adv Protein Chem* 62: 25–49.

Dunker, A. K., Garner, E., Guilliot, S., Romero, P., Albrecht, K., Hart, J., Obradovic, Z., Kissinger, C., and Villafranca, J. E. (1998). Protein disorder and the evolution of molecular recognition: theory, predictions and observations. *Pac Symp Biocomput*: 473–484.

Dunker, A. K., Lawson, J. D., Brown, C. J., Williams, R. M., Romero, P., Oh, J. S., Oldfield, C. J., Campen, A. M., Ratliff, C. M., Hipps, K. W., Ausio, J., Nissen, M. S., Reeves, R., Kang, C., Kissinger, C. R., Bailey, R. W., Griswold, M. D., Chiu, W., Garner, E. C., and Obradovic, Z. (2001). Intrinsically disordered protein. *J Mol Graph Model* 19: 26–59.

Dunker, A. K., and Obradovic, Z. (2001). The protein trinity-linking function and disorder. *Nat Biotechnol* 19: 805–806.

Duvoisin, R. C. (1992a). A brief history of parkinsonism. *Neurol Clin* 10: 301–316.

Duvoisin, R. C. (1992b). Overview of Parkinson's disease. *Ann N Y Acad Sci* 648: 187–193.

Dyson, H. J., and Wright, P. E. (2005). Intrinsically unstructured proteins and their functions. *Nat Rev Mol Cell Biol* 6: 197–208.

Eggers, D. K., and Valentine, J. S. (2001a). Crowding and hydration effects on protein conformation: a study with sol-gel encapsulated proteins. *J Mol Biol* 314: 911–922.

Eggers, D. K., and Valentine, J. S. (2001b). Molecular confinement influences protein structure and enhances thermal protein stability. *Protein Sci* 10: 250–261.

El-Agnaf, O. M., Bodles, A. M., Guthrie, D. J., Harriott, P., and Irvine, G. B. (1998a). The N-terminal region of non-A beta component of Alzheimer's disease amyloid is responsible for its tendency to assume beta-sheet and aggregate to form fibrils. *Eur J Biochem* 258: 157–163.

El-Agnaf, O. M., Jakes, R., Curran, M. D., Middleton, D., Ingenito, R., Bianchi, E., Pessi, A., Neill, D., and Wallace, A. (1998b). Aggregates from mutant and wild-type alpha-synuclein proteins and NAC peptide induce apoptotic cell death in human neuroblastoma cells by formation of beta-sheet and amyloid-like filaments. *FEBS Lett* 440: 71–75.

El-Agnaf, O. M., Jakes, R., Curran, M. D., and Wallace, A. (1998c). Effects of the mutations Ala30 to Pro and Ala53 to Thr on the physical and morphological properties of alpha-synuclein protein implicated in Parkinson's disease. *FEBS Lett* 440: 67–70.

Eliezer, D., Kutluay, E., Bussell, R., Jr., and Browne, G. (2001). Conformational properties of alpha-synuclein in its free and lipid-associated states. *J Mol Biol* 307: 1061–1073.

Ellis, R. J. (2001). Macromolecular crowding: obvious but underappreciated. *Trends Biochem Sci* 26: 597–604.

Fall, P. A., Fredrikson, M., Axelson, O., and Granerus, A. K. (1999). Nutritional and occupational factors influencing the risk of Parkinson's disease: a case-control study in southeastern Sweden. *Mov Disord* 14: 28–37.

Fandrich, M., Fletcher, M. A., and Dobson, C. M. (2001). Amyloid fibrils from muscle myoglobin. *Nature* 410: 165–166.

Farrer, M., Kachergus, J., Forno, L., Lincoln, S., Wang, D. S., Hulihan, M., Maraganore, D., Gwinn-Hardy, K., Wszolek, Z., Dickson, D., and Langston, J. W. (2004). Comparison of kindreds with parkinsonism and alpha-synuclein genomic multiplications. *Ann Neurol* 55: 174–179.

Feany, M. B., and Bender, W. W. (2000). A Drosophila model of Parkinson's disease. *Nature* 404: 394–398.

Fink, A. L. (2005). Natively unfolded proteins. *Curr Opin Struct Biol* 15: 35–41.

Floyd, R. A. (1990). Role of oxygen free radicals in carcinogenesis and brain ischemia. *FASEB J* 4: 2587–2597.

Forloni, G., Bertani, I., Calella, A. M., Thaler, F., and Invernizzi, R. (2000) Alpha-synuclein and Parkinson's disease: selective neurodegenerative effect of alpha-synuclein fragment on dopaminergic neurons *in vitro* and *in vivo*. *Ann Neurol* 47: 632–640.

Forno, L. S. (1996). Neuropathology of Parkinson's disease. *J Neuropathol Exp Neurol* 55: 259–272.

Fu, S., Dean, R., Southan, M., and Truscott, R. (1998). The hydroxyl radical in lens nuclear cataractogenesis. *J Biol Chem* 273: 28603–28609.

Fujiwara, H., Hasegawa, M., Dohmae, N., Kawashima, A., Masliah, E., Goldberg, M. S., Shen, J., Takio, K., and Iwatsubo, T. (2002). alpha-Synuclein is phosphorylated in synucleinopathy lesions. *Nat Cell Biol* 4: 160–164.

Fulton, A. B. (1982). How crowded is the cytoplasm? *Cell* 30: 345–347.

Furesz, S. (1970). Chemical and biological properties of rifampicin. *Antibiot Chemother* 16: 316–351.

Gai, W. P., Power, J. H., Blumbergs, P. C., and Blessing, W. W. (1998). Multiple-system atrophy: a new alpha-synuclein disease? *Lancet* 352: 547–548.

Gai, W. P., Power, J. H., Blumbergs, P. C., Culvenor, J. G., and Jensen, P. H. (1999). Alpha-synuclein immunoisolation of glial inclusions from multiple system atrophy brain tissue reveals multiprotein components. *J Neurochem* 73: 2093–2100.

Galvin, J. E., Lee, V. M., Schmidt, M. L., Tu, P. H., Iwatsubo, T., and Trojanowski, J. Q. (1999). Pathobiology of the Lewy body. *Adv Neurol* 80: 313–324.

Gao, Z., Huang, K., and Xu, H. (2001). Protective effects of flavonoids in the roots of Scutellaria baicalensis Georgi against hydrogen peroxide-induced oxidative stress in HS-SY5Y cells. *Pharmacol Res* 43: 173–178.

Gasser, T. (1998a). Genetics of Parkinson's disease. *Ann Neurol* 44: S53–57.

Gasser, T. (1998b). Genetics of Parkinson's disease. *Clin Genet* 54: 259–265.

Gasser, T. (2001). Genetics of Parkinson's disease. *J Neurol* 248: 833–840.

Gast, K., Damaschun, H., Eckert, K., Schulze-Forster, K., Maurer, H. R., Muller-Frohne, M., Zirwer, D., Czarnecki, J., and Damaschun, G. (1995). Prothymosin alpha: a biologically active protein with random coil conformation. *Biochemistry* 34: 13211–13218.

George, J. M., Jin, H., Woods, W. S., and Clayton, D. F. (1995). Characterization of a novel protein regulated during the critical period for song learning in the zebra finch. *Neuron* 15: 361–372.

Giasson, B. I., Duda, J. E., Murray, I. V., Chen, Q., Souza, J. M., Hurtig, H. I., Ischiropoulos, H., Trojanowski, J. Q., and Lee, V. M. (2000). Oxidative damage linked to neurodegeneration by selective alpha-synuclein nitration in synucleinopathy lesions. *Science* 290: 985–989.

Giasson, B. I., Duda, J. E., Quinn, S. M., Zhang, B., Trojanowski, J. Q., and Lee, V. M. (2002). Neuronal alpha-synucleinopathy with severe movement disorder in mice expressing A53T human alpha-synuclein. *Neuron* 34: 521–533.

Giasson, B. I., Uryu, K., Trojanowski, J. Q., and Lee, V. M. (1999). Mutant and wild type human alpha-synucleins assemble into elongated filaments with distinct morphologies *in vitro*. *J Biol Chem* 274: 7619–7622.

Glaser, C. B., Yamin, G., Uversky, V. N., and Fink, A. L. (2005). Methionine oxidation, alpha-synuclein and Parkinson's disease. *Biochim Biophys Acta* 1703: 157–169.

Goedert, M., Jakes, R., Spillantini, M. G., Hasegawa, M., Smith, M. J., and Crowther, R. A. (1996). Assembly of microtubule-associated protein tau into Alzheimer-like filaments induced by sulphated glycosaminoglycans. *Nature* 383: 550–553.

Goedert, M. (1999). Filamentous nerve cell inclusions in neurodegenerative diseases: tauopathies and alpha-synucleinopathies. *Philos Trans R Soc Lond B Biol Sci* 354: 1101–1118.

Goedert, M. (2001a). Alpha-synuclein and neurodegenerative diseases. *Nat Rev Neurosci* 2: 492–501.

Goedert, M. (2001b). Parkinson's disease and other alpha-synucleinopathies. *Clin Chem Lab Med* 39: 308–312.

Goers, J., Manning-Bog, A. B., McCormack, A. L., Millett, I. S., Doniach, S., Di Monte, D. A., Uversky, V. N., and Fink, A. L. (2003a). Nuclear localization of alpha-synuclein and its interaction with histones. *Biochemistry* 42: 8465–8471.

Goers, J., Uversky, V. N., and Fink, A. L. (2003b) Polycation-induced oligomerization and accelerated fibrillation of human alpha-synuclein *in vitro*. *Protein Sci* 12: 702–707.

Golbe, L. I. (1990). The genetics of Parkinson's disease: a reconsideration. *Neurology* 40: 7–14.

Goldberg, M. S., and Lansbury, P. T., Jr. (2000) Is there a cause-and-effect relationship between alpha-synuclein fibrillization and Parkinson's disease? *Nat Cell Biol* 2: E115–119.

Good, P. F., Olanow, C. W., and Perl, D. P. (1992). Neuromelanin-containing neurons of the substantia nigra accumulate iron and aluminum in Parkinson's disease: a LAMMA study. *Brain Res* 593: 343–346.

Gorell, J. M., Johnson, C. C., Rybicki, B. A., Peterson, E. L., and Richardson, R. J. (1998). The risk of Parkinson's disease with exposure to pesticides, farming, well water, and rural living. *Neurology* 50: 1346–1350.

Gorell, J. M., Johnson, C. C., Rybicki, B. A., Peterson, E. L., Kortsha, G. X., Brown, G. G., and Richardson, R. J. (1999a). Occupational exposure to manganese, copper, lead, iron, mercury and zinc and the risk of Parkinson's disease. *Neurotoxicology* 20: 239–247.

Gorell, J. M., Rybicki, B. A., Cole Johnson, C., and Peterson, E. L. (1999b) Occupational metal exposures and the risk of Parkinson's disease. *Neuroepidemiology* 18: 303–308.

Gosavi, N., Lee, H. J., Lee, J. S., Patel, S., and Lee, S. J. (2002). Golgi fragmentation occurs in the cells with prefibrillar alpha-synuclein aggregates and precedes the formation of fibrillar inclusion. *J Biol Chem* 277: 48984–48992.

Gourie-Devi, M., Ramu, M. G., and Venkataram, B. S. (1991). Treatment of Parkinson's disease in 'Ayurveda' (ancient Indian system of medicine): discussion paper. *J R Soc Med* 84: 491–492.

Greenbaum, E. A., Graves, C. L., Mishizen-Eberz, A. J., Lupoli, M. A., Lynch, D. R., Englander, S. W., Axelsen, P. H., and Giasson, B. I. (2005). The E46K mutation in alpha-synuclein increases amyloid fibril formation. *J Biol Chem* 280: 7800–7807.

Hageman, G., van der Hoek, J., van Hout, M., van der Laan, G., Steur, E. J., de Bruin, W., and Herholz, K. (1999). Parkinsonism, pyramidal signs, polyneuropathy, and cognitive decline after long-term occupational solvent exposure. *J Neurol* 246: 198–206.

Halfter, W., Schurer, B., Yip, J., Yip, L., Tsen, G., Lee, J. A., and Cole, G. J. (1997). Distribution and substrate properties of agrin, a heparan sulfate proteoglycan of developing axonal pathways. *J Comp Neurol* 383: 1–17.

Hamilton, B. A. (2004) alpha-Synuclein A53T substitution associated with Parkinson disease also marks the divergence of Old World and New World primates. *Genomics* 83: 739–742.

Han, H., Weinreb, P. H., and Lansbury, P. T., Jr. (1995). The core Alzheimer's peptide NAC forms amyloid fibrils which seed and are seeded by beta-amyloid: is NAC a common trigger or target in neurodegenerative disease? *Chem Biol* 2: 163–169.

Hartmann, G., Honikel, K. O., Knusel, F., and Nuesch, J. (1967). The specific inhibition of the DNA-directed RNA synthesis by rifamycin. *Biochim Biophys Acta* 145: 843–844.

Hashimoto, M., Rockenstein, E., Mante, M., Mallory, M., and Masliah, E. (2001). beta-Synuclein inhibits alpha-synuclein aggregation: a possible role as an anti-parkinsonian factor. *Neuron* 32: 213–223.

Hashimoto, M., Yoshimoto, M., Sisk, A., Hsu, L. J., Sundsmo, M., Kittel, A., Saitoh, T., Miller, A., and Masliah, E. (1997). NACP, a synaptic protein involved in Alzheimer's disease, is differentially regulated during megakaryocyte differentiation. *Biochem Biophys Res Commun* 237: 611–616.

Hemmings, H. C., Jr., Nairn, A. C., Aswad, D. W., and Greengard, P. (1984). DARPP-32, a dopamine- and adenosine 3':5'-monophosphate-regulated phosphoprotein enriched in dopamine-innervated brain regions. II. Purification and characterization of the phosphoprotein from bovine caudate nucleus. *J Neurosci* 4: 99–110.

Hertzman, C., Wiens, M., Snow, B., Kelly, S., and Calne, D. (1994). A case-control study of Parkinson's disease in a horticultural region of British Columbia. *Mov Disord* 9: 69–75.

Hirsch, E. C., Brandel, J. P., Galle, P., Javoy-Agid, F., and Agid, Y. (1991). Iron and aluminum increase in the substantia nigra of patients with Parkinson's disease: an X-ray microanalysis. *J Neurochem* 56: 446–451.

Hokenson, M. J., Uversky, V. N., Goers, J., Yamin, G., Munishkina, L. A., and Fink, A. L. (2004). Role of individual methionines in the fibrillation of methionine-oxidized alpha-synuclein. *Biochemistry* 43: 4621–4633.

Ikezoe, T., Chen, S. S., Heber, D., Taguchi, H., and Koeffler, H. P. (2001). Baicalin is a major component of PC-SPES which inhibits the proliferation of human cancer cells via apoptosis and cell cycle arrest. *Prostate* 49: 285–292.

Irizarry, M. C., Growdon, W., Gomez-Isla, T., Newell, K., George, J. M., Clayton, D. F., and Hyman, B. T. (1998). Nigral and cortical Lewy bodies and dystrophic nigral neurites in Parkinson's disease and cortical Lewy body disease contain alpha-synuclein immunoreactivity. *J Neuropathol Exp Neurol* 57: 334–337.

Ischiropoulos, H. (1998). Biological tyrosine nitration: a pathophysiological function of nitric oxide and reactive oxygen species. *Arch Biochem Biophys* 356: 1–11.

Ischiropoulos, H. (2003a). Biological selectivity and functional aspects of protein tyrosine nitration. *Biochem Biophys Res Commun* 305: 776–783.

Ischiropoulos, H. (2003b). Biological significance and clinical relevance of nitric oxide-mediated protein modifications. *Free Radic Res* 37: 15.

Ischiropoulos, H. (2003c). Oxidative modifications of alpha-synuclein. *Ann N Y Acad Sci* 991: 93–100.

Iwai, A., Masliah, E., Yoshimoto, M., Ge, N., Flanagan, L., de Silva, H. A., Kittel, A., and Saitoh, T. (1995a) The precursor protein of non-A beta component of Alzheimer's disease amyloid is a presynaptic protein of the central nervous system. *Neuron* 14: 467–475.

Iwai, A., Yoshimoto, M., Masliah, E., and Saitoh, T. (1995b). Non-A beta component of Alzheimer's disease amyloid (NAC) is amyloidogenic. *Biochemistry* 34: 10139–10145.

Jakes, R., Spillantini, M. G., and Goedert, M. (1994). Identification of two distinct synucleins from human brain. *FEBS Lett* 345: 27–32.

Jarrett, J. T., and Lansbury, P. T., Jr. (1992). Amyloid fibril formation requires a chemically discriminating nucleation event: studies of an amyloidogenic sequence from the bacterial protein OsmB. *Biochemistry* 31: 12345–12352.

Jarrett, J. T., and Lansbury, P. T., Jr. (1993) Seeding "one-dimensional crystallization" of amyloid: a pathogenic mechanism in Alzheimer's disease and scrapie? *Cell* 73: 1055–1058.

Jensen, P. H., Hojrup, P., Hager, H., Nielsen, M. S., Jacobsen, L., Olesen, O. F., Gliemann, J., and Jakes, R. (1997). Binding of Abeta to alpha- and beta-synucleins: identification of segments in alpha-synuclein/NAC precursor that bind Abeta and NAC. *Biochem J* 323 (Pt 2): 539–546.

Jensen, P. H., Nielsen, M. S., Jakes, R., Dotti, C. G., and Goedert, M. (1998). Binding of alpha-synuclein to brain vesicles is abolished by familial Parkinson's disease mutation. *J Biol Chem* 273: 26292–26294.

Ji, H., Liu, Y. E., Jia, T., Wang, M., Liu, J., Xiao, G., Joseph, B. K., Rosen, C., and Shi, Y. E. (1997). Identification of a breast cancer-specific gene, BCSG1, by direct differential cDNA sequencing. *Cancer Res* 57: 759–764.

Jo, E., McLaurin, J., Yip, C. M., St George-Hyslop, P., and Fraser, P. E. (2000). alpha-Synuclein membrane interactions and lipid specificity. *J Biol Chem* 275: 34328–34334.

Joseph, J. A., Shukitt-Hale, B., Denisova, N. A., Bielinski, D., Martin, A., McEwen, J. J., and Bickford, P. C. (1999). Reversals of age-related declines in neuronal signal transduction, cognitive, and motor behavioral deficits with blueberry, spinach, or strawberry dietary supplementation. *J Neurosci* 19: 8114–8121.

Kahle, P. J., Neumann, M., Ozmen, L., Muller, V., Jacobsen, H., Spooren, W., Fuss, B., Mallon, B., Macklin, W. B., Fujiwara, H., Hasegawa, M., Iwatsubo, T., Kretzschmar, H. A., and Haass, C. (2002). Hyperphosphorylation and insolubility of alpha-synuclein in transgenic mouse oligodendrocytes. *EMBO Rep* 3: 583–588.

Kanda, S., Bishop, J. F., Eglitis, M. A., Yang, Y., and Mouradian, M. M. (2000). Enhanced vulnerability to oxidative stress by alpha-synuclein mutations and C-terminal truncation. *Neuroscience* 97: 279–284.

Khurana, R., Ionescu-Zanetti, C., Pope, M., Li, J., Nielson, L., Ramirez-Alvarado, M., Regan, L., Fink, A. L., and Carter, S. A. (2003). A general model for amyloid fibril assembly based on morphological studies using atomic force microscopy. *Biophys J* 85: 1135–1144.

Kisilevsky, R. (2000). Review: amyloidogenesis-unquestioned answers and unanswered questions. *J Struct Biol* 130: 99–108.

Kitada, T., Asakawa, S., Hattori, N., Matsumine, H., Yamamura, Y., Minoshima, S., Yokochi, M., Mizuno, Y., and Shimizu, N. (1998). Mutations in the parkin gene cause autosomal recessive juvenile parkinsonism. *Nature* 392: 605–608.

Klawans, H. L., Stein, R. W., Tanner, C. M., and Goetz, C. G. (1982). A pure parkinsonian syndrome following acute carbon monoxide intoxication. *Arch Neurol* 39: 302–304.

Ko, L., Mehta, N. D., Farrer, M., Easson, C., Hussey, J., Yen, S., Hardy, J., and Yen, S. H. (2000). Sensitization of neuronal cells to oxidative stress with mutated human alpha-synuclein. *J Neurochem* 75: 2546–2554.

Kong, S. K., Yim, M. B., Stadtman, E. R., and Chock, P. B. (1996). Peroxynitrite disables the tyrosine phosphorylation regulatory mechanism: Lymphocyte-specific tyrosine kinase fails to phosphorylate nitrated cdc2(6–20)NH2 peptide. *Proc Natl Acad Sci USA* 93: 3377–3382.

Krishnan, S., Chi, E. Y., Wood, S. J., Kendrick, B. S., Li, C., Garzon-Rodriguez, W., Wypych, J., Randolph, T. W., Narhi, L. O., Biere, A. L., Citron, M., and Carpenter, J. F. (2003). Oxidative dimer formation is the critical rate-limiting step for Parkinson's disease alpha-synuclein fibrillogenesis. *Biochemistry* 42: 829–837.

Kruger, R., Kuhn, W., Muller, T., Woitalla, D., Graeber, M., Kosel, S., Przuntek, H., Epplen, J. T., Schols, L., and Riess, O. (1998). Ala30Pro mutation in the gene encoding alpha-synuclein in Parkinson's disease. *Nat Genet* 18: 106–108.

Langston, J. W., and Ballard, P. A., Jr. (1983). Parkinson's disease in a chemist working with 1-methyl-4-phenyl-1,2,5,6-tetrahydropyridine. *N Engl J Med* 309: 310.

Langston, J. W., Sastry, S., Chan, P., Forno, L. S., Bolin, L. M., and Di Monte, D. A. (1998) Novel alpha-synuclein-immunoreactive proteins in brain samples from the Contursi kindred, Parkinson's, and Alzheimer's disease. *Exp Neurol* 154: 684–690.

Lansbury, P. T., Jr. (1999). Evolution of amyloid: what normal protein folding may tell us about fibrillogenesis and disease. *Proc Natl Acad Sci USA* 96: 3342–3344.

Lashuel, H. A., Hartley, D., Petre, B. M., Walz, T., and Lansbury, P. T., Jr. (2002a). Neurodegenerative disease: amyloid pores from pathogenic mutations. *Nature* 418: 291.

Lashuel, H. A., Petre, B. M., Wall, J., Simon, M., Nowak, R. J., Walz, T., and Lansbury, P. T., Jr. (2002b). Alpha-synuclein, especially the Parkinson's disease-associated mutants, forms pore-like annular and tubular protofibrils. *J Mol Biol* 322: 1089–1102.

Lavedan, C. (1998). The synuclein family. *Genome Res* 8: 871–880.

Lavedan, C., Leroy, E., Dehejia, A., Buchholtz, S., Dutra, A., Nussbaum, R. L., and Polymeropoulos, M. H. (1998a). Identification, localization and characterization of the human gamma-synuclein gene. *Hum Genet* 103: 106–112.

Lavedan, C., Leroy, E., Torres, R., Dehejia, A., Dutra, A., Buchholtz, S., Nussbaum, R. L., and Polymeropoulos, M. H. (1998b). Genomic organization and expression of the human beta-synuclein gene (SNCB). *Genomics* 54: 173–175.

LaVoie, M. J., and Hastings, T. G. (1999). Dopamine quinone formation and protein modification associated with the striatal neurotoxicity of methamphetamine: evidence against a role for extracellular dopamine. *J Neurosci* 19: 1484–1491.

Lebeau, A., Esclaire, F., Rostene, W., and Pelaprat, D. (2001). Baicalein protects cortical neurons from beta-amyloid (25–35) induced toxicity. *Neuroreport* 12: 2199–2202.

Lee, M., Hyun, D., Halliwell, B., and Jenner, P. (2001). Effect of the overexpression of wild-type or mutant alpha-synuclein on cell susceptibility to insult. *J Neurochem* 76: 998–1009.

Lee, H. J., Choi, C., and Lee, S. J. (2002). Membrane-bound alpha-synuclein has a high aggregation propensity and the ability to seed the aggregation of the cytosolic form. *J Biol Chem* 277: 671–678.

Leeuwenburgh, C., Hansen, P., Shaish, A., Holloszy, J. O., and Heinecke, J. W. (1998). Markers of protein oxidation by hydroxyl radical and reactive nitrogen species in tissues of aging rats. *Am J Physiol* 274: R453–461.

Levine, H., 3rd. (1995). Soluble multimeric Alzheimer beta(1–40) pre-amyloid complexes in dilute solution. *Neurobiol Aging* 16: 755–764.

Levine, R. L., Mosoni, L., Berlett, B. S., and Stadtman, E. R. (1996). Methionine residues as endogenous antioxidants in proteins. *Proc Natl Acad Sci USA* 93: 15036–15040.

Lewy, F. H. (1912). Paralysis Agitans. Pathologische Anatomie. In *Handbuch der Neurologie*, ed. M. Lewandowski, pp. 920–933. Berlin: Springer.

Li, X., Romero, P., Rani, M., Dunker, A. K., and Obradovic, Z. (1999) Predicting protein disorder for N-, C-, and internal regions. *Genome Inform Ser Workshop Genome Inform* 10: 30–40

Li, B. Q., Fu, T., Gong, W. H., Dunlop, N., Kung, H., Yan, Y., Kang, J., and Wang, J. M. (2000). The flavonoid baicalin exhibits anti-inflammatory activity by binding to chemokines. *Immunopharmacology* 49: 295–306.

Li, J., Uversky, V. N., and Fink, A. L. (2001). Effect of familial Parkinson's disease point mutations A30P and A53T on the structural properties, aggregation, and fibrillation of human alpha-synuclein. *Biochemistry* 40: 11604–11613.

Li, J., Uversky, V. N., and Fink, A. L. (2002). Conformational behavior of human alpha-synuclein is modulated by familial Parkinson's disease point mutations A30P and A53T. *Neurotoxicology* 23: 553–567.

Li, J., Zhu, M., Manning-Bog, A. B., Di Monte, D. A., and Fink, A. L. (2004a). Dopamine and L-dopa disaggregate amyloid fibrils: implications for Parkinson's and Alzheimer's disease. *FASEB J* 18: 962–964.

Li, J., Zhu, M., Rajamani, S., Uversky, V. N., and Fink, A. L. (2004b). Rifampicin inhibits alpha-synuclein fibrillation and disaggregates fibrils. *Chem Biol* 11: 1513–1521.

Lindersson, E., and Jensen, P. H. (2004). Alpha-synuclein binding proteins. In *Molecular Mechanisms of Parkinson's Disease*, ed. P. J. Kahle, and C. Haass: Landes Bioscience.

Lindersson, E., Lundvig, D., Petersen, C., Madsen, P., Nyengaard, J. R., Hojrup, P., Moos, T., Otzen, D., Gai, W. P., Blumbergs, P. C., and Jensen, P. H. (2005). p25alpha Stimulates alpha-synuclein aggregation and is co-localized with aggregated alpha-synuclein in alpha-synucleinopathies. *J Biol Chem* 280: 5703–5715.

Lindersson, E. K., Hojrup, P., Gai, W. P., Locker, D., Martin, D., and Jensen, P. H. (2004). alpha-Synuclein filaments bind the transcriptional regulator HMGB-1. *Neuroreport* 15: 2735–2739.

Liou, H. H., Tsai, M. C., Chen, C. J., Jeng, J. S., Chang, Y. C., Chen, S. Y., and Chen, R. C. (1997). Environmental risk factors and Parkinson's disease: a case-control study in Taiwan. *Neurology* 48: 1583–1588.

Lo Bianco, C., Ridet, J. L., Schneider, B. L., Deglon, N., and Aebischer, P. (2002). alpha -Synucleinopathy and selective dopaminergic neuron loss in a rat lentiviral-based model of Parkinson's disease. *Proc Natl Acad Sci USA* 99: 10813–10818.

Lomakin, A., Teplow, D. B., Kirschner, D. A., and Benedek, G. B. (1997). Kinetic theory of fibrillogenesis of amyloid beta-protein. *Proc Natl Acad Sci USA* 94: 7942–7947.

Lowe, R., Pountney, D. L., Jensen, P. H., Gai, W. P., and Voelcker, N. H. (2004). Calcium(II) selectively induces alpha-synuclein annular oligomers via interaction with the C-terminal domain. *Protein Sci* 13: 3245–3252.

Lucking, C. B., and Brice, A. (2000). Alpha-synuclein and Parkinson's disease. *Cell Mol Life Sci* 57: 1894–1908.

Lundvig, D., Lindersson, E., and Jensen, P. H. (2005). Pathogenic effects of alpha-synuclein aggregation. *Brain Res Mol Brain Res* 134: 3–17.

Lyras, L., Perry, R. H., Perry, E. K., Ince, P. G., Jenner, A., Jenner, P., and Halliwell, B. (1998). Oxidative damage to proteins, lipids, and DNA in cortical brain regions from patients with dementia with Lewy bodies. *J Neurochem* 71: 302–312.

MacMillan-Crow, L. A., and Thompson, J. A. (1999). Tyrosine modifications and inactivation of active site manganese superoxide dismutase mutant (Y34F) by peroxynitrite. *Arch Biochem Biophys* 366: 82–88.

Maguire-Zeiss, K. A., and Federoff, H. J. (2003). Convergent pathobiologic model of Parkinson's disease. *Ann N Y Acad Sci* 991: 152–166.

Manning-Bog, A. B., McCormack, A. L., Li, J., Uversky, V. N., Fink, A. L., and Di Monte, D. A. (2002). The herbicide paraquat causes up-regulation and aggregation of alpha-synuclein in mice: paraquat and alpha-synuclein. *J Biol Chem* 277: 1641–1644.

Marder, K., Logroscino, G., Alfaro, B., Mejia, H., Halim, A., Louis, E., Cote, L., and Mayeux, R. (1998). Environmental risk factors for Parkinson's disease in an urban multiethnic community. *Neurology* 50: 279–281.

Maries, E., Dass, B., Collier, T. J., Kordower, J. H., and Steece-Collier, K. (2003). The role of alpha-synuclein in Parkinson's disease: insights from animal models. *Nat Rev Neurosci* 4: 727–738.

Markesbery, W. R., and Carney, J. M. (1999). Oxidative alterations in Alzheimer's disease. *Brain Pathol* 9: 133–146.

Maroteaux, L., Campanelli, J. T., and Scheller, R. H. (1988). Synuclein: a neuron-specific protein localized to the nucleus and presynaptic nerve terminal. *J Neurosci* 8: 2804–2815.

Masliah, E., Rockenstein, E., Veinbergs, I., Mallory, M., Hashimoto, M., Takeda, A., Sagara, Y., Sisk, A., and Mucke, L. (2000). Dopaminergic loss and inclusion body formation in alpha-synuclein mice: implications for neurodegenerative disorders. *Science* 287: 1265–1269.

McCormack, A. L., Thiruchelvam, M., Manning-Bog, A. B., Thiffault, C., Langston, J. W., Cory-Slechta, D. A., and Di Monte, D. A. (2002). Environmental risk factors and Parkinson's disease: selective degeneration of nigral dopaminergic neurons caused by the herbicide paraquat. *Neurobiol Dis* 10: 119–127.

Mcgeer, P. L., Harada, N., Kimura, H., Mcgeer, E. G., and Schulzer, M. (1992). Prevalence of dementia amongst elderly Japanese with leprosy-apparent effect of chronic drug-therapy. *Dementia* 3: 146–149.

McLaurin, J., Franklin, T., Kuhns, W. J., and Fraser, P. E. (1999). A sulfated proteoglycan aggregation factor mediates amyloid-beta peptide fibril formation and neurotoxicity. *Amyloid* 6: 233–243.

McLaurin, J., Yang, D., Yip, C. M., and Fraser, P. E. (2000). Review: modulating factors in amyloid-beta fibril formation. *J Struct Biol* 130: 259–270.

Miller, J. D., Cummings, J., Maresh, G. A., Walker, D. G., Castillo, G. M., Ngo, C., Kimata, K., Kinsella, M. G., Wight, T. N., and Snow, A. D. (1997). Localization of perlecan (or a perlecan-related macromolecule) to isolated microglia *in vitro* and to microglia/macrophages following infusion of beta-amyloid protein into rodent hippocampus. *Glia* 21: 228–243.

Miller, D. W., Hague, S. M., Clarimon, J., Baptista, M., Gwinn-Hardy, K., Cookson, M. R., and Singleton, A. B. (2004). Alpha-synuclein in blood and brain from familial Parkinson disease with SNCA locus triplication. *Neurology* 62: 1835–1838.

Minton, A. P. (2000a). Effect of a concentrated "inert" macromolecular cosolute on the stability of a globular protein with respect to denaturation by heat and by chaotropes: a statistical-thermodynamic model. *Biophys J* 78: 101–109.

Minton, A. P. (2000b). Implications of macromolecular crowding for protein assembly. *Curr Opin Struct Biol* 10: 34–39

Minton, A. P. (2001). The influence of macromolecular crowding and macromolecular confinement on biochemical reactions in physiological media. *J Biol Chem* 276: 10577–10580.

Morano, A., Jimenez-Jimenez, F. J., Molina, J. A., and Antolin, M. A. (1994). Risk-factors for Parkinson's disease: case-control study in the province of Caceres, Spain. *Acta Neurol Scand* 89: 164–170.

Morar, A. S., Olteanu, A., Young, G. B., and Pielak, G. J. (2001) Solvent-induced collapse of alpha-synuclein and acid-denatured cytochrome c. *Protein Sci* 10: 2195–2199.

Moskovitz, J., Jenkins, N. A., Gilbert, D. J., Copeland, N. G., Jursky, F., Weissbach, H., and Brot, N. (1996). Chromosomal localization of the mammalian peptide-methionine sulfoxide reductase gene and its differential expression in various tissues. *Proc Natl Acad Sci USA* 93: 3205–3208.

Munishkina, L. A., Phelan, C., Uversky, V. N., and Fink, A. L. (2003). Conformational behavior and aggregation of alpha-synuclein in organic solvents: modeling the effects of membranes. *Biochemistry* 42: 2720–2730.

Munishkina, L. A., Henriques, J., Uversky, V. N., and Fink, A. L. (2004b) .Role of protein-water interactions and electrostatics in alpha-synuclein fibril formation. *Biochemistry* 43: 3289–3300.

Munishkina, L. A., Cooper, E. M., Uversky, V. N., and Fink, A. L. (2004a). The effect of macromolecular crowding on protein aggregation and amyloid fibril formation. *J Mol Recognit* 17: 456–464.

Murray, I. V., Giasson, B. I., Quinn, S. M., Koppaka, V., Axelsen, P. H., Ischiropoulos, H., Trojanowski, J. Q., and Lee, V. M. (2003) Role of alpha-synuclein carboxy-terminus on fibril formation *in vitro*. *Biochemistry* 42: 8530–8540.

Naiki, H., Higuchi, K., Hosokawa, M., and Takeda, T. (1989). Fluorometric determination of amyloid fibrils *in vitro* using the fluorescent dye, thioflavin T1. *Anal Biochem* 177: 244–249.

Naiki, H., Higuchi, K., Matsushima, K., Shimada, A., Chen, W. H., Hosokawa, M., and Takeda, T. (1990). Fluorometric examination of tissue amyloid fibrils in murine senile amyloidosis: use of the fluorescent indicator, thioflavine T. *Lab Invest* 62: 768–773.

Nakajo, S., Shioda, S., Nakai, Y., and Nakaya, K. (1994). Localization of phosphoneuroprotein 14 (PNP 14) and its mRNA expression in rat brain determined by immunocytochemistry and *in situ* hybridization. *Brain Res Mol Brain Res* 27: 81–86.

Nakajo, S., Tsukada, K., Kameyama, H., Furuyama, Y., and Nakaya, K. (1996). Distribution of phosphoneuroprotein 14 (PNP 14) in vertebrates: its levels as determined by enzyme immunoassay. *Brain Res* 741: 180–184.

Nakajo, S., Tsukada, K., Omata, K., Nakamura, Y., and Nakaya, K. (1993). A new brain-specific 14-kDa protein is a phosphoprotein. Its complete amino acid sequence and evidence for phosphorylation. *Eur J Biochem* 217: 1057–1063.

Namba, Y., Kawatsu, K., Izumi, S., Ueki, A., and Ikeda, K. (1992). Neurofibrillary tangles and senile plaques in brain of elderly leprosy patients. *Lancet* 340: 978.

Narhi, L., Wood, S. J., Steavenson, S., Jiang, Y., Wu, G. M., Anafi, D., Kaufman, S. A., Martin, F., Sitney, K., Denis, P., Louis, J. C., Wypych, J., Biere, A. L., and Citron, M. (1999). Both familial Parkinson's disease mutations accelerate alpha-synuclein aggregation. *J Biol Chem* 274: 9843–9846.

Neumann, M., Kahle, P. J., Giasson, B. I., Ozmen, L., Borroni, E., Spooren, W., Muller, V., Odoy, S., Fujiwara, H., Hasegawa, M., Iwatsubo, T., Trojanowski, J. Q., Kretzschmar, H. A., and Haass, C. (2002). Misfolded proteinase K-resistant hyperphosphorylated alpha-synuclein in aged transgenic mice with locomotor deterioration and in human alpha-synucleinopathies. *J Clin Invest* 110: 1429–1439.

Nielsen, M. S., Vorum, H., Lindersson, E., and Jensen, P. H. (2001). Ca2+ binding to alpha-synuclein regulates ligand binding and oligomerization. *J Biol Chem* 276: 22680–22684.

Ninkina, N. N., Alimova-Kost, M. V., Paterson, J. W., Delaney, L., Cohen, B. B., Imreh, S., Gnuchev, N. V., Davies, A. M., and Buchman, V. L. (1998). Organization, expression and polymorphism of the human persyn gene. *Hum Mol Genet* 7: 1417–1424.

Ninkina, N. N., Privalova, E. M., Pinon, L. G., Davies, A. M., and Buchman, V. L. (1999). Developmentally regulated expression of persyn, a member of the synuclein family, in skin. *Exp Cell Res* 246: 308–311.

Nishie, M., Mori, F., Fujiwara, H., Hasegawa, M., Yoshimoto, M., Iwatsubo, T., Takahashi, H., and Wakabayashi, K. (2004). Accumulation of phosphorylated alpha-synuclein in the brain and peripheral ganglia of patients with multiple system atrophy. *Acta Neuropathol (Berl)* 107: 292–298.

Norris, E. H., Giasson, B. I., Ischiropoulos, H., and Lee, V. M. (2003). Effects of oxidative and nitrative challenges on alpha-synuclein fibrillogenesis involve distinct mechanisms of protein modifications. *J Biol Chem* 278: 27230–27240.

Okazaki, H., Lipkin, L. E., and Aronson, S. M. (1961). Diffuse intracytoplasmic ganglionic inclusions (Lewy type) associated with progressive dementia and quadriparesis in flexion. *J Neuropathol Exp Neurol* 20: 237–244.

Okochi, M., Walter, J., Koyama, A., Nakajo, S., Baba, M., Iwatsubo, T., Meijer, L., Kahle, P. J., and Haass, C. (2000). Constitutive phosphorylation of the Parkinson's disease associated alpha-synuclein. *J Biol Chem* 275: 390–397.

Olanow, C. W., and Tatton, W. G. (1999) Etiology and pathogenesis of Parkinson's disease. *Annu Rev Neurosci* 22: 123–144.

Oldfield, C. J., Cheng, Y., Cortese, M. S., Brown, C. J., Uversky, V. N., and Dunker, A. K. (2005). Comparing and combining predictors of mostly disordered proteins. *Biochemistry* 44: 1989–2000.

Orth, M., and Tabrizi, S. J. (2003). Models of Parkinson's disease. *Mov Disord* 18: 729–737.

Paik, S. R., Lee, J. H., Kim, D. H., Chang, C. S., and Kim, J. (1997). Aluminum-induced structural alterations of the precursor of the non-A beta component of Alzheimer's disease amyloid. *Arch Biochem Biophys* 344: 325–334.

Paik, S. R., Shin, H. J., Lee, J. H., Chang, C. S., and Kim, J. (1999). Copper(II)-induced self-oligomerization of alpha-synuclein. *Biochem J* 340 (Pt 3): 821–828.

Papp, M. I., Kahn, J. E., and Lantos, P. L. (1989). Glial cytoplasmic inclusions in the CNS of patients with multiple system atrophy (striatonigral degeneration, olivopontocerebellar atrophy and Shy-Drager syndrome). *J Neurol Sci* 94: 79–100.

Parkinson, J. (1817) *An essay on shaking palsy*. London: Sherwood, Neely and Jones.

Paxinou, E., Chen, Q., Weisse, M., Giasson, B. I., Norris, E. H., Rueter, S. M., Trojanowski, J. Q., Lee, V. M., and Ischiropoulos, H. (2001). Induction of alpha-synuclein aggregation by intracellular nitrative insult. *J Neurosci* 21: 8053–8061.

Perrin, R. J., Woods, W. S., Clayton, D. F., and George, J. M. (2000). Interaction of human alpha-Synuclein and Parkinson's disease variants with phospholipids. Structural analysis using site-directed mutagenesis. *J Biol Chem* 275: 34393–34398.

Perry, E. K., Pickering, A. T., Wang, W. W., Houghton, P. J., and Perry, N. S. (1999). Medicinal plants and Alzheimer's disease: from ethnobotany to phytotherapy. *J Pharm Pharmacol* 51: 527–534.

Pertinhez, T. A., Bouchard, M., Tomlinson, E. J., Wain, R., Ferguson, S. J., Dobson, C. M., and Smith, L. J. (2001). Amyloid fibril formation by a helical cytochrome. *FEBS Lett* 495: 184–186.

Pezzoli, G., Strada, O., Silani, V., Zecchinelli, A., Perbellini, L., Javoy-Agid, F., Ghidoni, P., Motti, E. D., Masini, T., Scarlato, G., Agid, Y., and Hirsch, E. C. (1996). Clinical and pathological features in hydrocarbon-induced parkinsonism. *Ann Neurol* 40: 922–925.

Polymeropoulos, M. H., Lavedan, C., Leroy, E., Ide, S. E., Dehejia, A., Dutra, A., Pike, B., Root, H., Rubenstein, J., Boyer, R., Stenroos, E. S., Chandrasekharappa, S., Athanassiadou, A., Papapetropoulos, T., Johnson, W. G., Lazzarini, A. M., Duvoisin, R. C., Di Iorio, G., Golbe, L. I., and Nussbaum, R. L. (1997). Mutation in the alpha-synuclein gene identified in families with Parkinson's disease. *Science* 276: 2045–2047.

Pountney, D. L., Lowe, R., Quilty, M., Vickers, J. C., Voelcker, N. H., and Gai, W. P. (2004). Annular alpha-synuclein species from purified multiple system atrophy inclusions. *J Neurochem* 90: 502–512.

Pountney, D. L., Voelcker, N. H., and Gai, W. P. (2005). Annular alpha-synuclein oligomers are potentially toxic agents in alpha-synucleinopathy. Hypothesis. *Neurotox Res* 7: 59–67.

Pronin, A. N., Morris, A. J., Surguchov, A., and Benovic, J. L. (2000). Synucleins are a novel class of substrates for G protein-coupled receptor kinases. *J Biol Chem* 275: 26515–26522.

Ragonese, P., Salemi, G., Morgante, L., Aridon, P., Epifanio, A., Buffa, D., Scoppa, F., and Savettieri, G. (2003). A case-control study on cigarette, alcohol, and coffee consumption preceding Parkinson's disease. *Neuroepidemiology* 22: 297–304.

Riederer, P., Sofic, E., Rausch, W. D., Schmidt, B., Reynolds, G. P., Jellinger, K., and Youdim, M. B. (1989). Transition metals, ferritin, glutathione, and ascorbic acid in parkinsonian brains. *J Neurochem* 52: 515–520.

Rochet, J. C., Conway, K. A., and Lansbury, P. T., Jr. (2000). Inhibition of fibrillization and accumulation of prefibrillar oligomers in mixtures of human and mouse alpha-synuclein. *Biochemistry* 39: 10619–10626.

Romero, P., Obradovic, Z., and Dunker, A. K. (2004). Natively disordered proteins: functions and predictions. *Appl Bioinformatics* 3: 105–113.

Romero, P., Obradovic, Z., Li, X., Garner, E. C., Brown, C. J., and Dunker, A. K. (2001). Sequence complexity of disordered protein. *Proteins* 42: 38–48.

Ross, G. W., Abbott, R. D., Petrovitch, H., Morens, D. M., Grandinetti, A., Tung, K. H., Tanner, C. M., Masaki, K. H., Blanchette, P. L., Curb, J. D., Popper, J. S., and White, L. R. (2000). Association of coffee and caffeine intake with the risk of Parkinson disease. *JAMA* 283: 2674–2679.

Ross, G. W., and Petrovitch, H. (2001). Current evidence for neuroprotective effects of nicotine and caffeine against Parkinson's disease. *Drugs Aging* 18: 797–806.

Rybicki, B. A., Johnson, C. C., Uman, J., and Gorell, J. M. (1993). Parkinson's disease mortality and the industrial use of heavy metals in Michigan. *Mov Disord* 8: 87–92.

Saito, Y., Kawashima, A., Ruberu, N. N., Fujiwara, H., Koyama, S., Sawabe, M., Arai, T., Nagura, H., Yamanouchi, H., Hasegawa, M., Iwatsubo, T., and Murayama, S. (2003). Accumulation of phosphorylated alpha-synuclein in aging human brain. *J Neuropathol Exp Neurol* 62: 644–654.

Schroeter, H., Williams, R. J., Matin, R., Iversen, L., and Rice-Evans, C. A. (2000). Phenolic antioxidants attenuate neuronal cell death following uptake of oxidized low-density lipoprotein. *Free Radic Biol Med* 29: 1222–1233.

Segrest, J. P., De Loof, H., Dohlman, J. G., Brouillette, C. G., and Anantharamaiah, G. M. (1990) Amphipathic helix motif: classes and properties. *Proteins* 8: 103–117.

Segrest, J. P., Jones, M. K., De Loof, H., Brouillette, C. G., Venkatachalapathi, Y. V., and Anantharamaiah, G. M. (1992). The amphipathic helix in the exchangeable apolipoproteins: a review of secondary structure and function. *J Lipid Res* 33: 141–166.

Seidler, A., Hellenbrand, W., Robra, B. P., Vieregge, P., Nischan, P., Joerg, J., Oertel, W. H., Ulm, G., and Schneider, E. (1996). Possible environmental, occupational, and other etiologic factors for Parkinson's disease: a case-control study in Germany. *Neurology* 46: 1275–1284.

Semchuk, K. M., Love, E. J., and Lee, R. G. (1992). Parkinson's disease and exposure to agricultural work and pesticide chemicals. *Neurology* 42: 1328–1335.

Semchuk, K. M., Love, E. J., and Lee, R. G. (1993). Parkinson's disease: a test of the multifactorial etiologic hypothesis. *Neurology* 43: 1173–1180.

Serpell, L. C., Berriman, J., Jakes, R., Goedert, M., and Crowther, R. A. (2000). Fiber diffraction of synthetic alpha-synuclein filaments shows amyloid-like cross-beta conformation. *Proc Natl Acad Sci USA* 97: 4897–4902.

Sherer, T. B., Kim, J. H., Betarbet, R., and Greenamyre, J. T. (2003). Subcutaneous rotenone exposure causes highly selective dopaminergic degeneration and alpha-synuclein aggregation. *Exp Neurol* 179: 9–16.

Shibayama-Imazu, T., Okahashi, I., Omata, K., Nakajo, S., Ochiai, H., Nakai, Y., Hama, T., Nakamura, Y., and Nakaya, K. (1993). Cell and tissue distribution and developmental change of neuron specific 14 kDa protein (phosphoneuroprotein 14). *Brain Res* 622: 17–25.

Shieh, D. E., Liu, L. T., and Lin, C. C. (2000). Antioxidant and free radical scavenging effects of baicalein, baicalin and wogonin. *Anticancer Res* 20: 2861–2865.

Shtilerman, M. D., Ding, T. T., and Lansbury, P. T., Jr. (2002). Molecular crowding accelerates fibrillization of alpha-synuclein: could an increase in the cytoplasmic protein concentration induce Parkinson's disease? *Biochemistry* 41: 3855–3860.

Singleton, A., Gwinn-Hardy, K., Sharabi, Y., Li, S. T., Holmes, C., Dendi, R., Hardy, J., Crawley, A., and Goldstein, D. S. (2004). Association between cardiac denervation and parkinsonism caused by alpha-synuclein gene triplication. *Brain* 127: 768–772.

Singleton, A. B., Farrer, M., Johnson, J., Singleton, A., Hague, S., Kachergus, J., Hulihan, M., Peuralinna, T., Dutra, A., Nussbaum, R., Lincoln, S., Crawley, A., Hanson, M., Maraganore, D., Adler, C., Cookson, M. R., Muenter, M., Baptista, M., Miller, D., Blancato, J., Hardy, J., and Gwinn-Hardy, K. (2003). alpha-Synuclein locus triplication causes Parkinson's disease. *Science* 302: 841.

Smith, C. D., Carney, J. M., Starke-Reed, P. E., Oliver, C. N., Stadtman, E. R., Floyd, R. A., and Markesbery, W. R. (1991). Excess brain protein oxidation and enzyme dysfunction in normal aging and in Alzheimer disease. *Proc Natl Acad Sci USA* 88: 10540–10543.

Souza, J. M., Giasson, B. I., Chen, Q., Lee, V. M., and Ischiropoulos, H. (2000). Dityrosine cross-linking promotes formation of stable alpha -synuclein polymers. Implication of nitrative and oxidative stress in the pathogenesis of neurodegenerative synucleinopathies. *J Biol Chem* 275: 18344–18349.

Spencer, J. P., Jenner, A., Butler, J., Aruoma, O. I., Dexter, D. T., Jenner, P., and Halliwell, B. (1996). Evaluation of the pro-oxidant and antioxidant actions of L-DOPA and dopamine *in vitro*: implications for Parkinson's disease. *Free Radic Res* 24: 95–105.

Spillantini, M. G., Schmidt, M. L., Lee, V. M., Trojanowski, J. Q., Jakes, R., and Goedert, M. (1997) Alpha-synuclein in Lewy bodies. *Nature* 388: 839–840.

Spillantini, M. G., Crowther, R. A., Jakes, R., Cairns, N. J., Lantos, P. L., and Goedert, M. (1998a). Filamentous alpha-synuclein inclusions link multiple system atrophy with Parkinson's disease and dementia with Lewy bodies. *Neurosci Lett* 251: 205–208.

Spillantini, M. G., Crowther, R. A., Jakes, R., Hasegawa, M., and Goedert, M. (1998b). alpha-Synuclein in filamentous inclusions of Lewy bodies from Parkinson's disease and dementia with lewy bodies. *Proc Natl Acad Sci USA* 95: 6469–6473.

Stachowiak, O., Dolder, M., Wallimann, T., and Richter, C. (1998). Mitochondrial creatine kinase is a prime target of peroxynitrite-induced modification and inactivation. *J Biol Chem* 273: 16694–16699.

Stadtman, E. R. (1993). Oxidation of free amino acids and amino acid residues in proteins by radiolysis and by metal-catalyzed reactions. *Annu Rev Biochem* 62: 797–821.

Stadtman, E. R., Oliver, C. N., Starke-Reed, P. E., and Rhee, S. G. (1993). Age-related oxidation reaction in proteins. *Toxicol Ind Health* 9: 187–196.

Sun, H., Gao, J., Ferrington, D. A., Biesiada, H., Williams, T. D., and Squier, T. C. (1999). Repair of oxidized calmodulin by methionine sulfoxide reductase restores ability to activate the plasma membrane Ca-ATPase. *Biochemistry* 38: 105–112.

Sveinbjornsdottir, S., Hicks, A. A., Jonsson, T., Petursson, H., Gugmundsson, G., Frigge, M. L., Kong, A., Gulcher, J. R., and Stefansson, K. (2000) Familial aggregation of Parkinson's disease in Iceland. *N Engl J Med* 343: 1765–1770.

Syme, C. D., Blanch, E. W., Holt, C., Jakes, R., Goedert, M., Hecht, L., and Barron, L. D. (2002). A Raman optical activity study of rheomorphism in caseins, synucleins and tau. New insight into the structure and behaviour of natively unfolded proteins. *Eur J Biochem* 269: 148–156.

Takahashi, M., Kanuka, H., Fujiwara, H., Koyama, A., Hasegawa, M., Miura, M., and Iwatsubo, T. (2003). Phosphorylation of alpha-synuclein characteristic of synucleinopathy lesions is recapitulated in alpha-synuclein transgenic Drosophila. *Neurosci Lett* 336: 155–158.

Takeda, A., Mallory, M., Sundsmo, M., Honer, W., Hansen, L., and Masliah, E. (1998). Abnormal accumulation of NACP/alpha-synuclein in neurodegenerative disorders. *Am J Pathol* 152: 367–372.

Tanner, C. M. (1989). The role of environmental toxins in the etiology of Parkinson's disease. *Trends Neurosci* 12: 49–54.

Tanner, C. M., Chen, B., Wang, W., Peng, M., Liu, Z., Liang, X., Kao, L. C., Gilley, D. W., Goetz, C. G., and Schoenberg, B. S. (1989). Environmental factors and Parkinson's disease: a case-control study in China. *Neurology* 39: 660–664.

Tanner, C. M., Ottman, R., Goldman, S. M., Ellenberg, J., Chan, P., Mayeux, R., and Langston, J. W. (1999). Parkinson disease in twins: an etiologic study. *JAMA* 281: 341–346.

Tieu, K., Ischiropoulos, H., and Przedborski, S. (2003). Nitric oxide and reactive oxygen species in Parkinson's disease. *IUBMB Life* 55: 329–335.

Tobe, T., Nakajo, S., Tanaka, A., Mitoya, A., Omata, K., Nakaya, K., Tomita, M., and Nakamura, Y. (1992). Cloning and characterization of the cDNA encoding a novel brain-specific 14-kDa protein. *J Neurochem* 59: 1624–1629.

Tomiyama, T., Asano, S., Suwa, Y., Morita, T., Kataoka, K., Mori, H., and Endo, N. (1994). Rifampicin prevents the aggregation and neurotoxicity of amyloid beta protein *in vitro*. *Biochem Biophys Res Commun* 204: 76–83.

Tomiyama, T., Kaneko, H., Kataoka, K., Asano, S., and Endo, N. (1997). Rifampicin inhibits the toxicity of pre-aggregated amyloid peptides by binding to peptide fibrils and preventing amyloid-cell interaction. *Biochem J* 322 (Pt 3): 859–865.

Tomiyama, T., Shoji, A., Kataoka, K., Suwa, Y., Asano, S., Kaneko, H., and Endo, N. (1996). Inhibition of amyloid beta protein aggregation and neurotoxicity by rifampicin. Its possible function as a hydroxyl radical scavenger. *J Biol Chem* 271: 6839–6844.

Tompa, P. (2002). Intrinsically unstructured proteins. *Trends Biochem Sci* 27: 527–533.

Trojanowski, J. Q. (2003). Rotenone neurotoxicity: a new window on environmental causes of Parkinson's disease and related brain amyloidoses. *Exp Neurol* 179: 6–8.

Trojanowski, J. Q., Goedert, M., Iwatsubo, T., and Lee, V. M. (1998). Fatal attractions: abnormal protein aggregation and neuron death in Parkinson's disease and Lewy body dementia. *Cell Death Differ* 5: 832–837.

Trojanowski, J. Q., and Lee, V. M. (2003). Parkinson's disease and related alpha-synucleinopathies are brain amyloidoses. *Ann N Y Acad Sci* 991: 107–110.

Tsen, G., Halfter, W., Kroger, S., and Cole, G. J. (1995). Agrin is a heparan sulfate proteoglycan. *J Biol Chem* 270: 3392–3399.

Ueda, K., Fukushima, H., Masliah, E., Xia, Y., Iwai, A., Yoshimoto, M., Otero, D. A., Kondo, J., Ihara, Y., and Saitoh, T. (1993). Molecular cloning of cDNA encoding an unrecognized component of amyloid in Alzheimer disease. *Proc Natl Acad Sci USA* 90: 11282–11286.

Uitti, R. J., Snow, B. J., Shinotoh, H., Vingerhoets, F. J., Hayward, M., Hashimoto, S., Richmond, J., Markey, S. P., Markey, C. J., and Calne, D. B. (1994). Parkinsonism induced by solvent abuse. *Ann Neurol* 35: 616–619.

Uversky, V. N. (1993) Use of fast protein size-exclusion liquid chromatography to study the unfolding of proteins which denature through the molten globule. *Biochemistry* 32: 13288–13298.

Uversky, V. N., Gillespie, J. R., and Fink, A. L. (2000). Why are "natively unfolded" proteins unstructured under physiologic conditions? *Proteins* 41: 415–427.

Uversky, V. N., Lee, H. J., Li, J., Fink, A. L., and Lee, S. J. (2001a). Stabilization of partially folded conformation during alpha-synuclein oligomerization in both purified and cytosolic preparations. *J Biol Chem* 276: 43495–43498.

Uversky, V. N., Li, J., and Fink, A. L. (2001b). Evidence for a partially folded intermediate in alpha-synuclein fibril formation. *J Biol Chem* 276: 10737–10744.

Uversky, V. N., Li, J., and Fink, A. L. (2001c). Metal-triggered structural transformations, aggregation, and fibrillation of human alpha-synuclein. A possible molecular NK between Parkinson's disease and heavy metal exposure. *J Biol Chem* 276: 44284–44296.

Uversky, V. N., Li, J., and Fink, A. L. (2001d). Pesticides directly accelerate the rate of alpha-synuclein fibril formation: a possible factor in Parkinson's disease. *FEBS Lett* 500: 105–108.

Uversky, V. N. (2002a). Natively unfolded proteins: a point where biology waits for physics. *Protein Sci* 11: 739–756.

Uversky, V. N. (2002b). What does it mean to be natively unfolded? *Eur J Biochem* 269: 2–12.

Uversky, V. N., and Fink, A. L. (2002a). Amino acid determinants of alpha-synuclein aggregation: putting together pieces of the puzzle. *FEBS Lett* 522: 9–13.

Uversky, V. N., and Fink, A. L. (2002b). Biophysical properties of human alpha-synuclein and its role in Parkinson's disease. In *Recent Research Developments in Proteins*, ed. S. G. Pandalai, pp. 153–186. Kerala, India: Transworld Research Network.

Uversky, V. N., E, M. C., Bower, K. S., Li, J., and Fink, A. L. (2002a). Accelerated alpha-synuclein fibrillation in crowded milieu. *FEBS Lett* 515: 99–103.

Uversky, V. N., Li, J., Souillac, P., Millett, I. S., Doniach, S., Jakes, R., Goedert, M., and Fink, A. L. (2002c). Biophysical properties of the synucleins and their propensities to fibrillate: inhibition of alpha-synuclein assembly by beta- and gamma-synucleins. *J Biol Chem* 277: 11970–11978.

Uversky, V. N. (2003a). Protein folding revisited. A polypeptide chain at the folding-misfolding-nonfolding cross-roads: which way to go? *Cell Mol Life Sci* 60: 1852–1871.

Uversky, V. N. (2003b). A protein-chameleon: conformational plasticity of alpha-synuclein, a disordered protein involved in neurodegenerative disorders. *J Biomol Struct Dyn* 21: 211–234.

Uversky, V. N. (2004). Neurotoxicant-induced animal models of Parkinson's disease: understanding the role of rotenone, maneb and paraquat in neurodegeneration. *Cell Tissue Res* 318: 225–241.

Uversky, V. N., and Fink, A. L. (2004). Conformational constraints for amyloid fibrillation: the importance of being unfolded. *Biochim Biophys Acta* 1698: 131–153.

Uversky, V. N., Yamin, G., Munishkina, L. A., Karymov, M. A., Millett, I. S., Doniach, S., Lyubchenko, Y. L., and Fink, A. L. (2005) Effects of nitration on the structure and aggregation of alpha-synuclein. *Brain Res Mol Brain Res* 134: 84–102.

Uversky, V. N., Li, J., Bower, K., and Fink, A. L. (2002b). Synergistic effects of pesticides and metals on the fibrillation of alpha-synuclein: implications for Parkinson's disease. *Neurotoxicology* 23: 527–536.

Uversky, V. N., Yamin, G., Souillac, P. O., Goers, J., Glaser, C. B., and Fink, A. L. (2002d). Methionine oxidation inhibits fibrillation of human alpha-synuclein *in vitro*. *FEBS Lett* 517: 239–244.

van der Putten, H., Wiederhold, K. H., Probst, A., Barbieri, S., Mistl, C., Danner, S., Kauffmann, S., Hofele, K., Spooren, W. P., Ruegg, M. A., Lin, S., Caroni, P., Sommer, B., Tolnay, M., and Bilbe, G. (2000). Neuropathology in mice expressing human alpha-synuclein. *J Neurosci* 20: 6021–6029.

Vanacore, N., Nappo, A., Gentile, M., Brustolin, A., Palange, S., Liberati, A., Di Rezze, S., Caldora, G., Gasparini, M., Benedetti, F., Bonifati, V., Forastiere, F., Quercia, A., and Meco, G. (2002). Evaluation of risk of Parkinson's disease in a cohort of licensed pesticide users. *Neurol Sci* 23 Suppl 2: S119–120.

Verbeek, M. M., Otte-Holler, I., van den Born, J., van den Heuvel, L. P., David, G., Wesseling, P., and de Waal, R. M. (1999).. Agrin is a major heparan sulfate proteoglycan accumulating in Alzheimer's disease brain. *Am J Pathol* 155: 2115–2125.

Vila, M., Vukosavic, S., Jackson-Lewis, V., Neystat, M., Jakowec, M., and Przedborski, S. (2000). Alpha-synuclein up-regulation in substantia nigra dopaminergic neurons following administration of the parkinsonian toxin MPTP. *J Neurochem* 74: 721–729.

Vogt, W. (1995). Oxidation of methionyl residues in proteins: tools, targets, and reversal. *Free Radic Biol Med* 18: 93–105.

Volles, M. J., and Lansbury, P. T., Jr. (2002). Vesicle permeabilization by protofibrillar alpha-synuclein is sensitive to Parkinson's disease-linked mutations and occurs by a pore-like mechanism. *Biochemistry* 41: 4595–4602.

Volles, M. J., and Lansbury, P. T., Jr. (2003). Zeroing in on the pathogenic form of alpha-synuclein and its mechanism of neurotoxicity in Parkinson's disease. *Biochemistry* 42: 7871–7878.

Wakabayashi, K., Matsumoto, K., Takayama, K., Yoshimoto, M., and Takahashi, H. (1997). NACP, a presynaptic protein, immunoreactivity in Lewy bodies in Parkinson's disease. *Neurosci Lett* 239: 45–48.

Wakabayashi, K., Yoshimoto, M., Tsuji, S., and Takahashi, H. (1998). Alpha-synuclein immunoreactivity in glial cytoplasmic inclusions in multiple system atrophy. *Neurosci Lett* 249: 180–182.

Weinreb, P. H., Zhen, W., Poon, A. W., Conway, K. A., and Lansbury, P. T., Jr. (1996). NACP, a protein implicated in Alzheimer's disease and learning, is natively unfolded. *Biochemistry* 35: 13709–13715.

Witko-Sarsat, V., Friedlander, M., Nguyen Khoa, T., Capeillere-Blandin, C., Nguyen, A. T., Canteloup, S., Dayer, J. M., Jungers, P., Drueke, T., and Descamps-Latscha, B. (1998a). Advanced oxidation protein products as novel mediators of inflammation and monocyte activation in chronic renal failure. *J Immunol* 161: 2524–2532.

Witko-Sarsat, V., Khoa, T. N., Jungers, P., Drueke, T., and Descamps-Latscha, B. (1998b). Advanced oxidation protein products: oxidative stress markers and mediators of inflammation in uremia. *Adv Nephrol Necker Hosp* 28: 321–341.

Wood, N. W. (1998). Genetic risk factors in Parkinson's disease. *Ann Neurol* 44: S58–62.

Wood, S. J., Wypych, J., Steavenson, S., Louis, J. C., Citron, M., and Biere, A. L. (1999). alpha-synuclein fibrillogenesis is nucleation-dependent. Implications for the pathogenesis of Parkinson's disease. *J Biol Chem* 274: 19509–19512.

Wright, P. E., and Dyson, H. J. (1999). Intrinsically unstructured proteins: re-assessing the protein structure-function paradigm. *J Mol Biol* 293: 321–331.

Wu, J. A., Attele, A. S., Zhang, L., and Yuan, C. S. (2001). Anti-HIV activity of medicinal herbs: usage and potential development. *Am J Chin Med* 29: 69–81.

Xu, J., Kao, S. Y., Lee, F. J., Song, W., Jin, L. W., and Yankner, B. A. (2002). Dopamine-dependent neurotoxicity of alpha-synuclein: a mechanism for selective neurodegeneration in Parkinson disease. *Nat Med* 8: 600–606.

Yamin, G., Glaser, C. B., Uversky, V. N., and Fink, A. L. (2003a). Certain metals trigger fibrillation of methionine-oxidized alpha-synuclein. *J Biol Chem* 278: 27630–27635.

Yamin, G., Uversky, V. N., and Fink, A. L. (2003b). Nitration inhibits fibrillation of human alpha-synuclein *in vitro* by formation of soluble oligomers. *FEBS Lett* 542: 147–152.

Yamin, G., Munishkina, L. A., Karymov, M. A., Lyubchenko, Y. L., Uversky, V. N., and Fink, A. L. (2005). Forcing the non-amyloidogenic beta-synuclein to fibrillate. *Biochemistry* 44: 9096–9107.

Yancey, P. H., Clark, M. E., Hand, S. C., Bowlus, R. D., and Somero, G. N. (1982). Living with water stress: evolution of osmolyte systems. *Science* 217: 1214–1222.

Yasui, M., Kihira, T., and Ota, K. (1992). Calcium, magnesium and aluminum concentrations in Parkinson's disease. *Neurotoxicology* 13: 593–600.

Yoshimoto, M., Iwai, A., Kang, D., Otero, D. A., Xia, Y., and Saitoh, T. (1995). NACP, the precursor protein of the non-amyloid beta/A4 protein (A beta) component of Alzheimer disease amyloid, binds A beta and stimulates A beta aggregation. *Proc Natl Acad Sci USA* 92: 9141–9145.

Zarranz, J. J., Alegre, J., Gomez-Esteban, J. C., Lezcano, E., Ros, R., Ampuero, I., Vidal, L., Hoenicka, J., Rodriguez, O., Atares, B., Llorens, V., Gomez Tortosa, E., del Ser, T., Munoz, D. G., and de Yebenes, J. G. (2004). The new mutation, E46K, of alpha-synuclein causes Parkinson and Lewy body dementia. *Ann Neurol* 55: 164–173.

Zayed, J., Ducic, S., Campanella, G., Panisset, J. C., Andre, P., Masson, H., and Roy, M. (1990). Environmental factors in the etiology of Parkinson's disease. *Can J Neurol Sci* 17: 286–291.

Zhou, W., Hurlbert, M. S., Schaack, J., Prasad, K. N., and Freed, C. R. (2000). Overexpression of human alpha-synuclein causes dopamine neuron death in rat primary culture and immortalized mesencephalon-derived cells. *Brain Res* 866: 33–43.

Zhu, M., Rajamani, S., Kaylor, J., Han, S., Zhou, F., and Fink, A. L. (2004). The flavonoid baicalein inhibits fibrillation of alpha-synuclein and disaggregates existing fibrils. *J Biol Chem* 279: 26846–26857.

Zimmerman, S. B., and Trach, S. O. (1991). Estimation of macromolecule concentrations and excluded volume effects for the cytoplasm of Escherichia coli. *J Mol Biol* 222: 599–620.

5

Cell Biology of α-Synuclein: Implications in Parkinson's Disease and Other Lewy Body Diseases

Seung-Jae Lee and Yoon Suk Kim

Abstract

Missense mutations and gene multiplication in the α-synuclein gene cause inherited forms of Parkinson's disease (PD), and deposition of amyloid fibrils composed of wild-type α-synuclein in Lewy bodies and Lewy neurites is a pathologic hallmark of sporadic PD and other related neurodegenerative diseases such as dementia with Lewy bodies. These findings, along with the studies in animal models, strongly suggest that misfolding and aggregation of α-synuclein is a critical component in the pathogenesis of these disorders. Though the physiologic function of α-synuclein has not been completely defined, there has been a body of evidence supporting its roles in synaptic transmission, synaptic vesicle biogenesis, and lipid transport and metabolism. More importantly, based on the lack of neurodegenerative phenotypes in synuclein knockout animals, it has been suggested that loss of normal function of this protein might not be the direct cause of PD. α-Synuclein forms various types of nonfibrillar and fibrillar aggregates. PD-linked mutations accelerate the aggregation process in one or more steps in the fibrillation. Although some oligomeric aggregates seem to be toxic to cell, leading to functional abnormalities such as proteasomal and lysosomal dysfunctions, mitochondrial deficits, and Golgi fragmentation and transport defects, cells have defense mechanisms against these potentially toxic oligomeric aggregates. The free oligomers are transported to and deposited in the pericentriolar region by the microtubule system, resulting in the sequestration of toxic aggregates. After deposition, the transition from oligomers to fibrils occurs, forming Lewy body–like inclusion bodies and perhaps transforming toxic species into inert aggregates. Moreover, we have also shown that cells can degrade the preformed oligomers by autophagy. Elucidation of the mechanisms of the cellular aggregation process and the handling of preformed toxic aggregates should greatly enhance our understanding of the disease and provide rational targets for therapeutic intervention.

5.1. Introduction

Parkinson's disease (PD) is an age-related progressive neurodegenerative disease whose defining pathologic features are selective loss of dopaminergic neurons in substantia nigra pars compacta and occurrence of proteinaceous inclusion bodies called Lewy bodies (LBs) (Dauer and Przedborski, 2003). The cause of the disease is unknown. However, genetic defects that are linked to inherited forms of parkinsonism might help us understand the pathogenic mechanism of more prevalent sporadic PD. One such example came from the discovery of the mutations in α-synuclein (α-syn), which so far include three missense mutations and gene multiplication mutations (Bonifati *et al.*, 2004). A direct role of α-syn and its aggregation in PD and other LB diseases is further supported by the findings that fibrillar aggregates of this protein are major components of LBs (Spillantini *et al.*, 1998) and that transgenic expression in animal models leads to neurodegenerative phenotypes (Maries *et al.*, 2003). In this chapter, we review the current state of our knowledge on normal and pathobiology of α-syn with the emphasis on the cell biological aspect of the matter.

5.2. Normal Biology

5.2.1. Identification and Expression Pattern

α-Syn was originally identified as a neuron-specific protein that is immunoreactive to an antiserum against purified cholinergic synaptic vesicles from *Torpedo* (Maroteaux *et al.*, 1988), followed by identification of homologues from other species including human (Ueda *et al.*, 1993; Jakes *et al.*, 1994). α-Syn, a member of the synuclein family, which also includes β-syn, γ-syn, and synoretin (George, 2001), is a natively unfolded protein of 140 amino acids (Weinreb *et al.*, 1996) and comprises an N-terminal amphipathic region with 7 imperfect 11-residue repeats and C-terminal acidic region.

α-Syn is expressed predominately in central nervous system, especially in hippocampus, olfactory bulb, striatum, and cerebellum, but is also expressed at low levels in other tissues as well (Ueda *et al.*, 1993; Iwai *et al.*, 1995). At the cellular level, α-syn is enriched in presynaptic nerve terminals (Iwai *et al.*, 1995). In brain homogenates, it is mainly found in the cytosolic fractions and is also present in various vesicular fractions, suggesting that the protein may be associated with membrane structures including possibly synaptic vesicles (George *et al.*, 1995; Irizarry *et al.*, 1996; Lee *et al.*, 2002a). However, localization of α-syn in the synaptic vesicles has not been unequivocally demonstrated.

Expression of α-syn is regulated during development. It has been shown that α-syn levels increase throughout development in mouse (Hsu *et al.*, 1998) and rat (Petersen *et al.*, 1999), and the timing of α-syn expression varies in discrete neuronal populations during human development (Raghavan *et al.*, 2004). In hippocampal culture, subcellular translocation of α-syn from soma to synaptic terminal occurs in later stages in neuronal development (Withers *et al.*, 1997). These results suggest that both proper expression and subcellular localization of α-syn is regulated temporally.

5.2.2. Physiologic Function

5.2.2.1. Synaptic Plasticity/Neurotransmitter Release

Because α-syn is enriched in presynaptic nerve terminals, studies for characterizing the physiologic function of α-syn have been mainly focused on its role in synaptic events.

It has been shown that expression of α-syn mRNA in the song control circuit is altered during critical period for song learning in Canary, thus α-syn may have a role in the regulation of neuronal plasticity (George *et al.*, 1995). Generation of mouse strains with homozygous deletion of α-syn gene has been instrumental in understanding the role of α-syn in synaptic function. These mice are viable with intact brain architecture and exhibit no sign of neurodegeneration. However, detailed characterization showed some changes in electrophysiologic and ultrastructural properties at the synapses. In one study, loss of α-syn led to an increased release of dopamine (DA) with paired stimuli (Abeliovich *et al.*, 2000), suggesting α-syn may be an activity-dependent negative regulator of DA neurotransmission. The same mouse also showed a modest decline in levels of striatal dopamine, but studies with two other lines of α-syn null mouse failed to find a significant change in dopamine levels (Cabin *et al.*, 2002; Schluter *et al.*, 2003). On the other hands, Cabin *et al.* (2002) showed a selective reduction in number of undocked vesicles without affecting docked vesicles and significant impairment in synaptic response to a prolonged train of repetitive stimulation in the hippocampal synapses of α-syn null mice (Cabin *et al.*, 2002). This indicates α-syn may be involved in the regulation of the biosynthesis/maintenance of synaptic vesicles. Recently, Liu *et al.* (2004) reported that hippocampal neurons that are derived from α-syn null mice or treated with antisense oligonucleotides showed a reduction in glutamate-induced neurotransmitter release (Liu *et al.*, 2004). These results suggest that α-syn has a role in modulation of neurotransmitter release, thereby regulating neuronal plasticity, learning, and synaptic transmission.

The fact that all the knockout lines show only subtle phenotypes triggered the suspicion that loss of α-syn function might be compensated by β-syn, another abundant neuronal protein. Interestingly, α- and β-syn double knockout mice also show overall normal neural development and viability with only a minor reduction in striatal dopamine levels (Chandra *et al.*, 2004). Therefore, synucleins appear to be involved in fine tuning of neural functions rather than basic neuronal activities or viability.

5.2.2.2. Lipid Metabolism

As mentioned above, in brain homogenates, some α-syn is present in vesicle fractions. Structural modeling predicts that the N-terminal region of α-syn is composed of the amphipathic α-helical arrangement typical of the lipid-binding domain in apolipoproteins and thus is capable of interacting with lipids (Clayton and George, 1998). Indeed, it has been demonstrated that α-syn binds to acidic phospholipid vesicles through the N-terminal region (Perrin *et al.*, 2000) and that upon binding, elongated α-helical conformation is stabilized (Davidson *et al.*, 1998; Bussell and Eliezer, 2003; Chandra *et al.*, 2003; Jao *et al.*, 2004). The exact physiologic meaning of α-syn/lipid interaction is unknown. Some studies, however, hinted that this interaction might have effects on lipid metabolism. Cole *et al.* (2002) has shown that α-syn accumulated on phospholipid monolayers surrounding triglyceride-rich lipid droplets and protected stored lipids from hydrolysis (Cole *et al.*, 2002). It was also shown that over expression of α-syn in a neuronal cell line and homozygous deletion of α-syn in mice led to changes in cellular fatty acid composition (Sharon *et al.*, 2003). α-Syn has been also shown to inhibit activity of phospholipase D_2 *in vitro* (Jenco *et al.*, 1998), implicating its role in the regulation of lipid second messengers.

5.2.2.3. Molecular Chaperone Function

Molecular chaperones are characterized by the following properties. First, they bind to denatured proteins and protect them from aggregation. Second, molecular chaperones are induced in response to denaturing conditions such as heat shock and oxidative stress. Finally, they protect cells from protein-denaturing stresses. α-Syn has been shown to possess all the above-mentioned characteristics and thus might function as a molecular chaperone.

It has been shown that α-syn suppresses the aggregation of thermally and chemically denatured proteins in *in vitro* assay (Kim *et al.*, 2000; Souza *et al.*, 2000). Moreover, α-syn-expressing *Escherichia coli* has higher survival rate than control *E. coli* in response to heat and oxidative stress, and the C-terminal region of α-syn is critical for this protective effect (Kim *et al.*, 2002, 2004). Importantly, the C-terminal region is also known to be critical for the chaperone-like activity

(Kim *et al.*, 2002). Park *et al.* (2002) proposed that the N-terminal region binds to substrate proteins to form high-molecular-weight complexes, whereas the C-terminal acidic region appears to be primarily responsible for solubilizing the complexes (Park *et al.*, 2002).

α-Syn is upregulated in response to certain stresses. For example, it has been shown that administration of 1-methyl-4-phenyl-1,2,3,6 tetrahydropyridine (MPTP) or the herbicide paraquat in mice induces its upregulation in substantia nigra (Vila *et al.*, 2000; Manning-Bog *et al.*, 2002). Also, α-syn upregulation occurred in 1-methyl-4-phenylpyridinium (MPP+)-treated neuroblastoma cells (Gomez-Santos *et al.*, 2002; Kalivendi *et al.*, 2004). Furthermore, α-syn overexpression protects against paraquat-induced dopaminergic neurodegeneration in mice (Manning-Bog *et al.*, 2003). In addition, direct delivery of recombinant α-syn fusion protein into cells showed protective effect against oxidative stresses at nanomolar concentrations (Albani *et al.*, 2004). In contrast, C-terminally truncated α-syn failed to protect the cells from oxidative stresses.

It is known that α-syn is capable of interacting with various proteins (Lee, 2003). This was further supported by a recent study in which more than 250 α-syn-interacting proteins were identified by proteomic approach (Zhou *et al.*, 2004). The notion that α-syn is a molecular chaperone may explain why there are so many α-syn binding proteins that are seemingly unrelated to each other.

5.2.3. Degradation

α-Syn is a long-lived protein whose half-life can be as long as a few days and is prolonged with aging (Li *et al.*, 2004). The precise mechanism of α-syn breakdown is yet to be elucidated. However, studies show that both proteasome and autophagy pathways can hydrolyze this protein (Bennett *et al.*, 1999; Tofaris *et al.*, 2001; Webb *et al.*, 2003; Cuervo *et al.*, 2004). Interestingly, for proteasomal degradation, α-syn does not appear to require ubiquitylation (Tofaris *et al.*, 2001; Liu *et al.*, 2003). As for the autophagic degradation, the chaperone-mediated autophagy seems to be responsible for the breakdown of wild-type α-syn, but not for the PD-linked mutant forms (Cuervo *et al.*, 2004). Instead, these mutants can be degraded by macroautophagy (Webb *et al.*, 2003). Other studies show that α-syn is also a substrate of calpain (Kim *et al.*, 2003; Mishizen-Eberz *et al.*, 2003) and neurosin (Iwata *et al.*, 2003).

5.3. Pathobiology

Genetic mutations in α-syn gene represent only a small fraction of Parkinson's cases. However, the molecular and cellular consequences caused by these genetic mutations may also occur in sporadic cases. Thus, common effects of these mutations may provide insights into how this protein might cause the disease. The most notable phenotype that these mutations all share at the molecular level is increased aggregation of α-syn (Conway *et al.*, 2000b; Choi *et al.*, 2004). Pathologic examination of PD and other related disorders indeed showed α-syn-positive inclusion body structures in neurons and glial cells of affected brain areas (Goedert, 2001) and high-molecular-weight α-syn aggregates from the tissue homogenates (Baba *et al.*, 1998; Langston *et al.*, 1998; Campbell *et al.*, 2001; Tofaris *et al.*, 2003). These studies strongly support the hypothesis that conformational defects and aggregation of α-syn play critical roles in the pathogenesis of PD and other α-synucleinopathies.

5.3.1. Determinants of Aggregation

Protein aggregates are highly ordered polymeric structures formed by alternative, off-pathway folding steps. This off-pathway folding involves self-assembly of partly folded folding intermediates that are accumulated due to their relative kinetic stability (Horwich, 2002). Interestingly, aggregation of these folding intermediates is sequence-specific, allowing only homotypic associations (London *et al.*, 1974). Recently, Ron Kopito and co-workers demonstrated that aggregation of misfolded proteins occurs through highly specific self-association in living cells that co-express two unrelated misfolded proteins (Rajan *et al.*, 2001). This specificity in assembly process provides theoretical basis for the concentration dependency of protein aggregation and emphasizes the importance of folding intermediates.

In solution, α-syn does not have well-defined structure and hence belongs to a group of "natively unfolded" proteins (Uversky *et al.*, 2000). Amyloidogenic aggregation of this protein, however, appears to require partly folded intermediates (Uversky *et al.*, 2001a). This partly folded form is unstable and readily reversible, but concentration of this conformer leads to dimerization of the protein, which in turn stabilize the partial conformation (Uversky *et al.*, 2001c). Therefore, at least *in vitro*, stabilization and accumulation of partly folded intermediates are critical initial steps of α-syn aggregation.

Unlike test tube conditions where conformers of single polypeptide are in equilibrium state without further metabolism, cellular proteins are situated in a more dynamic environment that fosters molecular interactions, covalent modifications of side chains, and enzyme-mediated peptide hydrolysis. These cellular events determine the levels of partly folded, or misfolded, α-syn and thus regulate the aggregation. It is indeed shown that in both cultured cells and animal models, all the conditions that are known to increase the levels of misfolded proteins, such as overexpression (Feany and Bender, 2000; Giasson *et al.*, 2002; Gosavi *et al.*, 2002), oxidative stresses (Paxinou *et al.*, 2001; Norris *et al.*, 2003), inhibition of proteolytic degradation (Rideout *et al.*, 2001; McNaught *et al.*, 2002; Hyun *et al.*, 2003), and mitochondrial dysfunction (Betarbet *et al.*, 2000; Lee *et al.*, 2002b), promote the aggregation of α-syn. This provides a strong argument for the importance of partly folded conformer of α-syn in its aggregation in cells (Fig. 5.1). It is also plausible that other known promoters of cellular α-syn oligomerization/aggregation also favor partly folded conformation. These include interaction with lipid (Perrin *et al.*, 2001; Cole *et al.*, 2002; Lee *et al.*, 2002a), interaction with metal ions (Uversky *et al.*, 2001b), interaction with other proteins (Engelender *et al.*, 1999; Alim *et al.*, 2002), phosphorylation (Fujiwara *et al.*,

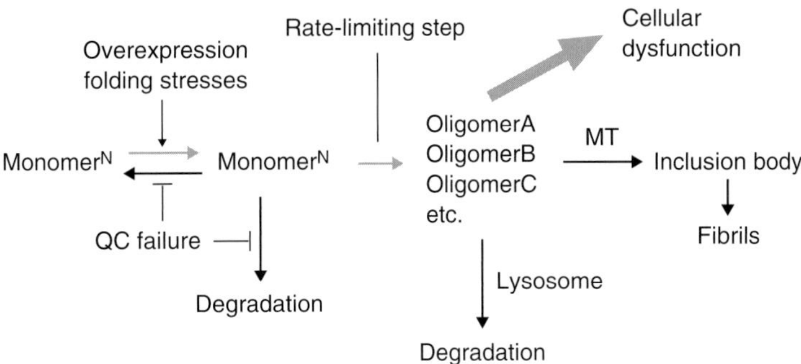

Figure 5.1. Working model for cellular α-syn aggregation. Native monomer (Monomer^N) is in equilibrium with partly folded monomer (Monomer^{PF}). Monomer^{PF} accumulates with overexpression of the protein or in the presence of various folding stresses, such as oxidative stresses, leading to self-assembly to produce various forms of oligomers. Oligomeric aggregates can be removed from the cytoplasm by autophagic degradation and microtubule (MT)-dependent sequestration in inclusion bodies. Failure of these clearance mechanisms results in the cytoplasmic accumulation of aggregates and leads to cellular dysfunction and death.

2002), and proteolytic truncation of C-terminus (Crowther *et al.*, 1998; Kim *et al.*, 2002). Whether all these modifications lead to a common defined conformation, and if so, what the conformation is, remains unanswered.

5.3.2. Biochemical and Cell Biological Dissection of Aggregates

In vitro, purified recombinant α-syn forms amyloid fibrils at high concentrations (Conway *et al.*, 2000a). During the course of the fibrillation, nonfibrillar intermediates, termed protofibrils, are observed in various sizes and shapes (Volles and Lansbury, 2003). Although it has not been unequivocally demonstrated, the transient appearance of the protofibrils suggests that they are the intermediates of the fibrillation process (Hoyer *et al.*, 2004).

When overexpressed in mammalian cells, α-syn also forms stable aggregates (Lee and Lee, 2002). Extraction of cellular proteins with a non-ionic detergent reveals "detergent-soluble" and "detergent-insoluble" aggregates based on their sedimentation properties in midspeed centrifugation [e.g., see Lee *et al.* (2004)]. Subsequent centrifugation of the detergent-soluble aggregates at ultraspeed shows various assembly states ranging from dimer to high-molecular-weight oligomers (Kim and Lee, unpublished data).

Microscopically, α-syn overexpression leads to the formation of two distinct types of aggregates. One of these aggregates is relatively large (several micrometers in diameter), thus easily visible at low magnification under microscope and always positioned at one end of the nucleus (Lee and Lee, 2002). These large inclusion bodies contain proteins other than α-syn that are frequently found in human LBs, such as ubiquitin, heat shock proteins, proteasome subunits, cytoskeletal proteins, and so forth. Like LBs, cellular inclusion bodies are also stained with thioflavin S, which shows elevated fluorescence when bound to amyloid fibrils. Electron microscopic analysis of the isolated inclusion bodies has demonstrated the presence of filamentous aggregates that were immunolabeled with α-syn antibodies (Lee and Lee, 2002). The width of these filaments is about 8–13 nm, which is similar to both typical amyloid fibrils formed *in vitro* and the fibrils isolated from LBs. These results clearly demonstrate that overexpression of α-syn leads to the fibrillation of this protein and the formation of inclusion bodies that recapitulate major characteristics of LBs. Further characterization of the inclusion bodies showed that their formation depends on the microtubule-dependent transport. This indicates that the α-syn inclusion bodies are aggresomes (Kopito, 2000).

The other type of aggregate is much smaller, thus visible only at high magnification, and occurs throughout the cytoplasm with granular appearance (Lee and Lee, 2002). These aggregates do not contain the Lewy body–associated proteins, such as ubiquitin, heat shock proteins, and proteasome subunits. EM analysis of these small aggregates showed a spherical morphology. Nonfibrillar structure of these small aggregates was confirmed by the lack of thioflavin S binding both in cells and in isolation. The striking differences in composition and structure between these small spherical aggregates and the large juxtanuclear inclusion bodies suggest that simple growth of the former would not ensure the formation of the latter.

Are these aggregates related or do they represent two independent aggregation pathways? Two experiments suggest that the small aggregates are the precursors of the inclusion bodies (Lee and Lee, 2002). First, time-course analysis showed that the small aggregates appeared before the formation of the inclusion bodies. Second, inhibition of inclusion formation with a microtubule-disrupting reagent, nocodazole, resulted in a concomitant increase in the levels of the spherical aggregates. Interestingly, blocking the inclusion body formation led to the increase in size of these spherical aggregates. These enlarged aggregates are still nonfibrillar and devoid of proteins that are found in the inclusion bodies. In fact, the fractionation study showed that fibrils are present only within the inclusion bodies. Therefore, cellular α-syn fibrillation is tightly linked to the microtubule-dependent deposition of precursor aggregates and other components of inclusion bodies (Fig. 5.1).

In a broad sense, "protein aggregate" refers to any quaternary assembly of abnormally structured proteins. For any given protein, there can be multiple aggregation pathways, and each pathway can involve multiple intermediate steps. In fact, as exemplified by the findings with α-syn, cytoplasmic aggregates include multicomponent, fibrillar inclusion bodies, microscopically visible, large nonfibrillar aggregates, and submicroscopic oligomers. Because of the heterogeneity of protein aggregates, use of the term "aggregates" has caused a substantial amount of confusion among the investigators. Distinction of different aggregate species in cell extracts and living cells is a critical step toward understanding the mechanism of aggregation process and the biological effects of individual aggregate species.

5.3.3. Effects of α-Syn Aggregation

5.3.3.1. Is α-Syn Aggregation Cytotoxic?

The cytotoxic effect of α-syn has been a subject of intense investigation, most of which employed exogenous overexpression of wild-type or mutant forms of α-syn in cultured cells. These studies generated a wide spectrum of data: some showed cytotoxic effects for all α-syn variants (Ostrerova *et al.*, 1999; Saha *et al.*, 2000; Iwata *et al.*, 2001; Xu *et al.*, 2002; Zhou *et al.*, 2002), whereas others found toxicity either only with the mutant forms (Zhou *et al.*, 2000; Lee *et al.*, 2001) or only in the presence of additional stress (Kanda *et al.*, 2000; Ko *et al.*, 2000; Zhou *et al.*, 2000; Junn and Mouradian, 2002). Some of these studies reported dopaminergic neuron-selective neurotoxicity with α-syn expression (Zhou *et al.*, 2000; Xu *et al.*, 2002; Zhou *et al.*, 2002). On the other hand, some studies showed that wild-type, but not mutant, α-synuclein has protective effects against cytotoxic insults (da Costa *et al.*, 2000; Lee *et al.*, 2001; Hashimoto *et al.*, 2002). Recent study by Albani *et al.* showed that nanomolar concentrations of cell-permeable TAT-α-syn fusion protein had a protective effect against oxidative stresses, whereas micromolar concentrations were toxic to cells (Albani *et al.*, 2004). This suggests that quantitative difference in α-syn expression could lead to opposite results. It is noteworthy that protein aggregation occurs as a function of concentration. Therefore, the aggregation of α-syn might be responsible for the toxicity of this protein, which otherwise has protective function to the cells.

In a recent study, our group has shown that in COS-7 cells, cytotoxicity of α-syn correlated with the amount of aggregates, whereas the increase in monomer levels did not show direct correlation with cell viability (Gosavi *et al.*, 2002). In addition, α-syn aggregation was also associated with fragmentation of the Golgi apparatus and impairment of protein trafficking through the biosynthetic pathway (Gosavi *et al.*, 2002). These results support the notion that α-syn gains pathogenic function through forming higher order quaternary structures. However, this interpretation is mostly based on correlative studies and needs to be further demonstrated. Use of specific blockers of α-syn aggregation, once available, would help resolve this issue. Interestingly, the same study showed that both cell death and Golgi fragmentation tightly correlated with the production of prefibrillar aggregates and actually preceded the formation of fibrillar inclusions, suggesting that the preinclusion state aggregates might be responsible for the cellular impairment (Fig. 5.1). In support of this, Tanaka *et al.* showed that mature inclusion bodies were much more frequent in nonapoptotic cells than in apoptotic cells in MG132-treated culture (Tanaka *et al.*, (2004). This is consistent with studies on expanded-polyglutamine proteins in which the toxicity of the proteins is dissociated from inclusion formation (Klement *et al.*, 1998; Saudou *et al.*, 1998; Cummings *et al.*, 1999; Faber *et al.*, 1999; Kazemi-Esfarjani and Benzer, 2000). Some of these studies even suggest that the inclusion body formation is a protective process (Taylor *et al.*, 2003;

Arrasate *et al.*, 2004). Again, in order to address whether aggregation is required for α-syn-mediated toxicity and which specific aggregate species are responsible for the toxicity, one has to be able to define individual aggregate species and manipulate the aggregation process.

5.3.3.2. Mechanism of α-Syn-Mediated Toxicity

It is not clear whether the aggregates themselves are toxic agents and what is the underlying mechanism by which α-syn aggregates lead to cellular dysfunction and death. Based on pore-like morphology of annular protofibrils and their ability to permeabilize synthetic liposomes, Peter Lansbury and colleagues proposed a model in which α-syn protofibrils act like nonspecific pores in the cellular membranes, thereby disrupting transport barriers for small molecules in membranous compartments (Volles and Lansbury, 2003). This protofibril-induced permeabilization shows strong size-selectivity, favoring low-molecular-mass molecules, consistent with the pore-like mechanism (Volles and Lansbury, 2002). Pathogenic mutations, A53T and A30P, promote the formation of annular protofibrils (Lashuel *et al.*, 2002) and membrane permeabilization activity (Volles and Lansbury, 2002), suggesting that the pore-like activity of α-syn protofibrils might be involved in the pathogenesis of PD. However, this model awaits further validation in cells.

Another line of studies suggests that α-syn aggregation leads to proteasomal dysfunction. *In vitro* studies showed that aggregated α-syn inhibited the proteolytic activity of the proteasomes (Snyder *et al.*, 2003; Lindersson *et al.*, 2004). α-Syn binds to specific subunits of proteasome complex, such as S6′ (Snyder *et al.*, 2003) and Tat binding protein 1 (Ghee *et al.*, 2000) of 19S regulatory complex, but fibrillar aggregates seem to have higher affinity to the 20S catalytic complex (Lindersson *et al.*, 2004). When overexpressed in cells, PD-linked mutant forms of α-syn reduced cellular proteasomal degradation, while wild-type protein showed little effect (Stefanis *et al.*, 2001; Tanaka *et al.*, 2001; Petrucelli *et al.*, 2002). It has been shown that the mutations accelerate oligomeric aggregation *in vitro* (Conway *et al.*, 2000b). However, whether oligomerization is responsible for the proteasome inhibition remains to be demonstrated.

Other studies showed lysosomal dysfunction (Stefanis *et al.*, 2001), mitochondrial defects (Hsu *et al.*, 2000), and vesicular trafficking defects (Gosavi *et al.*, 2002) in cells overexpressing α-syn. Considering the wide range of findings, it is likely that there are multiple mechanisms by which α-syn exerts its toxic effects. Whether these various effects of α-syn are related to one another is not known. A big challenge ahead is to identify pathogenic α-syn species that lead to these cellular dysfunctions and to dissect them into sequential and independent events.

5.3.4. How do Cells Handle Aggregates?

The extent of aggregate accumulation is likely determined by a dynamic equilibrium between the production and clearance of aggregates. Therefore, given the association between α-syn aggregation and various cellular dysfunctions, the aggregate clearance process may be critical for cell viability. In a recent study, we found that cells, including those of neuronal origin, were capable of clearing preformed α-syn aggregates via autophagic pathway (Lee *et al.*, 2004). Importantly, the same study showed that only the oligomeric intermediates, but not the mature fibrillar inclusion bodies, were susceptible to the aggregate-clearing mechanism, and that inhibition of the clearance exacerbated α-syn toxicity (Fig. 5.1). There are distinct autophagic mechanisms depending on how substrates are delivered to the lysosomes: macroautophagy, microautophagy, and chaperone-mediated autophagy. Our study hints that the clearance of α-syn aggregates is mediated by a non-macroautophagic mechanism, but further characterization is needed to reach a conclusion.

Another unanswered question is where the clearance occurs within cells. Does it occur wherever the aggregates form throughout the cytoplasm? or do the aggregates have to be delivered to a specific cytoplasmic location before they are processed by the lysosomes? With regard to this question, Kopito and co-workers speculated that autophagic clearance of misfolded proteins might be one of the purposes of inclusion body formation, which they suggested brings misfolded protein aggregates and autophagic machineries together (Kopito, 2000). Although we showed that compact, mature inclusion bodies were not susceptible to autophagic clearance, it is possible that, before they become condensed and inert structures, inclusion bodies may have a loose and dynamic stage where the autophagic degradation of aggregates occurs actively. In other words, autophagic degradation and inclusion body formation might in fact be along the line of the same process in which there are multiple stages, spanning from metabolically active and dynamic depositions to condensed clumps of inert organelles and protein aggregates.

5.4. Conclusion

Knockout animal studies clearly indicate that α-syn is not essential for neuronal survival but might be involved in fine-tuning of neuronal activities. Although loss of α-syn function cannot be completely ruled out as a component of disease process, it appears that pathogenic function of α-syn is an acquired property of the protein, which is otherwise harmless or even protective to neurons. In that regard, it is noteworthy that PD and other LB diseases are associated with abnormal aggregation of α-syn. In fact, it is increasingly evident that α-syn aggregation is not simply a pathologic marker of LB disorders but contributes to disease pathogenesis. However, it is not well understood how α-syn aggregation occurs and how the cells respond to them once aggregates form. From the tissue culture studies, we learned that the aggregation process in cells is not a diffusion-limited reaction and is therefore likely to be different from cell-free aggregation in solution. Also, it is reasonable to assume that the initial conformational defects that lead to aggregation have to occur within the context of the normal life cycle of α-syn in cells. Therefore, more emphasis should be placed on the research on normal biology of α-syn in terms of its interactions with other molecules and the impact of these interactions on conformation. For the past several years, a great deal of progress has been made, which not only advanced our knowledge but also generated a number of tissue culture and animal models. Using these models, future studies will bring us a more comprehensive picture combining biochemistry and cell biology of normal and pathogenic behavior of α-syn.

References

Abeliovich, A., Schmitz, Y., Farinas, I., Choi-Lundberg, D., Ho, W. H., Castillo, P. E., Shinsky, N., Verdugo, J. M., Armanini, M., Ryan, A., Hynes, M., Phillips, H., Sulzer, D. and Rosenthal, A. (2000). Mice lacking alpha-synuclein display functional deficits in the nigrostriatal dopamine system. *Neuron* 25: 239–252.

Albani, D., Peverelli, E., Rametta, R., Batelli, S., Veschini, L., Negro, A. and Forloni, G. (2004). Protective effect of TAT-delivered alpha-synuclein: relevance of the C-terminal domain and involvement of HSP70. *FASEB J* 18: 1713–1715.

Alim, M. A., Hossain, M. S., Arima, K., Takeda, K., Izumiyama, Y., Nakamura, M., Kaji, H., Shinoda, T., Hisanaga, S. and Ueda, K. (2002). Tubulin seeds alpha-synuclein fibril formation. *J Biol Chem* 277: 2112–2117.

Arrasate, M., Mitra, S., Schweitzer, E. S., Segal, M. R. and Finkbeiner, S. (2004). Inclusion body formation reduces levels of mutant huntingtin and the risk of neuronal death. *Nature* 431: 805–810.

Baba, M., Nakajo, S., Tu, P. H., Tomita, T., Nakaya, K., Lee, V. M., Trojanowski, J. Q. and Iwatsubo, T. (1998). Aggregation of alpha-synuclein in Lewy bodies of sporadic Parkinson's disease and dementia with Lewy bodies. *Am J Pathol* 152: 879–884.

Bennett, M. C., Bishop, J. F., Leng, Y., Chock, P. B., Chase, T. N. and Mouradian, M. M. (1999). Degradation of alpha-synuclein by proteasome. *J Biol Chem* 274: 33855–33858.

Betarbet, R., Sherer, T. B., MacKenzie, G., Garcia-Osuna, M., Panov, A. V. and Greenamyre, J. T. (2000). Chronic systemic pesticide exposure reproduces features of Parkinson's disease. *Nat Neurosci* 3: 1301–1306

Bonifati, V., Oostra, B. A. and Heutink, P. (2004). Unraveling the pathogenesis of Parkinson's disease—the contribution of monogenic forms. *Cell Mol Life Sci* 61: 1729–1750.

Bussell, R., Jr. and Eliezer, D. (2003). A structural and functional role for 11-mer repeats in alpha-synuclein and other exchangeable lipid binding proteins. *J Mol Biol* 329: 763–778.

Cabin, D. E., Shimazu, K., Murphy, D., Cole, N. B., Gottschalk, W., McIlwain, K. L., Orrison, B., Chen, A., Ellis, C. E., Paylor, R., Lu, B. and Nussbaum, R. L. (2002). Synaptic vesicle depletion correlates with attenuated synaptic responses to prolonged repetitive stimulation in mice lacking alpha-synuclein. *J Neurosci* 22: 8797–8807.

Campbell, B. C., McLean, C. A., Culvenor, J. G., Gai, W. P., Blumbergs, P. C., Jakala, P., Beyreuther, K., Masters, C. L. and Li, Q. X. (2001). The solubility of alpha-synuclein in multiple system atrophy differs from that of dementia with Lewy bodies and Parkinson's disease. *J Neurochem* 76: 87–96.

Chandra, S., Chen, X., Rizo, J., Jahn, R. and Sudhof, T. C. (2003). A broken alpha-helix in folded alpha-Synuclein. *J Biol Chem* 278: 15313–15318.

Chandra, S., Fornai, F., Kwon, H. B., Yazdani, U., Atasoy, D., Liu, X., Hammer, R. E., Battaglia, G., German, D. C., Castillo, P. E. and Sudhof, T. C. (2004). Double-knockout mice for alpha- and beta-synucleins: effect on synaptic functions. *Proc Natl Acad Sci USA* 101: 14966–14971.

Choi, W., Zibaee, S., Jakes, R., Serpell, L. C., Davletov, B., Crowther, R. A. and Goedert, M. (2004). Mutation E46K increases phospholipid binding and assembly into filaments of human alpha-synuclein. *FEBS Lett* 576: 363–368.

Clayton, D. F. and George, J. M. (1998). The synucleins: a family of proteins involved in synaptic function, plasticity, neurodegeneration and disease. *Trends Neurosci* 21: 249–254.

Cole, N. B., Murphy, D. D., Grider, T., Rueter, S., Brasaemle, D. and Nussbaum, R. L. (2002). Lipid droplet binding and oligomerization properties of the Parkinson's disease protein alpha-synuclein. *J Biol Chem* 277: 6344–6352.

Conway, K. A., Harper, J. D. and Lansbury, P. T., Jr. (2000a). Fibrils formed *in vitro* from alpha-synuclein and two mutant forms linked to Parkinson's disease are typical amyloid. *Biochemistry* 39: 2552–2563.

Conway, K. A., Lee, S.-J., Rochet, J. C., Ding, T. T., Williamson, R. E. and Lansbury, P. T., Jr. (2000b). Acceleration of oligomerization, not fibrillization, is a shared property of both alpha-synuclein mutations linked to early-onset Parkinson's disease: implications for pathogenesis and therapy. *Proc Natl Acad Sci USA* 97: 571–576.

Crowther, R. A., Jakes, R., Spillantini, M. G. and Goedert, M. (1998). Synthetic filaments assembled from C-terminally truncated alpha-synuclein. *FEBS Lett* 436: 309–312.

Cuervo, A. M., Stefanis, L., Fredenburg, R., Lansbury, P. T. and Sulzer, D. (2004). Impaired degradation of mutant alpha-synuclein by chaperone-mediated autophagy. *Science* 305: 1292–1295.

Cummings, C. J., Reinstein, E., Sun, Y., Antalffy, B., Jiang, Y., Ciechanover, A., Orr, H. T., Beaudet, A. L. and Zoghbi, H. Y. (1999). Mutation of the E6-AP ubiquitin ligase reduces nuclear inclusion frequency while accelerating polyglutamine-induced pathology in SCA1 mice. *Neuron* 24: 879–892.

da Costa, C. A., Ancolio, K. and Checler, F. (2000). Wild-type but not Parkinson's disease-related ala-53 → Thr mutant alpha-synuclein protects neuronal cells from apoptotic stimuli. *J Biol Chem* 275: 24065–24069.

Dauer, W. and Przedborski, S. (2003). Parkinson's disease: mechanisms and models. *Neuron* 39: 889–909.

Davidson, W. S., Jonas, A., Clayton, D. F. and George, J. M. (1998). Stabilization of alpha-synuclein secondary structure upon binding to synthetic membranes. *J Biol Chem* 273: 9443–9449.

Engelender, S., Kaminsky, Z., Guo, X., Sharp, A. H., Amaravi, R. K., Kleiderlein, J. J., Margolis, R. L., Troncoso, J. C., Lanahan, A. A., Worley, P. F., Dawson, V. L., Dawson, T. M. and Ross, C. A. (1999). Synphilin-1 associates with alpha-synuclein and promotes the formation of cytosolic inclusions. *Nat Genet* 22: 110–114.

Faber, P. W., Alter, J. R., MacDonald, M. E. and Hart, A. C. (1999). Polyglutamine-mediated dysfunction and apoptotic death of a Caenorhabditis elegans sensory neuron. *Proc Natl Acad Sci USA* 96: 179–184.

Feany, M. B. and Bender, W. W. (2000). A Drosophila model of Parkinson's disease. *Nature* 404: 394–398.

Fujiwara, H., Hasegawa, M., Dohmae, N., Kawashima, A., Masliah, E., Goldberg, M. S., Shen, J., Takio, K. and Iwatsubo, T. (2002). alpha-Synuclein is phosphorylated in synucleinopathy lesions. *Nat Cell Biol* 4: 160–164.

George, J. M. (2002). The synucleins. *Genome Biol* 3: reviews 3002.1–3002.6.

George, J. M., Jin, H., Woods, W. S. and Clayton, D. F. (1995). Characterization of a novel protein regulated during the critical period for song learning in the zebra finch. *Neuron* 15: 361–372.

Ghee, M., Fournier, A. and Mallet, J. (2000). Rat alpha-synuclein interacts with Tat binding protein 1, a component of the 26S proteasomal complex. *J Neurochem* 75: 2221–2224.

Giasson, B. I., Duda, J. E., Quinn, S. M., Zhang, B., Trojanowski, J. Q. and Lee, V. M.-Y. (2002). Neuronal alpha-synucleinopathy with severe movement disorder in mice expressing A53T human alpha-synuclein. *Neuron* 34: 521–533.

Goedert, M. (2001). Alpha-synuclein and neurodegenerative diseases. *Nat Rev Neurosci* 2: 492–501.

Gomez-Santos, C., Ferrer, I., Reiriz, J., Vinals, F., Barrachina, M. and Ambrosio, S. (2002). MPP+ increases alpha-synuclein expression and ERK/MAP-kinase phosphorylation in human neuroblastoma SH-SY5Y cells. *Brain Res* 935: 32–39.

Gosavi, N., Lee, H.-J., Lee, J. S., Patel, S. and Lee, S.-J. (2002). Golgi fragmentation occurs in the cells with prefibrillar alpha-synuclein aggregates and precedes the formation of fibrillar inclusion. *J Biol Chem* 277: 48984–48992.

Hashimoto, M., Hsu, L. J., Rockenstein, E., Takenouchi, T., Mallory, M. and Masliah, E. (2002). alpha-Synuclein protects against oxidative stress via inactivation of the C- jun N-terminal kinase stress signaling pathway in neuronal cells. *J Biol Chem* 14: 14.

Horwich, A. (2002). Protein aggregation in disease: a role for folding intermediates forming specific multimeric interactions. *J Clin Invest* 110: 1221–1232.

Hoyer, W., Cherny, D., Subramaniam, V. and Jovin, T. M. (2004). Rapid self-assembly of alpha-synuclein observed by *in situ* atomic force microscopy. *J Mol Biol* 340: 127–139.

Hsu, L. J., Mallory, M., Xia, Y., Veinbergs, I., Hashimoto, M., Yoshimoto, M., Thal, L. J., Saitoh, T. and Masliah, E. (1998). Expression pattern of synucleins (non-Abeta component of Alzheimer's disease amyloid precursor protein/alpha-synuclein) during murine brain development. *J Neurochem* 71: 338–344.

Hsu, L. J., Sagara, Y., Arroyo, A., Rockenstein, E., Sisk, A., Mallory, M., Wong, J., Takenouchi, T., Hashimoto, M. and Masliah, E. (2000). alpha-Synuclein promotes mitochondrial deficit and oxidative stress. *Am J Pathol* 157: 401–410.

Hyun, D. H., Lee, M., Halliwell, B. and Jenner, P. (2003). Proteasomal inhibition causes the formation of protein aggregates containing a wide range of proteins, including nitrated proteins. *J Neurochem* 86: 363–373.

Irizarry, M. C., Kim, T. W., McNamara, M., Tanzi, R. E., George, J. M., Clayton, D. F. and Hyman, B. T. (1996). Characterization of the precursor protein of the non-A beta component of senile plaques (NACP) in the human central nervous system. *J Neuropathol Exp Neurol* 55: 889–895.

Iwai, A., Masliah, E., Yoshimoto, M., Ge, N., Flanagan, L., de Silva, H. A., Kittel, A. and Saitoh, T. (1995). The precursor protein of non-A beta component of Alzheimer's disease amyloid is a presynaptic protein of the central nervous system. *Neuron* 14: 467–475.

Iwata, A., Maruyama, M., Kanazawa, I. and Nukina, N. (2001). alpha-Synuclein affects the MAPK pathway and accelerates cell death. *J Biol Chem* 276: 45320–45329.

Iwata, A., Maruyama, M., Akagi, T., Hashikawa, T., Kanazawa, I., Tsuji, S. and Nukina, N. (2003). Alpha-synuclein degradation by serine protease neurosin: implication for pathogenesis of synucleinopathies. *Hum Mol Genet* 12: 2625–2635.

Jakes, R., Spillantini, M. G. and Goedert, M. (1994). Identification of two distinct synucleins from human brain. *FEBS Lett* 345: 27–32.

Jao, C. C., Der-Sarkissian, A., Chen, J. and Langen, R. (2004). Structure of membrane-bound alpha-synuclein studied by site-directed spin labeling. *Proc Natl Acad Sci USA* 101: 8331–8336.

Jenco, J. M., Rawlingson, A., Daniels, B. and Morris, A. J. (1998). Regulation of phospholipase D2: selective inhibition of mammalian phospholipase D isoenzymes by alpha- and beta-synucleins. *Biochemistry* 37: 4901–4909.

Junn, E. and Mouradian, M. M. (2002). Human alpha-synuclein overexpression increases intracellular reactive oxygen species levels and susceptibility to dopamine. *Neurosci Lett* 320: 146–150.

Kalivendi, S. V., Cunningham, S., Kotamraju, S., Joseph, J., Hillard, C. J. and Kalyanaraman, B. (2004). Alpha-synuclein up-regulation and aggregation during MPP+-induced apoptosis in neuroblastoma cells: intermediacy of transferrin receptor iron and hydrogen peroxide. *J Biol Chem* 279: 15240–15247.

Kanda, S., Bishop, J. F., Eglitis, M. A., Yang, Y. and Mouradian, M. M. (2000). Enhanced vulnerability to oxidative stress by alpha-synuclein mutations and C-terminal truncation. *Neuroscience* 97: 279–284.

Kazemi-Esfarjani, P. and Benzer, S. (2000). Genetic suppression of polyglutamine toxicity in Drosophila. *Science* 287: 1837–1840.

Kim, S. J., Sung, J. Y., Um, J. W., Hattori, N., Mizuno, Y., Tanaka, K., Paik, S. R., Kim, J. and Chung, K. C. (2003). Parkin cleaves intracellular alpha-synuclein inclusions via the activation of calpain. *J Biol Chem* 278: 41890–41899.

Kim, T. D., Paik, S. R. and Yang, C. H. (2002). Structural and functional implications of C-terminal regions of alpha-synuclein. *Biochemistry* 41: 13782–13790.

Kim, T. D., Paik, S. R., Yang, C. H. and Kim, J. (2000). Structural changes in alpha-synuclein affect its chaperone-like activity *in vitro*. *Protein Sci* 9: 2489–2496.

Kim, T. D., Choi, E., Rhim, H., Paik, S. R. and Yang, C. H. (2004). Alpha-synuclein has structural and functional similarities to small heat shock proteins. *Biochem Biophys Res Commun* 324: 1352–1359.

Klement, I. A., Skinner, P. J., Kaytor, M. D., Yi, H., Hersch, S. M., Clark, H. B., Zoghbi, H. Y. and Orr, H. T. (1998). Ataxin-1 nuclear localization and aggregation: role in polyglutamine- induced disease in SCA1 transgenic mice [see comments]. *Cell* 95: 41–53.

Ko, L., Mehta, N. D., Farrer, M., Easson, C., Hussey, J., Yen, S., Hardy, J. and Yen, S. H. (2000). Sensitization of neuronal cells to oxidative stress with mutated human alpha-synuclein. *J Neurochem* 75: 2546–2554.

Kopito, R. R. (2000). Aggresomes, inclusion bodies and protein aggregation. *Trends Cell Biol* 10: 524–530.

Langston, J. W., Sastry, S., Chan, P., Forno, L. S., Bolin, L. M. and Di Monte, D. A. (1998). Novel alpha-synuclein-immunoreactive proteins in brain samples from the Contursi kindred, Parkinson's, and Alzheimer's disease. *Exp Neurol* 154: 684–690.

Lashuel, H. A., Petre, B. M., Wall, J., Simon, M., Nowak, R. J., Walz, T. and Lansbury, P. T., Jr. (2002). Alpha-synuclein, especially the Parkinson's disease-associated mutants, forms pore-like annular and tubular protofibrils. *J Mol Biol* 322: 1089–1102.

Lee, M., Hyun, D., Halliwell, B. and Jenner, P. (2001). Effect of the overexpression of wild-type or mutant alpha-synuclein on cell susceptibility to insult. *J Neurochem* 76: 998–1009.

Lee, H.-J. and Lee, S.-J. (2002). Characterization of cytoplasmic alpha-synuclein aggregates. Fibril formation is tightly linked to the inclusion-forming process in cells. *J Biol Chem* 277: 48976–48983.

Lee, H.-J., Choi, C. and Lee, S.-J. (2002a). Membrane-bound alpha-synuclein has a high aggregation propensity and the ability to seed the aggregation of the cytosolic form. *J Biol Chem* 277: 671–678.

Lee, H.-J., Shin, S. Y., Choi, C., Lee, Y. H. and Lee, S.-J. (2002b). Formation and removal of alpha-synuclein aggregates in cells exposed to mitochondrial inhibitors. *J Biol Chem* 277: 5411–5417.

Lee, S. J. (2003). alpha-Synuclein aggregation: a link between mitochondrial defects and Parkinson's disease? *Antioxid Redox Signal* 5: 337–348.

Lee, H.-J., Khoshaghideh, F., Patel, S. and Lee, S.-J. (2004). Clearance of alpha-synuclein oligomeric intermediates via the lysosomal degradation pathway. *J Neurosci* 24: 1888–1896.

Li, W., Lesuisse, C., Xu, Y., Troncoso, J. C., Price, D. L. and Lee, M. K. (2004). Stabilization of alpha-synuclein protein with aging and familial parkinson's disease-linked A53T mutation. *J Neurosci* 24: 7400–7409.

Lindersson, E., Beedholm, R., Hojrup, P., Moos, T., Gai, W., Hendil, K. B. and Jensen, P. H. (2004). Proteasomal inhibition by alpha-synuclein filaments and oligomers. *J Biol Chem* 279: 12924–12934.

Liu, C. W., Corboy, M. J., DeMartino, G. N. and Thomas, P. J. (2003). Endoproteolytic activity of the proteasome. *Science* 299: 408–411.

Liu, S., Ninan, I., Antonova, I., Battaglia, F., Trinchese, F., Narasanna, A., Kolodilov, N., Dauer, W., Hawkins, R. D. and Arancio, O. (2004). alpha-Synuclein produces a long-lasting increase in neurotransmitter release. *EMBO J* 23: 4506–4516.

London, J., Skrzynia, C. and Goldberg, M. E. (1974). Renaturation of Escherichia coli tryptophanase after exposure to 8 M urea. Evidence for the existence of nucleation centers. *Eur J Biochem* 47: 409–415.

Manning-Bog, A. B., McCormack, A. L., Li, J., Uversky, V. N., Fink, A. L. and Di Monte, D. A. (2002). The herbicide paraquat causes up-regulation and aggregation of alpha- synuclein in mice: paraquat and alpha-synuclein. *J Biol Chem* 277: 1641–1644.

Manning-Bog, A. B., McCormack, A. L., Purisai, M. G., Bolin, L. M. and Di Monte, D. A. (2003). Alpha-synuclein overexpression protects against paraquat-induced neurodegeneration. *J Neurosci* 23: 3095–3099.

Maries, E., Dass, B., Collier, T. J., Kordower, J. H. and Steece-Collier, K. (2003). The role of alpha-synuclein in Parkinson's disease: insights from animal models. *Nat Rev Neurosci* 4: 727–738.

Maroteaux, L., Campanelli, J. T. and Scheller, R. H. (1988). Synuclein: a neuron-specific protein localized to the nucleus and presynaptic nerve terminal. *J Neurosci* 8: 2804–2815.

McNaught, K. S., Mytilineou, C., Jnobaptiste, R., Yabut, J., Shashidharan, P., Jennert, P. and Olanow, C. W. (2002). Impairment of the ubiquitin-proteasome system causes dopaminergic cell death and inclusion body formation in ventral mesencephalic cultures. *J Neurochem* 81: 301–306.

Mishizen-Eberz, A. J., Guttmann, R. P., Giasson, B. I., Day, G. A., 3rd, Hodara, R., Ischiropoulos, H., Lee, V. M., Trojanowski, J. Q. and Lynch, D. R. (2003). Distinct cleavage patterns of normal and pathologic forms of alpha-synuclein by calpain I *in vitro*. *J Neurochem* 86: 836–847.

Norris, E. H., Giasson, B. I., Ischiropoulos, H. and Lee, V. M. (2003). Effects of oxidative and nitrative challenges on alpha-synuclein fibrillogenesis involve distinct mechanisms of protein modifications. *J Biol Chem* 278: 27230–27240.

Ostrerova, N., Petrucelli, L., Farrer, M., Mehta, N., Choi, P., Hardy, J. and Wolozin, B. (1999). alpha-Synuclein shares physical and functional homology with 14-3-3 proteins. *J Neurosci* 19: 5782–5791.

Park, S. M., Jung, H. Y., Kim, T. D., Park, J. H., Yang, C. H. and Kim, J. (2002). Distinct roles of the N-terminal-binding domain and the C-terminal solubilizing domain of alpha-synuclein, a molecular chaperone. *J Biol Chem* 277: 28512–28520.

Paxinou, E., Chen, Q., Weisse, M., Giasson, B. I., Norris, E. H., Rueter, S. M., Trojanowski, J. Q., Lee, V. M. and Ischiropoulos, H. (2001). Induction of alpha-synuclein aggregation by intracellular nitrative insult. *J Neurosci* 21: 8053–8061.

Perrin, R. J., Woods, W. S., Clayton, D. F. and George, J. M. (2000). Interaction of human alpha-synuclein and Parkinson's disease variants with phospholipids. Structural analysis using site-directed mutagenesis. *J Biol Chem* 275: 34393– 34398.

Perrin, R. J., Woods, W. S., Clayton, D. F. and George, J. M. (2001). Exposure to long-chain polyunsaturated fatty acids triggers rapid multimerization of synucleins. *J Biol Chem* 11: 11.

Petersen, K., Olesen, O. F. and Mikkelsen, J. D. (1999). Developmental expression of alpha-synuclein in rat hippocampus and cerebral cortex. *Neuroscience* 91: 651–659.

Petrucelli, L., O'Farrell, C., Lockhart, P. J., Baptista, M., Kehoe, K., Vink, L., Choi, P., Wolozin, B., Farrer, M., Hardy, J. and Cookson, M. R. (2002). Parkin protects against the toxicity associated with mutant alpha-synuclein: proteasome dysfunction selectively affects catecholaminergic neurons. *Neuron* 36: 1007–1019.

Raghavan, R., Kruijff, L., Sterrenburg, M. D., Rogers, B. B., Hladik, C. L. and White, C. L., 3rd (2004). alpha-Synuclein expression in the developing human brain. *Pediatr Dev Pathol* 7: 506–516.

Rajan, R. S., Illing, M. E., Bence, N. F. and Kopito, R. R. (2001). Specificity in intracellular protein aggregation and inclusion body formation. *Proc Natl Acad Sci USA* 98: 13060–13065.

Rideout, H. J., Larsen, K. E., Sulzer, D. and Stefanis, L. (2001). Proteasomal inhibition leads to formation of ubiquitin/alpha-synuclein-immunoreactive inclusions in PC12 cells. *J Neurochem* 78: 899–908.

Saha, A. R., Ninkina, N. N., Hanger, D. P., Anderton, B. H., Davies, A. M. and Buchman, V. L. (2000). Induction of neuronal death by alpha-synuclein [In Process Citation]. *Eur J Neurosci* 12: 3073–3077.

Saudou, F., Finkbeiner, S., Devys, D. and Greenberg, M. E. (1998). Huntingtin acts in the nucleus to induce apoptosis but death does not correlate with the formation of intranuclear inclusions. *Cell* 95: 55–66.

Schluter, O. M., Fornai, F., Alessandri, M. G., Takamori, S., Geppert, M., Jahn, R. and Sudhof, T. C. (2003). Role of alpha-synuclein in 1-methyl-4-phenyl-1,2,3,6-tetrahydropyridine-induced parkinsonism in mice. *Neuroscience* 118: 985–1002.

Sharon, R., Bar-Joseph, I., Mirick, G. E., Serhan, C. N. and Selkoe, D. J. (2003). Altered fatty acid composition of dopaminergic neurons expressing alpha-synuclein and human brains with alpha-synucleinopathies. *J Biol Chem* 278: 49874– 49881.

Snyder, H., Mensah, K., Theisler, C., Lee, J., Matouschek, A. and Wolozin, B. (2003). Aggregated and monomeric alpha-synuclein bind to the S6′ proteasomal protein and inhibit proteasomal function. *J Biol Chem* 278: 11753–11759.

Souza, J. M., Giasson, B. I., Lee, V. M. and Ischiropoulos, H. (2000). Chaperone-like activity of synucleins. *FEBS Lett* 474: 116–119.

Spillantini, M. G., Crowther, R. A., Jakes, R., Hasegawa, M. and Goedert, M. (1998). alpha-Synuclein in filamentous inclusions of Lewy bodies from Parkinson's disease and dementia with lewy bodies. *Proc Natl Acad Sci USA* 95: 6469–6473.

Stefanis, L., Larsen, K. E., Rideout, H. J., Sulzer, D. and Greene, L. A. (2001). Expression of A53T mutant but not wild-type alpha-synuclein in PC12 cells induces alterations of the ubiquitin-dependent degradation system, loss of dopamine release, and autophagic cell death. *J Neurosci* 21: 9549–9560.

Tanaka, Y., Engelender, S., Igarashi, S., Rao, R. K., Wanner, T., Tanzi, R. E., Sawa, A., V, L. D., Dawson, T. M. and Ross, C. A. (2001). Inducible expression of mutant alpha-synuclein decreases proteasome activity and increases sensitivity to mitochondria-dependent apoptosis. *Hum Mol Genet* 10: 919–926.

Tanaka, M., Kim, Y. M., Lee, G., Junn, E., Iwatsubo, T., Mouradian, M. M. (2004) Aggresomes formed by alpha-synuclein and synphilin-1 are cytoprotective. *J Biol Chem* 279: 4625–4631.

Taylor, J. P., Tanaka, F., Robitschek, J., Sandoval, C. M., Taye, A., Markovic-Plese, S. and Fischbeck, K. H. (2003). Aggresomes protect cells by enhancing the degradation of toxic polyglutamine-containing protein. *Hum Mol Genet* 12: 749–757.

Tofaris, G. K., Layfield, R. and Spillantini, M. G. (2001). alpha-Synuclein metabolism and aggregation is linked to ubiquitin-independent degradation by the proteasome. *FEBS Lett* 509: 22–26.

Tofaris, G. K., Razzaq, A., Ghetti, B., Lilley, K. S. and Spillantini, M. G. (2003). Ubiquitination of alpha-synuclein in Lewy bodies is a pathological event not associated with impairment of proteasome function. *J Biol Chem* 278: 44405–44411.

Ueda, K., Fukushima, H., Masliah, E., Xia, Y., Iwai, A., Yoshimoto, M., Otero, D. A., Kondo, J., Ihara, Y. and Saitoh, T. (1993). Molecular cloning of cDNA encoding an unrecognized component of amyloid in Alzheimer disease. *Proc Natl Acad Sci USA* 90: 11282–11286.

Uversky, V. N., Gillespie, J. R. and Fink, A. L. (2000). Why are "natively unfolded" proteins unstructured under physiologic conditions? *Proteins* 41: 415–427.

Uversky, V. N., Li, J. and Fink, A. L. (2001a). Evidence for a partially folded intermediate in alpha-synuclein fibril formation. *J Biol Chem* 276: 10737–10744.

Uversky, V. N., Li, J. and Fink, A. L. (2001b). Metal-triggered structural transformations, aggregation, and fibrillation of human alpha-synuclein. A possible molecular NK between Parkinson's disease and heavy metal exposure. *J Biol Chem* 276: 44284–44296.

Uversky, V. N., Lee, H.-J., Li, J., Fink, A. L. and Lee, S.-J. (2001c). Stabilization of partially folded conformation during {alpha}-synuclein oligomerization in both purified and cytosolic preparations. *J Biol Chem* 5: 5.

Vila, M., Vukosavic, S., Jackson-Lewis, V., Neystat, M., Jakowec, M. and Przedborski, S. (2000). Alpha-synuclein up-regulation in substantia nigra dopaminergic neurons following administration of the parkinsonian toxin MPTP. *J Neurochem* 74: 721–729.

Volles, M. J. and Lansbury, P. T., Jr. (2002). Vesicle permeabilization by protofibrillar alpha-synuclein is sensitive to Parkinson's disease-linked mutations and occurs by a pore-like mechanism. *Biochemistry* 41: 4595–4602.

Volles, M. J. and Lansbury, P. T., Jr. (2003). Zeroing in on the pathogenic form of alpha-synuclein and its mechanism of neurotoxicity in Parkinson's disease. *Biochemistry* 42: 7871–7878.

Webb, J. L., Ravikumar, B., Atkins, J., Skepper, J. N. and Rubinsztein, D. C. (2003). alpha-Synuclein is degraded by both autophagy and the proteasome. *J Biol Chem* 278: 25009–25013.

Weinreb, P. H., Zhen, W., Poon, A. W., Conway, K. A. and Lansbury, P. T., Jr. (1996). NACP, a protein implicated in Alzheimer's disease and learning, is natively unfolded. *Biochemistry* 35: 13709–13715.

Withers, G. S., George, J. M., Banker, G. A. and Clayton, D. F. (1997). Delayed localization of synelfin (synuclein, NACP) to presynaptic terminals in cultured rat hippocampal neurons. *Brain Res Dev Brain Res* 99: 87–94.

Xu, J., Kao, S. Y., Lee, F. J., Song, W., Jin, L. W. and Yankner, B. A. (2002). Dopamine-dependent neurotoxicity of alpha-synuclein: a mechanism for selective neurodegeneration in Parkinson disease. *Nat Med* 8: 600–606.

Zhou, W., Hurlbert, M. S., Schaack, J., Prasad, K. N. and Freed, C. R. (2000). Overexpression of human alpha-synuclein causes dopamine neuron death in rat primary culture and immortalized mesencephalon-derived cells. *Brain Res* 866: 33–43.

Zhou, W., Schaack, J., Zawada, W. M. and Freed, C. R. (2002). Overexpression of human alpha-synuclein causes dopamine neuron death in primary human mesencephalic culture. *Brain Res* 926: 42–50.

Zhou, Y., Gu, G., Goodlett, D. R., Zhang, T., Pan, C., Montine, T. J., Montine, K. S., Aebersold, R. H. and Zhang, J. (2004). Analysis of alpha-synuclein-associated proteins by quantitative proteomics. *J Biol Chem* 279: 39155–39164.

6

Pathogenesis of Prion Diseases

Giuseppe Legname, Stephen J. DeArmond, Fred E. Cohen, and Stanley B. Prusiner

Abstract

Prion diseases are invariably fatal neurodegenerative disorders affecting humans and many other mammals. Here we discuss the current understanding of prion biology and the pathogenesis of this group of illnesses. We introduce several aspects of prion biology, from the primary structure of the cellular form of the prion protein (PrP^C) to the conformational changes that occur during the conversion to the pathologic form, PrP^{Sc}. Moreover, we provide an overview of the various prion diseases. We then conclude with the recent discovery of mammalian synthetic prions and the implications that such findings may have to the future of prion research.

6.1. Introduction

Prion diseases are a group of incurable, fatal neurodegenerative maladies that afflict mammals. Examples of such devastating diseases are Creutzfeldt-Jakob disease (CJD), Gerstmann-Sträussler-Scheinker (GSS) disease, and kuru in humans; bovine spongiform encephalopathy (BSE) in cattle; scrapie in sheep; and chronic wasting disease (CWD) in cervids (Table 6.1) (Prusiner, 2001).

The existence of prion diseases has been known for centuries. In fact, scrapie in sheep was first documented in England nearly 300 years ago, but the unusual nature of the agent responsible for this pathology and the mechanism underlying its propagation has only more recently witnessed intense research and interest. In 1920, H.G. Creutzfeldt reported on the spongiform change in brain of a young woman who died after a neurologic illness that included seizures (Creutzfeldt, 1920). The next year, A. Jakob described five patients with what is thought to have been a similar disorder (Jakob, 1921). For more than 50 years, the disorder was called Jakob-Creutzfeldt disease until Clarence Joseph Gibbs and Daniel Carleton Gajdusek reported in 1968 that the disease was transmissible to apes and monkeys (Gibbs *et al.*, 1968). Gibbs decided to reverse the commonly used acronym of Jakob-Creutzfeldt disease (Kirschbaum, 1968) and call the disease Creutzfeldt-Jakob disease, or CJD; he liked the abbreviation CJD because his initials were CJ (Gibbs, 1992). The impact of these transmission studies on the investigations of neurodegenerative diseases has been immense.

The transmission of CJD prions into apes was prompted by earlier studies on kuru (Klatzo *et al.*, 1959). Kuru is a progressive debilitating central nervous system (CNS) disease found in New Guinea highlands tribes. In remote villages, women and children prepared the brains of dead relatives for cannibalistic feast in the hope of granting immortality to their relatives (Alpers, 1968). The neuropathologic similarities between kuru and scrapie prompted the suggestion that

Table 6.1. The pathogenesis of prion diseases

Disease	Host	Mechanism of pathogenesis
Kuru	Humans	Infection through ritualistic cannibalism
iCJD	Humans	Infection from prion-contaminated human tissue
vCJD	Humans	Infection from bovine prions
fCJD	Humans	Germ-line mutations in PrP gene
GSS	Humans	Germ-line mutations in PrP gene
FFI	Humans	Germ-line mutation in PrP gene (D178N, M129)
sCJD	Humans	Somatic mutation or spontaneous conversion of PrP^{C} into PrP^{Sc}
sFI	Humans	Somatic mutation or spontaneous conversion of PrP^{C} into PrP^{Sc}
Scrapie	Sheep	Infection in genetically susceptible sheep
BSE	Cattle	Infection with prion-contaminated MBM
TME	Mink	Infection with prions from sheep or cattle
CWD	Mule deer, elk	Unknown
FSE	Cats	Infection with prion-contaminated MBM
EUE	Greater kudu, nyala, oryx	Infection with prion-contaminated MBM

CJD, Creutzfeldt-Jakob disease; sCJD, sporadic CJD; fCJD, familial CJD; iCJD, iatrogenic CJD; vCJD, variant CJD; EUE, Exotic ungulate encephalopathy; FFI, fatal familial insomnia; sFI, sporadic fatal insomnia; GSS, Gerstmann-Sträussler-Scheinker disease; HGH, human growth hormone. BSE, bovine spongiform encephalopathy; CWD, chronic wasting disease; FSE, feline spongiform encephalopathy; MBM, meat and bone meal; TME, transmissible mink encephalopathy (Prusiner, 2001).

kuru like scrapie might be transmissible (Hadlow, 1959). Seven years later, the transmission of kuru prions into apes was reported (Gajdusek *et al.*, 1966).

In the mid-1980s, dairy cattle were identified that suffered from a new disease that was labeled bovine spongiform encephalopathy (BSE), or "mad cow disease" (Wells and Wilesmith, 2004). Cows suffering from BSE exhibited ataxia as well as apprehensive behavior. By the mid-1990s, the first reports appeared describing teenagers and young adults with variant (v) CJD, the human counterpart of BSE in cattle (Will *et al.*, 1996). Much evidence argues that vCJD prions were transmitted from tainted beef and beef products to young people (Scott *et al.*, 1999). To date, more than 190 people have died of vCJD.

In the 1960s and 1970s, many hypotheses were set forth to reconcile the resistance of the scrapie and CJD pathogens to inactivation by ionizing and ultraviolet radiations (Alper *et al.*, 1966, 1967; Gibbs *et al.*, 1978). These pathogens were often referred to as slow viruses, but there was no evidence that they contained a nucleic acid core, which carried the genetic blueprint for the production of progeny viruses. In 1982, one of us (S.B.P.) introduced the prion concept in order to explain a vast body of scientific data, much of which argued the pathogen causing scrapie is devoid of nucleic acid but contains a protein that is essential for infectivity (Prusiner, 1982). The term *prion* was derived from *pro*teinaceous and *in*fectious.

Although controversial for many years, the prion concept is now widely accepted (Prusiner, 1998). Prions are unprecedented infectious pathogens that give rise to invariably fatal neurodegenerative diseases via an entirely novel mechanism of disease. Prions are devoid of nucleic acid and are composed exclusively of disease-causing prion protein, PrP^{Sc}. The normal, cellular form of PrP, denoted PrP^{C}, is converted into PrP^{Sc} through a process whereby a portion of its α-helical and coil structure is refolded into β-sheet (Pan *et al.*, 1993). This structural transition is accompanied by profound changes in the physicochemical properties of PrP (see Fig. 6.1) (Govaerts *et al.*, 2004).

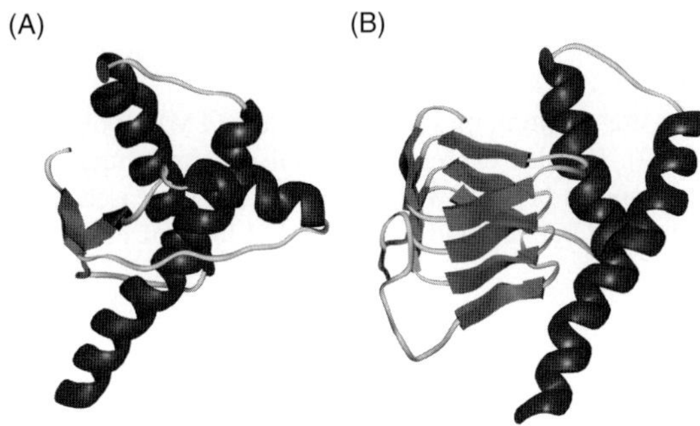

Figure 6.1. Two alternative conformations for the prion protein. The conversion of PrP^{C} into PrP^{Sc} is the fundamental event in prion diseases. (A) The NMR solution structure of recombinant PrP appears to be similar to that of normal cellular form of the prion protein, PrP^{C} (James *et al.*, 1997). (B) Model of PrP^{Sc} from electron crystallographic studies (Govaerts *et al.*, 2004).

6.1.1. Prion Biology and Diseases

Prion diseases are unique among all illnesses in that they can manifest as sporadic, genetic, or infectious maladies. Like many other neurodegenerative diseases, the sporadic form of prion disease accounts for about 80–90% of cases whereas the genetic forms account for 10–20% (Masters *et al.*, 1978; Will *et al.*, 2004). Infection by exogenous prions seems to be responsible for <1% of all human cases of prion disease, but this statement must be tempered by the caveat that the ubiquitous nature of bovine products in the food supply could make difficult the detection of infrequent prion infection (Asante *et al.*, 2002). The familial prion diseases are often subclassified according to their clinicopathologic phenotypes; they are familial CJD (fCJD), GSS disease, and fatal familial insomnia (FFI) (Masters *et al.*, 1981; Hsiao *et al.*, 1989; Prusiner, 1991; Medori *et al.*, 1992; Goldfarb *et al.*, 1994). These inherited prion diseases are autosomal dominant disorders, and each is caused by a point mutation or insert in the PrP gene. More than 30 different mutations have been recorded to segregate with familial prion disease phenotypes (Kong *et al.*, 2004).

For many years, the existence of prion strains was enigmatic. A wealth of data indicated the existence of many different prion strains, but the absence of a nucleic genome within the prion presented a conundrum. Eventually, evidence began to accumulate arguing that the strain-specific properties of prions are enciphered in the tertiary structure of PrP^{Sc} (Bessen and Marsh, 1994; Telling *et al.*, 1996b; Prusiner, 1997). The amino acid sequence of PrP^{Sc} corresponds with that encoded by the PrP gene of the mammalian host in which it last replicated. Transgenetic studies argue that PrP^{Sc} acts as a template upon which PrP^{C} is refolded into a nascent PrP^{Sc} molecule through a process most likely facilitated by an as-yet unidentified protein or set of proteins.

Three different pathogenic mechanisms have been found involving PrP metabolism in naturally occurring and experimentally manipulated diseases. The first group is the prion diseases; PrP^{C} is converted into PrP^{Sc} in caveolae-like domains (CLDs) (Gorodinsky and Harris, 1995; Taraboulos *et al.*, 1995; Vey *et al.*, 1996; Kaneko *et al.*, 1997a) and subsequently traffics to endosomes and lysosomes (Caughey, 1991; McKinley *et al.*, 1991b). The accumulation of PrP^{Sc} in membranes is thought to cause CNS dysfunction, but the mechanism remains unclear.

The second group is disorders similar to storage diseases. Deletion of either α helices B or C from PrP^{C} resulted in an accumulation of the mutant protein in the CNS (Muramoto *et al.*, 1997). As yet, this disease has been seen only in mice expressing mutant PrP transgenes; in these mice, a massive proliferation of the endoplasmic reticulum (ER) was found. Somewhat less severe alterations were found in the CNS and muscle of mice expressing foreign wild-type PrP transgenes at a very high level (Westaway *et al.*, 1994a).

The third group of disorders involves the adoption of a transmembrane (TM) topology by PrP (Hegde *et al.*, 1998; Hegde *et al.*, 1999). In mice expressing several different mutant PrP transgenes, a TM form of PrP accumulated, and in humans with GSS(A117V), increased levels of a similar TM form of PrP were found. In both humans and mice, this TM form of PrP, designated ^{Ctm}PrP, is thought to be pathogenic.

6.1.2. Prion Protein and Prion Structure

A single exon in the *Prnp* gene contains the open reading frame encoding the prion protein in rodents (Basler *et al.*, 1986; Westaway *et al.*, 1987). The primary structure of the protein consists of 254 amino acids in Syrian hamster and mouse. Two signal sequences are present in the preproprotein: an amino terminal ER translocation signal peptide and a glycosylphosphatidylinositol (GPI) signal sequence at the carboxy terminus. The mature protein of 209 residues has a single disulfide bridge between Cys179 and Cys213 (mouse PrP numbering) and two sites for Asn-linked glycosylation within the carboxy-terminal region (Fig. 6.2).

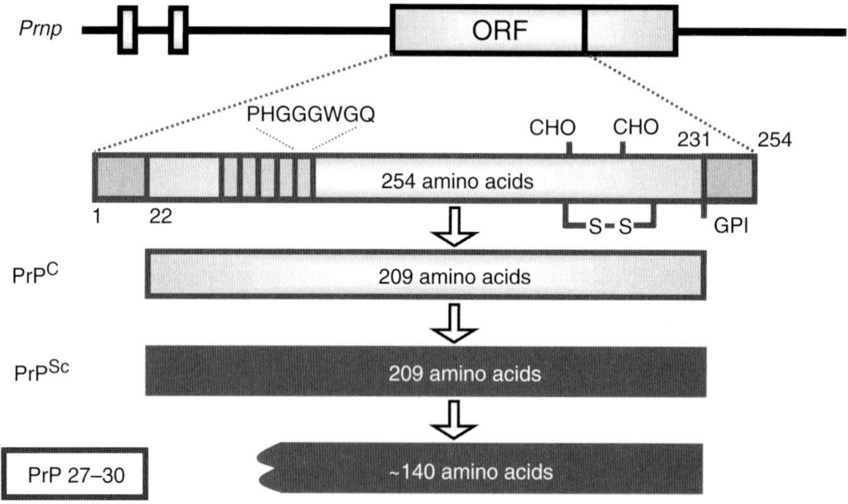

Figure 6.2. SHa *Prnp* gene and prion protein structures. The complete open reading frame (ORF) for the prion protein is contained in a single exon in the *Prnp* gene. The ORF codes for a preproprotein of 254 amino acids in rodents. The amino terminal signal sequence composed of 22 amino acids is necessary for the endoplasmic reticulum translocation. The carboxy terminal signal sequence is required for the post-translational glycosylphosphatidylinositol anchoring at the cell membrane. The mature sequence is composed of 209 amino acids between residue 23 to 231 in PrP^{C} and PrP^{Sc}. After limited treatment with protease, PrP^{Sc} is truncated to PrP 27-30, a ragged-ended protein composed of ~140 amino acids.

Mice with an ablated PrP gene (*Prnp*$^{0/0}$) are resistant to infection with prions (Büeler *et al.*, 1993; Prusiner *et al.*, 1993b), and their development and behavior appear normal (Büeler *et al.*, 1992). Although some aged *Prnp*$^{0/0}$ mice have been reported to develop ataxia and demyelination of cerebellar Purkinje cells (Sakaguchi *et al.*, 1996), this was not a consequence of the loss of PrP^{C} but rather due to the upregulation of the PrP-like protein doppel (Moore *et al.*, 1999). Attempts to define a physiologic role of PrP^{C} have focused on identifying interacting molecules. PrP^{C} has been found to interact with a number of proteins, including bacterial HSP-60 (Edenhofer *et al.*, 1996), the 37-kDa/67-kDa laminin receptor (Rieger *et al.*, 1997), laminin (Graner *et al.*, 2000), neural cell adhesion molecule (NCAM) (Schmitt-Ulms *et al.*, 2001), and the Grb2 protein, which is central in many signal transduction pathways (Spielhaupter and Schätzl, 2001). However, these studies have failed to identify a function for PrP^{C}. In search of interacting ligands, we showed the staining of several regions of the brain using novel PrP-Fc fusion proteins. In particular, the granule cell layer of the cerebellum showed intense staining, suggesting the presence of one or more PrP^{C}-interacting ligands in this region (Legname *et al.*, 2002). A PrP-Fc fusion protein was shown to increase moderately neurite extension and neuronal survival in primary cultures of mouse cerebellar granule cells (Chen *et al.*, 2003). Incubation of primary cultures of rat hippocampal neurons with recombinant PrP also demonstrated increased synaptic growth (Kanaani *et al.*, 2005).

Although the function of PrP^{C} has not yet been elucidated, one prominent finding is its interaction with copper (Brown *et al.*, 1997a). At the N-terminal, unstructured region of the protein, the presence of octarepeats is an interesting feature. These repeats [consensus sequence: P(H/Q)GGG(G)WGQ] are only found in PrP molecules across many different species of mammals, and they show an affinity for copper (II) ions (Cu^{++}) compared with other metals (Stöckel *et al.*, 1998). The coordination of Cu^{++} in PrP^{C} through His residues within the octarepeats raises the possibility that PrP^{C} functions as a copper-binding protein (Viles *et al.*, 1999; Burns *et al.*, 2003). The role of Cu^{++} in prion diseases remains to be established (Brown *et al.*, 1997b; Waggoner *et al.*, 2000).

Analysis of the primary structures of PrP^{C} and PrP^{Sc} does not account for their different biological properties (Stahl *et al.*, 1993). Moreover, no unique post-translation modification is found in PrP^{Sc} compared with PrP^{C} (Stahl *et al.*, 1993). Indeed, comparative analysis of PrP^{C} and PrP^{Sc} secondary structures revealed major differences in the conformation of these protein isoforms. While PrP^{C} contains ~40% α-helical structure, the α-helical content of PrP^{Sc} is only ~30%. In contrast to PrP^{C} with negligible β-sheet content, PrP^{Sc} consists of ~45% β-sheet (Pan *et al.*, 1993). Therefore, these differences in secondary structure argue that a major structural rearrangement occurs when PrP^{C} is converted into PrP^{Sc}, and this change in conformation is responsible for the strikingly different biological properties of the two protein isoforms.

Nuclear magnetic resonance (NMR) solution studies of the tertiary structure of recombinant PrP from different species have divided the protein into an amino-terminal moiety without apparent secondary structure and carboxy-terminal domain composed by a short stretch of β-sheet and three α-helices (Riek *et al.*, 1996; James *et al.*, 1997). The high α-helical content of recombinant PrP is in agreement with early studies on PrP^{C} purified from brain (Pan *et al.*, 1993; Pergami *et al.*, 1996).

During formation of PrP^{Sc}, PrP^{C} undergoes profound structural rearrangement as described above. α-Helix A is unfolded and is refolded into a β-sheet while α-helix C remains intact; the N-terminal portion of α-helix B might also adopt a β-sheet conformation. A recent model of PrP^{Sc}, constrained by electron crystallography data, proposes the formation of a β-helix in order to accommodate the large amount of β-sheet within a compact structure (Govaerts *et al.*, 2004) (see Fig. 6.1). This structural transition is the basis for the formation of nascent PrP^{Sc} and, thus, the replication of prions. As discussed above, prions consist solely of the disease-causing isoform, PrP^{Sc}.

Limited proteolysis of PrP^{Sc} leads to the formation of a smaller protein product of ~140 amino acids, denoted as PrP 27-30 (see Fig. 6.2). In the same experimental conditions, PrP^{C} is completely hydrolyzed. Addition of detergents to the digested fragment leads to the formation of prion amyloid rods, which are similar to PrP amyloid deposits seen in amyloid plaques in the CNS of animals and people afflicted with prion disease (McKinley *et al.*, 1991a).

6.1.3. Prion Replication

The mechanism of prion replication is not well understood. The conversion of α-helical, monomeric PrP^{C} into insoluble, protease-resistant, β-sheet-rich PrP^{Sc} is a process that seems to occur in CLDs (Gorodinsky and Harris, 1995; Taraboulos *et al.*, 1995; Vey *et al.*, 1996; Kaneko *et al.*, 1997a) and the resulting PrP^{Sc} subsequently traffics to other membranous compartments such as endosomes and lysosomes (Caughey, 1991; McKinley *et al.*, 1991b). The membranes of CLDs seem to be composed of cholesterol-rich rafts and presumably provide a milieu for the formation of PrP^{Sc}. Two models of conversion and replication have been suggested: (i) a nucleation-polymerization reaction and (ii) a templated-assisted process. In the first model, the rate-limiting step is the formation of critical amount of PrP^{Sc} to form a "seed" for the polymerization of PrP^{Sc}. It has been postulated that the two conformations, PrP^{C} and PrP^{Sc}, exist in equilibrium before the thermodynamically more stable PrP^{Sc} forms a "seed" and recruits more monomeric PrP^{Sc} to begin the polymerization of the infectious prion, including PrP^{Sc} amyloid (Aguzzi *et al.*, 2001).

In the template-assisted model, PrP^{C} must first undergo conversion toward a transition state (PrP*) that presumably corresponds with a partially destabilized structure (Fig. 6.3). This transition could be mediated by an auxiliary protein or proteins collectively denoted as protein X. The

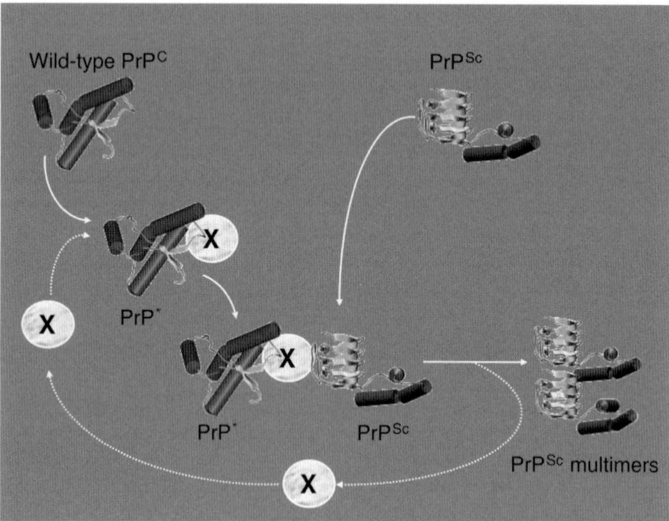

Figure 6.3. Hypothetical scheme for the initiation and replication of PrP^{Sc} formation in the template-assisted model. Initiation of the replication process in sporadic or infectious prion disease: Exogenous or endogenous PrP^{Sc} initiates PrP^{Sc} formation by binding to a PrP*/protein X complex. Facilitated by protein X and directed by the PrP^{Sc} template, PrP* changes conformation and forms a nascent molecule of PrP^{Sc}. When nascent PrP^{Sc} is formed, the heteromultimeric complex dissociates, yielding recycled protein X and two molecules of PrP^{Sc}. The two PrP^{Sc} molecules can now stimulate two conversion cycles, leading to an exponential increase in PrP^{Sc} formation.

PrP*/protein X complex can interact with available PrP^{Sc} molecules, which can be exogenous (infectious disease) or endogenous (sporadic disease or inherited disease) (Fig. 6.3). This interaction leads to the creation of new PrP^{Sc} molecules through templating existing PrP^{Sc}, which transmits its structural characteristics to the newly formed PrP^{Sc}. The latter is then available as a template for converting neighboring PrP*/protein X complexes, leading to an amplification cycle of the conversion (Fig. 6.3).

Many transgenic (Tg) mouse studies have provided evidence for the existence of protein X and that the latter might be a limiting factor in the conversion cycle. Indeed, Tg mice expressing human (Hu) PrP (HuPrP) on the wild-type (wt) mouse genetic background ($Prnp^{+/+}$), hereafter denoted Tg(HuPrP)$Prnp^{+/+}$ mice, are resistant to infection to Hu prions, which is attributed to the greater affinity of endogenous MoPrP compared with HuPrP to mouse protein X (Telling *et al.*, 1995). Notably, while Tg(HuPrP)$Prnp^{0/0}$ mice are susceptible to infection by Hu prions, they require an expression level 10-fold higher than that of Tg(MHu2M)$Prnp^{0/0}$ mice, which carry a chimeric Hu/Mo PrP transgene. This finding argues that the prion replication rate depends on the affinity of PrP for an endogenous auxiliary molecule (i.e., protein X). The results of recent transgenetic studies suggest that dominant-negative inhibition of PrP^{Sc} formation may occur through increased affinity of mutant PrP for protein X (Perrier *et al.*, 2002).

6.1.4. Transmission Barriers

In the mid-1960s, the concept of the "species barrier" was introduced (Pattison, 1965). The species barrier has been used to explain prolonged incubation times that were observed when prions were passaged from one species to another. With the application of transgenetics to the study of prion disease, the species barrier was correlated with the degree of similarity between the amino acid sequences of PrP^{C} in the host and of PrP^{Sc} in the inoculum (Prusiner *et al.*, 1990). Prions derive their PrP^{Sc} sequence from the mammal in which they were last passaged (Scott *et al.*, 1997). The importance of sequence similiarity between the host and donor PrP argues that PrP^{C} directly interacts with PrP^{Sc} in the prion-conversion process; PrP^{Sc} seems to function as a template for the tertiary structure of nascent PrP^{Sc} molecules as they are formed from PrP^{C}.

The term *species barrier* became obsolete once it was appreciated that the prion strains as well as the interaction of PrP^{C} with protein X influenced transmission of prions from one host to another. That being the case, we introduced the term *transmission barrier* to encompass all aspects of prion passage (Scott *et al.*, 2005). While human vCJD prions inefficiently infect Tg(HuPrP)$Prnp^{0/0}$ and Tg(MHu2M)$Prnp^{0/0}$ mice (Hill *et al.*, 1997; Asante *et al.*, 2002; Korth *et al.*, 2003), vCJD strains transmit readily into transgenic mice expressing bovine (Bo) PrP, designated Tg(BoPrP) mice (Scott *et al.*, 1999, 2005). This indicates that human vCJD prions strains, which are believed to originate from BSE-infected cattle, are more compatible with proteins comprised of the bovine PrP sequence than the human PrP sequence.

The interaction of PrP with protein X seems to be yet another component of the transmission barrier. In Tg mice, $HuPrP^{C}$ seems to bind much less efficiently to mouse protein X than does the chimeric protein MHu2M PrP^{C} (Telling *et al.*, 1995). After inoculation with human prions, equivalent incubation times were seen in Tg mice expressing $1 \times$ MHu2M PrP or $10 \times$ HuPrP on a $Prnp^{0/0}$ background. Earlier studies demonstrated that the level of PrP expression is inversely related to the length of the incubation time (Prusiner *et al.*, 1990).

6.1.5. Prion Strains

As discussed above, the strain of prion can greatly influence the interspecies transmissibility of prions. The existence of prion strains raised the question of how heritable biological information can be enciphered in a molecule other than nucleic acid (Dickinson *et al.*, 1968; Bruce and Dickinson, 1987; Ridley and Baker, 1996). Prion strains have been defined by incubation times, the distribution of neuronal vacuolation (Dickinson *et al.*, 1968), and patterns of PrP^{Sc} deposition (Bruce *et al.*, 1989; Hecker *et al.*, 1990). There is a wealth of evidence to support the thesis that prion diversity is enciphered in the conformation of PrP^{Sc} (Prusiner, 1991); thus, prion strains seem to represent distinct conformers of PrP^{Sc} (Bessen and Marsh, 1994; Telling *et al.*, 1996a; Scott *et al.*, 1997; Safar *et al.*, 1998; Peretz *et al.*, 2002). Persuasive evidence that strain-specific information is enciphered in the tertiary and quaternary structure of PrP^{Sc} comes from transmission of two different inherited human prion diseases, FFI and fCJD, to mice expressing the chimeric MHu2M PrP transgene (Telling *et al.*, 1996a). The deglycosylated PrP 27-30 fragment has a molecular size of 19 kDa in FFI and of 21 kDa in fCJD(E200K) and most sporadic prion diseases (Table 6.2) (Parchi *et al.*, 1996). This difference in molecular size was shown to be due to different sites of proteolytic cleavage at the N-termini of two human PrP^{Sc} molecules, which reflects different tertiary structures. These distinct conformations were not unexpected because the amino acid sequences of the PrPs differ. Using Tg(MHu2M) mice, extracts from the brains of FFI patients transmitted disease approximately 200 days after inoculation and induced formation of 19-kDa PrP^{Sc}, whereas homogenates from fCJD(E200K) and sporadic CJD (sCJD) patients produced the 21-kDa PrP^{Sc} (Telling *et al.*, 1996a). On second passage, Tg mice inoculated with FFI prions showed an incubation time of ~130 days and a 19-kDa PrP^{Sc} fragment, whereas Tg mice inoculated with fCJD(E200K) prions exhibited an incubation time of ~170 days and a 21-kDa PrP^{Sc} fragment (Table 6.2) (Prusiner, 1998). The experimental data demonstrate that chimeric PrP^{Sc} can exist in two different conformations based on the sizes of the PrP 27-30 fragments. This analysis was extended when patients with sporadic fatal insomnia (sFI) were found. Although they did not carry a PrP gene mutation, sFI patients had a clinical and pathologic phenotype that was indistinguishable from that of FFI patients. Furthermore, a 19-kDa PrP^{Sc} fragment was found in their brains, which was also found when prion disease was passaged to Tg(MHu2M) mice (Mastrianni *et al.*, 1999). These findings contend that the disease phenotype is dictated by the conformation of PrP^{Sc} and not by the amino acid sequence. The results also demand that PrP^{Sc} acts as a template for the conversion of PrP^{C} into nascent PrP^{Sc}.

In another set of studies, two prion strains, Sc237 and DY, that had been serially passaged in Syrian hamsters (SHa) were inoculated into Tg mice expressing a transgene encoding a chimeric SHa/Mo PrP, designated MH2M (Peretz *et al.*, 2002). On first passage in Tg(MH2M)*Prnp*$^{0/0}$ mice, the SHa(Sc237) prions exhibited prolonged incubation times, followed by a profound shortening on second passage. Moreover, the PrP^{Sc} of the MH2M(Sc237) strain possesses structural properties that differ from those of the SHa(Sc237) strain, as demonstrated

Table 6.2. Distinct prion strains generated in humans with inherited prion diseases and transmitted to transgenic mice

Inoculum	Host	PrP sequence	Incubation time (days ± SEM) [n/n^0]*	PrP^{Sc} (kDa)
None	Human	FFI (M129, D178N)		19
FFI	Mouse	MHu2M	206 ± 7 [7/7]	19
FFI → Tg(MHu2M)	Mouse	MHu2M	136 ± 1 [6/6]	19
None	Human	fCJD (E200K)		21
fCJD	Mouse	MHu2M	170 ± 2 [10/10]	21
fCJD → Tg(MHu2M)	Mouse	MHu2M	167 ± 3 [15/15]	21

* Number of sick animals/number of inoculated animals.

by relative conformational stability measurements (Peretz *et al.*, 2001). Conversely, transmission of SHa(DY) prions to Tg(MH2M)*Prnp*$^{0/0}$ mice showed no change in the incubation time, and the MH2M(DY) strain retained the conformational and phenotypic properties of SHa(DY) prions.

6.2. Prion Diseases of Animals

The prion diseases of animals include scrapie of sheep and goats, BSE, transmissible mink encephalopathy (TME), CWD of mule deer and elk, feline spongiform encephalopathy, and exotic ungulate encephalopathy (see Table 6.1).

6.2.1. PrP Polymorphisms in Sheep, Cattle, and Elk

Studies of natural scrapie in the United States have shown that 85% of afflicted sheep are of the Suffolk breed. Only those Suffolk sheep homozygous for Gln (Q) at codon 171 developed scrapie although healthy controls with Gln/Gln, Gln/Arg, and Arg/Arg genotypes were also found (Goldmann *et al.*, 1994; Westaway *et al.*, 1994b; Belt *et al.*, 1995; Clousard *et al.*, 1995; Hunter *et al.*, Ikeda *et al.*, 1995; O'Rourke *et al.*, 1997). These results argue that susceptibility in Suffolk sheep is governed by the PrP codon 171 polymorphism. As with the Suffolk breed, the PrP codon 171 polymorphism in Cheviot sheep has a profound influence on susceptibility to scrapie, and codon 136 also modulates susceptibility but less so than codon 171 (Hunter *et al.*, 1991; Goldmann *et al.*, 1991a).

In contrast with sheep, different breeds of cattle have no specific PrP polymorphisms. The only polymorphism recorded in cattle is a variation in the number of octarepeats: most cattle, like humans, have five octarepeats but some have six (Goldmann *et al.*, 1991b; Prusiner *et al.*, 1993a). However, the presence of six octarepeats does not seem to be overrepresented in BSE (Hunter *et al.*, 1994).

In studies of CWD, the susceptibility of elk, but not deer, seems to be modulated by codon 132, which corresponds with codon 129 in humans. Elk with CWD consistently express Met/Met at position 132; no elk with CWD expressing leucine at this residue have been found (O'Rourke *et al.*, 1999).

6.2.2. Bovine Spongiform Encephalopathy

Prion strains and the transmission barriers are of great importance in understanding the BSE epidemic in Britain, in which it is estimated that between 1 million and 2 million cattle were infected with prions (Anderson *et al.*, 1996; Nathanson *et al.*, 1997). The mean incubation time for BSE is approximately 5 years. Therefore, most cattle did not manifest disease because they were slaughtered between 2 and 3 years of age (Stekel *et al.*, 1996). Nevertheless, more than 180,000 cattle, primarily dairy cows, have died of BSE over the past decade (Anderson *et al.*, 1996; Phillips *et al.*, 2000). BSE is a massive common source epidemic caused by meat and bone meal (MBM) fed primarily to dairy cows (Wilesmith *et al.*, 1991; Nathanson *et al.*, 1997). MBM was prepared from the offal of sheep, cattle, pigs, and chickens as a high-protein nutritional supplement. In the late 1970s, the hydrocarbon-solvent extraction method used in the rendering of offal began to be abandoned, resulting in MBM with a much higher fat content (Wilesmith *et al.*, 1991). It is now thought that this change allowed scrapie prions from sheep to survive the rendering process and to be passed into cattle. Alternatively, bovine prions may have been present at low levels prior to modification of the rendering process and with the processing change, survived in sufficient numbers to initiate the BSE epidemic when re-introduced into cattle through ingestion of MBM (Phillips *et al.*, 2000). Perhaps a particular conformation of BoPrP^{Sc} selected for heat resistance during the rendering process and then res-elected multiple times as cattle infected by ingesting prion-contaminated MBM were slaughtered and their offal rendered into more MBM. Against the latter hypothesis is the widespread geographic distribution of the initial 17 cases of BSE throughout England, which occurred almost simultaneously (Wilesmith, 1991; Kimberlin, 1996; Nathanson *et al.*, 1997). In July 1988, the practice of feeding MBM to sheep and cattle was banned. Although statistical analyses demonstrate that the epidemic is disappearing as a result of this ruminant feed ban (Anderson *et al.*, 1996; Phillips *et al.*, 2000), it is unclear how many current cases of BSE are due to infection and how many arise spontaneously.

The origin of BSE prions cannot be determined by examining the amino acid sequence of PrP^{Sc} from cattle with BSE because it has the bovine sequence regardless of whether the initial prions originated from sheep or cattle. The bovine and sheep PrP sequences differ at seven or eight positions (Goldmann *et al.*, 1990, 1991b; Prusiner *et al.*, 1993a). Brain extracts from cattle with BSE cause disease in cattle, sheep, mice, pigs, and mink after intracerebral inoculation (Fraser *et al.*, 1988; Dawson *et al.*, 1990a, b; Bruce *et al.*, 1993, 1997; Wells *et al.*, 2003), but prions in brain extracts from sheep with scrapie fed to cattle produced an illness substantially different from BSE (Robinson *et al.*, 1995). However, no exhaustive effort has been made to test different strains of sheep prions or to examine the disease after bovine-to-bovine passage.

6.2.3. Chronic Wasting Disease

Mule deer, white-tailed deer, and elk have been reported to develop CWD, which is a prion disease unique among other prion diseases because it seems to be far more communicable than scrapie, BSE, or TME; moreover, it is the only prion disease known in free-ranging animals. CWD was first described in 1967 and was reported to be a spongiform encephalopathy in 1978 based on histopathology in the brain. CWD has been found in the United States, Canada, and South Korea. In the United States, CWD has been reported in Colorado, Wyoming, South Dakota, Nebraska, Oklahoma, Montana, New Mexico, Minnesota, Wisconsin, Utah, Illinois, Kansas, New York, West Virginia, and Michigan. In captive cervid herds, up to 90% of mule deer have been reported to be positive for prions (Williams and Young, 1980; Miller *et al.*, 1998; Peters *et al.*, 2000; Williams and Miller, 2002) and up to 60% of elk in Colorado and Wyoming develop CWD. Moreover, the incidence of CWD in cervids living in the wild has been estimated to be as high as 15% (Miller *et al.*, 2000). The mode of transmission of CWD prions among mule deer, white-tailed deer, and elk is unresolved, but contamination of grass with prions excreted in fecal matter seems to be a likely source (Williams and Miller, 2002). The high content of PrP^{Sc} in the intestinal lymphoid tissue of cervids with CWD supports such a scenario (Sigurdson *et al.*, 1999).

Brain homogenates from mule deer with CWD have transmitted disease to 4 of 13 cattle after intracerebral inoculation (Hamir *et al.*, 2001). These findings are particularly important because there is great concern that CWD prions might be transmitted to cattle grazing in contaminated pastures. In addition, CWD has been transmitted to ferrets, mink, squirrel monkeys, goats, and mice after intracerebral inoculation (Williams and Miller, 2002); however, only mule deer demonstrate efficient transmission of CWD prions by intracerebral inoculation. Recently, Tg mice expressing cervid PrP protein have been shown to be susceptible to CWD prions (Browning *et al.*, 2004; Tamgüney *et al.*, 2006).

6.3. Human Prion Diseases

6.3.1. Sporadic and Genetic Neurodegenerative Diseases: Prion Diseases

Initiation of sporadic disease may follow from a somatic mutation and thus develop in a manner similar to that for germ-line mutations in inherited disease. In this situation, the mutant PrP^{Sc} must be capable of recruiting wt PrP^{C}, a process known to be possible for some mutations (e.g., E200K, D178N) but less likely for others (e.g., P102L) (Telling *et al.*, 1995). More than 30 different mutations in *PRNP* resulting in nonconservative substitutions, missense mutation, and expansions in the octapeptide repeat region of the gene have been found to segregate with inherited human prion diseases (Kong *et al.*, 2004).

Although phenotypes may vary dramatically within families, there is a tendency for certain phenotypes to associate with specific mutations (Kong *et al.*, 2004). A clinical phenotype indistinguishable from sCJD is usually seen with substitutions at codons 180, 183, 200, 208, 210, or 232. Substitutions at codons 117, 102, 105, 198, or 217 are associated with GSS disease. The normal human PrP sequence contains five repeats of an octapeptide sequence. Insertions of two to nine extra octapeptide repeats are associated with variable phenotypes, ranging from a condition indistinguishable from CJD to a slowly progressive dementing illness of many years' duration. A mutation at codon 178, resulting in the substitution of asparagine for aspartate, produces FFI if a methionine is encoded at the polymorphic 129 residue on the same allele (Goldfarb *et al.*, 1992); typical CJD is seen if a valine is encoded at position 129 of the mutant allele. Particularly puzzling are the factors that determine the disease phenotype. As noted above, there is excellent evidence that the tertiary structure of PrP^{Sc} determines whether this disease-causing protein is deposited in subsets of thalamic neurons, producing fatal insomnia, or deposited more widely resulting in CJD. As more cases of inherited prion disease have been studied, multiple exceptions have been recorded, such as the discovery that a mutation that produces GSS with ataxia also results in a dementing illness and vice versa (Hsiao *et al.*, 1990; Mastrianni *et al.*, 1995). Within a single family carrying a PrP mutation, some patients develop peripheral neuropathies whereas others do not. Elucidating all of the factors that govern clinical and neuropathologic phenotypes will be challenging.

6.3.2. PrP Gene Polymorphisms and Dominant-Negative Inhibition

Polymorphisms influence the susceptibility to sporadic, inherited, and infectious forms of prion disease. The methionine/valine polymorphism at position 129 not only determines the clinical phenotype as noted above but also modulates the age of onset of some inherited prion diseases (Gambetti *et al.*, 1999; Palmer *et al.*, 1991). The finding that homozygosity at codon 129 predisposes a patient to sCJD supports a model of prion production that favors homologous PrP interactions as appears to occur in Tg mice expressing SHaPrP inoculated with either hamster prions or mouse prions (Prusiner *et al.*, 1990) as well as in Tg mice expressing MH2M PrP inoculated with novel prions. A lysine at 219 has been found in 12% of the Japanese population who seem to be resistant to prion disease (Shibuya *et al.*, 1998). In scrapie-infected neuroblastoma (ScN2a) cells and Tg mice, substitution of lysine at position 219 produced dominant-negative inhibition of prion replication and prevented PrP^{C} from being converted into PrP^{Sc} (Kaneko *et al.*, 1997b; Perrier *et al.*, 2002; Zulianello *et al.*, 2000). Dominant-negative inhibition of prion replication has been found in sheep with the substitution of basic residue arginine at position 171; these sheep are resistant to scrapie (Westaway *et al.*, 1994b; Hunter *et al.*, 1997b). Dominant-negative inhibition by PrP^{C}(171R) has been demonstrated in both ScN2a cells and Tg mice.

6.3.3. Infectious Prion Diseases

Although infectious prion diseases constitute less than 1% of all cases, the circumstances surrounding these infectious illnesses are often dramatic (see Table 6.1) (Will *et al.*, 1999a). The ritualistic cannibalism–involved transmission of kuru among the Fore people of New Guinea, the industrial cannibalism responsible for "mad cow disease" in Europe, and the growing group of patients with vCJD, contracted from prion-tainted beef products, are all examples of infectious prion diseases.

6.3.3.1. Kuru

Kuru has disappeared, but the impact of this illness on medical science remains immense. Kuru was the first human prion disease to be transmitted to experimental animals (Gajdusek *et al.*, 1966). Based on the similarities of the neuropathologic lesions of scrapie and kuru (Hadlow, 1959), transmission studies to apes were performed and similar investigations followed with CJD (Klatzo *et al.*, 1959; Gibbs *et al.*, 1968). The transmissibility of kuru suggested that the disease resulted from ritualistic cannibalism among the Fore people living in the highlands of New Guinea (Alpers, 1968). Presumably, kuru began with a person who had developed a sporadic case of prion disease; at death, the brain was removed and ingested by relatives in order to immortalize the spirit of the deceased.

6.3.3.2. Iatrogenic Prion Disease

Accidental transmission of CJD to humans appears to have occurred with corneal transplantation (Duffy *et al.*, 1974) and contaminated electroencephalograph (EEG) electrode implantation (Bernoulli *et al.*, 1977). Corneas were removed from donors who unknowingly had CJD and were transplanted to apparently healthy recipients who developed CJD after prolonged incubation periods. Improperly decontaminated EEG electrodes that caused CJD in two young patients with intractable epilepsy were experimentally implanted in a chimpanzee and were found to cause CJD 18 months later (Gibbs *et al.*, 1994). Surgical procedures may have resulted in accidental prion infection in patients (Gajdusek *et al.*, 1977; Will and Matthews, 1982; Collins *et al.*, 1999), presumably because some instrument or apparatus in the operating theater became contaminated when a CJD patient underwent surgery. Although the epidemiology of these studies is highly suggestive, no proof for such episodes exists.

6.3.3.3. Dura Mater Grafts

More than 100 cases of CJD after implantation of dura mater grafts have been recorded (Centers for Disease Control, 1997). All of the grafts were thought to have been acquired from a single manufacturer whose preparative procedures were inadequate to inactivate human prions. One case of CJD occurred after repair of an eardrum perforation with a pericardium graft (Tange *et al.*, 1989).

6.3.3.4. Human Growth Hormone Therapy

The possibility of transmission of CJD from contaminated human growth hormone (HGH) preparations derived from human pituitaries has been raised by the occurrence of fatal cerebellar disorders with dementia in more than 150 patients ranging in age from 10 to 41 years (Fradkin *et al.*, 1991; Mills *et al.*, 2004; Will *et al.*, 2004). These patients received injections of HGH every 2 to 4 days for 4 to 12 years (PHS, 1997). Assuming these patients developed CJD from injections of prion-contaminated HGH preparations, the possible incubation periods range from

4 to 30 years. Even though several investigations argue for the efficacy of inactivating prions in HGH fractions prepared from human pituitaries using 6 M urea, it seems doubtful that such protocols will be used for purifying HGH because recombinant HGH is available. Four cases of CJD have occurred in women receiving human pituitary gonadotropin (Cochius *et al.*, 1990).

6.3.3.5. Variant Creutzfeldt-Jakob Disease

The restricted geographic occurrence and chronology of vCJD have raised the possibility that BSE prions have been transmitted to humans. More than 190 vCJD cases have been recorded, but the incidence has been too low to be useful in establishing the origin of the disease (Will *et al.*, 1999b; Balter, 2000). No set of dietary habits distinguishes vCJD patients from apparently healthy people. Moreover, there is no explanation for the predilection of vCJD for teenagers and young adults. It is noteworthy that epidemiologic studies over the past three decades have failed to find evidence for transmission of sheep prions to humans (Cousens *et al.*, 1990). Attempts to predict the future number of cases of vCJD, assuming exposure to bovine prions prior to the offal ban, have been questioned because so few cases of vCJD have occurred (Ghani *et al.*, 2000). Are we at the beginning of a human prion disease epidemic in Great Britain like those seen for BSE and kuru, or will the number of vCJD cases remain small as seen with iCJD caused by cadaveric HGH? Although the mechanism of infection of vCJD has not been established yet, mounting evidence argues for transmission of bovine prions to humans. This conclusion is based on multiple lines of inquiry including (i) the spatial-temporal clustering of vCJD (Will *et al.*, 1996; Zeidler *et al.*, 1997); (ii) the successful transmission of BSE to macaques with induction of PrP plaques similar to those seen in vCJD (Lasmézas *et al.*, 1996); (iii) the similarity of the glycoform pattern of brain but not tonsil PrP^{Sc} in vCJD to that in cattle, mice, domestic cats, and macaques infected with BSE prions (Hill *et al.*, 1997, 1999); and (iv) transmission studies in non-Tg mice suggesting that vCJD and BSE represent the same prion strain (Bruce *et al.*, 1997). The prolonged incubation periods and inefficient transmission of prions that are seen after inoculation of foreign prions into non-Tg mice can readily confuse the interpretation of the findings (Lasmézas *et al.*, 1997; Prusiner, 1998).

The most compelling evidence that vCJD is caused by BSE prions has come from experiments using mice expressing the BoPrP transgene (Scott *et al.*, 1999, 2005). The incubation times, neuropathologic profiles, and patterns of PrP^{Sc} deposition in these Tg(BoPrP) mice were indistinguishable whether the inoculum originated from the brains of cattle with BSE or of humans with vCJD (Scott *et al.*, 1999, 2005). Neither sCJD nor fCJD(E200K) prions have transmitted disease to Tg(BoPrP) mice. In contrast with sCJD and fCJD(E200K) prions, which transmit disease efficiently to mice expressing a chimeric human-mouse PrP transgene, vCJD prions do not. Interestingly, none of the Tg mice expressing chimeric human-mouse PrP and inoculated with BSE prions from cattle developed disease after more than 600 days.

Because Tg(BoPrP) mice are also excellent hosts for transmission of natural sheep scrapie, it raises the possibility that sheep carry several strains of prions, including the BSE strain (Scott *et al.*, 2005). But the BSE strain might replicate more slowly than many scrapie strains in sheep and thus is present at low levels. If the scrapie strains were more heat-labile than the BSE strain, then the rendering process used in the late 1970s and most of the next decade might have selected for the BSE strain. This BSE strain then might have been reselected multiple times as cattle were infected through ingestion of prion-contaminated MBM. In this situation, the infected cattle were slaughtered and their offal rendered into more MBM, which was subsequently fed to more cattle.

The foregoing scenario suggests that the BSE strain may be widely distributed in sheep, and these prions composed of sheep PrP^{Sc} are non-pathogenic to humans. Once the BSE prions are passaged through cattle, they acquire bovine PrP^{Sc} and become pathogenic to humans.

6.4. Mammalian Synthetic Prions

Besides establishing beyond reasonable doubt that prions are infectious proteins, the production of synthetic prions opens many new approaches to the investigation of prion structure and biology. Prior to the production of wt synthetic prions using the 142-residue protein recMoPrP(89–230) (Legname *et al.*, 2004), we used a 55mer synthetic peptide carrying the GSS mutation at MoPrP residue 101 to create synthetic prions (Kaneko *et al.*, 2000; Tremblay *et al.*, 2004). The strains of mutant synthetic prions were protease-sensitive and required Tg196 mice expressing low levels of MoPrP(P101L) to demonstrate prion infectivity. Moreover, the mutant prions in Tg196 mice could not be passaged into wt mice. Development of a system using recMoPrP(89–230) to produce synthetic prions has facilitated investigations because these prions are protease-resistant and can be passaged in wt mice.

The use of amyloid fibrils formed from recMoPrP(89–230) mirrors work with fungal prions where recombinant proteins were polymerized into amyloid fibrils and introduced into fungi (King and Diaz-Avalos, 2004; Maddelein *et al.*, 2002; Sparrer *et al.*, 2000; Tanaka *et al.*, 2004). The ability to assay rapidly for prion infectivity in fungi greatly facilitated these studies.

The mouse synthetic prion strain 1 (MoSP1) differs from all strains of prions that we have studied to date. Incubation times, neuropathologic profiles of vacuolation, and conformational stability studies showed that MoSP1 prions differ from 20 other strains examined (Legname *et al.*, 2005). Defining the range of strain-specified characteristics using synthetic prions will be of considerable interest.

The foregoing investigations were directed at producing wt mouse synthetic prions using the formation of PrP amyloid as a surrogate marker for the folding of MoPrP(89–230) into a biologically active conformation. The rapidity and ease of measuring thioflavine T binding that reflects amyloid formation (Come *et al.*, 1993; LeVine, 1993) permitted us to determine conditions under which recMoPrP(89–230) would assemble into amyloid fibrils (Baskakov *et al.*, 2002). The finding that such fibrils harbor detectable levels of prion infectivity allows us to draw a series of conclusions about mammalian prions that were previously elusive. First, PrP is both necessary and sufficient for infectivity; prions are infectious proteins. Second, neither the Asn-linked oligosaccharides nor the GPI anchor are required for prion infectivity because the recMoPrP(89–230) used in the experiments contains neither of these post-translational modifications (Taraboulos *et al.*, 1990; DeArmond *et al.*, 1997; Gambetti and Parchi, 1999; Mastrianni *et al.*, 1999). Whether the Asn-linked oligosaccharides or the GPI anchor increase the efficiency of PrP^{Sc} formation for particular prion strains is unknown (Collinge *et al.*, 1996). Third, the biological information carried by distinct strains of prions resides in PrP^{Sc} as previously suggested (Prusiner, 1991; Bessen and Marsh, 1994; Telling *et al.*, 1996a; Caughey *et al.*, 1998; Peretz *et al.*, 2002). Fourth, the spontaneous formation of prions, which is thought to be responsible for sporadic forms of prion disease in livestock and humans, can occur in any mammal expressing PrP^{C} (Prusiner, 1989; Cohen and Prusiner, 1998).

6.5. Conclusion

The formation of prions from a 55mer synthetic peptide and recMoPrP(89–230) demonstrates that PrP^{C} is sufficient for the spontaneous formation of prions, and thus, no exogenous molecule is required for prions to form in any mammal. These findings provide an explanation for sporadic CJD, for which the spontaneous formation of prions has been hypothesized (Prusiner, 1989). Understanding sporadic prion disease is particularly relevant to controlling the exposure of humans to bovine prions (Biacabe *et al.*, 2004; Casalone *et al.*, 2004; Yamakawa *et al.*, 2003). That bovine prions are pathogenic for humans is well documented in the cases of more than 190 teenagers and young adults who have already died from prion-tainted beef derived from cattle with BSE (Will *et al.*, 1996, 2004; Scott *et al.*, 1999). Moreover, the sporadic forms of Alzheimer's and Parkinson's diseases as well as amyotrophic lateral sclerosis and the frontal temporal dementias are the most frequent forms of these age-dependent disorders as is the case for the prion diseases (Hebert *et al.*, 2003). Important insights into the etiologic events that feature in these more common neurodegenerative disorders, all of which are caused by the aberrant processing of proteins in the nervous system, are likely to emerge as more is learned about the molecular pathogenesis of sporadic prion diseases (Prusiner, 2001).

Acknowledgments

The authors wish to thank Dr. Kurt Giles for helpful discussions. The authors also wish to thank Dr. Cedric Govaerts for kindly providing Figures 6.1 and 6.3. This work was supported by grants from the National Institutes of Health (AG02132, AG10770, and AG021601) as well as by a gift from the G. Harold and Leila Y. Mathers Charitable Foundation.

References

Aguzzi, A., Montrasio, F., and Kaeser, P. S. (2001). Prions: health scare and biological challenge. *Nat Rev Mol Cell Biol* 2: 118–126.

Alper, T., Cramp, W. A., Haig, D. A., and Clarke, M. C. (1967). Does the agent of scrapie replicate without nucleic acid? *Nature* 214:764–766.

Alper, T., Haig, D. A., and Clarke, M. C. (1966). The exceptionally small size of the scrapie agent. *Biochem Biophys Res Commun* 22:278–284.

Alpers, M. P. (1968). Kuru: implications of its transmissibility for the interpretation of its changing epidemiological pattern In The Central Nervous System: Some Experimental Models of Neurological Diseases, O. T. Bailey, and D. E. Smith, eds. (Baltimore: Williams and Wilkins Company), pp. 234–251.

Anderson, R. M., Donnelly, C. A., Ferguson, N. M., Woolhouse, M. E. J., Watt, C. J., Udy, H. J., MaWhinney, S., Dunstan, S. P., Southwood, T. R. E., Wilesmith, J. W., *et al.* (1996). Transmission dynamics and epidemiology of BSE in British cattle. *Nature* 382:779–788.

Asante, E. A., Linehan, J. M., Desbruslais, M., Joiner, S., Gowland, I., Wood, A. L., Welch, J., Hill, A. F., Lloyd, S. E., Wadsworth, J. D., and Collinge, J. (2002). BSE prions propagate as either variant CJD-like or sporadic CJD-like prion strains in transgenic mice expressing human prion protein. *EMBO* J 21:6358–6366.

Balter, M. (2000). Tracking the human fallout from "mad cow disease." *Science* 289:1452–1454.

Baskakov, I. V., Legname, G., Baldwin, M. A., Prusiner, S. B., and Cohen, F. E. (2002). Pathway complexity of prion protein assembly into amyloid. *J Biol Chem* 277:21140–21148.

Basler, K., Oesch, B., Scott, M., Westaway, D., Wälchli, M., Groth, D. F., McKinley, M. P., Prusiner, S. B., and Weissmann, C. (1986). Scrapie and cellular PrP isoforms are encoded by the same chromosomal gene. *Cell* 46:417–428.

Belt, P. B., Muileman, I. H., Schreuder, B. E. C., Ruijter, J. B., Gielkens, A. L. J., and Smits, M. A. (1995). Identification of five allelic variants of the sheep PrP gene and their association with natural scrapie. *J Gen Virol* 76:509–517.

Bernoulli, C., Siegfried, J., Baumgartner, G., Regli, F., Rabinowicz, T., Gajdusek, D. C., and Gibbs, C. J., Jr. (1977). Danger of accidental person-to-person transmission of Creutzfeldt-Jakob disease by surgery. *Lancet* 1: 478–479.

Bessen, R. A., and Marsh, R. F. (1994). Distinct PrP properties suggest the molecular basis of strain variation in transmissible mink encephalopathy. *J Virol* 68: 7859–7868.

Biacabe, A. G., Laplanche, J. L., Ryder, S., and Baron, T. (2004). Distinct molecular phenotypes in bovine prion diseases. *EMBO Rep* 5: 110–115.

Brown, D. R., Qin, K., Herms, J. W., Madlung, A., Manson, J., Strome, R., Fraser, P. E., Kruck, T., von Bohlen, A., Schulz-Schaeffer, W., *et al.* (1997a). The cellular prion protein binds copper *in vivo*. *Nature* 390: 684–687.

Brown, D. R., Schulz-Schaeffer, W. J., Schmidt, B., and Kretzschmar, H. A. (1997b). Prion protein-deficient cells show altered response to oxidative stress due to decreased SOD-1 activity. *Exp Neurol* 146: 104–112.

Browning, S. R., Mason, G. L., Seward, T., Green, M., Eliason, G. A., Mathiason, C., Miller, M. W., Williams, E. S., Hoover, E., and Telling, G. C. (2004). Transmission of prions from mule deer and elk with chronic wasting disease to transgenic mice expressing cervid PrP. *J Virol* 78: 13345–13350.

Bruce, M., Chree, A., McConnell, I., Foster, J., and Fraser, H. (1993). Transmissions of BSE, scrapie and related diseases to mice (Abstr.). Proceedings of the IXth International Congress of Virology, Glasgow, Scotland, Aug. 9–12, p. 93.

Bruce, M. E., and Dickinson, A. G. (1987). Biological evidence that the scrapie agent has an independent genome. *J Gen Virol* 68: 79–89.

Bruce, M. E., McBride, P. A., and Farquhar, C. F. (1989). Precise targeting of the pathology of the sialoglycoprotein, PrP, and vacuolar degeneration in mouse scrapie. *Neurosci Lett* 102:1–6.

Bruce, M. E., Will, R. G., Ironside, J. W., McConnell, I., Drummond, D., Suttie, A., McCardle, L., Chree, A., Hope, J., Birkett, C., *et al.* (1997). Transmissions to mice indicate that 'new variant' CJD is caused by the BSE agent. *Nature* 389: 498–501.

Burns, C. S., Aronoff-Spencer, E., Legname, G., Prusiner, S. B., Antholine, W. E., Gerfen, G. J., Peisach, J., and Millhauser, G. L. (2003). Copper coordination in the full-length, recombinant prion protein. *Biochemistry* 42:6794–6803.

Büeler, H., Aguzzi, A., Sailer, A., Greiner, R.-A., Autenried, P., Aguet, M., and Weissmann, C. (1993). Mice devoid of PrP are resistant to scrapie. *Cell* 73:1339–1347.

Büeler, H., Fisher, M., Lang, Y., Bluethmann, H., Lipp, H.-P., DeArmond, S. J., Prusiner, S. B., Aguet, M., and Weissmann, C. (1992). Normal development and behavior of mice lacking the neuronal cell-surface PrP protein. *Nature* 356: 577–582.

Casalone, C., Zanusso, G., Acutis, P., Ferrari, S., Capucci, L., Tagliavini, F., Monaco, S., and Caramelli, M. (2004). Identification of a second bovine amyloidotic spongiform encephalopathy: Molecular similarities with sporadic Creutzfeldt-Jakob disease. *Proc Natl Acad Sci USA* 101:3065–3070.

Caughey, B. (1991). Cellular metabolism of PrP. Prion Diseases in Humans and Animals Conference, London, UK.

Caughey, B., Raymond, G. J., and Bessen, R. A. (1998). Strain-dependent differences in β-sheet conformations of abnormal prion protein. *J Biol Chem* 273:32230–32235.

Centers for Disease Control (1997). Creutzfeldt-Jakob disease associated with cadaveric dura mater grafts—Japan, January 1979–May 1996. *MMWR Morb Mortal Wkly Rep* 46:1066–1069.

Chen, S., Mange, A., Dong, L., Lehmann, S., and Schachner, M. (2003). Prion protein as trans-interacting partner for neurons is involved in neurite outgrowth and neuronal survival. *Mol Cell Neurosci* 22:227–233.

Clousard, C., Beaudry, P., Elsen, J. M., Milan, D., Dussaucy, M., Bounneau, C., Schelcher, F., Chatelain, J., Launay, J.-M., and Laplanche, J.-L. (1995). Different allelic effects of the codons 136 and 171 of the prion protein gene in sheep with natural scrapie. *J Gen Virol* 76:2097–2101.

Cochius, J. I., Mack, K., Burns, R. J., Alderman, C. P., and Blumbergs, P. C. (1990). Creutzfeldt-Jakob disease in a recipient of human pituitary-derived gonadotrophin. *Aust N Z J Med* 20:592–593.

Cohen, F. E., and Prusiner, S. B. (1998). Pathologic conformations of prion proteins. *Annu Rev Biochem* 67:793–819.

Collinge, J., Sidle, K. C. L., Meads, J., Ironside, J., and Hill, A. F. (1996). Molecular analysis of prion strain variation and the aetiology of "new variant" CJD. *Nature* 383:685–690.

Collins, S., Law, M. G., Fletcher, A., Boyd, A., Kaldor, J., and Masters, C. L. (1999). Surgical treatment and risk of sporadic Creutzfeldt-Jakob disease: a case-control study. *Lancet* 353:693–697.

Come, J. H., Fraser, P. E., and Lansbury, P. T., Jr. (1993). A kinetic model for amyloid formation in the prion diseases: importance of seeding. *Proc Natl Acad Sci USA* 90:5959–5963.

Cousens, S. N., Harries-Jones, R., Knight, R., Will, R. G., Smith, P. G., and Matthews, W. B. (1990). Geographical distribution of cases of Creutzfeldt-Jakob disease in England and Wales 1970–84. *J Neurol Neurosurg Psychiatry* 53, 459–465.

Creutzfeldt, H. G. (1920). Über eine eigenartige herdförmige Erkrankung des Zentralnervensystems. *Z Gesamte Neurol Psychiatrie* 57:1–18.

Dawson, M., Wells, G. A. H., and Parker, B. N. J. (1990a). Preliminary evidence of the experimental transmissibility of bovine spongiform encephalopathy to cattle. *Vet Rec* 126:112–113.

Dawson, M., Wells, G. A. H., Parker, B. N. J., and Scott, A. C. (1990b). Primary parenteral transmission of bovine spongiform encephalopathy to the pig. *Vet Rec* 127:338.

DeArmond, S. J., Sánchez, H., Yehiely, F., Qiu, Y., Ninchak-Casey, A., Daggett, V., Camerino, A. P., Cayetano, J., Rogers, M., Groth, D., *et al.* (1997). Selective neuronal targeting in prion disease. *Neuron* 19:1337–1348.

Dickinson, A. G., Meikle, V. M. H., and Fraser, H. (1968). Identification of a gene which controls the incubation period of some strains of scrapie agent in mice. *J Comp Pathol* 78:293–299.

Duffy, P., Wolf, J., Collins, G., DeVoe, A. G., Streeten, B., and Cowen, D. (1974). Possible person-to-person transmission of Creutzfeldt-Jakob disease. *N Engl J Med* 290:692–693.

Edenhofer, F., Rieger, R., Famulok, M., Wendler, W., Weiss, S., and Winnacker, E.-L. (1996). Prion protein PrP interacts with molecular chaperones of the Hsp60 family. *J Virol* 70:4724–4728.

Fradkin, J. E., Schonberger, L. B., Mills, J. L., Gunn, W. J., Piper, J. M., Wysowski, D. K., Thomson, R., Durako, S., and Brown, P. (1991). Creutzfeldt-Jakob disease in pituitary growth hormone recipients in the United States. *JAMA* 265: 880–884.

Fraser, H., McConnell, I., Wells, G. A. H., and Dawson, M. (1988). Transmission of bovine spongiform encephalopathy to mice. *Vet Rec* 123:472.

Gajdusek, D. C., Gibbs, C. J., Jr., and Alpers, M. (1966). Experimental transmission of a kuru-like syndrome to chimpanzees. *Nature* 209:794–796.

Gajdusek, D. C., Gibbs, C. J., Jr., Asher, D. M., Brown, P., Diwan, A., Hoffman, P., Nemo, G., Rohwer, R., and White, L. (1977). Precautions in medical care of, and in handling materials from, patients with transmissible virus dementia (Creutzfeldt-Jakob disease). *N Engl J Med* 297:1253–1258.

Gambetti, P., and Parchi, P. (1999). Insomnia in prion diseases: sporadic and familial. *N Engl J Med* 340:1675–1677.

Gambetti, P., Peterson, R. B., Parchi, P., Chen, S. G., Capellari, S., Goldfarb, L., Gabizon, R., Montagna, P., Lugaresi, E., Piccardo, P., and Ghetti, B. (1999). Inherited prion diseases, In Prion Biology and Diseases, S. B. Prusiner, ed. (Cold Spring Harbor: Cold Spring Harbor Laboratory Press), pp. 509–583.

Ghani, A. C., Ferguson, N. M., Donnelly, C. A., and Anderson, R. M. (2000). Predicted vCJD mortality in Great Britain. *Nature* 406:583–584.

Gibbs, C. J., Jr. (1992). Spongiform encephalopathies—slow, latent, and temperate virus infections—in retrospect, In Prion Diseases of Humans and Animals, S. B. Prusiner, J. Collinge, J. Powell, and B. Anderton, eds. (London: Ellis Horwood), pp. 53–62.

Gibbs, C. J., Jr., Asher, D. M., Kobrine, A., Amyx, H. L., Sulima, M. P., and Gajdusek, D. C. (1994). Transmission of Creutzfeldt-Jakob disease to a chimpanzee by electrodes contaminated during neurosurgery. *J Neurol Neurosurg Psychiatry* 57:757–758.

Gibbs, C. J., Jr., Gajdusek, D. C., Asher, D. M., Alpers, M. P., Beck, E., Daniel, P. M., and Matthews, W. B. (1968). Creutzfeldt-Jakob disease (spongiform encephalopathy): transmission to the chimpanzee. *Science* 161:388–389.

Gibbs, C. J., Jr., Gajdusek, D. C., and Latarjet, R. (1978). Unusual resistance to ionizing radiation of the viruses of kuru, Creutzfeldt-Jakob disease. *Proc Natl Acad Sci USA* 75:6268–6270.

Goldfarb, L. G., Brown, P., and Gajdusek, D. C. (1992). The molecular genetics of human transmissible spongiform encephalopathy, In Prion Diseases of Humans and Animals, S. B. Prusiner, J. Collinge, J. Powell, and B. Anderton, eds. (London: Ellis Horwood), pp. 139–153.

Goldfarb, L. G., Brown, P., Cervenakova, L., and Gajdusek, D. C. (1994). Genetic analysis of Creutzfeldt-Jakob disease and related disorders. *Philos Trans R Soc Lond B* 343:379–384.

Goldmann, W., Hunter, N., Manson, J., and Hope, J. (1990). The PrP gene of the sheep, a natural host of scrapie. Proceedings of the VIIIth International Congress of Virology, Berlin, Aug. 26–31, p. 284.

Goldmann, W., Hunter, N., Benson, G., Foster, J. D., and Hope, J. (1991a). Different scrapie-associated fibril proteins (PrP) are encoded by lines of sheep selected for different alleles of the Sip gene. *J Gen Virol* 72:2411–2417.

Goldmann, W., Hunter, N., Martin, T., Dawson, M., and Hope, J. (1991b). Different forms of the bovine PrP gene have five or six copies of a short, G-C-rich element within the protein-coding exon. *J Gen Virol* 72:201–204.

Goldmann, W., Hunter, N., Smith, G., Foster, J., and Hope, J. (1994). PrP genotype and agent effects in scrapie: change in allelic interaction with different isolates of agent in sheep, a natural host of scrapie. *J Gen Virol* 75:989–995.

Gorodinsky, A., and Harris, D. A. (1995). Glycolipid-anchored proteins in neuroblastoma cells form detergent-resistant complexes without caveolin. *J Cell Biol* 129:619–627.

Govaerts, C., Wille, H., Prusiner, S. B., and Cohen, F. E. (2004). Evidence for assembly of prions with left-handed β-helices into trimers. *Proc Natl Acad Sci USA* 101:8342–8347.

Graner, E., Mercadante, A. F., Zanata, S. M., Martins, V. R., Jay, D. G., and Brentani, R. (2000). Laminin-induced PC-12 cell differentiation is inhibited following laser inactivation of cellular prion protein. *FEBS Lett* 482:257–260.

Hadlow, W. J. (1959). Scrapie and kuru. *Lancet* 2:289–290.

Hamir, A. N., Cutlip, R. C., Miller, J. M., Williams, E. S., Stack, M. J., Miller, M. W., O'Rourke, K. I., and Chaplin, M. J. (2001). Preliminary findings on the experimental transmission of chronic wasting disease agent of mule deer to cattle. *J Vet Diagn Invest* 13:91–96.

Hebert, L. E., Scherr, P. A., Bienias, J. L., Bennett, D. A., and Evans, D. A. (2003). Alzheimer disease in the US population: prevalence estimates using the 2000 census. *Arch Neurol* 60:1119–1122.

Hecker, R., Stahl, N., Baldwin, M., Hall, S., McKinley, M. P., and Prusiner, S. B. (1990). Properties of two different scrapie prion isolates in the Syrian hamster. VIIIth International Congress of Virology, Berlin, Aug. 26–31, p. 284.

Hegde, R. S., Mastrianni, J. A., Scott, M. R., DeFea, K. A., Tremblay, P., Torchia, M., DeArmond, S. J., Prusiner, S. B., and Lingappa, V. R. (1998). A transmembrane form of the prion protein in neurodegenerative disease. *Science* 279, 827–834.

Hegde, R. S., Tremblay, P., Groth, D., Prusiner, S. B., and Lingappa, V. R. (1999). Transmissible and genetic prion diseases share a common pathway of neurodegeneration. *Nature* 402:822–826.

Hill, A. F., Desbruslais, M., Joiner, S., Sidle, K. C. L., Gowland, I., Collinge, J., Doey, L. J., and Lantos, P. (1997). The same prion strain causes vCJD and BSE. *Nature* 389:448–450.

Hill, A. F., Butterworth, R. J., Joiner, S., Jackson, G., Rossor, M. N., Thomas, D. J., Frosh, A., Tolley, N., Bell, J. E., Spencer, M., *et al.* (1999). Investigation of variant Creutzfeldt-Jakob disease and other human prion diseases with tonsil biopsy samples. *Lancet* 353:183–189.

Hsiao, K. K., Doh-ura, K., Kitamoto, T., Tateishi, J., and Prusiner, S. B. (1989). A prion protein amino acid substitution in ataxic Gerstmann-Sträussler syndrome. *Ann Neurol* 26:137.

Hsiao, K., Cass, C., Conneally, P. M., Dloughy, S. R., Hodes, M. E., Farlow, M. R., Ghetti, B., and Prusiner, S. B. (1990). Atypical Gerstmann-Sträussler-Scheinker syndrome with neurofibrillary tangles: no mutation in the prion protein open-reading-frame in a patient of the Indiana kindred. *Neurobiol Aging* 11:302.

Hunter, N., Foster, J. D., Benson, G., and Hope, J. (1991). Restriction fragment length polymorphisms of the scrapie-associated fibril protein (PrP) gene and their association with susceptiblity to natural scrapie in British sheep. *J Gen Virol* 72:1287–1292.

Hunter, N., Goldmann, W., Benson, G., Foster, J. D., and Hope, J. (1993). Swaledale sheep affected by natural scrapie differ significantly in PrP genotype frequencies from healthy sheep and those selected for reduced incidence of scrapie. *J Gen Virol* 74:1025–1031.

Hunter, N., Goldmann, W., Smith, G., and Hope, J. (1994). Frequencies of PrP gene variants in healthy cattle and cattle with BSE in Scotland. *Vet Rec* 135:400–403.

Hunter, N., Cairns, D., Foster, J. D., Smith, G., Goldmann, W., and Donnelly, K. (1997a). Is scrapie solely a genetic disease? *Nature* 386:137.

Hunter, N., Goldmann, W., Foster, J. D., Cairns, D., and Smith, G. (1997b). Natural scrapie and PrP genotype: case-control studies in British sheep. *Vet Rec* 141:137–140.

Hunter, N., Moore, L., Hosie, B. D., Dingwall, W. S., and Greig, A. (1997c). Association between natural scrapie and PrP genotype in a flock of Suffolk sheep in Scotland. *Vet Rec* 140:59–63.

Ikeda, T., Horiuchi, M., Ishiguro, N., Muramatsu, Y., Kai-Uwe, G. D., and Shinagawa, M. (1995). Amino acid polymorphisms of PrP with reference to onset of scrapie in Suffolk and Corriedale sheep in Japan. *J Gen Virol* 76:2577–2581.

Jakob, A. (1921). Über eigenartige Erkrankungen des Zentralnervensystems mit bemerkenswertem anatomischen Befunde (spastische Pseudosklerose-Encephalomyelopathie mit disseminierten Degenerationsherden). *Z Gesamte Neurol Psychiatrie* 64:147–228.

James, T. L., Liu, H., Ulyanov, N. B., Farr-Jones, S., Zhang, H., Donne, D. G., Kaneko, K., Groth, D., Mehlhorn, I., Prusiner, S. B., and Cohen, F. E. (1997). Solution structure of a 142-residue recombinant prion protein corresponding to the infectious fragment of the scrapie isoform. *Proc Natl Acad Sci USA* 94:10086–10091.

Kanaani, J., Prusiner, S.B., Diacovo, J., Baekkeskov, S., and Legname, G. (2005). Recombinant prion protein induces rapid polarization and development of synapses in embryonic rat hippocampal neurons *in vitro*. *J Neurochem* 95: 1373–1386.

Kaneko, K., Vey, M., Scott, M., Pilkuhn, S., Cohen, F. E., and Prusiner, S. B. (1997a). COOH-terminal sequence of the cellular prion protein directs subcellular trafficking and controls conversion into the scrapie isoform. *Proc Natl Acad Sci USA* 94:2333–2338.

Kaneko, K., Zulianello, L., Scott, M., Cooper, C. M., Wallace, A. C., James, T. L., Cohen, F. E., and Prusiner, S. B. (1997b). Evidence for protein X binding to a discontinuous epitope on the cellular prion protein during scrapie prion propagation. *Proc Natl Acad Sci USA* 94:10069–10074.

Kaneko, K., Ball, H. L., Wille, H., Zhang, H., Groth, D., Torchia, M., Tremblay, P., Safar, J., Prusiner, S. B., DeArmond, S. J., *et al.* (2000). A synthetic peptide initiates Gerstmann-Sträussler-Scheinker (GSS) disease in transgenic mice. *J Mol Biol* 295:997–1007.

Kimberlin, R. H. (1996). Speculations on the origin of BSE and the epidemiology of CJD, In Bovine Spongiform Encephalopathy: The BSE Dilemma, C. J. Gibbs, Jr., ed. (New York: Springer Verlag), pp. 155–175.

King, C. Y., and Diaz-Avalos, R. (2004). Protein-only transmission of three yeast prion strains. *Nature* 428:319–323.

Kirschbaum, W. R. (1968). Jakob-Creutzfeldt Disease (Amsterdam: Elsevier).

Klatzo, I., Gajdusek, D. C., and Zigas, V. (1959). Pathology of kuru. *Lab Invest* 8:799–847.

Kong, Q., Surewicz, W. K., Petersen, R. B., Zou, W., Chen, S. G., Gambetti, P., Parchi, P., Capellari, S., Goldfarb, L., Montagna, P., *et al.* (2004). Inherited prion diseases, In Prion Biology and Diseases, S. B. Prusiner, ed. (Cold Spring Harbor: Cold Spring Harbor Laboratory Press), pp. 673–775.

Korth, C., Kaneko, K., Groth, D., Heye, N., Telling, G., Mastrianni, J., Parchi, P., Gambetti, P., Will, R., Ironside, J., *et al.* (2003). Abbreviated incubation times for human prions in mice expressing a chimeric mouse—human prion protein transgene. *Proc Natl Acad Sci USA* 100:4784–4789.

Lasmézas, C. I., Deslys, J.-P., Demaimay, R., Adjou, K. T., Lamoury, F., Dormont, D., Robain, O., Ironside, J., and Hauw, J.-J. (1996). BSE transmission to macaques. *Nature* 381:743–744.

Lasmézas, C. I., Deslys, J.-P., Robain, O., Jaegly, A., Beringue, V., Peyrin, J.-M., Fournier, J.-G., Hauw, J.-J., Rossier, J., and Dormont, D. (1997). Transmission of the BSE agent to mice in the absence of detectable abnormal prion protein. *Science* 275:402–405.

Legname, G., Nelken, P., Guan, Z., Kanyo, Z. F., DeArmond, S. J., and Prusiner, S. B. (2002). Prion and doppel proteins bind to granule cells of the cerebellum. *Proc Natl Acad Sci USA* 99: 16285–16290.

Legname, G., Baskakov, I. V., Nguyen, H.-O. B., Riesner, D., Cohen, F. E., DeArmond, S. J., and Prusiner, S. B. (2004). Synthetic mammalian prions. *Science* 305:673–676.

Legname, G., Nguyen, H.-O. B., Baskakov, I. V., Cohen, F. E., DeArmond, S. J., and Prusiner, S. B. (2005). Strain-specified characteristics of mouse synthetic prions. *Proc Natl Acad Sci USA* 102:2168–2173.

LeVine, H. (1993). Thioflavine T interaction with synthetic Alzheimer's disease β-amyloid peptides: detection of amyloid aggregation in solution. *Protein Sci* 2:404–410.

Maddelein, M. L., Dos Reis, S., Duvezin-Caubet, S., Coulary-Salin, B., and Saupe, S. J. (2002). Amyloid aggregates of the HET-s prion protein are infectious. *Proc Natl Acad Sci USA* 99:7402–7407.

Masters, C. L., Gajdusek, D. C., and Gibbs, C. J., Jr. (1981). Creutzfeldt-Jakob disease virus isolations from the Gerstmann-Sträussler syndrome. *Brain* 104:559–588.

Masters, C. L., Harris, J. O., Gajdusek, D. C., Gibbs, C. J., Jr., Bernouilli, C., and Asher, D. M. (1978). Creutzfeldt-Jakob disease: patterns of worldwide occurrence and the significance of familial and sporadic clustering. *Ann Neurol* 5: 177–188.

Mastrianni, J. A., Curtis, M. T., Oberholtzer, J. C., Da Costa, M. M., DeArmond, S., Prusiner, S. B., and Garbern, J. Y. (1995). Prion disease (PrP-A117V) presenting with ataxia instead of dementia. *Neurology* 45:2042–2050.

Mastrianni, J. A., Nixon, R., Layzer, R., Telling, G. C., Han, D., DeArmond, S. J., and Prusiner, S. B. (1999). Prion protein conformation in a patient with sporadic fatal insomnia. *N Engl J Med 340*:1630–1638.

McKinley, M. P., Meyer, R. K., Kenaga, L., Rahbar, F., Cotter, R., Serban, A., and Prusiner, S. B. (1991a). Scrapie prion rod formation *in vitro* requires both detergent extraction and limited proteolysis. *J Virol* 65:1340–1351.

McKinley, M. P., Taraboulos, A., Kenaga, L., Serban, D., Stieber, A., DeArmond, S. J., Prusiner, S. B., and Gonatas, N. (1991b). Ultrastructural localization of scrapie prion proteins in cytoplasmic vesicles of infected cultured cells. *Lab Invest* 65:622–630.

Medori, R., Tritschler, H.-J., LeBlanc, A., Villare, F., Manetto, V., Chen, H. Y., Xue, R., Leal, S., Montagna, P., Cortelli, P., *et al.* (1992). Fatal familial insomnia, a prion disease with a mutation at codon 178 of the prion protein gene. *N Engl J Med* 326:444–449.

Miller, M. W., Wild, M. A., and Williams, E. S. (1998). Epidemiology of chronic wasting disease in captive Rocky Mountain elk. *J Wildl Dis* 34:532–538.

Miller, M. W., Williams, E. S., McCarty, C. W., Spraker, T. R., Kreeger, T. J., Larsen, C. T., and Thorne, E. T. (2000). Epizootiology of chronic wasting disease in free-ranging cervids in Colorado and Wyoming. *J Wildl Dis* 36:676–690.

Mills, J. L., Schonberger, L. B., Wysowski, D. K., Brown, P., Durako, S. J., Cox, C., Kong, F., and Fradkin, J. E. (2004). Long-term mortality in the United States cohort of pituitary-derived growth hormone recipients. *J Pediatr* 144: 430–436.

Moore, R. C., Lee, I. Y., Silverman, G. L., Harrison, P. M., Strome, R., Heinrich, C., Karunaratne, A., Pasternak, S. H., Chishti, M. A., Liang, Y., *et al.* (1999). Ataxia in prion protein (PrP)-deficient mice is associated with upregulation of the novel PrP-like protein doppel. *J Mol Biol* 292:797–817.

Muramoto, T., DeArmond, S. J., Scott, M., Telling, G. C., Cohen, F. E., and Prusiner, S. B. (1997). Heritable disorder resembling neuronal storage disease in mice expressing prion protein with deletion of an α-helix. *Nat Med* 3:750–755.

Nathanson, N., Wilesmith, J., and Griot, C. (1997). Bovine spongiform encephalopathy (BSE): cause and consequences of a common source epidemic. *Am J Epidemiol* 145:959–969.

O'Rourke, K. I., Holyoak, G. R., Clark, W. W., Mickelson, J. R., Wang, S., Melco, R. P., Besser, T. E., and Foote, W. C. (1997). PrP genotypes and experimental scrapie in orally inoculated Suffolk sheep in the United States. *J Gen Virol* 78:975–978.

O'Rourke, K. I., Besser, T. E., Miller, M. W., Cline, T. F., Spraker, T. R., Jenny, A. L., Wild, M. A., Zebarth, G. L., and Williams, E. S. (1999). PrP genotypes of captive and free-ranging Rocky Mountain elk (Cervus elaphus nelsoni) with chronic wasting disease. *J Gen Virol* 80:2765–2769.

Palmer, M. S., Dryden, A. J., Hughes, J. T., and Collinge, J. (1991). Homozygous prion protein genotype predisposes to sporadic Creutzfeldt-Jakob disease. *Nature* 352:340–342.

Pan, K.-M., Baldwin, M., Nguyen, J., Gasset, M., Serban, A., Groth, D., Mehlhorn, I., Huang, Z., Fletterick, R. J., Cohen, F. E., and Prusiner, S. B. (1993). Conversion of α-helices into β-sheets features in the formation of the scrapie prion proteins. *Proc Natl Acad Sci USA* 90:10962–10966.

Parchi, P., Castellani, R., Capellari, S., Ghetti, B., Young, K., Chen, S. G., Farlow, M., Dickson, D. W., Sima, A. A. F., Trojanowski, J. Q., *et al.* (1996). Molecular basis of phenotypic variability in sporadic Creutzfeldt-Jakob disease. *Ann Neurol* 39:767–778.

Pattison, I. H. (1965). Experiments with scrapie with special reference to the nature of the agent and the pathology of the disease, In Slow, Latent and Temperate Virus Infections, NINDB Monograph 2, D. C. Gajdusek, C. J. Gibbs, Jr., and M. P. Alpers, eds. (Washington, D.C: U.S. Government Printing), pp. 249–257.

Peretz, D., Scott, M., Groth, D., Williamson, A., Burton, D., Cohen, F. E., and Prusiner, S. B. (2001). Strain-specified relative conformational stability of the scrapie prion protein. *Protein Sci* 10:854–863.

Peretz, D., Williamson, R. A., Legname, G., Matsunaga, Y., Vergara, J., Burton, D., DeArmond, S. J., Prusiner, S. B., and Scott, M. R. (2002). A change in the conformation of prions accompanies the emergence of a new prion strain. *Neuron* 34:921–932.

Pergami, P., Jaffe, H., and Safar, J. (1996). Semipreparative chromatographic method to purify the normal cellular isoform of the prion protein in nondenatured form. *Anal Biochem* 236:63–73.

Perrier, V., Kaneko, K., Safar, J., Vergara, J., Tremblay, P., DeArmond, S. J., Cohen, F. E., Prusiner, S. B., and Wallace, A. C. (2002). Dominant-negative inhibition of prion replication in transgenic mice. *Proc Natl Acad Sci USA* 99:13079–13084.

Peters, J., Miller, J. M., Jenny, A. L., Peterson, T. L., and Carmichael, K. P. (2000). Immunohistochemical diagnosis of chronic wasting disease in preclinically affected elk from a captive herd. *J Vet Diagn Invest* 12:579–582.

Phillips, N. A., Bridgeman, J., and Ferguson-Smith, M. (2000). In The BSE Inquiry. The report (Volumes 1–16) London, England: The Stationery Office 2:1–274.

PHS (1997). Report on Human Growth Hormone and Creutzfeldt-Jakob Disease (Public Health Service Interagency Coordinating Committee), pp. 1–11.

Prusiner, S. B. (1982). Novel proteinaceous infectious particles cause scrapie. *Science* 216:136–144.

Prusiner, S. B. (1989). Scrapie prions. *Annu Rev Microbiol* 43:345–374.

Prusiner, S. B., Scott, M., Foster, D., Pan, K.-M., Groth, D., Mirenda, C., Torchia, M., Yang, S.-L., Serban, D., Carlson, G. A., *et al.* (1990). Transgenetic studies implicate interactions between homologous PrP isoforms in scrapie prion replication. *Cell* 63:673–686.

Prusiner, S. B. (1991). Molecular biology of prion diseases. *Science* 252:1515–1522.

Prusiner, S. B., Füzi, M., Scott, M., Serban, D., Serban, H., Taraboulos, A., Gabriel, J.-M., Wells, G. A. H., Wilesmith, J. W., Bradley, R., *et al.* (1993a). Immunologic and molecular biologic studies of prion proteins in bovine spongiform encephalopathy. *J Infect Dis* 167:602–613.

Prusiner, S. B., Groth, D., Serban, A., Koehler, R., Foster, D., Torchia, M., Burton, D., Yang, S.-L., and DeArmond, S. J. (1993b). Ablation of the prion protein (PrP) gene in mice prevents scrapie and facilitates production of anti-PrP antibodies. *Proc Natl Acad Sci USA* 90:10608–10612.

Prusiner, S. B. (1997). Biology of prions, In The Molecular and Genetic Basis of Neurological Disease, 2nd Edition, R. N. Rosenberg, S. B. Prusiner, S. DiMauro, and R. L. Barchi, eds. (Stoneham, MA: Butterworth Heinemann), pp. 103–143.

Prusiner, S. B. (1998). Prions. *Proc Natl Acad Sci USA* 95:13363–13383.

Prusiner, S. B. (2001). Shattuck Lecture—Neurodegenerative diseases and prions. *N Engl J Med* 344:1516–1526.

Ridley, R. M., and Baker, H. F. (1996). To what extent is strain variation evidence for an independent genome in the agent of the transmissible spongiform encephalopathies? *Neurodegeneration* 5:219–231.

Rieger, R., Edenhofer, F., Lasmézas, C. I., and Weiss, S. (1997). The human 37-kDa laminin receptor precursor interacts with the prion protein in eukaryotic cells. *Nat Med* 3:1383–1388.

Riek, R., Hornemann, S., Wider, G., Billeter, M., Glockshuber, R., and Wüthrich, K. (1996). NMR structure of the mouse prion protein domain PrP(121–231). *Nature* 382:180–182.

Robinson, M. M., Hadlow, W. J., Knowles, D. P., Huff, T. P., Lacy, P. A., Marsh, R. F., and Gorham, J. R. (1995). Experimental infection of cattle with the agents of transmissible mink encephalopathy and scrapie. *J Comp Pathol* 113:241–251.

Safar, J., Wille, H., Itri, V., Groth, D., Serban, H., Torchia, M., Cohen, F. E., and Prusiner, S. B. (1998). Eight prion strains have PrP^{Sc} molecules with different conformations. *Nat Med 4*:1157–1165.

Sakaguchi, S., Katamine, S., Nishida, N., Moriuchi, R., Shigematsu, K., Sugimoto, T., Nakatani, A., Kataoka, Y., Houtani, T., Shirabe, S., *et al.* (1996). Loss of cerebellar Purkinje cells in aged mice homozygous for a disrupted PrP gene. *Nature* 380:528–531.

Schmitt-Ulms, G., Legname, G., Baldwin, M. A., Ball, H. L., Bradon, N., Bosque, P. J., Crossin, K. L., Edelman, G. M., DeArmond, S. J., Cohen, F. E., and Prusiner, S. B. (2001). Binding of neural cell adhesion molecules (N-CAMs) to the cellular prion protein. *J Mol Biol* 314:1209–1225.

Scott, M. R., Groth, D., Tatzelt, J., Torchia, M., Tremblay, P., DeArmond, S. J., and Prusiner, S. B. (1997). Propagation of prion strains through specific conformers of the prion protein. *J Virol* 71:9032–9044.

Scott, M. R., Will, R., Ironside, J., Nguyen, H.-O. B., Tremblay, P., DeArmond, S. J., and Prusiner, S. B. (1999). Compelling transgenetic evidence for transmission of bovine spongiform encephalopathy prions to humans. *Proc Natl Acad Sci USA* 96:15137–15142.

Scott, M. R., Peretz, D., Nguyen, H.-O. B., DeArmond, S. J., and Prusiner, S. B. (2005). Transmission barriers for bovine, ovine, and human prions in transgenic mice. *J Virol* 79:5259–5271.

Shibuya, S., Higuchi, J., Shin, R.-W., Tateishi, J., and Kitamoto, T. (1998). Codon 219 Lys allele of PRNP is not found in sporadic Creutzfeldt-Jakob disease. *Ann Neurol* 43:826–828.

Sigurdson, C. J., Williams, E. S., Miller, M. W., Spraker, T. R., O'Rourke, K. I., and Hoover, E. A. (1999). Oral transmission and early lymphoid tropism of chronic wasting disease PrPres in mule deer fawns (Odocoileus hemionus). *J Gen Virol* 80:2757–2764.

Sparrer, H. E., Santoso, A., Szoka, F. C., Jr., and Weissman, J. S. (2000). Evidence for the prion hypothesis: induction of the yeast [PSI+] factor by *in vitro*-converted Sup35 protein. *Science* 289:595–599.

Spielhaupter, C., and Schätzl, H. M. (2001). PrP^C directly interacts with proteins involved in signaling pathways. *J Biol Chem* 276:44604–44612.

Stahl, N., Baldwin, M. A., Teplow, D. B., Hood, L., Gibson, B. W., Burlingame, A. L., and Prusiner, S. B. (1993). Structural analysis of the scrapie prion protein using mass spectrometry and amino acid sequencing. *Biochemistry* 32:1991–2002.

Stekel, D. J., Nowak, M. A., and Southwood, T. R. E. (1996). Prediction of future BSE spread. *Nature* 381:119.

Stöckel, J., Safar, J., Wallace, A. C., Cohen, F. E., and Prusiner, S. B. (1998). Prion protein selectively binds copper (II) ions. *Biochemistry* 37:7185–7193.

Tamgüney, G., Giles, K., Bouzamondo-Bernstein, E., Bosque, P. J., Miller, M. W., Safar, J., DeArmond, S. J., and Prusiner, S. B. (2006). Transmission of elk and deer prions to transgenic mice. *J Virol* 80: 9104–9114.

Tanaka, M., Chien, P., Naber, N., Cooke, R., and Weissman, J. S. (2004). Conformational variations in an infectious protein determine prion strain differences. *Nature* 428:323–328.

Tange, R. A., Troost, D., and Limburg, M. (1989). Progressive fatal dementia (Creutzfeldt-Jakob disease) in a patient who received homograft tissue for tympanic membrane closure. *Eur Arch Otorhinolaryngol* 247:199–201.

Taraboulos, A., Rogers, M., Borchelt, D. R., McKinley, M. P., Scott, M., Serban, D., and Prusiner, S. B. (1990). Acquisition of protease resistance by prion proteins in scrapie-infected cells does not require asparagine-linked glycosylation. *Proc Natl Acad Sci USA* 87:8262–8266.

Taraboulos, A., Scott, M., Semenov, A., Avrahami, D., Laszlo, L., and Prusiner, S. B. (1995). Cholesterol depletion and modification of COOH-terminal targeting sequence of the prion protein inhibits formation of the scrapie isoform. *J Cell Biol* 129:121–132.

Telling, G. C., Scott, M., Mastrianni, J., Gabizon, R., Torchia, M., Cohen, F. E., DeArmond, S. J., and Prusiner, S. B. (1995). Prion propagation in mice expressing human and chimeric PrP transgenes implicates the interaction of cellular PrP with another protein. *Cell* 83:79–90.

Telling, G. C., Parchi, P., DeArmond, S. J., Cortelli, P., Montagna, P., Gabizon, R., Mastrianni, J., Lugaresi, E., Gambetti, P., and Prusiner, S. B. (1996a). Evidence for the conformation of the pathologic isoform of the prion protein enciphering and propagating prion diversity. *Science* 274:2079–2082.

Telling, G. C., Scott, M., and Prusiner, S. B. (1996b). Deciphering prion diseases with transgenic mice, In Bovine Spongiform Encephalopathy: The BSE Dilemma, C. J. Gibbs, Jr., ed. (New York: Springer Verlag), pp. 202–231.

Tremblay, P., Ball, H. L., Kaneko, K., Groth, D., Hegde, R. S., Cohen, F. E., DeArmond, S. J., Prusiner, S. B., and Safar, J. G. (2004). Mutant PrP^{Sc} conformers induced by a synthetic peptide and several prion strains. *J Virol* 78:2088–2099.

Vey, M., Pilkuhn, S., Wille, H., Nixon, R., DeArmond, S. J., Smart, E. J., Anderson, R. G., Taraboulos, A., and Prusiner, S. B. (1996). Subcellular colocalization of the cellular and scrapie prion proteins in caveolae-like membranous domains. *Proc Natl Acad Sci USA 93*:14945–14949.

Viles, J. H., Cohen, F. E., Prusiner, S. B., Goodin, D. B., Wright, P. E., and Dyson, H. J. (1999). Copper binding to the prion protein: structural implications of four identical cooperative binding sites. *Proc Natl Acad Sci USA* 96:2042–2047.

Waggoner, D. J., Drisaldi, B., Bartnikas, T. B., Casareno, R. L., Prohaska, J. R., Gitlin, J. D., and Harris, D. A. (2000). Brain copper content and cuproenzyme activity do not vary with prion protein expression level. *J Biol Chem* 275:7455–7458.

Wells, G. A., Hawkins, S. A., Austin, A. R., Ryder, S. J., Done, S. H., Green, R. B., Dexter, I., Dawson, M., and Kimberlin, R. H. (2003). Studies of the transmissibility of the agent of bovine spongiform encephalopathy to pigs. *J Gen Virol* 84: 1021–1031.

Wells, G. A. H., and Wilesmith, J. W. (2004). Bovine spongiform encephalopathy and related diseases, In Prion Biology and Diseases, S. B. Prusiner, ed. (Cold Spring Harbor: Cold Spring Harbor Laboratory Press), pp. 595–628.

Westaway, D., Goodman, P. A., Mirenda, C. A., McKinley, M. P., Carlson, G. A., and Prusiner, S. B. (1987). Distinct prion proteins in short and long scrapie incubation period mice. *Cell* 51:651–662.

Westaway, D., DeArmond, S. J., Cayetano-Canlas, J., Groth, D., Foster, D., Yang, S.-L., Torchia, M., Carlson, G. A., and Prusiner, S. B. (1994a). Degeneration of skeletal muscle, peripheral nerves, and the central nervous system in transgenic mice overexpressing wild-type prion proteins. *Cell* 76:117–129.

Westaway, D., Zuliani, V., Cooper, C. M., Da Costa, M., Neuman, S., Jenny, A. L., Detwiler, L., and Prusiner, S. B. (1994b). Homozygosity for prion protein alleles encoding glutamine-171 renders sheep susceptible to natural scrapie. *Genes Dev* 8:959–969.

Wilesmith, J. W. (1991). The epidemiology of bovine spongiform encephalopathy. *Semin Virol* 2:239–245.

Wilesmith, J. W., Ryan, J. B. M., and Atkinson, M. J. (1991). Bovine spongiform encephalopathy: epidemiologic studies on the origin. *Vet Rec* 128:199–203.

Will, R. G., and Matthews, W. B. (1982). Evidence for case-to-case transmission of Creutzfeldt-Jakob disease. *J Neurol Neurosurg Psychiatry* 45:235–238.

Will, R. G., Ironside, J. W., Zeidler, M., Cousens, S. N., Estibeiro, K., Alperovitch, A., Poser, S., Pocchiari, M., Hofman, A., and Smith, P. G. (1996). A new variant of Creutzfeldt-Jakob disease in the UK. *Lancet* 347:921–925.

Will, R. G., Alpers, M. P., Dormont, D., Schonberger, L. B., and Tateishi, J. (1999a). Infectious and sporadic prion diseases, In Prion Biology and Diseases, S. B. Prusiner, ed. (Cold Spring Harbor: Cold Spring Harbor Laboratory Press), pp. 465–507.

Will, R. G., Cousens, S. N., Farrington, C. P., Smith, P. G., Knight, R. S. G., and Ironside, J. W. (1999b). Deaths from variant Creutzfeldt-Jakob disease. *Lancet* 353:979.

Will, R. G., Alpers, M. P., Dormont, D., and Schonberger, L. B. (2004). Infectious and sporadic prion diseases, In Prion Biology and Diseases, S. B. Prusiner, ed. (Cold Spring Harbor: Cold Spring Harbor Laboratory Press), pp. 629–671.

Williams, E. S., and Miller, M. W. (2002). Chronic wasting disease in deer and elk in North America. *Rev Sci Tech* 21: 305–316.

Williams, E. S., and Young, S. (1980). Chronic wasting disease of captive mule deer: a spongiform encephalopathy. *J Wildl Dis* 16:89–98.

Yamakawa, Y., Hagiwara, K., Nohtomi, K., Nakamura, Y., Nishijima, M., Higuchi, Y., Sato, Y., and Sata, T. (2003). Atypical proteinase K-resistant prion protein (PrPres) observed in an apparently healthy 23-month-old Holstein steer. *Jpn J Infect Dis* 56:221–222.

Zeidler, M., Stewart, G. E., Barraclough, C. R., Bateman, D. E., Bates, D., Burn, D. J., Colchester, A. C., Durward, W., Fletcher, N. A., Hawkins, S. A., *et al.* (1997). New variant Creutzfeldt-Jakob disease: neurological features and diagnostic tests. *Lancet* 350:903–907.

Zulianello, L., Kaneko, K., Scott, M., Erpel, S., Han, D., Cohen, F. E., and Prusiner, S. B. (2000). Dominant-negative inhibition of prion formation diminished by deletion mutagenesis of the prion protein. *J Virol* 74:4351–4360.

7

Mammalian Prion Protein

Ilia V. Baskakov

Abstract

The discoveries of prion disease transmission in mammals and a non-Mendelian type of inheritance in yeast has led to the establishment of a new concept in biology, "the prion hypothesis." This hypothesis postulates that an abnormal protein conformation propagates itself in an autocatalytic manner by recruiting the cellular isoform of the same protein, and, therefore, acts either as transmissible agent of disease (in mammals) or as heritable determinant of phenotype (in yeast and fungus). While the prion biology of yeast and fungus supports this idea strongly, the direct proof of the "protein-only" hypothesis in mammals, the generation of PrP^{Sc} *in vitro, de novo* from noninfectious PrP, has been difficult to achieve despite many years of effort. The current review discusses potential strategies for generation of prion infectivity *de novo* and summarizes our current knowledge about biophysical mechanisms of prion conversion.

7.1. Introduction

Several severe neurodegenerative diseases are associated with the misfolding and aggregation of prion protein (PrP). They include Creutzfeldt-Jakob disease (CJD), Gerstmann-Sträussler-Scheinker disease, kuru, and sporadic and familial insomnia in humans; scrapie in sheep; bovine spongiform encephalopathy (BSE) in cattle; and chronic wasting diseases in deer and elk (Prusiner, 1997). Prion maladies manifest in infectious, familial, and sporadic forms. A key event in all three forms of prion diseases is the conversion of the normal prion protein isoform, PrP^C, into the abnormal pathologic isoform, PrP^{Sc}. *Familial* forms are caused by single point mutations in PrP, with about 20 disease-inducing mutations identified so far (Prusiner and Scott, 1997). *Sporadic* forms occur spontaneously in about one person per million each year (Masters *et al.*, 1978). However, it is the *infectious* form that raises major public concern. Public awareness of the infectious form has gained attention due to the epidemic of bovine spongiform encephalopathy in the United Kingdom (Ghani *et al.*, 2000). Unlike other infectious agents, prions are not completely destroyed by sterilization, radiation, or cooking.

7.2. Structural Studies of Mammalian Prion Protein

7.2.1. PrP^C: Cellular Isoform of Prion Protein

Prion protein is predominately expressed in the neurons and glia of the brain and spinal cord; however, it was also found in peripheral tissues including muscles (Harris, 1999; Bosque *et al.*, 2002). Prion protein is translocated at the endoplasmic reticulum and then proteolytically processed to remove a 22-residue N-terminal signal peptide and 23 C-terminal amino acids to create a cellular isoform (PrP^C) with just 219 amino acids (Turk *et al.*, 1988; Bolton *et al.*, 1987) (Fig. 7.1). PrP^C contains a disulfide bridge connecting residues 179 and 214, two Asn-linked sugars at positions 181 and 197 (Endo *et al.*, 1989; Rudd *et al.*, 2001), and a glycophosphatidylinositol anchor (GPI) attached to Ser-231 (Stahl *et al.*, 1987). After post-translational modifications, PrP^C is exported to the cell surface and accumulates in the caveolae space (Borchelt *et al.*, 1994; Shyng *et al.*, 1994; Vey *et al.*, 1996; Harris, 1999).

Biophysical characterization of the unglycosylated recombinant PrP (rPrP) refolded into α-monomeric form resembling PrP^C reveals two domains (Fig. 7.1). The first is an N-terminal region (residues 23–123), which contains octarepeat region (57–91) that binds up to four Cu^{2+} ions per polypeptide chain (Hornshaw *et al.*, 1995a; Hornshaw *et al.*, 1995b; Stöckel *et al.*, 1998; Viles *et al.*, 1999; Whittal *et al.*, 2000; Aronoff-Spencer *et al.*, 2000; Garnett and Viles, 2003), and conservative hydrophobic region (residues 106–126). Octarepeat region is composed of multiple tandem copies of the eight-residue sequence. An additional Cu^{2+} binding site that binds up to two Cu^{2+} ions was identified recently within residues 92 and 111 (Jackson *et al.*, 2001; Burns *et al.*, 2003; Jones *et al.*, 2004). The N-terminal region is highly flexible and lacks any identifiable secondary structure in the absence of Cu^{2+} (Donne *et al.*, 1997; Riek *et al.*, 1997). The second is a C-terminal region (124–231) containing three α-helices and two short antiparallel β-strands (Riek *et al.*, 1996; Donne *et al.*, 1997; James *et al.*, 1997).

7.2.2. PrP^{Sc}: Pathologic Isoform of Prion Protein

Structural information about PrP^{Sc} is limited due to its insolubility, heterogeneity, and highly aggregated state (Gabizon and Prusiner, 1990; Wille *et al.*, 2002). In contrast with PrP^C, PrP^{Sc} exists in multiple conformations, which are referred to as *strains* of PrP^{Sc}. Biochemical

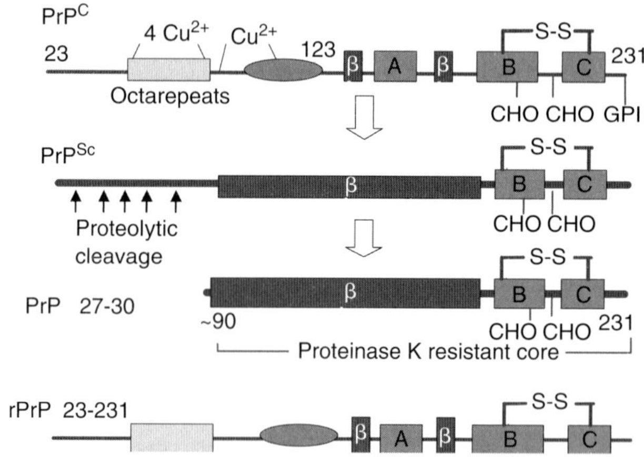

Figure 7.1. Prion protein isoforms: PrP^C, PrP^{Sc}, PrP^{Sc} 27-30, and recombinant PrP 23-231. After processing of the amino and carboxyl termini, both PrP^C and PrP^{Sc} consist of 209 amino acid residues. After limited proteolysis, the amino terminus of PrP^{Sc} is truncated to form PrP 27-30. Three α-helices (A–C, gray rectangles), the two β-strands of PrP^C (dark gray rectangles), the conservative hydrophobic region (gray oval), and the octarepeats (light gray rectangles) are highlighted. Attached carbohydrates (CHO) and a glycosylphosphatidylinositol (GPI) anchor are indicated. According to the molecular model by Govaerts *et al.* (2004) the region spanning residues ~90–170 forms β-helix upon conversion of PrP^C into PrP^{Sc}, as indicated by red box. Recombinant PrP 23-231 lacks carbohydrates and GPI anchor.

assays have been described that distinguish conformers of PrP^{Sc} by the extent of proteinase K (PK) resistance, the size of the PK-resistant core, thermodynamic stability, epitope presentation, and the relative amount of β-sheet-rich structure (Bessen and Marsh, 1994; Telling *et al.*, 1996; Caughey *et al.*, 1998; Safar *et al.*, 1998; Peretz *et al.*, 2001a). Conformationally different subtypes of PrP^{Sc} have also been found in patients with sporadic CJD (Parchi *et al.*, 1999; Hill *et al.*, 2003). These PrP^{Sc} *subtypes* have variations in the molecular mass of the PK-resistant core and exhibit different ratios of di-, mono-, and unglycosylated PrP^{Sc}. Conformational diversity of PrP^{Sc} subtypes is linked to the methionine/valine polymorphism at codon 129 and may also be affected by unidentified host factors (Telling *et al.*, 1995; Parchi *et al.*, 2000).

To distinguish different conformers and to classify subtypes of PrP^{Sc}, limited digestion with PK has been widely used (McKinley *et al.*, 1983; Bessen and Marsh, 1994; Safar *et al.*, 1998; Parchi *et al.*, 1999; Hill *et al.*, 2003). Treatment of PrP^{Sc} with PK generates a C-terminal PK-resistant core, referred to as PrP 27-30 (McKinley *et al.*, 1983; Oesch *et al.*, 1985) (Fig. 7.1). Depending on the subtype of PrP^{Sc}, PrP 27-30 exhibits minor variations in gel mobility due to differences in the PK cleavage site between residues 79 and 103 (Parchi *et al.*, 2000). Because PrP 27-30 is capable of transmitting prion disease (Prusiner *et al.*, 1983), and because transgenic mice expressing only PrP 90-231 but not the full-length PrP^{C} develop prion disease (Fischer *et al.*, 1996), the N-terminus region including residues 23–89 is believed to be unnecessary for propagation of prions. Multiple lines of evidence indicate that the central region of PrP, encompassing residues 90–141, is important for prion propagation (Tagliavini *et al.*, 1993; Fischer *et al.*, 1996; Peretz *et al.*, 1997; Chabry *et al.*, 1998). The current molecular model of PrP^{Sc} postulates that the region 90–174 adopts left-handed β-helical fold upon conversion of PrP^{C} into PrP^{Sc}, while the region spanning residues 177–227 remains in predominately α-helices conformation (Govaerts *et al.*, 2004) (Fig. 7.1).

In addition to classical PrP^{Sc}, two novel PK-resistant fragments have been identified recently in patients with sporadic CJD (Zou *et al.*, 2003). Deglycosylated forms of these novel PK-resistant fragments migrate at 13 and 12 kDa by SDS-PAGE and are generated by cleavage at residues 154/156 and 162/167, respectively, retaining the C-terminal region intact.

7.3. Reconstitution of Prion Infectivity *In Vitro*

To explain the infectious form of prion diseases, the "protein-only" hypothesis postulates that PrP^{Sc} acts as a transmissible agent and that it self-propagates its pathologic conformation in an autocatalytic manner using PrP^{C} as a substrate (Prusiner, 1982). The direct proof of the protein-only hypothesis, the reconstitution of PrP^{Sc} *in vitro* from noninfectious PrP, has been the ultimate goal of prion research for many years. In an attempt to generate infectious form of PrP, two different strategies have been pursued: (i) amplification of PrP^{Sc} using PrP^{C} as a substrate and (ii) conversion of recombinant PrP into abnormal β-sheet-rich conformation in the absence of PrP^{Sc} template.

There are several technical challenges related to reconstitution of infectivity *in vitro*. Despite substantial conformational differences between PrP^{C} and PrP^{Sc} (PrP^{C} is a PK-sensitive α-helical monomer, whereas PrP^{Sc} is an assembled multimer characterized by enhanced resistance toward PK-digestion and increased amount of β-structure (Caughey *et al.*, 1991; Pan *et al.*, 1993), it is still unclear what physical property can be used as a valid probe to monitor formation of PrP^{Sc} *de novo* in a cell-free system. For example, it has been shown that prion disease can be transmitted in the absence of the PK-resistant form of PrP (Lasmézas *et al.*, 1997; Manuelidis *et al.*, 1997; Barron *et al.*, 2001). On the other hand, generation of PK-resistant PrP *in vitro* did not correlate with amplification of infectivity (Hill *et al.*, 1999; Jackson *et al.*, 1999; Bieschke *et al.*, 2004). The fact that PrP^{Sc} exists in multiple forms referred to as strains and subtypes, which are characterized by a broad range of PK-resistance and diverse physical properties, creates additional challenges for reconstitution of prion infectivity *in vitro*.

7.3.1. Amplification of PrP^{Sc}

Several *in vitro* amplification protocols have been developed, in which PrP^{Sc} was used as a template for conversion of PrP^{C} into nascent PrP^{Sc} (Bessen *et al.*, 1995; Saborio *et al.*, 2001; Deleault *et al.*, 2003). Because difference in PK-resistance has been used historically to distinguish PrP^{C} from PrP^{Sc} isoforms, all these *in vitro* amplification protocols exploited enhancement in PK-resistance of PrP to track the conversion.

In the protocol developed by Caughey and co-workers, approximately 20% of PrP^{C} was converted into PK-resistant form, referred to as PrP-res, upon addition of 50-fold molar excess of purified PrP^{Sc} (Caughey *et al.*, 1995; Horiuchi *et al.*, 1999). Conversion of PrP^{C} into PrP-res was found to be species- and strain-specific (Bessen *et al.*, 1995; Kocisko *et al.*, 1995; Horiuchi *et al.*, 2000). Soto and co-workers reported that approximately 30-fold excess of PrP^{C} can be converted into PrP-res form upon repetitive cycles of sonication of brain homogenate containing small amounts of PrP^{Sc} (Saborio *et al.*, 2001). In most recent studies, unlimited amplification of PrP-res was achieved upon subsequent serial dilution-amplification cycles of miniscule amounts of initial PrP^{Sc} seeds (Castilla *et al.*, 2005). Amplification of PrP-res was accompanied by amplification of prion infectivity. This work illustrates that PrP-res molecules generated *in vitro* were able to catalyze the formation of new PrP-res molecules confirming the autocatalytic properties of *de novo* generated PrP-res. Without sonication, substantially lower level of amplification of PrP-res was achieved in crude brain homogenate suggesting that sonication generates active replication centers (Deleault *et al.*, 2003; Lucassen *et al.*, 2003). The mechanisms for fragmentation of PrP^{Sc} aggregates and generation of active replication centers *in vivo* are currently unknown.

7.3.2. *In Vitro* Conversion of Recombinant PrP

An alternative strategy for the reconstitution of PrP^{Sc} *in vitro* includes cell-free conversion of recombinant or synthetic PrP in the absence of PrP^{Sc}. Recombinant PrP (rPrP) was proved to be a valuable biophysical model for elucidating structural aspects of prion replication. In its α-helical conformation, rPrP resembles PrP^{C}, as both proteins were shown to have similar secondary and tertiary structure (Hornemann *et al.*, 2004) (Fig. 7.1). Cell-free conversions of rPrPs produced a large diversity of abnormal β-sheet-rich isoforms, many of which possess some but not all PrP^{Sc}-like properties (Jackson *et al.*, 1999; Swietnicki *et al.*, 2000; Baskakov *et al.*, 2001; Cordeiro *et al.*, 2001; Rezaei *et al.*, 2002; Kazlauskaite *et al.*, 2003; Lee and Eisenberg, 2003; Sokolowski *et al.*, 2003; Torrent *et al.*, 2003). To monitor conversion *in vitro*, most investigators used an enhancement of PK-resistance and/or change in secondary structure of rPrPs. Despite many years of efforts and broad conformational diversity of abnormal isoforms generated *in vitro*, reconstitution of the infectious form of PrP has been difficult to achieve (May *et al.*, 2004).

Instead of exploiting PK-resistance as a biochemical marker of cell-free conversion, an alternative approach for reconstitution of infectivity *in vitro* involved the generation of self-propagating amyloid isoform. In our previous studies using mouse rPrP encompassing residues 89-231 (rPrP 89-231), we developed a cell-free conversion system that produces amyloid fibrils (Baskakov *et al.*, 2002). This *in vitro* conversion system exhibited peculiar features of the autocatalytic process and was limited by a species barrier, two key aspects of prion propagation (Baskakov, 2004). When inoculated into transgenic mice expressing only PrP 89-231, these amyloid fibrils

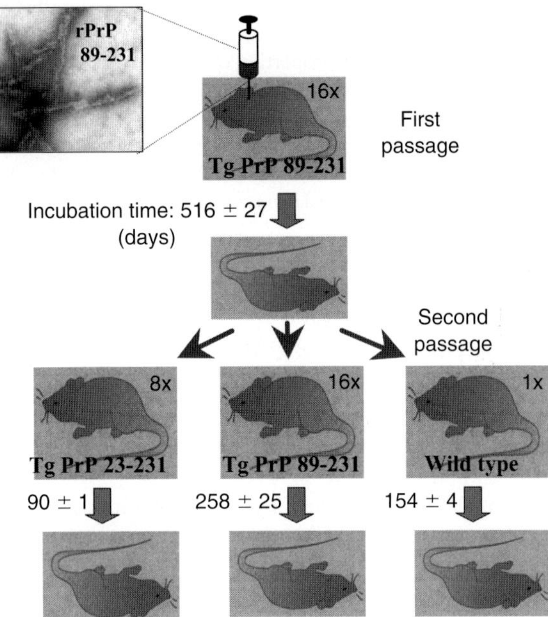

Figure 7.2. Transmission of synthetic prions. Amyloid fibrils consisting of rPrP 89-230 were produced *in vitro* and inoculated into transgenic (Tg) mice expressing PrP 89-231. The animals inoculated with the fibrils developed neurologic clinical signs of prion disease and died between days 380 and 620 after inoculation. The brain tissues of these mice were injected into wild-type mice and two groups of Tg mice expressing PrP 89-231 and PrP 23-231. All three groups of animals developed scrapie with much shorter incubation time.

induced prion disease that could be efficiently transmitted to both wild-type and transgenic mice (Legname *et al.*, 2004) (Fig. 7.2). Biochemical and neuropathologic features indicated that inoculation of amyloid fibrils produced a novel strain of prion disease in experimental animals (Legname *et al.*, 2005). When inoculated into mice, amyloid fibrils displayed a long incubation time compared with known strains of PrP^{Sc}, a possible indication of low infectivity titer (Fig. 7.2). Are there any other molecules that are involved in the production of infections prions? One may speculate that yet unidentified cellular cofactors may be required to produce fibrils with higher infectivity titers *in vitro*.

7.3.3. Search for Cofactors Involved in Prion Replication

Yet unidentified cellular cofactors may be required for efficient conversion of PrP^{C} or recombinant PrP into PrP^{Sc}. Available data suggested that additional species-specific factor, designated as *protein X*, is involved in prion replication (Telling *et al.*, 1995; Kaneko *et al.*, 1997). Despite substantial efforts to identify *protein X* using genetic, biochemical, and immunologic approaches, *protein X* remains elusive (Schmitt-Ulms *et al.*, 2001).

Several classes of macromolecules were shown to bind to PrP and, specifically, to the N-terminal region of PrP encompassing residues 23–90 (Cordeiro *et al.*, 2001; Gonzalez-Iglesias *et al.*, 2002; Warner *et al.*, 2002; Gabus *et al.*, 2004). While the N-terminal region is not critical for transmission of prions (Prusiner *et al.*, 1983; Fischer *et al.*, 1996), this region impacts conformational diversity of PrP^{Sc} conformers (Wadsworth *et al.*, 1999; Lawson *et al.*, 2004), and, therefore, may also influence structural properties of fibrils generated *in vitro*. In particular, RNA, DNA, and heparan sulfate were shown to bind to PrP^{C} with high affinities (Cordeiro *et al.*, 2001; Gonzalez-Iglesias *et al.*, 2002; Warner *et al.*, 2002; Gabus *et al.*, 2004). Furthermore, RNA and sulfated glycans were found to stimulate cell-free conversion of PrP^{C} into PrP-res form *in vitro* (Wong *et al.*, 2001; Deleault *et al.*, 2003). Whether or not binding of the above cofactors is helpful for reconstitution of infectivity *in vitro* remains to be determined.

7.4. Kinetic Models of Prion Propagation

7.4.1. Nucleation-Polymerization and Template-Assisted Models

Two kinetic models have been proposed to explain prion propagation: (1) *the nucleation-polymerization model* (NPM) (Harper and Lansbury, Jr., 1997) and (2) *the template-assisted model* (TAM) (Cohen *et al.*, 1994). NPM was originally introduced in 1975 to describe polymerization of proteins (Oosawa and Asakura, 1975) and was later applied to amyloid formation (Harper and Lansbury, Jr., 1997). On the other hand, TAM was introduced specifically to model prion replication. NPM postulates that the rate-limiting step is the formation of a nucleus: an oligomeric aggregate of PrP. The nucleus is thermodynamically unstable and this makes spontaneous process a very rare event. However, as the nucleus is formed, further polymerization is facile. Exogenous addition of PrP^{Sc} bypasses the formation of nucleus. TAM, on the other hand, argues that PrP^{C} is separated from PrP^{Sc} by a substantial energy barrier. The high barrier precludes the formation of PrP^{Sc} under normal conditions. However, the process of conversion is facilitated by an exogenous administration of PrP^{Sc}, which acts as a catalyst and lowers the energy barrier. According to TAM, in addition to its catalytic role, each strain of PrP^{Sc} acts as a unique template, providing conformational constraints for the formation of nascent PrP^{Sc}, and determines the conformation of nascent PrP^{Sc} (Cohen *et al.*, 1994). The presence of a nucleus or a template accelerates the process of conversion, either by avoiding a rate-limiting step in NPM or by lowering of the energy barrier in TAM. Both models predict low occurrences of spontaneous form of disease, because spontaneous formation of the first catalytic center is inaccessible either thermodynamically or kinetically. However, neither model explicitly specifies the mechanism of multiplication of the catalytic centers during the time course of the disease.

7.4.2. Autocatalytic Model that Uses the Mechanism of Branched-Chain Reactions

While studying the kinetics of fibril formation, we were surprised to discover that *in vitro* conversion of rPrP into amyloid fibrils displays several unusual kinetic features: (1) the weak dependence of the lag-phase on the concentration of rPrP; (2) the dramatic effect of reaction volume on the length of the lag-phase; (3) volume-dependent threshold effect; and (4) the highly cooperative sigmoidal kinetic curves of polymerization (Baskakov and Bocharova, 2005). These kinetic features are not typical for the polymerization kinetics displayed by other amyloidogenic proteins and could not be rationalized within NPM.

To explain these unique kinetic features, we propose an autocatalytic model, in which the conversion reaction is regulated by the dynamic balance between the rates of multiplication and deactivation of self-propagating fibrillar rPrP isoforms. This model exploits a framework developed for branched-chain reactions (Semenov, 1957). A common characteristic of autocatalytic and branched-chain reactions is the multiplication of catalytic or active centers in the time course of the reactions (Fig. 7.3). In a simplified mathematical equation, the reaction velocity and threshold effect are determined by the multiplication coefficient (r), which is proportional to the probability of generating new active/catalytic centers divided by the probability of their loss or quenching in the time course of the process. Depending upon the rate of multiplication of catalytic centers versus the rate of their deactivation, the autocatalytic reactions may switch between autoacceleration mode and decay mode (Table 7.1). When multiplication of the active centers exceeds their quenching ($r > 1$; Table 7.1), the conversion reaction proceeds with self-acceleration. If the rate of quenching is higher than the rate of multiplication ($r < 1$; Table 7.1), the reaction decays. Correspondingly, it is expected that PrP^{Sc} will be cleared throughout an animal's lifetime if the process of prion replication is slower than the rate of clearance. When r is close to 1, the new model postulates that apparently negligible changes in experimental parameters may switch the *in vitro* reaction from decay mode to autoacceleration mode, and vice versa, causing stochastic behavior, in which the reaction may follow the "all or nothing" rule. Correspondingly, if the rate of PrP^{Sc} propagation is equal to or only slightly exceeds the rate of its clearance, prion disease will never progress during the animal/human's lifetime, however the disease may exist in subclinical form (Table 7.1).

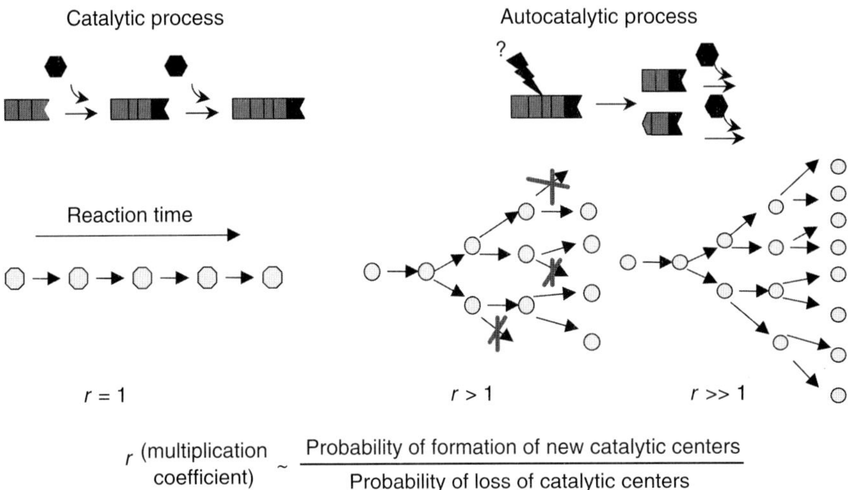

Figure 7.3. Schematic illustration of the autocatalytic model of prion propagation. The autocatalytic model proposes that prion propagation is controlled by the relative rate of multiplication versus deactivation of active centers (circles). Multiplication of active centers *in vitro* may occur through fibril fragmentation. Notably, the elongation reaction by itself does not generate new active centers unless fibril fragmentation occurs.

Table 7.1. Autocatalytic model predicts three possible outcomes of prion disease

	$r > 1$	$r \approx 1$	$r < 1$
Kinetics follow formal mechanism of	Branched-chain reactions	Enzyme catalyzes	Apparent first-order kinetics
Clinical form of disease	Progression of prion disease	Subclinical form of disease	Clearance of prions

What factors determine the rate of multiplication and clearance? For the majority of chemical processes following a branched-chain mechanism, the multiplication coefficient *r* depends on the ratio of surface to volume of the reaction vessel (Semenov, 1957). Vessel surfaces may either catalyze or deactivate active centers, thus having a significant impact on the lag-phase and final yield of the reactions. Remarkably, we found that surface plays an important role in polymerization of rPrP in the cell-free conversion system. Amyloid formation by rPrP 90–231 displayed apparent volume-dependent threshold—the conversion was observed only in reaction volume > 0.3 mL (Baskakov, 2004; Baskakov and Bocharova, 2005). The volume-dependent threshold is consistent with the scenario that self-propagating isoforms of rPrP are adsorbed and deactivated by the vessel surface. As the reaction volume decreases, the surface-to-volume ratio grows. Therefore, the threshold may be reached when the rate of surface-dependent deactivation exceeds the rate of multiplication of self-propagating isoforms. Indeed, we found that amyloid fibrils have high propensity to adsorb to walls of different reaction vessels. Binding of fibrillar rPrP to surfaces is reminiscent of that of PrP^{Sc}. It is known that prion diseases can be efficiently transmitted through wires and surgical instruments contaminated with PrP^{Sc} (Bernouilli *et al.*, 1977; Gibbs, Jr. *et al.*, 1994; Zobeley *et al.*, 1999; Weissmann *et al.*, 2002).

The mechanism responsible for the multiplication of catalytic centers of PrP^{Sc} *in vivo* is currently unknown and needs to be explored in the nearest future. Fragmentation of yeast prion aggregates *in vivo* was shown to involve cellular chaperones (Shorter and Lindquist, 2004). We recently found that fragmentation of PrP fibrils accounts for the multiplication of the active centers of polymerization *in vitro*, while surface-dependent sorption of fibrils is responsible for their deactivation (Baskakov and Bocharova, 2005). Similar mechanism of multiplication of active centers through fibril fragmentation was described for polymerization of the yeast prion protein Sup 35 (Collins *et al.*, 2004). Studies on modeling of *in vitro* polymerization argue that fragmentation of fibrils is an important step of polymerization and may account for the highly cooperative kinetics of conversion (Masel *et al.*, 1999; Poschel *et al.*, 2003; Collins *et al.*, 2004). If the intrinsic fragility of PrP^{Sc} aggregates does dictate the rate of prion propagation, this property could account for substantial differences in the incubation times produced by different strains of PrP^{Sc}. The ability to control multiplication of the catalytic isoform of PrP would be a powerful approach to treat prion diseases.

The novel mechanism postulates that the dynamic balance between multiplication and clearance of PrP^{Sc} determines the progression and final outcome of the prion disease (Table 7.1). This model helps to explain numerous experimental observations, which otherwise would be difficult to rationalize. For example, the concentration of PrP^{Sc} in the brain of experimental animals drops substantially in the first week after intracerebral inoculation (Bolton *et al.*, 1991; Bueler *et al.*, 1993), indicating that the rate of clearance may exceed the rate of multiplication during the initial stage of prion transmission. Surprisingly, the lifetime of PrP^{Sc} was found to be relatively short despite substantial resistance to proteolytic digestion (Enari *et al.*, 2001; Peretz *et al.*, 2001b). Under conditions that inhibit the production of nascent PrP^{Sc}, the half-life of PrP^{Sc} in ScN2a cells was shown to be only 28 h (Peretz *et al.*, 2001b). Multiple factors may influence the clearance rate of PrP^{Sc} *in vivo*: strain-specific intrinsic stability of PrP^{Sc} (Kuczius and Groschup, 1999; Peretz *et al.*, 2001a); an intensity of proteolytic processing (Luhr *et al.*, 2004; Yadavalli *et al.*, 2004); an interaction of PrP^{Sc} with stabilizing cofactors such as glycosaminoglycans (Shaked *et al.*, 2001; Wong *et al.*, 2001; Ben-Zaken *et al.*, 2003).

7.5. Conformational Diversity of Self-Propagating Aggregates

7.5.1. Strains of PrP^{Sc}

The existence of different prion *strains* has been recognized for a long time in the history of prion diseases (Dickinson *et al.*, 1968; Fraser and Dickinson, 1973). Each prion strain displays distinct neuropathologic features, such as the length of incubation time and the distribution of neuronal vacuolation in the brain. These features are stable during serial transmission of particular PrP^{Sc} strain in a given host species (Scott *et al.*, 2004), illustrating high fidelity in the propagation of prion strains. To explain strain phenomenon, the protein-only hypothesis postulates that the same PrP amino acid sequence must be capable of adopting separate states when it is converted into pathologic isoforms, referred to as strains of PrP^{Sc}.

During the past several years, considerable evidence has been found indicating that the properties of prion strains are enciphered in the conformation of PrP^{Sc} (Bessen and Marsh, 1994; Telling *et al.*, 1996; Caughey *et al.*, 1998; Safar *et al.*, 1998; Peretz *et al.*, 2001a; Thomzig *et al.*, 2004). The questions of fundamental importance are (i) what factors determine the conformational diversity of prion strains, and, specifically, what is the role of the primary structure of PrP versus template and environmental factors in determining conformational diversity? (ii) How do new strains arise? Understanding these issues is essential for the development of an efficient strategy to prevent the pandemic spread of new strains of prions in human populations in the future. Notably, the effectiveness of potential anti-prion inhibitors is often strongly dependent on particular strain of PrP^{Sc}. For example, amphotericin B was effective against 263K strain of scrapie in hamsters but not against 139H or DY strains (McKenzie *et al.*, 1994; Demaimay *et al.*, 1999). PPI dendrimers were also found to be active against RML and 139A strains of mouse scrapie but ineffective against Me7 and 87V strains (Supattapone *et al.*, 2001).

Besides generation of new strains, selection is another factor in strain evolution of which we must be aware. Transmission of BSE to humans is believed to occur after selection of the most thermodynamically stable strain of BSE that resisted heat treatment during the rendering process. It is quite possible that the novel strains are the results of selection and evolution of different conformational subtypes of PrP^{Sc} of sporadic origin. As an example, kuru is believed to have originated from a single case of sporadic Creutzfeldt-Jakob disease and spread through ritualistic cannibalism (Glasse and Lindenbaum, 1992).

7.5.2. Complexity of *In Vitro* Misfolding Pathways

Recent biophysical studies revealed the complexity of plausible misfolding pathways and the diversity of abnormal β-sheet-rich forms that could be generated from full-length recombinant PrP *in vitro* (Bocharova *et al.*, 2005a). An α-monomeric form of rPrP 23-230 was shown to exist in a slow equilibrium with β-sheet-rich oligomeric isoform (Fig. 7.4). This equilibrium is shifted toward the β-oligomer at acidic pH but favors α-rPrP at neutral pH. The β-oligomer was found to be off the kinetic pathway to the amyloid form, which is formed at neutral or slightly acidic pH. While the β-oligomer accumulates predominately at acidic pH and at high protein concentrations (>10 μM), the amyloid fibrils are formed at physiologic pH values and at much lower protein concentration (~1 μM), similar to those found in normal brains (Safar *et al.*, 1998). Oxidation of methionine residues interferes with the formation of mature rPrP fibrils leading to accumulation of intermediate prefibrillar products (Breydo *et al.*, 2005).

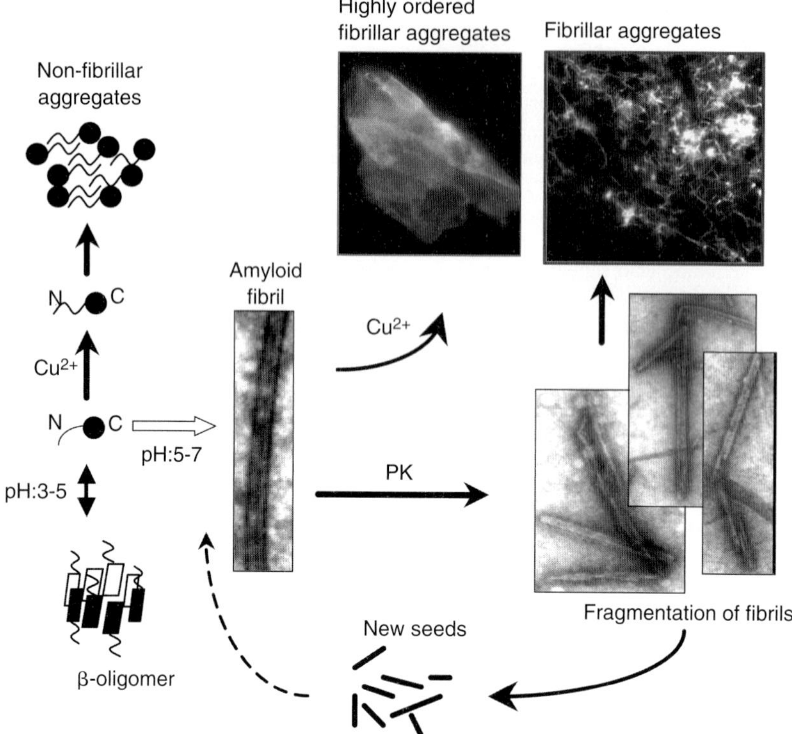

Figure 7.4. Schematic diagram illustrating complexity of *in vitro* conversion pathways. The β-oligomic isform of rPrP 23-230 is formed at acidic pH, whereas the amyloid fibrils are generated at neutral pH. Upon PK-digestion, the amyloid fibrils break into short fragments creating new active centers for propagation and also aggregate into large amyloid clumps. Cu^{2+} inhibits conversion of α-rPrP 23-230 into amyloid fibrils by stabilizing non-amyloidogenic PK-resistant form of α-rPrP 23-230. PK-resistant α-rPrP 23-230 slowly forms nonfibrillar aggregates. Treatment of preformed amyloid fibrils with Cu^{2+} induces further assembly of fibrils into highly ordered amyloid structures and enhances PK-resistance of the amyloid aggregates.

PK treatment of the amyloid fibrils results in cleavage of the N-terminal region encompassing residues 23–90, while the C-terminal region preserves β-rich fibrillar structure (Bocharova *et al.*, 2005a). After digestion of the N-terminal part, the fibrils tend to fragment into short pieces and co-aggregate laterally to form large clumps (Fig. 7.4). Both aggregation of the fibrils and their fragmentation may have physiologic implications for the progression of prion disease. Aggregation of PrP^{Sc} in the brain may interfere with the normal function of neurons. On the other hand, fibril fragmentation may create new active centers and, therefore, accelerates the rate of prion replication. Remarkably, fibrils treated with PK did not lose the ability to seed the conversion reactions in fresh reaction mixtures, suggesting that proteolytic enzymes may stimulate prion propagation (Bocharova *et al.*, 2005a). Consistent with this hypothesis are recent studies by Telling and co-authors, who demonstrated that inhibitors of cellular Ca^{2+}-dependent proteases reduced the rate of PrP^{Sc} accumulation in cultured cells (Yadavalli *et al.*, 2004). As proteolytic digestion may accelerate both clearance and propagation of prions, potential therapeutic strategies that target cellular proteases should be examined with great caution (Checler and Vincent, 2002).

Harris and co-workers showed that Cu^{2+} converts PrP^C into a protease K-resistant species, which are distinct from scrapie isoform (Quaglio *et al.*, 2001; Chiesa *et al.*, 2003). Consistent with these studies, binding of Cu^{2+} to α-rPrP was shown to induce aggregation and formation of non-amyloidogenic PK-resistant form of α-rPrP (Bocharova *et al.*, 2005b) (Fig. 7.4). On the other hand, Cu^{2+} also enhanced PK-resistance of the fibrillar form and triggered the assembly of fibrils into highly ordered macromolecular aggregates (Bocharova *et al.*, 2005b). Higher order assembly is presumably driven by interfibrillar coordination of Cu^{2+} by the octarepeat regions. Several active binding modes are possible for coordination of Cu^{2+} by the octarepeats, which depend on copper occupancy and solvent conditions (Viles *et al.*, 1999; Aronoff-Spencer *et al.*, 2000; Burns *et al.*, 2002, 2003). Moreover, different types of Cu^{2+} site geometries were observed depending on whether coordination occurs via interrepeat or intrarepeat binding modes, where interrepeat binding can be both intramolecular and intermolecular (Morante *et al.*, 2004).

Observation of further higher order assembly of rPrP 23-230 fibrils into ordered aggregates is important in light of the fact that progression of prion disease is accompanied by the accumulation of amyloid prion plaques rather than separately laying fibrils. Prion plaques vary dramatically with respect to their sizes and shapes and include kuru-type "spiked ball" plaques, "florid plaques," small punctate deposits, and others (DeArmond *et al.*, 2004). Importantly, electron microscopy studies revealed that the amyloid prion plaques have a filamentous ultrastructure similar to amyloid fibrils (DeArmond *et al.*, 1985). Considering that amyloid fibrils generated *in vitro* are prone to further assembly into highly ordered aggregates and that the shape of these aggregates is very sensitive to experimental conditions, it is understandable why the morphology of PrP^{Sc} deposits formed in brain under pathologic conditions can also be highly variable.

7.6. Can the Prion Hypothesis Be Expanded to Non-Prion Proteins?

The discoveries of prion disease transmission in mammals and a non-Mendelian type of inheritance in yeast has led to the establishment of a new concept in biology, "the prion hypothesis." Strong support for the idea that certain proteins are capable of self-propagating their abnormal conformation came from the prion biology of yeast and fungus (Wickner *et al.*, 1995; Maddelein *et al.*, 2002). *In vitro*–made amyloid fibrils of Sup35 and HET-s—prion protein of yeast and fungus—caused prion propagation *in vivo* and changed the phenotype of cells (Sparrer *et al.*, 2000; Maddelein *et al.*, 2002). In all the above cases, abnormal protein conformation acts either as transmissible agent of disease (in mammals) or as heritable determinant of phenotype (in yeast and fungus).

Despite differences in the primary and tertiary structures of yeast and mammalian prions, the yeast prions display all peculiar features of the replication of mammalian prions, such as (i) the ability to propagate an alternative protein conformation in an autocatalytic manner and (ii) high-fidelity propagation. Fidelity can be defined as the ability to maintain particular conformational properties of the original self-propagating aggregates/strains in the course of serial passages. Yeast and mammalian prion proteins are capable of adopting conformationally distinct self-propagating states—strains within the same amino acid sequence. To maintain high fidelity, efficient propagation of prion strains requires identical or high homologous amino acid sequences of normal and self-propagating isoform, implying high specificity of their interaction (Santoso *et al.*, 2000; Chien and Weissman, 2001). A transmission barrier is observed when normal and self-propagating isoforms have even minor differences in their primary structures.

Recent studies demonstrated that the ability to form amyloid structure seems to be a general property of proteins and is determined by the chemical nature of the polypeptide backbone (Dobson, 2002). Many proteins and polypeptides not related to any known conformational diseases have been found to adopt alternative amyloid fold *in vitro* and *in vivo*. Medin (a proteolytic fragment of lactadherin) is an example of amyloidogenic polypeptides that form amyloid deposits in aorta of virtually all individuals older than 60 years of age (Haggqvist *et al.*, 1999). Amyloidogenic proteins are now found in a variety of organisms including prokaryotes, plants, and mammals (Konno, 2001; Podrabsky *et al.*, 2001; Chapman *et al.*, 2002). No sequence consensuses that would predetermine the ability to form amyloid have been identified in this class of proteins. At the same time, amino acid side chains may influence amyloidogenic propensity of each individual protein. For instance, the alternating patterns of polar and nonpolar amino acids were shown to be intrinsically amyloidogenic (West *et al.*, 1999; Broome and Hecht, 2000).

If the process of amyloid formation is a general property of the polypeptide backbone, how many proteins are capable of self-propagating conformational transition in prion-like manner? A few studies argue that the amyloid formation by non-prion proteins can also display some peculiar properties of prions. The systemic amyloidosis caused by the amyloid deposition of *the serum amyloid A protein* can be transmitted from animal to animal in a prion-like mechanism (Lundmark *et al.*, 2002). The process of amyloid formation by non-prion proteins also requires high homology of amino acid sequences of a precursor and products of fibrillization. For instance, the fibrillization of mouse α-synuclein was inhibited by human α-synuclein, which differs from mouse protein at seven positions (Rochet *et al.*, 2000). Recent studies indicate that conversion of neuronal CPEB protein, responsible for the regulation of translation, into alternative self-perpetuating prion-like state may help to maintain long-term synaptic changes associated with memory storage (Si *et al.*, 2003). The question of whether peculiar features, which are believed to be unique for prion propagation, are indeed inherent properties of the amyloidogenic process remains to be answered.

References

Aronoff-Spencer, E., C. S. Burns, N. I. Avdievich, G. J. Gerfen, J. Peisach, W. E. Antholine, H. L. Ball, F. E. Cohen, S. B. Prusiner, and G. L. Millhauser. 2000. Identification of the Cu^{2+} binding sites in the N-terminal domain of the prion protein by EPR and CD spectroscopy. *Biochemistry* 39:13760–13771.

Barron, R. M., V. Thomson, E. Jameison, D. W. Melton, J. Ironside, R. Will, and J. C. Manson. 2001. Changing a single amino acid in the N-terminus of murine PrP alters TSE incubation time across three species barrier. *EMBO J.* 20:5070–5078.

Baskakov, I. V., G. Legname, S. B. Prusiner, and F. E. Cohen. 2001. Folding of prion protein to its native α-helical conformation is under kinetic control. *J. Biol. Chem.* 276:19687–19690.

Baskakov, I. V., G. Legname, M. A. Baldwin, S. B. Prusiner, and F. E. Cohen. 2002. Pathway complexity of prion protein assembly into amyloid. *J. Biol. Chem.* 277:21140–21148.

Baskakov, I. V. 2004. Autocatalytic conversion of recombinant prion proteins displays a species barrier. *J. Biol. Chem.* 279:586–595.

Baskakov, I. V. and O. V. Bocharova. 2005. *In vitro* conversion of mammalian prion protein into amyloid fibrils is not consistent with the nucleation polymerization mechanism. *Biochemistry* 44: 2339–2348.

Ben-Zaken, O., S. Tzaban, Y. Tal, Horonchik. L., J. D. Esko, I. Vlodavsky, and A. Taraboulos. 2003. Cellular heparan sulfate participates in the metabolism of prions. *J. Biol. Chem.* 41:40041–40049.

Bernouilli, C., J. Siegfried, G. Baumgartner, F. Regli, T. Rabinowicz, D. C. Gajdusek, and C. J. Gibbs, Jr. 1977. Danger of accidental person to person transmission of Creutzfeldt-Jakob disease by surgery. *Lancet* 1:478–479.

Bessen, R. A., D. A. Kocisko, G. J. Raymond, S. Nandan, P. T. Lansbury, and B. Caughey. 1995. Non-genetic propagation of strain-specific properties of scrapie prion protein. *Nature* 375:698–700.

Bessen, R. A. and R. F. Marsh. 1994. Distinct PrP properties suggest the molecular basis of strain variation in transmissible mink encephalopathy. *J. Virol.* 68:7859–7868.

Bieschke, J., P. Weber, N. Sarafoff, M. Beekes, A. Giese, and H. Kretzschmar. 2004. Autocatalytic self-propagation of misfolded prion protein. *Proc. Natl. Acad. Sci. USA* 101:12207–12211.

Bocharova, O. V., L. Breydo, A. S. Parfenov, V. V. Salnikov, and I. V. Baskakov. 2005a. *In vitro* conversion of full length mammalian prion protein produces amyloid form with physical property of PrPSc. *J. Mol. Biol.* 346:645–659.

Bocharova, O. V., L. Breydo, V. V. Salnikov, and I. V. Baskakov. 2005b. Cu(II) inhibits *in vitro* conversion of prion protein into amyloid fibrils. *Biochemistry* 44:6776–6787.

Bolton, D. C., P. E. Bendheim, A. D. Marmorstein, and A. Potempska. 1987. Isolation and structural studies of the intact scrapie agent protein. *Arch. Biochem. Biophys.* 258:579–590.

Bolton, D. C., S. J. Seligman, G. Bablanian, D. Windsor, L. J. Scala, K. S. Kim, C. J. Chen, R. J. Kascsak, and P. E. Bendheim. 1991. Molecular location of a species-specific epitope on the hamster scrapie agent protein. *J. Virol.* 65:3667–3675.

Borchelt, D. R., V. E. Koliatsis, M. Guarnieri, C. A. Pardo, S. S. Sisodia, and D. L. Price. 1994. Rapid anterograde axonal transport of the cellular prion glycoprotein in the peripheral and central nervous systems. *J. Biol. Chem.* 269:14711–14714.

Bosque, P. J., C. Ryou, G. Telling, D. Peretz, G. Legname, S. DeArmond, and S. B. Prusiner. 2002. Prions in skeletal muscle. *Proc. Natl. Acad. Sci. USA* 99:3812–3817.

Breydo, L., O. V. Bocharova, N. Makarava, V. V. Salnikov, M. Anderson, and I. V. Baskakov. 2005. Methionine oxidation interferes with conversion of the prion protein into the fibrillar proteinase K-resistant conformation. *Biochemistry* 15534–15543.

Broome, B. M. and M. H. Hecht. 2000. Nature disfavors sequences of alternating polar and non-polar amino acids: implications for amyloidogenesis. *J. Mol. Biol.* 296:961–968.

Bueler, H., A. Aguzzi, A. Sailer, R. A. Greiner, P. Autenried, M. Aguet, and C. Weissmann. 1993. Mice Devoid of PrP are resistant to scrapie. *Cell* 73:1339–1347.

Burns, C. S., E. Aronoff-Spencer, C. M. Dunham, P. Lario, N. I. Avdievich, W. E. Antholine, M. M. Olmstead, A. Vrielink, G. J. Gerfen, J. Peisach, W. G. Scott, and G. L. Millhauser. 2002. Molecular features of the copper binding sites in the octarepeat domain of the prion protien. *Biochemistry* 41:4001.

Burns, C. S., E. Aronoff-Spencer, G. Legname, S. B. Prusiner, W. E. Antholine, G. J. Gerfen, J. Peisach, and G. L. Millhauser. 2003. Copper coordination in the full-length, recombinant prion protein. *Biochemistry* 42:6794–6803.

Castilla, J., P. Saa, C. Hetz, and C. Soto. 2005. *In vitro* generation of infectious scrapie prions. *Cell* 121:195–206.

Caughey, B., D. A. Kocisko, G. J. Raymond, and P. T. Lansbury, Jr. 1995. Aggregates of scrapie-associated prion protein induce the cell-free conversion of protease-sensitive prion protein to the protease-resistant state. *Chem. Biol.* 2:807–817.

Caughey, B., G. J. Raymond, and R. A. Bessen. 1998. Strain-dependent differences in β-sheet conformations of abnormal prion protein. *J. Biol. Chem.* 273:32230–32235.

Caughey, B. W., A. Dong, K. S. Bhat, D. Ernst, S. F. Hayes, and W. S. Caughey. 1991. Secondary structure analysis of the scrapie-associated protein PrP 27–30 in water by infrared spectroscopy. *Biochemistry* 30:7672–7680.

Chabry, J., B. Caughey, and B. Chesebro. 1998. Specific inhibition of *in vitro* formation of proteinase-resistant prion protein by synthetic peptides. *J. Biol. Chem.* 273:13203–13207.

Chapman, M. R., L. S. Robinson, J. S. Pinkner, R. Roth, J. Heuser, M. Hammar, S. Normark, and S. J. Hultgren. 2002. Role of Escherichia coli curli operones in directing amyloid fiber formation. *Science* 295:851–855.

Checler, F. and B. Vincent. 2002. Alzheimer's and prion diseases: distinct pathologies, common proteolytic denominators. *Trends Neurosci.* 25:616–620.

Chien, P. and J. S. Weissman. 2001. Conformational diversity in a yeast prion dictates its seeding specificity. *Nature* 410:223–227.

Chiesa, R., P. Piccardo, E. Quaglio, B. Drisaldi, S. L. Si-Hoe, M. Takao, B. Ghetti, and D. A. Harris. 2003. Molecular distinction between pathogenic and infectious properties of the prion protein. *J. Virology* 77:7622.

Cohen, F. E., K.-M. Pan, Z. Huang, M. Baldwin, R. J. Fletterick, and S. B. Prusiner. 1994. Structural clues to prion replication. *Science* 264:530–531.

Collins, S. R., A. Douglass, R. D. Vale, and J. S. Weissman. 2004. Mechanism of prion propagation: amyloid growth occurs by monomer addition. *PLOS Biol.* 2:e321.

Cordeiro, Y., F. Machado, L. Juliano, M. A. Juliano, R. R. Brentani, D. Foguel, and J. L. Silva. 2001. DNA converts cellular prion protein into the beta-sheet conformation and inhibits prion peptide aggregation. *J. Biol. Chem.* 276:49400–49409.

DeArmond, S. J., J. W. Ironside, E. Bouzamondo, D. Peretz, and J. R. Fraser. 2004. Neuropathology of prion diseases. *In* Prion Biology and Diseases. S. B. Prusiner, editor. Cold Spring Harbor Laboratory Press, Cold Spring Harbor, NY. 777–856.

DeArmond, S. J., M. P. McKinley, R. A. Barry, M. B. Braunfeld, J. R. McColloch, and S. B. Prusiner. 1985. Identification of prion amyloid filaments in scrapie-infected brain. *Cell* 41:221–235.

Deleault, N. R., R. W. Lucassen, and S. Supattapone. 2003. RNA molecules stimulate prion protein conversion. *Nature* 425:717–720.

Demaimay, R., R. Race, and B. Chesebro. 1999. Effectiveness of polyene antibiotics in treatment of transmissible spongiform encephalopathy in transgenic mice expressing Syrian hamster PrP only in neurons. *J. Virol.* 73:3513.

Dickinson, A. G., V. M. H. Meikle, and H. Fraser. 1968. Identification of a gene which controls the incubation period of some strains of scrapie agent in mice. *J. Comp. Pathol.* 78:293–299.

Dobson, C. M. 2002. Protein misfolding, evolution and disease. *Trends Biochem. Sci.* 24:329–332.

Donne, D. G., J. H. Viles, D. Groth, I. Mehlhorn, T. L. James, F. E. Cohen, S. B. Prusiner, P. E. Wright, and H. J. Dyson. 1997. Structure of the recombinant full-length hamster prion protein PrP(29–231): the N terminus is highly flexible. *Proc. Natl. Acad. Sci. USA* 94:13452–13457.

Enari, M., E. Flechsig, and C. Weissmann. 2001. Scrapie prion protein accuulation by scrapie-infected neuroblastoma cells abrogated by exosure to a prion protein antibody. *Proc. Natl. Acad. Sci. USA* 98:9295–9299.

Endo, T., D. Groth, S. B. Prusiner, and A. Kobata. 1989. Diversity of oligosaccharide structures linked to asparagines of the scrapie prion protein. *Biochemistry* 28:8380–8388.

Fischer, M., T. Rülicke, A. Raeber, A. Sailer, M. Moser, B. Oesch, S. Brandner, A. Aguzzi, and C. Weissmann. 1996. Prion protein (PrP) with amino-proximal deletions restoring susceptibility of PrP knockout mice to scrapie. *EMBO J.* 15:1255–1264.

Fraser, H. and A. G. Dickinson. 1973. Scrapie in mice. Agent-strain differences in the distribution and intensity of grey matter vacuolation. *J. Comp. Pathol.* 83:29–40.

Gabizon, R. and S. B. Prusiner. 1990. Prion liposomes. *Biochem. J.* 266:1–14.

Gabus, C., E. Derrington, P. Leblanc, J. Chnaiderman, D. Dormont, W. Swietnicki, M. Morillas, W. K. Surewicz, D. Marc, P. Nandi, and J. L. Darlix. 2004. The prion protein has RNA binding chaperoning properties characteristic of nucleocasid protein NCP7 of HIV-1. *J. Biol. Chem.* 276:19301–19309.

Garnett, A. P. and J. H. Viles. 2003. Copper binding to the octarepeats of the prion protein. *J. Biol. Chem.* 278:6795–6802.

Ghani, A. C., N. M. Ferguson, C. A. Donnelly, and R. M. Anderson. 2000. Predicted vCJD mortality in Great Britain. *Nature* 406:583–584.

Gibbs, C. J., Jr., D. M. Asher, A. Kobrine, H. L. Amyx, M. P. Sulima, and D. C. Gajdusek. 1994. Transmission of Creutzfeldt-Jakob disease to a chimpanzee by electrodes contaminated during neurosurgery. *J. Neurol. Neurosurg. Psychiatry* 57:757–758.

Glasse, R. and S. Lindenbaum. 1992. Fieldwork in the South Fore: The process of ethnographic inquiry. In Prion Diseases of Human and Animals. S. B. Prusiner and *et al.*, editors. Ellis Horwood, New York. 77–91.

Gonzalez-Iglesias, R., M. A. Pajares, C. O. J. C. Espinosa, B. Oesch, and M. Gasset. 2002. Prion protein interaction with glycosaminoglycan occurs with the formation of oligomeric complexes stabilized by Cu(II) bridges. *J. Mol. Biol.* 319:527–540.

Govaerts, C., H. Wille, S. B. Prusiner, and F. E. Cohen. 2004. Evidance for assembly of prions with left-handed b-helices into trimers. *Proc. Natl. Acad. Sci. USA* 101:8342–8347.

Haggqvist, B., J. Naslund, K. Sletten, G. Westermark, G. Mucchiano, L. O. Tjernberg, C. Nordstedt, U. Engstrom, and P. Westermark. 1999. Medin: An integral fragment of aortic amooth muscle cell-produced lactadherin forms the most common human amyloid. *Proc. Natl. Acad. Sci. USA* 96:8669–8674.

Harper, J. D. and P. T. Lansbury, Jr. 1997. Models of amyloid seeding in Alzheimer's disease and scrapie: mechanistic truths and physiological consequences of the time-dependent solubility of amyloid proteins. *Annu. Rev. Biochem.* 66:385–407.

Harris, D. A. 1999. Cellular biology of prion diseases. *Clin. Microbiol. Rev.* 12:429–444.

Hill, A. F., M. Antoniou, and J. Collinge. 1999. Protease-resistant prion protein produced *in vitro* lacks detectable infectivity. *J. Gen. Virol.* 80:11–14.

Hill, A. F., S. Joiner, J. D. F. Wadsworth, K. C. L. Sidle, J. E. Bell, H. Budka, J. W. Ironside, and J. Collinge. 2003. Molecular classification of sporadic Crutzfeldt-Jakob disease. *Brain* 126:1333–1346.

Horiuchi, M., J. Chabry, and B. Caughey. 1999. Specific binding of normal prion protein to the scrapie form via a localized domain initiates its conversion to the protease-resistant state. *EMBO J.* 18:3193–3203.

Horiuchi, M., S. A. Priola, J. Chabry, and B. Caughey. 2000. Interactions between heterologous forms of prion protein: binding, inhibition of conversion, and species barriers. *Proc. Natl. Acad. Sci. USA* 97:5836–5841.

Hornemann, S., C. Schorn, and K. Wuthrich. 2004. NMR structure of bovine prion protein isolated from healthy calf brain. *EMBO Rep.* 5:1159–1164.

Hornshaw, M. P., J. R. McDermott, and J. M. Candy. 1995a. Copper binding to the N-terminal tandem repeat regions of mammalian and avian prion protein. *Biochem. Biophys. Res. Commun.* 207:621–629.

Hornshaw, M. P., J. R. McDermott, J. M. Candy, and J. H. Lakey. 1995b. Copper binding to the N-terminal tandem repeat region of mammalian and avian prion protein: structural studies using synthetic peptides. *Biochem. Biophys. Res. Commun.* 214:993–999.

Jackson, G. S., L. L. P. Hosszu, A. Power, A. F. Hill, J. Kenney, H. Saibil, C. J. Craven, J. P. Waltho, A. R. Clarke, and J. Collinge. 1999. Reversible conversion of monomeric human prion protein between native and fibrilogenic conformations. *Science* 283:1935–1937.

Jackson, G. S., I. Murray, L. Hosszu, N. Gibbs, J. P. Waltho, A. R. Clarke, and J. Collinge. 2001. Location and properties of metal-binding sites on the human prion protein. *Proc. Natl. Acad. Sci. USA* 98:8531–8535.

James, T. L., H. Liu, N. B. Ulyanov, S. Farr-Jones, H. Zhang, D. G. Donne, K. Kaneko, D. Groth, I. Mehlhorn, S. B. Prusiner, and F. E. Cohen. 1997. Solution structure of a 142-residue recombinant prion protein corresponding to the infectious fragment of the scrapie isoform. *Proc. Natl. Acad. Sci. USA* 94:10086–10091.

Jones, C. E., S. R. Abdelraheim, D. R. Brown, and J. H. Viles. 2004. Preferential Cu^{2+} coordination by His96 and His111 induces b-sheet formation in the unstructured amyloidogenic region of the prion protein. *J. Biol. Chem.* 279: 32018–32027.

Kaneko, K., L. Zulianello, M. Scott, C. M. Cooper, A. C. Wallace, T. L. James, F. E. Cohen, and S. B. Prusiner. 1997. Evidence for protein X binding to a discontinuous epitope on the cellular prion protein during scrapie prion propagation. *Proc. Natl. Acad. Sci. USA* 94:10069–10074.

Kazlauskaite, J., N. Sanghera, I. Sylvester, C. Venien-Bryan, and T. J. Pinheiro. 2003. Structural changesof the prion protein in lipid membranes leading to aggregation and fibrillization. *Biochemistry* 42:3295–3304.

Kocisko, D. A., S. A. Priola, G. J. Raymond, B. Chesebro, P. T. Lansbury, Jr., and B. Caughey. 1995. Species specificity in the cell-free conversion of prion protein to protease-resistant forms: a model for the scrapie species barrier. *Proc. Natl. Acad. Sci. USA* 92:3923–3927.

Konno, T. 2001. Multistep nucleus formation and a separate subunit contribution of the amyloidogenesis of heat-denatured monellin. *Prot. Sci.* 10:2093–2101.

Kuczius, T. and M. H. Groschup. 1999. Differences in proteinase K resistance and neuronal deposition of abnormal prion proteins characterize bovine spongioform encephalopathy (BSE) and scrapie strains. *Mol. Med.* 5:406–418.

Lasmézas, C. I., J.-P. Deslys, O. Robain, A. Jaegly, V. Beringue, J.-M. Peyrin, J.-G. Fournier, J.-J. Hauw, J. Rossier, and D. Dormont. 1997. Transmission of the BSE agent to mice in the absence of detectable abnormal prion protein. *Science* 275:402–405.

Lawson, V. A., S. A. Priola, K. Meade-White, M. Lawton, and B. Chesebro. 2004. Flexible N-terminal region of prion protein influences conformation of protease resistant prion protein isoforms associated with cross-species scrapie infection *in vivo* and *in vitro*. *J. Biol. Chem.* 279:13689–13695.

Lee, S. and D. Eisenberg. 2003. Seeded conversion of recombinant prion protein to a disulfide-bonded oligomer by a reduction-oxidation process. *Nat. Struc. Biol.* 10:725–730.

Legname, G., I. V. Baskakov, H.-O. B. Nguyen, D. Riesner, F. E. Cohen, S. J. DeArmond, and S. B. Prusiner. 2004. Synthetic mammalian prions. *Science* 305:673–676.

Legname, G., H.-O. B. Nguyen, I. V. Baskakov, F. E. Cohen, S. J. DeArmond, and S. B. Prusiner. 2005. Strain-specified characteristics of mouse synthetic prion. *Proc. Natl. Acad. Sci. USA* 102:2168–2173.

Lucassen, R., K. Nishina, and S. Supattapone. 2003. *In vitro* amplification of protease-resistant prion protein requires free sulfhydryl groups. *Biochemistry* 42:4127–4135.

Luhr, K. M., E. K. Nordstromm, P. Low, H. G. Ljunggren, A. Taraboulos, and K. Kristensson. 2004. Scrapie protein degradation by cysteine protease in CD11c+ dendritic cells and GT1-1 neuronal cells. *J. Virol.* 78:4776–4782.

Lundmark, K., G. T. Westermark, S. Nystrom, C. L. Murphy, A. Solomon, and P. Westermark. 2002. Transmissibility of systemic amyloidosis by a prion-like mechanism. *Proc. Natl. Acad. Sci. USA* 99:6979–6984.

Maddelein, M. L., S. Dos Reis, S. Duvezin-Caubet, B. Coulary-Salin, and S. J. Saupe. 2002. Amyloid aggregates of the HET-s prion protein are infectious. *Proc. Natl. Acad. Sci. USA* 99:7402–7407.

Manuelidis, L., W. Fritch, and Y.-G. Xi. 1997. Evolution of a strain of CJD that induces BSE-like plaques. *Science* 277:94–98.

Masel, J., V. A. A. Jansen, and M. A. Nowak. 1999. Quantifying the kinetic parameters of prion replication. *Biophys. Chem.* 77:139–152.

Masters, C. L., J. O. Harris, D. C. Gajdusek, C. J. Gibbs, Jr., C. Bernouilli, and D. M. Asher. 1978. Creutzfeldt-Jakob disease: patterns of worldwide occurrence and the significance of familial and sporadic clustering. *Ann. Neurol.* 5:177–188.

May, B. C. H., C. Govaerts, S. B. Prusiner, and F. E. Cohen. 2004. Prions: so many fibers, so little infectivity. *Trends Biochem. Sci.* 29:162–165.

McKenzie, D., J. Kaczkowski, R. Marsh, and J. Aiken. 1994. Amphotericin B delays both scrapie agent replication and PrP-res accumulation early in infection. *J. Virol.* 68:7534–7536.

McKinley, M. P., D. C. Bolton, and S. B. Prusiner. 1983. A protease-resistant protein is a structural component of the scrapie prion. *Cell* 35:57–62.

Morante, S., R. Gonzalez-Iglesias, C. Potrich, C. Meneghini, W. Meyer-Klaucke, G. Menestrina, and M. Gasset. 2004. Inter- and intra-octarepeat Cu(II) site geometries in the prion protein: implications in Cu(II) binding cooperativity and Cu(II)-mediated assemblies. *J. Biol. Chem.* 279:11753–11759.

Oesch, B., D. Westaway, M. Wälchli, M. P. McKinley, S. B. H. Kent, R. Aebersold, R. A. Barry, P. Tempst, D. B. Teplow, L. E. Hood, S. B. Prusiner, and C. Weissmann. 1985. A cellular gene encodes scrapie PrP 27-30 protein. *Cell* 40:735–746.

Oosawa, F. and S. Asakura. 1975. Thermodynamics of Polymerization of Protein. Academic Press, London.

Pan, K.-M., M. Baldwin, J. Nguyen, M. Gasset, A. Serban, D. Groth, I. Mehlhorn, Z. Huang, R. J. Fletterick, F. E. Cohen, and S. B. Prusiner. 1993. Conversion of α-helices into β-sheets features in the formation of the scrapie prion proteins. *Proc. Natl. Acad. Sci. USA* 90:10962–10966.

Parchi, P., A. Giese, S. Capellari, P. Brown, W. Schulz-Schaeffer, O. Windl, I. Zerr, H. Budka, N. Kopp, P. Piccardo, S. Poser, A. Rojiani, N. Streichemberger, J. Julien, C. Vital, B. Ghetti, P. Gambetti, and H. Kretzschmar. 1999. Classification of sporadic Creutzfeldt-Jakob disease based on molecular and phenotypic analysis of 300 subjects. *Ann. Neurol.* 46:224–233.

Parchi, P., W. Zou, W. Wang, P. Brown, S. Capellari, B. Ghetti, N. Kopp, W. J. Schulz-Schaeffer, H. A. Kretzschmar, M. W. Head, J. W. Ironside, P. Gambetti, and S. G. Chen. 2000. Genetic Influence on the structural variations of the abnormal prion protein. *Proc. Natl. Acad. Sci. USA* 97:10168–10172.

Peretz, D., R. A. Williamson, Y. Matsunaga, H. Serban, C. Pinilla, R. B. Bastidas, R. Rozenshteyn, T. L. James, R. A. Houghten, F. E. Cohen, S. B. Prusiner, and D. R. Burton. 1997. A conformational transition at the N terminus of the prion protein features in formation of the scrapie isoform. *J. Mol. Biol.* 273:614–622.

Peretz, D., M. Scott, D. Groth, A. Williamson, D. Burton, F. E. Cohen, and S. B. Prusiner. 2001a. Strain-specified relative conformational stability of the scrapie prion protein. *Protein Sci.* 10:854–863.

Peretz, D., R. A. Williamson, K. Kaneko, J. Vergara, E. Leclerc, G. Schmitt-Ulms, I. R. Mehlhorn, G. Legname, M. R. Wormald, P. M. Rudd, R. A. Dwek, D. R. Burton, and S. B. Prusiner. 2001b. Antibodies inhibit prion propagation and clear cell cultures of prion infectivity. *Nature* 412:739–743.

Podrabsky, J. E., J. E. Carpenter, and S. C. Hand. 2001. Survival of water stress in annual fish embryos: dehydration avoidance and egg envelope amyloid fibers. *Am. J. Physiol.* 280:R123–R131.

Poschel, T., N. V. Brilliantov, and C. Frommel. 2003. Kinetics of prion growth. *Biophys. J.* 85:3460–3474.

Prusiner, S. B. 1982. Novel proteinaceous infectious particles cause scrapie. *Science* 216:136–144.

Prusiner, S. B. 1997. Prion diseases and the BSE crisis. *Science* 278:245–251.

Prusiner, S. B., M. P. McKinley, K. A. Bowman, D. C. Bolton, P. E. Bendheim, D. F. Groth, and G. G. Glenner. 1983. Scrapie prions aggregate to form amyloid-like birefringent rods. *Cell* 35:349–358.

Prusiner, S. B. and M. R. Scott. 1997. Genetics of prions. *Annu. Rev. Genet.* 31:139–175.

Quaglio, E., B. Chiesa, and D. Harris. 2001. Copper converts the cellular prion protein into a protease-resistant species that is distinct from the scrapie isoform. *J. Biol. Chem.* 276:11432–11438.

Rezaei, H., Y. Choiset, F. Eghiaian, E. Treguer, P. Mentre, P. Debey, J. Grosclaude, and T. Haertle. 2002. Amyloidogenic unfolding intermediates differentiate sheep prion protein variants. *J. Mol. Biol.* 322:799–814.

Riek, R., S. Hornemann, G. Wider, M. Billeter, R. Glockshuber, and K. Wüthrich. 1996. NMR structure of the mouse prion protein domain PrP(121-231). *Nature* 382:180–182.

Riek, R., S. Hornemann, G. Wider, R. Glockshuber, and K. Wüthrich. 1997. NMR characterization of the full-length recombinant murine prion protein, *m*PrP(23-231). *FEBS Lett.* 413:282–288.

Rochet, J. C., K. A. Conway, and P. T. Lansbury, Jr. 2000. Inhibition of fibrillization and accumulation of prefibrillar oligomers in mixtures of human and mouse alpha-synuclein. *Biochemistry* 39:10619–10626.

Rudd, P. M., M. R. Wormald, D. R. Wing, S. B. Prusiner, and R. A. Dwek. 2001. Prion glycoprotein: structure, dynamics, and roles for the sugars. *Biochemistry* 40:3759–3766.

Saborio, G. P., B. Permanne, and C. Soto. 2001. Sensitive detection of pathological prion protein by cyclic amplification of protein misfolding. *Nature* 411:810–813.

Safar, J., H. Wille, V. Itri, D. Groth, H. Serban, M. Torchia, F. E. Cohen, and S. B. Prusiner. 1998. Eight prion strains have PrP^{Sc} molecules with different conformations. *Nat. Med.* 4:1157–1165.

Santoso, A., P. Chien, L. Z. Osherovich, and J. S. Weissman. 2000. Molecular basis of a yeast prion species barrier. *Cell* 100:277–288.

Schmitt-Ulms, G., G. Legname, M. A. Baldwin, H. L. Ball, N. Bradon, P. J. Bosque, K. L. Crossin, G. M. Edelman, S. J. DeArmond, F. E. Cohen, and S. B. Prusiner. 2001. Binding of neural cell adhesion molecules (N-CAMs) to the cellular prion protein. *J. Mol. Biol.* 314:1209–1225.

Scott, M., D. Peretz, R. M. Ridley, H. F. Baker, S. J. DeArmond, and S. B. Prusiner. 2004. Trangenic investigations of the species barrier and prion strains. In Prion Biology and Diseases. S. B. Prusiner, editor. Cold Spring Harbor Laboratory Press, Cold Spring Harbor, NY. 435–82.

Semenov, N. N. 1957. The Nobel Prize Lecture: Einige Probleme der Kettenreaktionen und der Verbrennungstheorie. *Angew. Chem.* 69:767–777.

Shaked, G. M., Z. Meiner, I. Avraham, A. Taraboulos, and R. Gabizon. 2001. Reconstitution of Prion Infectivity from Solubolozed Protease-resistant PrP and nonprotein components of prion rods. *J Biol. Chem.* 276:14324–14328.

Shorter, J. and S. Lindquist. 2004. Hsp104 catalyzes formation and elimination of self-replicating Sup35 prion conformers. *Science* 304:1793–1797.

Shyng, S.-L., J. E. Heuser, and D. A. Harris. 1994. A glycolipid-anchored prion protein is endocytosed via clathrin-coated pits. *J. Cell Biol.* 125:1239–1250.

Si, K., S. Lindquist, and E. R. Kandel. 2003. A Neuronal Isoform of the Aplysia CPEB has prion-like properties. *Cell* 115:879–891.

Sokolowski, F., A. J. Modler, R. Masuch, D. Zirwer, M. Baier, G. Lutsch, M. D. A., K. Gast, and D. Naumann. 2003. Formation of critical oligomers is a key event during conformational transition of recombinant syrian hamster prion protein. *J. Biol. Chem.* 278:40481–40492.

Sparrer, H. E., A. Santoso, F. C. Szoka, Jr., and J. S. Weissman. 2000. Evidence for the prion hypothesis: induction of the yeast [*PSI*$^+$] factor by *in vitro*-converted Sup35 protein. *Science* 289:595–599.

Stahl, N., D. R. Borchelt, K. Hsiao, and S. B. Prusiner. 1987. Scrapie prion protein contains a phosphatidylinositol glycolipid. *Cell* 51:229–240.

Stöckel, J., J. Safar, A. C. Wallace, F. E. Cohen, and S. B. Prusiner. 1998. Prion protein selectively binds copper (II) ions. *Biochemistry* 37:7185–7193.

Supattapone, S., H. Wille, L. Uyechi, J. Safar, P. Tremblay, F. C. Szoka, F. E. Cohen, S. B. Prusiner, and M. R. Scott. 2001. Branched polyamines cure prion-infected neuroblastoma cells. *J. Virol.* 75:3453–3461.

Swietnicki, W., M. Morillas, S. G. Chen, P. Gambetti, and W. K. Surewicz. 2000. Aggregation and fibrillization of the recombinant human prion protein huPrP 90-231. *Biochemistry* 39:424–431.

Tagliavini, F., F. Prelli, L. Verga, G. Giaccone, R. Sarma, P. Gorevic, B. Ghetti, F. Passerini, E. Ghibaudi, G. Forloni, M. Salmona, O. Bugiani, and B. Frangione. 1993. Synthetic peptides homologous to prion protein residues 106–147 form amyloid-like fibrils *in vitro*. *Proc. Natl. Acad. Sci. USA* 90:9678–9682.

Telling, G. C., M. Scott, J. Mastrianni, R. Gabizon, M. Torchia, F. E. Cohen, S. J. DeArmond, and S. B. Prusiner. 1995. Prion propagation in mice expressing human and chimeric PrP transgenes implicates the interaction of cellular PrP with another protein. *Cell* 83:79–90.

Telling, G. C., P. Parchi, S. J. DeArmond, P. cortelli, P. Montagna, R. Gabizon, J. Mastrianni, E. Lugaresi, P. Gambetti, and S. B. Prusiner. 1996. Evidence for the conformation of the pathologic isoform of the prion protein enciphering and propagating prion diversity. *Science* 274:2079–2082.

Thomzig, A., S. Spassov, M. Friedrich, D. Naumann, and M. Beekes. 2004. Discriminating scrapie and bovine spongiform encephalopathy isolates by infrared spectroscopy of pathological prion protein. *J. Biol. Chem.* 279:33854.

Torrent, J., M. T. Alvarez-Martinez, F. Heitz, J. P. Liautard, C. Balny, and R. Lange. 2003. Alternative prion structural changes revealed by high pressure. *Biochemistry* 42:1318–1325.

Turk, E., D. B. Teplow, L. E. Hood, and S. B. Prusiner. 1988. Purification and properties of the cellular and scrapie hamster prion proteins. *Eur. J. Biochem.* 176:21–30.

Vey, M., S. Pilkuhn, H. Wille, R. Nixon, S. J. DeArmond, E. J. Smart, R. G. Anderson, A. Taraboulos, and S. B. Prusiner. 1996. Subcellular colocalization of the cellular and scrapie prion proteins in caveolae-like membranous domains. *Proc. Natl. Acad. Sci. USA* 93:14945–14949.

Viles, J. H., F. E. Cohen, S. B. Prusiner, D. B. Goodin, P. E. Wright, and H. J. Dyson. 1999. Copper binding to the prion protein: structural implications of four identical cooperative binding sites. *Proc. Natl. Acad. Sci. USA* 96:2042–2047.

Wadsworth, J. D. F., A. F. Hill, S. Joiner, G. S. Jackson, A. R. Clarke, and J. Collinge. 1999. Strain-specific prion-protein conformation determined by metal ions. *Nat. Cell Biol.* 1:55–59.

Warner, R. G., C. Hundt, S. Weiss, and J. E. Turnbull. 2002. Identification of the heparan sulfate binding sites in the cellular prion protein. *J Biol. Chem.* 277:18421–18430.

Weissmann, C., M. Enari, P. C. Klohn, D. Rossi, and E. Flechsig. 2002. Transmission of prions. *J. Infect. Dis.* 186(Suppl 2):S157–S165.

West, M. W., W. Wang, J. Patterson, J. D. Mancias, J. R. Beasley, and M. H. Hecht. 1999. De novo amyloid proteins from designed combinatorial libraries. *Proc. Natl. Acad. Sci. USA* 96:11211–11216.

Whittal, R. M., H. L. Ball, F. E. Cohen, A. L. Burlingame, S. B. Prusiner, and M. A. Baldwin. 2000. Copper binding to octarepeat peptides of the prion protein monitored by mass spectrometry. *Protein Sci.* 9:332–343.

Wickner, R. B., D. C. Masison, and H. K. Edskes. 1995. [PSI] and [URE3] as yeast prions. *Yeast* 11:1671–1685.

Wille, H., M. D. Michelitsch, V. Guenebaut, S. Supattapone, A. Serban, F. E. Cohen, D. A. Agard, and S. B. Prusiner. 2002. Structural studies of the scrapie prion protein by electron crystallography. *Proc. Natl. Acad. Sci. USA* 99:3563–3568.

Wong, C., L. W. Xiong, M. Horiuchi, L. Raymond, K. Wehrly, B. Chesebro, and B. Caughey. 2001. Sulfated glycans and elevated temperature stimulate PrP(Sc)-dependent cell-free formation of protease-resistant prion protein. *EMBO J.* 20:377–386.

Yadavalli, R., R. P. Guttmann, T. Seward, A. P. Centers, R. A. Williamson, and G. C. Telling. 2004. Calpain-dependent endoproteolytic cleavage of PrPSc modulates scrapie prion. *J. Biol. Chem.* 279:21948–21956.

Zobeley, E., E. Flechsig, A. Cozzio, M. Enari, and C. Weissmann. 1999. Infectivity of scrapie prions bound to a stainless steel surface. *Mol. Med.* 5:240–243.

Zou, W. Q., S. Capellari, P. Parchi, M. S. Sy, P. Gambetti, and S. G. Chen. 2003. Identification of novel proteinase K-resistant C-terminal fragments of PrP in Creutzfeldt-Jakob disease. *J. Biol. Chem.* 278:40429–40436.

8

The Yeast Prion Proteins Sup35p and Ure2p

Joanna Krzewska and Ronald Melki

Abstract

In the yeast *Saccharomyces cerevisiae*, two genetic elements, [PSI$^+$] and [URE3], were discovered nearly 40 years ago (Cox, 1965; Lacroute, 1971). These traits do not obey Mendel's laws. They were first considered to be due to a nonchromosomal nucleic acid. However, [PSI$^+$] and [URE3] behavior differs significantly from that of DNA plasmids, RNA viruses, the mitochondrial genome, or RNA replicons. In addition, the genes encoding these traits are located in the nucleus of yeast cells (Schoun and Lacroute, 1969; Cox *et al.*, 1988). The very unusual properties of [PSI$^+$] and [URE3] led only 10 years ago to extend the prion hypothesis to the yeast *Saccharomyces cerevisiae* (Wickner, 1994). The term *prion* is meant for infectious protein that can lose for unknown reasons its normal function and that converts the functional polypeptide into a nonfunctional form. This chapter defines the genetic criteria that define a prion in yeast and relates the discovery of *S. cerevisiae* prions. The structural features of the soluble and insoluble forms of *S. cerevisiae* prions are detailed, and the mechanistic models for aggregation are presented. Finally, the mechanisms of maintenance and propagation of yeast prions, in particular the role played by molecular chaperones and the potential role of yeast prions, are described.

8.1. Genetic Criteria for the Prions in Yeast

Three genetic criteria need to be fulfilled in order to establish that an infectious agent is a prion and not a virus, plasmid, or other nucleic acid replicon: (i) reversible curability, (ii) induction of prion formation by overexpression of the normal protein, and (iii) a unique phenotypic relationship between the prion state and mutations in the gene of the protein (Wickner, 1994).

8.1.1. Reversible Curability

When a yeast strain is cured from a virus or plasmid, the nucleic acid is not expected to arise again in the cells unless it is reintroduced from another cell or by transformation. Prion traits reappear spontaneously in cured cells with a low frequency. This is due to the fact that the functional protein is present in the cells and can undergo the spontaneous change that leads to the nonfunctional prion form.

8.1.2. The Frequency with which the Prion Phenotype Arises Increases upon Overexpression of Infectious Proteins

Nucleic acid replicons do not arise *de novo* when proteins involved in their propagation are overproduced. In contrast, the frequency with which a prion phenotype arises increases upon overexpression of an infectious protein. This is due to the fact that the probability of proteins with prion properties to switch from a functional to a nonfunctional form, although low, increases when the overall cellular content of the normal form of the protein is increased, regardless of the mechanism of prion propagation.

8.1.3. Phenotypic Relationship Between the Prion State and Mutations in the Gene of the Protein

A prion state can only be maintained in cells where the gene of the protein is expressed. In *S. cerevisiae*, prion phenotypes mimic the loss of function of prion proteins as do mutations or deletions of the chromosomal genes encoding these proteins. This is because the nonfunctional prion form depletes the functional proteins. However, when the loss of function phenotype is due to mutations or deletions of the chromosomal genes encoding these proteins, the wild-type phenotypes are not restored when yeast cells are grown in the presence, for example, of millimolar concentrations of the protein denaturant guanidinium chloride or under high osmotic strength conditions and cannot reappear without introduction of new DNA in contrast with what happens in the case of authentic prion phenotypes (Aigle and Lacroute, 1975; Singh *et al.*, 1979; Lund and Cox, 1981; Tuite *et al.*, 1981).

8.2. The [PSI$^+$], [PIN], and [URE3] Traits

8.2.1. The [PSI$^+$] and [PIN] Traits

[PSI$^+$] is associated with Sup35p, an essential component of the translation termination machinery (Hawthorne and Mortimer, 1968; Inge-Vechtomov and Andrianova, 1970). Sup35p, known also as eRF3 (eukaryotic release factor 3), and Sup45p constitute the translation release factor that recognizes stop codons and releases nascent polypeptides from the final t-RNA. As the termination event is in competition with the misreading of the termination codon and the continuation of a polypeptide chain, the loss of the function of Sup35p reduces the fidelity with which ribosomes terminate translation at stop codons. In [PSI$^+$] cells, the effect of the weak suppressor tRNA *SUQ5* specific to the stop codon ochre (UAA) is increased. In [psi$^-$] cells carrying an *ADE2* gene interrupted by an ochre stop codon (*ade2-1* nonsense allele), the adenine biosynthesis pathway (Fig. 8.1A) is interrupted and a red precursor of adenine accumulates. The cells are unable to grow on a medium lacking adenine while the colonies appear red when adenine is supplemented (Fig. 8.2). The *ade2-1* nonsense allele is suppressed in [PSI$^+$] cells. The adenine biosynthesis pathway is restored and the colonies grow on media lacking adenine and are of normal white color (Fig. 8.2).

In *S. cerevisiae*, Sup35p (Swiss-Prot P05453) is a large polypeptide made of 685 amino acid residues that has a calculated molecular mass of 76,551 Da (Kushnirov *et al.*, 1988). One can distinguish three regions in Sup35p (Fig. 8.3). The N-terminal region (*N*) extends from amino acid residues 1 to 122 and is rich in Q (glutamine), N (Asparagine), and G (Glycine) residues (47%). This region is not essential for the role of the protein in translation termination (Ter-Avanesyan *et al.*, 1993; 1994; Derkatch *et al.*, 1996). It is called prion domain (PrD) as it plays a critical role in

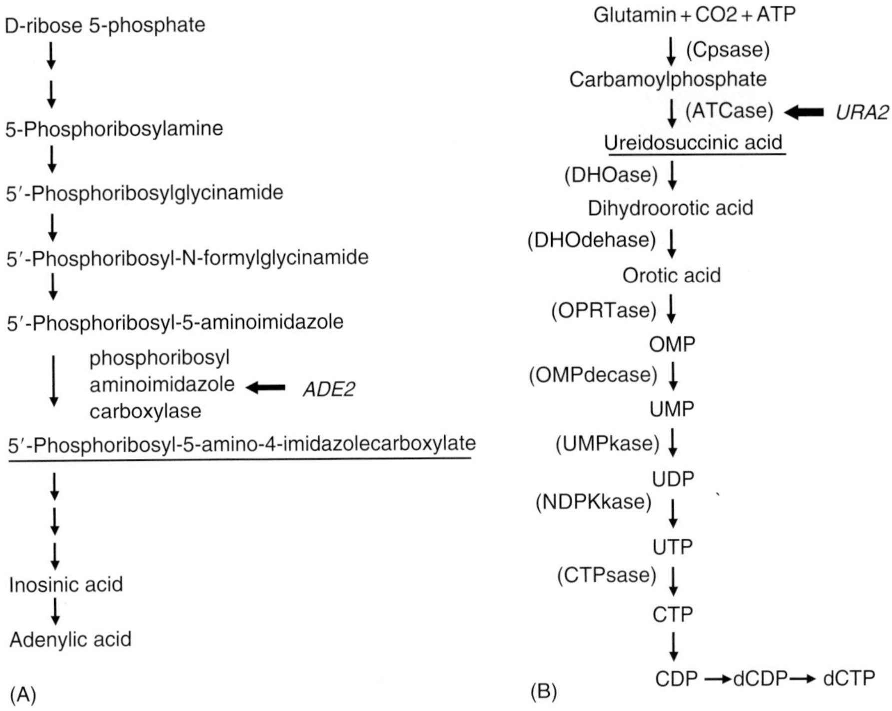

Figure 8.1. (A) Adenine and (B) pyrimidine biosynthesis pathways in *S. cerevisiae*. The steps under the control of *ADE2* and *URA2* gene products are indicated.

prion propagation. The middle region (*M*) has probably a structural function. It extends from amino acid residues 123 to 253. The C-terminal region is the functional domain of the protein providing translation-termination activity (Ter-Avanesyan *et al.*, 1993; Zhouravleva *et al.*, 1995). Sup35p is soluble and functional in [psi$^-$] cells while it is mainly insoluble and nonfunctional in [PSI$^+$] cells (Stansfield *et al.*, 1995; Patino *et al.*, 1996).

The overexpression of Sup35p increases the frequency of [PSI$^+$] appearance. This is also what is observed upon the overexpression of Rnq1p (Swiss-Prot P25367), a 42.5-kDa polypeptide with unknown function, rich in Q and N residues, in particular in its part extending from amino acid residue 153 to 405 (Fig. 8.3). Rnq1p exists as a soluble protein as well as in insoluble aggregates (Sondheimer and Lindquist, 2000). The amount of insoluble Rnq1p increases when the protein is overproduced (Derkatch *et al.*, 2001). In its insoluble state, Rnq1p allows the conversion of [psi$^-$] cells into [PSI$^+$] cells and is at the origin of the phenotype [PIN$^+$] that comes from "[PSI$^+$] inducibility" (Derkatch *et al.*, 1997). The overexpression of many other G- and N-rich polypeptides has [PIN$^+$]-like effects (Derkatch *et al.*, 2001; Osherovich and Weissman, 2001). Nevertheless, only [PIN$^+$] is cured when yeast cells are grown on guanidinium chloride and can reappear in cured cells suggesting it is a prion (Derkatch *et al.*, 2000).

Growth of [psi-] *ade2-1* mutant *Saccharomyces cerevisiae* cell

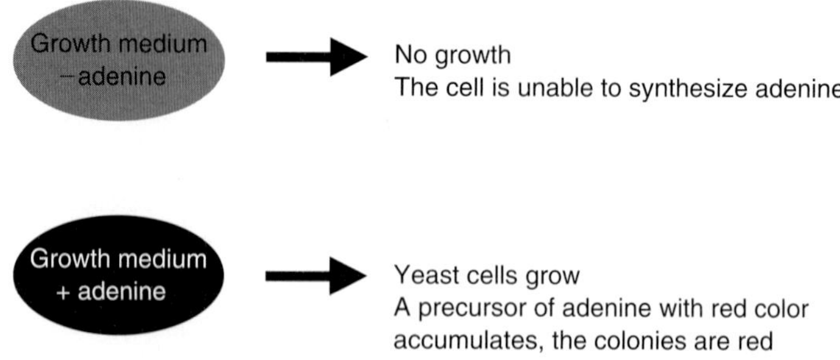

When an *ade2-1* mutant *S. cerevisiae* cell acquires the [PSI+] phenotype

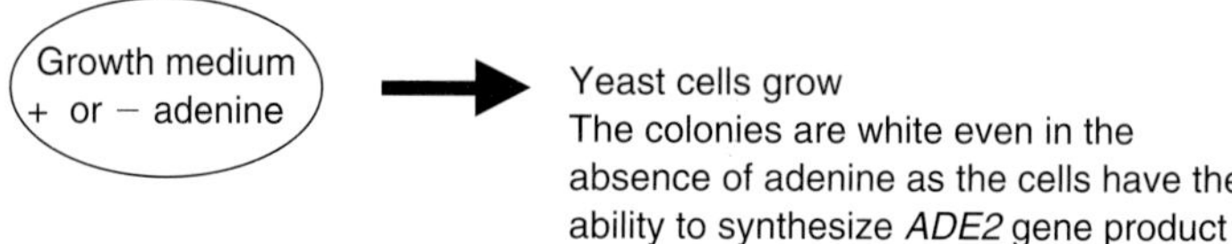

Figure 8.2. Synthetic scheme summarizing the behavior of [psi⁻] and [PSI⁺] *S. cerevisiae* cells bearing an *ade2-1* allele on different growth media. In a [psi⁻] context, Sup35p is functional, and a truncated, nonfunctional *ADE2* gene product is synthesized. The biosynthesis of adenine is interrupted, and a red precursor accumulates in the cells (shown as black oval). Adenine is needed for growth. In a [PSI⁺] context, Sup35p is not functional, and translation termination is not efficient. The partial suppression of the stop codon in *ade2-1* allele restores the function of *ADE2* gene product. The biosynthesis of adenine is restored. *S. cerevisiae* colonies are white (shown as white oval).

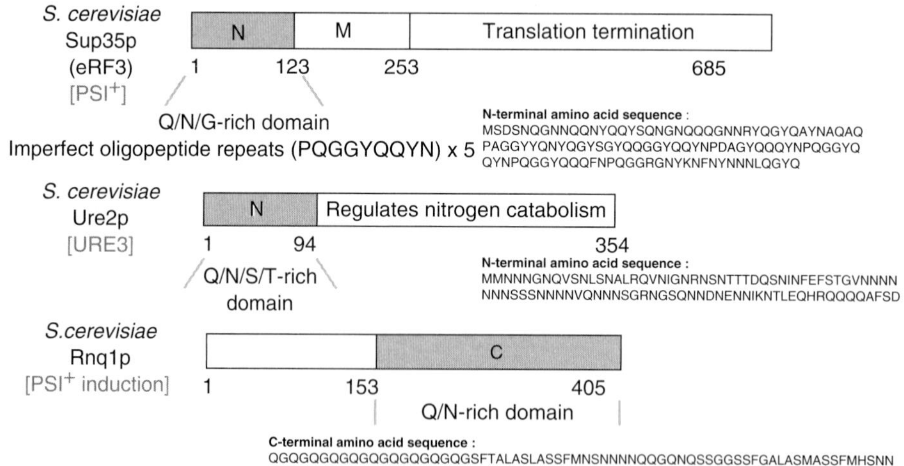

Figure 8.3. Comparison of the characteristics of the yeast prions Ure2p, Sup35p, and Rnq1p. The boundaries between the different domains in each prion are indicated. The prion domains are outlined (in gray) and their amino acid composition or characteristics presented. The consensus sequence of the oligopeptide repeat in Sup35p is shown and underlined in the N-terminal amino acid sequence.

8.2.2. The [URE3] Trait

In the presence of nitrogen sources that can be used with ease, such as ammonia or glutamine, yeast cells turn off the synthesis of enzymes and transporter involved in the use of complex nitrogen sources such as proline, allantoate, or urea (Cooper, 1982; Magasanik, 1992). While studying mutant yeast cells unable to synthesize ureidosuccinate (USA), a precursor of uracile, because of a mutation in *URA2* gene, the gene encoding aspartate transcarbamylase (Fig. 8.1B), François Lacroute (1971) found that USA was not taken up when ammonium is supplied in the medium and the cells were unable to grow. In contrast, when a complex source of nitrogen such as proline is present in the medium, USA is taken up and the cells grow (Fig. 8.4). Thus, USA uptake appears regulated by the nitrogen source. To understand this regulation, Lacroute screened for *ure2* mutants that are able to grow on a medium containing ammonium and USA. Among the isolated mutants, a genetic element that segregates in a non-Mendelian manner was found and called [URE3].

[URE3] is associated with Ure2p, a polypeptide whose exact function is unknown, which acts as a negative regulator of nitrogen metabolism (Mitchell and Magasanik, 1984; Courchesne and Magasanik, 1988; Coschigano and Magasanik, 1991). Functional Ure2p is believed to interact with the GATA transcription factor Gln3p in the cytoplasm thus preventing the entry of Gln3p into the nucleus where it allows the transcription of a number of genes, among which is *DAL5*, the gene encoding the allantoate transporter Dal5p (Rai *et al.*, 1987). When Ure2p is not functional, Dal5p is expressed in a constitutive manner, and allantoate, a complex nitrogen source, is taken up (Fig. 8.5). Allantoate resembles structurally USA. Dalp5p can therefore transport this precursor of uracil and the cells can grow on media containing ammonium and USA (see Fig. 8.4).

Growth of *URA2* mutant *Saccharomyces cerevisiae cell*

Minimal medium
+ ammonium
+ USA

Ammonium is a good nitrogen source

No growth
The import of ureidosuccinate is impossible as a particular permease is not expressed.

Minimal medium
+ proline + USA

Proline is a poor nitrogen source

Yeast cells grow
The permease Dal5p is expressed and ureidosuccinate is imported into the cells.

When a *URA2* mutant *S. cerevisiae* cell acquires the [URE3] phenotype

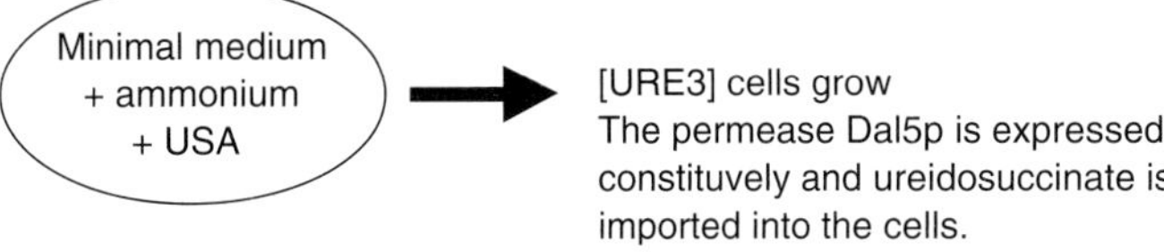

Figure 8.4. Synthetic scheme summarizing the behavior of *URA2* mutant *S. cerevisiae* cells in [wild-type] and [URE3] contexts. In a [wild type] context, Ure2p is functional. The uptake of ureidosuccinate is repressed in the presence of ammonium, and the *URA2* cells that are unable to synthesize ureidosuccinate, a precursor of pyrimidines, from ammonium via glutamine are unable to grow. In the absence of ammonium and the presence of a complex nitrogen source such as proline, the synthesis of enzymes and transporters involved in the use of complex nitrogen sources is turned on. Ureidosuccinate is incorporated and the cells grow. In a [URE3] context, Ure2p is not functional. The regulation of nitrogen metabolism is compromised. *URA2* mutant cells incorporate ureidosuccinate even in the presence of ammonium and grow.

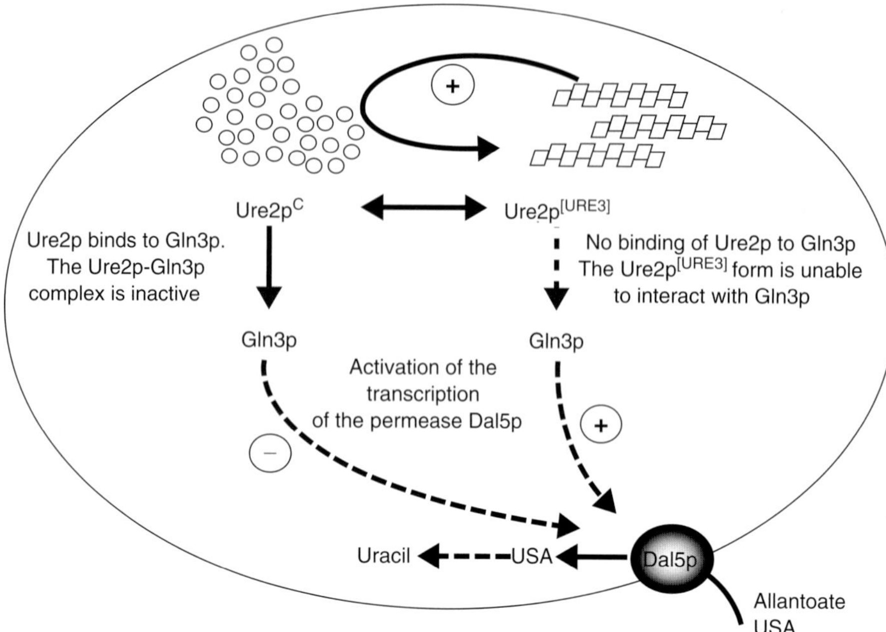

Figure 8.5. The [URE3] phenotype and the role of Ure2p in the regulation of nitrogen catabolism. In wild-type cells, functional Ure2p is believed to inactivate the transcription factor Gln3p through the formation of an inactive Ure2p-Gln3p complex. The allantoate permease, Dal5p, is not synthesized and the cells are unable to incorporate allantoin and its structural analogue ureidosuccinate. In a [URE3] context, Ure2p is unable to interact with Gln3p. Gln3p triggers constitutively the expression of Dal5p, and the cells take up ureidosuccinate. The inactive form of Ure2p in [URE3] cells converts catalytically the active form of the protein in an inactive form. *URA2* cells are unable to grow in the presence of ammonium and ureidosuccinate, unless they have acquired the [URE3] phenotype (see Figs. 8.1 and 8.4).

In *S. cerevisiae*, Ure2p (Swiss-Prot P23202) has a calculated molecular mass of 40,271 Da (Dalton) and is composed of 354 amino acid residues (see Fig. 3). The N-terminal part of Ure2p extending from amino acid residue 1 to 93 (Thual *et al.*, 1999) has an unusual composition. It is very rich in Q, N, S (serine), and T (threonine) residues (62%). This region is required for the propagation of the [URE3] phenotype (Masison and Wickner, 1995) and is referred to as the prion domain (PrD) of the protein. The C-terminal domain extends from amino acid residues 94 to 354 (Thual *et al.*, 1999). It complements *URE2* gene deletion and therefore constitutes the functional domain of the protein (Masison and Wickner, 1995).

8.3. Other Potential Prions

Any chromosomally encoded protein that undergoes a change into a modified form that is necessary for its own modification is a potential prion. Such proteins are detected if the prion form is toxic or if it confers selective advantages or disadvantages. In yeast and fungi, prions do not kill the organism but instead allow cells to read through stop codons, to use poor nitrogen sources, or to forbid fusion between cells, which can be advantageous under particular conditions but disadvantageous under normal growth conditions.

A number of surveys have been conducted using the unusual content and distribution of Q and N amino acid residues to identify novel yeast prions. Several proteins with Q- and N-rich domains have been revealed (Michelisch and Weissman, 2000; Sondheimer and Lindquist, 2000). Novel prions have also been created by fusing the prion domain of Sup35p to polypeptides such as the rat glucocorticoid receptor (Li and Lindquist, 2000; True and Lindquist, 2000). All the polypeptides with Q- and N-rich domains do not possess prion properties. This is the case, for example, with the protein Huntingtin (associated with Huntington disease) that exhibits no infectious activity. Furthermore, the mammalian prion PrP and the *P. anserina* prion HET-s are neither rich in Q nor N residues. These facts strongly suggest that the presence of Q- and N-rich domains in a polypeptide is not sufficient to make a protein a prion.

8.4. The "Prion Domains" of Yeast Prions

Yeast prions possess a "prion domain" (PrD) crucial for prion propagation that can be either N- or C-terminal. Indeed, truncated forms of yeast prions lacking their PrD cannot convert into the prion forms. In addition, the overexpression of the PrDs of Sup35p or Ure2p in wild-type cells elevates very significantly the rate of appearance of [PSI$^+$] and [URE3] (Masison and Wickner, 1995; Derkatch *et al.*, 1996). The minimal PrDs of yeast prions have been defined by deletion studies (Maddelein and Wickner, 1999; Sondheimer and Lindquist, 2000; Parhan *et al.*, 2001). However, regions outside the PrDs of Sup35p and Ure2p affect markedly [PSI$^+$] and [URE3] propagation as well (Maddelein and Wickner, 1999; Liu *et al.*, 2002). While the only remarkable feature in Ure2p PrD are two poly-asparagine stretches, that of Sup35p includes five imperfect repeats of the oligopeptide PQGGYQQ-YN that resemble to some extent the octarepeat PHGGG-WGQ of mammalian PrP (Tuite, 2000). Changes in the number of repeat influences [PSI$^+$] propagation (TerAvanesyan *et al.*, 1994; Liu and Lindquist, 1999). The role of such repeats in prion propagation is unclear as the PrDs of Ure2p and Rnq1p lack such repeats.

8.5. Structural Features

The PrDs of yeast prions are flexible while the functional domains are compactly folded. Proteolytic studies indicate that the N-terminal parts of Sup35p, Ure2p and the C-terminal parts of Rnq1p are highly exposed to the solvent (Thual *et al.*, 1999; Serio *et al.*, 2000; Derkatch *et al.*, 2001). In the case of Ure2p, the physical boundary between the N- and C-terminal domains was determined using limited proteolysis (Thual *et al.*, 1999). The three-dimensional structure of Ure2p lacking the N-terminal domain extending from amino acid residues 1 to 93 (Bousset *et al.*, 2001; Umland *et al.*, 2001) reveals the compact folding of the functional part of the protein.

Purified, full-length Sup35p and Ure2p are soluble and helical (Glover *et al.*, 1997; Thual *et al.*, 1999). The soluble forms can convert *in vitro* into high-molecular-weight oligomers at neutral pH (Glover *et al.*, 1997; Thual *et al.*, 1999). The high-molecular-weight oligomers are either soluble and spherical or insoluble and fibrillar (Figs. 8.6A and 8.6B) (Thual *et al.*, 1999; Serio *et al.*, 2000). The formation of the fibrils can be followed by a variety of spectroscopic methods and is greatly accelerated by seeding with preformed fibrils (Figs. 8.6C and 8.6D) (Glover *et al.*, 1997; Thual *et al.*, 1999, 2001). Furthermore, there is evidence that *in vitro*–assembled fibrils made of recombinant Sup35pNM fragment are infectious as their introduction in yeast cells induces the *de novo* appearance of the prion phenotype (Tanaka *et al.*, 2004; King and Diaz-Avalos, 2004).

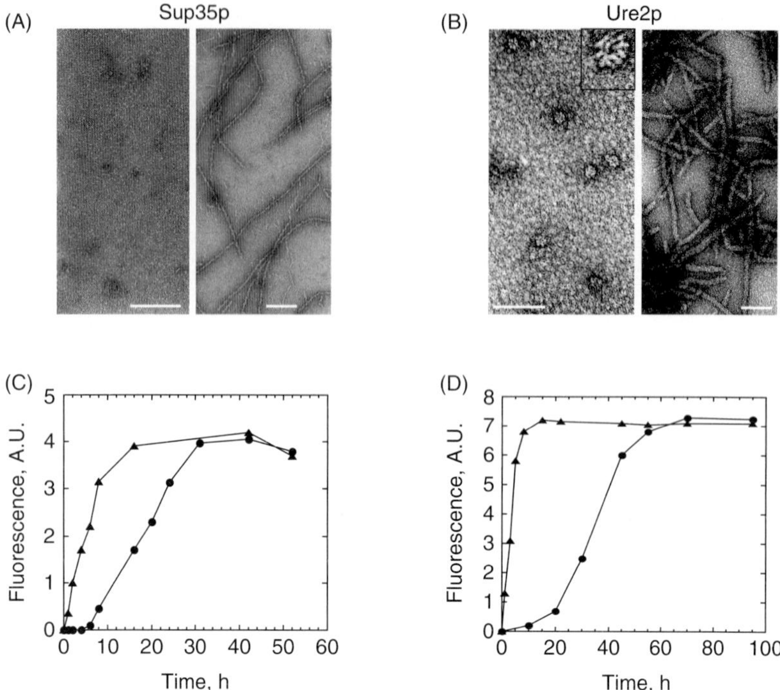

Figure 8.6. Oligomeric states of Sup35p and Ure2p and *in vitro* assembly reactions. Electron micrographs of negatively stained soluble (left panel) and fibrillar (right panel) Sup35p (A) and Ure2p (B) oligomers at pH 7.5. Bar = 0.2 μm in (A), 0.1 μm in (B), and 20 nm in the inset in (B). Time courses of Sup35p (C) and Ure2p (D) assembly into fibrils monitored by thioflavine T binding in the absence (●) and the presence (▲) of preformed fibrils.

8.6. *In Vitro* Assembly Process of Yeast Prions

While there is a general agreement on the molecular mechanism underlying the formation of the fibrillar form of prion proteins, the nature and extent of the conformational changes and the properties of the oligomeric species remain subject to great debate.

The self-assembly of Sup35NM fragment is a cooperative process where one can distinguish a lag phase where nucleation occurs followed by an elongation phase where the assembly accelerates preceding the onset of a plateau. The lag phase is shortened significantly upon increasing Sup35NM fragment concentration or seeding the reaction with preformed Sup35NM fibrils further demonstrating the cooperative character of the assembly reaction (Glover *et al.*, 1997). In the lag phase preceding assembly, variably sized oligomers containing between 20 and 80 Sup35NM molecules form and convert later on into nucleating units that have been proposed to aggregate in a manner similar to colloidal particles giving rise to fibrils (Serio *et al.*, 2000; Xu *et al.*, 2001). These fibrils have been shown to possess an intrinsic polarity (Inoue *et al.*, 2001) that may be the consequence of a structural diversity within the fibrillar scaffold that might account for some of the aspects related to prion strains *in vivo* (DePace and Weissman, 2002).

It is widely believed that Sup35p assembles into β-sheet-rich aggregates through the establishment of an array of hydrogen bonds between the side-chain and the main-chain amides of its polyglutamine/asparagine-containing region (Perutz, 1999; Perutz *et al.*, 1994, 2002). While this is certainly the case in solution when unfolded Sup35NM fragment is diluted from denaturant or when a peptide reproducing the amino acid stretch 7–13 of Sup35p, unrelated to the repeats that modulate the stability of [PSI$^+$], is used (Balbirnie *et al.*, 2001), no data is available for the assembly reaction of full-length Sup35p. It is therefore not clear whether the assembly of full-length Sup35p follows closely that of its NM fragment, and the size of the stable nucleus and the structure of full-length Sup35p in its fibrillar form are unknown.

The assembly reaction of full-length Ure2p is also cooperative and follows a lag phase (Thual *et al.*, 1999). The polymerization of Ure2p is greatly accelerated by addition of minute amounts of preformed Ure2p fibrils (Thual *et al.*, 1999), thus indicating that the formation of nuclei that requires the interaction of many Ure2p molecules is the limiting step in the assembly reaction. The dissociation of the native dimeric form of Ure2p to its constituent monomers is a prerequisite for assembly into fibrils (Bousset *et al.*, 2002). The minimal stable nucleus that is the precursor of Ure2p fibrils is made of six Ure2p molecules (Bousset *et al.*, 2002; Fay *et al.*, 2003). Ure2p fibrils grow in a polar manner (Fay *et al.*, 2003).

The assembly of Ure2p is certainly not solely due to its polyglutamine/asparagine extensions. Indeed, the assembly process of a variant Ure2p truncated of its amino acids 15–42 was recently described. While the Q and N proportion in the N-terminal domain of full-length Ure2p is 46% (43 residues out of 93), this proportion increases to 54% (35 residues out of 65) in the Ure2pΔ15-42 variant. In marked contrast with what one would expect if the assembly of Ure2p was driven by the polyglutamine/asparagine extensions, the Ure2pΔ15-42 variant does not assemble under conditions where full-length Ure2p does (Jiang *et al.* 2004). This clearly indicates either that assembly is not driven by the polyglutamine/asparagine extensions of this prion or that the secondary structure elements of the N-terminal domain of Ure2p revealed recently by limited proteolysis studies (Bousset *et al.*, 2004) that are crucial for assembly are lost upon removal of amino acid residues 15 to 42.

8.7. Nature of the Fibrillar Forms of Sup35p and Ure2p

The fibrillar forms of Sup35p and Ure2p have an increased resistance to proteolytic treatments and bind the dye Congo red (Glover *et al.*, 1997; Thual *et al.*, 1999). Ure2p fibrils bind as well thioflavine T and exhibit yellow-green birefringence upon binding of Congo red in polarized light (Thual *et al.*, 2001). These properties are not unique to prion fibrils as biological polymers such as actin filaments and microtubules exhibit increased resistance to proteolysis and yellow-green birefringence in polarized light upon Congo red binding (Bousset *et al.*, 2004). It is claimed that prion fibrils are amyloid as the assembly reaction is accompanied by a change in the circular dichroism spectrum that resembles an α-helical to β-sheet transition (Glover *et al.*, 1997; Schlumpberger *et al.*, 2000). Circular dichroism (CD) is not an adequate method to measure conformational changes occurring during the assembly of a polypeptide into protein fibrils. Indeed, although the optical activity in the region 190–230 nm is dominated by the peptide backbone, the CD of an α-helix or a β-sheet is a function of its length. This is not adequately taken into account when homopolymeric structures are used as references to compute the experimental spectra. Suitable proteins with known three-dimensional structures should instead be used to compute the experimental data. In addition, when a protein assembles into highly ordered polymers, individual regions of secondary structure that are exposed to the solvent are packed next to each other.

It is nearly impossible to adequately compute the contribution to the overall CD spectrum of packed helical and sheet regions even by the use of model proteins with known three-dimensional structures instead of the widely used homopolymeric structures (Cantor and Schimmel, 2001).

Amyloids are defined as compact extracellular deposits with inherent birefringence that increases intensely upon binding of the dye Congo red. Amyloids are associated with various pathologies and made of fibrillar materials that exhibit a typical cross-ß structure in x-ray fiber diffraction images (Sipe and Cohen, 2000; Westermark *et al.*, 2002). This is a characteristic x-ray diffraction pattern observed with fibrils aligned and oriented with their main axes perpendicular to the x-ray beam. At least two reflections must be present in the x-ray diffraction pattern to define it as that of a cross-β structure: the main chain reflection, a sharp meridional reflection at 4.7 Å, corresponding with the separation of two hydrogen-bonded chains; and an equatorial reflection at 10 Å, also called side-chain reflection, corresponding with the packing distance between two juxtaposed β-sheets. The typical anisotropy of the diffraction patterns of amyloid fibrils implies a β-sheet orientation in which hydrogen bonds and the plane of the sheet are parallel to the main fibrils axis, whereas backbone and side chains extend perpendicular to the fibrils axis (Sunde *et al.*, 1997). Fourier transform infrared (FTIR) spectroscopy is another technique ideally suited for analyzing β-sheet content, the strong hydrogen bonds of amyloid giving rise to specific bands in the IR spectrum between 1623 and 1618 cm^{-1} that can be uniquely assigned as amyloid (Zurdo *et al.*, 2001).

The nature of fibrils made of full-length Sup35p has never been accessed. Fibrils made *in vitro* after dilution from denaturant of Sup35p NM fragment that seed the aggregation of full-length Sup35p (Glover *et al.*, 1997) are of amyloid nature (Serio *et al.*, 2000). Similarly, fibrils made after dilution from denaturant of a synthetic polypeptide reproducing part of the N-terminal domain of Ure2p are probably of amyloid nature (Taylor *et al.*, 1999). However, fibrils assembled *in vitro* of native full-length Ure2p under physiologically relevant conditions are not built on the cross-β framework of amyloid (Bousset *et al.*, 2002, 2003), thus indicating that subtle conformational changes (Carrell and Gooptu, 1998; Huntington *et al.*, 1999; Liu *et al.*, 2001) such as that occuring upon assembly of lithostathine (Grégoire *et al.*, 2001) and the members of the serpin superfamily (Lomas and Carrell, 2002) may govern the assembly reaction. It is worth nothing that when the latter fibrils are heated at 60 °C or incubated at pH 2.0, they acquire the characteristics of amyloids but lose the propensity to seed the assembly of native Ure2p (Bousset *et al.*, 2003).

8.8. Mechanistic Models for Prion Propagation

The mechanism of prion propagation is not yet understood at the molecular level. It is widely accepted though that a protein-folding event, where a partial unfolding or folding of a region in the protein with prion properties takes place, plays a critical role in the acquisition by native prion of infectious properties.

Two working models have been proposed. The first is called the template-assistance model (Griffith, 1967; Gajdusek, 1988; Prusiner, 1991), the second the seeded-polymerization model (Jarret and Lansbury, 1993).

In the template-assistance model (Fig. 8.7A), the native soluble form of a prion protein interacts with the infectious form. As a consequence of this physical interaction, the native form is transformed into the infectious form. The infectious homodimer may then dissociate allowing the interaction of each infectious polypeptide with a native prion protein and the generation of additional infectious forms. In this model, the infectious form acts somehow as a template facilitating the occurrence of an energetically unfavorable protein-folding event. A variant of this

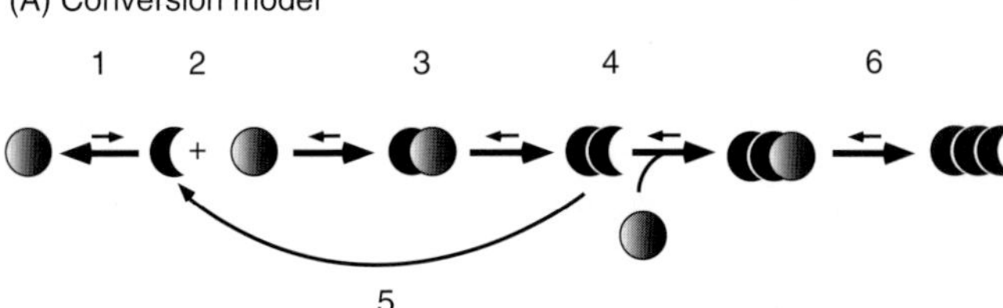

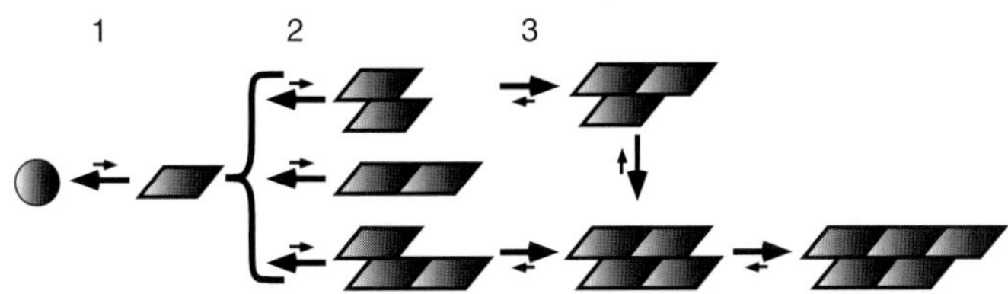

Figure 8.7. Theoretical models for prion propagation. (A) Template-assistance model. (1) The normal form of the prion protein is in equilibrium with the rare abnormal form that is the precursor of the aggregated form. The abnormal form can be of exogenous origin in iatrogenic or dietary infections. (2) The prion protein in its abnormal form interacts with the normal form of the prion protein. (3) Following this interaction, (4) the normal prion protein converts into the abnormal form. (5) The homodimer can then either dissociate, each abnormal monomer converting the cellular pool of native prion protein into the abnormal form through cycles of interaction and dissociation or 6) elongate by interacting with native prion proteins and converting them to the abnormal form. (B) Seeded polymerization model. (1) The normal form of the prion protein is in equilibrium with the rare abnormal form of the prion protein that is the precursor of the aggregated form. (2) Prion proteins in their abnormal form can interact with each other. The interaction is unstable because the intermolecular interactions are not strong enough to outweigh the entropic cost of binding. Thus, the low-molecular-weight oligomers that are formed dissociate until a stable nucleus is formed. (3) This nucleus or seed can grow indefinitely from one or both ends depending on the structural properties of the abnormal form of prion protein. It can also break into smaller stable fibrils that can elongate by incorporation of the abnormal form of the prion protein. The seed can be of exogenous origin in iatrogenic or dietary infections.

scenario has been proposed where the interaction between an unidentified enzyme or chaperone and native prion proteins could favor the protein folding event at the origin of the prion form (Telling *et al.*, 1995).

In the seeded-polymerization model (Fig. 8.7B), the native monomeric prion molecule P is in equilibrium with a rare and unstable conformational isoform P^*. P^* can be stabilized by complementary association with another P^* molecule. Because P^* is unstable, its concentration must be very low and the formation of low-molecular-weight oligomers of P^* are not favored because the energy gained from intermolecular interactions does not outweigh the entropic cost of binding until a stable nucleus P^n is formed. A number of inherited mutations that destabilize yeast prions predispose them to convert to their unstable P^* form (Fernandez-Bellot *et al.*, 2000; Uptain *et al.*, 2001), thus increasing the concentration of the latter form and favoring their oligomerization. Once stable oligomers are formed, they can grow by incorporation of the P^* form at their ends. Such polymers can break into smaller units, each of which would behave as a seed. The requirement for stable nuclei to form before conversion is stable accounts for the low frequency of occurrence of the prion phenotypes. However, the high efficiency of incorporation of P^* into oligomers and polymers made of P^* accounts for the efficiency of transmission from mother to daughter cells.

8.9. Maintenance and Inheritance

The efficiency of transmission of prion disease between species is governed by a phenomenon known as the "species barrier" (Prusiner, 1998). This barrier is due to differences in primary structure of PrP from different species and to the ability of PrP to adopt a number of different conformational states that stably propagate the same structural forms giving rise to specific neuropathologies and different rates of disease progression that are termed prion "strains" (Bruce, 1993; Bessen and Marsh, 1992; Caughey *et al.*, 1998).

In yeast, different prion "strains" also called "variants" have been described (Derkatch *et al.*, 1996; Chien and Weissman, 2001; Schlumpberger *et al.*, 2001; Uptain *et al.*, 2001). Indeed, a prion phenotype can disappear spontaneously from a yeast strain with a frequency comparable to chromosomal gene mutations (Lund and Cox, 1981). The frequency with which a prion phenotype is lost upon cell division defines the stability of a prion strain. A relationship between prion protein primary structure and prion strains has been established (Liu and Lindquist, 1999; King, 2001; Chien *et al.*, 2003). It is reasonable to envisage that differences in primary structure allow the assembly of prion protein into different fibrillar forms with distinct morphologies, growth rates, and polarity (Chien and Weissman, 2001).

When yeast prion strains were first observed, a difference in the efficiency of curing these strains upon overexpression of Hsp104 was described (Chernoff *et al.*, 1995; Derkatch *et al.*, 1996). The molecular chaperone Hsp104 is strictly required for yeast prion propagation as strains carrying *HSP104* gene deletion are unable to propagate the [PSI$^+$] (Chernoff *et al.*, 1995) and the [URE3] (Moriyama *et al.*, 2000) traits. Hsp104 is not the only cellular factor necessary for the continued propagation of prions in yeast cells. The expression levels of other molecular chaperones are also of great importance. *SSA1* gene expression has been reported to be important for [PSI$^+$] stability (Jones and Masison, 2003) as are other members of the Hsp70 family (Chernoff *et al.*, 1999; Newnam *et al.*, 1999).

The overexpression of Ydj1p (a member of the Hsp40 family) and Ssa1p in exponentially growing [URE3] cells (Moriyama *et al.*, 2000; Jones and Masison, 2003) leads to the destabilization of [URE3] phenotypes. Finally, Sis1p (another member of the Hsp40 family) is required for [PIN$^+$] propagation (Sondheimer *et al.*, 2001).

A number of models that account for the role played by molecular chaperones in the maintenance or the destabilization of the prion phenotypes have been published. The first model hypothesizes that the elevated levels of molecular chaperones facilitate the disaggregation of the high-molecular-weight species of prion proteins that act as seeds (Glover and Lindquist, 1998; Kushnirov and Ter-Avanesyan, 1998). In the second model, the molecular chaperones sequester either the folding intermediate(s) that assemble into prion aggregates or the cellular factor(s) that are required for the generation of the high-molecular-weight species of prion proteins that act as seeds (Chernoff *et al.*, 1999). The third model hypothesizes that the faithful transmission of the high-molecular-weight species of prion proteins that act as seeds from mother to daughter cells is compromised upon the overexpression of molecular chaperones (Cox *et al.*, 2003). Additional scenarios represented in Fig. 8.8 can be proposed. A direct *in vitro* approach, where the effect of each purified molecular chaperone on the assembly reaction of soluble Sup35p, Rnq1p, and Ure2p and on *in vitro*–assembled fibrils such as those carried out very recently (Krzewska and Melki, 2006; Shorter and Lindquist, 2006), will allow in the future the identification of the exact role of each molecular chaperone in yeast prion propagation.

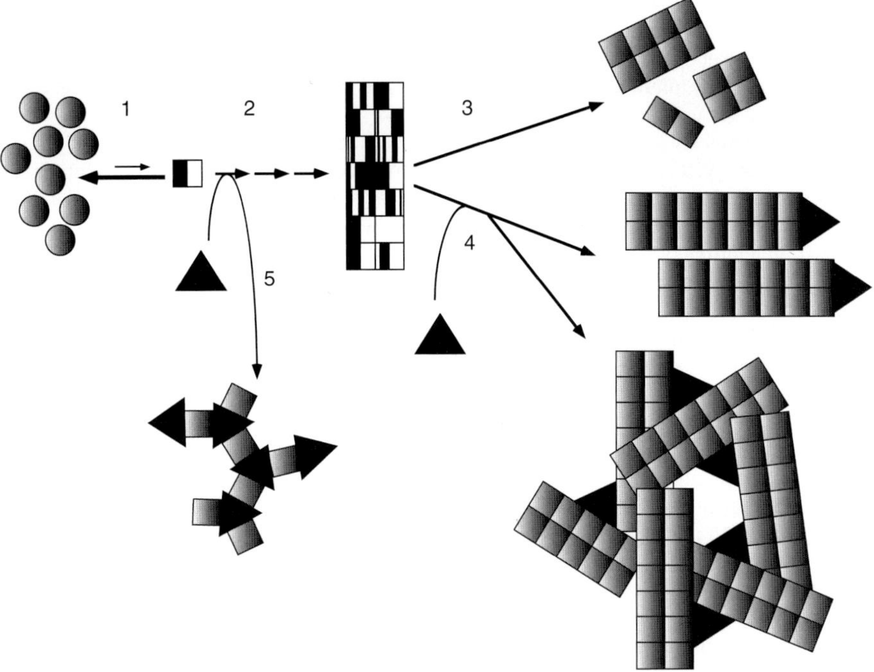

Figure 8.8. Molecular chaperones can modulate the aggregation of prion proteins. (1) Molecular chaperones (triangle-shaped molecule) can displace the equilibrium between the normal form of the prion protein (molecule with a circular shape) and its abnormal form (molecule with a square shape) either toward the normal form (i.e., refold misfolded prion protein) or toward the abnormal form (i.e., unfold native prion protein). In the first case, molecular chaperones will have an unfavorable effect on oligomerization while in the second molecular chaperones will favor oligomerization. (2) The abnormal prion protein oligomerizes into fibrils that can grow from one or both ends. (3) Molecular chaperones can shear the fibrils, thus generating additional ends (i.e., incorporation sites for abnormal prion protein). This will favor assembly and decrease the limiting lag phase preceding assembly. (4) Molecular chaperones can cap one or the two ends of the fibrils. This will either inhibit the assembly reaction or have no effect depending on the intrinsic polarity of the fibrillar scaffold and whether the capped end is the end where abnormal prion protein is incorporated. Molecular chaperones can interact with fibril walls and favor the bundling or the disordered aggregation of the fibrils into very large aggregates. Such an aggregation may lead to the neutralization of prion fibrils, which would reflect in a negative effect on assembly. (5) Finally, molecular chaperones can interact specifically with the abnormal soluble form of prion protein and sequester the precursors of the fibrils into structured or amorphous aggregates. The latter possibility would restrain assembly into fibrils.

8.10. Potential Role of Yeast Prions

Yeast prions do not spread from cell to cell. They are inherited by daughter cells from mothers or passed between partners during mating. The frequencies of appearance and loss of a yeast prion phenotype are 10^{-5} and 10^{-7}, respectively. It is not trivial to imagine a beneficial function for the loss of function of a protein via an irreversible aggregation process. A plausible function for the behavior of yeast prions has been proposed (Li and Lindquist, 2000; True and Lindquist, 2000). These epigenetic and metastable traits would provide a selective advantage for yeast cells in fluctuating environments. Whether the prion phenotype mimics the loss or gain of a function, the genes of these proteins are present in an unaltered form in the cell, and the loss or gain of function can be considered as reversible provided that the aggregated form of prion protein that perpetuates the structure-based mode of inheritance can be eliminated, counterbalanced, or neutralized. Thus, prions might constitute a class of proteins whose activity can be modulated without requirement for genetic mutations.

Protein synthesis is an energy-consuming process in the cell. It is therefore very difficult to accept the idea that a cell would have designed a process where the activity of a number of proteins would be regulated by their irreversible aggregation after synthesis, in particular as the transcription and translation of the genes encoding these proteins (i.e., their synthesis) is not turned off. In addition, while the regulation of inducible genes is very efficient, immediate, and costless, the regulation of the activity of a protein through its irreversible aggregation is inefficient, slow, and energy-consuming. The efficiency of the prion inactivation depends on the amount of soluble form of the prion protein left in yeast cells. In cells exhibiting the [PSI$^+$] phenotype for example, the essential protein Sup35 is in part functional as witnessed by translation termination. This is probably due to the fact that the critical concentration for aggregation of prion proteins is not zero but in the nanomolar range. *In vitro*, prion proteins assemble into fibrils within a timescale (hours when the solutions are not shaken) incompatible with shutting off an activity after changes in the environment. Finally, the synthesis of prion proteins in cells exhibiting prion phenotypes are not shut off, that is, neo-synthesized prion proteins are destined to aggregate, which is pretty illogical and energy consuming.

The only case where the inactivation of a protein via its aggregation may have indirect beneficial aspects is the case of Sup35p. The inactivation of Sup35p leads to an enhanced suppression of non-sense codons and non-sense codon mutations (i.e., a reduction in translational fidelity). The [PSI$^+$]-mediated read-through of naturally occurring stop codons, or open reading frames (ORFs) that have acquired inactivating stop codon mutations, surely alters the function and/or the stability of encoded proteins. It may though allow the expression of proteins that are not naturally expressed, thus contributing to phenotypic changes. In the case the new phenotype is advantageous, the population will grow and mutations will arise to fix the trait by eliminating stop codons that are relevant to the phenotype and require [PSI$^+$] for read-through, and the cell will occupy new niches. If this scenario occurs, the protein profiles of wild-type and yeast cells exhibiting the prion phenotype should differ significantly when compared using proteomic tools.

8.11. Conclusions, Perspectives, and Limitations of *S. cerevisiae* Prions as Model Systems for Mammalian Prions and Polyglutamine Diseases

Important steps have been made in understanding the biology of yeast prions during the past decade. This includes the characterization of the structure and folding dynamics of yeast prions. Although a number of precursors have been identified, the folding intermediates that assemble into the low- and high-molecular-weight oligomers involved in prion propagation are still unidentified. Their identification is crucial for developing tools that inhibit prion assembly. A better characterization of the structural properties of prion fibrils is a prerequisite for developing tools that allow either fibrillar deposits disassembly or lowering of their toxicity. The identification of partners of the different prions will also allow a better comprehension of the genesis of prion phenotype.

It is also crucial to better understand the biological and evolutionary significance of *S. cerevisiae* prions and determine whether this epigenetic phenomenon is widespread or restricted to this yeast strain. Indeed, the vast majority of Ure2p and Sup35p homologues that are expressed in various yeast strains possess asparagine/glutamine-rich N-terminal extensions. Apart from a unique exception, none induce [URE3] or [PSI^+] traits in their respective yeast strains. In contrast, a number of these homologues induce the prion phenotype when expressed in *S. cerevisiae*. The reason for this is unclear and suggests either that the prion behavior of these proteins is limited to *S. cerevisiae* or that the observed phenotype is not due to Ure2p or Sup35p aggregation but to indirect physiologic events occuring within *S. cerevisiae* and only in this yeast strain. A careful analysis of the physiologic changes that accompany the propagation of the prion phenotype by comparing the proteomes of wild-type and cells exhibiting the prion phenotypes should allow a better comprehension of the molecular events at the origin of these phenotypes in *S. cerevisiae*.

Finally, although mammalian and yeast prions and their respective propagation share similarities, they differ in a number of features, in particular the involvement of molecular chaperones. The same comment applies to the *in vitro* aggregation process of yeast prions and mammalian polyglutamine-rich polypeptides. However, a constant feature of the many polyglutamine diseases is the intranuclear localization of the inclusion bodies (Davies *et al.*, 1998) in contrast with the cytosolic inclusion bodies in the case of yeast prions. While the formation of an intranuclear inclusion has obvious deleterious consequences, it is not clear how a cytosolic inclusion would influence the growth rate of a cell as observed in *S. cerevisiae*. Thus, caution must be taken in directly extrapolating findings with yeast prions to prions and proteins rich in glutamine and asparagine residues from mammals.

Acknowledgment

This work was supported by the Centre National de la Recherche Scientifique and the Fondation pour la Recherche Médicale.

References

Aigle, M., and Lacroute, F. (1975). Genetic aspects of [URE3] a non-Mendelian cytoplasmically inherited mutation in yeast. *Mol. Gen. Genet.* 136:327–335.

Balbirnie, M., Grothe, R., and Eisenberg, D.S. (2001). An amyloid-forming peptide from the yeast prion Sup35 reveals a dehydrated β-sheet structure for amyloid. *Proc. Natl. Acad. Sci. USA* 98:2375–2380.

Bessen, R.A., and Marsh, R.F. (1992). Identification of two biologically distinct strains of transmissible mink encephalopathy in hamsters. *J. Gen. Virol.* 73:329–334.

Bousset, L., Belrhali, H., Janin, J., Melki, R., and Morera, S. (2001). Structure of the globular region of the prion protein Ure2 from the yeast Saccharomyces cerevisiae. *Structure* 9:39–46.

Bousset, L., Thomson, N.H., Radford, S.E., and Melki, R. (2002). The yeast prion Ure2p retains its native alpha-helical conformation upon assembly into protein fibrils *in vitro*. *EMBO J.* 21:2903–2911.

Bousset, L., Briki, F., Doucet, J., and Melki, R. (2003). The native-like conformation of Ure2p in fibrils assembled under physiologically relevant conditions switches to an amyloid-like conformation upon heat-treatment of the fibrils. *J. Struct. Biol.* 141:132–142.

Bousset, L., Redeker, V., Decottignies, P., Dubois, S., Le Marechal, P., and Melki, R. (2004). Structural characterization of the fibrillar form of the yeast Saccharomyces cerevisiae prion Ure2p. *Biochemistry* 43:5022–5032.

Bruce, M.E. (1993). Scrapie strain variation and mutation. *Br. Med. Bull.* 49:822–838.

Cantor, C.R., and Schimmel, P.R. (2001). Biophysical Chemistry, twelfth printing. WH Freeman and Co. New York. pp. 409–431.

Carrell, R.W., and Gooptu, B. (1998). Conformational changes and diseaseserpins, prions and Alzheimer's. *Curr. Opin. Struct. Biol.* 8:799–809.

Caughey, B., Raymond, G.J., and Bessen, R.A. (1998). Strain dependent differences in beta-sheet conformations of abnormal prion protein. *J. Biol. Chem.* 273:32230–32235.

Chernoff, Y.O., Derkatch, I.L., and Inge-Vechtomov, S.G. (1993). Multicopy SUP35 gene induces de-novo appearance of psi-like factors in the yeast *Saccharomyces cerevisiae*. *Curr. Genet.* 24:268–270.

Chernoff, Y.O., Lindquist, S.L., Ono, B., Inge-Vechtomov, S.G., and Liebman, S.W. (1995). Role of the chaperone protein Hsp104 in propagation of the yeast prion-like factor [psi+]. *Science* 268:880–884.

Chernoff, Y.O., Newnam, G.P., Kumar, J., Allen, K., and Zink, A.D. (1999). Evidence for a protein mutator in yeast: Role of Hsp70-related chaperone Ssb in formation, stability and toxicity of the [PSI+] prion. *Mol. Cell. Biol.* 19:8103–8112.

Chien, P., and Weissman, J.S. (2001). Conformational diversity in a yeast prion dictates its seeding specificity. *Nature* 410:223–227.

Chien, P., DePace, A.H., Collins, S.R., and Weissman, J.S. (2003). Generation of prion transmission barriers by mutational control of amyloid conformations. *Nature* 424:948–951.

Cooper, T.G. (1982). in The Molecular and Cellular Biology of the Yeast Saccharomyces: Metabolism and Gene Expression (Strathern, JN, Jones EW, Broach JR, eds), Vol 2, pp. 39–99, Cold Spring Harbor Laboratory, Cold Spring Harbor, NY.

Coschigano, P.M., and Magasanik, B. (1991). The URE2 gene product of S. cerevisiae plays an important role in the cellular response to the nitrogen source and has homology to glutathione-S-transferases. *Mol. Cell. Biol.* 11:822–832.

Courchesne, W.E., and Magasanik, B. (1988). Regulation of nitrogen assimilation in S. cerevisiae: Roles of the URE2 and GLN3 genes. *J. Bacteriol.* 170:708–713.

Cox, B.S. (1965). PSI, a cytoplasmic suppressor of super-suppressor in yeast. *Heredity* 20:505–521.

Cox, B.S., Tuite, M.F., and McLaughlin, C.S. (1988). The Psi factor of yeast: A problem in inheritance. *Yeast* 4:159–179.

Cox, B.S., Ness, F., and Tuite, M.F. (2003). Analysis of the generation and segregation of propagons: entities that propagate the [PSI+] prion in yeast. *Genetics* 165:23–33.

Davies, S.W., Beardsall, K., Turmaine, M., DiFiglia, M., Aronin, N., and Bates, G.P. (1998). Are neuronal intranuclear inclusions the common neuropathology of triplet-repeat disorders with polyglutamine-repeat expansions? *Lancet* 351:131–133.

DePace, A.H., and Weissman, J.S. (2002). Origins and kinetic consequences of diversity in Sup35 yeast prion fibers. *Nat. Struct. Biol.* 9:389–396.

Derkatch, I.L., Chernoff, Y.O., Kushnirov, V.V., Inge-Vechtomov, S.G., and Liebman, S.W. (1996). Genesis and variability of [PSI] prion factors in Saccharomyces cerevisiae. *Genetics* 144:1375–1386.

Derkatch, I.L., Bradley, M.E., Zhou, P., Chernoff, Y.O., and Liebman, S.W. (1997). Genetic and environmental factors affecting the de novo appearance of the [PSI+] prion in *Saccharomyces cerevisiae*. *Genetics* 147:507–519.

Derkatch, I.L., Bradley, M.E., Masse, S.V., Zadorsky, S.P., Polozkov, G.V., Inge-Vechtomov, S.G., and Liebman, S.W. (2000). Dependence and independence of [PSI(+)] and [PIN(+)]: a two-prion system in yeast? *EMBO J.* 19:1942–1952.

Derkatch, I.L., Bradley, M.E., Hong, J.Y., and Liebman, S.W. (2001). Prions affect the appearance of other prions: the story of [PIN(+)]. *Cell* 106:171–182.

Dobson, C.M. (1999). Protein misfolding, evolution and disease. *Trends. Biochem. Sci.* 24:329–332.

Fay, N., Inoue, Y., Bousset, L., Taguchi, H., and Melki, R. (2003). Assembly of the yeast prion Ure2p into protein fibrils: Thermodynamic and kinetic characterization. *J. Biol. Chem.* 278:30199–30205.

Fernandez-Bellot, E., Guillemet, E., and Cullin, C. (2000). The yeast prion [URE3] can be greatly induced by a functional mutated URE2 allele. *EMBO J.* 19:3215–3222.

Gajdusek, D.C. (1988). Transmissible and non-transmissible amyloidoses: Autocatalyticv post-translational conversion of host precursor proteins to β-pleated conformations. *J. Neuroimmunol.* 20:95–110.

Glover, J.R., and Lindquist, S. (1998). Hsp104, Hsp70, and Hsp40: A novel chaperone system that rescues previously aggregated proteins. *Cell* 94:73–82.

Glover, J.R., Kowal, A.S., Schirmer, E.C., Patino, M.M., Liu, J.J., and Lindquist, S. (1997). Self-seeded fibers formed by Sup35, the protein determinant of [PSI+], a heritable prion-like factor of *S. cerevisiae*. *Cell* 89:811–819.

Gregoire, C., Marco, S., Thimonier, J., Duplan, L., Laurine, E., Chauvin, J.P., Michel, B., Peyrot, V., and Verdier, J.M. (2001). Three-dimensional structure of the lithostathine protofibril, a protein involved in Alzheimer's disease. *EMBO J.* 20:3313–3321.

Griffith, J.S. (1967). Self-replication and scrapie. *Nature* 215:1043–1044.

Hawthorne, D.C., and Mortimer, R.K. (1968) Genetic mapping of nonsense suppressors in yeast. *Genetics* 60:735–742.

Huntington, J.A., Pannu, N.S., Hazes, B., Read, R.J., Lomas, D.A., and Carrell, R.W. (1999). A 2.6Å structure of a serpin polymer and implications for conformational disease. *J. Mol. Biol.* 293:449–455.

Inge-Vechtomov, S.G., and Andrianova, V.M. (1970). Recessive super-suppressors in yeast. *Genetika* 6:103–115.

Inoue, Y., Kishimoto, A., Hirao, J., Yoshida, M., and Taguchi, H. (2001). Strong growth polarity of yeast prion fiber revealed by single fiber imaging. *J. Biol. Chem.* 276:35227–35230.

Jarret, J.T., and Lansbury, P.T. (1993) Seeding "one-dimensional-crystallization" of amyloid: a pathogenic mechanism in Alzheimer's disease and scrapie? *Cell* 73:1055–1058.

Jiang, Y., Li, H., Zhu, L., Zhou, J.M., and Perrett, S. (2004). Amyloid nucleation and hierarchical assembly of Ure2p fibrils: Role of asparagine/glutamine repeat and nonrepeat regions of the prion domain. *J. Biol. Chem.* 279:3361–3369.

Jones, G.W., and Masison, D.C. (2003). Saccharomyces cerevisiae Hsp70 mutations affect [PSI+] prion propagation and cell growth differently and implicate Hsp40 and tetratricopeptide repeat cochaperones in impairment of [PSI+]. *Genetics* 163:495–506.

King, C.Y. (2001). Supporting the structural basis of prion strains: Induction and identification of [PSI] variants. *J. Mol. Biol.* 307:1247–1260.

King, C.Y. and Diaz-Avalos, R. (2004). Protein-Only transmission of three yeast Prion strains. *Nature* 428:319–323.

Krzewska, J. and Melki, R. (2006). Molecular chaperones and the assembly of the prion Sup35p, an *in vitro* study. *EMBO J.* 25:822–833.

Kushnirov, V.V., and Ter-Avanesyan, M.D. (1998). Structure and replication of yeast prions. *Cell* 94:13–16.

Kushnirov, V.V., Ter-Avanesyan, M.D., Telckov, M.V., Surguchov, A.P., Smirnov, V.N., and Inge-Vechtomov, S.G. (1988). Nucleotide sequence of the SUP2 (SUP35) gene of Saccharomyces cerevisiae. *Gene* 66:45–54.

Lacroute, F. (1971). Non-Mendelian mutation allowing ureidosuccinic acid uptake in yeast. *J. Bacteriol.* 106:519–522.

Li, L., and Lindquist, S.L. (2000). Creating a protein-based element of inheritance. *Science* 287:661–664.

Liu, J.J., and Lindquist, S.L. (1999). Oligopeptide-repeat expansions modulate "protein-only" inheritance in yeast. *Nature* 400:573–576.

Liu, J.J., Sondheimer, N., and Lindquist, S.L. (2002). Changes in the middle region of Sup35 profoundly alter the nature of epigenetic inheritance for the yeast prion [PSI+]. *Proc. Natl. Acad. Sci. USA* 99:16446–16453.

Lomas, D.A., and Carrell, R.W. (2002). Serpinopathies and the conformational dementias. *Nat. Rev. Genet.* 3:759–768.

Lund, P.M., and Cox, B.S. (1981). Reversion analysis of [psi] mutations in Saccharomyces cerevisiae. *Genet. Res.* 37:173–182.

Maddelein, M.L., and Wickner, R.B. (1999). Two prion inducing regions of Ure2p are non-overlapping. *Mol. Cell. Biol.* 19:4516–4524.

Magasanik, B. (1992). in The Molecular and Cellular Biology of the Yeast Saccharomyces cerevisiae (Jones, EW, Pringle JR, and Broach JR, eds), Vol 2, 2nd Ed., pp. 283–317, Cold Spring Harbor Laboratory, Cold Spring Harbor, NY.

Masison, D.C., and Wickner, R.B. (1995). Prion-inducing domain of yeast Ure2p and protease resistance of Ure2p in prion-containing cells. *Science* 270:93–95.

Michelitsch, M.D., and Weissman, J.S. (2000). A census of glutamine/asparagine-rich regions: Implications for their conserved function and the prediction of novel prions. *Proc. Natl. Acad. Sci. USA* 97:11910–11915.

Mitchell, A.P., and Magasanik, B. (1984). Regulation of glutamine-repressible gene products by GLN3 function in *S. cerevisiae*. *Mol. Cell. Biol.* 4:2758–2766.

Moriyama, H., Edskes, H.K., and Wickner, R.B. (2000). [URE3] prion propagation in *Saccharomyces cerevisiae*: Requirement for chaperone Hsp104 and curing by overexpressed chaperone Ydj1p. *Mol. Cell. Biol.* 20:8916–8922.

Newnam, G.P., Wegrzyn, R.D., Lindquist, S.L., and Chernoff, Y.O. (1999). Antagonistic interaction between yeast chaperones Hsp104 and Hsp70 in prion curing. *Mol. Cell. Biol.* 19:1325–1333.

Osherovich, L.Z., and Weissman, J.S. (2001). Multiple Gln/Asn-rich prion domains confer susceptibility to induction of the yeast [PSI(+)] prion. *Cell* 106:183–194.

Parham, S.N., Resende, C.G. and Tuite, M.F. (2001). Oligopeptide repeats in the yeast protein Sup35p stabilize intermolecular prion interactions. *EMBO J.* 20:2111–2119.

Patino, M.M., Liu, J.J., Glover, J.R., and Lindquist, S. (1996). Support for the prion hypothesis for inheritance of a phenotypic trait in yeast. *Science* 273:622–626.

Perutz, M.F. (1999). Glutamine repeats and neurodegenerative diseases: Molecular aspects. *Trends. Biochem. Sci.* 24:58–63.

Perutz, M.F., Pope, B.J., Owen, D., Wanker, E.E., and Scherzinger, E. (2002). Aggregation of proteins with expanded glutamine and alanine repeats of the glutamine-rich and asparagine-rich domains of Sup35 and of the amyloid β-peptide of amyloid plaques. *Proc. Natl. Acad. Sci. USA* 99:5596–5600.

Prusiner, S.B. (1991). Molecular biology of prion diseases. *Science* 252:1515–1522.

Rai, R., Genbauffe, F., Lea, H.Z., and Cooper, T.G. (1987). Transcriptional regulation of the *DAL5* gene in *S. cerevisiae. J. Bacteriol.* 169:3521–3524.

Schlumpberger, M., Wille, H., Baldwin, M.A., Butler, D.A., Herskowitz, I., and Prusiner, S.B. (2000). The prion domain of yeast Ure2p induces autocatalytic formation of amyloid fibers by a recombinant fusion protein. *Prot. Sci.* 9:440–451.

Schlumpberger, M., Prusiner, S.B., and Herskowitz, I. (2001). Induction of distinct [URE3] yeast prion strains. *Mol. Cell. Biol.* 21:7035–7046.

Schoun, J., and Lacroute, F. (1969). Etude physiologique d'une mutation permettant l'incorporation d'acide ureidosuccinique chez la levure. *C. R. Acad. Sci.* 269:1412–1414.

Schubert, U., Anton, L.C., Gibbs, J., Norbury, C.C., Yewdell, J.W., and Bennink, J.R. (2000). Rapid degradation of a large fraction of newly synthesized proteins by proteasomes. *Nature* 404:770–774.

Serio, T.R., Cashikar, A.G., Kowal, A.S., Sawicki, G.J., Moslehi, J.J., Serpell, L., Arnsdorf, M.F., and Lindquist, S.L. (2000). Nucleated conformational conversion and the replication of conformational information by a prion determinant. *Science* 289: 1317–1321.

Shorter, J., and Lindquist, S. (2006). Destruction or potentiation of different prions catalyzed by similar Hsp104 remodeling activities. *Mol. Cell.* 23:425–438.

Singh, A., Helms, C., and Sherman, F. (1979). Mutation of the non-Mendelian suppressor, Psi+, in yeast by hypertonic media. *Proc. Natl. Acad. Sci. USA* 76:1952–1956.

Sipe, J.D., and Cohen, A.S. (2000). History of the amyloid fibril. *J. Struct. Biol.* 130:88–98.

Sondheimer, N., and Lindquist, S. (2000). Rnq1: An epigenetic modifier of protein function in yeast. *Mol. Cells* 5:63–172.

Sondheimer, N., Lopez, N., Craig, E.A., and Lindquist, S. (2001). The role of Sis1 in the maintenance of the [RNQ+] prion. *EMBO J.* 20:2435–2442.

Stansfield, I., Jones, K.M., Kushnirov, V.V., Dagkesamanskaya, A.R., Poznyakovski, A.I., Paushkin, S.V., Nierras, C.R., Cox, B.S., Ter-Avanesyan, M.D., and Tuite, M.F. (1995). The products of the SUP45 (eRF1) and *SUP35* genes interact to mediate translation termination in Saccharomyces cerevisiae. *EMBO J.* 14:4365–4373.

Sunde, M., Serpell, L.C., Bartlam, M., Fraser, P.E., Pepys, M.B., and Blake, C.C. (1997). Common core structure of amyloid fibrils by synchrotron X-ray diffraction. *J. Mol. Biol.* 273:729–739.

Tanaka, M., Chien, P., Naber, N., Cooke, R., and Weissman, J.S. (2004). Conformational variations in an infections protein determine prion strain differences. *Nature* 428:323–328.

Taylor, K.L., Cheng, N., Williams, R.W., Steven, A.C., and Wickner, R.B. (1999). Prion domain initiation of amyloid formation *in vitro* from native Ure2p. *Science* 283:1339–1343.

Telling, G.C., Scott, M., Mastrianni, J., Gabizon, R., Torchia, M., Cohen, F.E., DeArmond, S.J., and Prusiner, S.B. (1995). Prion propagation in mice expressing human and chimeric PrP transgenes implicates the interaction of cellular PrP with another protein. *Cell* 83:79–90.

Ter-Avanesyan, M.D., Kushnirov, V.V., Dagkesamanskaya, A.R., Didichenko, S.A., Chernoff, Y.O., Inge-Vechtomov, S.G., and Smirnov, V.N. (1993). Deletion analysis of the *SUP35* gene of the yeast *Saccharomyces cerevisiae* reveals two non-overlapping functional regions in the encoded protein. *Mol. Microbiol.* 7:683–692.

Ter-Avanesyan, M.D., Dagkesamanskaya, A.R., Kushnirov, V.V., and Smirnov, V.N. (1994). The SUP35 omnipotent suppressor gene is involved in the maintenance of the non-mendelian determinant [psi+] in the yeast *saccharomyces cerevisiae. Genetics* 137:1339–1343.

Thual, C., Komar, A.A., Bousset, L., Fernandez-Bellot, E., Cullin, C., and Melki, R. (1999). Structural characterization of *saccharomyces cerevisiae* prion-like protein Ure2. *J. Biol. Chem.* 274:13666–13674.

Thual, C., Bousset, L., Komar, A. A, Walter, S., Buchner, J., Cullin, C., and Melki, R. (2001). Stability, folding, dimerization, and assembly properties of the yeast prion Ure2p. *Biochemistry* 40:1764–1773.

True, H.L., and Lindquist, S.L. (2000). A yeast prion provides a mechanism for genetic variation and phenotypic diversity. *Nature* 407:477–483.

Tuite, M.F., Mundy, C.R., and Cox, B.S. (1981). Agents that cause a high frequency of genetic change from [psi+] to [psi–] in *S. cerevisiae*. *Genetics* 98:691–711.

Tuite, M.F. (2000). Yeast prions and their prion-forming domain. *Cell* 100:289–292.

Umland, T.C., Taylor, K.L., Rhee, S., Wickner, R.B., and Davies, D.R. (2001). The crystal structure of the nitrogen regulation fragment of the yeast prion protein Ure2p. *Proc. Natl. Acad. Sci. USA* 98:1459–1464.

Uptain, S.M., Sawicki, G.J., Caughey, B., and Lindquist, S. (2001). Strains of [PSI+] are distinguished by their efficiencys of prion-mediated conformational conversion. *EMBO J.* 20:6236–6245.

Westermark, P., Benson, M.D., Buxbaum, J.N., Cohen, A.S., Frangione, B., Ikeda, S., Masters, C.L., Merlini, G., Saraiva, M.J., and Sipe, J.D. (2002). Amyloid fibril protein nomenclature2002. *Amyloid* 9:197–200.

Wickner, R.B. (1994). Evidence for a prion analog in *S. cerevisiae*: the [URE3] non-Mendelian genetic element as an altered URE2 protein. *Science* 264:566–569.

Xu, S., Bevis, B., and Arnsdorf, M.F. (2001). The assembly of amyloidogenic yeast Sup35 as assessed by scanning (atomic) force microscopy: An analogy to linear colloidal aggregation? *Biophys. J.* 81:446–454.

Zhouravleva, G., Frolova, L., Le Goff, X., Le Guellec, R., Inge-Vechtomov, S., Kisselev, L., and Philippe, M. (1995). Termination of translation in eukaryotes is governed by two interacting polypeptide chain release factors, eRF1 and eRF3. *EMBO J.* 14:4065–4072.

Zurdo, J., Guijarro, J.I., and Dobson, C.M. (2001). Preparation and characterization of purified amyloid fibrils. *J. Am. Chem. Soc.* 123:8141–8142.

9

Immunoglobulin Light Chain and Systemic Light-Chain Amyloidosis

Marina Ramirez-Alvarado, Janelle K. De Stigter, Elizabeth M. Baden, Laura A. Sikkink, Richard W. McLaughlin, and Anya L. Taboas

Abstract

Light-chain amyloidosis (AL) is characterized by the clonal expansion of plasma B cells that secrete large amounts of monoclonal immunoglobulin light chains. The free light chains circulate in serum and form amyloid fibrils on vital organs such as the kidney, heart, and liver causing organ failure and eventually death. Multiple myeloma (MM) is another type of B-cell malignancy. MM results in bone lesions, hypercalcemia, renal failure, and anemia. Free light chains from MM patients are non-amyloidogenic. There is a small subset of MM patients (10–15%) that also develop light-chain amyloidosis complications. Amyloid fibrils are derived from the N-terminal region of the light-chain variable domain (VL). Due to the antigen-driven selection process, there is a large degree of mutational variability, thus each patient has a unique VL sequence. In some cases, the organ involvement has some correlation between VL subtype and germ-line donor sequence gene. To understand the amyloidogenicity of AL proteins, their thermodynamic properties have been compared with non-amyloidogenic MM proteins. Generally, AL proteins have lower thermodynamic stability than MM proteins. Current and future research is focusing on understanding the mutational diversity and organ tropism associated with AL, as well as understanding which species along the amyloid fibril formation pathway causes cellular toxicity. The current treatment for AL targets the plasma cell clone. Chemotherapy and peripheral stem cell transplantation are commonly used. Future therapies to treat this disease could involve small molecules that stabilize the folded state and inhibit amyloid fibril formation and molecules that bind to amyloid fibrils, destabilize the fibrils, thus reducing the amyloid burden. In this chapter, we discuss the structure, truncations, mutational diversity, and organ tropism of immunoglobulin light chains, thermodynamics, and fibril formation studies using AL and MM proteins and the current treatment options and also discuss directions in the study and treatment of AL.

9.1. Light-Chain Amyloidosis and Multiple Myeloma as Related Hematologic Malignancies

9.1.1. Light-Chain Amyloidosis

Light-chain amyloidosis (AL) is among a group of devastating misfolding disorders called amyloidoses in which normally soluble proteins misfold and aggregate to form insoluble amyloid fibrils in the extracellular space. The formation of amyloid fibrils leads to cell death and tissue degeneration. To date, 20 different proteins and polypeptides have been identified in disease-associated amyloid deposits. These proteins include the amyloid beta (Aβ) peptide in Alzheimer's disease (AD), the prion protein in transmissible spongiform encephalopathies (TSE), and the islet-associated polypeptide (IAPP) in type 2 diabetes, among others (Buxbaum, 2003; Ross and Poirier, 2004). No sequence or structural similarities are apparent between any of the proteins that form amyloid fibrils. Amyloid fibrils, however, present a number of common features. All amyloid fibrils are long, unbranched filaments 40–120 Å in diameter. They bind to histologic dyes such as Congo red and thioflavine T and are resistant to protease degradation. X-ray fibril diffraction reveals a cross-β structure, in which the long axis of the helix is parallel to the helical array of β-sheets and perpendicular to the β-strands (Merlini and Bellotti, 2003).

AL is the only known hematologic malignancy associated with a misfolding disorder. The incidence of AL is 8 cases per 1 million persons per year (Gertz *et al.*, 1999), the same incidence as Hodgkin's lymphoma. The median age of patients with amyloidosis is 65 years with a marginal male dominance (Kyle and Greipp, 1983; Kyle *et al.*, 1992). AL is characterized by an abnormal proliferation of monoclonal plasma cells in the bone marrow. These plasma cells secrete high amounts of monoclonal immunoglobulin light chains. These light-chain dimers are secreted into circulation and excreted in large amounts in urine. While they are in circulation, immunoglobulin light chains misfold and form amyloid fibrils composed of the immunoglobulin light chain N-terminus variable domain (in most AL cases) (Olsen *et al.*, 1998). The amyloid fibrils affect multiple vital organs where the deposition causes organ failure leading to death. The kidney is the most frequently affected organ (Gertz and Kyle, 1990), followed by the heart (Kyle and Gertz, 1995) and the liver (Gertz and Kyle, 1997). Peripheral nerve, gastrointestinal, and pulmonary involvement are also observed. In addition to visceral involvement, AL amyloid can infiltrate the tongue, salivary glands, skeletal muscle, joints, ligaments, and skin (Sezer *et al.*, 2000). The disease is insidious, relentlessly progressive, and uniformly fatal with a median survival of approximately 12–18 months after diagnosis and only 5% survival rate after 10 years (Kyle *et al.*, 1986; Gertz and Kyle, 1989; Kyle *et al.*, 1999).

9.1.2. Multiple Myeloma

Multiple myeloma (MM) is a B-cell malignancy characterized by abnormal proliferation of monoclonal plasma cells in the bone marrow causing bone lesions, hypercalcemia, renal failure, and anemia. The bone marrow plasma cells secrete large amounts of monoclonal immunoglobulin light chain into circulation. As with AL, these light-chain dimers are secreted into circulation and excreted in large amounts in urine. Contrary to what happens during AL, no amyloid deposits are found in MM patients, so MM light chains are considered non-amyloidogenic and are used as controls for AL studies. However, there is a subset of MM patients (10–15%) that present further complications with AL either at the initial MM diagnosis or at a later time (Rajkumar *et al.*, 1998). Proteins derived from patients with both MM and AL are treated as a special third class of protein in our laboratory. We have included data from one of these proteins in Section 9.6.3.

9.2. Immunoglobulin Light-Chain Structure

Plasma cells are stimulated by antigen binding to secrete antibodies, also known as immunoglobulins, into the bloodstream. All immunoglobulin molecules are composed of two identical light chains and two identical heavy chains linked together by disulfide bonds. A schematic diagram of an immunoglobulin G (IgG) molecule is shown in Figure 9.1A.

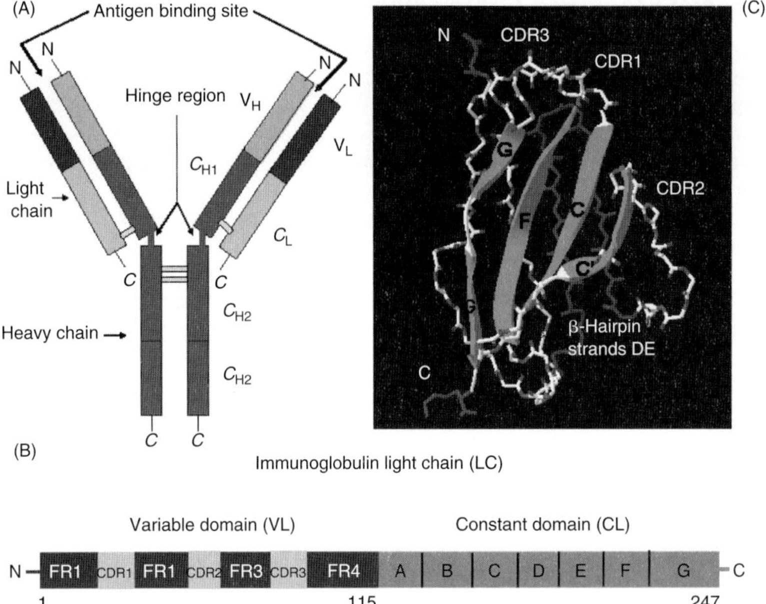

Figure 9.1. Panel (A) Immunoglobulin G IgG basic structure. Immunoglobulin light chains are bound to heavy chains via a disulfide bonds. (B) Schematic representation of an immunoglobulin light chain. FR=Framework regions. The variable domain is in the N-terminus. Most of the proteolysis occurs between the variable domain and the constant domain. (C) VL structure (1BRE.pdb) showing the N-terminus and C-terminus of the domain (N and C), β-hairpin between strands D and E and the heavy chain/light chain dimer interface in ribbons.

Each light chain is composed of an N-terminal variable domain (VL) and a C-terminal constant domain (CL) (Fig. 9.1B). VLs are not uniformly variable throughout their lengths. Three small regions, called hypervariable or complementarity determining regions (CDRs), show much more variability than the rest of the domain. They vary both in size and in sequence among different isotypes and are responsible for determining the specificity of the antigen-antibody interactions. The remaining parts of the VL, the four framework regions (FRs), have quite similar amino acid sequences. The overall structure of the VL is an immunoglobulin fold of about 105 amino acids with two antiparallel β-sheets packed tightly against each other and joined together by a disulfide bridge (Fig. 9.1C). The topology could be described as a form of a Greek key β-barrel formed of 9 β-strands (A, B, C, C′, C″, D, E, F, G) where strands A and G are the N- and C-termini, respectively. Strands C, C′, F, and G form the β-sheet that interacts with the heavy chain. The three light-chain CDR loops combine with the heavy-chain variable domain (VH) CDR loops to form the antigen binding site (Branden and Tooze, 1999).

The primary source of sequence variability in light chains comes from combinatorial pairing of the V genes (30 λ and 40 κ), which encode for the first 90 residues of the VL, and the J genes coding for residues 91–105. This pairing allows the possibility to generate about 3000 different light-chain sequences. In addition, further sequence variation appears from somatic mutations, which affect CDR regions more frequently than FR regions. Somatic mutation is a normal process in the immune system, as the accumulated mutations function to improve the affinity of the antibody for the antigen.

9.2.1. Crystal Structures of Amyloidogenic Light-Chain Proteins

A large number of high-resolution structures of AL proteins for both full-length light chain and VL have been previously reported (Epp *et al.*, 1975; Huang *et al.*, 1994; Schormann *et al.*, 1995; Alim *et al.*, 1999; Pokkuluri *et al.*, 1999; Wall *et al.*, 2004). Examples of the structural comparison between VL of amyloidogenic and non-amyloidogenic proteins are discussed here. In general terms, the structure of the VL domain from AL patients is very well conserved, with differences in only a few particular regions of the protein.

The crystal structure of BRE, a κ1 VL from an AL protein, was reported at 2 Å resolution (Schormann *et al.*, 1995). Structural comparison between the AL protein BRE and the non-amyloidogenic protein REI (also a κ1 VL) shows the difference to be in the 40–44 loop, which forms part of the dimer interface with the heavy chain, located between strand C and C′. BRE crystal packing resembled a pseudohexagonal spiral structure consistent with dimensions of amyloid fibrils. The comparison between two λVI VL domains (Wil from an AL protein, Jto from a MM protein) revealed the absence of the loop extension in the CDR1 region for Wil. This loop extension included an electrostatic interaction between D29 and R68 that increased the stability of the MM protein (Pokkuluri *et al.*, 1999; Wall *et al.*, 2004). The dimeric face interaction of two VλII light-chain crystal structures have also been compared (Bourne *et al.*, 2002). This work proposed that the symmetry derived from dimerizing variable regions of two light chains instead of variable regions between a light chain and a heavy chain greatly affects some key regions of the protein, namely the CDR regions.

9.3. Unique Properties of AL

9.3.1. Light-Chain Isotypes in MM and AL, Truncations, and Mutational Diversity

Normal human repertoire of immunoglobulin light chains and immunoglobulin light chains associated with MM have a preference to be of the κ subtype (λ/κ 1:2). In contrast, AL immunoglobulin light chains are preferentially λ (λ/κ 3:1) (Abraham *et al.*, 2003).

AL amyloid fibrils are derived from the N-terminal region VL of the monoclonal immunoglobulin light chain, and proteolysis is likely involved by one of two mechanisms: (i) The full-length light chain may suffer proteolytic cleavage that destabilizes the protein and induces amyloid formation. (ii) Another possibility is that full-length light chain is involved in fibril formation and that the VL is the core of the amyloid fibril. Subsequent proteolysis of the C-terminal domain of the light chain leaves the VL as the only portion of the light chain in the fibril.

AL is a disease with a large degree of mutational variability derived from the antigen-driven selection process that immunoglobulin VL and VH undergo. Each patient has a unique VL protein sequence. General aspects associated with the location and the nature of the mutations will be discussed in Sections 9.4.1 and 9.6.1.

9.3.2. Association with Heavy Chain

The normal function of the light chain is to associate with immunoglobulin heavy chain to form antibodies such as IgG, IgA, and IgM. The antigen binding site consists of the variable regions of both the heavy and light chains (VH and VL). Free light chain can be secreted, but not free heavy chain (Dul and Argon, 1990). There is, however, a rare condition called heavy chain deposition disease (HCDD), in which amyloid deposits are formed by the heavy chain. Heavy chains alone are highly insoluble and tend to be degraded in plasma cells if the interaction between light and heavy chains does not occur. It has been suggested that nonsecreted heavy chains may be toxic to plasma B cells (Morrison and Scharff, 1979; Kohler, 1980; Haas and Wabl, 1984; Chou and Morrison, 1993).

Monoclonal light chains can be produced and secreted in large amounts in patients with a variety of B-cell malignancies, including AL. As a consequence, plasma cells secrete free light chains in the form of dimers, also known as Bence Jones proteins, in which the interaction between the heavy chain and the light chain is substituted by the interaction between two light chains. The heavy and light chains interact strongly with an association constant ($K_A > 10^{10} M^{-1}$) (Klein *et al.*, 1979; Alexandru *et al.*, 1980). This interaction is the combination of the interaction between VH and VL ($K_A \sim 10^6 M^{-1}$), and the constant region 1 on the heavy chain (CH1) and CL ($K_A \sim 10^7 M^{-1}$). Based on these results, the interaction is slightly stronger between the constant domains than between VL domains. It is possible that there is competition between a heavy chain and a light chain to form a dimer with another light chain.

The loss of association as a possible source of instability has been seen in other amyloid diseases. Transthyretin amyloid fibril formation has been observed in familial amyloid polyneuropathy and senile systemic amyloidosis. Transthyretin is normally present in human plasma as a tetramer; however, mutations that disrupt the tetramer association are amyloidogenic and give rise to monomers that are unstable and rapidly form amyloid fibrils (Koo *et al.*, 1999). Small molecules have been shown to stabilize the folded tetramer structure and inhibit amyloid formation (Sacchettini and Kelly, 2002).

9.3.3. Systemic Nature and Mortality

In systemic amyloidosis, the deposits may be present in the parenchyma of the viscera of all tissues except the brain, as well as in the walls of blood vessels throughout the body. A recent study has indicated that 63% of AL patients had two or more organs affected by severe amyloid deposition (Abraham *et al.*, 2003). Systemic amyloidosis is uniformly fatal, although the prognosis has been improved by the process of removing the protein source or by organ transplantation. Early diagnosis becomes an essential component to improve the prognosis of the patients. The survival of the patient has been correlated with the organ involved as AL cardiac patients have a decreased survival rate compared with renal patients. It was also shown that AL female patients live longer than males (Abraham *et al.*, 2003).

9.3.4. AL Organ Tropism

AL patients present a predominately dysfunctional organ. Some correlation between the light chain subtype and germ-line donor sequence gene and organ involvement has been established (Comenzo *et al.*, 1999). Patients whose monoclonal light chain derived from the VλVI (also described as IGLV6S1/Vλ6a) germ-line donor uniformly presented with dominant renal involvement, while those with other Vλ genes or unknown donors often had dominant cardiac or other organ involvement. A more recent study (Abraham *et al.*, 2003) confirms the correlation between λVI 6a and renal involvement. Moreover, this report finds that proteins corresponding with the λIII 3r germ-line gene are usually involved with soft tissue AL, and the λII 2b2 germ-line is associated with cardiac amyloidosis and reduced survival. Light chains corresponding with λ6a, λ3r, λ2b2, κIV, and κI account for 60% of all amyloid cases. The molecular mechanism describing why certain light chains affect primarily one or more organs is not understood.

9.3.5. Current Therapies

Therapeutic options for AL are limited. There is no available treatment that improves impaired organ function by induction of amyloid mobilization. AL and most amyloidoses are dynamic processes, so measures that even moderately reduce the supply of amyloid fibril precursor protein can result in a major regression of the deposits. The most effective treatments thus far are those that target the plasma cell clone such as chemotherapy and peripheral stem cell transplantation (Khan and Falk, 2001). Unfortunately, many patients with AL are too sick to withstand the rigors of peripheral stem cell transplantation. In cases limited to cardiac amyloidosis, heart transplantation has been performed, but progression of the disease and recurrence in the transplanted organ limits this approach.

9.4. Thermodynamics Studies of AL Proteins

9.4.1. Possible Causes of Instability for AL Proteins

One important step toward understanding the amyloidogenic properties of AL proteins is to look at their thermodynamic characteristics, especially in comparison with non-amyloidogenic counterparts. Most AL proteins have been shown, by means of thermal and chaotropic denaturation (using urea and guanidinum hydrochloride), to have lower melting temperatures (T_m), chemical melting (C_m), and lower free energies of unfolding (ΔG_{unf}) than MM proteins.

Studies have shown that one potential cause of thermodynamic instability is mutations within AL proteins. The Wetzel group used the MM protein REI and created mutant constructs using the most frequent AL mutations identified from a panel of 36 AL proteins (Hurle *et al.*, 1994). All of the mutants showed a lower energy of unfolding and a decrease in C_m as compared with the wild-type REI protein. These mutations have a global destabilizing effect, thus requiring less energy to unfold the protein (Hurle *et al.*, 1994; Wetzel, 1997; Stevens *et al.*, 1999). The propensity to form amyloid fibrils *in vitro* for some VL domains appears to be inversely correlated with their free energy of unfolding, suggesting that stabilizing interactions within the VL domain can influence the kinetics of amyloid formation (Wetzel, 1997; Wall *et al.*, 1999b; Kim *et al.*, 2000).

Other studies have also identified sequence characteristics that may contribute to the instability of AL proteins, including mutations that disrupt critical ionic interactions, mutations that remove prolines that anchor β-sheets (Stevens, 2000), or mutations that allow for glycosylation in the framework regions (Omtvedt *et al.*, 2000). Stevens has analyzed more than 100 AL protein sequences from the κ1 light-chain family (Stevens, 2000). He identified four structural risk factors for κ1 VL domains that may enhance the amyloidogenicity of light chains. These risk factors are mutations in the isoleucine residue in position 27b (CDR1), mutations of the amino acid in position 31 to aspartic acid (CDR1), mutations in arginine 61 (located in strand D, part of β-hairpin DE), loss of proline residues in β-turns, and the creation of glycosylation sites (asparagine-X-serine/threonine) anywhere in the protein sequence. The Stevens (2000) report is an important sequence analysis work to establish the role of the nature and the location of the mutations in AL proteins. Amyloid formation due to protein destabilization could be local or global. In the first case, the destabilization in a local region of the protein promotes a conformational change that would trigger amyloid formation. A global effect is considered when the positions of the mutations are not as important as their free energy effects (Wetzel, 1997; Ramirez-Alvarado and Regan, 2002).

9.4.2. Thermodynamic Stability Differences Between MM and AL Proteins

Light-chain proteins Wil and Jto have frequently been compared as they are both λ6 proteins with more than 90% identity; however, Wil is amyloidogenic and Jto is non-amyloidogenic (Wall *et al.*, 1999b). A primary difference seen in the crystal structures of both proteins is one salt-bridge between D29 and R68 located in CDR1 loop and CDR2 loop, respectively, which is believed to contribute stability to Jto. The protein Wil unfolds at a lower concentration of guanidine hydrochloride and at a lower temperature as measured by an increase in tryptophan fluorescence.

Another pair of proteins, BIF (AL) and GAL (MM), both κ1, have been extensively compared thermodynamically. Once again, the amyloidogenic protein BIF was found to have a lower T_m, C_m, and ΔG_{unf} compared with GAL (Kim *et al.*, 2000). Interestingly, by using a nonspecific thermodynamic stabilizer, 1 M sucrose, BIF thermodynamic properties were equal to those from GAL in the absence of the stabilizer. This study indicates that increasing the thermodynamic stability of AL proteins may render them less amyloidogenic.

9.5. Fibril Formation Studies of AL Proteins

Formation of AL amyloid fibrils is favored under conditions that destabilize the LC protein, potentially due to some of the thermodynamic factors discussed above. Fibril formation for AL proteins is typically characterized as a nucleation-dependent mechanism, in which the formation of an oligomeric structure is necessary to start forming fibrils. This nucleation-dependent mechanism is detected by a lag time before a rapid increase in fibrillation. This lag time can be abolished by adding "seeds" of previously formed fibrils that can act as a nucleus to trigger fibril formation. Seeding experiments with the AL protein SMA have confirmed this characteristic; addition of 5% seeds of SMA to a reaction was sufficient to produce a maximal amount of fibrils in half the time as an unseeded reaction (Davis *et al.*, 2000). SMA seeds were also able to nucleate soluble MM protein LEN, suggesting that LEN is capable of undergoing conformational changes that allow it to elongate SMA fibrils, despite the fact that LEN is more thermodynamically stable than SMA (Khurana *et al.*, 2003). Studies with mutant forms of LEN pointed out that one mutational difference (P40L) between SMA and LEN contributed most significantly to LEN being able to form fibrils. It was shown by decreasing the pH of these proteins that SMA populates a distinct unstable intermediate that may be responsible for fibril formation. In contrast, LEN does not form such an intermediate.

Most studies of the kinetics of AL fibril formation rely on altering *in vitro* conditions in an attempt to destabilize the LC protein so that fibril formation occurs in a reasonable timescale. The kinetics of AL fibril formation varies greatly depending on the protein. One frequently altered condition is the temperature at which the fibril formation assay is carried out. One report suggests that the key to fibril formation is to maximize the amount of unfolded conformations during the unfolding transition of the protein, which is found to be the melting temperature of the protein (Ramirez-Alvarado *et al.*, 2000). Another frequently altered condition is the pH of the buffer solution. Studies with SMA have shown that the more acidic the pH, the more the stability decreases and fibril formation/general aggregation increase (Khurana *et al.*, 2001). Other factors that play significant roles in fibril formation assays include ionic strength, agitation/stirring, concentration of protein (Wall *et al.*, 1999a), addition of sodium sulfate (Ramirez-Alvarado *et al.*, 2000), and pressure (Kim *et al.*, 2002).

Another notable feature of light-chain proteins is that they can also form nonfibrillar aggregates such as in light chain deposition disease (LCDD). LCDD is another light-chain disorder characterized by amorphous aggregates in the kidney rather than fibrillar deposits. Amyloidogenic light-chain SMA has been shown to form both fibrillar and amorphous deposits, depending on the conditions of the assay, indicating that the protein can aggregate by way of two different intermediates (Khurana *et al.*, 2001). SMA and AL-09 have also been shown to form oligomeric species, including annular or spherical intermediates (Zhu *et al.*, 2004; McLaughlin *et al.*, 2006). SMA has also been reported to form a granular aggregate upon exposure to copper *in vitro* and *in vivo*, under conditions that would normally favor forming fibrils (Davis *et al.*, 2001). Gaining a better understanding of the different protein morphologies and the conditions under which they form fibrils or aggregates may prove useful in determining the pathway of fibril formation (Fig. 9.2).

One factor that has been considered in AL fibril formation is whether fibrils are formed in solution or on a surface. The Fink group reported SMA fibrils on a mica surface at pH 5, whereas usually it forms amorphous aggregates in solution (Zhu *et al.*, 2002). Fibrils also grew at increased rates with lower concentrations if a surface was present. This may indicate some physiologic relevance as the organ surfaces *in vivo* may play a role in the fibril deposits of AL proteins.

It has been shown that even proteins deemed non-amyloidogenic by their *in vivo* behavior may have an ability to form fibrils *in vitro* under the right conditions (Dobson, 2003). This somewhat general feature can be studied to give more insight into what may occur in fibril formation

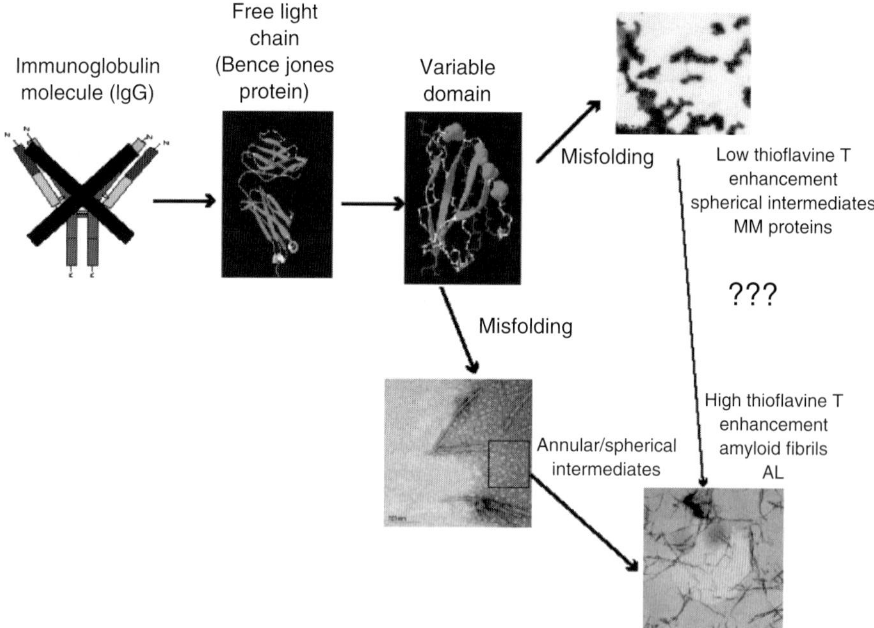

Figure 9.2. Proposed misfolding pathway followed by immunoglobulin light chains. Loss of association with the heavy chain results in the secretion of free light chains (also called Bence Jones proteins), a proteolytic cleavage occurs (we are assuming that cleavage is occurring prior to the aggregation, although this has not been tested) followed by misfolding, which gives rise to different kinds of species.

pathways. This is true of the MM proteins Jto and LEN, which are both non-amyloidogenic *in vivo*. Jto forms fibrils *in vitro*, however, it displays much slower kinetics of fibril formation than the AL protein Wil, having a lag time an order of magnitude greater (Wall *et al.*, 1999b). The MM protein LEN also forms fibrils *in vitro* under mild urea denaturation, and as the concentration of urea increases, so does the rate of fibrillation (Souillac *et al.*, 2002a). Another interesting feature of LEN is that the rate of fibril formation is also concentration dependent; however, in contrast with most AL proteins, as the concentration of the protein increases, the rate and propensity to form fibrils decreases (Souillac *et al.*, 2002b).

As we have mentioned before, the final protein sequence of each light chain is a combination of the use of a specific germ-line donor sequence gene and the accumulation of somatic mutations. It has been shown that a protein corresponding with the "pure" germ-line Vλ λ6a with no accumulated somatic mutations forms amyloid fibrils if incubated for an extended period of time under the same conditions in which Wil and Jto form fibrils (del Pozo and Becerril, personal communication).

9.6. Recent Findings

9.6.1. Large Mutational Diversity

AL is a disease with a large degree of mutational variability. In order to design rational strategies for AL treatment, it is imperative that the mechanisms of AL amyloid formation are understood. This can only be accomplished through the study of a wide number of mutationally diverse AL proteins. We have mapped the different regions of mutational diversity for more than 20 VL domains from AL proteins (Abraham *et al.*, 2004). We have classified mutational regions critical for the overall stability of the VL domain into four groups, including the mutational regions of MM proteins for comparison.

1. Mutations are distributed throughout the structure in the top and/or bottom of the immunoglobulin Greek key β-barrel. VL protein SMA, Wil, GAL, LEN (MM), and Jto (MM) belong to this mutational group. These mutations will affect the overall stability of the domain.
2. Mutations are located in the N-terminus and/or the C-terminus β-strands. Interestingly, most of our MM VL protein sequences belong to this group. Loss of interactions between these two strands could enhance conformational flexibility in this region of the protein. uMM-01 belongs to this mutational group.
3. Mutations are located in the β-hairpin formed by strands D and E (Fig. 9.1C). It has been previously reported that peptides with a sequence corresponding with this region of the protein inhibit amyloid formation *in vitro* (Davis *et al.*, 2000). Increased flexibility of this "hinge" region may cause a conformational change that exposes hydrophobic regions of the domain that could trigger amyloid formation. AL-12 belongs to this mutational group.
4. Mutations are present in the dimer interface where heavy chain and light chain dimerize (strands C, C′, F, and G). Mutations in the external face of the antiparallel β-sheet corresponding with the heavy chain/light chain interface may impair the light chain to dimerize with the heavy chain. MM VL BIF and AL-09 belong to this mutational group.

It is entirely possible that we will find other mutational regions that are affected as we proceed with our sequence/structure analysis. This is a first step in our efforts to try to classify the mutational diversity in AL.

9.6.2. Addressing Organ Tropism by Studying the Effect of Glycosaminoglycans

Amyloid fibrils have been found to interact with different cofactors such as metals, serum amyloid P, apolipoprotein E, and glycosaminoglycans (GAGs) (Ancsin, 2003). GAGs, along with collagens and proteoglycans, are the main components found in the extracellular matrix. Structurally, they are a group of negatively charged heterogeneous polysaccharides made from assembling repeating disaccharide units. The disaccharide units consist of either glucuronate or iduronate linked to a glucosamine or galactosamine. The interaction between immunoglobulin light chains and glycosaminoglycans has been tested using size-exclusion chromatography and sodium dodecyl sulfate–polyacrylamide gel electrophoresis (SDS-PAGE). It was shown that heparin 16000, chondroitin sulfate B and C precipitated full-length LC and recombinant VL (Jiang *et al.*, 1997). A recent study has shown that incubation of fibroblast growth factor–treated cardiac fibroblasts with urine-derived light chains caused an internalization of the light chains to perinuclear localization and the translocation of heparin sulfate into the nucleus. The authors show that the presence of heparan sulfate slightly changed the conformation of light-chain proteins and slightly decreased their thermal stability (Trinkaus-Randall *et al.*, 2005). The authors propose that light chains cause an injury-like response that alters the regulation of proteoglycans once they are internalized in the cell. The effect of glycosaminoglycans on the kinetics of amyloid formation had not been described until recently (McLaughlin *et al.*, 1996). Heparin and chondroitin sulfate B accelerated amyloid formation while chondroitin sulfate A stabilized spherical intermediates in the amyloid formation pathway.

9.6.3. Studying Protein Thermodynamics and Fibril Formation to Understand the Complication from MM to AL

Fifteen percent of MM patients will present complications associated with AL. It is not known whether these patients have a single monoclonal light-chain population or if the two diseases are associated with two different monoclonal populations. We have sequenced cDNA from bone marrow RNA samples taken at different stages of the progression between MM and AL. Surprisingly, a patient suffering from both MM and AL (MMAL-01) presents two distinct populations of λ1c and λ3r in plasma cell RNA molecules. Urine protein from this patient (uMMAL-01) presents a β-sheet structure and is less stable than usual MM or AL proteins according to the melting temperature (Fig. 9.3A). This protein is capable of forming amyloid fibrils (Fig. 9.3B) when incubated at its T_m. Further studies to identify the exact protein sequence present in the urine and to confirm the biclonal nature of this patient's disease will give insights into the mechanisms associated with these two hematologic malignancies.

9.6.4. Biochemical Characteristics of AL Proteins that Depend Largely on Their Mutational Diversity: Low pH is Not Always Destabilizing

In order to understand the role of electrostatic interactions in the stability of proteins, we can utilize low-pH and high-salt experimental conditions to screen the electrostatic interactions that are thought to be important for protein stability. As previously mentioned, studies with SMA protein have shown that acidic pH destabilized the protein and promoted amyloid formation (Khurana *et al.*, 2001). Studies with AL-09, AL-12, and uMM-01 proteins from our laboratory have shown that AL proteins do not follow a unique pattern of protein stability at different pHs (Table 9.1). AL-09 does not change conformation or protein stability in the presence of NaCl (Fig. 9.4A; **see color insert**). Rates of urea-induced unfolding for AL-09 have been measured at

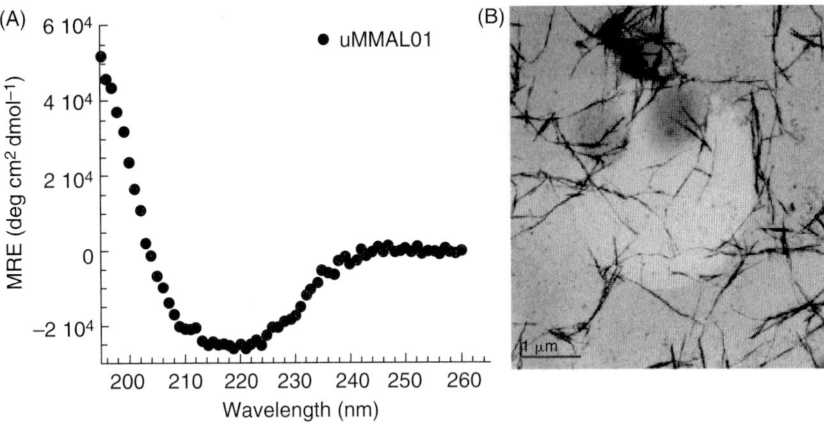

Figure 9.3. (A) Far UV-CD spectrum of uMMAL-01 in 10 mM Tris-HCl, pH 7.4, at 4°C. The T_m of this protein is 50.1°C at pH 7.4. (B) Electron micrograph of uMMAL-01 fibrils. These fibrils were formed during the thermal denaturation experiment in which the protein is heated up from 4°C to 90°C and cooled down back to 4°C. Scale bar is shown for the micrograph.

Table 9.1. Melting temperatures corresponding with AL-09, AL-12, and uMM-01 proteins followed by circular dichroism and tryptophan fluorescence

pH	AL-09 (°C)	AL-12 (f) (°C)	uMM-01 (°C)
5	37.5	54.8	49.2
7	37.1	47.5	51.0
9	37.4	30.9	55.1
11	33.6	n/d	49.8

n/d, not able to calculate T_m. No two-state folding occurs under these conditions. (f), determined by using fluorescence.

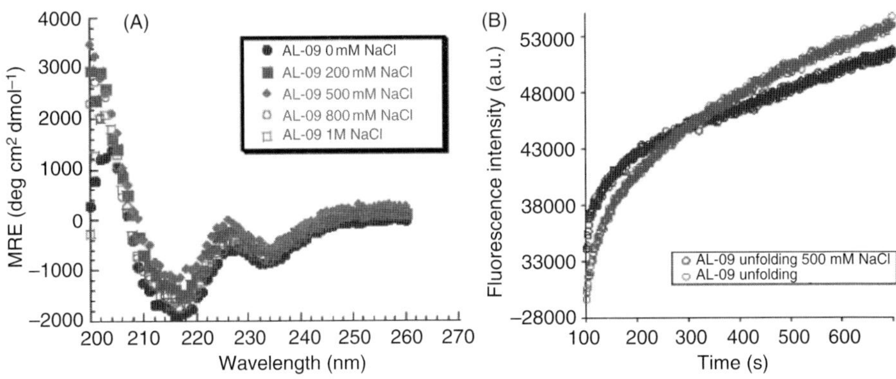

Figure 9.4. (A) The structure of AL-09 is not affected by NaCl. Far UV-CD spectra of AL-09 in the presence of different concentrations of NaCl. All reactions were done in the presence of 10 mM Tris-HCl at pH 7.4. (B) Unfolding kinetics of AL-09 in 10 mM Tris-HCl at pH 7.4 in the presence and the absence of 500 mM NaCl followed by the changes in fluorescence spectroscopy corresponding with Trp 35.

pH 7.4 in the presence and absence of 500 mM NaCl by monitoring changes in intrinsic protein fluorescence from tryptophan 35. The unfolding kinetics follow single exponential in the absence of NaCl and double exponential in the presence of 500 mM NaCl (Fig. 9.4B; **see color insert**). AL-09 unfolds about four-fold faster in the absence of NaCl. The folding kinetics result indicates that electrostatic effects could either affect the thermodynamic stability in equilibrium or the rates of unfolding. The presence of electrostatic interactions that could be destabilized at low pHs depends largely on the mutational diversity of the system.

9.7. Future Directions in the Study and Treatment of AL

9.7.1. Understand Mutational Diversity and Organ Tropism

The high mutational diversity of AL protein sequences could be one explanation for the type and number of organs/tissues affected, as well as the severity of the disease. It is therefore extremely important to understand how mutations affect protein stability and amyloid formation pathways. The role of the proteolysis in AL pathogenesis has not been addressed previously; therefore it is important to determine if the constant domain (absent in most AL fibrils) could confer any stability to the protein.

9.7.2. Determine the Toxic Species in AL Pathogenesis

It has been reported that the presence of soluble light chains impairs cardiomyocyte function by increasing cellular oxidant stress (Brenner *et al.*, 2004). More recently, it has been shown that cardiac fibroblasts in the presence of soluble light chains and fibroblast growth factor-2 caused heparan sulfate translocation to the nucleus and sulfation of the secreted glycosaminoglycans, mimicking cardiac injury (Trinkaus-Randall *et al.*, 2005). These studies suggest that soluble light chains can promote stress responses in the cell. Future studies should address the role of purified populations of different oligomeric intermediates and amyloid fibrils in cell stress and cell death.

9.7.3. Therapeutic Approaches

AL is a complex disease in which the destabilization that leads to amyloid formation could be caused by one or more of the following reasons: (i) somatic mutations are accumulated; (ii) proteolysis may play a role in destabilizing the structure; (iii) the loss of association with the heavy chain. Inhibition of fibril formation for AL has been previously reported (Davis *et al.*, 2000). In this study, the endoplasmic reticulum chaperone BiP and a peptide corresponding with strands D and E were used as inhibitors for amyloid formation. The use of conformation-specific antibodies that target the AL amyloid fibrils to enhance amyloid clearance has been reported (Solomon *et al.*, 2003). Another type of antibody targets oligomeric intermediates (Glabe, 2004). Both approaches could be potentially useful to treat AL. Once the toxic species in AL pathogenesis are defined, possible therapeutic approaches that could be used are (i) stabilization of the folded state using small molecules as has been done to inhibit amyloid formation by transthyretin mutants; (ii) enhanced clearance of amyloid fibrils by using molecules that bind to the fibrils and destabilize them or that bind to serum amyloid P-component (Sacchettini and Kelly, 2002).

References

Abraham, R. S., Geyer, S. M., Price-Troska, T. L., Allmer, C., Kyle, R. A., Gertz, M. A., and Fonseca, R. (2003). Immunoglobulin light chain variable (V) region genes influence clinical presentation and outcome in light chain-associated amyloidosis (AL). *Blood* 101: 3801–3808.

Abraham, R. S., Geyer, S. M., Ramirez-Alvarado, M., Price-Troska, T. L., Gertz, M. A., and Fonseca, R. (2004). Analysis of somatic hypermutation and antigen selection in the clonal B cell in immunoglobulin light chain amyloidosis (AL). *J. Clin. Immunol.* 24: 340–353.

Alexandru, I., Kells, D. I., Dorrington, K. J., and Klein, M. (1980). Non-covalent association of heavy and light chains of human immunoglobulin G: studies using light chain labelled with a fluorescent probe. *Mol. Immunol.* 17: 1351–1363.

Alim, M. A., Yamaki, S., Hossain, M. S., Takeda, K., Kozima, M., Izumi, T., Takashi, I., and Shinoda, T. (1999). Structural relationship of kappa-type light chains with AL amyloidosis: multiple deletions found in a VkappaIV protein. *Clin. Exp. Immunol.* 118: 344–348.

Ancsin, J. B. (2003). Amyloidogenesis: historical and modern observations point to heparan sulfate proteoglycans as a major culprit. *Amyloid Int. J. Exper. Clin. Invest.* 10: 67–69.

Bourne, P. C., Ramsland, P. A., Shan, L., Fan, Z. C., DeWitt, C. R., Shultz, B. B., Terzyan, S. S., Moomaw, C. R., Slaughter, C. A., Guddat, L. W., and Edmundson, A. B. (2002). Three-dimensional structure of an immunoglobulin light-chain dimer with amyloidogenic properties. *Acta Crystallogr. D Biol. Crystallogr.* 58: 815–823.

Branden, C., and Tooze, J. (1999). *Introduction to protein structure*. New York, Garland Publishing, Inc.

Brenner, D. A., Jain, M., Pimentel, D. R., Wang, B., Connors, L. H., Skinner, M., Apstein, C. S., and Liao, R. (2004). Human amyloidogenic light chains directly impair cardiomyocyte function through an increase in cellular oxidant stress. *Circ. Res.* 94: 1008–1010.

Buxbaum, J. N. (2003). Diseases of protein conformation: what do *in vitro* experiments tell us about *in vivo* diseases? *Trends Biochem. Sci.* 28: 585–592.

Chou, C. L., and Morrison, S. L. (1993). An insertion-deletion even in murine immunoglobulin kappa gene resembles mutations at heavy-chain disease loci. *Somat. Cell Mol. Genet.* 19: 131–139.

Comenzo, R. L., Wally, J., Kica, G., Murray, J., Ericsson, T., Skinner, M., and Zhang, Y. (1999). Clonal immunoglobulin light chain variable region germline gene use in AL amyloidosis: association with dominant amyloid-related organ involvement and survival after stem cell transplantation. *Br. J. Haematol.* 106: 744–751.

Davis, P. D., Raffen, R., Dul, L. J., Vogen, M. S., Williamson, K. E., Stevens, J. F., and Argon, Y. (2000). Inhibition of amyloid fiber assembly by both BiP and its target peptide. *Immunity* 13: 433–442.

Davis, D. P., Gallo, G., Vogen, S. M., Dul, J. L., Sciarretta, K. L., Kumar, A., Raffen, R., Stevens, F. J., and Argon, Y. (2001). Both the environment and somatic mutations govern the aggregation pathway of pathogenic immunoglobulin light chain. *J. Mol. Biol.* 313: 1021–1034.

Dobson, C. M. (2003). Protein folding and misfolding. *Nature* 426: 884–890.

Dul, J. L., and Argon, Y. (1990). A single amino acid substitution in the variable region of the light chain specifically blocks immunoglobulin secretion. *Proc. Natl. Acad. Sci. U.S.A.* 87: 8135–8139.

Epp, O., Lattman, E. E., Schiffer, M., Huber, R., and Palm, W. (1975). The molecular structure of a dimer composed of the variable portions of the Bence Jones protein REI refined at 2.0-A resolution. *Biochemistry* 14: 4943–4952.

Gertz, M. A., and Kyle, R. A. (1989). Primary systemic amyloidosis—a diagnostic primer. *Mayo Clin. Proc.* 64: 1505–1519.

Gertz, M. A., and Kyle, R. A. (1990). Prognostic value of urinary protein in primary systemic amyloidosis (AL). *Am. J. Clin. Pathol.* 94: 313–317.

Gertz, M. A., and Kyle, R. A. (1997). Hepatic amyloidosis: clinical appraisal in 77 patients. *Hepatology* 25: 118–121.

Gertz, M. A., Lacy, M. Q., and Dispenzieri, A. (1999). Amyloidosis. *Hematol. Oncol. Clin. North Am.* 13: 1211–1233.

Glabe, C. G. (2004). Conformation-dependent antibodies target diseases of protein misfolding. *Trends Biochem. Sci.* 29: 542–547.

Haas, I. G., and Wabl, M. R. (1984). Immunoglobulin heavy chain toxicity in plasma cells is neutralized by fusion by pre-B cells. *Proc. Natl. Acad. Sci. U.S.A.* 81: 7185–7188.

Huang, D. B., Chang, C. H., Ainsworth, C., Brunger, A. T., Eulitz, M., Solomon, A., Stevens, F. J., and Schiffer, M. (1994). Comparison of crystal structures of two homologous proteins: structural origin of altered domain interactions in immunoglobulin light-chain dimers. *Biochemistry* 33: 14848–14857.

Hurle, M. R., Helms, L. R., Li, L., Chan, W., and Wetzel, R. (1994). A role for destabilizing amino acid replacements in light-chain amyloidosis. *Proc. Natl. Acad. Sci. U.S.A.* 91: 5446–5450.

Jiang, X., Myatt, E., Lykos, P., and Stevens, F. J. (1997). Interaction between glycosaminoglycans and immunoglobulin light chains. *Biochemistry* 36: 13187–13194.

Khan, M. F., and Falk, R. H. (2001). Amyloidosis. *Postgrad. Med. J.* 77: 686–693.

Khurana, R., Gillespie, J. R., Talapatra, A., Minert, L. J., Ionescu-Zanetti, C., Millett, I., and Fink, A. L. (2001). Partially folded intermediates as critical precursors of light chain amyloid fibrils and amorphous aggregates. *Biochemistry* 40: 3525–3535.

Khurana, R., Souillac, P. O., Coats, A. C., Minert, L. J., Ionescu-Zanetti, C., Carter, S. A., Solomon, A., and Fink, A. L. (2003). A model for amyloid formation in immunoglobulin light chains based on comparison of amyloidogenic and benign proteins and specific antibody binding. *Amyloid Int. J. Exper. Clin. Invest.* 10: 97–109.

Kim, Y., Wall, J. S., Meyer, J., Murphy, C., Randolph, T. W., Manning, M. C., Solomon, A., and Carpenter, J. F. (2000). Thermodynamic modulation of light chain amyloid fibril formation. *J. Biol. Chem.* 275: 1570–1574.

Kim, Y.-S., Randolph, T. W., Stevens, F. J., and Carpenter, J. F. (2002). Kinetics and energetics of assembly, nucleation, and growth of aggregates and fibrils for an amyloidogenic protein. Insights into transition states from pressure, temperature, and co-solute studies. *J. Biol. Chem.* 277: 27240–27246.

Klein, M., Kortan, C., Kells, D. I., and Dorrington, K. J. (1979). Equilibrium and kinetic aspects of the interaction of isolated variable and constant domains of light chain with the Fd' fragment of immunoglobulin G. *Biochemistry* 18: 1473–1481.

Kohler, G. (1980). Immunoglobulin chain loss in hybridoma lines. *Proc. Natl. Acad. Sci. U.S.A.* 77: 2197–2199.

Koo, E. H., Lansbury, P. T., Jr., and Kelly, J. W. (1999). Amyloid diseases: abnormal protein aggregation in neurodegeneration. *Proc. Natl. Acad. Sci. U.S.A.* 96: 9989–9990.

Kyle, R. A., and Greipp, P. R. (1983). Amyloidosis (AL). Clinical and laboratory features in 229 cases. *Mayo Clin. Proc.* 58: 665–683.

Kyle, R. A., Greipp, P. R., and O'Fallon, W. M. (1986). Primary systemic amyloidosis: multivariate analysis for prognostic factors in 168 cases. *Blood* 68: 220–224.

Kyle, R. A., Linos, A., Beard, C. M., Linke, R. P., Gertz, M. A., O'Fallon, W. M., and Kurland, L. T. (1992). Incidence and natural history of primary systemic amyloidosis in Olmsted County, Minnesota, 1950 through 1989. *Blood* 79: 1817–1822.

Kyle, R. A., and Gertz, M. A. (1995). Primary systemic amyloidosis: clinical and laboratory features in 474 cases. *Semin. Hematol.* 32: 45–59.

Kyle, R. A., Gertz, M. A., Greipp, P. R., Witzig, T. E., Lust, J. A., Lacy, M. Q., and Therneau, T. M. (1999). Long-term survival (10 years or more) in 30 patients with primary amyloidosis. *Blood* 93: 1062–1066.

McLaughllin R.W., De Stigter J.K., Sikkink L.A, Baden E.M., and Ramirez-Alvarado M. (2006). The effects of sodium sulfate, glycosaminoglycans and Congo red on the structure, stability and amyloid formation of an immunoglobulin light chain protein. *Protein Sci.* 15: 1710–1722.

Merlini, G., and Bellotti, V. (2003). Molecular mechanisms of amyloidosis. *N. Engl. J. Med.* 349: 583–596.

Morrison, S. L., and Scharff, M. D. (1979). A mouse myeloma variant with a defect in light chain synthesis. *Eur. J. Immunol.* 9: 461–465.

Olsen, K. E., Sletten, K., and Westermark, P. (1998). Extended analysis of AL-amyloid protein from abdominal wall subcutaneous fat biopsy: kappa IV immunoglobulin light chain. *Biochem. Biophys. Res. Commun.* 245: 713–716.

Omtvedt, L. A., Bailey, D., Renouf, D. V., Davies, M. J., Paramonov, N. A., Haavik, S., Husby, G., Sletten, K., and Hounsell, E. F. (2000). Glycosylation of immunoglobulin light chains associated with amyloidosis. *Amyloid Int. J. Exper. Clin. Invest.* 7: 227–244.

Pokkuluri, P. R., Solomon, A., Weiss, D. T., Stevens, F. J., and Schiffer, M. (1999). Tertiary structure of human lambda 6 light chains. *Amyloid Int. J. Exper. Clin. Invest.* 6: 165–171.

Rajkumar, S. V., Gertz, M. A., and Kyle, R. A. (1998). Primary systemic amyloidosis with delayed progression to multiple myeloma. *Cancer* 82: 1501–1505.

Ramirez-Alvarado, M., Merkel, J. S., and Regan, L. (2000). A systematic exploration of the influence of the protein stability on amyloid fibril formation *in vitro*. *Proc. Natl. Acad. Sci. U.S.A.* 97: 8979–8984.

Ramirez-Alvarado, M., and Regan, L. (2002). Does the location of a mutation determine the ability to form amyloid fibrils? *J. Mol. Biol.* 323: 17–22.

Ross, C. A., and Poirier, M. A. (2004). Protein aggregation and neurodegenerative disease. *Nat. Med.* 10: S10-S17.

Sacchettini, J. C., and Kelly, J. W. (2002). Therapeutic strategies for human amyloid diseases. *Nat. Rev. Drug Discov.* 1: 267–275.

Schormann, N., Murrell, J. R., Liepnieks, J. J., and Benson, M. D. (1995). Tertiary structure of an amyloid immunoglobulin light chain protein: a proposed model for amyloid fibril formation. *Proc. Natl. Acad. Sci. U.S.A.* 92: 9490–9494.

Sezer, O., Eucker, J., Schmid, P., and Possinger, K. (2000). New therapeutic approaches in primary systemic AL amyloidosis. *Ann. Hematol.* 79: 1–6.

Solomon, A., Weiss, D. T., and Wall, J. S. (2003). Immunotherapy in systemic primary (AL) amyloidosis using amyloid-reactive monoclonal antibodies. *Cancer Biother. Radiopharm.* 18: 853–860.

Souillac, P. O., Uversky, V. N., Millett, I. S., Khurana, R., Doniach, S., and Fink, A. L. (2002a). Effect of association state and conformational stability on the kinetics of immunoglobulin light chain amyloid fibril formation at physiological pH. *J. Biol. Chem.* 277: 12657–12665.

Souillac, P. O., Uversky, V. N., Millett, I. S., Khurana, R., Doniach, S., and Fink, A. L. (2002b). Elucidation of the molecular mechanism during the early events in immunoglobulin light chain amyloid fibrillation. Evidence for an off- pathway oligomer at acidic pH. *J. Biol. Chem.* 277: 12666–12679.

Stevens,. F. J., Weiss, D. T., and Solomon, A. (1999). Structural bases of light chain-related pathology. In: Zanetti, M. and Capra, J.D., (Eds.), *The Antibodies*, Vol. 5. Harwood Academic Publishers, Langehorne, PA, pp. 175–208.

Stevens, F. J. (2000). Four structural risk factors identify most fibril-forming kappa light chains. *Amyloid Int. J. Exper. Clin. Invest.* 7: 200–211.

Trinkaus-Randall, V., Walsh, M. T., Steeves, S., Monis, G., Connors, L. H., and Skinner, M. (2005). Cellular response of cardiac fibroblasts to amyloidogenic light chains. *Am. J. Pathol.* 166: 197–208.

Wall, J., Murphy, C. L., and Solomon, A. (1999a). In vitro immunoglobulin light chain fibrillogenesis. *Methods Enzymol.* 309: 204–217.

Wall, J., Schell, M., Murphy, C., Hrncic, R., Stevens, F. J., and Solomon, A. (1999b). Thermodynamic instability of human lambda 6 light chains: correlation with fibrillogenicity. *Biochemistry* 38: 14101–14108.

Wall, J., Gupta, V., Wilkerson, M., Schell, M., Loris, R., Adams, P., Solomon, A., Stevens, F. J., and Dealwis, C. (2004). Structural basis of light chain amyloidogenicity: comparison of thermodynamic properties, fibrillogenic potential and tertiary structural features of four Vlambda6 proteins. *J. Mol. Recognit.* 17: 323–331.

Wetzel, R. (1997). Domain stability in immunoglobulin light chain deposition disorders. *Adv. Protein Chem.* 50: 183–242.

Zhu, M., Souillac, P. O., Ionescu-Zanetti, C., Carter, S. A., and Fink, A. L. (2002). Surface-catalyzed amyloid fibril formation. *J. Biol. Chem.* 277: 50914–50922.

Zhu, M., Han, S., Zhou, F., Carter, S. A., and Fink, A. L. (2004). Annular oligomeric amyloid intermediates observed by *in situ* atomic force microscopy. *J. Biol. Chem.* 279: 24452–24459.

10

Pancreatic Islet Amyloid and Diabetes

Anne Clark and Jenni Moffitt

Abstract

Deposition of amyloid in pancreatic islets is a feature of type 2 diabetes in man but the causal factors are unknown. Progressive amyloidosis results in loss of insulin-producing cells and increased severity of disease. The fibril component is a 37-amino-acid peptide, islet amyloid polypeptide (IAPP; amylin), which is a normal cellular peptide, cosecreted with insulin; the species-specific amino-acid sequence of IAPP is unaltered in diabetes. Peptide refolding in islet perivascular spaces to form β-sheets could be related to production of fragments of the precursor peptide, proIAPP. Islet amyloidosis is not a causal factor for hyperglycemia in man but contributes to onset of diabetes in non-human primates and transgenic rodent models. Because islet amyloidosis cannot be detected *in vivo*, clinical development of inhibitors is likely to be difficult.

10.1. Introduction

Hyaline deposits in the pancreas have been recognised for more than a century; Opie in 1900 described amorphous hyaline material in pancreatic islets from a diabetic patient (Opie, 1901). This descriptor for the "clear" or hyaline substance was subsequently changed to amyloid (Ehrlich and Ratner, 1961) when the glycosyl residues were recognized as being a ubiquitous component of proteinaceous amyloid deposits. Through these early years, careful observations by pancreatic pathologists (Cecil, 1909; Maclean and Ogilvie, 1955; Maloy *et al.*, 1981) indicated that there were different types of pancreatic islet disease that were collectively called diabetes; patients who were diagnosed with hyperglycemia in their first and second decades had little islet amyloid, whereas patients developing diabetes in middle age had extensive deposits (Cecil, 1909; Wright, 1927). However, it was not until more than 50 years later that these two groups of patients were recognized as belonging to two distinct forms of diabetes with very different etiologies and that amyloid was related only to disease developing in older subjects-now known as type 2 (non–insulin-dependent) diabetes; this form of diabetes contrasted with that of the younger age of onset where insulin-secreting cells were entirely lost as a result of autoimmune processes now known as type 1 (insulin-dependent diabetes) (Clark, 2004). Originally, amyloid deposits in any organ were considered to be a nonspecific feature of aging and with no particular relationship to the etiology or progression of disease. For Pancreatic, pathologic studies identified that islet amyloid was a specific feature of diabetes developing in older patients; however, as a result of the extreme insolubility of the amyloid fibrils, it was not until 1987–1988 that the peptide responsible for the deposits was identified (Clark *et al.*, 1987a; Westermark *et al.*, 1987) and some of the structural factors for fibril formation identified (Westermark *et al.*, 1990). However, the pathophysiologic features that result in refolding of the amyloid peptide to form fibrils in diabetes and some other conditions remain remarkably elusive.

Pancreatic islet amyloid is located only within the islets of Langerhans, which are the sites of production of insulin in the body. Pancreatic islets (average diameter 150–300 μm) are distributed throughout the exocrine pancreas (weight 90–150 g in man) (Fig. 10.1A; **see color insert**) and form 3–5% of the total organ weight (Clark, 2004). Therefore, like cerebral amyloidosis in Alzheimer's disease, these deposits are small and difficult to identify at postmortem. Currently, islet amyloidosis cannot be identified *in vivo*. Unlike the invasive nature of some forms of systemic amyloidosis, pancreatic amyloid remains inside the confines of the islet and does not spread to the exocrine pancreatic tissue; the islets do not increase in size as a result of the deposits, but endocrine cells are replaced (Fig. 10.1B; **see color insert**) (Westermark and Wilander, 1978; Clark *et al.*, 1988; Röcken *et al.*, 1992). Thus, the cytotoxic nature of islet amyloid has important implications for maintenance of insulin reserves in diabetes. Indeed, islet amyloid has been postulated to be an etiological factor for the inadequate insulin secretion that is characteristic of type 2 diabetes (Höppener *et al.*, 2002). Deposition of amyloid in pancreatic islets is one of the most common pathologic features of type 2 diabetes being found in at least one islet at postmortem in more than 90% of type 2 diabetic subjects (Schneider *et al.*, 1980; Westermark, 1994). Islet amyloid has also been identified at postmortem in 15% of elderly patients who were apparently nondiabetic *in vivo* (Westermark, 1994) and is found in islet cell tumors known as insulinomas in man (O'Brien *et al.*,

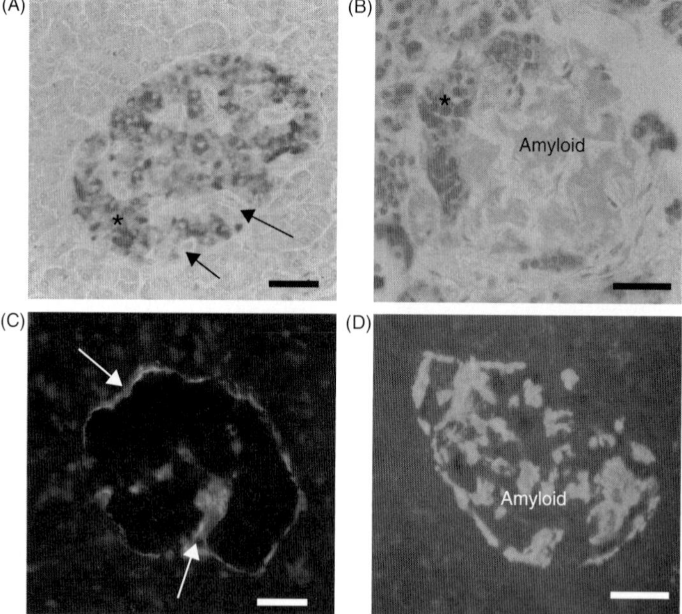

Figure 10.1. Distribution of islet amyloid and insulin-containing β-cells in the pancreas of nondiabetic (A) and type 2 diabetic patients (B, C, D). (A) Immunolabeling for insulin (brown) demonstrates that 60–80% of the islet mass contains β-cells. Capillaries (arrows) surround and penetrate into the islet centre. (B) Amyloid stained with Congo red (amorphous pink color) fills > 60% of the islet space. The remaining β-cells (*) occupy < 30% of the islet. (C) Islet from a subject diagnosed for > 10 years with type 2 diabetes; amyloid labeled with thioflavine S (green fluorescence) is perivascular, adjacent to capillaries around and within the islet. (D) Amyloid in a type 2 diabetic subject (diabetes duration > 10 years) (labeled with thioflavine S) occupies more than 80% of the islet. Scale bars = 20 μm.

1994). This suggests that at least some of the factors responsible for islet amyloid deposition could be unrelated to diabetes.

The relationship of islet amyloid to the onset and progression of type 2 diabetes is largely unknown; formation of amyloid cannot be directly related to the pathophysiology of the disease in man because the deposits are undetectable *in vivo*. The identification of the component peptide of islet amyloid deposits as islet amyloid polypeptide (IAPP; also known as "amylin") in the 1980s (Westermark *et al.*, 1986; Clark *et al.*, 1987a; Cooper *et al.*, 1987; Westermark *et al.*, 1987) was a breakthrough in islet amyloid research. This permitted measurements of the circulating peptide to be made in clinical observations (Sanke *et al.*, 1991), a synthetic analogue was produced for commercial exploitation (Butler *et al.*, 1990), and investigations continue using genetically modified (Fox *et al.*, 1993; D'Alessio *et al.*, 1994; de Koning *et al.*, 1994; Janson *et al.*, 1996) and other animal models (Howard, 1978; O'Brien *et al.*, 1986; de Koning *et al.*, 1993) to determine the relationship of amyloidosis to the pathophysiology of diabetes. In addition, this small, 37-amino-acid peptide has been used extensively in biophysical experiments *in vitro* to identify amyloidogenic folding processes and potential actions on cellular systems.

10.2. Islet Amyloid Polypeptide and Fibril Formation

Islet amyloid polypeptide (IAPP; amylin) is a 37-amino-acid peptide (Fig. 10.2) that was identified as the component peptide of islet amyloid deposits found in the pancreas of cats, humans, and in insulinomas (Westermark *et al.*, 1987; Cooper *et al.*, 1988). IAPP is structurally related to calcitonin-gene-related peptide (CGRP) having more than 40% amino-acid sequence homology with the CGRP peptides (Cooper, 1994). The gene for IAPP consists of three exons and is found on chromosome 12 (Fig. 10.3); chromosomes 11 (containing the CGRP family of genes) and 12 are thought to share some ancestral homologies. This similarity is also demonstrated in some of the proposed actions of IAPP such as calcium metabolism (Gilbey *et al.*, 1991) and shared receptor binding with the calcitonin gene family of peptides (Barth *et al.*, 2004). IAPP is costored with insulin in β-cell secretory granules (Clark *et al.*, 1989; Lukinius *et al.*, 1989) and cosecreted in response to agents that promote secretion of insulin (e.g., glucose) (Butler *et al.*, 1990; Sanke *et al.*, 1991; Kahn *et al.*, 1998). IAPP is synthesized as a larger precursor peptide, proIAPP (67 amino acids in man) (Sanke *et al.*, 1988). This peptide is proteolytically cleaved at the N- and C-terminal junctions by enzymes, prohormone convertase 1/3 and 2 within the β-cell secretory granule as the granule matures (Sanke *et al.*, 1988; Badman *et al.*, 1996; Higham *et al.*, 2000; Marzban *et al.*, 2005a). The normally soluble 37-amino-acid peptide is normally found in the circulation at 5–15 pmol/L concentration in man and is excreted by the kidney (Kautzky-Willer *et al.*, 1994). Although a multitude of physiologic roles has been attributed to IAPP, it remains a peptide without a clearly identified function *in vivo* (Gebre-Medhin *et al.*, 2000); the only notable feature in the IAPP knockout mouse was moderately increased insulin secretion in response to glucose (Gebre-Medhin *et al.*, 1998). In addition, islet amyloid is not formed in type 1 diabetic subjects where β-cells (and therefore insulin and IAPP) are absent, but these subjects do not have pathophysiologic features that could be attributed directly to IAPP deficiency. Unlike many other forms of amyloidosis, islet amyloid formation in type 2 diabetes cannot be directly related to any post-translational modification of the peptide or gene mutations that would confer increased amyloidogenicity to the peptide.

Pro-Islet amyloid polypeptide

```
              N-terminal                         IAPP                          C-terminal
                              1      8          20          29      37
Human    TPIES:::HQVEKR   KCNTATCATQRLANFLVHSSNNFGAILSSTNVGSNTY   GKRNAVEVLKREPLNYLPL
Monkey   -----:::------   -----------------R------T---------D--   -------------------
Cat      -----:::N-----   ----------------IR----L-----P--------   ---ST-DI-N-------F
Dog      ---K-:::N-----   -----------------RT---L-----P--------   ----TI-I-N-G------
Rat      --VG-GTNP--D--   -----------------R----L-PV-PP--------   ----VA-DPN-S-DF-L-
Mouse    --VR-GSNP-MD--   -----------------R----L-PV-PP--------   -----AGDPN-S-DF-KV
Hamster--VR-GTN--MD--     -------------------N--L-PV-P---------   ---S-A-IPDGDS-DLFL-
G-pig    -S-A-DTG---G--   -----------T-----R--H-L-A-LP-D-------   -----PQISD-LCH----
Degu     ---A-DTD-R-D--   -----------T-----R--H-L-A-PP-K-------   -R---:Q-VDV-L-H----
Rabbit                            --------I--------F-PPS
Hare                              --------I--------F-PP-
Cougar                    -*----*---------IRSS*------*
Pig                         -M------H-----DR-R-L:-T-F-P-K---
```

Figure 10.2. Amino acid sequences of proIAPP illustrated in single letter code. Three residues are missing (:) in human, monkey, cat, and dog when aligned with other sequences. Porcine IAPP is missing one amino acid. Amino acids common to the human sequence are indicated by (−). Asterisks (*) represent amino acids not yet determined. Proline residues in IAPP 20–29 (underlined) are present in rat and mouse IAPP, which prevent folding to form fibrils. Only partial sequences are available for rabbit, hare, cougar, and pig. G-pig = guinea-pig.

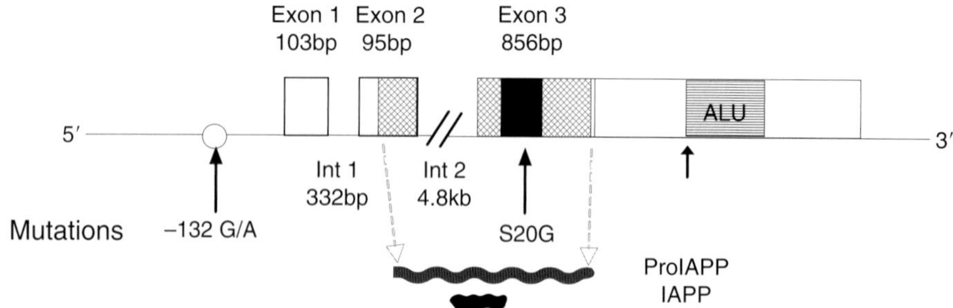

Figure 10.3. Schematic of the human IAPP gene. Exon 1 encodes 103 bp of the 5′ untranslated region; exon 2 encodes 15 base pairs of 5′ untranslated region, the signal peptide, and the first five residues of the N-terminal peptide of proIAPP (hatched); exon 3 encodes the remainder of the N-terminal region of proIAPP (hatched), the IAPP peptide sequence (black), the C-terminal peptide sequence of proIAPP (hatched), and the 3′ untranslated region of the mRNA. Two polyadenylation sites (polyA) have been identified at 390 bp and 1246 bp from the start of exon 3. An ALU repetitive sequence has been identified in human exon 3 (hashed box). Mutations have been identified in diabetic subjects (o) in the promoter sequence at −132 and at IAPP 20 serine to glycine.

10.2.1. Molecular Structure of IAPP and β-Sheet Formation

IAPP has been identified in all species in which it has been examined [reviewed in Jaikaran and Clark (2001)]. The primary structure of IAPP 1–37 is well conserved although there are some important species-specific substitutions that have implications for amyloid formation (Fig. 10.2). Diabetes-associated amyloid is found in non-human primates (Howard, 1978; de Koning *et al.*, 1993) and cats (O'Brien *et al.*, 1986) but not in rodents. The region

IAPP 20–29 was considered to be the primary amyloidogenic region of the peptide. Proline substitutions in this region are thought to be responsible for the lack of fibril formation in rats and mice and some other species (Betsholtz *et al.*, 1989; Westermark *et al.*, 1990; Moriarty and Raleigh, 1999; Green *et al.*, 2003). Peptide fragments of this central region of the molecule readily assume β-sheet conformation and form fibrils *in vitro*, and this domain was originally considered to be the only contributor to the amyloidogenic potential of the peptide, being primarily responsible for initiation of the refolding of the monomeric human IAPP and contributing to intermolecular β-sheet formation (Glenner *et al.*, 1988; Griffiths *et al.*, 1995; Kayed *et al.*, 1999; Goldsbury *et al.*, 2000; Tenidis *et al.*, 2000). However, two additional domains of the molecule, amino acids 30–37 and 8–20 (both of which are largely conserved in rodents and other species) have been shown experimentally to be amyloidogenic (Jaikaran *et al.*, 2001; Scrocchi *et al.*, 2003). This would allow formation of an intramolecular β-sheet (Fig. 10.4A; **see color insert**) with parallel and antiparallel association of strands and two turns at residues 17–19 and at 28–29 (Jaikaran *et al.*, 2001); these proposals are consistent with measurement of protofilament dimensions (Goldsbury *et al.*, 1997; Makin and Serpell, 2004). In rodents, the proline-rich central domain is a major stabilizing factor for the two potentially amyloidogenic molecular sequences, residues 30–37 and 8–20. Recent studies have shown, however, that other residues that differ between human and rat IAPP, (hIAPP 18, 23, 26), contribute to the amyloidogenicity (Green *et al.*, 2003) and when the human residues are substituted into the rat sequence, fibrils are formed.

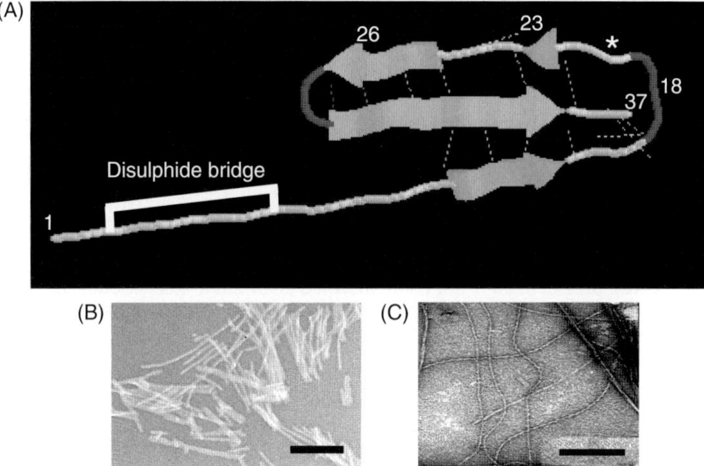

Figure 10.4. Molecular model of folded human IAPP and amyloid fibrils. (A) Hypothetical linear model of a folded β-sheet monomer of hIAPP generated using Insight II. Three major β-strands are predicted for hIAPP with hydrogen bonds (yellow dotted lines) between the strand regions; the region containing the cysteine bridge may not be incorporated into the fibril core. β-Turns in the region 17–19 and 28–29 are predicted. Serine 20 (asterisk) is replaced by glycine in a small number of Asian diabetic subjects. Replacement of residues 18, 23, 26 in rodent IAPP with the relevant human amino acids confers amyloidogenicity to rodent IAPP. (B) Amyloid fibrils extracted from the pancreas of a type 2 diabetic subject. Fibrils are short, straight, and do not show twisted fibril structure. (C) Fibrils formed *in vitro* from synthetic human IAPP demonstrating helical twisting of protofilaments. Scale bars = 200 nm.

The mechanisms that are involved in the initial stages of the refolding of IAPP and the intermediates of oligomeric assembly and protofilament formation are largely unknown. These intermediates are particularly difficult to identify *in vitro* because synthetic human IAPP 1–37 and many of its peptide fragments rapidly (within minutes) form insoluble aggregates in aqueous media compared with the 1–3 days required for Aβ peptides (Larson *et al.*, 2000; O'Nuallain *et al.*, 2004). Therefore, careful preparation is required for experimental examination of human IAPP (Higham *et al.*, 2000a). In common with other amyloid proteins, IAPP fibril formation *in vitro* involves transition from random structure to an aggregation-prone conformation with increased hydrophobicity (Kayed *et al.*, 1999; Kapurniotu, 2001); aggregation of peptides to create a nucleus is followed by oligomeric assembly and extension into protofilaments (Goldsbury *et al.*, 1999; Goldsbury *et al.*, 2000). However, the exact sequence of events that occurs *in vivo* is not known because the concentration of the peptide would be relatively low in comparison with that studied *in vitro* and would fluctuate. Indeed, fibrils extracted from islets of a type 2 diabetic subject were short, straight, and showing no ribbon or twisted features (Fig. 10.4B; **see color insert**), which are characteristic of fibrils formed *in vitro* (Fig. 10.4C; **see color insert**).

10.3. Islet Amyloid and the Pathophysiology of Type 2 Diabetes

The relationship of islet amyloid deposition to the characteristic features of the progression of type 2 diabetes in man (increased insulin resistance, onset of hyperglycemia, and β-cell dysfunction) is difficult to establish; there is no clinical parameter that can be related to the formation or extent of the deposits, and pancreatic biopsies are ethically unacceptable. Quantification of islet amyloid in postmortem human specimens has shown that extensive islet amyloidosis is found in patients with more severe islet dysfunction who required insulin replacement therapy rather than diet or oral hypoglycemic agents (Schneider *et al.*, 1980; Westermark, 1994). This indicates that the deposits have become so extensive that many islet cells have been replaced (Fig. 10.1D; **see color insert**) and insufficient numbers of functional β-cells are available to supply adequate insulin. However, the severity of the pathology is unrelated to the duration of disease in patients; this is consistent with the wide heterogeneity of the clinical features, progression of the syndrome of type 2 diabetes, and the diverse, often unidentified etiologies (DeFronzo *et al.*, 2004).

Although there is an association of the degree of amyloidosis with the later more severe stages of islet dysfunction, the role of amyloid in the onset and progression of diabetes is more complex. In patients with diabetes of long duration, the degree of islet amyloidosis at postmortem can vary from a low prevalence (<1% islets affected with small perivascular deposits) to up to 90% islets affected and a high degree of amyloid replacement of islet cells (up to 80% islet mass occupied with amyloid) (Westermark, 1972; Clark *et al.*, 1988; Zhao *et al.*, 2003; Clark *et al.*, 1990) (Figs. 10.1B and 10.1D). This indicates that islet amyloid and cellular replacement is not the sole precipitating factor for islet dysfunction and diabetes in man.

Longitudinal and cross-sectional studies on the development of islet amyloid are possible in animal models. Spontaneously developing diabetes in cats and monkeys is associated with progressively increasing islet amyloidosis (Howard, 1978; O'Brien *et al.*, 1986; de Koning *et al.*, 1993). These models of diabetes have a similar physiologic syndrome to that seen in man, including older age of onset, obesity, impaired glucose tolerance progressing to hyperglycemia, and dependence upon insulin therapy (Hansen and Bodkin, 1986; O'Brien *et al.*, 1993). Although these large animal models have verified what was thought to be occurring in man, they are not easily manipulated in the laboratory, and the disease process occurs over years rather than months. To facilitate laboratory observations, transgenic mice and rats expressing the gene for human IAPP have been created (hIAPP TM) where the human IAPP gene under the control of the rat insulin gene promoter has been inserted (Fox *et al.*, 1993; Höppener *et al.*, 1993; D'Alessio *et al.*, 1994; Couce *et al.*, 1996); this model allows targeting of the transgene to islet β-cells and regulation of human IAPP expression in parallel with insulin. In these models, amyloid is deposited only against a background of increased transgene expression (Couce *et al.*, 1996) or obesity (high-fat feeding, genetically determined obesity) (Verchere *et al.*, 1996; Soeller *et al.*, 1998; Höppener *et al.*, 1999); these rodents provide a useful laboratory model of islet amyloid with many features being similar to that of the spontaneous amyloidosis of primates and cats, but some aspects are very different, and all animal models differ from the pathology in humans.

In all non-human species and transgenic hIAPP, fat-fed mice, initial fibril formation accompanies or precedes obesity and/or glucose intolerance (Hull *et al.*, 2004). At first, only a few islets may be affected (low prevalence) with no significant change in islet cell mass (de Koning *et al.*, 1993; Wang *et al.*, 2001). At this stage, the very small bundles of amyloid fibrils are found adjacent to islet capillaries as in man (Fig. 10.1C; **see color insert)**, (Figs. 10.5A, and 10.5B). There is no apparent impact of the fibrils on glucose tolerance or on the proportion of islet β-cells in the animal models because they are normoglycemic suggesting that there is minimal cytotoxicity. However, these fine perivascular deposits are found in diabetic humans and, although not apparently cytotoxic, could compromise transfer of nutrients and insulin across the pericapillary space in affected islets and contribute to glucose intolerance. In transgenic mice and

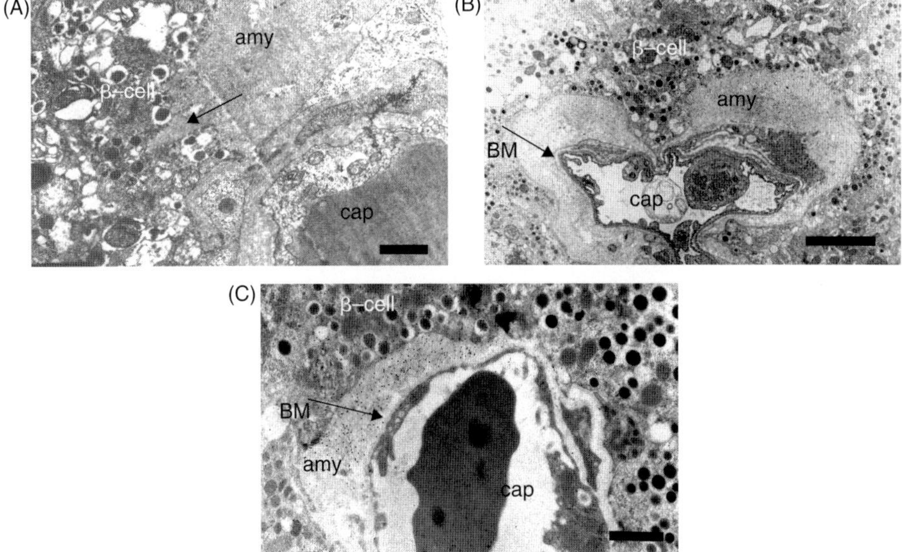

Figure 10.5. Amyloid deposits examined by electron microscopy: (A) a human diabetic subject, (B) a glucose-intolerant *Macaca mulatta* monkey, and (C) an old (aged 18 months), nondiabetic, non-obese, transgenic mouse expressing the gene for human IAPP. (A) Amyloid deposits are situated between an islet capillary and the β-cells with some fibril bundles in invaginations of the β-cell membrane (arrow). This could disrupt membrane signaling of glucose uptake and insulin secretion by interference with membrane fluidity. (B) Perivascular amyloid (arrow) surrounds this profile of the islet capillary in a monkey and fills the space between the cells and the basement membrane. (C) Amyloid deposits were infrequently found in this old nondiabetic transgenic mouse expressing hIAPP. The deposits were identified as IAPP by immunogold labeling (small dots). Amy, amyloid deposits; cap, capillary; BM, basement membrane. Scale bars (A) 2.0 μm, (B) 5.0 μm, (C) 1.0 μm.

monkeys, more islets become affected as the deposition progresses (increased islet prevalence), and when most islets are affected there is a substantial increase in severity (larger deposits in islets) (Fig. 10.6; **see color insert**). Measurements in hIAPP TM have shown that a prevalence of 80% was associated with 1.5% of islet space occupied by amyloid (Wang *et al.*, 2001). With moderate islet replacement with amyloid (~10% islet mass filled with amyloid and 100% islets affected), animals had impaired glucose-stimulated insulin secretion (Wang *et al.*, 2001) suggesting that islet amyloid (and the associated destruction of islet cells) was responsible for islet dysfunction. The progressive relationship between prevalence and severity has also been identified in monkey islets (Howard, 1986; Clark and Nilsson, 2004) where the severity of amyloidosis was generally less than 10% in glucose-intolerant animals and increased when almost 100% of islets were affected (Fig. 10.6). In cats, the degree of amyloidosis increased with worsening of glucose tolerance leading to hyperglycemia (O'Brien *et al.*, 1993). A transgenic rat model of islet amyloidosis appeared to have a more aggressive disease process than in mice, and amyloid appeared to be the major causative factor for onset of hyperglycemia (Butler *et al.*, 2004). All of these animal models had extensive amyloidosis when diabetic suggesting that the islet pathology was a causative factor for onset of hyperglycemia; this is not the case in man where the degree

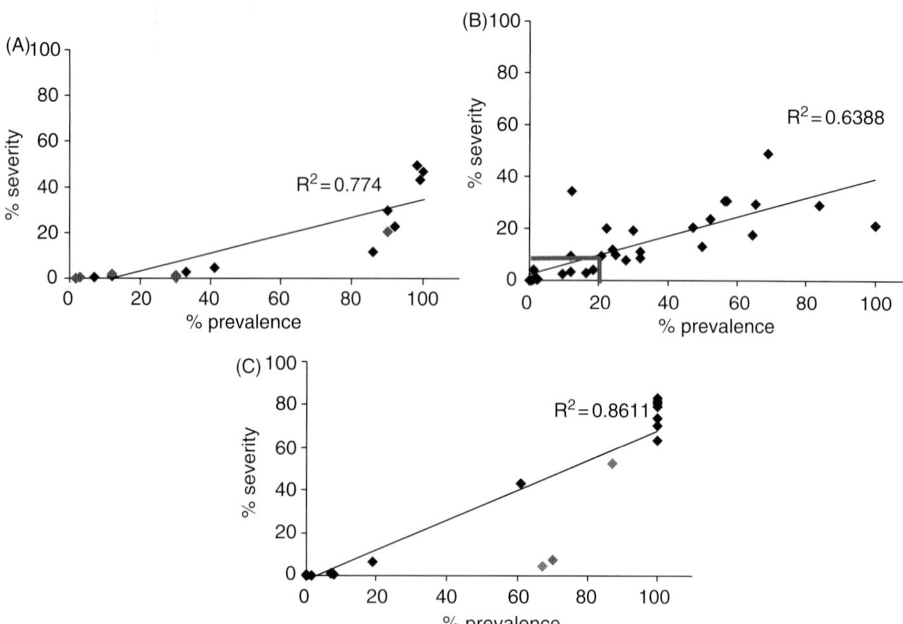

Figure 10.6. The relationship between islet amyloid prevalence (% islets affected) and amyloid severity (% islet area occupied) in human pancreas (A, B) and *Macaca mulatta* monkeys (C). (A) Data redrawn from Westermark *et al.* (1972) showing correlation between prevalence and severity in pancreatic tissue from 10 diabetic (blue) and 7 nondiabetic (red) patients. (B) Data was derived from studies on postmortem human diabetic pancreas (reported in Clark *et al.*, 1988); there was a significant relationship between prevalence and severity ($r^2 = 0.64$, $p < 0.05$); more than 50% of diabetic subjects had less than 10% islet area replaced in less than 20% of islets (red line). (C) Cross-sectional data from monkeys redrawn from Clark and Nilsson (2004) demonstrating a positive linear relationship of amyloid prevalence and severity in nondiabetic (red), glucose intolerant (green), and diabetic (blue) animals; in diabetes there was 100% prevalence with > 80% severity.

of amyloidosis after many years of diabetes is very variable. The positive relationship between the prevalence and severity of islet amyloidosis is shown in diabetic subjects; more than 50% of diabetic subjects have less than 20% prevalence and less than 10% severity (Figs. 10.6a and 10.6b). It is therefore unlikely that islet amyloidosis and the associated loss of β-cells are the primary causative factor for onset of diabetes in man.

10.4. Potential Fibrillogenic Factors for IAPP

10.4.1. Where Is Amyloid Found in the Islet?

In vivo, in animal models and humans, islet amyloid deposits are found largely outside the cells, but considerable controversy exists concerning the site of initiation of islet amyloidosis and the possibility of fibril assembly inside the cell (Couce *et al.*, 1996). However, in type 2 diabetes, fibrils are almost exclusively located at extracellular sites in the islets (Fig. 10.1). Small deposits in islets are located adjacent to the basement membrane of islet capillaries (Figs. 10.5A, 10.5B, and 10.5C). The basement membrane could therefore act as a permissive site for refolding of IAPP and the initiation of fibril formation in diabetes.

Heparan sulfate proteoglycans (HSPG), a component of the basement membrane, have been shown to play a role in synthetic IAPP fibrillogenesis (Snow *et al.*, 1987; Wisniewski and Frangione, 1992), and proIAPP has consensus sequences for HSPG binding (Park and Verchere, 2001). The role of the basement membrane could be to anchor aggregates of IAPP or proIAPP, which would form a "nucleus" or "nidus" for fibril formation (Rhoades *et al.*, 2000); once a nidus is formed adjacent to the basement membrane, IAPP monomers secreted from the islet β-cells after stimulation of insulin secretion could be converted rapidly to β-conformation at the nucleation site. This site would correspond with the perivascular location, adjacent to the basement membrane, where fibrils are first observed. As long as the nidus remains exposed acting as a nucleation site, the rate of amyloid deposition might be related to any changes in IAPP secretion. This hypothesis has been tested in a feline model of islet amyloidosis and diabetes (Hoenig *et al.*, 2000); increased production of IAPP in diabetic cats induced by drugs that stimulate β-cell secretion (sulfonylureas) resulted in islet amyloidosis; this was seen only in sulfonylurea-treated animals and not in those treated with insulin. These data suggest that prolonged stimulation of β-cell secretion with sulfonylureas coupled with hyperglycemia (as occurs in many patients with diabetes) can promote islet amyloid deposition.

Fibril-like accumulations of IAPP with tinctorial properties of amyloid have been identified at intracellular sites in β-cells of human insulinomas and of islets from hIAPP TM (de Koning *et al.*, 1994; O'Brien *et al.*, 1994; Jaikaran and Clark, 2001). This suggests that the conditions that promote peptide refolding and assembly of fibrils can occur at different sites under variable conditions. Intracellular bundles of fibrils without a boundary membrane are situated in the cytoplasm and some of these intracellular fibrils are immunoreactive for the N-terminal peptide of proIAPP (Jaikaran and Clark, 2001) and have been shown to be ubiquitinated (O'Brien *et al.*, 1994). If aberrant synthesis of IAPP occurs under some conditions (e.g., in β-cell tumors), it is possible that proIAPP could accumulate in the endoplasmic reticulum and form fibrils that are directed by ubiquitination toward the proteosome for degradation. This complex route is an unlikely common pathway for fibril formation in human diabetes as intracellular fibrils have not been detected.

10.4.2. What Normally Prevents IAPP Conversion into Fibrils?

Synthetic hIAPP prepared *in vitro* is rapidly converted to β-sheet and fibrils (Goldsbury *et al.*, 1999; Kayed *et al.*, 1999; Higham *et al.*, 2000a); this conversion is dependent upon peptide concentration, time, pH, and salt concentration of the buffer and the presence of a nidus. In comparison, insulin and IAPP are stored in secretory granules of β-cells, where IAPP is at a relatively high concentration; this has been estimated to be approximately 4 mM based on a ratio of 1:10 with insulin in pancreatic extracts and an estimated concentration of granular insulin of 40 mM (Hutton 1984). At this concentration, synthetic hIAPP *in vitro* would be rapidly converted into fibrils. This suggests that the peptide must be stabilized to prevent oligomerization and fibril formation *in vivo*. Many laboratories have demonstrated that insulin will inhibit IAPP fibril formation *in vitro* (Westermark *et al.*, 1996; Janciauskiene *et al.*, 1997); IAPP and insulin (but not proinsulin) form heteromolecular complexes *in vitro* (Jaikaran *et al.*, 2004; Larson and Miranker, 2004) suggesting that insulin stabilizes hIAPP in the β-cell granule. Proinsulin does not bind to IAPP or have a stabilizing effect (Jaikaran *et al.*, 2004).

10.4.3. What Factors Could Destabilize Solubility of Human IAPP?

The "environmental" factors that can destabilize a protein conformation include some situations that are present in diabetes. These include glycation, deamidation, and pH. Increased protein glycation is a complication of the sustained hyperglycemia of diabetes resulting in tissue damage as a result of formation of advanced glycation end-products (AGEs) (Furth, 1997). Glycation of human IAPP *in vitro* has been shown to promote fibrillogenesis (Kapourniotu *et al.*, 1998). However, such forms of protein glycation occur over period of days/weeks in the body, and the half-life of IAPP in the circulation has been estimated to be 30–40 mins (Kautzky-Willer *et al.*, 1997). Glycation of proteins in diabetes may affect cellular viability through binding to the so-called receptor for advanced glycation end-products, or RAGE. Monomeric nonglycated hIAPP does not bind to this receptor but fibrillar IAPP (and other amyloid proteins) have been shown to interact (Yan *et al.*, 2000; Deane *et al.*, 2003). If glycated monomeric IAPP binds to the RAGE receptor, this could act as a "nidus" for fibril formation and/or induce apoptosis, but there is no evidence from *in vivo* studies for this mechanism. Although glycation of established amyloid deposits almost certainly occurs, it is unlikely that glycation of IAPP monomers could initiate fibrillogenesis. Experiments of IAPP model peptides show that deamidation can dramatically accelerate amyloid formation (Nilsson *et al.*, 2002). Changes in pH and salt concentration are features of normal granule release; the granule has acid pH and high calcium concentration, which are both changed upon insulin secretory granule exocytosis, and the insulin crystal dissolves. These conditions do not promote fibril formation for biosynthetic IAPP *in vivo* or synthetic IAPP *in vitro* (Jaikaran *et al.*, 2001).

10.4.4. Genetically Determined Structural Changes: Mutations in the IAPP Gene

No genetically determined aberrant IAPP sequence has been found in the majority of populations with type 2 diabetes (Nishi *et al.*, 1990; Cook *et al.*, 1991). A missense heterozygous substitution in hIAPP at position 20 of a serine residue for a glycine (Figs. 10.2 and 10.3) has been identified in a very small number of patients in Japanese, Chinese, and Taiwanese studies (Seino, 2001); this mutation has not been found in large populations of Caucasian subjects with diabetes (Birch *et al.*, 1997) indicating that it is not a high-risk factor for the total population of diabetic subjects. A mutation in the IAPP gene promoter (-132G/A) (Fig. 10.3) has been identified and linked to type 2 diabetes in a Spanish population (Novials *et al.*, 2001) and in a small number of Polynesian subjects (Poa *et al.*, 2003). This mutation has the potential to increase expression of IAPP as a result of generation of a transcription factor binding site (Novials *et al.*,

2004). However, this variant was not linked to diabetes in Caucasian populations or to the degree of islet amyloidosis in man suggesting that it cannot be the only causative factor for islet amyloidosis (Esapa *et al.*, 2001).

10.4.5. Aberrant Production of IAPP

The role of inappropriate processing of proIAPP in type 2 diabetes remains high as a candidate factor for fibrillogenesis. In subjects with type 2 diabetes (Clark *et al.*, 1987) and in patients with insulinoma (both conditions associated with amyloid), there is increased secretion of incompletely processed proinsulin (Porte and Kahn, 1989; Marks and Teale, 1996; Halban and Kahn, 1997) as a result of incomplete processing by the proteolytic enzymes in granules. This could equally affect proIAPP processing by the same enzymes. Islet studies *in vitro* indicate increased production of N-terminal intact proIAPP under stimulated secretory conditions (Hou *et al.*, 1999), and immunoreactivity for the N-terminal peptide of proIAPP (see Fig. 10.2) has been identified in islet amyloid deposits in insulinomas and type 2 diabetes (Clark *et al.*, 1993; Westermark *et al.*, 2000). N-terminal intact proIAPP and proIAPP fragments have been identified in mouse models of deleted prohormone convertase enzymes (Wang *et al.*, 2001b; Marzban *et al.*, 2005a,b). This confirms that proIAPP could be cosecreted with proinsulin in type 2 diabetes and in insulinomas where aberrant protein processing of β-cell peptides occurs; this is currently one of the most likely hypotheses for islet amyloidosis.

10.4.6. IAPP Concentration

As with other amyloid diseases, the extent of islet amyloid formation is related to the production and local concentrations of IAPP; these are dependent upon secretion, clearance from the islet spaces, or by degradation. No correlation has been observed between an increase in circulating IAPP concentrations and either glucose intolerance or type 2 diabetes in man other than that which would be expected as a result of increased insulin resistance (Enoki *et al.*, 1992; Kahn *et al.*, 1998, 1999). In addition, insulin and IAPP are coordinately regulated in terms of gene expression in response to most stimuli (German *et al.*, 1992), and circulating concentrations in glucose intolerance and diabetes in man vary in parallel (Hartter *et al.*, 1991; Enoki *et al.*, 1992; Kahn *et al.*, 1998; Jorgensen *et al.*, 2000); proportions of the two β-cell products do however change during treatment with steroids when expression of insulin is reduced and IAPP enhanced (O'Brien *et al.*, 1991; Ludvik *et al.*, 1993; Mulder *et al.*, 1995).

Transgenic mice overexpressing the gene for human IAPP (hIAPP TM) have increased circulating human IAPP that can be up to 30 times higher than the endogenous mouse IAPP (Fox *et al.*, 1993; Höppener *et al.*, 1993; D'Alessio *et al.*, 1994; de Koning *et al.*, 1994; Janson *et al.*, 1996). However, these high levels are not necessarily associated with amyloidosis suggesting that increased concentration is not adequate alone to induce changes in peptide conformation (Soeller *et al.*, 1998; Höppener *et al.*, 1999). Increased local concentrations of secreted peptides are also influenced by clearance from the islet spaces. IAPP fibrils form between β-cells in islets from hIAPP TM maintained in culture (de Koning *et al.*, 1994; MacArthur *et al.*, 1999); the degree of extracellular fibril formation in these islets was increased by β-cell secretagogues suggesting that a combination of increased secretion and decreased clearance in the poorly perfused, extracellular spaces contribute to fibrillogenesis (MacArthur *et al.*, 1999).

10.5. Consequences of Fibrils in Islets

10.5.1. Fibril-Induced Cell Death

Fibrils formed from synthetic IAPP and other amyloidogenic proteins rapidly exhibit toxic properties to islet and other cells *in vitro* by an unknown mechanism (Janson *et al*., 1999; Caughey and Lansbury 2003; Scrocchi *et al*., 2003); insertion into the lipid bilayer and changes in cell membrane ion channel activity (Wagoner *et al*., 1993; Mirzabekov *et al*., 1996; Janson *et al*., 1998) leading to apoptosis (Lorenzo *et al*., 1994; Saafi *et al*., 2001), have been proposed. Not only mature fibrils but also small soluble or insoluble oligomers of hIAPP are thought to have cytotoxic properties (Harroun *et al*., 2001; Anguiano *et al*., 2002; Porat *et al*., 2003; Demuro *et al*., 2005). However, in cytotoxicity experiments with human IAPP, it is difficult to separate toxic effects of mature fibrils from that of smaller invisible oligomers because, *in vitro*, the process of fibrillogenesis is continuous. Although the relevance of these intermediates of peptide folding for understanding fibrils assembly is very high, their significance in cell death *in vivo* should be interpreted with caution. If nonfibrillar, small invisible hIAPP oligomers are the causative factor for death of β-cells in type 2 diabetes, it is surprising that there is not a severe reduction in islet mass in the absence of visible amyloidosis or at the first stages of small perivascular deposits; apoptotic cells are very infrequent in adult human islets (Kassem *et al*., 2000). Similarly, the degree of toxicity induced by fibrils formed from secreted, biosynthetic human IAPP in hIAPP TM islets in short-term culture is minimal (MacArthur *et al*., 1999); cells adjacent to fibrils show little evidence of cytotoxicity or apoptosis suggesting that the time course of synthetic fibril-induced toxicity is more rapid than that induced by biosynthetic hIAPP *in vitro* or *in vivo*.

10.5.2. Development of Inhibitors for Islet Amyloidosis

There is currently considerable interest in developing therapeutic agents directed against islet amyloid for use in the ever-growing populations affected by type 2 diabetes. Some of the proposed targets for therapeutic intervention include agents designed to prevent refolding of IAPP to form fibrils. Beta-sheet breaker peptides (Soto *et al*., 1998) bind specifically to amyloidogenic sequences and prevent assembly of β-sheets. Several differerent approaches have been made using small molecules that are dissimilar from the amyloidogenic protein or modified sequences of the same protein; N-methylation, (Kapurniotu *et al*., 2002) or modification of aromatic residues (Porat *et al*., 2004) in the peptide or short non-amyloidogenic sequences of IAPP have been shown to reduce fibril formation and the associated cytotoxicity.

However, there are some major problems in the final acceptance of such compounds for diabetes therapy. Because islet amyloid cannot be identified *in vivo*, the efficacy of any therapeutic agent designed to reduce or prevent islet amyloid formation would have to be demonstrated in transgenic mouse studies. However, for humans, it would be more difficult. Due to the heterogeneity of type 2 diabetes and the fact that the contribution that islet amyloid makes to the pathophysiology of the disease is unknown, the outcomes of such a clinical trial would be difficult to assess. Even then, postmortem evidence would be required to show reduced proportions of patients on therapy with demonstrable islet amyloidosis. The most useful aspect of development of inhibitors has been a greater understanding of the process of fibrillogenesis.

10.6. Conclusion

Islet amyloidosis is a common feature of pancreatic islets in type 2 diabetes in man. The degree of islet amyloidosis is very heterogeneous in diabetic humans. Progressive deposition of islet amyloid replaces insulin-secreting β-cells. The deposits are formed from the 37-amino-acid peptide IAPP that is cosecreted from β-cells with insulin. The conditions that provoke refolding of the peptide into β-conformation are largely unknown; the amino-acid sequence, the concentration of IAPP, and the proportion of proIAPP in the islet are likely to be important. Diabetes-related islet amyloidosis occurs in cats, monkeys, and in fat-fed hIAPP transgenic mice and rats. However, in these models, islet amyloidosis and replacement of islet cells are likely to be causal factors for development of diabetes. This is not similar to humans where the evidence suggests that islet amyloidosis is not a precipitating factor for hyperglycemia in most patients with type 2 diabetes but could result from complications of the disease (Clark and Nilsson, 2004).

Acknowledgments

We are grateful to Professor John Morris (Oxford) for his continued support for islet amyloid research and for use of the electron microscope in the Department of Human Anatomy and Genetics, which was purchased by the Wellcome Trust. We thank Dr. Melanie Nilsson, Professor Barbara Hansen, Michael Badman, Rashpal Bhogal, Claire Higham, Emma Jaikaran, and Eelco de Koning for contributing unpublished data from their theses. Funding was provided by the Wellcome Trust (A.C. and J.M.).

References

Anguiano, M., Nowak, R. J. and Lansbury, P. T., Jr. (2002). Protofibrillar islet amyloid polypeptide permeabilizes synthetic vesicles by a pore-like mechanism that may be relevant to type II diabetes. *Biochemistry* 41: 11338–11343.

Badman, M. K., Shennan, K. I., Jermany, J. L., Docherty, K. and Clark, A. (1996). Processing of pro-islet amyloid polypeptide (proIAPP) by the prohormone convertase PC2. *FEBS Lett* 378: 227–231.

Barth, S. W., Riediger, T., Lutz, T. A. and Rechkemmer, G. (2004). Peripheral amylin activates circumventricular organs expressing calcitonin receptor a/b subtypes and receptor-activity modifying proteins in the rat. *Brain Res* 997: 97–102.

Betsholtz, C., Christmansson, L., Engstrom, U., Rorsman, F., Svensson, V., Johnson, K. H. and Westermark, P. (1989). Sequence divergence in a specific region of islet amyloid polypeptide (IAPP) explains differences in islet amyloid formation between species. *FEBS Lett* 251: 261–264.

Birch, C. L., Fagan, L. J., Armstrong, M. J., Turnbull, D. M. and Walker, M. (1997). The S20G islet-associated polypeptide gene mutation in familial NIDDM (Letter). *Diabetologia* 40: 1113.

Butler, P. C., Chou, J., Carter, W. B., Wang, Y. N., Bu, B. H., Chang, D., Chang, J. K. and Rizza, R. A. (1990). Effects of meal ingestion on plasma amylin concentration in NIDDM and nondiabetic humans. *Diabetes* 39: 752–756.

Butler, A. E., Jang, J., Gurlo, T., Carty, M. D., Soeller, W. C. and Butler, P. C. (2004). Diabetes due to a progressive defect in beta-cell mass in rats transgenic for human islet amyloid polypeptide (HIP Rat): a new model for type 2 diabetes. *Diabetes* 53: 1509–1516.

Caughey, B. and Lansbury, P. T., Jr. (2003). Protofibrils, pores, fibrils, and neurodegeneration: separating the responsible protein aggregates from the innocent bystanders. *Annu Rev Neurosci* 26: 267–298.

Cecil, R. L. (1909). A study of the pathological anatomy of the pancreas in ninety cases of diabetes mellitus. *J Exp Med* 11: 266–290.

Clark, A. (2004). *International Textbook of Diabetes Mellitus*. Chichester, UK, John Wiley & Sons, Ltd.

Clark, A. and Nilsson, M. R. (2004). Islet amyloid: a complication of islet dysfunction or an aetiological factor in type 2 diabetes? *Diabetologia* 47: 157–169.

Clark, A., Cooper, G. J., Lewis, C. E., Morris, J. F., Willis, A. C., Reid, K. B. and Turner, R. C. (1987a). Islet amyloid formed from diabetes-associated peptide may be pathogenic in type-2 diabetes. *Lancet* 2: 231–234.

Clark, A., Matthews, D. R., Naylor, B. A., Wells, C. A., Hosker, J. P. and Turner, R. C. (1987b). Pancreatic islet amyloid and elevated proinsulin secretion in familial maturity-onset diabetes. *Diabetes Res* 4: 51–55.

Clark, A., Wells, C. A., Buley, I. D., Cruickshank, J. K., Vanhegan, R. I., Matthews, D. R., Cooper, G. J., Holman, R. R. and Turner, R. C. (1988). Islet amyloid, increased A-cells, reduced B-cells and exocrine fibrosis: quantitative changes in the pancreas in type 2 diabetes. *Diabetes Res* 9: 151–159.

Clark, A., Edwards, C. A., Ostle, L. R., Sutton, R., Rothbard, J. B., Morris, J. F. and Turner, R. C. (1989). Localisation of islet amyloid peptide in lipofuscin bodies and secretory granules of human beta-cells and in islets of type-2 diabetic subjects. *Cell Tissue Res* 257: 179–185.

Clark, A., Saad, M. F., Nezzer, T., Uren, C., Knowler, W. C., Bennett, P. H. and Turner, R. C. (1990). Islet amyloid polypeptide in diabetic and non-diabetic Pima Indians. *Diabetologia* 33: 285–289.

Clark, A., de Koning, E. J. P., Baker, C. A., Charge, S. and Morris, J. F. (1993). Localisation of N-terminal pr pro-islet amyloid polypeptide in beta cells of man and transgenic mice. *Diabetologia* 36: A136.

Cook, J. T., Patel, P. P., Clark, A., Hoppener, J. W., Lips, C. J., Mosselman, S., O'Rahilly, S., Page, R. C., Wainscoat, J. S. and Turner, R. C. (1991). Non-linkage of the islet amyloid polypeptide gene with type 2 (non-insulin-dependent) diabetes mellitus. *Diabetologia* 34: 103–108.

Cooper, G. J. S. (1994). Amylin compared with calcitonin-gene-related peptidestructure, biology, and relevance to metabolic disease. *Endocrine Rev* 15: 163–201.

Cooper, G. J., Willis, A. C., Clark, A., Turner, R. C., Sim, R. B. and Reid, K. B. (1987). Purification and characterization of a peptide from amyloid-rich pancreases of type 2 diabetic patients. *Proc Natl Acad Sci USA* 84: 8628–8632.

Cooper, G., Leighton, B., Dimitriadis, G., Parry-Billings, M., Kowalchuk, J., Howland, K., Rothbard, J., Willis, A. and Reid, K. (1988). Amylin found in amyloid deposits in human type 2 diabetes mellitus may be a hormone that regulates glycogen metabolism in skeletal muscle. *Proc. Nat. Acad. Sci. USA* 85: 7763–7766.

Couce, M., Kane, L. A., O'Brien, T. D., Charlesworth, J., Soeller, W., McNeish, J., Kreutter, D., Roche, P. and Butler, P. C. (1996). Treatment with growth hormone and dexamethasone in mice transgenic for human islet amyloid polypeptide causes islet amyloidosis and beta-cell dysfunction. *Diabetes* 45: 1094–1101.

D'Alessio, D. A., Verchere, C. B., Kahn, S. E., Hoagland, V., Baskin, D. G., Palmiter, R. D. and Ensinck, J. W. (1994). Pancreatic expression and secretion of human islet amyloid polypeptide in a transgenic mouse. *Diabetes* 43: 1457–1461.

de Koning, E. J., Bodkin, N. L., Hansen, B. C. and Clark, A. (1993). Diabetes mellitus in Macaca mulatta monkeys is characterised by islet amyloidosis and reduction in beta-cell population. *Diabetologia* 36: 378–384.

de Koning, E. J., Morris, E. R., Hofhuis, F. M., Posthuma, G., Höppener, J. W., Morris, J. F., Capel, P. J., Clark, A. and Verbeek, J. S. (1994). Intra- and extracellular amyloid fibrils are formed in cultured pancreatic islets of transgenic mice expressing human islet amyloid polypeptide. *Proc Natl Acad Sci USA* 91: 8467–8471.

Deane, R., Du Yan, S., Submamaryan, R. K., LaRue, B., Jovanovic, S., Hogg, E., Welch, D., Manness, L., Lin, C., Yu, J., Zhu, H., Ghiso, J., Frangione, B., Stern, A., Schmidt, A. M., Armstrong, D. L., Arnold, B., Liliensiek, B., Nawroth, P., Hofman, F., Kindy, M., Stern, D. and Zlokovic, B. (2003). RAGE mediates amyloid-beta peptide transport across the blood-brain barrier and accumulation in brain. *Nat Med* 9: 907–913.

DeFronzo, R. A., Mandarino, L. and Ferrannini, E. (2004). *International Textbook of Diabetes Mellitus*. Chichester, UK, John Wiley & Sons, Ltd.

Demuro, A., Mina, E., Kayed, R., Milton, S. C., Parker, I. and Glabe, C. G. (2005). Calcium dysregulation and membrane disruption as a ubiquitous neurotoxic mechanism of soluble amyloid oligomers. *J Biol Chem.* 280: 17294–300.

Ehrlich, J. and Ratner, M. (1961). Amyloidosis of the islets of Langerhans. A restudy of islet hyaline in diabetic and non-diabetic individuals. *Am J Pathol* 38: 49–59.

Enoki, S., Mitsukawa. T., Takemura, J., Nakazato, M., Aburaya, J., Tosgiori, H. and Matsukara, S. (1992). Plasma islet amyloid polypeptide levels in obesity, impaired glucose tolerance and non-insulin dependant diabetes mellitus. *Diabetes Res Clin Pract* 15: 97–102.

Esapa, C., Moffitt, J., McNamara, C., JC, J. and A, C. (2001). Prevalence of islet amyloid polypeptide promoter mutation (−132 G to A) in type 2 diabetes. *Diabetologia* 44: A89, 339.

Fox, N., Schrementi, J., Nishi, M., Ohagi, S., Chan, S. J., Heisserman, J. A., Westermark, G. T., Leckstrom, A., Westermark, P. and Steiner, D. F. (1993). Human islet amyloid polypeptide transgenic mice as a model of non-insulin-dependent diabetes mellitus (NIDDM). *FEBS Lett* 323: 40–44.

Furth, A. J. (1997).Glycated proteins in diabetes. *Br J Biomed Sci* 54: 192–200.

Gebre Medhin, S., Mulder, H., Pekny, M., Westermark, G., Tornell, J., Westermark, P., Sundler, F., Ahren, B. and Betsholtz, C. (1998). Increased insulin secretion and glucose tolerance in mice lacking islet amyloid polypeptide (amylin). *Biochem Biophys Res Commun* 250: 271–277.

Gebre-Medhin, S., Olofsson, C. and Mulder, H. (2000). Islet amyloid polypeptide in the islets of Langerhans: friend or foe? *Diabetologia* 43: 687–695.

German, M. S., Moss, L. G., Wang, J. and Rutter, W. J. (1992). The insulin and islet amyloid polypeptide genes contain similar cell-specific promoter elements that bind identical beta-cell nuclear complexes. *Mol Cell Biol* 12: 1777–1788.

Gilbey, S. G., Ghatei, M. A., Bretherton-Watt, D., Zaidi, M., Jones, P. M., Perera, T., Beacham, J., Girgis, S. and Bloom, S. R. (1991). Islet amyloid polypeptide: production by an osteoblast cell line and possible role as a paracrine regulator of osteoclast function in man. *Clin Sci (Lond)* 81: 803–808.

Glenner, G. G., Eanes, E. D. and Wiley, C. A. (1988). Amyloid fibrils formed from a segment of the pancreatic-islet amyloid protein. *Biochem Biophys Res Commun* 155: 608–614.

Goldsbury, C., Kistler, J., Aebi, U., Arvinte, T. and Cooper, G. J. (1999). Watching amyloid fibrils grow by time-lapse atomic force microscopy. *J Mol Biol* 285: 33–39.

Goldsbury, C., Goldie, K., Pellaud, J., Seelig, J., Frey, P., Muller, S. A., Kistler, J., Cooper, G. J. S. and Aebi, U. (2000). Amyloid fibril formation from full-length and fragments of amylin. *J Struct Biol* 130: 352–362.

Goldsbury, C. S., Cooper, G. J., Goldie, K. N., Muller, S. A., Saafi, E. L., Gruijters, W. T., Misur, M. P., Engel, A., Aebi, U. and Kistler, J. (1997). Polymorphic fibrillar assembly of human amylin. *J Struct Biol* 119: 17–27.

Green, J., Goldsbury, C., Mini, T., Sunderji, S., Frey, P., Kistler, J., Cooper, G. and Aebi, U. (2003). Full-length rat amylin forms fibrils following substitution of single residues from human amylin. *J Mol Biol* 326: 1147–1156.

Griffiths, J. M., Ashburn, T. T., Auger, M., Costa, P. R., Griffin, R. G. and Lansbury, P. T. (1995). Rotational resonance solid-state NMR elucidates a structural model of pancreatic amyloid. *J Am Chem Soc* 12: 3539–3546.

Halban, P. A. and Kahn, S. E. (1997). Release of incompletely processed proinsulin is the cause of the disproportionate proinsulineamia of NIDDM. *Diabetes* 46: 1725–1732.

Hansen, B. C. and Bodkin, N. L. (1986). Heterogeneity of insulin responses: phases leading to type 2 (non-insulin-dependent) diabetes mellitus in the rhesus monkey. *Diabetologia* 29: 713–719.

Harroun, T. A., Bradshaw, J. P. and Ashley, R. H. (2001). Inhibitors can arrest the membrane activity of human islet amyloid polypeptide independently of amyloid formation. *FEBS Lett* 507: 200–204.

Hartter, E., Svoboda, T., Ludvik, B., Schuller, M., Lell, B., Kuenburg, E., Brunnbauer, M., Woloszczuk, W. and Prager, R. (1991). Basal and stimulated plasma levels of pancreatic amylin indicate its co-secretion with insulin in humans. *Diabetologia* 34: 52–54.

Higham, C. E., Jaikaran, E. T. A. S., Fraser, P. E., Gross, M. and Clark, A. (2000a). Preparation of synthetic human islet amyloid polypeptide (IAPP) in a stable conformation to enable stucy of conversion to amyloid-like fibrils. *FEBS Lett* 470: 55–60.

Higham, C. E., Hull, R. L., Lawrie, L., Shennan, K. I. J., Morris, J. F., Birch, N. P., Docherty, K. and Clark, A. (2000b). Processing os synthetic pro-islet amyloid polypeptide (proIAPP) 'amylin' by recombinant prohormone convertase enzymes, PC2 and PC3, in vitro. *Eur J Biochem* 267: 4998–5004.

Hoenig, M., Hall, G., Ferguson, F., Jordan, K., Henderson, M., Johnson, K. and O'Brien, T. (2000). A feline model of experimentally induced amyloidosis. *Am J Pathol* 157: 2143–2150.

Höppener, J. W., Verbeek, J. S., de Koning, E. J., Oosterwijk, C., van Hulst, K. L., Visser Vernooy, H. J., Hofhuis, F. M., van Gaalen, S., Berends, M. J., Hackeng, W. H. and *et al.* (1993). Chronic overproduction of islet amyloid polypeptide/amylin in transgenic mice: lysosomal localization of human islet amyloid polypeptide and lack of marked hyperglycaemia or hyperinsulinaemia. *Diabetologia* 36: 1258–1265.

Höppener, J. W. M., Oosterwijk, C., Nieuwenhuis, M. G., Posthuma, G., Thijssen, J. H. H., Vroom, T. M., Ahren, B. and Lips, C. J. M. (1999). Extensive islet amyloid formation is induced by development of type II diabetes mellitus and contributes to its progression: pathogenesis of diabetes in a mouse model. *Diabetologia* 42: 427–434.

Höppener, J. W., Nieuwenhuis, M. G., Vroom, T. M., Ahren, B. and Lips, C. J. (2002). Role of islet amyloid in type 2 diabetes mellitus: consequence or cause? *Mol Cell Endocrinol* 197: 205–212.

Hou, X., Ling, Z., Quartier, E., Foriers, A., Schuit, F., Pipeleers, D. and VanSchravendijk, C. (1999). Prolonged exposure of pancreatic beta cells to raised glucose concentrations results in increased cellular content of islet amyloid polypeptide precursors. *Diabetologia* 42: 188–194.

Howard, C. (1978). Insular amyloidosis and diabetes mellitus in Macaca nigra. *Diabetes* 27: 357–364.

Howard, C. F., Jr. (1986).Longitudinal studies on the development of diabetes in individual Macaca nigra. *Diabetologia* 29: 301–306.

Hull, R. L., Westermark, G. T., Westermark, P. and Kahn, S. E. (2004). Islet amyloid: a critical entity in the pathogenesis of type 2 diabetes. *J Clin Endocrinol Metab* 89: 3629–3643.

Hutton, J. C. (1984). Secretory granules. *Experientia* 40: 1091–1098.

Jaikaran, E. T. A. S. and Clark, A. (2001). Islet amyloid and type 2 diabetes: from molecular misfolding to islet pathophysiology. *Biochim Biophys Acta* 1537: 179–203.

Jaikaran, E. T. A. S., Higham, C. E., Serpell, L. C., Zurdo, J., Gross, M., Clark, A. and Fraser, P. E. (2001). Identification of a novel human islet amyloid polypeptide beta-sheet domain and factors influencing fibrillogenesis. *J Mol Biol* 308: 515–525.

Jaikaran, E. T., Nilsson, M. R. and Clark, A. (2004). Pancreatic beta-cell granule peptides form heteromolecular complexes which inhibit islet amyloid polypeptide fibril formation. *Biochem J* 377: 709–716.

Janciauskiene, S., Eriksson, S., Carlemalm, E. and Ahren, B. (1997). B cell granule peptides affect human islet amyloid polypeptide (IAPP) fibril formation in vitro. *Biochem Biophys Res Commun* 236: 580–585.

Janson, J., Soeller, W. C., Roche, P. C., Nelson, R. T., Torchia, A. J., Kreutter, D. K. and Butler, P. C. (1996). Spontaneous diabetes mellitus in transgenic mice expressing human islet amyloid polypeptide. *Proc Natl Acad Sci U S A* 93: 7283–7288.

Janson, J., Ashley, R. and Butler, P. (1998). Human islet amyloid polypeptide (hIAPP) cytotoxicity is caused by membrane damage. *Diabetes* 47: 970.

Janson, J., Ashley, R. H., Harrison, D., McIntyre, S. and Butler, P. C. (1999). The mechanism of islet amyloid polypeptide toxicity is membrane disruption by intermediate-sized toxic amyloid particles. *Diabetes* 48: 491–498.

Jorgensen, J. O., Rosenfalck, A. M., Fisker, S., Nyholm, B., Fineman, M. S., Schmitz, O., Madsbad, S., Holst, J. J. and Christiansen, J. S. (2000). Circulating levels of incretin hormones and amylin in the fasting state and after oral glucose in GH-deficient patients before and after GH replacement: a placebo-controlled study. *Eur J Endocrinol* 143: 593–599.

Kahn, S. E., Verchere, C. B., Andrikopoulos, S., Asberry, P. J., Leonetti, D. L., Wahl, P. W., Boyko, E. J., Schwartz, R. S., Newell Morris, L. and Fujimoto, W. Y. (1998). Reduced amylin release is a characteristic of impaired glucose tolerance and type 2 diabetes in Japanese Americans. *Diabetes* 47: 640–645.

Kahn, S. E., Andrikopoulos, S. and Verchere, C. B. (1999). Islet amyloid: A long-recognized but underappreciated pathological feature of type 2 diabetes. *Diabetes* 48: 241–253.

Kapourniotu, A., Bernhagen, J., Greenfield, N., AlAbed, Y., Teichberg, S., Frank, R. W., Voelter, W. and Bucala, R. (1998). Contribution of advanced glycosylation to the amyloidogenicity of islet amyloid polypeptide. *Eur J Biochem* 251: 208–216.

Kapurniotu, A. (2001). Amyloidogenicity and cytotoxicity of islet amyloid polypeptide. *Biopolymers* 60: 438–459.

Kapurniotu, A., Schmauder, A. and Tenidis, K. (2002). Structure-based design and study of non-amyloidogenic, double N-methylated IAPP amyloid core sequences as inhibitors of IAPP amyloid formation and cytotoxicity. *J Mol Biol* 315: 339–350.

Kassem, S. A., Ariel, I., Thornton, P. S., Scheimberg, I. and Glaser, B. (2000). Beta-cell proliferation and apoptosis in the developing normal human pancreas and in hyperinsulinism of infancy. *Diabetes* 49: 1325–1333.

Kautzky Willer, A., Thomaseth, K., Pacini, G., Clodi, M., Ludvik, B., Streli, C., Waldhausl, W. and Prager, R. (1994). Role of islet amyloid polypeptide secretion in insulin-resistant humans. *Diabetologia* 37: 188–194.

Kautzky Willer, A., Thomaseth, K., Ludvik, B., Nowotny, P., Rabensteiner, D., Waldhausl, W., Pacini, G. and Prager, R. (1997). Elevated islet amyloid pancreatic polypeptide and proinsulin in lean gestational diabetes. *Diabetes* 46: 607–614.

Kayed, R., Bernhagen, J., Greenfield, N., Sweimeh, K., Brunner, H., Voelter, W. and Kapurniotu, A. (1999). Conformational transitions of islet amyloid polypeptide (IAPP) in amyloid formation in vitro. *J Mol Biol* 287: 781–796.

Larson, J. L., Ko, E. and Miranker, A. D. (2000). Direct measurement of islet amyloid polypeptide fibrillogenesis by mass spectrometry. *Protein Sci* 9: 427–431.

Larson, J. L. and Miranker, A. D. (2004). The mechanism of insulin action on islet amyloid polypeptide fiber formation. *J Mol Biol* 335: 221–231.

Lorenzo, A., Razzaboni, B., Weir, G. and Yankner, B. (1994). Pancreatic islet cell toxicity of amylin associated with type-2 diabetes mellitus. *Nature* 368: 756–760.

Ludvik, B., Clodi, M., Kautzky-Willer, A., Capek, M., Hartter, E., Pacini, G. and Prager, R. (1993). Effect of dexamethasone on insulin sensitivity, islet amyloid polypeptide and insulin secretion in humans. *Diabetologia* 36: 84–87.

Lukinius, A., Wilander, E., Westermark, G. T., Engstrom, U. and Westermark, P. (1989). Co-localization of islet amyloid polypeptide and insulin in the B cell secretory granules of the human pancreatic islets. *Diabetologia* 32: 240–244.

MacArthur, D. L. A., deKoning, E. J. P., Verbeek, J. S., Morris, J. E. and Clark, A. (1999). Amyloid fibril formation is progressive and correlates with beta-cell secretion in transgenic mouse isolated islets. *Diabetologia* 42: 1219–1227.

Maclean, N. and Ogilvie, R. F. (1955). Quantitative estimation of the pancreatic islet tissue in diabetic subjects. *Diabetes* 4: 367–376.

Maloy, A. L., Longnecker, D. S. and Greenberg, E. R. (1981). The relation of islet amyloid to the clinical type of diabetes. *Hum Pathol* 12: 917–922.

Marks, V. and Teale, J. D. (1996). Investigation of hypoglycaemia. *Clin Endocrinol (Oxf)* 44: 133–136.

Marzban, L., Soukhatcheva, G. and Verchere, C. B. (2005a). Role of carboxypeptidase E in processing of pro-islet amyloid polypeptide in {beta}-cells. *Endocrinology* 146: 1808–1817.

Marzban, L., Trigo-Gonzalez, G. and Verchere, C. B. (2005b). Processing of pro-islet amyloid polypeptide in the constitutive and regulated secretory pathways of beta cells. *Mol Endocrinol* 19: 2154–63.

Mirzabekov, T., Lin, M. and Kagan, B. (1996). Pore formation by the cytotixic islet amyloid peptide amyloin. *J Biol Chem* 271: 1988–1992.

Moriarty, D. F. and Raleigh, D. P. (1999). Effects of sequential proline substitutions on amyloid formation by human amylin(20–29). *Biochemistry* 38: 1811–1818.

Mulder, H., Ahren, B., Stridsberg, M. and Sundler, F. (1995). Non-parallelism of islet amyloid polypeptide (amylin) and insulin gene expression in rats islets following dexamethasone treatment. *Diabetologia* 38: 395–402.

Nilsson, M. R., Driscoll, M. and Raleigh, D. P. (2002). Low levels of asparagine deamidation can have a dramatic effect on aggregation of amyloidogenic peptides: implications for the study of amyloid formation. *Protein Sci* 11: 342–349.

Nishi, M., Bell, G. I. and Steiner, D. F. (1990). Islet amyloid polypeptide (amylin): no evidence of an abnormal precursor sequence in 25 type 2 (non-insulin-dependent) diabetic patients. *Diabetologia* 33: 628–630.

Novials, A., Rojas, I., Casamitjana, R., Usac, E. F. and Gomis, R. (2001). A novel mutation in islet amyloid polypeptide (IAPP) gene promoter is associated with Type II diabetes mellitus. *Diabetologia* 44: 1064–1065.

Novials, A., Mato, E., Lucas, M., Franco, C., Rivas, M., Santisteban, P. and Gomis, R. (2004). Mutation at position-132 in the islet amyloid polypeptide (IAPP) gene promoter enhances basal transcriptional activity through a new CRE-like binding site. *Diabetologia* 47: 1167–1174.

O'Brien, T., Hayden, D., Johnson, K. and Fletcher, T. (1986). Immuno-histochemical morphometry of pancreatic endocrine cells in diabetic normoglycaemic glucose-intolerant and normal cats. *J Comp Pathol* 96: 357–369.

O'Brien, T. D., Westermark, P. and Johnson, K. H. (1991). Islet amyloid polypeptide and insulin secretion from isolated perfused pancreas of fed, fasted, glucose-treated, and dexamethasone-treated rats. *Diabetes* 40: 1701–1706.

O'Brien, T. D., Butler, P. C., Westermark, P. and Johnson, K. H. (1993). Islet amyloid polypeptide: a review of its biology and potential roles in the pathogenesis of diabetes mellitus. *Vet Pathol* 30: 317–332.

O'Brien, T. D., Butler, A. E., Roche, P. C., Johnson, K. H. and Butler, P. C. (1994). Islet amyloid polypeptide in human insulinomas. Evidence for intracellular amyloidogenesis. *Diabetes* 43: 329–336.

O'Nuallain, B., Williams, A. D., Westermark, P. and Wetzel, R. (2004). Seeding specificity in amyloid growth induced by heterologous fibrils. *J Biol Chem* 279: 17490–17499.

Opie, E. (1901). The relation of diabetes mellitus to lesions of the pancreas. Hyaline degeneration of the islands of Langerhans. *J Exp Med* 5: 527–540.

Park, K. and Verchere, C. B. (2001).Identification of a heparin-binding domain in the N-terminal cleavage site of pro-islet amyloid polypeptide: implications for islet amyloid formation. *J Biol Chem* 276: 16611–16616.

Poa, N. R., Cooper, G. J. and Edgar, P. F. (2003). Amylin gene promoter mutations predispose to type 2 diabetes in New Zealand Maori. *Diabetologia* 46: 574–578.

Porat, Y., Kolusheva, S., Jelinek, R. and Gazit, E. (2003). The human islet amyloid polypeptide forms transient membrane-active prefibrillar assemblies. *Biochemistry* 42: 10971–10977.

Porat, Y., Mazor, Y., Efrat, S. and Gazit, E. (2004). Inhibition of islet amyloid polypeptide fibril formation: a potential role for heteroaromatic interactions. *Biochemistry* 43: 14454–14462.

Porte, D. and Kahn, S. E. (1989). Hyperproinsulinaemia and amyloid in NIDDM: clues to etiology of islet β-cell dysfunction. *Diabetes* 38: 1333–1336.

Rhoades, E., Agarwal, J. and Gafni, A. (2000). Aggregation of an amyloidogenic fragment of human islet amyloid polypeptide. *Biochim Biophys Acta* 1476: 230–238.

Röcken, C., Linke, R. P. and Saeger, W. (1992). Immunohistology of islet amyloid polypeptide in diabetes mellitus: semi-quantitative studies in a post-mortem series. *Virchows Arch A Pathol Anat Histopathol* 421: 339–344.

Saafi, E. L., Konarkowska, B., Zhang, S., Kistler, J. and Cooper, G. J. (2001). Ultrastructural evidence that apoptosis is the mechanism by which human amylin evokes death in RINm5F pancreatic islet beta-cells. *Cell Biol Int* 25: 339–350.

Sanke, T., Bell, G. I., Sample, C., Rubenstein, A. H. and Steiner, D. F. (1988). An islet amyloid peptide is derived from an 89-amino acid precursor by proteolytic processing. *J Biol Chem* 263: 17243–17246.

Sanke, T., Hanabusa, T., Nakano, Y., Oki, C., Okai, K., Nishimura, S., Kondo, M. and Nanjo, K. (1991). Plasma islet amyloid polypeptide (Amylin) levels and their responses to oral glucose in type 2 (non-insulin-dependent) diabetic patients. *Diabetologia* 34: 129–132.

Schneider, H. M., Storkel, S. and Will, W. (1980). Das Amyloid der Langerhansschen Inseln und seine Beziehung zum Diabetes mellitus. *Dtsch Med Wochenschr* 105: 1143–1147.

Scrocchi, L. A., Ha, K., Chen, Y., Wu, L., Wang, F. and Fraser, P. E. (2003). Identification of minimal peptide sequences in the (8–20) domain of human islet amyloid polypeptide involved in fibrillogenesis. *J Struct Biol* 141: 218–227.

Seino, S. (2001).S20G mutation of the amylin gene is associated with type II diabetes in Japanese. Study Group of Comprehensive Analysis of Genetic Factors in Diabetes Mellitus. *Diabetologia* 44: 906–909.

Snow, A. D., Willmer, J. and Kisilevsky, R. (1987). Sulfated glycosaminoglycans: a common constituent of all amyloids? *Lab Invest* 56: 120–123.

Soeller, W. C., Janson, J., Hart, S. E., Parker, J. C., Carty, M. D., Stevenson, R. W., Kreutter, D. K. and Butler, P. C. (1998). Islet amyloid-associated diabetes in obese A(vy)/a mice expressing human islet amyloid polypeptide. *Diabetes* 47: 743–750.

Soto, C., Sigurdsson, E. M., Morelli, L., Kumar, R. A., Castano, E. M. and Frangione, B. (1998). Beta-sheet breaker peptides inhibit fibrillogenesis in a rat brain model of amyloidosis: implications for Alzheimer's therapy [see comments]. *Nat Med* 4: 822–826.

Sumner Makin, O. and Serpell, L. C. (2004). Structural characterisation of islet amyloid polypeptide fibrils. *J Mol Biol* 335: 1279–1288.

Tenidis, K., Waldner, M., Bernhagen, J., Fischle, W., Bergmann, M., Weber, M., Merkle, M. L., Voelter, W., Brunner, H. and Kapurniotu, A. (2000). Identification of a penta- and hexapeptide of islet amyloid polypeptide (IAPP) with amyloidogenic and cytotoxic properties. *J Mol Biol* 295: 1055–1071.

Verchere, C. B., D'Alessio, D. A., Palmiter, R. D., Weir, G. C., Bonner Weir, S., Baskin, D. G. and Kahn, S. E. (1996). Islet amyloid formation associated with hyperglycemia in transgenic mice with pancreatic beta cell expression of human islet amyloid polypeptide. *Proc Natl Acad Sci USA* 93: 3492–3496.

Wang, F., Hull, R. L., Vidal, J., Cnop, M. and Kahn, S. E. (2001a). Islet amyloid develops diffusely throughout the pancreas before becoming severe and replacing endocrine cells. *Diabetes* 50: 2514–2520.

Wang, J., Xu, J., Finnerty, J., Furuta, M., Steiner, D. F. and Verchere, C. B. (2001b). The prohormone convertase enzyme 2 (PC2) is essential for processiong pro-islet amyloid polypeptide at the Nh2-terminal cleavage site. *Diabetes* 50: 534–539.

Westermark, P. (1972). Quantitiative studies on amyloid in the islets of Langerhans. *Ups J Med Sci* 77: 91–94.

Westermark, P. and Wilander, E. (1978). The influence of amyloid deposits on the islet volume in maturity onset diabetes mellitus. *Diabetologia* 15: 417–421.

Westermark, P., Wernstedt, C., Wilander, E. and Sletten, K. (1986). A novel peptide in the calcitonin gene related peptide family as an amyloid fibril protein in the endocrine pancreas. *Biochem Biophys Res Commun* 140: 827–831.

Westermark, P., Wernstedt, C., Wilander, E., Hayden, D. W., O'Brien, T. D. and Johnson, K. H. (1987). Amyloid fibrils in human insulinoma and islets of Langerhans of the diabetic cat are derived from a neuropeptide-like protein also present in normal islet cells. *Proc Natl Acad Sci USA* 84: 3881–3885.

Westermark, P., Engstrom, U., Johnson, K. H., Westermark, G. T. and Betsholtz, C. (1990). Islet amyloid polypeptide--pinpointing amino-acid-residues linked to amyloid fibril formation. *Proc Natl Acad Sci USA* 87: 5036–5040.

Westermark, P. (1994). Amyloid and polypeptide hormones: what is their interrelationship? *Int J Exp Clin Invest* 1: 47–60.

Westermark, P., Li, Z. C., Westermark, G. T., Leckstrom, A. and Steiner, D. F. (1996). Effects of beta cell granule components on human islet amyloid polypeptide fibril formation. *FEBS Lett* 379: 203–206.

Westermark, G. T., Steiner, D. F., Gebre-medhin, S., Engsrom, U. and Westermark, P. (2000). Pro Islet Amyloid Polypeptide (proIAPP) immunoreactivity in the islets of langerhans. *Ups J Med Sci* 105: 97–106.

Wisniewski, T. and Frangione, B. (1992).Apolipoprotein E: a pathological chaperone protein in patients with cerebral and systemic amyloid. *Neurosci Lett* 135: 235–238.

Wright, (1927). Hyaline degeneration of the islets of Langerhans in non-diabetics. *Am J Pathol* 3: 461–484.

Yan, S. D., Zhu, H., Zhu, A., Golabek, A., Du, H., Roher, A., Yu, J., Soto, C., Schmidt, A. M., Stern, D. and Kindy, M. (2000). Receptor-dependent cell stress and amyloid accumulation in systemic amyloidosis. *Nat Med* 6: 643–651.

Zhao, H. L., Lai, F. M., Tong, P. C., Zhong, D. R., Yang, D., Tomlinson, B. and Chan, J. C. (2003). Prevalence and clinicopathological characteristics of islet amyloid in chinese patients with type 2 diabetes. *Diabetes* 52: 2759–2766.

11

β_2-Microglobulin and Dialysis-Related Amyloidosis

Isobel J. Morten, Eric W. Hewitt, and Sheena E. Radford

Abstract

β_2-Microglobulin (β_2m) is the noncovalently bound component of the class I major histocompatibility complex (MHC) and is one of more than 20 proteins that are known to cause human amyloid disease. β_2m is degraded and excreted by the kidney as part of its normal catabolic cycle. In patients suffering renal failure, therefore, the principal mechanism of β_2m degradation is abrogated, with the consequence that the concentration of β_2m in the serum increases by up to 60-fold, which leads to the association of freely circulating β_2m into insoluble amyloid fibrils, which typically accumulate in the musculoskeletal system, causing dialysis-related amyloidosis (DRA). The unique environment of the synovial joint may also be an important feature in stimulating β_2m deposition in a fibrillar form, although the factors that play the key role in the development and deposition of β_2m amyloid in this compartment remain unclear. In this chapter, we outline the events involved in the development of DRA. We discuss biophysical and biochemical studies of β_2m fibrillogenesis *in vitro* and summarize the insights that these studies have provided into the structural molecular mechanism of its self-association into amyloid fibrils. We then discuss the role of different biological factors in influencing amyloid formation *in vitro* and *in vivo*, highlighting key questions about the mechanism of aggregation of β_2m into amyloid fibrils that have arisen from these studies. Finally we discuss routes forward in the search for potential new therapies for DRA patients.

11.1. Introduction

Amyloidosis is a class of conformational diseases that arises from the conversion of globular proteins into insoluble fibrils (Dobson, 1999; Rochet and Lansbury, 2000; Zerovnik, 2002). β_2-Microglobulin (β_2m) is one of more than 20 proteins that are known to cause human amyloid disease (Westermark *et al.*, 2002). The native structures of the known amyloidogenic proteins vary widely (Dobson, 1999; Rochet and Lansbury, 2000; Zerovnik, 2002), but the fibrils they produce exhibit a common cross-β structure that gives rise to a characteristic x-ray fiber diffraction pattern (Geddes *et al.*, 1968; Sunde and Blake, 1997). Amyloid formation from β_2m is unique in that the fibrils deposit in the musculoskeletal system of uremic patients who have been subjected to renal dialysis for 10–15 years, causing the disorder dialysis-related amyloidosis (DRA) (Gejyo *et al.*, 1985; Floege and Ehlerding, 1996; Bellotti *et al.*, 1998).

In this chapter, we outline the events involved in the development of DRA. These are summarized schematically in Figure 11.1. In addition, we discuss biophysical and biochemical studies of β_2m fibrillogenesis *in vitro* and summarize the insights that these studies have provided into the mechanism of aggregation of β_2m into amyloid fibrils *in vivo*. We then conclude with key questions about the molecular mechanism of aggregation of β_2m into amyloid fibrils that have arisen from these studies and discuss routes forward in the search for potential new therapies for DRA patients.

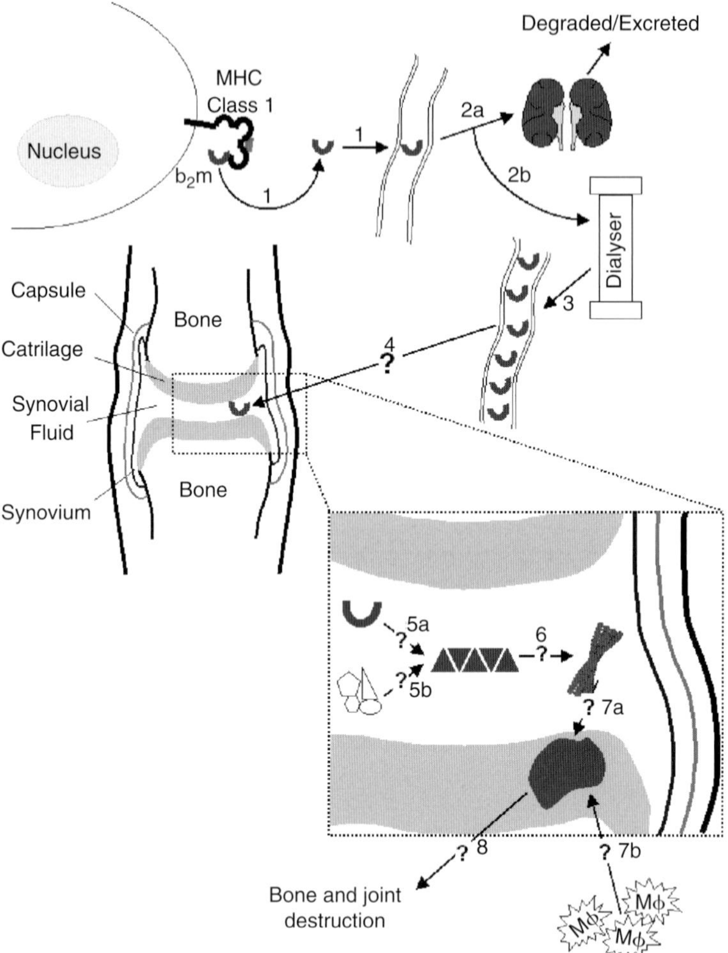

Figure 11.1. Overview of the stages of development of DRA. β_2m (gray cresent) dissociates from MHC class I molecules and enters the serum (1). In healthy individuals, β_2m is degraded and excreted by the kidney (2a). Uremic patients without functional kidneys undergo hemodialysis to filter their blood (2b). Dialysis membranes do not effectively remove β_2m from the blood. As a consequence, the serum concentration of β_2m increases (3). β_2m enters synovial joints by an unknown mechanism (4), where it forms a prefibrillar state (5a) possibly with the assistance of other biological factors (5b). The β_2m prefibrillar state then forms fully assembled amyloid fibrils (6). Fibrils deposit as amyloid plaques on the internal surfaces of the joint (7a), causing macrophage (Mϕ;) infiltration (7b). The activated macrophages release proinflammatory cytokines, which result in the bone and joint destruction (*8*), which results in the onset of the symptomatic phase of DRA.

11.2. Normal Cellular Role of β_2M

β_2m is a 99-residue soluble protein, which forms the noncovalently bound light chain of the major histocompatibility complex (MHC) class I molecule (Fig. 11.1) (Pamer and Cresswell, 1998). MHC class I molecules are expressed on the surface of all nucleated cells and function to present peptides, typically of 8 to 10 residues in length, to cytotoxic T lymphocytes (Pamer and Cresswell, 1998). The peptides presented by MHC class I molecules are generated from a cytosolic pool of proteins by proteosomal proteolysis (Rock and Goldberg, 1999). The peptides are then transported into the endoplasmic reticulum, where they are assembled noncovalently with β_2m and the heavy chain to form the fully assembled class I molecules that are then transported to the cell surface (Pamer and Cresswell, 1998).

In vivo, β_2m is continuously shed from the surface of cells that express MHC class I molecules, whereupon the protein is transported by the serum to the kidney where it is degraded and excreted (Floege and Ehlerding, 1996). The major site of β_2m clearance is the proximal tubule of the kidney (Nguyen-Simonnet *et al.*, 1982). In the proximal tubule, β_2m is endocytosed in a classic receptor-mediated endocytic pathway with recycling of the receptor and subsequent degradation of β_2m in the lysosomes (Sundin *et al.*, 1994). In patients suffering renal failure, therefore, the principal mechanism of β_2m degradation is abrogated with the consequence that the concentration of β_2m in the serum increases from 1 to 3 mg/L in a healthy individual to > 30 mg/L in patients suffering from renal failure and undergoing hemodialysis (Fig. 11.2; **see color insert**) (Gejyo *et al.*, 1986; Floege and Ehlerding, 1996; Floege and Ketteler, 2001). The increased β_2m serum concentration is a key initiating factor in the association of freely circulating β_2m into insoluble amyloid fibrils, which typically accumulate in the musculoskeletal system (Gorevic *et al.*, 1986). This accumulation of β_2m fibrils causes patients who have been subjected to long-term renal dialysis to develop DRA (Floege and Ehlerding, 1996; Bellotti *et al.*, 2001).

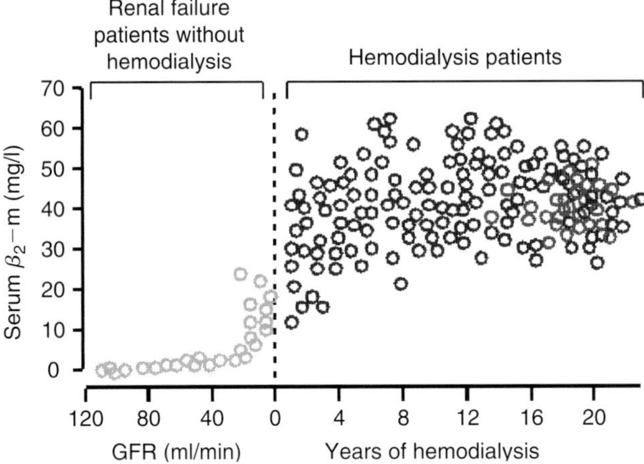

Figure 11.2. Serum β_2m concentrations in patients with chronic renal failure and those undergoing hemodialysis. The glomerular filtration rate (GFR) of patients in renal failure who are not undergoing hemodialysis is shown in green; patients undergoing hemodialysis who lack symptoms of DRA are shown in blue; and hemodialysis patients who have clinical symptoms of DRA are shown in red. The figure was kindly supplied by Professors Naiki and Gejyo (Department of Pathology, Fukui Medical University, Fukui, Japan, and Division of Clinical Nephrology and Rheumatology, Niigata University Graduate School of Medical and Dental Science, Niigata, Japan, respectively).

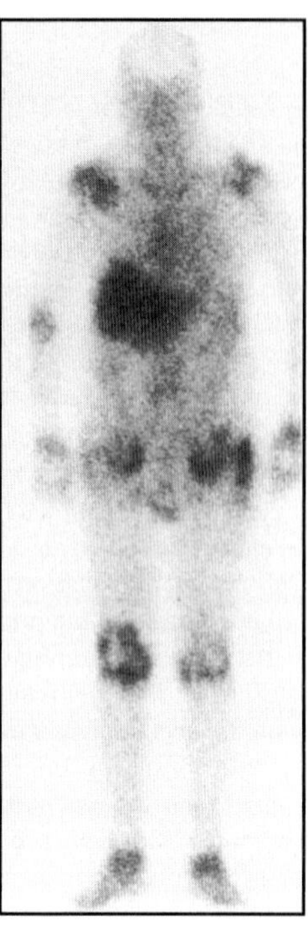

Figure 11.3. Anterior image of a ^{111}In-β_2m scintigraph of a 55-year-old male suffering from DRA. In this projection, local tracer accumulations are noted in both shoulders, elbows, wrists, hips, knees, and ankles. Tracer accumulation in the liver is a nonspecific phenomenon, which is observed with most ^{111}In-labeled radiodiagnostics and is related to uptake of tracer in the recticuloendothelial system. Reproduced from Floege J and Ehlerding G, *Nephron* 1996; 72:9–26, with permission from S. Karger AG, Basel.

Clinically, the first β_2m amyloid deposits appear in the cervical intervertebral disks in patients that have been subjected to renal dialysis for 3 years or more, a feature that has been attributed to the mechanical stress in these regions (White and Panjabi, 1978; Ohashi *et al.*, 1992). The disk deposits are asymptomatic, and the disease usually presents with carpel tunnel syndrome (CTS). Clinical symptoms rarely present before 5 years but increase to a nearly 100% prevalence after 15 years of renal dialysis (Fig. 11.2) (Koch, 1992; Floege and Ehlerding, 1996; Jadoul *et al.*, 1997; Miyata *et al.*, 1998). In the symptomatic phase, other joints (knee, hip, shoulder, and ankle joints) are affected with painful osteoarticluar complications that include chronic arthropathy, cystic bone lesions, destructive arthropathy, pathologic fractures, and scapulohumeral periarthritis (Garcia-Garcia *et al.*, 1999; Danesh and Ho, 2001). Figure 11.3 shows a scintigraph of a 55-year-old male who has undergone hemodialysis for 13 years (Floege and Ehlerding, 1996). The patient was injected with radiolabeled ^{111}Inidium-β_2m, which shows the accumulation of β_2m amyloid deposits in both shoulders, elbows, wrists, hips, knees, and ankles, highlighting the typical manifestation of DRA as a disorder associated with the collagen-rich regions of the musculoskeletal system.

11.3. Treatments for DRA

There are currently no small-molecule treatments for DRA, but studies have shown that a successful renal transplant leads to rapid symptomatic improvement (reduction in joint pain and swelling), because the circulating plasma levels of β_2m are reduced by the transplanted kidney

(Floege and Ehlerding, 1996). Although renal transplantation halts the progression of the disease, it is controversial whether transplantation leads to the regression of established β_2m amyloid deposits (Campistol, 2001). Moreover, in patients for whom transplantation is not an option, such as those with cardiovascular problems or advancing years, as well as individuals for whom transplantation is not permitted on religious or ethical grounds, the only hope for therapies to prevent DRA either involve the development of dialysis membranes with enhanced bioselectivity (Winchester *et al.*, 2003) or for the development of small-molecule therapies that halt or prevent the association of β_2m into amyloid fibrils.

Clearance of β_2m from the serum of patients with renal failure has been improved by the use of continuous ambulatory peritoneal dialysis (CAPD) or high-flux dialyzer membranes (van Ypersele and Drueke, 1996). Studies have shown that a single standard high-flux hemodialysis session reduces β_2m plasma levels by 50%, possibly by adsorption of β_2m to the synthetic dialyzer membrane, in addition to diffusive clearance (van Ypersele de Strihou, 1996; Lornoy *et al.*, 2000), and studies have confirmed the efficacy of hemodiafiltration (HDF) for the even more efficient removal of β_2m (Koda *et al.*, 1997). Methods to remove β_2m specifically from serum, in conjunction with hemodialysis columns, have also been developed and include a direct hemoperfusion β_2m adsorption column (Lixelle; Kaneka Corporation, Osaka, Japan) (Gejyo *et al.*, 1993; Abe *et al.*, 2003; Gejyo *et al.*, 2004), which reduces the concentration of circulating β_2m by a combination of hydrophobic interactions and adequate pore size (Gejyo *et al.*, 1993, 2004). The long-term benefits of such approaches, however, are not yet clear.

11.4. Components of *Ex Vivo* Amyloid

11.4.1. Modifications of β_2m in Amyloid Deposits in DRA Patients

Amyloid deposits extracted from DRA patients contain many different constituents (Table 11.1), but the major component is full-length, unmodified wild-type β_2m (Gejyo *et al.*, 1985; Gorevic *et al.*, 1985). β_2m fibrils *ex vivo* have also been shown to retain the single disulfide bond in the oxidized form (Bellotti *et al.*, 1998). Consistent with these observations, maintenance of an intact disulfide bond has been shown to be required both for the formation of β_2m amyloid fibrils *de novo in vitro* (Smith and Radford, 2001), as well as for the extension of seeds formed from fibrils from DRA patients with monomeric β_2m (Ohhashi *et al.*, 2002).

A number of modified species of β_2m are also found in β_2m amyloid, including up to 25% of β_2m in which either the N-terminal six (ΔN6β_2m) or nineteen (ΔN19β_2m) amino acids have been removed (Linke *et al.*, 1987, 1989; Esposito *et al.*, 2000). A modified species of β_2m has also been isolated from *ex vivo* material in which Asn17 is deamidated to Asp (Odani *et al.*, 1990), although other studies have failed to recognize this modification in *ex vivo* material (Argiles *et al.*, 1995). Using site-directed mutagenesis to examine the effect of this chemical modification on amyloid formation *in vitro*, Kad *et al.* (2001) have shown that deamidation of Asn17 destabilizes β_2m but has little effect on the ability of the protein to from fibrils *in vitro* (Kad *et al.*, 2001).

An acidic isoform of β_2m that contains protein modified by advanced glycation end products (AGE) (Miyata *et al.*, 1994, 1996c; Degenhardt *et al.*, 1997; Niwa *et al.*, 1997) is found at low concentrations in the serum from healthy individuals. The concentration of this species is increased in the ultrafiltrate and amyloid deposits of dialysis patients (Odani *et al.*, 1990). AGE modification typically occurs at the N-terminus of the protein and at the ε-amino group of lysine residues (Brownlee *et al.*, 1988; Baynes *et al.*, 1989). The primary site of such modifications in β_2m occurs at the N-terminal isoleucine residue (Miyata *et al.*, 1994) and results in a number of

Table 11.1. Constituents of β_2m amyloid deposits *in vivo*

	Component	Reference
β_2m protein	Full-length, intact β_2m	(Gejyo *et al.*, 1985; Gorevic *et al.*, 1985)
	N-terminal truncated β_2m (ΔN6β_2m and ΔN19β_2m)	(Linke *et al.*, 1987, 1989)
	Deamidated β_2m (N17D)	(Odani *et al.*, 1990)
	Age-modified β_2m	(Odani *et al.*, 1990)
	Radical-modified β_2m	(Capeillere-Blandin *et al.*, 1991)
Other components	Glycosaminoglycans	(Athanasou *et al.*, 1995)
	Proteoglycans	(Aruga *et al.*, 1993)
	Collagen	(Coggi *et al.*, 1989)
	AGE-modified collagen	(Homma *et al.*, 1989; Hou *et al.*, 1997)
	α_2-Macroglobulin	(Argiles *et al.*, 1989)
	Serum amyloid P component (SAP)	(Campistol *et al.*, 1992a)
	Apolipoprotein E	(Campistol *et al.*, 1992a; Yamada *et al.*, 1994)
Cells	Macrophages	(Ohashi *et al.*, 1992)

different products (pentosidine, *N*-carboxymethyllysine, and imidazolone) in uremic sera and in β_2m amyloid deposits (Miyata *et al.*, 1996c; Degenhardt *et al.*, 1997; Niwa *et al.*, 1997). It is thought that AGE-modification of β_2m occurs after the deposition of the protein as amyloid in osteoarticular tissues (Niwa, 2001). AGE modification, therefore, is not thought to be a key initiating factor in the deposition of β_2m in amyloid fibrils. Nonetheless, AGE modification may play a role in inducing an inflammatory response that causes the onset of the symptoms of DRA (Miyata *et al.*, 1994; Garbar *et al.*, 1999; O'Neill *et al.*, 2003).

Dialysis therapy accelerates the generation of reactive oxygen species (Capeillere-Blandin *et al.*, 1991), which may also participate in fibril formation. Radicals decompose protein by modifying and cross-linking amino acid residues. *In vitro* studies have shown that hydroxyl radicals at low concentrations stimulate polymerization of β_2m, and at high concentrations these radicals destroy the β-sheet structure of the native protein, causing the protein to unfold (Capeillere-Blandin *et al.*, 1991). Although the generation and subsequent role of such modifications in β_2m fibril formation is not fully understood, the presence of these species in β_2m amyloid deposits formed *in vivo* leaves open the possibility that they could contribute to the ability of β_2m to form fibrils that deposit in osteoarticular tissues.

11.4.2. Other Components of β_2m Amyloid

One of the most fascinating features of DRA is the apparent high specificity of β_2m-containing amyloid fibrils to deposit in the joint areas, a feature typical and unique to DRA. Figure 11.4 shows a schematic diagram of a synovial joint that is surrounded by the capsule. The capsule is a fibrous membrane through which solutes can pass and has ligamentous regions to provide structural support to the joint (Simkin, 1997). Lining the inside of the joint is the synovium (also known as the synovial membrane), which covers all capsular structures other than cartilage (Simkin, 1997). The synovium consists of glycosaminoglycans (GAGs), which are linear polysaccharide chains composed of repeating disaccharide units and are abundant in all

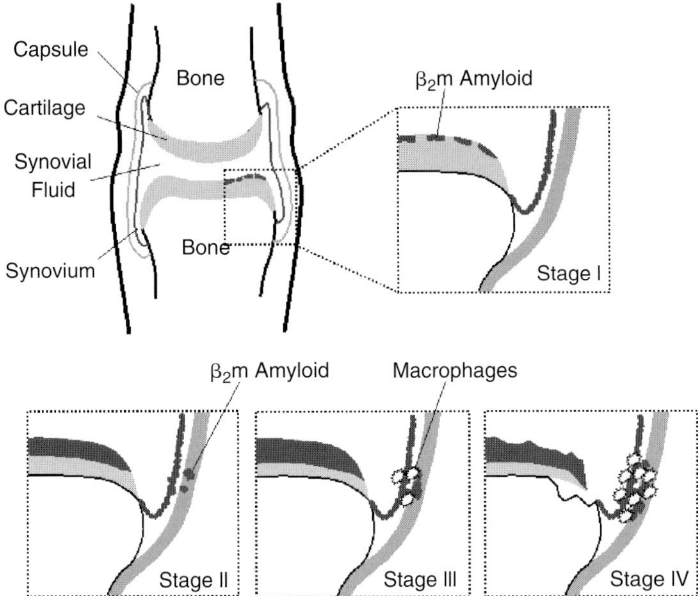

Figure 11.4. Schematic representation of the stages of deposition of β_2m amyloid fibrils in DRA. Stage I: β_2m amyloid deposits as a thin discontinuous layer on the articular cartilage. Stage II: The cartilaginous layer of β_2m amyloid thickens into a continuous layer and deposits appear in the capsule and/or synovium. Stage III: Deposits thicken and synovium is infiltrated by macrophages. Stage IV: Activated macrophages secrete proinflammatory cytokines, which cause bone and joint destruction. Adapted and redrawn from Garbar *et al.* (1999).

structures of the joint. Interestingly, GAGs are not only found in β_2m amyloid deposits *ex vivo* but are also found in many other forms of amyloid (Ohishi *et al.*, 1990; Snow *et al.*, 1991; Athanasou *et al.*, 1995; Morita *et al.*, 1995). GAGs attach to a protein core to form proteoglycan (PG), which is also found in deposits of β_2m *ex vivo* (Ohashi *et al.*, 1995). In addition to GAGs, the synovium contains matrix proteins and synovial fibroblasts that secrete synovial fluid that fills the joint space. Synovial fluid is composed principally of hyaluronan and sulfated GAGs and recirculates into the bloodstream via the lymphatic system. Finally, the joint is rich in cartilage, a matrix of type II collagen, GAGs, and aggrecan, a proteoglycan that is secreted by chondrocytes, raising the possibility that one or more of these components could be involved in the development of β_2m amyloid deposits *in vivo*.

Although GAGs and PGs have been found associated with β_2m fibrils *ex vivo*, the precise roles of these molecules in β_2m fibrillogenesis is poorly understood. *In vitro* studies have shown that the GAGs heparin, dermatan sulfate, and heparan sulfate exhibit a strong inhibitory effect on the depolymerization of β_2m fibrils (Yamaguchi *et al.*, 2003), suggesting that these components could enhance the deposition of β_2m amyloid fibrils *in vivo* by binding directly to the fibril surface (Yamaguchi *et al.*, 2003; Yamamoto *et al.*, 2004b). This stabilizes the conformation of β_2m in the fibrils and, by laterally aggregating the fibrils, protects them from proteolysis *in vivo* (Yamaguchi *et al.*, 2003). A similar effect of these molecules is also found for fibrils formed from the Aβ peptide in Alzheimer's disease (Gupta-Bansal *et al.*, 1995), suggesting that GAG binding may be a generic and important feature of amyloid deposition disorders.

Other components that could contribute to the stability and hence to enhance deposition of β_2m amyloid fibrils include the antiprotease α_2-macroglobulin, as well as the serum proteins serum amyloid P component (SAP) and apolipoprotein E (Apo E). α_2-Macroglobulin is often found associated with soluble β_2m and β_2m amyloid deposits, although it is unknown what effect this protein has upon amyloid deposition *in vivo* (Motomiya *et al.*, 2003). *In vitro*, β_2m fibrils have been shown to be stabilized by binding SAP (Ono and Uchino, 1994) and by association with Apo E (Yamaguchi *et al.*, 2001), making both proteins potentially contributing factors to the evolution of DRA. Finally, the fact that β_2m amyloid deposits do not occur in the shafts of long bones indicates that the synovial joint is a necessary prerequisite for deposition (Moe and Chen, 2001), possibly because of the abundance of type II collagen fibrils therein, to which β_2m in known to bind (Coggi *et al.*, 1989).

11.5. Role of the Synovial Cellular Environment in DRA

While amyloid deposition occurs within 3 years after renal dysfunction, the pathologic symptoms of the disorder usually present 5–10 years after the onset of renal therapy (Fig. 11.2) (Gejyo *et al.*, 1986; Koch, 1992; Floege and Ehlerding, 1996; Jadoul *et al.*, 1997; Miyata *et al.*, 1998). Several factors have been implicated in the evolution of the disease including macrophage infiltration and subsequent expression of proinflammatory cytokines. Macrophages have been shown to be associated with β_2m amyloid deposits *in vivo* (Ohashi *et al.*, 1992) and are known to be the main cell type to infiltrate deposits (Argiles *et al.*, 1994). In light chain derived (AL) and serum amyloid A (AA) amyloidosis, the precursor protein has been shown to be proteolysed by lysosomal proteases in macrophages prior to their deposition as amyloid fibrils (Durie *et al.*, 1982; Takahashi *et al.*, 1989). Macrophage-derived proteolysis was also proposed as a key initiating factor in DRA based on a number of observations: (i) β_2m amyloid filaments have been observed in the cytoplasm of macrophage-like cells surrounding amyloid deposits (Morita *et al.*, 1985); (ii) β_2m has been shown to be synthesized and released by macrophages after cytokine or lipopolysaccharide stimulation (Knudsen *et al.*, 1990); and (iii) polymerization of β_2m into fibrils was detected in the supernatant of cultures of peripheral blood mononuclear cells harvested from dialysis patients (Campistol *et al.*, 1992b). By contrast, a study of β_2m amyloid deposits obtained from the carpel tunnel of patients suffering from DRA suggested that intracellular β_2m is almost exclusively intralysosomal and there were no signs of synthesis of either β_2m or amyloid fibrils in human macrophage cells (Garcia-Garcia *et al.*, 1999).

In a large, more direct, study using sternoclavicular joints from postmortem samples of dialyzed uremic patients, Garbar *et al.* (1999) investigated the involvement of macrophages in the deposition of β_2m amyloid in synovial joints. These authors demonstrated that the presence of macrophages is not a prerequisite for β_2m amyloid deposition, showing instead that macrophage infiltration occurs later in the development of amyloid plaques and correlates with the symptomatic stage of DRA (Fig. 11.4) (Garbar *et al.*, 1999). β_2m amyloid first deposits on the surface of the cartilage of the synovial joint in a thin patchy layer (Fig. 11.4, stage I). In stage II, the deposits on the cartilage thicken into a continuous layer and appear in the capsule and the synovium. Macrophages are then recruited and migrate to the deposits of the capsule and synovium (stage III) (Garbar *et al.*, 1999). The activated macrophages are thought to express proinflammatory cytokines and chemokines, which cause articular inflammation and typical bone erosions (stage IV) (Miyata *et al.*, 1998; Garbar *et al.*, 1999). The expressed proinflammatory cytokines include interleukin (IL)-1β, tumor necrosis factor (TNF)-β, and IL-6 (Sargent *et al.*, 1989; Ohashi *et al.*, 1992; Inoue *et al.*, 1995). Expressed chemokines include monocyte chemoattractant protein-1 and macrophage inflammatory protein-1, which cause the migration and activation of monocytes

(Hou and Owen, 2002). In the inflamed synovium, macrophage infiltration occurs mainly around blood vessels or adjacent to the deposition of the amyloid protein (Sargent *et al.*, 1989; Inoue *et al.*, 1995), suggesting that the macrophages are recruited from the peripheral blood system.

As mentioned previously, AGE-modified β_2m is present in *ex vivo* deposits of amyloid from DRA patients (Odani *et al.*, 1990). AGE-modified β_2m enhances chemotaxis of monocytes and stimulates macrophages to release bone-resorbing cytokines, such as IL-1β, TNF-α, and IL-6 (Iida *et al.*, 1994; Miyata *et al.*, 1994, 1996b). The amount of cytokines secreted from macrophages is sufficient to stimulate the synthesis of collagenase in cultured human synovial cells (Miyata *et al.*, 1994) and might contribute to progressive bone loss and the development of bone cysts. Monocyte chemotaxis and TNF-α production induced by AGE-modified β_2m can be inhibited by AGE receptor (RAGE) blockade, suggesting that the biological effects of AGE-modified β_2m on monocytes/macrophages are mediated by RAGE (Miyata *et al.*, 1996a).

A recent study, using live cell imaging, has shown that cellular uptake of β_2m and AGE-modified β_2m by monocytes/macrophages and synovial fibroblasts differs (O'Neill *et al.*, 2003). In monocytes/macrophages, AGE-modified β_2m was found to be more potent than unmodified β_2m at inducing a cellular response, whereas in synovial fibroblasts the opposite occurred. Based on these results, the authors suggest that the onset of β_2m amyloid formation may be triggered by the binding of β_2m to the MHC class I molecule of surrounding cells. This induces protein production (tissue inhibitor metalloproteases-1, vascular adhesion molecule-1, and cyclooxygenase-2) by the synovial fibroblasts, which leads to the degradation of the joint surface, inducing local amyloid deposition. The deposits are subsequently glycated, which induces the infiltration of macrophages, compounding the inflammatory response and leading to the symptoms of DRA (O'Neill *et al.*, 2003).

Taken together, these data suggest that the involvement of macrophages and synovial fibroblasts is in the development of the symptoms of DRA, caused by the expression of numerous proinflammatory cytokines and chemokines. Although opinions on the precise role of these factors differ (Morita *et al.*, 1985; Knudsen *et al.*, 1990; Campistol *et al.*, 1992b; Garbar *et al.*, 1999; Garcia-Garcia *et al.*, 1999; O'Neill *et al.*, 2003), these studies concur that the participation of these cells is a postdeposition event and is not a prerequisite for amyloid formation.

11.6. Structure of β_2m

11.6.1. MHC Class I-Associated β_2m

The crystal structure of the MHC class I complex is shown in (Fig. 11.5A; **see color insert**). This structure, representing more than 80 structures of the MHC class I complex currently deposited in the structural databank, shows β_2m, a 99-residue protein with a seven-stranded β-sandwich fold typical of the immunoglobulin superfamily (Fig. 11.5B; **see color insert**), which is noncovalently bound to the MHC heavy chain (Saper *et al.*, 1991). β_2m is a crucial component of the MHC class I complex, it stabilizes the heavy chain and is essential for the folding and assembly of the whole complex, which is necessary for the stable surface expression of the class I molecule (Pamer and Cresswell, 1998). β_2m contains two β-sheets, one comprising β-strands A, B, D, and E, the other β-strands C, F, and G. The protein is stabilized by a single disulfide bond between Cys25 and Cys80 that links β-strands B and F, across the two β-sheets (Saper *et al.*, 1991; Smith and Radford, 2001; Ohhashi *et al.*, 2002). When complexed with the MHC class I heavy chain, residues 50–56 of β_2m form two short (two-residue) β-strands (D1 and D2), separated by a two-residue β-bulge, which facilitates its binding to the surface of the heavy chain. β_2m thus binds to the heavy chain α1 domain via residues in strands D (53 and 55-56), α2 domain via the D–E loop (residue 60), and the α3 domain via strand A (residues 8 and 10-13) (Bjorkman *et al.*, 1987).

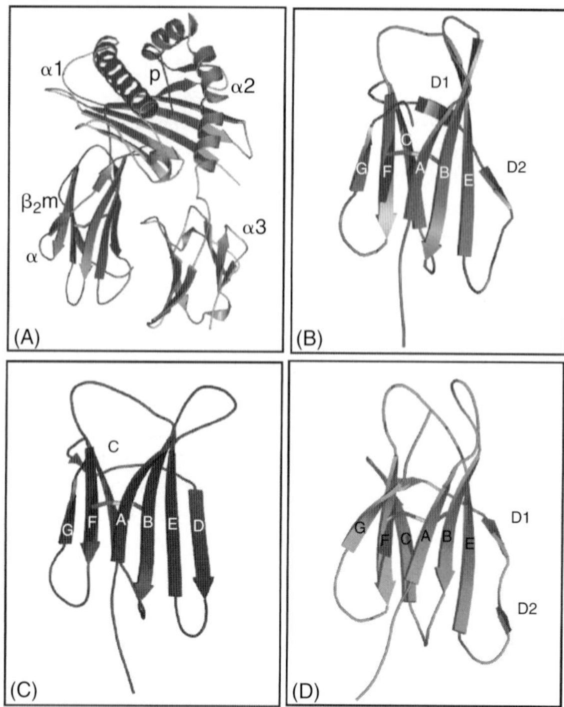

Figure 11.5. Structures of β_2m. (A) Ribbon diagram of the MHC class I molecule showing the extracellular domains: β_2m (purple), the heavy-chain subunits (α1–3), and the bound peptide (p) (red) are shown. (B) Enlargement of the structure of MHC bound β_2m. The disulfide bond that links strands B and F is shown in yellow. Structures were taken from PDB code 1UVQ (Siebold *et al.*, 2004). (C) Crystal structure of monomeric human β_2m. The single disulfide bond is shown in yellow. The structure was created using the coordinates 1LDS (Trinh *et al.*, 2002). Solution structure of monomeric human β_2m, determined by ^{1}H NMR spectroscopy, structure taken from PDB code 1JNJ (Verdone *et al.*, 2002). Each β-strand A–G is labeled. Note that the structure of the D strand differs significantly in (C) compared with (B) and (D). Diagrams were drawn using the program PyMol (DeLano, 2002).

11.6.2. Monomeric β_2m

The first crystal structure of monomeric human β_2m was published in 2002 (Fig. 11.5C; **see color insert**) (Trinh *et al.*, 2002), revealing that the structure of the monomeric protein is very similar to its MHC class I–bound counterpart (rms deviation of 3.1 Å over all atoms or 0.5 Å when omitting the largest deviations involving residues 12–20, 49–60, and 71–77) (Trinh *et al.*, 2002). The most significant difference in the crystal structures of the free and bound protein involves residues in the β-strand D (50–56) and the succeeding loop (Trinh *et al.*, 2002). In the crystal structure of monomeric human β_2m, the β-bulge observed between strands D1 and D2 in the MHC-bound conformation is no longer present. Instead, residues 51–56 form a continuous six-residue β-strand stabilized by six main-chain–main-chain hydrogen bonds with β-strand E, forming a regular antiparallel β-sheet. Loss of the β-bulge, together with changes in the positions of three residues in the loop linking strands C and D, causes structural rearrangements of the N-terminal section of strand D. These result in a slip of $i + 1$ in the hydrogen-bonding partners of residues 50–53 relative to their positions in the MHC-bound conformation, with concomitant reorganization of their side chains (including His51, which changes its orientation by 180 degrees) (Trinh *et al.*, 2002). These conformational changes, although rarely populated in solution (Trinh *et al.*, 2002; Verdone *et al.*, 2002), have been proposed to enhance the aggregation potential of β_2m *via* the edge strand D. In a more recent crystal structure of the β_2m variant H31Y, however, the β-bulge is retained in the x-ray structure, while conformational changes are observed in the A strand and have been suggested to reflect early events in the β_2m amyloidogenic transformation (Rosano *et al.*, 2004). Thus, while the overall structure of β_2m is maintained in the monomeric forms studied to date, one or more of the conformational changes observed in the crystal structures could be important in initiating the aggregation of the protein.

The solution structure of monomeric human β_2m has also been solved using ^{1}H NMR spectroscopy (Fig. 11.5D; **see color insert**). This structure also resembles that of MHC-bound β_2m (Okon *et al.*, 1992; Verdone *et al.*, 2002). The region encompassing residues 50–57 is highly ordered in the crystal structure (Trinh *et al.*, 2002) but appears to be dynamic in solution (Okon *et al.*, 1992; Verdone *et al.*, 2002). Studies have shown that residues 55 and 56, which lie in β-strand D2 of the crystal structure of MHC-bound β_2m (Fig. 11.5B) and in the C-terminal region of strand D in the crystal structure of monomeric β_2m (Fig. 11.5C), are in intermediate exchange between different conformations in solution, one of which could include the continuous β-strand involving residues 51–56 observed in the crystal structure of monomeric β_2m (Okon *et al.*, 1992; Trinh *et al.*, 2002; Verdone *et al.*, 2002). The regions of β_2m which vary the most in the structures described (strands A and D) are involved in contacts with the α3 and α1 heavy-chain domains of the MHC class I molecule, possibly explaining why β_2m does not self-assemble in its native MHC-bound environment.

11.7. β_2m Amyloid Fibril Formation *In Vitro*

A number of studies have now demonstrated the assembly of β_2m into amyloid-like fibrils *in vitro*. Below we review the conformational properties of the amyloid precursor conformation observed in these *in vitro* experiments. We also discuss current knowledge of the mechanism of assembly; the role of biological factors in modulating fibril formation; and current insights into the structure of the amyloid fibril itself. These data are summarized schematically in Figure 11.6A.

11.7.1. β_2m Fibrillogenesis Under Acidic Conditions

11.7.1.1. Partially Unfolded β_2m and Curved Nodular Fibrils

Monomeric β_2m at pH 3.0–4.0 adopts a partially unfolded conformation that has been shown to form fibrils rapidly *in vitro* (McParland *et al.*, 2000; Kad *et al.*, 2001; Hong *et al.*, 2002; Smith *et al.*, 2003). Using NMR and other spectroscopic methods, it has been shown that this partially unfolded species lacks many of the fixed tertiary interactions of the native protein, yet it retains significant stable structure, specifically in regions corresponding to the native strands B, C, E, and F (McParland *et al.*, 2002). Incubation of wild-type β_2m under these conditions and in the presence of $>$100 mM NaCl results in the formation of short (200 to 600 nm) fibrils, with a curved nodular morphology, which form rapidly without an observable lag phase (Fig. 11.6B) (McParland *et al.*, 2000; Kad *et al.*, 2001; Hong *et al.*, 2002; Smith *et al.*, 2003). These fibrils have a repeat distance between consecutive nodules of approximately 30 nm and are 3.5 nm in height, as measured by atomic force microscopy (AFM) (Kad *et al.*, 2001, 2003). Although these

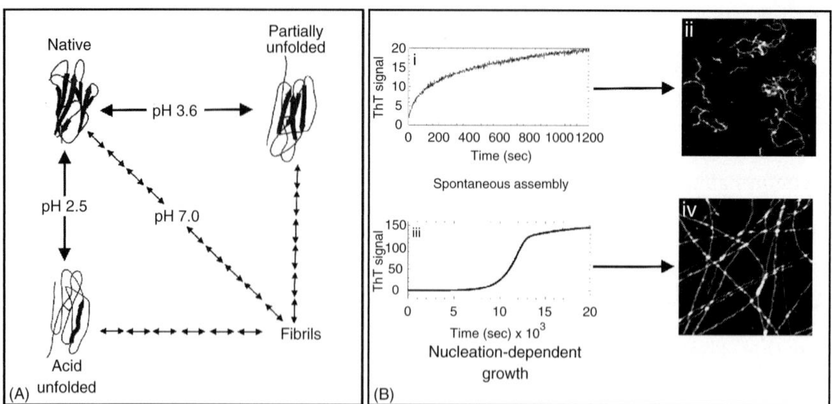

Figure 11.6. (A) β_2m forms fibrils with different morphologies commencing with partially unfolded β_2m (pH 3.6), acid unfolded β_2m (pH 2.5), and native β_2m (pH 7.0). (B) Growth kinetics of fibrils of β_2m at 37°C monitored by ThT fluorescence, at (i) pH 3.6, 0.4 M ionic strength, no agitation, and (iii) pH 2.5, 50 mM ionic strength, agitation at 1400 rpm. Tapping mode AFM images of β_2m fibrils formed at (ii) pH 3.6 and (iv) pH 2.5 dried onto mica. Each image is 1 μm^2. The molecular mechanism by which β_2m spontaneously assembles (without seeds) into fibrils at pH 7.0 is currently unknown, although a number of methods that result in fibrillogenesis at this pH in the presence of seeds formed from fibrils from DRA deposits have been reported (see text).

fibrils have morphologic properties similar to the protofibrils seen in other amyloid proteins (Rochet and Lansbury, 2000) and give rise to a fibre diffraction pattern consistent with cross-β structure (Smith *et al.*, 2003), there is no evidence that these species are able to assemble further into higher ordered structures (Kad *et al.*, 2003; Gosal *et al.*, 2005).

11.7.1.2. Acid Unfolded β_2m and Long, Straight Fibrils

At low pH (3.0–1.0), monomeric β_2m is highly unfolded (McParland *et al.*, 2000; Kad *et al.*, 2001; Katou *et al.*, 2002). Investigations into the conformational properties of β_2m under these conditions using NMR relaxation methods have shown that acid unfolded β_2m is a rapidly fluctuating species that is more highly unfolded than the partially unfolded state at pH 3.0–4.0 (McParland *et al.*, 2000; Kad *et al.*, 2001; Katou *et al.*, 2002; McParland *et al.*, 2002; Platt *et al.*, 2004). NMR data, including measurements of the chemical shift and T_2 relaxation properties, have shown that at pH values of 2.5 and below, β_2m adopts a random coil-like structure in the N-and C-terminal regions (residues 1–20 and 86–99, respectively), but forms significant non-native hydrophobic clusters involving both aliphatic and aromatic side chains in the region 21–85 (Platt *et al.*, 2004). Further studies using NMR dispersion methods (Platt *et al.*, 2004) and electrospray ionization mass spectrometry (ESI MS) (Borysik *et al.*, 2004) have highlighted the complexity of this ensemble, suggesting that it contains two or more distinct conformational states in equilibrium, one or more of which could be involved in the onset of fibrillogenesis. Together these data demonstrate that formation of the cross β-structure of amyloid-like fibrils at pH 2.5 must involve substantial refolding events (Platt *et al.*, 2004).

At pH 2.5, β_2m assembles into amyloid-like fibrils with a long, straight morphology in a nucleation-dependent manner (Naiki *et al.*, 1997) (Fig. 11.6B). These fibrils also give rise to a characteristic cross-β x-ray diffraction pattern (Smith *et al.*, 2003). At these pH values, and dependent on the agitation conditions used, aggregates may be detected in the lag phase of assembly, including amorphous species, rings, and toriods (Kad *et al.*, 2003). Similar species have been observed in the assembly of many proteins into amyloid fibrils, suggesting commonalities in the mechanism of assembly of different proteins to the cross-β structure of amyloid (Lashuel *et al.*, 2002). The fibrils formed under these conditions are of varied morphology including straight aperiodic fibrils, 8 ± 1 nm in height, and fibrils displaying a left-handed twist with regular periodicity of approximately 60 nm that contain at least four protofilaments. An additional twisted-ribbon fibril morphology is also observed, which is formed from four or more protofilaments that are associated laterally with a regular periodicity of 70 nm and are 5 ± 1 nm in height (Kad *et al.*, 2003).

Although a high-resolution structure of β_2m amyloid fibrils, or indeed for any protein amyloid fibril formed to date has not yet been obtained, recent studies of fibrils formed *in vitro* from β_2m at pH 2.5 and pH 3.6 have begun to reveal important information about the environment of individual amino acids in the fibrillar state. Studies of fibrils formed *in vitro* using hydrogen exchange monitored by NMR, for example, have revealed that the N-terminal ~20 residues of the polypeptide chain are predominately unprotected from hydrogen exchange in the fibril, while the remainder of the protein is involved in the formation of stable hydrogen-bonded structure, including residues that form loops or unstructured regions in the native structure (Hoshino *et al.*, 2002; Yamaguchi *et al.*, 2004) (Fig. 11.7A; **see color insert**). In accord with these results, limited proteolysis experiments have shown that only the N-terminal 10–20 residues of β_2m are exposed to proteases in fibrils formed at this pH *in vitro* (Esposito *et al.*, 2000; Monti *et al.*, 2002; Myers *et al.*, 2006a). The nodular fibrils formed at pH 3.6 show very similar hydrogen exchange protection factors to fibrils formed at pH 2.5 (Fig. 11.7A). The core of the fibril (strands B–F) is highly protected

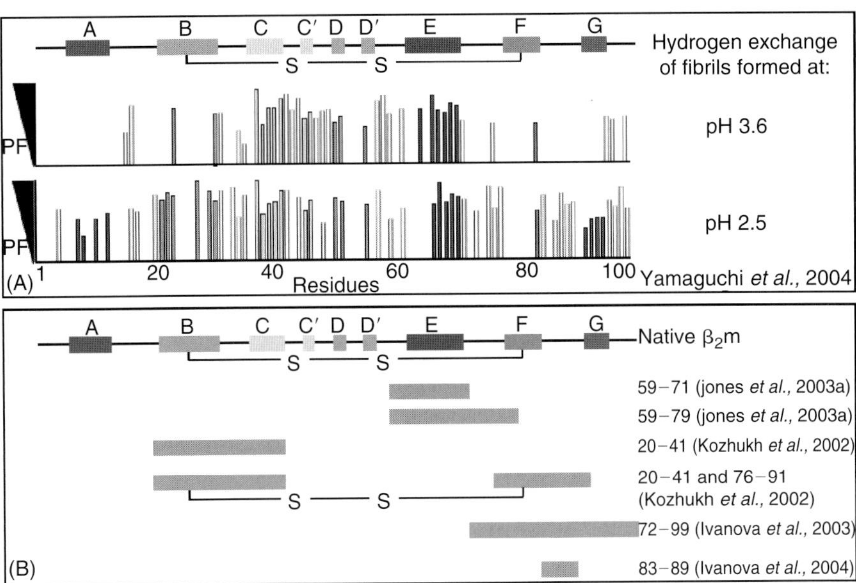

Figure 11.7. (A) Protection factors (PF) of residues in β_2m fibrils formed at pH 3.6 and pH 2.5 elucidated by hydrogen exchange. Hydrogen exchange data were kindly supplied by Professor Goto, Institute for Protein Research, Osaka University, Osaka, Japan (Yamaguchi *et al.*, 2004). (B) Secondary structural elements of human β_2m colored from the N-terminus (red) to the C-terminus (violet). Synthetic peptides shown to form fibrils *in vitro* are shown in blue with the residues involved in each peptide shown alongside (Kozhukh *et al.*, 2002; Ivanova *et al.*, 2003; Jones *et al.*, 2003a; Ivanova *et al.*, 2004).

from hydrogen exchange in both fibril morphologies, indicating that both fibril types are built around a rigid core involving residues 20–86. However, the N- and C-terminal regions of the polypeptide chain in fibrils formed at pH 3.6 exchange much more rapidly, obviating measurement of their protection factors (Fig. 11.7A) (Yamaguchi *et al.*, 2004).

11.7.2. Regions of β_2m Involved in Fibril Formation

An important objective in the study of protein assembly is to identify regions involved in intermolecular recognition events. To determine if specific regions of β_2m are responsible for fibril formation, peptides corresponding to different regions of the β_2m structure have been designed and their aggregation properties investigated (Fig. 11.7B; **see color insert**) (Kozhukh *et al.*, 2002; Ivanova *et al.*, 2003; Jones *et al.*, 2003a; Ivanova *et al.*, 2004).

Jones *et al.* (2003a) systematically studied the ability of seven peptides of β_2m to assemble into amyloid-like fibrils *in vitro*, each of which corresponded with a β-strand in the native monomer, including several residues from each adjacent loop region (the peptides were named A to G, corresponding with their equivalent β-strands in the full length protein). An eighth peptide was included (E′), corresponding to the most stable region (residues 59–79) in partially unfolded β_2m pH 3.6 (McParland *et al.*, 2000). These data showed that only two of the eight peptides studied at pH 2.5 and 3.6 formed amyloid-like fibrils *de novo in vitro*: peptides E (residues 59–71) and E′ (residues 59–79) (Fig. 11.7B) (Jones *et al.*, 2003a). The only significant difference between peptides E and E′ and the other peptides studied is that they have an unusually high content of aromatic residues (6 and 7 of the total 13 and 21 residues in each sequence, respectively). Interactions between aromatic residues have been proposed to facilitate ordered assembly of peptides and proteins by favourable π-π stacking that aligns adjacent β-strands in the polymer (Aggeli *et al.*, 1997; Azriel and Gazit, 2001; Gazit, 2002; Porat *et al.*, 2004). Jones *et al.* (2003a) suggest that the ability to undergo favorable π-π stacking may be critically important in determining the ability of these peptides (and possibly even the intact protein) to form amyloid fibrils *in vitro*. Interestingly, predictions of regions of β_2 m with high aggregation propensity using TANGO (Fernandez-Escamilla *et al.*, 2004) also point to the E strand of β_2m as the only region with high aggregation propensity (de la Paz and Serrano, 2004).

Kozhukh *et al.* (2002) used a different approach in which *Achromobacter* protease I was used to obtain a series of peptides from recombinant human β_2m. Out of the series of peptides obtained, one peptide (K7) that encompasses residues 20–41 (strands B and C) and 76–91 (strand F) linked by a disulfide bond formed fibrils at pH 2.5 (Fig. 11.7B) (Kozhukh *et al.*, 2002). Upon reduction of the disulfide bond, only one (K3, residues 20–41) of the two peptides generated retained the ability to form fibrils both *de novo in vitro* and in the presence of seeds (Fig. 11.7B). These data highlight the context dependence of different sequences in determining their fibrillogenic properties, as peptides B (residues 17–29) and C (residues 30–41) in isolation or in a pair-wise combination were unable to form fibrils *in vitro* (Jones *et al.*, 2003a), while a single peptide encompassing residues 20–41 (K3) rapidly formed fibrils under similar conditions (Kozhukh *et al.*, 2002).

Finally, Ivanova *et al.* (2003) showed that a peptide encompassing the F and G strands of human β_2m (residues 72–99) (Fig. 11.7B) forms fibrils with a long straight morphology when incubated at pH 2.0 in the presence of 1.5 M NaCl (Ivanova *et al.*, 2003). Interestingly, this region encompasses a sequence seven residues in length (83–89) that differs in six residues between the mouse and human sequence. The human 7mer peptide also formed fibrils at pH 2.0 in 0.25 M NaCl (Fig. 11.7B), but the mouse peptide does not (Ivanova *et al.*, 2004). Based on these data, Ivanova *et al.* (2004) concluded that residues 83–89 are important in fibril formation in human β_2m and may explain the very different aggregation properties of the mouse and human proteins.

Other regions of β_2m that possibly play an important role in fibril assembly are the N- and C-terminal strands (A and G, respectively) that display significant destabilization at pH 3.6

(McParland *et al.*, 2002) and at pH 2.5 are highly unstructured (Katou *et al.*, 2002; Platt *et al.*, 2004). Point mutations made in strands A (I7A and V9A) or G (V93A) destabilize β_2m to different extents (denaturation midpoints of 2.7 M, 1.6 M, and 3.4 M urea, respectively, compared with 4.7 M for wild-type-β_2m at pH 7.0, 37°C) (Jones *et al.*, 2003b). Not only did these variants form amyloid fibrils rapidly at acidic pH, but they also formed a small, but significant amount of fibrillar material at pH 7.0, in the absence of seed (Jones *et al.*, 2003b). Based on these results, Jones *et al.* (2003b) propose that the enhanced amyloidogenicity of these variants results from a specific decrease in the local stability of the A and G strands, caused by the truncation of hydrophobic residues that dock these strands into the core of the native protein (Jones *et al.*, 2003b). A recent study involving the introduction of proline residues (the least favored amino acid in β-strands) (Minor and Kim, 1994a, 1994b) in different regions of β_2m indicated that the introduction of prolines in the N- and C-terminal regions of the protein decrease the stability of native β_2m but have no effect on amyloid formation (Chiba *et al.*, 2003). Together these data suggest that the fraying of the A and G strands is an important initiating event in β_2m amyloid formation (Esposito *et al.*, 2000; De Lorenzi *et al.*, 2002; McParland *et al.*, 2002; Jones *et al.*, 2003b).

The studies described above have highlighted many regions of β_2m that could be involved in fibril formation. Six different peptides encompassing strands B–G have been shown to form fibrils under acidic conditions *in vitro* and mutagenesis studies have shown that the fraying of strands A and G may also be associated with fibril formation. The ensuing rapid dynamics of these regions may then expose the hydrophobic clusters in the central region of the polypeptide chain in the acid unfolded or partially unfolded states, facilitating the recognition process and stimulating self assembly (McParland *et al.*, 2002; Platt *et al.*, 2004). This is supported by independent observations that destabilizing mutations in the N- and C-terminal regions of the polypeptide chain accelerate fibril formation at pH 2.5 (Jones *et al.*, 2003b).

11.7.3. β_2m Fibril Formation at pH 7.0

β_2m does not form fibrils spontaneously *in vitro* at pH 7.0, but fibril assembly can be induced under these conditions by a variety of approaches, including extension of *ex vivo* seeds from DRA deposits with (i) wild-type β_2m or ΔN6β_2m (Esposito *et al.*, 2000), as well as the addition of monomeric β_2m in (ii) 20% (vol/vol) trifluoroethanol (Yamamoto *et al.*, 2004b), (iii) low concentrations of sodium dodecyl sulfate (SDS) (Yamamoto *et al.*, 2004a), or (iv) the presence of GAGs (Yamamoto *et al.*, 2004b). β_2m fibrils have also been formed at pH 7.0 in the presence of seeds made in vitro, stabilized by the addition of heparin (Mayers *et al.*, 2006b) and in the absence of seed by (i) concentration and drying of the protein on a dialysis membrane (Connors *et al.*, 1985), (ii) specific destabilization of the N- or C-terminal strands by mutagenesis (Jones *et al.*, 2003b), (iii) the addition of copper (II) ions in the presence of 100 mM urea (Morgan *et al.*, 2001), or (iv) the addition of SAP (Ono and Uchino, 1994).

N-terminally truncated β_2m has been found in *ex vivo* β_2m amyloid plaques (Linke *et al.*, 1987; Linke *et al.*, 1989). Indeed, 25% of β_2m extracted from fibrils *ex vivo* lacks six residues from the N-terminal region (ΔN6β_2m) (Bellotti *et al.*, 1998). *In vitro* studies have shown that recombinant ΔN6β_2m forms fibrils at pH 7.0 in the presence of *ex vivo* seeds (Esposito *et al.*, 2000). Studies of this β_2m variant using NMR have demonstrated that the loss of the first six residues not only modifies the three-dimensional structure of β_2m but also reduces its stability, facilitates fibril formation (by extension) at pH 7.0, and enhances its susceptibility to proteolytic digestion (Esposito *et al.*, 2000). Although ΔN6β_2m is found in *ex vivo* fibrils, it has not been isolated from the serum of DRA patients, suggesting that proteolysis may be a postdeposition event (Linke *et al.*, 1989; Esposito *et al.*, 2000; Stoppini *et al.*, 2000). Molecular dynamics simulations of ΔN6β_2m found that the main effect of the truncation is to increase the conformational dynamics of strand A and the probability of strand A separation (Ma and Nussinov, 2003), which has been proposed by many to initiate intermolecular association (Esposito *et al.*, 2000; Ma and Nussinov, 2003; Jones *et al.*, 2003b).

Divalent copper ions (Cu^{2+}) have been suggested to play a role in the evolution of a number of amyloid diseases including Alzheimer's disease (Miura *et al.*, 2000), Parkinson's disease (Uversky *et al.*, 2001), and Creutzfeldt-Jakob disease (CJD) (Brown *et al.*, 1997; Cereghetti *et al.*, 2001). Investigations have shown that Cu^{2+} can also initiate β_2m fibril formation *in vitro* at pH 7.0 in the presence of 100 mM urea, and these authors have suggested that exposure to Cu^{2+} during the dialysis procedure could initiate amyloidogenesis (Morgan *et al.*, 2001; Eakin *et al.*, 2002). Native β_2m is destabilized by Cu^{2+} binding (Morgan *et al.*, 2001; Eakin *et al.*, 2002), His13 (A–B loop), and His31 (B–C loop) forming the ligand binding sites at pH 6.5 (Verdone *et al.*, 2002) and His13, His31, and His51 at pH 7.0 (Villanueva *et al.*, 2004). Villanueve *et al.* (2004) showed that when Cu^{2+} binds to the histidine residues, especially His51, mobility is increased in the adjacent regions by weakening the hydrogen bonds along the backbone and subtly altering hydrophobic interactions, making the thermodynamics and the kinetics more favorable for the formation of the amyloidogenic β_2m precursor necessary for fibril formation (Villanueva *et al.*, 2004). In fact, the conformational changes in the D strand observed in the crystal structure of human β_2m cause a change in the orientation of the side chain of His51 by 180 degrees (Trinh *et al.*, 2002), which could facilitate Cu^{2+} binding (Villanueva *et al.*, 2004). Alternatively, Eakin *et al.* (2004) propose that Cu^{2+} is necessary for the generation and stabilization of oligomeric intermediates of β_2m, suggesting that Cu^{2+} acts as an initiating factor of amyloidogenesis by a mechanism involving domain swapping (Eakin *et al.*, 2004). Finally, very recent studies have implicated the role of cis/trans proline isomerization in initiating amyloid formation at pH 7.0, specifically by the trapping of an intermediate with a non-native trans proline at residue 32 (Jahn *et al.*, 2006; Eakin *et al.*, 2006). These data highlight the fact that only very minor changes can tip the balance between a stable, soluble fold and protein self-assembly.

11.7.4. Kinetic Mechanisms of Fibril Formation *In Vitro*

Many groups have now shown that amyloid-like fibrils form *in vitro* with nucleation-dependent kinetics (Rochet and Lansbury, 2000). This model was tested, using thioflavine T (ThT) fluorescence, for the assembly of β_2m fibrils at acidic pH, by analyzing the rate of elongation of seeds formed from β_2m fibrils from DRA deposits (Naiki *et al.*, 1997). The extension of β_2m seeds proceeds by pseudo-first-order reaction kinetics with a rate that is maximal at pH 2.5. Based on these studies, Naiki *et al.* (1997) propose that the extension of β_2m proceeds via the consecutive association of β_2m monomers onto the ends of existing fibrils (Naiki *et al.*, 1997). More recent studies using ThT fluorescence detected using total internal reflection fluorescence microscopy (TIRFM) showed that extension is predominately unidirectional and that some seeds must be polar (Ban *et al.*, 2003). The extension reaction for individual fibrils could be fitted to a single exponential curve, consistent with the extension rates shown by Naiki *et al.* (1997), yielding a fibril growth rate of approximately 47.4 ± 15.0 nm min^{-1} (Ban *et al.*, 2003). Interestingly, and by contrast with the mechanism of fibril assembly at pH 2.5, the formation of the curved nodular fibrils of β_2m formed at pH 3.0–4.5 does not include a lag phase even in the absence of seed, suggesting that these assemblies do not form in a nucleation-dependent manner (Fig. 11.6B) (McParland *et al.*, 2000; Smith *et al.*, 2003).

11.8. Conclusion

Considerable progress has been made in understanding the molecular events that lead to the formation of fibrils from monomeric β_2m both *in vitro* and *in vivo*. It is now clear that β_2m accumulates in the blood of uremic patients due to the inability of dialysis membranes to remove the free monomer from the circulating system (Fig. 11.1). The high concentration of β_2m in the serum of uremic patients is one key initiating event in DRA. The unique environment of the synovial joint is also an important feature in stimulating β_2m deposition in a fibrillar form, although the key factors in this compartment remain unclear. *Ex vivo* β_2m deposits have been shown to contain a variety of components in addition to full-length and truncated versions of wild-type β_2m, which include GAGs and PGs, both of which are abundant in the synovial environment, raising the possibility that these species may be critical components for fibril deposition. In accordance with this view, *in vitro* studies have shown that GAGs and PGs might enhance deposition of β_2m by stabilizing fibrils and protecting them from proteolysis (Yamaguchi *et al.*, 2003; Yamamoto *et al.*, 2004b). Other cellular factors including α_2-macroglobulin, SAP, and Apo E could also play a similar role (Ono and Uchino, 1994; Yamaguchi *et al.*, 2001; Motomiya *et al.*, 2003).

In vitro studies have also shown that destabilization of monomeric β_2m facilitates fibril formation, such that lowering pH or altering the amino acid sequence enhances the rate of fibril formation *in vitro* (Ono and Uchino, 1994; McParland *et al.*, 2000; Kad *et al.*, 2001; Yamaguchi *et al.*, 2001; Hong *et al.*, 2002; Katou *et al.*, 2002; Jones *et al.*, 2003b; Smith *et al.*, 2003). Similarly, the addition of copper (II) ions also destabilizes β_2m and enhances fibrillogenesis (Morgan *et al.*, 2001). Whether such features play a role in fibrillogenesis *in vivo* remains unclear.

The assembly of monomeric β_2m into fibrils is a complex process that involves the formation of amorphous aggregates, the rapid development of curved, nodular fibrils, and the generation of higher order fibrillar assemblies, the later with nucleation-dependent kinetics. Despite elegant studies involving EM, AFM, proteolysis, and NMR, which have led to one of the most detailed maps of the role of individual residues in different fibrillar forms, how β_2m subunits stack in the cross-β structure of amyloid is yet to be fully understood. The fibrils formed under different acidic conditions *in vitro* give rise to fibrils with distinct morphologies, but how these species relate to amyloid *in vivo* is unknown. Finally, although *in vitro* studies have given extensive information about the mechanisms of fibril formation, how fibrils form *in vivo* remains elusive.

The detailed knowledge of fibrils formed under acidic conditions and insights into the role of cellular factors that stabilize β_2m in the fibrillar state are now beginning to offer new avenues for therapeutics to treat DRA. There are currently no treatments for DRA except transplantation, which has been shown to halt the disease, but it is controversial as to whether it leads to fibril regression. In the absence of a ready supply of replacement organs, other approaches to treatment must be developed. One route to prevent the disease would be to develop dialysis membrane technology such that the membranes used effectively remove β_2m from the blood of uremic patients. Future therapeutic options for DRA could be aimed at preventing the deposition of β_2m into fibrils, destabilizing the deposited fibrillar plaques, stimulating fibril degradation, or blocking the inflammatory response that leads to bone and joint destruction. For these avenues to become a reality, further research into the structural molecular mechanism of β_2m assembly *in vitro*, as well as to the role of cellular factors in aiding deposition *in vivo* is paramount. Finally, and equally importantly, the discovery of the molecular mechanism of β_2m fibril formation may aid in the development of therapies for many other amyloid diseases.

Acknowledgments

We thank members of the S.E.R. group, especially Thomas Jahn for help with Figure 11.5, and Geoff Platt, Walraj Gosal, and Sue Jones for helpful discussions. We also thank Professors Gejyo and Naiki for kindly providing Figure 11.2 and Professor Goto for kindly providing the data used to create Figure 11.7A. The AFM images in Figure 11.6 were kindly provided by Walraj Gosal, and David Smith obtained the ThT traces shown in Figure 11.6. I.J.M. is funded by Kidney Research UK, and S.E.R. is a BBSRC Professorial Fellow.

References

Abe, T., Uchita, K., Orita, H., Kamimura, M., Oda, M., Hasegawa, H., Kobata, H., Fukunishi, M., Shimazaki, M., Akizawa, T. and Ahmad, S. (2003). Effect of beta-2-microglobulin adsorption column on dialysis-related amyloidosis. *Kidney Int* 64: 1522–1528.

Aggeli, A., Bell, M., Boden, N., Keen, J. N., Knowles, P. F., McLeish, T. C., Pitkeathly, M. and Radford, S. E. (1997). Responsive gels formed by the spontaneous self-assembly of peptides into polymeric beta-sheet tapes. *Nature* 386: 259–262.

Argiles, A., Garcia-Garcia, M., Derancourt, J., Mourad, G. and Demaille, J. G. (1995). Beta-2-microglobulin isoforms in healthy individuals and in amyloid deposits. *Kidney Int* 48: 1397–1405.

Argiles, A., Mourad, G., Axelrud-Cavadore, C., Watrin, A., Mion, C. and Cavadore, J. C. (1989). High-molecular-mass proteins in hemodialysis-associated amyloidosis. *Clin Sci (Lond)* 76: 547–552.

Argiles, A., Mourad, G., Kerr, P. G., Garcia, M., Collins, B. and Demaille, J. G. (1994). Cells surrounding hemodialysis-associated amyloid deposits are mainly macrophages. *Nephrol Dial Transplant* 9: 662–667.

Aruga, E., Ozasa, H., Teraoka, S. and Ota, K. (1993). Macromolecules that are colocalized with deposits of beta-2-microglobulin in hemodialysis-associated amyloidosis. *Lab Invest* 69: 223–230.

Athanasou, N. A., Puddle, B. and Sallie, B. (1995). Highly sulfated glycosaminoglycans in articular cartilage and other tissues containing beta-2-microglobulin amyloid deposits. *Nephrol Dial Transplant* 10: 1672–1678.

Azriel, R. and Gazit, E. (2001). Analysis of the minimal amyloid-forming fragment of the islet amyloid polypeptide. An experimental support for the key role of the phenylalanine residue in amyloid formation. *J Biol Chem* 276: 34156–34161.

Ban, T., Hamada, D., Hasegawa, K., Naiki, H. and Goto, Y. (2003). Direct observation of amyloid fibril growth monitored by thioflavin T fluorescence. *J Biol Chem* 278: 16462–16465.

Baynes, J. W., Watkins, N. G., Fisher, C. I., Hull, C. J., Patrick, J. S., Ahmed, M. U., Dunn, J. A. and Thorpe, S. R. (1989). The Amadori product on protein: structure and reactions. *Prog Clin Biol Res* 304: 43–67.

Bellotti, V., Gallieni, M., Giorgetti, S. and Brancaccio, D. (2001). Dynamic of beta-2-microglobulin fibril formation and reabsorption: the role of proteolysis. *Semin Dial* 14: 117–122.

Bellotti, V., Stoppini, M., Mangione, P., Sunde, M., Robinson, C., Asti, L., Brancaccio, D. and Ferri, G. (1998). Beta-2-microglobulin can be refolded into a native state from *ex vivo* amyloid fibrils. *Eur J Biochem* 258: 61–67.

Bjorkman, P. J., Saper, M. A., Samraoui, B., Bennett, W. S., Strominger, J. L. and Wiley, D. C. (1987). Structure of the human class I histocompatibility antigen, HLA-A2. *Nature* 329: 506–512.

Borysik, A. J., Radford, S. E. and Ashcroft, A. E. (2004). Co-populated conformational ensembles of beta-2-microglobulin uncovered quantitatively by electrospray ionization mass spectrometry. *J Biol Chem* 279: 27069–27077.

Brown, D. R., Qin, K., Herms, J. W., Madlung, A., Manson, J., Strome, R., Fraser, P. E., Kruck, T., von Bohlen, A., Schulz-Schaeffer, W., Giese, A., Westaway, D. and Kretzschmar, H. (1997). The cellular prion protein binds copper *in vivo*. *Nature* 390: 684–687.

Brownlee, M., Cerami, A. and Vlassara, H. (1988). Advanced glycosylation end products in tissue and the biochemical basis of diabetic complications. *N Engl J Med* 318: 1315–1321.

Campistol, J. M. (2001). Dialysis-related amyloidosis after renal transplantation. *Semin Dial* 14: 99–102.

Campistol, J. M., Shirahama, T., Abraham, C. R., Rodgers, O. G., Sole, M., Cohen, A. S. and Skinner, M. (1992a). Demonstration of plasma proteinase inhibitors in beta-2-microglobulin amyloid deposits. *Kidney Int* 42: 915–923.

Campistol, J. M., Sole, M., Bombi, J. A., Rodriguez, R., Mirapeix, E., Munoz-Gomez, J. and Revert, O. W. (1992b). *In vitro* spontaneous synthesis of beta-2-microglobulin amyloid fibrils in peripheral blood mononuclear cell culture. *Am J Pathol* 141: 241–247.

Capeillere-Blandin, C., Delaveau, T. and Descamps-Latscha, B. (1991). Structural modifications of human beta-2-microglobulin treated with oxygen-derived radicals. *Biochem J* 277 (Pt 1): 175–182.

Cereghetti, G. M., Schweiger, A., Glockshuber, R. and Van Doorslaer, S. (2001). Electron paramagnetic resonance evidence for binding of Cu(2+) to the C-terminal domain of the murine prion protein. *Biophys J* 81: 516–525.

Chiba, T., Hagihara, Y., Higurashi, T., Hasegawa, K., Naiki, H. and Goto, Y. (2003). Amyloid fibril formation in the context of full-length protein: effects of proline mutations on the amyloid fibril formation of beta-2-microglobulin. *J Biol Chem* 278: 47016–47024.

Coggi, G., Dell'Orto, P., Braidotti, P., Coggi, A. and Viale, G. (1989). Coexpression of intermediate filaments in normal and neoplastic human tissues: a reappraisal. *Ultrastruct Pathol* 13: 501–514.

Connors, L. H., Shirahama, T., Skinner, M., Fenves, A. and Cohen, A. S. (1985). *In vitro* formation of amyloid fibrils from intact beta-2-microglobulin. *Biochem Biophys Res Commun* 131: 1063–1068.

Danesh, F. and Ho, L. T. (2001). Dialysis-related amyloidosis: history and clinical manifestations. *Semin Dial* 14: 80–85.

de la Paz, M. L. and Serrano, L. (2004). Sequence determinants of amyloid fibril formation. *Proc Natl Acad Sci USA* 101: 87–92.

De Lorenzi, E., Grossi, S., Massolini, G., Giorgetti, S., Mangione, P., Andreola, A., Chiti, F., Bellotti, V. and Caccialanza, G. (2002). Capillary electrophoresis investigation of a partially unfolded conformation of beta-2-microglobulin. *Electrophoresis* 23: 918–925.

Degenhardt, T. P., Grass, L., Reddy, S., Thorpe, S. R., Diamandis, E. P. and Baynes, J. W. (1997). Technical note. The serum concentration of the advanced glycation end-product N epsilon-(carboxymethyl)lysine is increased in uremia. *Kidney Int* 52: 1064–1067.

DeLano, W. (2002). The PyMOL Molecular Graphics System. San Carlos, CA, USA, DeLano Scientific.

Dobson, C. M. (1999). Protein misfolding, evolution and disease. *Trends Biochem Sci* 24: 329–332.

Durie, B. G., Persky, B., Soehnlen, B. J., Grogan, T. M. and Salmon, S. E. (1982). Amyloid production in human myeloma stem-cell culture, with morphologic evidence of amyloid secretion by associated macrophages. *N Engl J Med* 307: 1689–1692.

Eakin, C. M., Attenello, F. J., Morgan, C. J. and Miranker, A. D. (2004). Oligomeric assembly of native-like precursors precedes amyloid formation by beta-2-microglobulin. *Biochemistry* 43: 7808–7815.

Eakin, C. M., Knight, J. D., Morgan, C. J., Gelfand, M. A. and Miranker, A. D. (2002). Formation of a copper specific binding site in non-native states of beta-2-microglobulin. *Biochemistry* 41: 10646–10656.

Eaken, C. M., Bermin, A. J. and Miranker, A. D. (2006). A native to amyloidogenic transition regulated by a backbone trigger. *Nat Struct Mol Biol* 13: 202–208.

Esposito, G., Michelutti, R., Verdone, G., Viglino, P., Hernandez, H., Robinson, C. V., Amoresano, A., Dal Piaz, F., Monti, M., Pucci, P., Mangione, P., Stoppini, M., Merlini, G., Ferri, G. and Bellotti, V. (2000). Removal of the N-terminal hexapeptide from human beta-2-microglobulin facilitates protein aggregation and fibril formation. *Protein Sci* 9: 831–845.

Fernandez-Escamilla, A. M., Rousseau, F., Schymkowitz, J. and Serrano, L. (2004). Prediction of sequence-dependent and mutational effects on the aggregation of peptides and proteins. *Nat Biotechnol* 22: 1302–1306.

Floege, J. and Ehlerding, G. (1996). Beta-2-microglobulin-associated amyloidosis. *Nephron* 72: 9–26.

Floege, J. and Ketteler, M. (2001). Beta-2-microglobulin-derived amyloidosis: an update. *Kidney Int Suppl* 78: S164–171.

Garbar, C., Jadoul, M., Noel, H. and van Ypersele de Strihou, C. (1999). Histological characteristics of sternoclavicular beta-2-microglobulin amyloidosis and clues for its histogenesis. *Kidney Int* 55: 1983–1990.

Garcia-Garcia, M., Argiles, Gouin-Charnet, A., Durfort, M., Garcia-Valero, J. and Mourad, G. (1999). Impaired lysosomal processing of beta-2-microglobulin by infiltrating macrophages in dialysis amyloidosis. *Kidney Int* 55: 899–906.

Gazit, E. (2002). A possible role for pi-stacking in the self-assembly of amyloid fibrils. *FASEB J* 16: 77–83.

Geddes, A. J., Parker, K. D., Atkins, E. D. and Beighton, E. (1968). "Cross-beta" conformation in proteins. *J Mol Biol* 32: 343–358.

Gejyo, F., Homma, N., Hasegawa, S. and Arakawa, M. (1993). A new therapeutic approach to dialysis amyloidosis: intensive removal of beta-2-microglobulin with adsorbent column. *Artif Organs* 17: 240–243.

Gejyo, F., Homma, N., Suzuki, Y. and Arakawa, M. (1986). Serum levels of beta-2-microglobulin as a new form of amyloid protein in patients undergoing long-term hemodialysis. *N Engl J Med* 314: 585–586.

Gejyo, F., Kawaguchi, Y., Hara, S., Nakazawa, R., Azuma, N., Ogawa, H., Koda, Y., Suzuki, M., Kaneda, H., Kishimoto, H., Oda, M., Ei, K., Miyazaki, R., Maruyama, H., Arakawa, M. and Hara, M. (2004). Arresting dialysis-related amyloidosis: a prospective multicenter controlled trial of direct hemoperfusion with a beta-2-microglobulin adsorption column. *Artif Organs* 28: 371–380.

Gejyo, F., Yamada, T., Odani, S., Nakagawa, Y., Arakawa, M., Kunitomo, T., Kataoka, H., Suzuki, M., Hirasawa, Y. and Shirahama, T. (1985). A new form of amyloid protein associated with chronic hemodialysis was identified as beta-2-microglobulin. *Biochem Biophys Res Commun* 129: 701–706.

Gorevic, P. D., Casey, T. T., Stone, W. J., DiRaimondo, C. R., Prelli, F. C. and Frangione, B. (1985). Beta-2-microglobulin is an amyloidogenic protein in man. *J Clin Invest* 76: 2425–2429.

Gorevic, P. D., Munoz, P. C., Casey, T. T., DiRaimondo, C. R., Stone, W. J., Prelli, F. C., Rodrigues, M. M., Poulik, M. D. and Frangione, B. (1986). Polymerization of intact beta-2-microglobulin in tissue causes amyloidosis in patients on chronic hemodialysis. *Proc Natl Acad Sci USA* 83: 7908–7912.

Gosal, W. S., Morten, I. J., Hewitt, H. W., Smith, D. A., Thomson, N. H., and Radford, S. E. (2005). Competing pathways determine fibril morphology in the self-assembly of beta-2-microglobulin into amyloid. *J Mol Biol* 351: 850–864.

Gupta-Bansal, R., Frederickson, R. C. and Brunden, K. R. (1995). Proteoglycan-mediated inhibition of A beta proteolysis. A potential cause of senile plaque accumulation. *J Biol Chem* 270: 18666–71.

Homma, N., Gejyo, F., Isemura, M. and Arakawa, M. (1989). Collagen-binding affinity of beta-2-microglobulin, a preprotein of hemodialysis-associated amyloidosis. *Nephron* 53: 37–40.

Hong, D. P., Gozu, M., Hasegawa, K., Naiki, H. and Goto, Y. (2002). Conformation of beta-2-microglobulin amyloid fibrils analyzed by reduction of the disulfide bond. *J Biol Chem* 277: 21554–21560.

Hoshino, M., Katou, H., Hagihara, Y., Hasegawa, K., Naiki, H. and Goto, Y. (2002). Mapping the core of the beta-2-microglobulin amyloid fibril by H/D exchange. *Nat Struct Biol* 9: 332–336.

Hou, F. F., Chertow, G. M., Kay, J., Boyce, J., Lazarus, J. M., Braatz, J. A. and Owen, W. F., Jr. (1997). Interaction between beta-2-microglobulin and advanced glycation end products in the development of dialysis related-amyloidosis. *Kidney Int* 51: 1514–1519.

Hou, F. F. and Owen, W. F., Jr. (2002). Beta-2-microglobulin amyloidosis: role of monocytes/macrophages. *Curr Opin Nephrol Hypertens* 11: 417–421.

Iida, Y., Miyata, T., Inagi, R., Sugiyama, S. and Maeda, K. (1994). Beta-2-microglobulin modified with advanced glycation end products induces interleukin-6 from human macrophages: role in the pathogenesis of hemodialysis-associated amyloidosis. *Biochem Biophys Res Commun* 201: 1235–1241.

Inoue, H., Saito, I., Nakazawa, R., Mukaida, N., Matsushima, K., Azuma, N., Suzuki, M. and Miyasaka, N. (1995). Expression of inflammatory cytokines and adhesion molecules in hemodialysis-associated amyloidosis. *Nephrol Dial Transplant* 10: 2077–2082.

Ivanova, M. I., Gingery, M., Whitson, L. J. and Eisenberg, D. (2003). Role of the C-terminal 28 residues of beta-2-microglobulin in amyloid fibril formation. *Biochemistry* 42: 13536–13540.

Ivanova, M. I., Sawaya, M. R., Gingery, M., Attinger, A. and Eisenberg, D. (2004). An amyloid-forming segment of beta-2-microglobulin suggests a molecular model for the fibril. *Proc Natl Acad Sci USA* 101: 10584–10589.

Jadoul, M., Drueke, T., Zingraff, J. and van Ypersele de Strihou, C. (1997). Does dialysis-related amyloidosis regress after transplantation? *Nephrol Dial Transplant* 12: 655–657.

Jahn, T. R., Parker, M. J., Homans, S. W. and Radford, S. E. (2006). Amyloid formation under physiological conditions proceeds via a native-like folding intermediate. *Nat Struct Mol Biol* 13: 195–201.

Jones, S., Manning, J., Kad, N. M. and Radford, S. E. (2003a). Amyloid-forming peptides from beta-2-microglobulin-Insights into the mechanism of fibril formation *in vitro*. *J Mol Biol* 325: 249–257.

Jones, S., Smith, D. P. and Radford, S. E. (2003b). Role of the N- and C-terminal strands of beta-2-microglobulin in amyloid formation at neutral pH. *J Mol Biol* 330: 935–941.

Kad, N. M., Myers, S. L., Smith, D. P., Smith, D. A., Radford, S. E. and Thomson, N. H. (2003). Hierarchical assembly of beta-2-microglobulin amyloid *in vitro* revealed by atomic force microscopy. *J Mol Biol* 330: 785–797.

Kad, N. M., Thomson, N. H., Smith, D. P., Smith, D. A. and Radford, S. E. (2001). Beta-2-microglobulin and its deamidated variant, N17D form amyloid fibrils with a range of morphologies *in vitro*. *J Mol Biol* 313: 559–571.

Katou, H., Kanno, T., Hoshino, M., Hagihara, Y., Tanaka, H., Kawai, T., Hasegawa, K., Naiki, H. and Goto, Y. (2002). The role of disulfide bond in the amyloidogenic state of beta-2-microglobulin studied by heteronuclear NMR. *Protein Sci* 11: 2218–2229.

Knudsen, P. J., Shaldon, S., Floege, J. and Koch, M. (1990). Hemodialysis-related induction of beta-2-microglobulin synthesis and release by mononuclear phagocytes. *Int J Artif Organs* 13: 73–76.

Koch, K. M. (1992). Dialysis-related amyloidosis. *Kidney Int* 41: 1416–1429.

Koda, Y., Nishi, S., Miyazaki, S., Haginoshita, S., Sakurabayashi, T., Suzuki, M., Sakai, S., Yuasa, Y., Hirasawa, Y. and Nishi, T. (1997). Switch from conventional to high-flux membrane reduces the risk of carpal tunnel syndrome and mortality of hemodialysis patients. *Kidney Int* 52: 1096–1101.

Kozhukh, G. V., Hagihara, Y., Kawakami, T., Hasegawa, K., Naiki, H. and Goto, Y. (2002). Investigation of a peptide responsible for amyloid fibril formation of beta-2-microglobulin by *Achromobacter* protease I. *J Biol Chem* 277: 1310–1315.

Lashuel, H. A., Hartley, D., Petre, B. M., Walz, T. and Lansbury, P. T., Jr. (2002). Neurodegenerative disease: amyloid pores from pathogenic mutations. *Nature* 418: 291.

Linke, R. P., Hampl, H., Bartel-Schwarze, S. and Eulitz, M. (1987). Beta-2-microglobulin, different fragments and polymers thereof in synovial amyloid in long-term hemodialysis. *Biol Chem Hoppe Seyler* 368: 137–144.

Linke, R. P., Hampl, H., Lobeck, H., Ritz, E., Bommer, J., Waldherr, R. and Eulitz, M. (1989). Lysine-specific cleavage of beta-2-microglobulin in amyloid deposits associated with hemodialysis. *Kidney Int* 36: 675–681.

Lornoy, W., Becaus, I., Billiouw, J. M., Sierens, L., Van Malderen, P. and D'Haenens, P. (2000). On-line hemodiafiltration. Remarkable removal of beta-2-microglobulin. Long-term clinical observations. *Nephrol Dial Transplant* 15 Suppl 1: 49–54.

Ma, B. and Nussinov, R. (2003). Molecular dynamics simulations of the unfolding of beta-2-microglobulin and its variants. *Protein Eng* 16: 561–575.

McParland, V. J., Kad, N. M., Kalverda, A. P., Brown, A., Kirwin-Jones, P., Hunter, M. G., Sunde, M. and Radford, S. E. (2000). Partially unfolded states of beta-2-microglobulin and amyloid formation *in vitro*. *Biochemistry* 39: 8735–8746.

McParland, V. J., Kalverda, A. P., Homans, S. W. and Radford, S. E. (2002). Structural properties of an amyloid precursor of beta-2-microglobulin. *Nat Struct Biol* 9: 326–331.

Minor, D. L., Jr. and Kim, P. S. (1994a). Context is a major determinant of beta-sheet propensity. *Nature* 371: 264–267.

Minor, D. L., Jr. and Kim, P. S. (1994b). Measurement of the beta-sheet-forming propensities of amino acids. *Nature* 367: 660–663.

Miura, T., Suzuki, K., Kohata, N. and Takeuchi, H. (2000). Metal binding modes of Alzheimer's amyloid beta-peptide in insoluble aggregates and soluble complexes. *Biochemistry* 39: 7024–7031.

Miyata, T., Hori, O., Zhang, J., Yan, S. D., Ferran, L., Iida, Y. and Schmidt, A. M. (1996a). The receptor for advanced glycation end products (RAGE) is a central mediator of the interaction of AGE-beta-2-microglobulin with human mononuclear phagocytes *via* an oxidant-sensitive pathway. Implications for the pathogenesis of dialysis-related amyloidosis. *J Clin Invest* 98: 1088–1094.

Miyata, T., Iida, Y., Ueda, Y., Shinzato, T., Seo, H., Monnier, V. M., Maeda, K. and Wada, Y. (1996b). Monocyte/macrophage response to beta-2-microglobulin modified with advanced glycation end products. *Kidney Int* 49: 538–550.

Miyata, T., Inagi, R., Iida, Y., Sato, M., Yamada, N., Oda, O., Maeda, K. and Seo, H. (1994). Involvement of beta-2-microglobulin modified with advanced glycation end products in the pathogenesis of hemodialysis-associated amyloidosis. Induction of human monocyte chemotaxis and macrophage secretion of tumor necrosis factor-alpha and interleukin-1. *J Clin Invest* 93: 521–528.

Miyata, T., Jadoul, M., Kurokawa, K. and Van Ypersele de Strihou, C. (1998). Beta-2-microglobulin in renal disease. *J Am Soc Nephrol* 9: 1723–1735.

Miyata, T., Taneda, S., Kawai, R., Ueda, Y., Horiuchi, S., Hara, M., Maeda, K. and Monnier, V. M. (1996c). Identification of pentosidine as a native structure for advanced glycation end products in beta-2-microglobulin-containing amyloid fibrils in patients with dialysis-related amyloidosis. *Proc Natl Acad Sci USA* 93: 2353–2358.

Moe, S. M. and Chen, N. X. (2001). The role of the synovium and cartilage in the pathogenesis of beta-2-microglobulin amyloidosis. *Semin Dial* 14: 127–130.

Monti, M., Principe, S., Giorgetti, S., Mangione, P., Merlini, G., Clark, A., Bellotti, V., Amoresano, A. and Pucci, P. (2002). Topological investigation of amyloid fibrils obtained from beta-2-microglobulin. *Protein Sci* 11: 2362–2369.

Morgan, C. J., Gelfand, M., Atreya, C. and Miranker, A. D. (2001). Kidney dialysis-associated amyloidosis: a molecular role for copper in fiber formation. *J Mol Biol* 309: 339–345.

Morita, H., Cai, Z., Shinzato, T., David, G., Mizutani, A., Itano, N., Habuchi, H., Yoneda, M., Maeda, K. and Kimata, K. (1995). Glycosaminoglycans in dialysis-related amyloidosis. *Contrib Nephrol* 112: 83–89.

Morita, T., Suzuki, M., Kamimura, A. and Hirasawa, Y. (1985). Amyloidosis of a possible new type in patients receiving long-term hemodialysis. *Arch Pathol Lab Med* 109: 1029–1032.

Motomiya, Y., Ando, Y., Haraoka, K., Sun, X., Iwamoto, H., Uchimura, T. and Maruyama, I. (2003). Circulating level of alpha2–macroglobulin-beta-2-microglobulin complex in hemodialysis patients. *Kidney Int* 64: 2244–2252.

Myers, S. L., Thomson, N. H., Radford, S. E. and Ashcroft, A. E. (2006a). Investigating the structural properties of amyloid-like fibrils formed from beta-2-microglobulin using limited proteolysis and electrospray ionization mass spectrometry. *Rapid Communications in Mass Spectrometry* 20: 1628–1636.

Myers, S. L., Jones, S., Jahn, T. R., Morten, I. J., Tennent, G. A., Hewitt, E. W. and Radford, S. E. (2006b). A systematic study of the effect of physiological factors on beta-2-microgloblin amyloid formation at neutral pH. *Biochemistry* 45: 2274–2282.

Naiki, H., Hashimoto, N., Suzuki, S., Kimura, H., Nakakuki, K. and Gejyo, F. (1997). Establishment of a kinetic model of dialysis-related amyloid fibril extension *in vitro. Amyloid-Int J Exp Clin Invest* 4: 223–232.

Nguyen-Simonnet, H., Vincent, C., Gauthier, C., Revillard, J. P. and Pellet, M. V. (1982). Turnover studies of human beta-2-microglobulin in the rat: evidence for a beta-2-microglobulin-binding plasma protein. *Clin Sci (Lond)* 62: 403–410.

Niwa, T., Katsuzaki, T., Miyazaki, S., Momoi, T., Akiba, T., Miyazaki, T., Nokura, K., Hayase, F., Tatemichi, N. and Takei, Y. (1997). Amyloid beta-2-microglobulin is modified with imidazolone, a novel advanced glycation end product, in dialysis-related amyloidosis. *Kidney Int* 51: 187–194.

Niwa, T. (2001). Dialysis-related amyloidosis: pathogenesis focusing on AGE modification. *Semin Dial* 14: 123–126.

Odani, H., Oyama, R., Titani, K., Ogawa, H. and Saito, A. (1990). Purification and complete amino acid sequence of novel beta-2-microglobulin. *Biochem Biophys Res Commun* 168: 1223–1229.

Ohashi, K., Hara, M., Kawai, R., Ogura, Y., Honda, K., Nihei, H. and Mimura, N. (1992). Cervical discs are most susceptible to beta-2-microglobulin amyloid deposition in the vertebral column. *Kidney Int* 41: 1646–1652.

Ohashi, K., Hara, M., Yanagishita, M., Kawai, R., Tachibana, S. and Ogura, Y. (1995). Proteoglycans in hemodialysis-related amyloidosis. *Virchows Arch* 427: 49–59.

Ohhashi, Y., Hagihara, Y., Kozhukh, G., Hoshino, M., Hasegawa, K., Yamaguchi, I., Naiki, H. and Goto, Y. (2002). The intra-chain disulfide bond of beta-2-microglobulin is not essential for the immunoglobulin fold at neutral pH, but is essential for amyloid fibril formation at acidic pH. *J Biochem (Tokyo)* 131: 45–52.

Ohishi, H., Skinner, M., Sato-Araki, N., Okuyama, T., Gejyo, F., Kimura, A., Cohen, A. S. and Schmid, K. (1990). Glycosaminoglycans of the hemodialysis-associated carpal synovial amyloid and of amyloid-rich tissues and fibrils of heart, liver, and spleen. *Clin Chem* 36: 88–91.

Okon, M., Bray, P. and Vucelic, D. (1992). 1H NMR assignments and secondary structure of human beta-2-microglobulin in solution. *Biochemistry* 31: 8906–8915.

O'Neill, K. D., Chen, N. X., Wang, M., Cocklin, R., Zhang, Y. and Moe, S. M. (2003). Cellular uptake of beta-2-microglobulin and AGE-beta-2-microglobulin in synovial fibroblasts and macrophages. *Nephrol Dial Transplant* 18: 46–53.

Ono, K. and Uchino, F. (1994). Formation of amyloid-like substance from beta-2-microglobulin *in vitro*. Role of serum amyloid P component: a preliminary study. *Nephron* 66: 404–407.

Pamer, E. and Cresswell, P. (1998). Mechanisms of MHC class I-restricted antigen processing. *Annu Rev Immunol* 16: 323–358.

Platt, G. W., McParland, V. J., Kalverda, A. P., Homans, S. W. and Radford, S. E. (2004). Dynamics in the unfolded state of beta-2-microglobulin studied by NMR. *J Mol Biol* 346: 279–294.

Porat, Y., Mazor, Y., Efrat, S. and Gazit, E. (2004). Inhibition of islet amyloid polypeptide fibril formation: a potential role for heteroaromatic interactions. *Biochemistry* 43: 14454–14462.

Rochet, J. C. and Lansbury, P. T., Jr. (2000). Amyloid fibrillogenesis: themes and variations. *Curr Opin Struct Biol* 10: 60–68.

Rock, K. L. and Goldberg, A. L. (1999). Degradation of cell proteins and the generation of MHC class I-presented peptides. *Annu Rev Immunol* 17: 739–779.

Rosano, C., Zuccotti, S., Mangione, P., Giorgetti, S., Bellotti, V., Pettirossi, F., Corazza, A., Viglino, P., Esposito, G. and Bolognesi, M. (2004). Beta-2-microglobulin H31Y variant 3D structure highlights the protein natural propensity towards intermolecular aggregation. *J Mol Biol* 335: 1051–1064.

Saper, M. A., Bjorkman, P. J. and Wiley, D. C. (1991). Refined structure of the human histocompatibility antigen HLA-A2 at 2.6 A resolution. *J Mol Biol* 219: 277–319.

Sargent, M. A., Fleming, S. J., Chattopadhyay, C., Ackrill, P. and Sambrook, P. (1989). Bone cysts and haemodialysis-related amyloidosis. *Clin Radiol* 40: 277–281.

Siebold, C., Hansen, B. E., Wyer, J. R., Harlos, K., Esnouf, R. E., Svejgaard, A., Bell, J. I., Strominger, J. L., Jones, E. Y. and Fugger, L. (2004). Crystal structure of HLA-DQ0602 that protects against type 1 diabetes and confers strong susceptibility to narcolepsy. *Proc Natl Acad Sci USA* 101: 1999–2004.

Simkin, P. A. (1997). The musculoskeletal system, A. Joints. *Primer on Rheumatic Diseases*. M. J. Favus. Philadelphia, Arthritis Foundation: 10–11.

Smith, D. P., Jones, S., Serpell, L. C., Sunde, M. and Radford, S. E. (2003). A systematic investigation into the effect of protein destabilisation on beta-2-microglobulin amyloid formation. *J Mol Biol* 330: 943–954.

Smith, D. P. and Radford, S. E. (2001). Role of the single disulphide bond of beta-2-microglobulin in amyloidosis *in vitro*. *Protein Sci* 10: 1775–1784.

Snow, A. D., Bramson, R., Mar, H., Wight, T. N. and Kisilevsky, R. (1991). A temporal and ultrastructural relationship between heparan sulfate proteoglycans and AA amyloid in experimental amyloidosis. *J Histochem Cytochem* 39: 1321–1330.

Stoppini, M. S., Arcidiaco, P., Mangione, P., Giorgetti, S., Brancaccio, D. and Bellotti, V. (2000). Detection of fragments of beta-2-microglobulin in amyloid fibrils. *Kidney Int* 57: 349–350.

Sunde, M. and Blake, C. (1997). The structure of amyloid fibrils by electron microscopy and X-ray diffraction. *Adv Protein Chem* 50: 123–159.

Sundin, D. P., Cohen, M., Dahl, R., Falk, S. and Molitoris, B. A. (1994). Characterization of the beta-2-microglobulin endocytic pathway in rat proximal tubule cells. *Am J Physiol* 267: F380–389.

Takahashi, M., Yokota, T., Kawano, H., Gondo, T., Ishihara, T. and Uchino, F. (1989). Ultrastructural evidence for intracellular formation of amyloid fibrils in macrophages. *Virchows Arch A Pathol Anat Histopathol* 415: 411–419.

Trinh, C. H., Smith, D. P., Kalverda, A. P., Phillips, S. E. and Radford, S. E. (2002). Crystal structure of monomeric human beta-2-microglobulin reveals clues to its amyloidogenic properties. *Proc Natl Acad Sci USA* 99: 9771–9776.

Uversky, V. N., Li, J. and Fink, A. L. (2001). Metal-triggered structural transformations, aggregation, and fibrillation of human alpha-synuclein. A possible molecular NK between Parkinson's disease and heavy metal exposure. *J Biol Chem* 276: 44284–44296.

van Ypersele, C. and Drueke, T. (1996). *Dialysis Amyloid*. Oxford, Oxford University Press.

van Ypersele de Strihou, C. (1996). beta-2-microglobulin amyloidosis: effect of ESRF treatment modality and dialysis membrane type. *Nephrol Dial Transplant* 11 Suppl 2: 147–149.

Verdone, G., Corazza, A., Viglino, P., Pettirossi, F., Giorgetti, S., Mangione, P., Andreola, A., Stoppini, M., Bellotti, V. and Esposito, G. (2002). The solution structure of human beta-2-microglobulin reveals the prodromes of its amyloid transition. *Protein Sci* 11: 487–499.

Villanueva, J., Hoshino, M., Katou, H., Kardos, J., Hasegawa, K., Naiki, H. and Goto, Y. (2004). Increase in the conformational flexibility of beta-2-microglobulin upon copper binding: a possible role for copper in dialysis-related amyloidosis. *Protein Sci* 13: 797–809.

Westermark, P., Benson, M. D., Buxbaum, J. N., Cohen, A. S., Frangione, B., Ikeda, S., Masters, C. L., Merlini, G., Saraiva, M. J. and Sipe, J. D. (2002). Amyloid fibril protein nomenclature. 2002. *Amyloid* 9: 197–200.

White, A. A. and Panjabi, M. M. (1978). The basic kinematics of the human spine. A review of past and current knowledge. *Spine* 3: 12–20.

Winchester, J. F., Salsberg, J. A. and Levin, N. W. (2003). Beta-2-microglobulin in ESRD: an in-depth review. *Adv Ren Replace Ther* 10: 279–309.

Yamada, T., Kakihara, T., Gejyo, F. and Okada, M. (1994). A monoclonal antibody recognizing apolipoprotein E peptides in systemic amyloid deposits. *Ann Clin Lab Sci* 24: 243–249.

Yamaguchi, I., Hasegawa, K., Takahashi, N., Gejyo, F. and Naiki, H. (2001). Apolipoprotein E inhibits the depolymerization of beta-2-microglobulin-related amyloid fibrils at a neutral pH. *Biochemistry* 40: 8499–8507.

Yamaguchi, I., Suda, H., Tsuzuike, N., Seto, K., Seki, M., Yamaguchi, Y., Hasegawa, K., Takahashi, N., Yamamoto, S., Gejyo, F. and Naiki, H. (2003). Glycosaminoglycan and proteoglycan inhibit the depolymerization of beta-2-microglobulin amyloid fibrils *in vitro. Kidney Int* 64: 1080–1088.

Yamaguchi, K., Katou, H., Hoshino, M., Hasegawa, K., Naiki, H. and Goto, Y. (2004). Core and heterogeneity of beta-2-microglobulin amyloid fibrils as revealed by H/D exchange. *J Mol Biol* 338: 559–51.

Yamamoto, S., Hasegawa, K., Yamaguchi, I., Tsutsumi, S., Kardos, J., Goto, Y., Gejyo, F. and Naiki, H. (2004a). Low concentrations of sodium dodecyl sulfate induce the extension of beta-2-microglobulin-related amyloid fibrils at a neutral pH. *Biochemistry* 43: 11075–11082.

Yamamoto, S., Yamaguchi, I., Hasegawa, K., Tsutsumi, S., Goto, Y., Gejyo, F. and Naiki, H. (2004b). Glycosaminoglycans enhance the trifluoroethanol-induced extension of beta-2-microglobulin-related amyloid fibrils at a neutral pH. *J Am Soc Nephrol* 15: 126–133.

Zerovnik, E. (2002). Amyloid-fibril formation. Proposed mechanisms and relevance to conformational disease. *Eur J Biochem* 269: 3362–3371.

12

Serum Amyloid A and AA Amyloidosis

Zafer Ali-Khan

Abstract

Amyloid-related diseases show marked extracellular deposition of nonbranching protein fibrils, called amyloid, in various soft organs. The mechanisms by which soluble amyloid precursor proteins transform into amyloid fibrils remain mainly obscure. Here we discuss data derived from alveolar hydatid cyst-infected (AHC) mouse model of inflammation-associated amyloid A (AA) amyloidosis and concepts related to the potential roles of oxidative stress–related factors and metabolism of monocytoid cell–associated serum amyloid A (SAA) precursor protein in AA fibril formation *in vivo*. Evidence shows that AA fibrils are generated intracellularly within the activated host monocytoid cells, which prior to and during AA fibril formation generate, among other oxidative stress–related derivatives, 4-hydroxynonenal, a lipid peroxidation product. We suggest that release of the AA fibrils formed intracellularly, either by exocytosis or cell death, could act as the "seed" for the nucleation-dependent expansion of extracellular AA fibril deposition.

12.1. Introduction

The term *amyloidosis* describes a heterogeneous collection of both localized and systemic chronic diseases, characterized by the extracellular deposition of proteinaceous insoluble fibrils, termed *amyloid*, of great chemical diversity in various organs and tissues (Glenner, 1980; Merlini and Westermark, 2004). Progressive tissue deposition of amyloid fibrils may lead to dysfunction of the affected organs and even death. Alzheimer's disease, type 2 diabetes mellitus, a number of polyneuropathies, the transmissible encephalopathies, and both the primary and secondary amyloidoses stand out as prominent amyloid-related diseases (Husby *et al.*, 1994; Buxbaum and Tagoe, 2000). To date, at least 21 precursor proteins have been identified that yield amyloid fibrils (Merlini and Westermark, 2004). Regardless of the chemical diversity, all amyloid fibrils tend to be insoluble under physiologic conditions, possess cross-β structure, appear indistinguishable under light and electron microscopy, and bind Congo red stain (Glenner, 1980; Merlini and Westermark, 2004). The core pathogenetic process involves conversion of structurally diverse soluble amyloid precursors into morphologically similar amyloid fibrils. This is an intriguing phenomenon. However, the conditions that lead to conformational changes in the precursor proteins, notably, from α-helix to β-pleated sheet amyloid fibrils, remain mostly speculative.

Kelly (1996) proposed "acid denaturation pathway" as a common mechanism of amyloidogenesis. A number of *in vitro* findings, originally proposed by Glenner *et al.* (1971) and Glenner (1980), support this proposal. For example, the light chain of immunoglobulin, various fragments of the Alzheimer amyloid beta (Aβ) peptide, prion protein, and serum amyloid A (SAA) and transthyretin, exposed to acid conditions *in vitro*, yielded amyloid fibrils (Glenner *et al.*, 1971; Fraser *et al.*, 1991; Westermark *et al.*, 1992; Pan *et al.*, 1993; Yamada *et al.*, 1994; Kelly, 1996; Kelly and Lansbury, 1996; Kirschner *et al.*, 1998). Recently, Guijarro *et al.* (1998) and Chiti *et al.* (1999) replicated these experiments and were able to generate amyloid fibrils from proteins unrelated to any of the amyloid-related diseases. These findings led to the concept that amyloid fibril formation may not be restricted to a unique subset of proteins, but under "appropriate conditions" many proteins possess the tendency to yield amyloid fibrils *in vitro*. Some of the proposed appropriate predisposing conditions may include acidic pH, partial proteolysis of the precursor protein (although not a universal phenomenon), and structural modification of the precursor protein leading to its aggregation or fibril formation through oxidative stress (OX-St)-generated factors (Dyrks *et al.*, 1992; Friedrlich and Butcher, 1994; Schweers *et al.*, 1995; Sayer *et al.*, 1997). How, when, and under what conditions these factors intersect to precipitate amyloid fibril formation remain the central issue in amyloidosis research.

A portion of the concept, presented here, emanates from our experience using the alveolar hydatid cyst (AHC)-mouse model of amyloid A (AA) amyloidosis (Ali-Khan *et al.*, 1983, 1996). AHC is the larval stage of *Echinococcus multilocularis*, a cestode parasite that infects both humans and a host of rodent species (Ali-Khan and Rausch, 1987; Ali-Khan *et al.*, 1996). In sum, AHC grows like a solid tumor and induces chronic inflammation, massive hepato-splenomegaly, systemic lymphadenitis, and multiple-organ AA fibril deposition in 100% of mice infected intraperitoneally with 250 AHC vesicles, each measuring approximately 2 mm in diameter (Treves and Ali-Khan, 1984; Ali-Khan *et al.*, 1988; Du and Ali-Khan, 1990; Ali-Khan *et al.*, 1996). Some of these conditions, however, contrast with the frequency of AA amyloidosis seen in patients with various inflammatory disorders. For example, only 1–2% of rheumatoid arthritis (RA) patients in the United States and 10–15% of patients in Japan develop AA amyloidosis (Yamada *et al.*, 2003). This disparity in the frequency of AA amyloidosis between the RA patients and the AHC-infected mice is unclear. Both the duration and intensity of chronic inflammation may partly explain the high incidence of AA amyloidosis in the AHC-infected mice (Treves and Ali-Khan, 1984).

Using the AHC-mouse model, we were able to localize both SAA and its end product (AA fibrils) to lysosomes (LY) in the inflammatory monocytoid cells [macrophages (MΦ) and reticuloendothelial (RE) cells] (Chronopoulos *et al.*, 1992, 1994, 1995; Chan *et al.*, 1997; Bell *et al.*, 1999). These findings, as discussed below, led us to revive the concept that dysfunction of LY and abnormal degradation of SAA within LY in inflammatory monocytoid cells may set the stage for intracellular nascent AA fibril formation (Shirahama and Cohen, 1975).

Evidence suggests that secondary AA amyloidosis is the prototype of a family of chronic diseases, singularly associated with protein misfolding disorders. Contextually, it is important to recall that chronic inflammation is central to the pathogenesis of AA amyloidosis. Thus, the focus of this chapter is an attempt to highlight current information on AA amyloidosis, particularly the possible mechanisms of ingress of SAA into activated monocytoid cells, and OX-St-derived factors that may influence abnormal metabolism of SAA within the LY and promote induction of nascent AA fibril formation.

12.2. AA Amyloidosis

Reactive amyloidosis, characterized by the tissue deposition of AA amyloid fibrils, is a rare but a serious complication of certain chronic inflammatory diseases. Patients with tuberculosis, leprosy, RA, osteomyelitis, familial Mediteranian fever, and a host of other chronic inflammatory disorders, including parasitic diseases, develop multiple organ AA amyloidosis (Glenner, 1980; Ali-Khan and Rausch, 1987; Ali-Khan *et al.*, 1996; Gillmore *et al.*, 2001). Kidney involvement leading to kidney failure is the major complication of AA amyloidosis, and fatality may reach up to 60% (Gillmore *et al.*, 2001). Currently, unlike patients with leprosy, the incidence of AA amyloidosis in treatable tuberculosis cases is virtually rare. Chronic inflammation in conjunction with chronic elevation in circulating acute-phase serum amyloid A (A-SAA; see below) are believed to be the two principal predisposing conditions associated with this disease. The AA fibrils, deposited systemically in various visceral tissues, are the N-terminal cleaved products of A-SAA, ranging in size from 5.5 to 8 kDa (Husby *et al.*, 1994). After cleavage, the N-terminally intact A-SAA fragments appear to be intrinsically unstable and prone to polymerize into AA fibrils (Stevens, 2004). Results of recent studies, using amyloid P component scintography as a monitoring device, show that suppression of inflammatory activity combined with the maintenance of serum A-SAA concentration below 10 mg/L in patients with systemic AA amyloidosis caused significant regression of tissue deposited AA amyloid (Gillmore *et al.*, 2001).

12.3. The Apolipoprotein Serum Amyloid A and AA Amyloid Protein

SAA is a family of apolipoproteins consisting of multiple isoforms, found predominatly associated with high-density lipoprotein (HDL) (Benditt and Ericksen, 1977; Benditt *et al.*, 1979). Both SAA genes and SAA have been described from a host of mammalian species but studied comprehensively only from mice and humans (Uhlar *et al.*, 1994; Uhlar and Whitehead, 1999). The data show that both SAA genes and the SAA isoforms have been well conserved during evolution, indicating that these apolipoproteins may be physiologically important in homeostasis. Based on the induction profile, SAA may be constitutive or the sequela of an acute-phase SAA response (A-SAA) (Gaby and Kushner, 1999). During this phase, the plasma concentration of A-SAA may increase up to 1000-fold (normal value ~5 μg/mL) (Steel and Whitehead 1994; Gaby and Kushner, 1999; Thorn *et al.*, 2004). The liver is the primary source of this protein (Benson and Kleiner, 1980; Meek *et al.*, 1986). However, extrahepatic expression of A-SAA also occurs during the acute-phase response (APR); detectable levels of SAA mRNA and A-SAA have been described in monocytoid and smooth muscle cells in atherosclerotic lesions (Meek *et al.*, 1994), and in Alzheimer brain tissues, indicating inflammation as an associated factor in the genesis of Alzheimer dementia (Liang *et al.*, 1997).

Multiple isoforms of A-SAA are produced by many mammalian species in response to tissue insult including infection or inflammation. Two human genes, SAAl and SAA2, which are ~90% identical, encode 12–14 kDa A-SAA1 and A-SAA2 proteins, respectively (Uhlar and Whitehead, 1999). A-SAA1 and A-SAA2 isoforms, each consisting of 104 amino acids, differ only at positions 60, 68, 69, 84, and 90 but contribute differentially to tissue AA amyloid. A-SAA1 manifests greater propensity to yield AA fibrils and is the predominant protein constituting the tissue AA amyloid substance (Yamada *et al.*, 1994; Liepnieks *et al.*, 1995).

Thus far, five and three alleles have been described for SAA1 and SAA2, respectively; SAA1γ was found to be the main constituent of AA amyloid substance in Japanese patients (Baba *et al.*, 1995). Of considerable interest is the finding that a single nucleotide polymorphism at position −13 in the SAA1 5′ flanking region acts as a risk factor and, thus, may explain differing incidence of AA amyloidosis between different ethnic groups afflicted with RA (Yamada *et al.*, 2003). How this risk factor might influence the differing incidence of AA amyloidosis in different ethnic groups, however, remains to be determined.

In mice, four SAA genes (Saa1, Saa2, Saa3, and Saa4) and a pseudogene, clustered on chromosome 7, have been described (Lowell *et al.*, 1986; Uhlar and Whitehead, 1999). A-SAA1, A-SAA2, the two major murine A-SAA isoforms and the products of Saa1 and Saa2 genes, respectively, predominate in the acute-phase plasma (Hoffman and Benditt, 1982; Yamamoto and Migita, 1985). Mice stimulated with bacterial lipopolysaccharides (LPS), casein, or infection with AHC generate high levels of A-SAA1 and A-SAA2. These two hepatocyte-secreted proteins bind to HDL, circulate in approximately equal quantities, and share ~95% structural identity (Yamamoto and Migita, 1985; Kluv-Beckerman *et al.*, 1997). Mouse A-SAA1, similar to its human counterpart, is the main precursor of AA fibrils, although, a small fraction, less than 1% of the total AA fibril protein, in amyloid susceptible C57BL/6 is also contributed by A-SAA2 (Bell *et al.*, 1996).

Evidence shows that both human and mouse AA proteins, purified from amyloidotic tissues, are heterogeneous in size (5.5 to 8 kDa) and represent the N-terminal two-thirds of the ~12-kDa A-SAA molecule (Husby *et al.*, 1994). However, in rare instances complete A-SAA with intact C-terminus has also been found in both human and mouse AA proteins (Westermark and Sletten, 1982). These rare findings, however, raise the question whether proteolytic removal of the C-terminus of the molecule is a prerequisite for amyloidogenesis? By contrast, under acid conditions *in vitro*, both full-length human recombinant A-SAA1 molecules and its synthetic ~12 N-terminal amino acid residue, the hydrophobic HDL-binding portion, but not the C-terminus, yielded AA fibrils, structurally similar to that of native AA amyloid substance (Westermark *et al.*, 1992; Yamada, *et al.*, 1994). In sum, the amyloidogenic property of the A-SAA1 molecule appears to reside at its amino terminus.

12.4. Regulation of Serum Amyloid A

Both exogenous and numerous host-derived endogenous factors appear to upregulate A-SAA synthesis. They include bacterial LPS, several proinflammatory cytokines [interleukin (IL)-1, IL-6, tumor necrosis factor (TNF)-α, interferon gamma (IFN-γ), IL-18] and glucocorticoids; differential activities of these factors have been comprehensively reviewed recently (Uhlar and Whitehead, 1999). During inflammation, IL-1, IL-6, and TNF-α, derived mostly from activated monocytoid cells, are the principal stimulators of A-SAA synthesis by inducing SAA mRNA expression in hepatocytes (Ganapathi *et al.*, 1991). Evidence shows that the activity of IL-1 and TNF-α is amplified synergistically by both IL-6 and glucocorticoids and the latter was shown to have a dual effect on A-SAA synthesis during the APR (Thorn *et al.*, 2004). Studies in mice further demonstrated that IL-6 was a key cytokine for the induction of AA amyloidosis; amyloidogen-treated mice, injected with anti-IL-6 receptor antibody, failed to develop AA amyloidosis, indicating a promising method for the management of AA amyloidosis (Shiina *et al.*, 2004). More recently, elevated levels of both IL-18, a proinflammatory cytokine, and A-SAA were detected in synovial fluid from RA patients, and synovial cells treated *in vitro* with IL-18 expressed A-SAA in a dose-dependent manner (Tanaka *et al.*, 2004). In addition to IL-18-mediated local synthesis of A-SAA, Maury *et al.* (2000) found increased humoral level of IL-18 in RA patients with AA amyloidosis compared with those without amyloidosis. Taken together, it appears that several members of IL-1 cytokine family regulate a widespread A-SAA response in a variety of mammalian cells and tissues.

At least five transcriptional factors, the nuclear factor-kappa B (NF-kB) family, cytosine-cytosine-adenine-adenine-thymine/enhancer binding protein family, nuclear factor for IL-6 and Sp1, and SAA activating sequence factor, have been shown to play a pivotal role in the proinflammatory cytokine–induced upregulation of A-SAA expression (Uhlar and Whitehead, 1999). These factors have been shown to be differentially expressed under different biological conditions; SAF and NF-kB activities, however, have been associated with chronic elevation in A-SAA (Ray and Ray, 1999). In sum, the regulation of A-SAA gene expression is complex and multiple factors are involved in the induction process. For more comprehensive information on cytokine-mediated expression of A-SAA mRNA and A-SAA and related events, the readers are referred to a recent review on this subject (Uhlar and Whitehead, 1999).

12.5. Serum and Monocytoid Cell-Associated Pools of Serum Amyloid A

The metabolism of A-SAA has been studied extensively both *in vivo* and *in vitro*. Mice stimulated with an amyloidogen (LPS, casein, or azocasein) or infected with AHC display at least

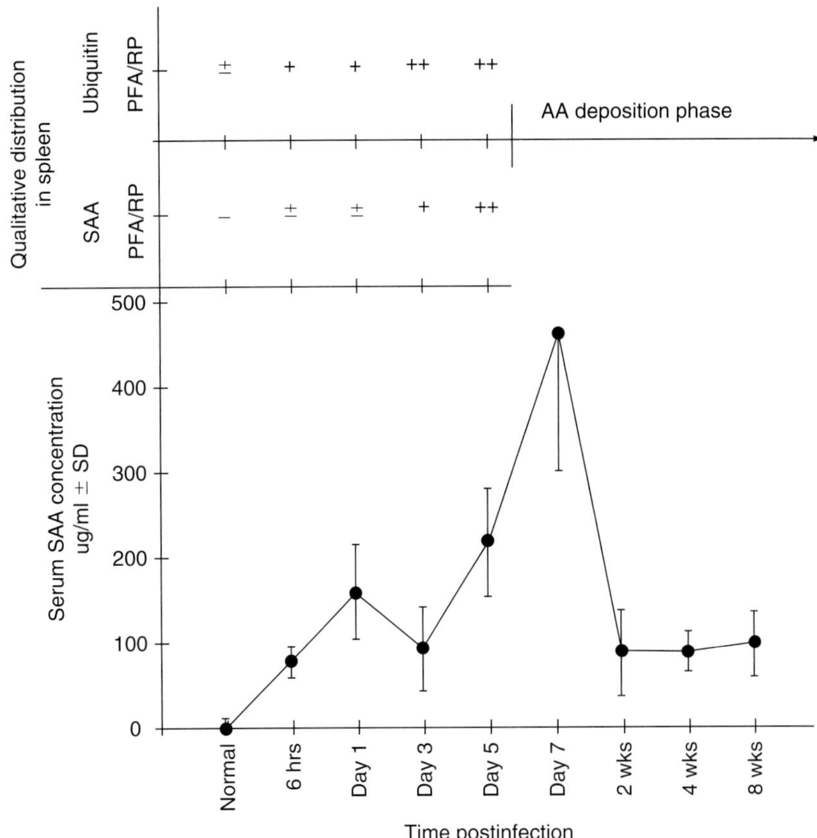

Figure 12.1. Determination of serum amyloid A protein (SAA) concentration by enzyme-linked immunosorbent assay (lower panel) and qualitative estimation of SAA and ubiquitin levels by strepavidin-biotin-peroxidase immunocytochemistry (upper panel) in splenic red pulp (RP) and perfollicular areas (PFA). The spleen sections were stained using rabbit anti-bovine ubiquitin IgG or rabbit anti-mouseAA amyloid IgG antibodies and graded from negative (-) or patchy (±) to maximum (++) staining. Serum and spleen sections were obtained from normal or alveolar hydatid cyst infected mice. Reproduced from Chronopoulos S. *et al.*, 1992. Journal of Pathology. 167: 249–259. © Pathological Society of Great Britain and Ireland. Permission is granted by John Wiley & Sons Ltd on behalf of PathSoc.

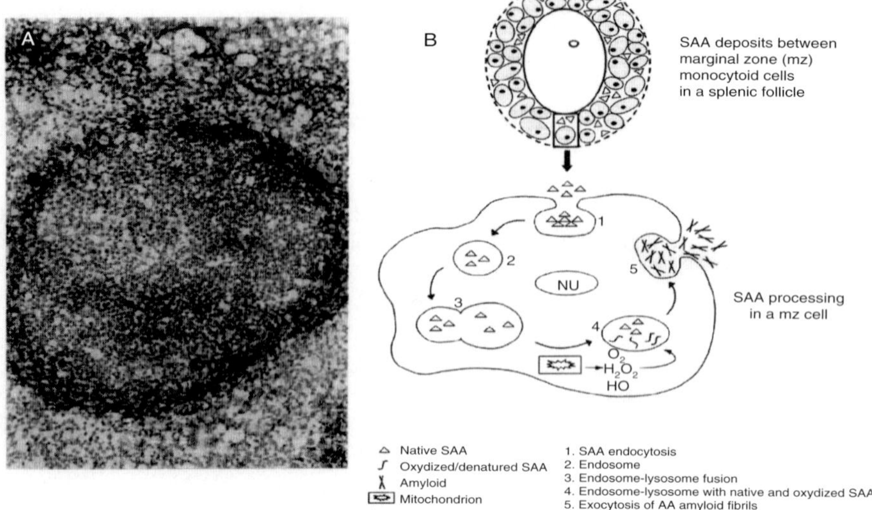

Figure 12.2. (A) Cryostat spleen section from a pre-amyloidotic mouse reacted with rabbit anti-mouse AA amyloid IgG-goat anti-rabbit Ig conjugated to peroxidase-labeled dextran polymer purchased from DAKO; the color reaction was developed with diaminonobenzidine. Note serum amyloid A (SAA) deposition in the splenic marginal zones (SMZ). (B) Schematic representation of the steps leading to intracellular nascent AA amyloid fibril formation by inflammatory SMZ monocytoid cells after SAA deposition in the interstitium. SAA is endocytosed, possibly through specific receptors, by the SMZ cells (1); fusion of gravid endosomes to lysosomes (2, 3); partial degradation and oxidation of SAA by mitochonria-generated reactive oxygen radicals (4); polymerization of partially degraded SAA into AA fibrils and its exocytosis into the interstitium (5). Reproduced from Ali-Khan Z. 2002. *Journal of Alzheimer's Disease* 4:105–114, with permission from IOS Press.

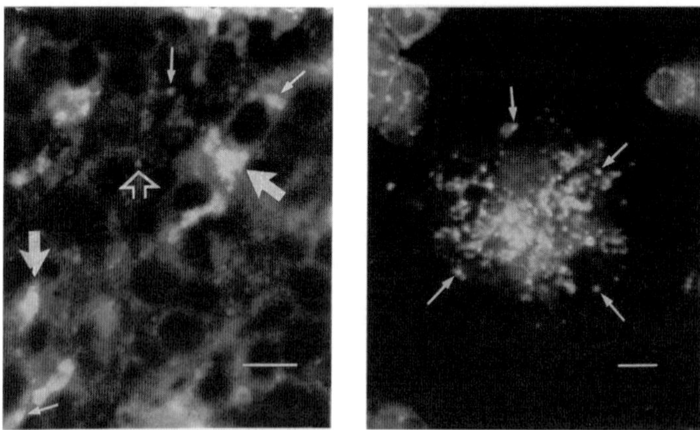

Figure 12.3. Confocal microscopic photomicrograph of serum amyloid A (SAA) and lysosome-associated membrane proteins (LAMP-1 and LAMP-2): a double-labeled spleen section (top), and a cytocentrifuged macrophage from amyloidotic mice. Fluorescein-labeled SAA and Texas red–labeled LAMP-1/LAMP-2 were colocalized, evident by the yellow color resulting from image overlays; small arrows indicate lysosomes and because of tight cell packing in the spleen section made the resolution of individual perikaryal (large arrows) lysosomes difficult. Bar = 10 μm. Reproduced from Chronopoulos S. *et al.*, 1994. *Journal of Pathology* 173: 361–369. © Pathological Society of Great Britain and Ireland. Permission is granted by John Wiley & Sons Ltd on behalf of PathSoc.

two distinct pools of A-SAA: increasing levels of HDL bound A-SAA in the serum (Benditt and Ericksen, 1977; Benditt *et al.*, 1979; Benson and Kleiner, 1980) and A-SAA localized to LY in the inflammatory monocytoid cells (Chronopoulos *et al.*, 1992, 1994, 1995). Results of plasma clearance studies in mice, using ^{35}S-labeled and HDL-associated murine rSAA1 and rSAA2, revealed that when these apolipoproteins were bound to acute-phase HDL, compared with normal HDL, their clearance was significantly retarded (Kluve-Beckerman *et al.*, 1997). This clearance pattern, particularly during inflammation, was proposed to be conducive to AA amyloidogenesis. Indeed, Kupffer cells isolated from C57B/L6 mice, chronically stimulated with casein, compared with the unstimulated controls, demonstrated defective A-SAA degradation and generation of soluble AA amyloid-size A-SAA derivatives in the culture (Fuks and Zucker-Franklin, 1985). We examined the metabolism of A-SAA in the AHC-infected C57B/L6 mouse model. The peak serum A-SAA level, approximately 480 μg/mL, was detected at 7 days postinfection, and after which its concentration decreased threefold and corresponded with systemic AA fibril deposition (Chronopoulos *et al.*, 1992) (Fig. 12.1). Intracellular A-SAA was detected starting 6 h postinfection in the peritoneal leukocytes, including macrophages. At 5 days, prior to splenic AA fibril deposition, A-SAA was localized interstitially to the splenic marginal zones (SMZ) and as punctate deposits to LY in the SMZ cells and peritoneal MΦ (Fig. 12.2A) and (Fig. 12.3; **see color insert**) (Chronopoulos *et al.*, 1992, 1994). Interestingly, both the temporal and spatial deposition of A-SAA in both the MΦ and the SMZ sites paralleled with ubiquitin expression (Fig. 12.1). To evaluate the fate of MΦ-associated A-SAA, the cells were lysed and A-SAA was affinity purified and characterized by immunoblotting and N-terminal sequencing. We found 5.5- to 12-kDa A-SAA derivatives, similar in mass to those of AA protein purified from amyloidotic tissues, indicating that these derivatives could serve as precursors of intracellular and/or the tissue-deposited AA fibrils (Bell *et al.*, 1999). In sum, evidence shows that inflammatory murine monocytoid cells internalize A-SAA/rSAA, with or without HDL, but unlike the unstimulated controls the sequestered A-SAA is only partially degraded. Questions pertaining to the progression of A-SAA from the SMZ interstitium to LY in the SMZ cells and whether A-SAA endocytosis is a receptor-mediated event are discussed below.

12.6. Do Murine Monocytoid Cells Possess Serum Amyloid A Receptor?

Internalization of A-SAA by the inflammatory monocytoid cells is a seminal event involved in nascent intralysosomal AA fibril formation (Shirahama and Cohen 1975). At least three sequential steps appear to be involved in this process (Chronopoulos *et al.*, 1992, 1994): (a) deposition of circulating A-SAA in the SMZ (Fig. 12.2A), (b) binding of A-SAA to the plasma membrane of SMZ cells, and (c) trafficking of A-SAA to the LY (Fig. 12.3, right). A series of *in vitro* studies supports these basic findings (Kluve-Beckerman *et al.*, 1999, 2001, 2002). Rocken and Kisilevsky (1997), using colloidal gold-tagged HDL-SAA, localized the ligand in coated pits/vesicles and endo-lysosomes of murine macrophages and proposed that both the binding and uptake of A-SAA may be mediated by a specific A-SAA receptor. The putative A-SAA receptor, however, remains to be identified and biochemically characterized.

In the past 4 years, three chemically dissimilar A-SAA-binding proteins have been described. They include receptors for advanced glycation end products on macrophage-like BV-2 cells (Yan *et al.*, 2000), two 72- and 100-kDa membrane proteins on macrophage J774 cells (Artl *et al.*, 2000), and Tanis, a 189-amino-acid protein on the plasma membranes of hepatocytes, myocytes, and adepocytes (Walder *et al.*, 2002; Gao *et al.*, 2003; Karlsson *et al.*, 2004). BV-2 cells interacted with both murine AA fibrils and A-SAA1, but not with non-amyloidogenic

A-SAA2, whereas, using ligand blotting, the two J774 cell plasma membrane proteins were shown to interact with both the normal and acute-phase HDL. Surface plasmon resonance analysis was used to establish interaction between human SAA1β and Tanis protein; the latter has been identified as a diabetes-associated gene product in *Psamommys obesus* (Israeli sand rat), a polygenic animal model of type 2 diabetes. So far, none of these putative receptors have been functionally connected with the uptake of A-SAA. Although a body of evidence supports the contention that A-SAA is internalized by and processed within monocytoid cells, it is therefore important to understand the biology and the chemical nature of this receptor.

12.7. Metabolism of Serum Amyloid A: Role of Lysosomal Cathepsins

Three proteolytic pathways have been proposed in the metabolism of A-SAA: (a) an endosomal-lysosomal pathway (Yamada *et al.*, 1995a, b, 1996), (b) protease(s) bound to monocytoid cell plasma membranes (Lavie *et al.*, 1978), and (c) proteases secreted by MΦ (Skogen *et al.*, 1980). Existing evidence favors the first possibility. As mentioned, Glenner *et al.* (1971) generated immunoglobulin light chain–derived AL amyloid fibrils *in vitro* and implicated both lysosomal enzymes and its acidic environment in amyloidogenesis. Subsequently, AA fibrils were localized to LY in murine monocytoid cells and LY were proposed to be the *in vivo* sites for nascent AA fibril formation (Shirahama and Cohen, 1975). We confirmed these findings and went on to localize both A-SAA and AA fibrils to the LY (Chronopoulos *et al.*, 1992, 1994).

Furthermore, recent findings, based on both *in vitro* and *in vivo* studies, have directly implicated lysosomal cathepsins (Cat), the major mammalian monocytoid cell proteases, in the metabolism of mouse A-SAA. Cat B partially degraded mouse rSAA *in vitro* and generated soluble 8-kDa fragments that were similar in mass and structure to the tissue derived AA fibril protein (Yamada *et al.*, 1995a). Cat D, on the other hand, generated non-amyloidogenic rSAA derivative (Yamada *et al.*, 1995b). Significantly, administration of pepstatin (an inhibitor of Cat D) to mice, exposed to repeated injections with an amyloidogen (Yamada *et al.*, 1996), or addition of pepstatin to murine MΦ-rSAA *in vitro* (Kluve-Beckerman *et al.*, 1999), hastened AA fibril deposition/formation in each of these experimental models. It was concluded that pepstatin-mediated blockage of lysosomal Cat D may have enhanced the amyloidogenic activities of Cat B. These data, nonetheless, bring into focus two important points. First, involvement of LY cathepsins in A-SAA metabolism, and second, elevation in the concentration of Cat B over Cat D may promote generation of amyloidogenic A-SAA derivatives. How such a switch might occur, possibly a prelude to AA amyloidogenesis *in vivo*, remains to be elucidated?

Possible connections between aberrant LY functions and disorders in A-SAA degradation by inflammatory monocytoid cells have been suggested for many years. Evidence shows that resident monocytoid cells degrade endocytosed A-SAA completely (Gonnerman *et al.*, 1996), whereas inflammatory monocytoid cells show faulty A-SAA clearance, generating 8-kDa A-SAA fragments, similar in molecular mass to AA fibril protein (Fuks and Zucker-Franklin, 1985; Bell *et al.*, 1999). Furthermore, normal $CD14^+$ monocytes were shown to degrade A-SAA completely, but pretreatment of these cells with IL-1β or IFN-γ led to the generation of A-SAA fragments (Migita *et al.*, 2001). Lah *et al.* (1995) also reported alteration in both the concentration and proteolytic activity of LY cathepsins in monocytoid cells exposed to IFN-γ, indicating a proinflammatory cytokine–mediated effect. Interestingly, Crabb *et al.* (2002) were able to show that covalent interaction between Cat B and 4-hydroxy-2-nonenal (HNE), an OX-St-derived diffusible aldehyde, produced intracellularly by the oxidative cleavage of arachidonic acid and linoleic acid, altered the proteolytic activity of Cat B. In sum, intracellular proteins, including LY proteases, appear to be vulnerable to OX-St-generated factors.

12.8. Proinflammatory Functions of Serum Amyloid A

Normal HDL has been shown to manifest antioxidant properties whereas the acute-phase HDL (i.e., HDL-SAA) proinflammatory activities (Van Lenten *et al.*, 1995; Furlaneto *et al.*, 2000; Yavin *et al.*, 2000). As such, these two apolipoproteins particles appear to be functionally distinct (Cabana *et al.*, 2003). Normal HDL was shown to block the cytotoxic activity of oxidized low-density lipoprotein (LDL) against endothelial and smooth muscle cells *in vitro* (Hessler *et al.*, 1979; Parthasarathy *et al.*, 1990). This activity appeared to be conferred by paraoxonase, one of the two biologically potent antioxidant enzymes, found linked to apolipoprotein (apo) A-1, the major protein component of normal HDL (Mackness *et al.*,1991; Cabana *et al.*, 2003). Conversely, the proinflammatory activity of acute-phase HDL was attributed to A-SAA (Van Lenten *et al.*, 1995; Furlaneto *et al.*, 2000; Yavin *et al.*, 2000; O'Hara *et al.*, 2004). Evidence shows that A-SAA, following its secretion by hepatocytes during the APR, binds primarily to HDL by displacing apo A-1 where A-SAA comprises up to 87% of the total apolipoprotein (Marhaug *et al.*, 1982; Clifton *et al.*, 1985). These major events during the APR help explain the altered functions of HDL-SAA.

Based on *in vivo* and *in vitro* findings, several distinct proinflammatory and pro-oxidant functions have been associated with A-SAA, rSAA, and AA fibrils. Both A-SAA and rSAA were shown to possess chemotactic activities for neutrophils and monocytes, mediated by formyl peptide receptor-like 1 expressed by the target cells (Badolato *et al.*, 1994; Su *et al.*, 1999). Evidence shows that A-SAA, apparently unassociated with HDL, is generated locally (Urieli-Shoval *et al.*, 2000), and cleaved products of A-SAA have been detected in the circulation during inflammatory conditions (Kiernan *et al.*, 2003). These events raise the possibility that the products generated due to the interaction of A-SAA or its derivatives with certain host cells or inflammatory leukocytes could adversely affect the pathophysiology of the disease process. Indeed, interaction of A-SAA with fibroblast-like synovocytes in inflamed human synovial tissue was shown to cause production of cartilage-degrading matrix metalloproteinase and bone matrix degeneration in RA patients (O'Hara *et al.*, 2004). A-SAA was also shown to induce a dramatic increase in the release of TNF-α (2746–3112 pg/mL), IL-1β (158.7–426.3 pg/mL), and IL-8 (10,313–16,738 pg/mL) by neutrophils; particularly, the amount of TNF-α was found to be ~200-fold higher than that from LPS-stimulated neutrophils (Furlaneto and Campa, 2000). As to the other critical proinflammatory activities, interaction between murine A-SAA1 and/or AA fibrils with various leukocytes (monocytes, macrophage, and T lymphocytes) led to increased expression/release of TNF-α, macrophage colony-stimulating factor (M-CSF), IFN-γ, cyclooxygenase metabolites, and heme oxygenase-1 (HO-1), the latter a sensitive marker of oxidative stress (Malle *et al.*, 1997; Furlaneto *et al.*, 2000; Yan *et al.*, 2000; Yavin *et al.*, 2000). Conceivably, both A-SAA and its blood-borne fragments, following their tissue deposition, could act as autocatalytic factors triggering locally and/or systemically a sequence of inflammatory events. These may include release of proinflammatory cytokines by activated phagocytes, local or hepatocyte-mediated synthesis of A-SAA, and induction of OX-St response. Significantly, septic patients or patients undergoing AA amyloidosis showed increased serum concentration of TNF-α, IL-6, and M-CSF with a corresponding increase in A-SAA concentration (Fratelli *et al.*, 1997; Rysava *et al.*, 1999). Potentially, as discussed below, both A-SAA and proinflammatory cytokines, acting singly or in concert, could induce generation of OX-St-related factors and affect intracellular aggregation of proteins including monocytoid cell–associated A-SAA. It is quite likely that these events could occur in the SMZ and, thus, provide a link between abnormal A-SAA clearance and AA fibril formation (Figs. 12.2A and 12.2B).

12.9. Possible Links Among Oxidative Stress, Defective Intralysosomal Serum Amyloid A Metabolism, and Nascent AA Fibril Formation

Proinflammatory cytokine–mediated activation of NADPH oxidase, a plasma membrane enzyme, is known to generate various reactive oxygen/nitrogen radicals (H_2O_2, O_2^-, NO, HO), and OX-St response in phagocytes (MΦ, RE cells, and granulocytes) (Sato *et al.*, 1998, Barbior, 2000). These reactants have been implicated in the (a) modification of cellular lipoproteins and proteins (Jessup *et al.*, 1992; Mander *et al.*, 1994; Dean *et al.*, 1997), (b) damage to lysosome membrane (Kalra *et al.*, 1988; Zdolsek *et al.*, 1993), and (c) cross-linking/unfolding of cellular proteins, resulting into secondary and tertiary structural modifications (Dean *et al.*, 1997).

Although, the precise mechanism remains to be fully understood, OX-St-generated factors appear to contribute to the pathogenesis of Alzheimer's disease (AD) (Smith *et al.*, 1998; Sayer *et al.*, 2001), including Aβ fibril (Dyrks *et al.*, 1992), tau filament (Schweers *et al.*, 1995), and α-synuclein–derived amyloid fibril formation (Paik *et al.*, 2003), the cardinal markers associated with AD and Parkinson's neuropathology, respectively. Notably, interaction of oxidized form of glutathione was required for the conversion of the full-length α-synuclein into amyloid fibrils. In addition, the combined data of Ichiropoulos (2003) and Krishnan *et al.* (2003) suggested that dimerization of α-synuclein, through oxidative/nitrative modifications, is a prerequisite for the relatively rapid fibrillogenetic event. Accumulating data further suggest that highly oxidized proteins, through the formation of protein cross-links, become remarkably resistant to proteolysis: Aβ fibrils, neurofibrillary tangles, α-synuclein–derived amyloid fibrils, and oxidized apo B, found in atherosclerotic foam cells (Merten and Holvoet, 2001), fall in this category. For example, Rapala-Kozik *et al.* (1998) reported that substitution of the methionine-596 residue with methionine sulfoxide dramatically impeded protease-mediated cleavage of a model peptide matching the 587–606 sequence fragment of Alzheimer precursor protein. Similarly, oxidized apo B was shown to accumulate in LY in the monocytoid cells without undergoing proteolysis, indicating dysfunction of lysosomal proteases (Jessup *et al.*, 1992; Mander *et al.*, 1994).

The events described above are consistent with the findings that HNE and carboxymethyllysine (CML) adducts (Sayer *et al.*, 2001; and other references therein) and pyronitrite anion (Ischiropoulos, 2003), the three potent OX-St-generated oxidants, each through their protein-modifying properties play critical roles in protein aggregation and/or amyloidogenesis. Some of these oxidants have been localized to chemically distinct amyloid fibrils that include familial amyloidotic polyneuropathy, light-chain amyloidosis (AL) and AA fibrils (Ando *et al.*, 1997; Miyata *et al.*, 1997; Fugimoto *et al.*, 2002; Rocken *et al.*, 2003), AD plaques (Pappola *et al.*, 1998; Smith *et al.*, 1998; Sayer *et al.*, 2001), and α-synuclein fibrils (Ischiropoulos and Beckman, 2003). Thus, OX-St response appears to be a common denominator linking various clinical and chemical forms of amyloidoses.

To our knowledge, Meydani *et al.* (1981) were the first to implicate OX-St response in the induction of AA amyloidosis in the casein-mouse model. We extended their findings and showed that mice undergoing AA amyloidosis manifest systemic HO-1 response, and the tissue-deposited AA fibrils become targets of HNE and CML (Kamalvand *et al.*, 2003; Kamalvand and Ali-Khan, 2004), including 3-nitrotyrosine (unpublished). We were also able to immunolocalize both HNE and CML to the A-SAA-positive SMZ cells as well as interstitially in the SMZ. More importantly, the A-SAA-loaded peritoneal MΦs from the AHC-mice demonstrated relatively large HNE- and CML-positive perikaryal vesicles (Kamalvand and Ali-Khan, 2004). Thus, the possibility of HNE-mediated structural modification in Cat B and the endocytosed A-SAA in the inflammatory monocytoid cells cannot be ruled out. Indeed, using the Oxiblot kit, we found oxidized A-SAA in the spleen extract from mice undergoing AA amyloidosis (unpublished). More recently, we found relatively high levels of humoral IFN-γ in mice exposed to the azocesein regimen for the induction of AA amyloidosis; the peak IFN-γ response appeared to correspond with a significant decrease in the concentration of cathepsin B in favor of Cat D in the monocytoid cells (Phipps-Yonas *et al.*, 2004). Whether this shift is an IFN-γ-mediated event, as shown by Lah *et al.* (1995), is unclear at this time. However, given the fact that SMZ, the major sites for A-SAA clearance in mice, contain relatively high concentration of A-SAA, which is targeted for clearance by the inflammatory SMZ monocytoid cells, it is reasonable to consider that OX-St-generated factors could compromise the A-SAA clearance. Thus, prolonged persistence of A-SAA fragments and their exposure to both low pH conditions and OX-St-generated factors could tip the balance in favor of intracellular AA fibril formation. A hypothetical concept showing convergence of multiple OX-St-related factors affecting AA fibril formation in the SMZ is shown in Figure 12.2B.

12.10. Conclusion

It is plausible that during chronic inflammation, an incessant unidirectional flow of A-SAA into inflammatory monocytoid cells may overburden the LY causing faulty degradation of A-SAA. Concurrently, OX-St-related factors generated in A-SAA-loaded monocytoid cells, through the interaction with proinflammatory cytokines and/or A-SAA or its fragments, could oxidatively modify the amyloidogeneic A-SAA fragments, render them resistant to LY proteases, and under low pH conditions induce their transformation into intralysosomal AA fibrils. Such fibrils after their release either by exocytosis or toxic-cell death, caused by H_2O_2 generated through the AA fibrils (Behl *et al.*, 1994; Zhang *et al.*, 1997), could act as "seeds" for the nucleation-dependent expansion of interstitial AA fibril deposition in the SMZ (Figs. 12.2A and 12.2B).

Acknowledgments

This work was supported by a research grant (MOP-42526) from the Canadian Institutes of Health Research. I thank Dr. M. Baines for his help in the preparation of the illustrations and Savannah DeVarney for excellent editorial services.

References

Ali-Khan, Z., Jothy. S., and Alkarmi, T. O. (1983). Murine alveolar hydatidosis: A potential experimental model for the study of AA amyloidosis. *Br. J. Exp. Pathol.* 64: 599–611.

Ali-Khan, Z., and Rauch, R. L. (1987). Demonstration of amyloid and immune complex deposits in renal and hepatic parenchyma of Alaskan alveolar hydatid disease patients. *Ann. Trop. Med. Parasitol.* 81: 381–392.

Ali-Khan, Z., Sipe, J. D., Du, T., and Rimal, H. (1988). Echinococcus multilocularis: Relationship between persistent inflammation, serum amyloid A protein and amyloidosis in four mouse strains. *Exp. Parasitol.* 67: 334–345.

Ali-Khan, Z., Li, W., and Chan, S. L. (1996). Animal model for the pathogenesis of reactive amyloidosis. *Parasitol Today.* 12: 297–302.

Ando, Y., Nyhlin, N., Suhr, O., Holmgren, G., Uchida, K., El Sahly, M., Yamashita, T., Terasaki, H., Nakamura, M., Uchino, M., and Ando, M. (1997). Oxidative stress is found in amyloid deposits in systemic amyloidosis. *Biochem. Biophys. Res. Commun.* 232: 497–502.

Artl, A., Marsche, G., Lestavel, S., Sattler, W., and Malle, E. (2000). Role of serum amyloid A during metabolism of acute-phase HDL by macrophages. *Arterioscler. Thromb. Vasc. Biol.* 20: 763–772.

Baba, S., Masago, T., Takahashi, T., Kasama, T., Sugimura, H., Tsugane, S., Tsutsui, Y., and Shirasawa, H. (1995). A novel alleic variant of serum amyloid A, SAA1γ: genomic evidence, evolution, frequency, and implication as a risk factor for reactive systemic AA amyloidosis. *Hum. Mol. Genet.* 4: 1083–1087.

Babior, B. (2000). Phagocytes and oxidative stress. *Am. J. Med.* 109: 33–44.

Badolato, R., Wang, J. K., Murphy J. M., Murphy, W. J., Lloyd, A. R., Michiel, D. F., Bausserman, L. L., Kelvin, D. J., and Oppenheim, J. J. (1994). Serum amyloid A is a chemoattractant: induction of migration, adhesion, and tissue infiltration of monocytes and polymorphonuclear leukocytes. *J. Exp. Med.* 180: 203–209.

Behl, C., Davis, J. B., Lesley, R., and Schubert, D. (1994). Hydrogen peroxide mediates amyloid B protein toxicity. *Cell.* 77: 817–827.

Bell, A., Chan, S. L., Marcantonio, D., and Ali-Khan, Z. (1996). Both murine SAA1 and SAA2 yield AA amyloid in alveolar hydatid cyst-infected mice. *Scand. J. Immunol.* 43: 173–180.

Bell, A., Chan, S. L., and Ali-Khan, Z. (1999). N-terminal sequence analysis of SAA-derivatives purified from murine inflammatory macrophages. *Amyloid Int. J. Exp. Clin. Invest.* 6: 31–36.

Benditt, E. P., and Ericksen N. (1977). Amyloid protein SAA is associated with high density lipoprotein from human serum. *Proc. Natl. Acad. Sci. USA.* 74: 4025–4028.

Benditt, E. P., Eriksen, N., and Hanson, R. H. (1979). Amyloid protein SAA is an apoprotein of mouse plasma high density lipoprotein. *Proc. Natl. Acad. Sci. USA.* 66: 4092–4096.

Benson, M. D., and Kleiner, L. (1980). Synthesis and secretion of serum amyloid protein A (SAA) by hepatocytes in mice treated with casein. *J. Immunol.* 124: 495–499.

Buxbaum, J. N., and Tagoe, C. E. (2000). The genetics of the amyloidoses. *Ann. Rev. Med.* 51: 543–569.

Cabana, V. G., Reardon, C. A., Feng, N., Neath, S., Lukens, J., and Getz, G. S. (2003). Serum paraoxonase: effect of the apolipoprotein composition of HDL and the acute phase response. *J. Lipid Res.* 44: 780–792.

Chan, S. L., Chronopoulos, S., Murray, J., Laird, D. W., and Ali-Khan., Z. (1997). Selective localization of murine ApoSAA1/SAA2 in endosomes-lysosomes of activated macrophages and their degradation products. *Amyloid Int. J. Exp. Clin. Invest.* 4: 40–48.

Chiti, F., Webster, P., Taddei, N., Clark, A., Stefani, M., Ramponi, G., and Dobson, C. M. (1999). Designing conditions for *in vitro* formation of amyloid protofilaments and fibrils. *Proc. Natl. Acad. Sci. U. S.A.* 96: 3590–3594.

Chronopoulos, S., Lembo, P., Alizadeh-Khiavi, K., and Ali-Khan, Z. (1992). Ubiquitin: Its potential significance in murine AA amyloidogenesis. *J. Pathol.* 167: 249–259.

Chronopoulos, S., Laird, D. W., and Ali-Khan, Z. (1994). Immunolocalization of serum amyloid A and AA amyloid in lysosomes in murine monocytoid cells: confocal and immunogold electron microscopic studies. *J. Pathol.* 173: 361–369.

Chronopoulos, S., Chan, S., Ratcliffe, M. J. H., and Ali-Khan. Z. (1995). Colocalization of ubiquitin and serum amyloid A and ubiquitin-bound AA in the endosomes-lysosomesa double immunogold electron microscopic study. *Amyloid Int. J. Exp. Clin. Invest.* 2: 191–194.

Clifton, P. M., Mackinnon, A, M., and Barter, P. J. (1985). Effects of serum amyloid A protein (SAA) on composition, size, and density of high density lipoprotein in subjects with myocardial infarction. *J. Lipid Res.* 26: 1389–1398.

Crabb, J. W., O'Neil, J., Miyagi, M., West, K., and Hoff, H. F. (2002). Hydroxynonenal inactivates cathepsin B by forming Michel adducts with active site residues. *Protein Sci.* 11: 831–840.

Dean, R., Fu, S., Stocker, R., and Davies, M. (1997). Biochemistry and pathology of radical-mediated protein oxidation. *Biochem. J.* 324: 1–18.

Du, T., and Ali-Khan, Z. (1990). Pathogenesis of secondary amyloidosis in an alveolar hydatid cyst mouse model: Histopathology and immuno/enzyme-histochemical analysis of splenic marginal zone cells during amyloidogenesis. *J. Exp. Pathol.* 71: 313–335.

Dyrks, T., Dyrks, E., Hartmann, T., Masters, C., and Beyreuther, K. (1992). Amyloidogenicity of beta A4 and beta A4-bearing amyloid protein precursor fragments by metal-catalyzed oxidation. *J. Biol. Chem.* 267: 18210–18217.

Fraser, P. E., Nguyen, J. T., Surewicz W. K., and Kirschner, D. A. (1991). pH-dependent structural transitions of Alzheimer amyloid peptides. *Biophys. J.* 60: 1190–1201.

Fratelli, M., Zinetti, M., Fantuzzi, G., Spina, C., Napoletano, G., Donatiello, G., Ravagnan, R., Sipe, J. D., Casey, C. A., and Ghezzi, P. (1997). Time course of circulating acute phase proteins and cytokines in septic patients. *Amyloid Int. J. Exp. Clin. Invst.* 4: 33–39.

Friedlich, A. L., and Butcher, L. L. (1994). Involvement of free oxygen radicals in B- amyloidosis: An hypothesis. *Neurobiol Aging.* 15, 443–455.

Fujimoto, N., Yajima, M., Ohnishi, Y., Tagima, S., Ishibashi, A., Hata, Y., Enomoto, U., Konohana, I., Wachi, H., and Seyama, Y. (2002). Advanced glycation end product-modified B2-microglobulin is a component of amyloid fibrils of primary localized cutaneous nodular amyloidosis. *J. Invest. Dermatol.* 118: 479–484.

Fuks, A., and Zucker-Franklin, D. (1985). Impaired Kupffer cell function precedes development of secondary amyloidosis. *J. Exp. Med.* 161: 1013–1028.

Furlaneto, C. J., and Campa A. (2000). A novel function of serum amyloid A: A potent stimulus for the release of TNF alpha, IL-1b and IL-8 by human blood neutrophils. *Biochem. Biophys. Res. Commun.* 268: 405–408.

Gabay, C., and Kushner, I. (1999). Acute-phase proteins and other systemic responses to inflammation. *N. Engl. J. Med.* 340: 448–454.

Ganapathi, M. K., Rzewnicki, D., Samols, D., Jiang, S. L., and Kushner, I. (1991). Effect of combinations of cytokines and hormones on synthesis of serum amyloid A and C-reactive protein in Hep 3B. *J. Immunol.* 147: 1261–1265.

Gonnerman, W. A., Lim, M., Sipe, J. D., Hayes, K. C., and Cathcart, E. S. (1996). The acute phase response in Syrian hamsters elevates apolipoprotein serum amyloid A (apo SAA) and disrupts lipoprotein metabolism. *Amyloid Int. J. Exp. Clin. Invest.* 3: 261–269.

Gao, Y., Wilder, K., Sunderland, T., Kantham, L., Feng, H. C., Quick, M., Bishara, N., de Silva, A., Augert, G., Tenne-Brown, J., and Collier, G. R. (2003). Elevation in Tanis expression alters glucose metabolism and insulin sensitivity in HAIIE cells. *Diabetes* 52: 929–934.

Gillmore, J. D., Lovat, L. B., Percy, M. R., Pepys, M. B., and Hawkins, P. N. (2001). Amyloid load and clinical outcome in AA amyloidosis in relation to circulating concentration of serum amyloid A protein. *Lancet* 358: 24–29.

Glenner, G. G. (1980). Amyloid deposits and amyloidosis: The β- Fibrillogeneses. *N. Engl J. Med.* 302: 1283–1292, 1333–1343.

Glenner, G. G., Ein, D., Eanes, E. D., Balden, H. A., Terry, W., and Page, D. L. (1971). Creation of "amyloid" fibrils from Bence Jones proteins *in vitro*. *Science* 174: 712–714.

Guijarro, J. I., Sunde, M., Jones, J. A., Cambell, I. D., and Dobson, C. M. (1998). Amyloid fibril formation by an SH3 domain. *Proc. Natl. Acad. Sci. USA.* 95: 4224–4228.

Hessler, J. R., Robertson, A. L., and Chisolm, G. M. (1979). LDL- induced cytotoxicity and its inhibition by HDL in human vascular smooth muscle and endothelial cells in culture. *Arteriosclerosis* 32: 213–229.

Hoffman, J. S, and Benditt, E. P. (1982). Changes in high density content following endotoxin administration in the mouse. *J. Biol. Chem.* 257: 10510–10517.

Husby, G., Marhaug, G., Dowton, B., Sletten, K., and Sipe, J. D. (1994). Serum amyloid A (SAA): biochemistry, genetics and the pathogenesis of AA amyloidosis. *Amyloid Int. J. Exp. Clin. Invest.* 1: 119–137.

Ischiropoulos, H. (2003). Oxidative modifications of alpha-synuclein. *Ann. N.Y. Acad. Sci.* 991: 93–100.

Ischiropoulos, H., and Beckman, J. S. (2003). Oxidative stress and nitration in neurodegeneration: cause, effect, or association? *J. Clin. Invest.* 111: 163–169.

Jessup, W., Mander, E. L., and Dean, R. T. (1992). The intracellular storage and turnover of apolipoprotein B of oxidized LDL in macrophages. *Biochim. Biophys. Acta.* 1126: 167–177.

Kamalvand, G., Pinard, G., and Ali-Khan, Z. (2003). Heme-oxygenase-1 response, a marker of oxidative stress, in a mouse model of AA amyloidosis. *Amyloid J. Protein Folding Disord.* 10:151–159.

Kamalvand, G, Z., and Ali-Khan, Z. (2004). Immunolocalization of lipid peroxidation/advanced glycation end products in amyloid A amyloidosis. *Free. Radic. Biol. Med.* 36: 657–664.

Karlsson, H. K. R., Tchuda, H., Lake, S., Koistenen, H. A., and Krook, A. (2004). Relationship between serum amyloid A level and Tanis/SelSmRna expression in skeletal muscle and adipose tissue from healthy and Type 2 diabetic subjects. *Diabetes* 53: 1424–1428.

Kelly, J. W. (1996). Alternative conformations of amyloidogenic proteins govern their behaviour. *Curr. Opin. Struct. Biol.* 6: 11–17.

Kelly, J. W., and Lansbury, P. T. (1996). A chemical approach to elucidate the mechanism of transthyretin and β-protein amyloid fibril formation. *Amyloid Int. J. Exp. Clin. Invest.* 1: 186–205.

Kluve-Beckerman, B., Yamada, T., Hardwick, J., Liepnieks, J. J., and Benson, M. D. (1997). Differential plasma clearance of murine acute-phase serum amyloid A proteins SAA_1 and SAA_2. *Biochem. J.* 332: 663–669.

Kluve-Beckerman, B., Leipnieks, J. J., Wang, L., and Benson, M. D. (1999). A culture system for the study of amyloid pathogenesis: Amyloid formation by peritoneal macrophages cultured with recombinant serum amyloid A. *Am. J. Pathol.* 155: 123–133.

Kluve-Beckerman, B., Manaloor, J., and Liepnieks J. J. (2001). Binding, trafficking and accumulation of serum amyloid A in peritoneal macrophages. *Scand. J. Immunol.* 53: 393–400.

Kluve-Beckerman, B., Manaloor, J., and Liepnicks, J. J. (2002). A pulse-chase study tracking the conversion of macrophage-endocytosed serum amyloid A into extracellular amyloid. *Arthritis Rheum.* 46: 1905–1913.

Kalra, J., Lautner, D., Massey, K., and Prasad, R. (1988). Oxygen free radicals induced release of lysosomal enzymes *in vitro*. *Mol. Cell. Biochem.* 84: 233–238.

Kirschner, D. A., Elliot-Bryant, R., Szumowski, K. E., Gonnerman, W. A., Kindy, M. S., and Cathcart, E. S. (1998). *In vitro* amyloid fibril formation by synthetic peptides corresponding to the amino-terminus of apoSAA isoforms from amyloid-susceptible and amyloid resistant mice. *J. Struct. Biol.* 124: 88–98.

Kiernan, U. A., Tubbs, K. A., Nedelkov, D., Niederkofler, E. E., and Nelson, R. W. (2003). Detection of truncated forms of human serum amyloid A protein in human plasma. *FEBS Lett.* 537: 166–170.

Krishnan. S., Chi, E. Y., Wood, S. J., Kendrick, B. S., Li, C., Garzon-Rodriguez, W., Wypych, J., Randolph, T. W., Narhi, L, O., Biere, A. L., Citron, M., and Carpenter, J. F. (2003). Oxidative dimer formation is the critical rate-limiting step for Parkinson's disease alpha-synclein fibrillogenesis. *Biochemistry.* 42**:** 829–837.

Lah, T. T., Hawley, M., Rock, K. L., and Goldberg, A. I. (1995). Gamma-interferon causes a selective induction of the lysosomal proteases, cathepsin B and L, in macrophages. *FEBS Lett.* 537: 85–89.

Lavie, G., Zucker-Franklin, D., and Franklin, E C. (1978). Degradation of serum amyloidA protein by surface-associated enzymes of human blood monocytes. *J. Exp. Med.* 148**:** 1020–1031.

Lowell, C. A., Potter, D. A, Stearman, R. S., and Marrow, J. F. (1986). Structure of the murine serum amyloid A gene family: gene conversion. *J. Biol. Chem.* 261: 8442–8452.

Liang, J. S., Sloane, J. A., Wells, J. M., Abraham, C. R., Fine, R. E., and Sipe, J. D. (1997). Evidence for local production of acute phase response apolipoptotein serum amyloid A in Alzheimer's disease brain. *Neurosci. Lett.* 225: 73–76.

Liepnieks, J. J., Kluve-Beckerman, B., and Benson, M. D. (1995). Characterization of amyloid A protein in human secondary amyloidosis: the predominant deposition of serum amyloid A1. *Biochim. Biophys. Acta* 1270: 81–86.

Mackness, M. L., Arrol, S., Abbott, C. A., and Durrington, P. N. (1993). Is paraoxanase related to atherosclerosis. *Chem-Biol. Interact.* 87: 161–171.

Malle, E., Bollman, A., Steinmetz, A., Gemsa, D., Leis, H. J., and Satler, W. (1997). Serum amyloid A (SAA) protein enhances formation of cyclooxygenase metabolites of activated human monocytes. *FEBS Lett.* 419, 215–219.

Marhaug, G., Sletten, K., and Husby, G. (1982). Characterization of amyloid related protein SAA complexed with serum lipoprotein (apoSAA). *Clin. Exp. Immunol.* 50: 382–389.

Mander, E. L., Dean, R. T., Stanley, K. K., and Jessup, W. (1994). Apolipoprotein B of oxidized LDL accumulates in the lysosomes of macrophages. *Biochim. Biophys. Acta* 1212: 80–92.

Maury, C. P., Tiitinen, S., Laiho, K., Kaarela, K., and Liljestrom, M. (2000). Raised circulating IL-18 levels in reactive AA amyloidosis. *Amyloid Int. J. Exp. Clin. Invest.* 9: 141–144.

Meek, R. L., Hoffman, J. S., and Benditt, E. P. (1986). Amyloidogenesis: One serum amyloid A isotype is selectively removed from the circulation. *J. Exp. Med.* 163: 499–510.

Meek, R. L., Uriiel-Shoval, S., and Benditt, E. P. (1994). Expression of apolipoprotein serum amyloid A mRNA in human atherosclerotic lesions and cultured vascular cells: Implications for serum amyloid A function. *Proc. Natl. Acad. Sci. USA.* 91: 3186–3190.

Merlini, G., and Westermark, P. (2004). The systemic amyloidoses: clear understanding of the molecular mechanisms offer hope for more effective therapies. *J. Int. Med.* 255: 159–178.

Mertens, A., and Holvoet, P. (2001). Oxidized LDL and HDL: Antagonists in atherothrombosis. *FASEB J.* 15: 2073–2084.

Meydani, S. N., Cathcart, E. S., and Hopkins R. E. (1981). Antioxidants in experimental amyloidosis of young and old mice. In: Glenner, G. G., Osserman, E. F., Benditt, E. P., Calkins, E., Cohen, A. S., and Zucker-Franklin, D. (eds), *Amyloidosis.* Plenum Press, New York, pp. 683–690.

Migita, K., Yamasaki, S., Shibatomi, K., Ida, H., Kita, M., Kawakami, A., and Eguchi, K. (2001). Impaired degradation of serum amyloid A (SAA) protein by cytokine-stimulated monocytes. *Clin. Exp. Immunol.* 123: 408–411.

Miyata, T., Wada, Y., Cai, Z., Iida, Y., Horie, K., Yasuda, Y., Maeda, K., Kurokawa, B., and van Ypersele de Strihou, C. (1997). Implication of an increased oxidative stress in the formation of advanced glycation end products in patients with end-stage renal failure. *Kidney Int.* 51: 1170–118.

O'Hara, R., Murphy, E. P., Whitehead, A. S., FitzGerald, O., and Bresnihan, B. (2004). Local expression of serum amyloid A and formyl peptide receptor-like 1 genes in synovial tissue is associated with matrix metalloproteinase production in patients with inflammatory arthritis. *Arthritis Rheum.* 50: 1788–1799.

Okada, K., Wangpoengrakul, C., Osawa, T., Toyokunit, S., Tanaka, K., and Uchida, K. (1999). 4-Hydroxy-2-nonenal-mediated impairment of intracellular proteolysis during oxidative stress. *J. Biol. Chem.* 274: 23787–23793.

Paik, S. R., Lee, D., Cho, H-J., Lee, E-N., and Chang, C-S. (2003). Oxidized glutathione stimulated the amyloid formation of α-synuclein. *FEBS Lett.* 537: 63–67.

Pan, K., Baldwin, M., Nguyen, J. T., Gasset, M., Serban, A., Groth, D., Mehlhorn, I., Huang, Z., Fletterick, R. J., and Prusiner, S. B. (1993). Conversion of alpha-helices into beta sheets features in the formation of scrapie prion proteins. *Proc. Natl. Acad. Sci. USA.* 90, 10962–10966.

Pappola, M. A., Chayan, Y-J., Omar, R. A., Hsiao, K., Perry, G., Smith, M. A., and Bozner, P. (1998). Evidence of oxidative stress and *in vivo* neurotoxicity of B-amyloid in a transgenic mouse model of Alzheimer's disease. *Am. J. Pathol.* 152: 871–877.

Parthasarathy, S., Barnett, J., and Fong, L. G. (1990). High-density lipoprotein inhibits the oxidative modification of low-density lipoprotein. *Biochim. Biophys. Acta* 1044: 275–283.

Phipps-Yonas, H., Pinnard, G., and Ali-Khan, Z. (2004). Humoral proinflammatory cytokine and SAA generation profiles and spatio-temporal relationship between SAA and lysosomal cathepsin B and D in murine splenic monocytoid cells during AA amyloidosis. *Scand. J. Immunol.* 59:168–176.

Rapala-Kozik, M., Kozik, A., and Traves, J. (1998). Effect of oxidation of B-amyloid precursor protein on its B-secretase cleavage. A model study with synthetic peptides and candidate B-secretase. *J. Peptide Res.* 52: 315–320.

Ray, A., and Ray, B. K. (1999). Persistent expression of serum amyloid A during experimentally induced chronic inflammatory condition in rabbit involves differential activation of SAF, NF(kappa)B, and C/EBP transcription factors. *J. Immunol.* 163: 2143–2150.

Rocken, C., and Kisilevsky, R. (1997). Binding and endocytosis of murine high density lipoprotein from healthy (HDL_{SAA}) by murine macrophages *in vitro*. A light and electronmicroscopic investigation. *Amyloid: Int. J. Exp. Clin. Invest.* 4: 259–273.

Rocken, C., Kientsch-Engel, R., Mansfield, S., Stix, B., Stubenrauch, K., Weigle, B., Buling, F., Schwan, M., and Saeger, W. (2003). Advanced glycation end products and receptor for advanced glycation end products in AA amyloidosis. *Am. J. Pathol.* 162: 1213–1220.

Rysava, R., Merta, M., Tesar, V., Jirsa, M., and Zima, T. (1999). Can serum amyloid A or macrophage colony stimulating factor serve as marker of amyloid formation process? *Biochem. Mol. Biol. Intern.* 47: 845–850.

Sato, K., Akaki, T., and Tomioka, H. (1998). Differential potentiation of anti-mycobacterial activity and reactive nitrogen intermediate-producing ability of murine peritoneal macrophages activated interferon-gamma (IFN-gamma) and tumor necrosis factor-alpha (TNF-alpha). *Clin. Exp. Immunol.* 112: 63–68.

Sayer, L. M., Zagorski, M. G., Surewicz, W. K., Krafft, G. A., and Perry, G. (1997). Mechanism of neurotoxicity associated with amyloid B deposition and the role of free radicals in the pathogenesis of Alzheimer's disease: A critical appraisal. *Chem. Res. Toxicol.* 10**:** 518–526.

Sayer, L. M., Smith, M. A., and Perry, G. (2001). Chemistry and biochemistry of oxidative sress in neurodegenerative disease. *Curr. Med. Chem.* 8: 721–738.

Schweers, O., Mandelkow, E. M., Blenart, J., and Mandelkow, E. (1995). Oxidation of cysteine-322 in the repeat domain of microtubule-associated protein T controls the *in vitro* assembly of paired helical filaments. *Proc. Natl. Acad. Sci. USA.* 92: 8463–8467.

Shiina, M. M., Nishimoto, N., Yoshizaki, K., Kishimoto, T., and Akamatsu, K. (2004). Anti-interleukin 6 receptor antibody inhibits murine AA amyloidosis. *J. Rheumatol.* 31: 1132–1138.

Shirahama, T., and Cohen, A. S. (1975). Intralysosomal formation of amyloid fibrils. *Am. J. Pathol.* 81: 101–116.

Skogen, B., Thorsteinsson, L., and Natvig J. B. (1980). Degradation of protein SAA to an AA-like fragment by enzymes of monocytic origin. *Scand. J. Immunol.* 11: 533–540.

Smith, M. A., Hirai, K., Hsiao, K., Pappola, M. A., Haris, P. L., Siedlak, S. L., Tabaton, M., and Perry, G. (1998). Amyloid beta deposition in Alzheimer transgenic mice is associated with oxidative stress. *J. Neurochem.* 70: 2212–2215.

Steel, D. M., and Whitehead A. S. (1994). The major acute phase reactants: C-reactive protein, serum amyloid P component and serum amyloid A protein. *Immunol. Today.* 15: 81–88.

Stevens, F. J. (2004). Hypothetical structure of human serum amyloid A. *Amyloid J. Protein Folding Disord* 11: 71–84.

Su, S. B., Gong, W., Gao, J-L., Shen, W., Murphy, P. M., Oppenheim, J. J., and Wang, J. M. (1999). A seven-transmembrane, G protein-coupled receptor, FPRL1, mediates the chemotactic activity of serum amyloid A for human phagocytic cells. *J. Exp. Med.* 189: 395–402.

Tanaka, F., Migita, K., Kawabe, Y., Aoyagi, T., Ida, H., Kawakami, A., and Eguchi, K. (2004). Interleukin-18 induces serum amyloid A (SAA) protein production from rheumatoid synovial fibroblasts. *Life Sci.* 74: 1671–1679.

Thorn, C. F., Lu, Z.-Y., and Whitehead A. S. (2004). Regulation of the human acute phase serum amyloid A genes by tumour necrosis factor-alpha, interleukin-6 and glucocorticoids in hepatic and epithelial cell lines. *Scand. J. Immunol.* 59: 152–158.

Treves, S., and Ali-Khan, Z. 1984. Characterization of the inflammatory cells in progressing tumour-like hydatid cysts. 1. Kinetics and composition of inflammatory infiltrates. *Tropenmed. Parasit.* 35, 183–188.

Uhlar, C. M., Burgess, C. J., Sharp, P. M., and Whitehead, A. S. (1994). Evolution of the serum amyloid A (SAA) protein superfamily. *Genomics* 19: 228–234.

Uhlar, C. M., and Whitehead, A. S. (1999). Serum amyloid A, the major vertebrate acute-phase reactant. *Eur. J. Biochem.* 265: 501–523.

Uriel-Shoval, S., Reinhold, L., and Yaacov, M. (2000). Expression and function of serum amyloid A, a major acute-phase protein, in normal and disease states. *Curr. Opin. Hematol.* 71: 64–69.

Van Lenten, B. J., Hama, S. Y., De Beer, F. C., Stafforini, D. M., McIntyre, T. M., Prescott, S. M., La Du, B. N., Fogelman, A. M., and Navab, M. (1995). Anti-inflammatory HDL becomes pro-inflammatory during the acute phase response. *J. Clin. Invest.* 96: 2758–2767.

Walder, K., Kantham, L., Millan, J. S., Trevaskis, J., Kerr, L., de Silva, A., Sunderland, T., Godde, N., Gao, Y., Bishara, N., Windmill, K., Tenne-Brown., Augert, G., Zimmet, P. Z, and Collier, G. R. (2002). Tanis: A link between Type 2 diabetes and inflammation? *Diabetes* 51:1859–1866.

Westermark, G. T., and Engstrom, U. (1982). A serum AA-like protein as a common constituent of secondary amyloid fibrils. *Clin. Exp. Immunol.* 49: 725–731.

Westermark, G. T., Engstrom, U., and Westermark, P. (1992). The N-terminal segment of protein AA determines its fibrillogenic property. *Biochem. Biophys. Res. Commun.* 182: 27–33.

Yamada, T., Kluve-Beckerman, B., Liepnieks, J. J., and Benson, M. D. (1994). Fibril formation from recombinant human serum amyloid A. *Biochim. Biophys. Acta* 1226: 323–329.

Yamada, T., Kluve-Beckerman, B., Liepnieks, J. J., and Benson, M. D. (1995). *In vitro* degradation of serum amyloid A by cathepsin D and other acid proteases: Possible protection against amyloid fibril formation. *Scand. J. Immunol.* 41: 570–574.

Yamada, T., Liepnieks, J. J., Kluve-Beckerman, B., and Benson, M. D. (1995). Cathepsin B generates the most common form of amyloid A (76 residues) as a degradation product from serum amyloid A. *Scand. J. Immunol.* 41: 94–97.

Yamada, T., Liepnieks, J. J., Benson, M. D., and Kluve-Beckerman, B. (1996). Accelerated amyloid deposition in mice treated with the aspartic protease inhibitor, pepstatin. *J. Immunol.* 157: 901–907.

Yamada, T., Okuda, Y., Takasugi, K., Wang, L., Marks, D., Benson M. D., and Kluve-Beckerman, B. (2003). An allele of serum amyloid A1 associated with amyloidosis in both Japanese and Caucasians. *Amyloid J. Protein Folding Disord.* 10: 7–11.

Yamamoto, K., and Migita, S. (1985). Complete primary structures of two major murine serum amyloid A proteins deduced from cDNA sequences. *Proc. Natl. Acad. Sci. USA.* 82: 2915–2919.

Yan, S. D., Zhu, H., Zhu, A., Glabek, A., Du, H., Roher, A., Yu, J., Soto, C., Schmidt, A. M., Stern, D., and Kindy, M. (2000). Receptor-dependent cell stress and amyloid accumulation in systemic amyloidosis. *Nat. Med.* 6: 643–651.

Yavin, E. J., Preciado-Patt, L., Rosen, O., Yaron, M., Suessmuth, R. D., Levartowsky, D., Jung, G., Lider, O., and Fridkin, M. (2000). Serum amyloid A protein-derived peptide, present in human rheumatic synovial fluids, induce the secretion of interferon-gamma by human CD + 4 T lymphocytes. *FEBS Lett.* 472: 259–262.

Zdolsek, J., Zhang, H., Roberg, K., and Brunk, U. (1993). H_2O_2-mediated damage to lysosomal membranes of J-774 cells. *Free. Radic. Res. Commun.* 18: 71–85.

Zhang, L., Zhao, B., Yew, D. T., Kusiak, J. W., and Roth, G. S. (1997). Processing of Alzheimer's amyloid precursor protein during H_2O_2-induced apoptosis in human neuronal cells. *Biochem Biophys. Res. Commun.* 235: 845–848.

II

Point Mutations and Enhanced Protein Deposition

13

Transthyretin and the Transthyretin Amyloidoses

Joel N. Buxbaum

Abstract

Transthyretin is a normal serum protein that carries the secondary thyroid hormone thyroxine and retinol binding protein when it is loaded with retinol. It is synthesized primarily in the liver but there is also significant production in the choroid plexus and the retina. Both message and protein are found in the kidney but that site does not appear to contribute to the serum level in any meaningful way. The protein is a homotetramer composed of the 14-kDa monomer with the thyroxine binding sites in the central groove. The structure is highly conserved particularly in the regions responsible for ligand binding, suggesting that its carrier function has been retained over the millennia. More than 80 mutations at 55 different positions in the gene encoding the protein are the primary cause of a set of human disorders collectively known as the familial amyloidotic polyneuropathies and cardiomyopathies. In these diseases, the soluble protein becomes insoluble under physiologic conditions resulting in functional compromise in the organs in which the protein is deposited. Peripheral nerves, heart, kidneys, gastrointestinal tract, and the leptomeninges have all been described as sites of deposition. It is possible that particular syndromes are associated with particular sets of mutations but, because of the rarity of some of the mutations, it is uncertain if the relationship between any mutation and any clinical syndrome is absolute. Further, it is also clear that the wild-type protein can deposit in tissues with subsequent dysfunction, particularly in the heart and carpal tunnel. Biophysical studies *in vitro* indicate that the process leading from soluble tetramer to insoluble aggregates involves monomer release, misfolding, oligomerization, and extension and lateral aggregation apparently as a downhill polymerization to a more stable lower energy state. How this is modulated *in vivo* is not known, although accessory molecules appear to be involved in the process. The early, albeit incomplete, understanding of the processes have led to potential therapies directed at stabilizing the tetramer or interfering with the interaction with other molecules as well as replacing the offending gene by liver transplantation.

13.1. Introduction

Human transthyretin (TTR) was originally called thyroxine-binding prealbumin, on the basis of observations made in the 1960s that the protein migrated anodally to serum albumin in electrophoresis and was a carrier of the thyroid hormone thyroxine (T4). Other studies defined its isoelectric point (4.7), its concentration in serum (28–35 mg/dL), and its capacity to bind plasma retinol binding protein (RBP) charged with retinol (Robbins, 2002). Its tetrameric structure, that is, four 14-kDa identical subunits, was surmised in 1971 (Gonzalez and Offord, 1971). Its amino acid sequence was determined in 1974 and the first crystal structures published in 1977 with a more-refined structure in 1978 (Kanda *et al.*, 1974; Blake and Oatley, 1977; Blake *et al.*, 1978). In 1980, on the basis of these data and the recognition that murine α_1-antitrypsin migrated in the prealbumin region, the nomenclature committee of the International Union of Biochemists changed the name to transthyretin; that is, *trans*porter of *thy*roxine and *retin*ol binding protein (Robbins, 2002). While the attempt at a functional nomenclature was well meaning, it may have presumed a more complete knowledge of the protein than was actually the case. When the murine gene was silenced by targeted disruption, the effects on both thyroid function and retinol metabolism were far less than expected, suggesting perhaps that the evolutionarily important function(s) of the molecule lay elsewhere (Palha, 2002). Alternatively, these functions may be so critical to the survival of the species that redundant mechanisms ensure that they are maintained, even in the absence of TTR (Schreiber, 2002).

The protein has acquired significance in clinical medicine in two settings. The serum concentration has been utilized as a marker of nutritional status in a variety of conditions, particularly subacute malnutrition (Ingenbleek and Young, 2002; Potter and Luxton, 2002). Further, the wild-type protein and a large of number of mutants have been shown to aggregate and deposit in tissues causing the disorders senile systemic amyloidosis (SSA) (wild type), familial amyloidotic polyneuropathy (FAP), and familial amyloidotic cardiomyopathy (FAC) (mutants) (Connors *et al.*, 2003). These aggregation diseases seem to be independent of the physiologic function of the protein, because, as a rule, the carriers of the mutant alleles do not exhibit any identifiable loss of function. The explanation for the lack of a phenotype representing the haploinsufficient state is the autosomal dominant nature of the disease. Tissue dysfunction is related to the aggregation and deposition of mutant molecules, that is, a gain of a pathologic function; the soluble, circulating protein functions normally to carry both T4 and RBP, although binding of RBP is diminished in mutants at position 84, and T4 binding is enhanced by substitutions at position 6 and 109 (Fitch *et al.*, 1991; Rosen *et al.*, 1993). The clinical disorders will be discussed later in this chapter.

13.1.1. The Transthyretin Gene

The single-copy transthyretin gene (*TTR*) (Fig. 13.1) is relatively conserved across most species (Schreiber and Richardson, 1997). In humans, it is located on chromosome 18 (18q11.2–q12.1), in the mouse on chromosome 4 (Whitehead *et al.*, 1984; Qiu *et al.*, 1992). Both contain four exons included within 7 kb of DNA. The first exon encodes the leader sequence and the first three amino acids. The majority of the protein is encoded in exons 2, 3, and 4. In addition to the sequence encoding the functional protein, the gene contains two open reading frames, one beginning in the second intron and one in the third (Tsuzuki *et al.*, 1985). Both are transcribed but not independently of the primary transcript and do not encode proteins (Soares *et al.*, 2003). Their functions, if any are unknown.

Six single nucleotide polymorphisms (SNPs) are present within the exon-intron structure of the gene (Yoshioka *et al.*, 1989). Recently additional flanking SNPs and microsatellite markers have extended the haplotypes encompassing the *TTR* gene well beyond the coding region (Soares *et al.*, 2004). As yet the noncoding region SNPs have no known function but are useful genetic markers for identifying the particular chromosome on which mutations have arisen, thus allowing the identification of founders and in some cases the geographic gene flow of the amyloidogenic mutations (e.g.,.V30M) (Reilly *et al.*, 1995a).

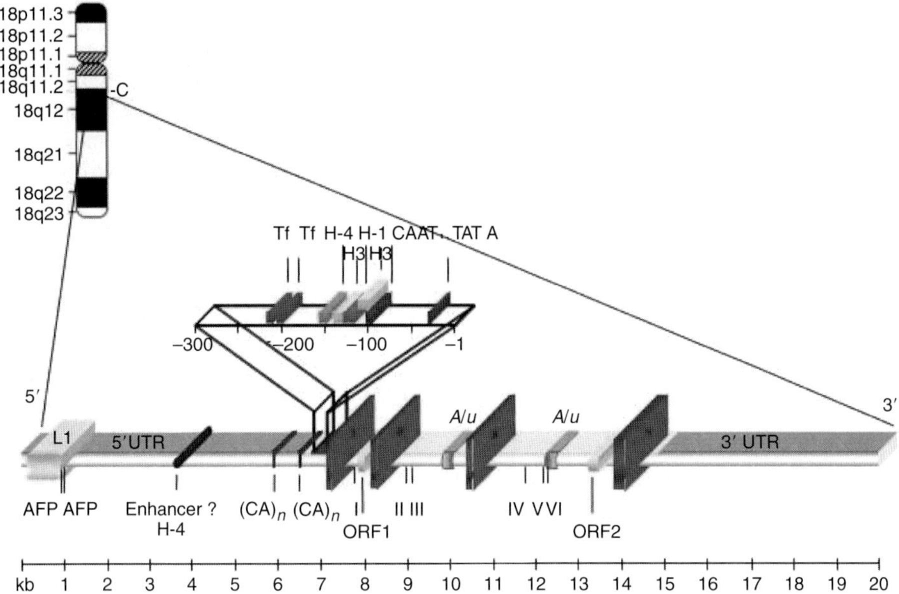

Figure 13.1. (Soares, 2002) Schematic representation of the genomic interval containing *TTR* and respective regulatory sequences. The interval corresponding with *TTR* cds (GenBank M11518) is depicted in light gray. TTR exons are numbered from 1 to 4; the positions of the intronic SNPs identified by Yoshioka *et al.* (1989) are numbered from I to VI. $(CA)_n$: CA-dinucleotide repeats. Alu and L1: respective sequences. ORF1 and ORF2: intronic open reading frames 1 and 2, respectively. TATA: TATA box. CAAT: CAAT box. H-1, H3, H-4, C/E: binding sites for hepatocyte nuclear factors HNF-1, HNF-3, and HNF-4, respectively. Tf: common motif to Tf-LF-1, Tf-LF-2, and LF-A1. Enhancer: a region highly momologous to a tissue-specific enhancer of the mouse TTR gene. AFP: binding site for AFP-1 factor. Adapted from Sakaki *et al.* (1989).

13.1.2. TTR Expression

The gene is highly expressed in the liver and the choroid plexus (Herbert *et al.*, 1986). The protein is also synthesized in the leptomeningeal and retinal pigment epithelia (Getz *et al.*, 1999). The concentration of TTR mRNA, when measured per gram of wet weight of tissue, is higher in choroid plexus epithelium than in liver, however this may be a function of the total number and amount of all mRNAs present in liver tissue rather than an absolute value, that is, there is more mRNA in hepatocytes and the TTR mRNA represents a small proportion of the total, while in choroid plexus TTR mRNA represents a larger fraction of the mRNAs because TTR is a major protein component of cerebrospinal fluid (CSF) (Dickson *et al.*, 1985).

Small amounts of TTR mRNA or protein have been identified in the pancreas and endocrine tumors of the gut, in embryonic sheep kidney, and in developing brain (Jacobsson, 1989; Jacobsson *et al.*, 1989). Several recent studies have demonstrated *TTR* transcription and protein production in various areas of adult murine brain including the cortex, hippocampus, and amygdyla, confirming earlier work in transgenic mice in which TTR mini-genes were expressed under the control of TTR promoter sequences and shown to be regulated differently in brain cells than in the choroid plexus (Yan *et al.*, 1990).

The regulatory elements required for hepatic expression of *TTR* have been well characterized. The promoter proximal sequences within 300 bp 5′of the initiation codon and a 100-bp enhancer approximately 2 kb upstream are sufficient for hepatocyte synthesis, however those necessary for choroid plexus expression are further upstream, within 3 kb 5′ of the translation start site, and have not been fully defined (Yan *et al.*, 1990). The regulatory sequences identified as functional in the hepatocyte also appear to be sufficient for transcription in brain cells outside the choroid plexus and leptomeninges. The *TTR* proximal promoter region contains binding sites for the transcription factors HNF 3 and 4 while the enhancer region has binding sites for C/EBP, HNF-1, and AP-1, all of which are present in hepatocytes. HNF 3 is composed of α, β, and γ polypeptides and is a member of the fork head transcription factor family. It is necessary but insufficient for *TTR* transcription (Costa *et al.*, 1990). In the liver TTR, like albumin, behaves as a negative acute phase reactant, that is, it is diminished in the course of inflammation. IL-6, perhaps in response to IL-1 and/or TNF, suppresses production of the HNF 3α that is required for *TTR* transcription. The suppression may be mediated by another transcription factor, HNF 6 (Samadani and Costa, 1996). Choroid plexus epithelial cells lack the transcription factors present in the liver so that, despite the presence of the same regulatory DNA sequences, *TTR* expression in the choroid is not influenced by systemic inflammation (Schreiber *et al.*, 1989). The factors that regulate *TTR* transcription in the choroid and leptomeninges have not been identified.

In the hepatocyte, the primary TTR transcript is spliced to yield the mature mRNA, which is translated on membrane-bound polyribosomes and trafficked through the cell for export. There is no evidence for alternatively spliced forms of the primary transcript. Apart from removal of the leader sequence, the protein does not undergo functionally significant post-translational modification (i.e., glycosylation, lipidation, or phosphorylation).

The TTR tetramer is assembled in the endoplasmic reticulum where it appears to bind RBP charged with retinol. Based on *in vitro* reconstitution studies, it has been suggested that different cells may have different capacities to assemble the tetramer (Melhus *et al.*, 1991; Bellovino *et al.*, 1996, 1998). Mice with the *RBP* gene disrupted have lower serum levels of TTR. RBP does not exit the hepatocyte if TTR is not present. In the absence of TTR, plasma RBP is rapidly cleared via the kidney. Mice with a nonfunctional *TTR* gene generated by targeted disruption have approximately 5% of the normal concentration of RBP in the serum. Thus, the serum concentration of RBP appears to be dependent on the TTR level and the availability of retinol. Nonetheless, the TTR knockout animals have no evidence of clinical retinol deficiency.

The role of thyroid hormones in hepatic TTR expression is less clear. Human hyperthyroidism has been associated with reduced levels of serum TTR, but the concentration appears to be normal in hypothyroid individuals (Ishida *et al.*, 1982). Results of studies on thyroidectomized rats from different laboratories are inconsistent, showing either normal or reduced hepatic TTR mRNA levels compared with controls. Administration of tri-iodothyronine (T3) restored the mRNA concentrations where it was reduced and had no effect when it was found to be equal to the control levels. Administration of growth hormone reduced serum TTR concentration and hepatic mRNA levels. Paradoxically, in those experiments hypophysectomy, with removal of the pituitary source of growth hormone, also reduced hepatic TTR mRNA concentrations, which were not restored by either growth hormone or T3 administration. The most comprehensive study showed no effect of thyroid status on TTR mRNA levels in hepatocytes, choroid plexus epithelium, astrocytes, or cerebral endothelial cells as measured by Northern blotting, but it is unclear if significant small differences in transcription would be detected by this technique. In sum, the experiments suggest that thyroid hormone availability *per se* does not directly affect TTR mRNA levels in any tissue, but this conclusion may not be the final word on the subject.

Animals with *TTR* silenced by targeted gene disruption do not have systemic thyroid hormone deficiency, presumably because there are other transport mechanisms available to compensate for the absent TTR. These two sets of observations, that is, the failure of thyroid status to affect TTR transcription rate and the absence of TTR to have any discernible effect on thyroid function, have raised questions regarding the reason for evolutionary conservation of its structure. Further they suggest that in the rodent, at least, a major physiologic role of TTR could lie elsewhere.

13.1.3. TTR Function

Quantitatively, the major known function of TTR is that of transporting retinol bound to RBP. The interaction of the two proteins within the hepatocyte and the rapid loss of RBP from the circulation via renal excretion in the absence of TTR indicate that in rodents the relationship is functional and important. The lack of a phenotype in the unstressed knockout was somewhat surprising and indicates that there may be other pathways utilized in meeting the baseline retinol needs of the animal.

It has been estimated that in humans, between 25% and 50% of serum TTR is complexed with RBP, the 21-kDa polypeptide bound to one molecule of retinol, and secreted by the hepatocyte. There is little interaction of TTR with empty RBP. In retinol deficiency, hepatic RBP synthesis is decreased. In contrast with the TTR knockout, silencing the RBP gene by targeted disruption results in a visual defect from birth and an inability to mobilize retinyl esters from hepatocytes. The former can be overcome by feeding retinol, hence there must be alternative pathways to either transport or generate retinoids in the retinal epithelium.

The RBP sequences involved in the interaction with TTR include residues 21–37, 54–59, and 77–84. When these were inserted into the structure of the epididymal retinoic acid binding protein, an intracellular molecule encoded by its own gene, which binds retinoids but does not interact with TTR, the construct containing residues 77–84 was sufficient to impart *in vitro* TTR binding capacity.

Complexes of human TTR with chicken RBP and human TTR with human RBP have been crystallized and analyzed at resolutions of 3.1 and 3.2 Å. Each TTR subunit has a single binding site for RBP involving residues Gly 83, Ile 84, and Ser 85, which interact with RBP residues Leu 35, Leu 97, and Phe 96. TTR residues Asp 99 and Ser 100 also interact with Lys 89, Lys 99, and Trp 91 of RBP. The involvement of Ile 84 was suspected because of the reduction of serum RBP in carriers of the TTR I84S mutation. The fully assembled tetramer has only two of the four potential RBP binding sites available. In the heterologous complex, the two RBP molecules react with subunits of the same dimer. In the homologous interaction, the most extensive interactions are with the opposite dimers. In addition, the two binding sites show negative cooperativity, with binding to the second site reduced when the first site is occupied. There is no impact of RBP binding on the accessibility of the T4 binding sites to their ligand (Raz, 1969).

In purified systems, the affinity of TTR is much greater for holo-RBP (RBP loaded with retinol) than the apo-protein (Noy *et al.*, 1992). The dissociation of retinol from the TTR-RBP complex is faster than that of the TTR/RBP complex itself, suggesting that in the circulation TTR has three possible states with respect to RBP. It may be unbound to RBP, bound to RBP loaded with retinol (as it is secreted from the liver), or bound to RBP that has given up its retinol. Any or all of these may be carrying thyroxine. Because these experiments were performed in the absence of cells, the interaction of TTR with receptors for retinol (unlikely), RBP, TTR, or the TTR-RBP could not be evaluated. Binding of RBP stabilizes TTR decreasing its propensity for fibril formation, hence the release of RBP in a specific location could determine the tendency of a given tissue to be a site of aggregation and subsequent deposition, a hypothesis that remains to be tested (White and Kelly, 2001).

TTR is only one of three potential thyroxine (T4) carrier proteins in the serum (Robbins, 2002). The bulk of plasma T4 is carried by thyroxine-binding globulin, present in the lowest concentrations (12–31 mg/L) but having the highest affinity. Albumin is present in the highest concentration (3–4 g/L) and has the lowest affinity. TTR serum concentrations range from 0.1 to 0.45 g/L and the affinity for T4 is intermediate. The molecule has two binding sites for T4 in the channel created by the tetramerization of the two dimers. Residues Lys 15 and Glu 54 are the major contact residues (Blake and Oatley, 1977; Wojtczak, 1996). The T4 sites display negative cooperativity ($K_1 = 10^8\ M^{-1}$, $K_2 = 9.5 \times 10^5\ M^{-1}$). It has been calculated, using the normal serum T4 concentration ($10–^7$ M), the normal TTR concentration (5 μM), and the fact that a maximum of 15% of T4 is TTR bound, that only 0.3% of the available TTR is carrying T4 (Neumann *et al.*, 2001).

13.1.4. TTR in Plasma

Normal plasma TTR has been noted to exhibit considerable heterogeneity. The bulk of the protein is a noncovalently linked tetramer, however a variable proportion of the protein is present as monomer, presumably reflecting the equilibrium between the two forms. In addition, variable amounts of glutathionylated, cysteinylated, and homocysteinylated TTR modified at the Cys 10 position are present. There has also been a report of the conversion of Cys 10 to gly *in vivo*. The significance of these changes in the pathogenesis of human TTR amyloidosis has not been established (*vide infra*).

The plasma TTR level shows a gradual increase up to 0.2 g/L from birth to age 7 or 8. It increases more sharply with puberty, peaking between ages 25 and 40 (range 0.2 to 0.35 g/L) and remaining stable until approximately age 50 when it declines, more sharply in males than females. Levels in adult males are a bit higher than those in females (0.29–0.3 2 g/L vs. 0.26–0.28 g/L) but are similar by age 80 (Ritchie *et al.*, 1999a, b).

Plasma turnover of TTR has been measured in the rat, mouse, and human. All utilized radiolabeled TTR. When wild-type and V30M variant TTR were purified from human plasma, iodinated with two different radioisotopes (I^{131} or I^{125}), and injected into either a normal subject or an FAP patient, the two proteins had different plasma disappearance curves with the variant protein having a $t_{1/2}$ of 8–11 h while the $t_{1/2}$ of the wild type was approximately 18 h (Hanes *et al.*, 1996). The fractional catabolic rate for the wild-type protein was 0.039 per hour while that for the variant averaged 0.070 per hour. The studies suggested a multicompartment distribution although the number of discrete pools was unclear. In rats the data for the wild-type endogenous protein suggested a three-compartment pool and a fractional catabolic rate in the plasma of 0.15 per hour (Makover *et al.*, 1988). Human and rat TTR had similar survivals in rats. There did not seem to be any difference in the decay of the human wild-type and V30M variant (Ando *et al.*, 1989).

When wild-type and variant human TTRs were injected into mice, the results best fit a distribution into two compartments, with only the very stable Met119 variant differing from the wild type and V30M (Longo, *et al.*, 1997). Interestingly, homozygous V30M showed no difference in plasma clearance from the wild-type protein in this system, but mixed wild-type and V30M tetramers were cleared more rapidly and Met 119 more slowly. The mélange of results to date suggest that further studies are necessary, particularly because the number of studies is small, were generally done at a single age, and were performed with TTR purified from its T4 and RBP ligands, either or both of which might exert an effect on turnover. Age-dependent changes in the capacity of humans or experimental animals to clear wild-type or variant TTRs could be partially responsible for the age dependence of the deposition disorders, hence require further investigation.

In experimental animals, most of the label is excreted into the urine as small peptides. The major sites of degradation appear to be the liver, muscle, skin, adipose tissue, and kidneys, with minor amounts seen in many tissues (Makover *et al.*, 1988). These results, as well as questions regarding the specificity of TTR in delivering its ligands (T4 and RBP) to tissues, stimulated studies designed to identify a cell receptor for TTR. Putative saturable binding sites specific for

TTR have been reported in human (Hep G2) and rat (FAO) hepatoma cells, chicken oocytes, and a transformed murine ependymyal cell line (TG-65B) (Divino and Schussler, 1990; Vieira *et al.*, 1995; Kuchler-Bopp *et al.*, 2000; Sousa and Saraiva, 2001). Scatchard analyses in the hepatoma cells from the two species were highly disparate with K_Ds of 14 nM and 4.3 nM in the human and rat, respectively. The candidate receptor molecules also differed considerably in size. In both cell lines, TTR binding was enhanced when it carried T4, a finding that may be related to either stabilization of the TTR conformation or reflect the hypothetical importance of TTR binding to T4 transport. In oocytes, RBP binding to TTR seemed to have no effect on TTR receptor interactions, while in the FAO cells the RBP-TTR complex was less avidly bound (70% decrease at 3 h relative to unliganded TTR). It was not determined whether this represented a quantitative or kinetic effect. On the other hand, the results in the HepG2 hepatoma and TG-65B ependymoma cells were comparable with the latter's receptor having a K_D of 18 nM and an estimated molecular size of 100 kDa. The heterogeneity of the results suggests either methodologic inconsistencies or considerable variability in the nature of a TTR receptor on different cells in different species. Alternatively, the evidence for a specific receptor could be artifactual, as none of the results have been precisely replicated, hence the question remains experimentally open. It has also been reported that megalin, a large membrane protein expressed in the proximal tubular cells of the kidney, binds TTR. The affinity is relatively low (K_D 5 μM at 20°C) hence the protein may be involved in tubular transport into the urine rather than binding a molecule necessary for cell function.

13.1.5. TTR in the Cerebrospinal Fluid

Transthyretin is synthesized and secreted into the cerebrospinal fluid (CSF) by choroid plexus (CP) epithelium, representing 12–25% of CP synthesized proteins but 5–43% of CSF proteins in various reports (Dickson, 1986). Some studies indicate that the TTR concentration in CSF decreases with increasing age. The CSF molecules show similar modifications to those seen in the serum protein, that is, cysteinylation and glycyl-cysteinylation. It has been suggested that, in contrast with the serum, the dominant form in the CSF is the monomer, however this seems unlikely because T4 binding requires a fully assembled tetramer (Marchi *et al.*, 2003). Further, the published studies of CSF TTR have generally used methods in which the molecules are exposed to denaturants prior to analysis. Because the tetramer is readily dissociated to dimers and monomers in sodium dodecyl sulfate and acetonitrile, reliable assessment of the presence of tetramers is difficult under these conditions.

It has been argued that TTR bound T4 in CSF is critical for the normal cerebral development of the fetus. However, there is general agreement that T4 is membrane soluble and enters cells without the benefit of a specific receptor. It has also been noted that there is relatively little transport into brain parenchyma after direct injection of T4 into the cerebral ventricles. Further, the absence of an apparent functional hypothyroid (as indicated by normal T3 levels) brain phenotype in murine TTR knockouts makes a strict requirement for TTR transport of T4 unlikely. However, the recent report of a behavioral phenotype in the knockouts suggests that absent cerebral TTR and a reduced brain T4 concentration may not be without consequence, either directly or via some downstream event (Sousa *et al.*, 2004). While the T3 content of the brain may be normal, its distribution may not. Alternatively, T4 might have some other function not requiring its conversion to T3.

13.1.6. Transthyretin, the Protein

The approximate molecular weight of isolated human serum TTR was determined to be 50,000 Da by sedimentation equilibrium analysis, however gel filtration in the presence of 5 M guanidine indicated that the protein could consist of a single smaller subunit assembled noncovalently. The nature of the subunit structure was initially defined by crystallography. The *2,2,2* symmetry was consistent with either four identical subunits, or two pairs of nonidentical subunits. Subsequent gel filtration and tryptic peptide analyses favored the identical subunit structure that was confirmed by the determination of the complete amino acid sequence of both human and rhesus monkey serum TTR that also established the molecular weight as 54,980 Da (Van Jaarsveld *et al.*, 1973; Kanda *et al.*, 1974).

Crystallographic studies, at 2.5 and 1.8 Å resolution, revealed that the subunits were assembled around a central channel containing the two T4 binding sites (Blake and Oatley, 1977; Blake *et al.*, 1978). The channel showed a constriction formed by the Ser 117, Thr 119 pairs and Lys 15 and Glu 54. Each subunit contains 8 β-strands (designated A to H) 6 to 9 amino acids long, except for the D strand (amino acids 53–55). They are all connected by β-bends except for strands E and F, which are connected by an α-helix. Thus the monomer structure is that of a β-barrel formed by the antiparallel arrangement of the β-strands, strands DAGH forming one sheet and CEBF forming the other. The 9 amino-terminal and the 7 carboxy-terminal amino acids appeared to be disordered. There are now multiple crystallographic analyses of wild-type and mutant TTR molecules in accessible databases. Interestingly, when a peptide extending from amino acid 1 to 51, spanning β-strands A to C was synthesized, in solution it adopted a random coil rather than a β-sheet conformation, a demonstration of the importance of context in protein structure (Wilce *et al.*, 2002). The intrinsic stability of the wild-type quaternary structure, with respect to denaturation by acid (requiring pH 2), urea, guanidine, or SDS, is striking considering the effects of multiple single amino acid substitutions scattered throughout the molecule on its sensitivity to denaturation and the tendency of the released misfolded monomer to aggregate (*vide infra*).

Sequence analyses of human and rhesus TTRs revealed only two differences, at positions 5 and 6 in the disordered amino-terminal tail. The amino acid sequences of TTR from 22 other species have largely been derived by DNA sequencing (Schreiber and Richardson, 1997) Examination of those sequences indicates that the most conserved regions encompass amino acids 10 through 22 and 106 to 121. Not surprisingly, those are the sites involved in forming the T4 binding sites. There is less conservation in the residues involved in RBP binding, that is, 83–85, 99, and 100. The most variable regions are the randomly ordered carboxy and amino termini (residues 1–10 and 123–127). There is also substantial variation in the region from amino acid 23 to 70, although among some species those regions are reasonably conserved, particularly from 41 to 60. The observation that the most highly conserved sequences are those that are involved in T4 and RBP binding are consistent with the notion that evolution retains functionally important structures.

Interestingly, a family of proteins has been described in which all the members are highly conserved and share about 35% homology with TTR (Eneqvist *et al.*, 2003). The proteins are found in bacteria as well as in *C. elegans* and do not appear to bind either T4 or RBP but do have a binding site similar to that of TTR for an as yet unknown ligand.

13.2. TTR and Human Disease

13.2.1. Plasma TTR as a Marker of Nutritional Status

Hepatic TTR synthesis is responsive to protein depletion, a characteristic it shares with several other proteins (transferrin, albumin, retinol binding protein, and insulin-like growth factor-1) (Ingenbleek and Young, 2002). It is more sensitive to dietary insufficiency and repletion than albumin, however it is also clear (*vide supra*) that TTR synthesis is suppressed by cytokine release during infectious or noninfectious inflammation. Hence its usefulness as a marker may be compromised when both inflammation and malnutrition are present, a common occurrence in children in tropical settings (Brugler *et al.*, 2002). Because both serum TTR and RBP are suppressed in that context, it has been suggested that the ratio between RBP and TTR be utilized, based on the notion that RBP synthesis by the liver is retinol dependent and it is secreted as an RBP-TTR complex (Filteau, 2000; Rosales, 1998). Because vitamin A deficiency will reduce the RBP available to interact with TTR, the ratio of RBP/TTR, normally about 0.5, would fall in the presence of inadequate vitamin A intake. Experimental caloric restriction (40%) in rats, whether or not accompanied by vitamin and mineral supplementation, results in markedly reduced plasma RBP levels, corresponding with lower hepatic synthesis (Chevalier *et al.*, 1999). Interestingly, while plasma TTR levels fall to a lesser extent than those of RBP, liver TTR content is unchanged, suggesting that hepatic TTR secretion or turnover may be more affected by the reduction in available RBP than its synthesis.

13.2.2. The TTR Amyloidoses as Disorders of Protein Conformation

13.2.2.1. Mutations

In general, the TTR amyloidoses typify the extracellular subset of disorders of protein conformation. At the time of this writing, 82 different amino acid substitutions at 51 different positions, as well as the wild-type protein have been associated with tissue amyloid deposition in humans. Eleven additional substitutions have not been associated with tissue deposits. The mutations are tabulated and summarized at two websites, which also provide references including the original description of the mutation.

The substitutions are spread throughout the molecule. The unstructured amino-terminus is spared, except for an apparent normal polymorphism at position 6 (Jacobson *et al.*, 1995). Only three of the substitutions are based on mutations at CpG dinucleotides, known to be mutational "hot spots" (Cooper and Youssoufian, 1988). Three positions (30, 33, 47) have been the sites of all the possible base substitutions yielding four different mutant amino acids at the same site. In at least one instance, both amyloidogenic and benign mutations have been reported at the same residue (Ala 109). It has been suggested that the D-loop is a particularly sensitive site for amyloidogenic substitutions, however it is difficult to support this statistically (Goldsteins *et al.*, 1997).

Two mutations are more stable than the wild type and one (TTR V119M) has been shown to be capable of kinetically stabilizing hybrid tetramers with the mutant protein (Hammarstrom *et al.*, 2003b). It also suppresses the expression of disease *in vivo* when the gene is present as the other allele (Coelho *et al.*, 1993). It is uncertain if the Arg104H stabilizes via the same mechanism, although it may have a similar protective effect *in vivo* (Almeida *et al.*, 2000).

13.2.3. *In Vitro* Analyses

There are extensive substantial *in vitro* experimental data exploring the process of fibril formation. While much of the evidence has been obtained under conditions that are far from physiologic, reproducibility in multiple laboratories suggests that the conclusions are valid (Fig. 13.2).

It appears likely that the first event is release of monomer from the tetramer followed by misfolding and aggregation (Kelly, 1998). Subunit exchange has been examined for recombinant homotetramers of wild-type and mutant proteins and for heterotetramers of V30M and V119M and V30M and wild-type (Hammarstrom *et al.*, 2001). Under the experimental conditions used, wild-type and V30M exchanged at the same rate at 37°C but the mutant protein exchanges more

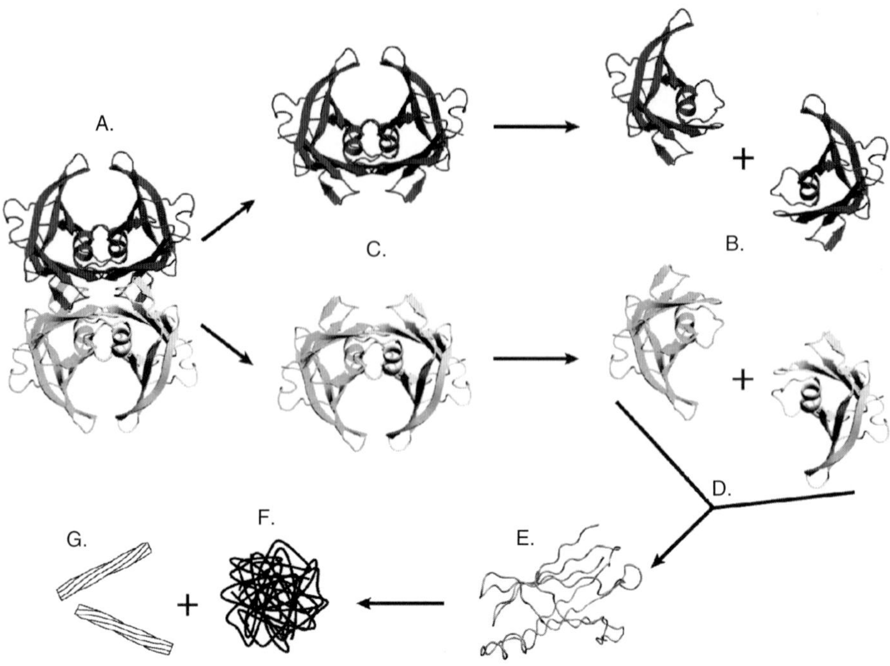

Figure 13.2. Under physiologic conditions, transthyretin exists as a tetramer in equilibrium with its monomers via a thus far hypothetical dimeric state. Release of monomer is followed rapidly by its misfolding. The propensity for the monomer to misfold is enhanced by the presence of amyloidogenic mutations. The misfolded monomer rapidly aggregates to form oligomers, which go on to form spherical aggregates, protofibrils, and fibrils. *In vivo*, the ultimate configuration may be facilitated by interactions with a variety of other proteins or metabolites (figure provided by Dr. T. Foss).

rapidly at 0°C. V119M shows little subunit exchange. Hybrid tetramers between V30M and V119M are extremely stable kinetically explaining the apparent protective effect of V119M when it is present in trans in V30M carriers. On the other hand, when hybrids between V30M and wild type dissociate to monomers, the V30M monomers rapidly misfold and aggregate pulling the tetramer-monomer equilibrium toward monomer and away from reconstituting the tetramer. Hence both thermodynamic and kinetic considerations determine the tendency of TTR tetramers to give rise to the misfolded monomer that is predisposed to oligomerize forming aggregates, protofibrils, prefibrils, and fibrils. Energetically, it appears that the sequence is a downhill process with the protofibril/fibril constituting the most stable form of the misfolded monomer (Hurshman *et al.*, 2004). It is only the last that is responsible for the tissue binding of Congo red with birefringence under polarized light, binding of serum amyloid P-component (SAP), and apolipoprotein E (Apo E) (Gallo *et al.*, 1988, 1994).

Recombinant forms of many of the mutant proteins have been crystallized and examined to very high orders of resolution. The results have been interesting, although disappointing with respect to explaining the propensity of the mutations to oligomerize and form fibrils. Most interesting was a study suggesting that replacing the tyrosine at residue 78 with a phenylalanine would enhance fibrillogenicity (Redondo *et al.*, 2000). A recombinant was produced containing the substitution. The protein formed fibrils *in vitro*.

13.2.4. Transthyretin Amyloidosis

The clinical syndromes related to mutant and wild-type TTR deposition are primarily a function of the site of the deposits. The initial description of a disorder, later found to be produced by TTR V30M amyloid deposition, included peripheral neuropathy affecting both sensory and motor functions, first affecting the lower extremities, autonomic neuropathy with constipation and subsequent loss of bladder control. Peripheral nerve conduction studies are abnormal and the diagnosis can frequently be made by sural nerve biopsy (Hanyu *et al.*, 1989; Ii *et al.*, 1992). Recent observations have suggested that mild early symptoms are associated with nonfibrillar TTR deposition in the affected nerves and biochemical evidence of tissue reactivity to the aggregates (Sousa *et al.*, 2001).

Cardiomyopathy with congestive heart failure and arrhythmia was common, with sudden death sometimes the cause of death. It has become clear that the heart disease is initially restrictive with diastolic dysfunction and a normal ejection fraction. Late in the course of the disease, left ventricular failure may occur. The diagnosis of the heart disease was revolutionized with the availability of ultrasound analysis of cardiac function and the recognition of a characteristic pattern of findings including thickening of the interventricular septum and posterior wall, evidence of cardiac noncompliance, particularly rapid ventricular filling, and a granular pattern of the myocardial echoes well before heart failure ensued (Falk *et al.*, 1987).

Vitreous opacities (composed of TTR fibrils) with marked visual compromise was noted later (Kaufman and Thomas, 1959). It has now also become clear that chronic kidney disease related to TTR amyloid deposition occurs in some kindreds and may require renal transplantation (Lobato *et al.*, 2003).

In TTR mutation carriers the serum TTR concentration is often low prior to the clinical manifestation of disease and becomes lower when disease becomes symptomatic (Benson and Dwulet, 1983; Skinner *et al.*, 1985). In the heterozygous state, the normal and mutant alleles appear to be transcribed and the mRNAs translated at equal rates. While there are data regarding the rate of subunit exchange between wild-type and mutant tetramers made as recombinant proteins (*vide infra*), there are none regarding the relative proportions of wild-type and mutant subunits in the *in vivo*–assembled tetramers. Hence the circulating TTR pool may consist of a mixture of tetramers of varying monomer composition or may contain two populations, one consisting of all wild type while the other contains all variant, although the latter seems less likely, given the stabilizing effects of some variants.

The serum TTR level is in the normal range until the disease becomes symptomatic, when it decreases. It has been assumed, although not proved, that the reduction in serum concentration is related to TTR being removed from the circulation by the process of deposition. The proportion of total serum TTR as monomer is increased with the majority representing the normal subunit (Sekijima *et al.*, 2001). The implication of the analysis was that the mutant monomer is less stable in serum. Whether this is true throughout life or only after the appearance of symptomatic disease is not known, because longitudinal data have not been reported.

13.2.5. The Relationship Between Specific Mutations and Clinical Disease

TTR tetramers are stabilized by their ligands (T4 and RBP) (White and Kelly, 2001). However, because a maximum of 50% of serum TTR has bound RBP and a much smaller proportion contains T4, at any moment at least half of circulating TTR is available to contribute to the pool of monomer at risk for misfolding and oligomerization *in vivo*. It is clear from the *in vitro* work and animals transgenic for human TTR that fibrillogenesis is concentration dependent and facilitated by denaturing conditions, however it is uncertain how these phenomena translate into human disease with respect to its age dependence, the phenotypic variability of the same mutation, and the apparent differences in the hierarchy of tissue deposition among particular mutations. The last is particularly perplexing because virtually all the experimental and clinical evidence supports the notion that the major sites of tissue pathology and significant clinical disease with most mutations are at a distance from the site of synthesis of the protein. That having been said, there are exceptions. Many mutations are associated with collections of TTR fibrils in the vitreous humor of the eye (Skinner *et al.*, 1992; Salvi *et al.*, 1993; Zeldenrust *et al.*, 1994; Ando *et al.*, 1999; Yazaki *et al.*, 2002; Imasawa *et al.*, 2004). The protein is presumably secreted by the retinal epithelial cells and aggregates in a concentration-dependent manner in the vitreous fluid. Similarly, there are autopsy reports of choroid plexus TTR deposits, without associated symptoms.

It has been proposed that the mutants D18G and A25T (as well as all the other mutants associated with a similar phenotype) display significant leptomeningeal deposition because they are particularly unstable, never getting out of the liver in sufficient concentrations to produce systemic deposition and being secreted by choroid plexus cells only because there is sufficient T4 available in choroid plexus epithelium to chaperone the molecule into the CSF where it rapidly aggregates to deposit in the leptomeninges (Hammarstrom *et al.*, 2003a). This mechanism predicts that all the leptomeningeal mutants would have late onset, little systemic deposition and some evidence of hepatic or perihepatic deposition. In fact, only one individual, a TTR L12P carrier with leptomeningeal disease, has been reported to have deposits in the liver, although these were assumed on the basis of an serum amyloid P-component (SAP) scan, rather than confirmed by autopsy or biopsy (Brett *et al.*, 1999).

An index derived from the analysis of 23 TTR variants (prepared as recombinant proteins) with respect to their thermodynamic and kinetic properties coupled with their rates of secretion suggests that clinical phenotypes can be generally predicted, in particular those proteins associated with leptomeningeal disease (Sekijima *et al.*, 2005). These observations and the resultant hypothesis are provocative. Is it possible to draw such firm conclusions based on the clinical phenotypes described to date?

In examining the TTR mutation databases, it is apparent that only 15 mutations have been clinically and pathologically characterized in more than one or two individuals or kindreds, hence the reproducibility of the tissue phenotype of a given mutation is problematic without further instances in which the mutation is found in other than a single biological context. The best example of this is TTR V30M. The original description, by Andrade, an eminent Portuguese neurologist, resulted in naming the disorder familial amyloidotic polyneuropathy (Andrade, 1952). Examination of Andrade's published descriptions indicates that all his patients had significant cardiac disease, suggesting that, had Andrade been a cardiologist, the nomenclature might have been different. The peripheral neuropathic phenotype may only reflect the fact that peripheral neuron function may be more easily disrupted with the same degree of tissue deposition than cardiac compromise, and that tissue specificity may only be apparent, related to our inability to sensitively and accurately ascertain deposition in some tissues.

The V30M mutation has now been identified and studied clinically and pathologically in several hundred individuals in three ethnic groups, Portuguese, Swedish, and Japanese. With increasing numbers of subjects, it is now clear that the apparent homogeneous clinical phenotype was a misperception. While penetrance of the disease in Portuguese and Japanese is similarly high, it is quite low in Swedish V30M carriers (Sousa *et al.*, 1993). Further, among Japanese and Portuguese gene carriers there is extensive phenotypic variability, most readily appreciated in the age of clinical onset (Ikeda *et al.*, 2002). V30M can present with primarily leptomeningeal, nephropathic, and vitreous phenotypes as well as the more clinically evident neuropathy and cardiomyopathy. Hence the biological variability in the presence of the same mutation makes a simple structural explanation for any disease phenotype unlikely. Preliminary genetic studies examining the age of onset in Portuguese V30M carriers suggest that gene interactions among alleles at the RBP, SAP, Apo E, Perlecan, and (CE1) Serum Amyloid A (SAA) loci may influence clinical penetrance (Soares, 2005).

The lack of pathologic specificity is confirmed in autopsy studies of other mutations, which show rather generalized deposition in blood vessels, skin, gut, and other viscera as well as in autonomic nerves in patients characterized as having FAP or FAC. However, deposition in the liver, the main site of synthesis, apart from hepatic vessels has not been described, except as noted above in the single L12P patient. While the single patient with A25T had primarily leptomeningeal disease, the kindred with D18G displayed significant systemic deposition in addition to the leptomeningeal findings (Harkany *et al.*, 2002; Sekijima *et al.*, 2002). A leptomeningeal phenotype has also been reported in some V30M carriers (Herrick *et al.*, 1996).

Three mutations, other than V30M, have been reported in a sufficient number of individuals or kindreds to suggest an association with a particular clinical phenotype with some degree of statistical confidence. Ala 60, Met 111, and V122I have all had predominant cardiac deposition with congestive heart failure. However, each of these has been described in patients of a particular ethnic or national origin that may condition the site of deposition, hence greater variability may be seen if the allele appears in other populations. Met 111 is the most restrictive, with a striking frequency of heart disease. However, because it has only been reported in a single very large Danish kindred, it may be premature to exclusively ascribe the dominant cardiac phenotype to the properties of the protein before it is found on another genetic background with the same phenotype (Svendsen *et al.*, 1998).

Most individuals with Ala 60 (Reilly *et al.*, 1995b) have been of Irish extraction.

Similarly, cardiac deposition related to both homozygosity and heterozygosity of the V122I allele is found almost exclusively in individuals with traceable African ancestry with an allele prevalence of 0.016 in African Americans, which is even higher in some West African locales. While carpal tunnel syndrome and gastrointestinal infiltration have also been noted as has a recent instance of peripheral neuropathy, suggesting that the disorder may be more heterogeneous in this ethnic group, two kindreds with no apparent African ties have been identified with cardiac amyloid deposition, again suggesting that tissue specificity is related to the structure of the mutant protein. Autopsy data and recent clinical studies indicate that the anatomic and clinical penetrance of the allele is quite high in African Americans over age 60 (Jacobson *et al.*, 1997).

While synthesis of TTR by cardiomyocytes or other cardiac cells has not been reported, it is not clear whether this possibility has been ruled out as a mechanism underlying primary deposition in the heart as some microarray experiments have suggested a low level of myocardial transcription of the *TTR* gene. Further, a single case of inclusion body myositis, a common muscle disorder of the elderly associated with the intramyocytic accumulation of amyloid beta (Aβ) aggregates, had co-aggregation of TTR V122I fibrils intracellularly (Askanas *et al.*, 2000). The significance of this apparently extraordinary event is unclear because it implies that skeletal muscle cells either synthesize TTR or can take it up from the circulation.

In another instance L55P, reported in only two well-studied kindreds, but of very different genetic backgrounds (European and Chinese), had a similar pattern of diffuse deposition with a marked autonomic phenotype and aggressive anticipation in succeeding generations. Here the extreme thermodynamic instability of the protein could be associated with almost indiscriminant tissue affinity and the absence of anything resembling tissue tropism. However, it is difficult to account for the striking genetic anticipation (earlier onset and more severe disease in succeeding generations) in families carrying the L55P mutation on a purely protein structural basis (Jacobson *et al.*, 1992; Chou *et al.*, 1997).

13.2.6. Age of Onset

Studies of V30M kindreds showing anticipation did not reveal evidence for the trinucleotide repeat expansion seen in other neurologic disorders such as Huntington's disease and DRPLA where the so-called dynamic mutations (based on expansion of the number of trinucleotide repeats encoding strings of glutamine, for example) strongly impact on the age of onset of disease (Soares *et al.*, 1999).

It is not yet possible to fully account for the variable age of onset of the clinical disorders related to the *TTR* mutations. The *in vitro* properties of the recombinant TTR mutants would suggest that aggregation and tissue deposition should be occurring throughout life, as does atherosclerosis. However, in the instances where tissues from human gene carriers or transgenic animals that ultimately develop deposits have been examined (prior to the development of clinical disease in the former or early in life in the latter), deposition in the eventual affected sites has not been detected (Harats *et al.*, 1989). Hence age dependence is not merely an effect of the accumulation of deposited protein throughout life. However, once deposition occurs in a clinically sensitive site, it appears to be cumulative and symptomatic. These observations, as well as those from transgenic animals, where it has not been possible to experimentally increase deposition by the injection of preformed fibrils, argue against lifelong nucleated seeding as being responsible for the age dependence of clinical penetrance (Jarrett and Lansbury, 1993). These experiments are supported by *in vitro* studies that failed to show that seeding (i.e., the addition of a small amount of preformed fibrils) accelerates TTR aggregation and fibril formation (Hurshman *et al.*, 2004). Further, the analysis of a significant number of Portuguese individuals carrying the V30M mutation with onset before 40 and after 50 indicate that combinations of genetic factors encoding proteins involved in deposition (SAP, Apo E, Perlecan) or TTR function (RBP) may play a role in the age dependence of the clinical presentation (Soares, 2005).

13.2.7. Environmental Effects

Environmental factors are frequently invoked to explain the occurrence of diseases late in life, because the passage of time certainly increases the cumulative exposure to agents present in the environment. The strongest evidence in favor of an environmental influence is the discordance of disease occurrence in monozygotic twins. While concordance has been shown in most TTR V30M monozygotic twin pairs, several have been described in which this was not the case (Munar-Ques *et al.*, 1999; Holmgren *et al.*, 2004). In one instance in Sweden, the twin who did not develop disease described ingesting large quantities of fish oil, a practice not pursued by his affected brother. The anti-inflammatory effects of ω-3 fatty acids, a major component of some fish oils, are well-known and it is possible that this dietary habit may have had a favorable influence on either TTR aggregation or deposition. However, sufficient details were not available to draw such a conclusion. A general survey of potential environmental exposures failed to show any statistically significant occurrence of any occupational or other exposure among symptomatic gene carriers (Hardell *et al.*, 1995).

The notion that aging is associated with increased oxidative changes in macromolecules has frequently been cited as a factor in the development of the age-associated protein deposition diseases, particularly in those in which the deposits consist of wild-type proteins (Levine and Stadtman, 1996). The fact that TTR, as well as other serum proteins, turn over with relatively short half-lives, argues against age-association of cumulative post-translational changes in a single molecule as the cause of its propensity to aggregate. More likely is a decreased tendency for such proteins to be degraded with age, giving them a longer temporal window in which to sample the misfolded state. Studies have shown that bulk intercellular protein degradation is reduced with age, but there are few available studies concerning specific intracellular or circulating proteins and none investigating TTR (Ward, 2002). Several reports have indicated that oxidation of TTR enhances its potential for *in vitro* fibril formation (Tajiri *et al.*, 2002; Zhang and Kelly, 2003). In addition, circulating TTR from elderly individuals has been shown to display a greater degree of oxidative change by mass spectroscopy, however individuals with demonstrated wild-type cardiac TTR deposition did not show a greater degree of oxidation in the circulating TTR than age-matched controls without cardiac amyloidosis (Connors *et al.*, 2002). It is possible, however, that molecules having such changes were no longer in the circulating TTR pool, having been deposited in the myocardium. A comparative analysis of circulating and deposited TTR from the same individual has not been reported but should be very useful in defining whether oxidative changes play a role in potentiating fibrillogenesis *in vivo*.

Another possible explanation for the age-dependence of deposition might be a decreased ability of the host to clear circulating aggregates or prefibrillar deposits with increasing age. This possibility has not been explored experimentally hence remains another viable explanation.

13.2.8. Wild-Type TTR Deposition

Deposits of wild-type TTR have been found in the gut, carpal tunnel, and hearts of the elderly (Cornwell, III *et al.*, 1983; Westermark *et al.*, 1990; Kyle *et al.*, 1992; Rocken *et al.*, 1994). The gastrointestinal deposits have not been a clinical problem although they may be a source of protracted bleeding in the presence of other gastrointestinal pathology. The carpal tunnel syndrome can be both annoying and disabling but it is not life-threatening. Such is not the case with respect to cardiac deposition, which has been reported to occur in from 10% to 25% of individuals over age 80 and is associated with atrial fibrillation and congestive heart failure (Cornwell, III *et al.*, 1983; Hodkinson and Pomerance, 1977). In one autopsy study of subjects over 90, it was found in 20% of the subjects and was the cause of death in half of those (Lie and Hammond, 1988). It appears later in life than that associated with mutant forms of TTR, although there is significant overlap in the age of detection of the disease related to V122I with that of the wild type. A study of elderly Swedes found to have congestive heart failure related to wild-type TTR deposition indicated that their serum TTR concentrations were significantly lower that those of age- and gender-matched controls, a finding similar to that reported in symptomatic FAP patients (Westermark *et al.*, 1985).

It has recently been reported that cardiac deposits found in Swedish SSA subjects consist largely of truncated wild-type molecules. Based on these findings, it has been argued that the mechanism of deposition of the wild-type protein may differ from that of the various mutants (Bergstrom *et al.*, 2005). There have been several reports of TTR fragments beginning at position 49 being isolated from tissue deposits of both mutant and wild-type proteins (Gustavsson *et al.*, 1995). However, in mice transgenic for wild-type human TTR, neither the cardiac nor renal fibrillar deposits contained detectable amounts of the truncated human protein, an observation not totally consistent with the reports in human (Teng *et al.*, 2001). This may be a species difference or reflect the fact that in the transgenics the human protein was highly overexpressed and might not require digestion to aggregate, while in the intact human the protein is expressed at normal levels and digestion is required to predispose to aggregation and deposition. There are no reported *in vitro* analyses of the latter possibility. If that is the case, the question remains, does the partial digestion occur after deposition or is it actually part of the aggregation process? Further, it may be that the capacity to digest TTR to its pro-fibrillogenic state is genetically determined. There are no data regarding this possibility. There have been no analyses of the wild-type TTR found in the gut or carpal tunnel of the elderly to determine whether it is intact or truncated.

13.2.9. Experimental Models

13.2.9.1. Transgenic Mice

There are no naturally occurring animal models of the TTR amyloidoses. In fact, it appears that both rat and mouse TTR are much more stable with respect to misfolding and amyloid formation than the human protein. Nonetheless, several laboratories have produced mice transgenic for various forms of human TTR. The intial successes were achieved with a V30M construct driven by the murine metallothionein promoter. The number of integrated TTR gene copies was not determined. Expression was widespread and the animals developed amyloid deposits in the gut, heart, skin, and kidneys beginning at about 6 months of age (Yi *et al.*, 1991), but no deposits were seen in peripheral nerve. The same investigators produced a second line with approximately 30–60 copies of the gene driven by its own promoter, the construct containing 6 kb of upstream DNA. Expression appeared to be tissue specific, primarily in the liver with deposition. The onset of deposition was at about 11 months in the gut, skin, heart, and kidneys, but none in peripheral nerve (Kohno *et al.*, 1997). Breeding this construct on to a mouse strain in which the endogenous mouse TTR gene had been silenced by targeted disruption had no apparent effect on either the frequency or extent of tissue deposition.

These investigators noted that transfer of the transgenic mice to a specific-pathogen-free animal facility markedly reduced the frequency of tissue deposition. They subsequently provided some evidence that the nature of the gastrointestinal flora may have had some influence on the prevalence of tissue deposition (Noguchi *et al.*, 2002). The latter explanation was somewhat counterintuitive, as TTR production should be reduced in the presence of active inflammation and cytokine production.

A third transgenic strain was produced carrying approximately 90 copies of a 20-kb genomic wild-type TTR clone containing all the recognized regulatory elements required for tissue-specific TTR expression (Teng *et al.*, 2001). The sera of these animals contained high levels of the normal human protein, and the animals began to develop deposits at 1 year of age. The earliest and most extensive site of deposition was the kidney. Cardiac deposits were noted increasingly after 18 months of age. The deposits were initially nonfibrillar, being identified by anti-TTR staining. However in the older animals, the deposits bound Congo red with positive birefringence, stained for both SAP and Apo E, properties not shown by the nonfibrillar, presumably precursor, tissue deposits, and had the characteristic ultrastructure on electron microscopy. The animals of this strain showed no difference in deposition whether housed in a conventional or specific-pathogen-free animal facility (Tagoe *et al.*, 2003). In contrast with both murine (CE 2) Amyloid A (AA) and Apo AII amyloidosis, neither the prevalence, age of onset, or extent of deposition could be enhanced by the intravenous injection of sonicated preformed TTR fibrils, an observation subsequently confirmed by investigators studying the V30M transgenics (Tagoe *et al.*, 2004; Wei, 2004).

Three other TTR transgenic mouse strains have been described. Animals carrying one or two copies of the 20-kb construct containing the highly amyloidogenic L55P mutation showed little if any deposition of the human protein throughout life, however when the gene was crossed onto the mouse TTR knockout background, 25% of the animals displayed non-Congophilic human TTR deposits in the kidneys after 18 months of age (Sousa *et al.*, 2002). A subsequently derived strain utilizing the L55P cDNA driven by the murine metallothionein promoter and induced with zinc in their drinking water showed little human TTR deposition, all of it non-Congophilic. When that construct was transferred to the murine knockout background, the frequency of deposition increased and in the older animals the deposits were Congophilic, again consistent with murine TTR stabilizing hybrid tetramers. However, despite these suggestive findings, because of methodologic issues, that is, the antibodies used were not shown to distinguish between mouse and human TTR, the presence of mixed human mouse tetramers in these or any other animals has not yet been definitively demonstrated.

Recently, two additional strains were reported, one with a genomic V30M construct and the other with the same construct but having a base change resulting in the substitution of serine for cysteine 10 (Takaoka *et al.*, 2004). In these mice, the conventional V30M animals developed non-Congophilic deposits after 19 months of age, while those with Ala 10 showed no deposition whatever, a convincing demonstration of the enabling role of cysteine 10 in V30M aggregation.

In summary, the mice transgenic for various forms of human TTR have shown both similarities and differences with humans carrying the same mutations. They correspond in showing age-dependent deposition. They share the specificity for skin, gastrointestinal, renal, and cardiac deposition but fail to show deposition in peripheral nerves. Like humans, they do not show liver deposits. They provided the first demonstration, later confirmed in humans, that nonfibrillar deposits preceded the fibrils. They also demonstrated the potential stabilizing effect of the murine TTR molecule on the human protein, presumably by forming stabilized hybrid tetramers.

As models in which potential therapies could be tested, they have not fulfilled their promise, perhaps because the levels of human TTR that must be present to achieve deposition in the presence of the murine protein place substantial stoichiometric requirements on candidate compounds, narrowing the toxic therapeutic ratio. Some experiments have already shown that at least one compound given to animals carrying two copies of TTR L55P can stabilize the serum protein to urea denaturation (Buxbaum, unpublished). Clearly, animals in which the human genes are expressed moderately in the absence of the mouse protein and develop deposits represent much better tools for drug development.

13.2.10. Cell Culture Models

The traditional view of the pathogenesis of the amyloidoses has held that it was the fibril mass that compromised tissue function and that removal of that "inert" collection would allow normal tissue repair and functional restoration. The notion that it is not the fibrils but the process of fibril formation that produces tissue damage by generating toxic prefibrillar intermediates has arisen among those investigating the pathogenesis of Alzheimer's disease (AD), because a number of observations made on autopsy and biopsy specimens from patients with FAP and FAC and TTR transgenic animals resemble those seen in AD, for example, the presence of hydroxyl-nonenal in proximity to tissue deposits; activation of NFκB associated with binding of fibrils to the receptor for advanced glycation end products (RAGE) (Ando *et al.*, 1997; Sousa *et al.*, 2000). While there have not been a large number of such observations and some of them have not been consistent, it is possible that they provide clues regarding the pathogenesis of tissue damage. However, the *in vivo* analyses are difficult to perform in a systematic fashion controlling for all the potential variables.

Several attempts have been made to simplify the analysis of pathogenesis by utilizing recombinant wild-type and mutant TTR proteins in experiments with cells representative of the *in vivo* tissue targets. A highly amyloidogenic artificial three-amino-acid deletion mutant produced toxicity in cultured neuroblastoma cells, a result that was not seen when the investigators used the naturally occurring V30M mutant protein in the same system (Andersson *et al.*, 2002). The most reproducible results were obtained using both engineered and V30M recombinant proteins, showing consistent cytotoxic effects on cultured human neuroblastoma cells, a cultured rat cardiomyocyte cell line, and primary rat cardiomyocytes. Further, the analyses showed that fully formed fibrils, soluble aggregates greater than 100 kDa, and nondissociating tetramers were nontoxic. The only active species was relatively small in size, perhaps a misfolded monomer, but certainly no larger than hexamer, and transient in the aggregation process (Reixach *et al.*, 2004). Killing appeared to be apoptotic and could be inhibited by small molecules that stabilized the V30M tetramer.

These types of experiments suggest that if the same process, or surrogates for that process, can be identified *in vivo*, it would indicate that pathogenesis may be primarily a function of cytotoxicity or a combination of cytotoxicity and tissue compromise by fibril mass, rather than fibril mass alone. It must be emphasized, however, that to date, there is little definitive evidence that these events actually occur *in vivo*.

13.2.11. Treatment of the Human TTR Amyloidoses

At first glance, it would appear that patients with the TTR amyloidoses would not be good candidates for gene therapy because the disorders are the result of autosomal dominant gain of function mutations. A more reasonable choice would be an autosomal recessive loss of function disorder in which one would not have to achieve quantitative replacement of the lost function. Nonetheless, the results in FAP represent a significant achievement albeit utilizing a rather crude procedure with significant risk. The fact that the liver is the major source of systemic TTR led clinicians to consider removing the liver producing a mutant TTR and replacing it with an organ synthesizing only the wild-type molecule (i.e., gene therapy by organ transplant).

Approximately 1000 such procedures have been performed to date (Herlenius *et al.*, 2004). Sixty percent of the recipients are stabilized, 20% improve, and 20% do poorly. The best results appear to be in TTR V30M carriers who have not yet manifested the significant weight loss associated with severe autonomic gastrointestinal disease. Individuals with other mutations do not appear to do as well. Patients with cardiac disease fare particularly poorly, some going on to deposit wild-type TTR in addition to the residual mutant protein after the transplant. In order to circumvent this problem, some individuals have had combined heart liver transplants (Nardo *et al.*, 2004). An international registry of all the transplants carried out for this disorder is maintained. Without the existence of such a resource to monitor both process and outcomes, it would not be possible to sustain uniformly advanced technical development worldwide. Further, it allows the detection of systematic problems that can only become evident when there is disciplined sufficient and appropriate data collection over a long period of time (Herlenius *et al.*, 2001).

Several investigative groups are involved in developing other therapeutic modalities based on the increasing knowledge concerning the biochemistry and pathogenesis of the diseases. One approach has been to attack the problem using compounds that have the capacity to dissolve all amyloid fibrils. Compounds based on anionic sulfonates, which appear to interfere with the interaction between the fibril or fibril precursor and matrix and membrane proteoglycans that may be involved in the formation of tissue deposits, are currently in clinical trials for other forms of amyloidosis and, if successful, could be applied to the TTR diseases (Kisilevsky *et al.*, 2004).

The fact that SAP is found in all the fibrillar deposits and that the rate of deposition of inflammation-associated AA amyloid is reduced in murine SAP knockouts have led to the development of compounds that deplete total body SAP (Pepys *et al.*, 2002). The results of an early phase II trial in humans indicated that serum SAP was markedly reduced with few side effects in a group of mixed systemic amyloid patients, but clinical efficacy was not assessed and further results have not yet been reported. Should this approach be effective in other amyloidoses it would warrant a trial in the TTR diseases.

4′-iodo-4′-Deoxydoxorubicin, an iodinated derivative of the anticancer agent doxorubicin, was shown to mobilize L-chain amyloid deposits in a few patients but failed to be effective in a larger more systematic trial (Merlini *et al.*, 1999). In laboratory studies, the drug was shown to bind to TTR tissue deposits and dissolve fibrils (Cardoso *et al.*, 2003). Toxicity, particularly that affecting the heart, represents a major drawback to its use in TTR disease. A search for less toxic homologues is ongoing.

These approaches are based on the idea that the fibrillar deposits are responsible for the clinical manifestations of all of the amyloidoses and are directed at reducing the tissue fibril load and enhancing tissue function. Hence success in any form of clinical amyloidosis, as measured by the disappearance of deposits and increased tissue function, would result in the agent being applied to the TTR diseases.

Two current therapeutic strategies are directed specifically at the diseases resulting from TTR deposition. The success of active or passive Aβ-immunization in mice carrying genes responsible for genetic forms of Alzheimer's disease has stimulated a similar approach to the TTR amyloidoses (Schenk *et al.*, 1999). Whether it will work in the face of significant levels of circulating TTR, representing a potential risk of immune complex disease, remains to be determined. If the immunization can be directed at antigenic sites on the oligomeric, prefibrillar, misfolded TTR complexes, rather than at epitopes on the circulating native molecule, the potential for pathogenic immune complex formation may be avoided. As of yet, no results have been reported.

The second, more traditional, small-molecule strategy is based on the observation that T4 and molecules that bind in the T4 binding pocket of TTR stabilize the tetramer kinetically, reducing the availability of monomer to misfold and enter the oligomeric aggregation pathway (Adamski-Werner *et al.*, 2004). A number of such compounds have been identified and have been shown to both inhibit fibril formation in the test tube and oligomer-induced cell cytoxicity under more physiologic conditions (Reixach *et al.*, 2004). A small-molecule therapeutic has the advantage that, by reducing the amount of monomer available to misfold, it both inhibits the formation of potentially cytotoxic oligomers as well as fibrils and could shift the equilibrium from the deposits, ultimately reducing the fibril mass by increasing the off rate. One drug of this class of agents is just entering a clinical trial in FAP.

References

Adamski-Werner, S. L., Palaninathan, S. K., Sacchettini, J. C., and Kelly, J. W. (2004). Diflunisal analogues stabilize the native state of transthyretin. Potent inhibition of amyloidogenesis. *J. Med. Chem.* 47:355–374.

Almeida, M. R., Alves, I. L., Terazaki, H., Ando, Y., and Saraiva, M. J. (2000). Comparative studies of two transthyretin variants with protective effects on familial amyloidotic polyneuropathy: TTR R104H and TTR T119M. *Biochem. Biophys. Res. Commun.* 270:1024–1028.

Andersson, K., Olofsson, A., Nielsen, E. H., Svehag, S. E., and Lundgren, E. (2002). Only amyloidogenic intermediates of transthyretin induce apoptosis. *Biochem. Biophys. Res. Commun.* 294:309–314.

Ando, Y., Ikegawa, S., Miyazaki, A., Inoue, M., Morino, Y., and Araki, S. (1989). Role of variant prealbumin in the pathogenesis of familial amyloidotic polyneuropathy: fate of normal and variant prealbumin in the circulation. *Arch. Biochem. Biophys.* 274:87–93.

Ando, Y., Nyhlin, N., Suhr, O., Holmgren, G., Uchida, K., el, S., Yamashita, T., Terasaki, H., Nakamura, M., Uchino, M., and Ando, M. (1997). Oxidative stress is found in amyloid deposits in systemic amyloidosis. *Biochem. Biophys. Res. Commun.* 232:497–502.

Ando, Y., Ando, E., Ohlsson, P. I., Olofsson, A., Sandgren, O., Suhr, O., Terazaki, H., Obayashi, K., Lundgren, E., Ando, M., and Negi, A. (1999). Analysis of transthyretin amyloid fibrils from vitreous samples in familial amyloidotic polyneuropathy (Val30Met). *Amyloid Int. J. Exp. Clin. Invest.* 6:119–123.

Andrade, C. (1952). A peculiar form of peripheral neuropathy. Familial atypical generalized amyloidosis with special involvement of the peripheral nerves. *Brain* 75:408–427.

Askanas, V., Engel, W. K., Alvarez, R. B., Frangione, B., Ghiso, J., and Vidal, R. (2000). Inclusion body myositis, muscle blood vessel and cardiac amyloidosis, and transthyretin Val122Ile allele. *Ann. Neurol.* 47:544–549.

B.U.M.C. (Boston University Medical Campus), Boston (February 3, 2004). Available at http://www.bumc.bu.edu/Dept/Content.aspx?departmentid=354&PageID=5530.

Bellovino, D., Morimoto, T., Tosetti, F., and Gaetani, S. (1996). Retinol binding protein and transthyretin are secreted as a complex formed in the endoplasmic reticulum in HepG2 human hepatocarcinoma cells. *Exp. Cell Res.* 222:77–83.

Bellovino, D., Morimoto, T., Pisaniello, A., and Gaetani, S. (1998). *In vitro* and *in vivo* studies on transthyretin oligomerization. *Exp. Cell Res.* 243:101–112.

Benson, M. D. and Dwulet, F. E. (1983). Prealbumin and retinol binding protein serum concentrations in the Indiana type hereditary amyloidosis. *Arthritis Rheum.* 26:1493–1498.

Bergstrom, J., Gustavsson, A., Hellman, U., Sletten, K., Murphy, C. L., Weiss, D. T., Solomon, A., Olofsson, B. O., and Westermark, P. (2005). Amyloid deposits in transthyretin-derived amyloidosis: cleaved transthyretin is associated with distinct amyloid morphology. *J. Pathol.* 206(2): 224–232.

Blake, C. C. and Oatley, S. J. (1977). Protein-DNA and protein-hormone interactions in prealbumin: a model of the thyroid hormone nuclear receptor? *Nature* 268:115–120.

Blake, C. C. F., Geisow, M. J., and Oatley, S. J. (1978). Structure of prealbumin: secondary, tertiary and quaternary interactions determined by Fourier refinement at 1.8 Å. *J. Mol. Biol.* 121:339–356.

Brett, M., Persey, M. R., Reilly, M. M., Revesz, T., Booth, D. R., Booth, S. E., Hawkins, P. N., Pepys, M. B., and Morgan-Hughes, J. A. (1999). Transthyretin Leu12Pro is associated with systemic, neuropathic and leptomeningeal amyloidosis. *Brain* 122:183–190.

Brugler, L., Stankovic, A., Bernstein, L., Scott, F., and O'Sullivan-Maillet, J. (2002). The role of visceral protein markers in protein calorie malnutrition. *Clin. Chem. Lab. Med.* 40:1360–1369.

Cardoso, I., Merlini, G., and Saraiva, M. J. (2003). 4′-iodo-4′-deoxydoxorubicin and tetracyclines disrupt transthyretin amyloid fibrils *in vitro* producing noncytotoxic species: screening for TTR fibril disrupters. *FASEB J.* 17:803–809.

Chevalier, S., Blaner, W. S., Azais-Braesco, V., and Tuchweber, B. (1999). Dietary restriction alters retinol and retinol-binding protein metabolism in aging rats. *J. Gerontol. A Biol. Sci. Med. Sci.* 54:B384-B392.

Chou, C. T., Lee, C. C., Chang, D. M., Buxbaum, J. N., and Jacobson, D. R. (1997). Familial amyloidosis in one Chinese family: clinical, immunological, and molecular genetic analysis. *J. Intern. Med.* 241:327–331.

Coelho, T., Carvalho, M., Saraiva, M., Alves, I., Almeida, M. R., and Costa, P. P. (1993). A strikingly benign evolution of FAP in an individual found to be a compound heterozygote for two TTR mutations: TTR MET 30 and TTR MET 119. *J. Rheumatol.* 20:179.

Connors, L. H., Karbassi, J., Lim, A., Costello, C. E., and Skinner, M. (2002). Senile cardiac amyloidosis: serum levels of S-sulfated transthyretin (abs). The 5th International Symposium on Familial Amyloidotic Polyneuropathy and other Transthyretin Related Disorders. 23.9-24-2002. Boston University. The 5th International Symposium on Familial Amyloidotic Polyneuropathy and Other Transthyretin Related Disorders.

Connors, L. H., Lim, A., Prokaeva, T., Roskens, V. A., and Costello, C. E. (2003). Tabulation of human transthyretin (TTR) variants, 2003. *Amyloid* 10:160–184.

Cooper, D. N. and Youssoufian, H. (1988). The CpG dinucleotide and human genetic disease. *Hum. Genet.* 78:151–155.

Cornwell, G. G., III, Murdoch, W. L., Kyle, R. A., Westermark, P., and Pitkanen, P. (1983). Frequency and distribution of senile cardiovascular amyloid. A clinicopathologic correlation. *Am. J. Med.* 75:618–623.

Costa, R. H., Van Dyke, T. A., Yan, C., Kuo, F., and Darnell, J. E., Jr. (1990). Similarities in transthyretin gene expression and differences in transcription factors: liver and yolk sac compared to choroid plexus. *Proc. Natl. Acad. Sci USA* 87:6589–6593.

Dickson, P. W., Howlett, G. J., and Schreiber, G. (1985). Rat transtyhretin (prealbumin). *J. Biol. Chem.* 260:8214–8219.

Divino, C. M. and Schussler, G. C. (1990). Receptor-mediated uptake and internalization of transthyretin. *J. Biol Chem.* 265:1425–1429.

Eneqvist, T., Lundberg, E., Nilsson, L., Abagyan, R., and Sauer-Eriksson, A. E. (2003). The transthyretin-related protein family. *Eur. J. Biochem.* 270:518–532.

Falk, R. H., Plehn, J. F., Deering, T., Shick, E. C. J., Bosnay, P., Rubinow, A., Skinner, M., and Cohen, A. S. (1987). Sensitivity and specificity of the echocardiographic features of cardiac amyloidosis. *Am. J. Cardiol.* 59:418–422.

Filteau, S. M. (2000). Use of the retinol-binding protein: transthyretin ratio for assessment of vitamin A status during the acute-phase response. *Br. J. Nutr.* 83:513–520.

Fitch, N. J. S., Akbari, M. T., and Ramsden, D. B. (1991). An inherited non-amyloidogenic transthyretin variant, [Ser6]-TTR, with increased thyroxine-binding affinity, characterized by DNA sequencing. *J. Endocrinol.* 129:309–313.

Gallo, G., Picken, M., Frangione, B., and Buxbaum, J. (1988). Nonamyloidotic monoclonal immunoglobulin deposits lack amyloid P component. *Mod. Pathol.* 1:453–456.

Gallo, G., Wisniewski, T., Choi-Miura, N. H., Ghiso, J., and Frangione, B. (1994). Potential role of apolipoprotein-E in fibrillogenesis. *Am. J. Pathol.* 145:526–530.

Getz, R. K., Kennedy, B. G., and Mangini, N. J. (1999). Transthyretin localization in cultured and native human retinal pigment epithelium. *Exp. Eye Res.* 68:629–636.

Goldsteins, G., Andersson, K., Olofsson, A., Dacklin, I., Edvinsson, A., Baranov, V., Sandgren, O., Thylen, C., Hammarstrom, S., and Lundgren, E. (1997). Characterization of two highly amyloidogenic mutants of transthyretin. *Biochemistry* 36:5346–5352.

Gonzalez, G. and Offord, R. E. (1971). The subunit structure of prealbumin. *J. Biochem.* 125:309–317.

Gustavsson, Å., Jahr, H., Tobiassen, R., Jacobson, D. R., Sletten, K., and Westermark, P. (1995). Amyloid fibril composition and transthyretin gene structure in senile systemic amyloidosis. *Lab. Invest.* 73:703–708.

H.G.M.D. (the Human Gene Mutation Database), 1984, Cardiff (May 18, 2005). Available at http://archive.uwcm.ac.uk/uwcm/mg/search/119471.html.

Hammarstrom, P., Schneider, F., and Kelly, J. W. (2001). Trans-suppression of misfolding in an amyloid disease. *Science* 293:2459–2462.

Hammarstrom, P., Sekijima, Y., White, J. T., Wiseman, R. L., Lim, A., Costello, C. E., Altland, K., Garzuly, F., Budka, H., and Kelly, J. W. (2003a). D18G transthyretin is monomeric, aggregation prone, and not detectable in plasma and cerebrospinal fluid: a prescription for central nervous system amyloidosis? *Biochemistry* 42:6656–6663.

Hammarstrom, P., Wiseman, R. L., Powers, E. T., and Kelly, J. W. (2003b). Prevention of transthyretin amyloid disease by changing protein misfolding energetics. *Science* 299:713–716.

Hanes, D., Zech, L. A., Murrell, J., and Benson, M. D. (1996). Metabolism of normal and MET30 transthyretin. *Advanced Food Nutr. Res.* 40:149–153.

Hanyu, N., Ikeda, S., Nakadai, A., Yanagisawa, N., and Powell, H. C. (1989). Peripheral nerve pathological findings in familial amyloid polyneuropathy: A correlative study of proximal sciatic nerve and sural nerve lesions. *Ann. Neurol.* 25:340–350.

Harats, N., Worth, R. M., and Benson, M. D. (1989). Hereditary amyloidosis: Evidence against early amyloid deposition. *Arthritis Rheum.* 32:1474–1476.

Hardell, L., Holmgren, G., Steen, L., Fredrikson, M., and Axelson, O. (1995). Occupational and other risk factors for clinically overt familial amyloid polyneuropathy. *Epidemiology* 6:598–601.

Harkany, T., Garzuly, F., Csanaky, G., Luiten, P. G., Nyakas, C., Linke, R. P., and Viragh, S. (2002). Cutaneous lymphatic amyloid deposits in "Hungarian-type" familial transthyretin amyloidosis: a case report. *Br. J. Dermatol.* 146:674–679.

Herbert, J., Wilcox, J. N., Pham, K. C., Fremeau, R. T., Zaviani, M., Dwork, A., Soprano, D., Makover, A., Goodman, D. S., Zimmerman, E. Z., Roberts, J. L., and Schon, E. A. (1986). Transthyretin: a choroid plexus-specific transport protein in human brain. *Neurology* 36:900–911.

Herlenius, G., Larsson, M., and Ericzon, B. G. (2001). FAP World Transplant Register and domino/sequential register update. *Transplant. Proc.* 33:1367.

Herlenius, G., Wilczek, H. E., Larsson, M., and Ericzon, B. G. (2004). Ten years of international experience with liver transplantation for familial amyloidotic polyneuropathy: results from the Familial Amyloidotic Polyneuropathy World Transplant Registry. *Transplantation* 77:64–71.

Herrick, M. K., DeBruyne, K., Horoupian, D. S., Skare, J., Vanefsky, M. A., and Ong, T. (1996). Massive leptomeningeal amyloidosis associated with a Val30Met transthyretin gene. *Neurology* 47:988–992.

Hodkinson, H. M. and Pomerance, A. (1977). The clinical significance of senile cardiac amyloidosis: a prospective clinicopathological study. *Q. J. Med.* 46:381–387.

Holmgren, G., Wikstrom, L., Lundgren, H. E., and Suhr, O. B. (2004). Discordant penetrance of the trait for familial amyloidotic polyneuropathy in two pairs of monozygotic twins. *J. Intern. Med.* 256:453–456.

Hurshman, A. R., White, J. T., Powers, E. T., and Kelly, J. W. (2004). Transthyretin aggregation under partially denaturing conditions is a downhill polymerization. *Biochemistry* 43:7365–7381.

Ii, K., Kyle, R. A., and Dyck, P. J. (1992). Immunohistochemical characterization of amyloid proteins in sural nerves and clinical associations in amyloid neuropathy. *Am. J. Pathol.* 141:217–226.

Ikeda, S., Nakazato, M., Ando, Y., and Sobue, G. (2002). Familial transthyretin-type amyloid polyneuropathy in Japan: clinical and genetic heterogeneity. *Neurology* 58:1001–1007.

Imasawa, M., Toda, Y., Sakurada, Y., Imai, M., and Iijima, H. (2004). Vitreous opacities in a case of familial amyloidotic polyneuropathy associated with a transthyretin Lys 54. *Acta Ophthalmol. Scand.* 82:635–636.

Ingenbleek, Y. and Young, V. R. (2002). Significance of transthyretin in protein metabolism. *Clin. Chem. Lab. Med.* 40:1281–1291.

Ishida, M., Kajita, Y., Ochi, Y., Hachiya, T., Miyazaki, T., Yoshimura, M., and Ijichi, H. (1982). Measurement of serum thyroxine binding prealbumin in various thyroidal states by radioimmunoassay. *Endocrinol. Jpn.* 29:607–613.

Jacobson, D. R., McFarlin, D. E., Kane, I., and Buxbaum, J. N. (1992). Transthyretin Pro55, a variant associated with early-onset, aggressive, diffuse amyloidosis with cardiac and neurologic involvement. *Hum. Genet.* 89:353–356.

Jacobson, D. R., Alves, I. L., Saraiva, M. J., Thibodeau, S. N., and Buxbaum, J. N. (1995). Transthyretin Ser 6 gene frequency in individuals without amyloidosis. *Hum. Genet.* 95:308–312.

Jacobson, D. R., Pastore, R. D., Yaghoubian, R., Kane, I., Gallo, G., Buck, F. S., and Buxbaum, J. N. (1997). Variant-sequence transthyretin (isoleucine 122) in late-onset cardiac amyloidosis in black Americans. *N. Engl. J. Med.* 336:466–473.

Jacobsson, B. (1989). Localization of transthyretin-mRNA and of immunoreactive transthyretin in the human fetus. *Virchows Arch. A Pathol. Anat.* 415:259–263.

Jacobsson, B., Collins, V. P., Grimelius, L., Pettersson, T., Sandstedt, B., and Carlstrom, A. (1989). Transthyretin immunoreactivity in human and porcine liver, choroid plexus, and pancreatic islets. *J. Histochem. Cytochem.* 37:31–37.

Jarrett, J. T. and Lansbury, P. T., Jr. (1993). Seeding "one-dimensional crystallization" of amyloid: a pathogenic mechanism in Alzheimer's disease and scrapie. *Cell* 73:1055–1058.

Kanda, Y., Goodman, D. S., Canfield, R. E., and Morgan, F. J. (1974). The amino acid sequence of human plasma prealbumin. *J Biol. Chem.* 249:6796–6805.

Kaufman, H. E. and Thomas, L. B. (1959). Vitreous opacities diagnostic of familial primary amyloidosis. *N. Engl. J. Med.* 261:1267–1271.

Kelly, J. W. (1998). The alternative conformations of amyloidogenic proteins and their multi-step assembly pathways. *Curr. Opin. Struct. Biol.* 8:101–106.

Kisilevsky, R., Szarek, W. A., Ancsin, J., Vohra, R., Li, Z., and Marone, S. (2004). Novel glycosaminoglycan precursors as antiamyloid agents: part IV. *J. Mol. Neurosci.* 24:167–172.

Kohno, K., Palha, J. A., Miyakawa, K., Saraiva, M. J., Ito, S., Mabuchi, T., Blaner, W. S., Iijima, H., Tsukahara, S., Episkopou, V., Gottesman, M. E., Shimada, K., Takahashi, K., Yamamura, K., and Maeda, S. (1997). Analysis of amyloid deposition in a transgenic mouse model of homozygous familial amyloidotic polyneuropathy. *Am. J. Pathol.* 150:1497–1508.

Kuchler-Bopp, S., Dietrich, J. B., Zaepfel, M., and Delaunoy, J. P. (2000). Receptor-mediated endocytosis of transthyretin by ependymoma cells. *Brain Res.* 870:185–194.

Kyle, R. A., Gertz, M. A., and Linke, R. P. (1992). Amyloid localized to tenosynovium at carpal tunnel release. Immunohistochemical identification of amyloid type. *Am. J. Clin. Pathol.* 97:250–253.

Levine, R. L. and Stadtman, E. R. (1996). Protein modifications with aging. In: E. L. Schneider & J. W. Rowe (Eds.), *Handbook of the biology of aging* (pp. 184–197). San Diego, Calif.; London: Academic Press.

Lie, J. T. and Hammond, P. I. (1988). Pathology of the senescent heart: anatomic observations on 237 autopsy studies of patients 90 to 105 years old. *Mayo Clinic Proc.* 63:552–564.

Lobato, L., Ventura, A., Beirao, I., Miranda, H. P., Seca, R., Henriques, A. C., Teixeira, M., Sarmento, A. M., and Pereira, M. C. (2003). End-stage renal disease in familial amyloidosis ATTR Val30Met: a definitive indication to combined liver-kidney transplantation. *Transplant. Proc.* 35:1116–1120.

Longo, A. I., Hays, M. T., and Saraiva, M. J. (1997). Comparative stability and clearance of [Met30]transthyretin and [Met119]transthyretin. *Eur. J. Biochem.* 249:662–668.

Makover, A., Moriwaki, H., Ramakrishnan, R., Saraiva, M. J., Blaner, W. S., and Goodman, D. S. (1988). Plasma transthyretin. Tissue sites of degradation and turnover in the rat. *J. Biol. Chem.* 263:8598–8603.

Marchi, N., Fazio, V., Cucullo, L., Kight, K., Masaryk, T., Barnett, G., Vogelbaum, M. A., Kinter, M., Rasmussen, P., Mayberg, M. R., and Janigro, D. (2003). Serum transthyretin monomer as a possible marker of blood-to-CSF barrier disruption. *J. Neurosci.* 23:1949–1955.

Melhus, H., Nilsson, T., Peterson, P. A., and Rask, L. (1991). Retinol-binding protein and transthyretin expressed in HeLa cells form a complex in the endoplasmic reticulum in both the absence and presence of retinol. *Exp. Cell Res.* 197:119–124.

Merlini, G., Anesi, E., Garini, P., Perfetti, V., Obici, L., Ascari, E., Lechuga, M. H., Capri, G., and Gianni, L. (1999). Treatment of AL amyloidosis with 4′-iodo-4′-deoxydoxorubicin: an update [letter]. *Blood* 93:1112–1113.

Munar-Ques, M., Pedrosa, J. L., Coelho, T., Gusmao, L., Seruca, R., Amorim, A., and Sequeiros, J. (1999). Two pairs of proven monozygotic twins discordant for familial amyloid neuropathy (FAP) TTR Met 30. *J. Med. Genet.* 36:629–632.

Nardo, B., Beltempo, P., Bertelli, R., Montalti, R., Vivarelli, M., Cescon, M., Grazi, G. L., Salvi, F., Magelli, C., Grigioni, F., Arpesella, G., Martinelli, G., and Cavallari, A. (2004). Combined heart and liver transplantation in four adults with familial amyloidosis: experience of a single center. *Transplant. Proc.* 36:645–647.

Neumann, P., Cody, V., and Wojtczak, A. (2001). Structural basis of negative cooperativity in transthyretin. *Acta Biochim. Pol.* 48:867–875.

Noguchi, H., Ohta, M., Wakasugi, S., Noguchi, K., Nakamura, N., Nakamura, O., Miyakawa, K., Takeya, M., Suzuki, M., Nakagata, N., Urano, T., Ono, T., and Yamamura, K. (2002). Effect of the intestinal flora on amyloid deposition in a transgenic mouse model of familial amyloidotic polyneuropathy. *Exp. Anim.* 51:309–316.

Noy, N., Slosberg, E., and Scarlata, S. (1992). Interactions of retinol with binding proteins: Studies with retinol-binding protein and with transthyretin. *Biochemistry* 31:11118–11124.

Palha, J. A. (2002). Transthyretin as a thyroid hormone carrier: function revisited. *Clin. Chem. Lab. Med.* 40:1292–1300.

Pepys, M. B., Herbert, J., Hutchinson, W. L., Tennent, G. A., Lachmann, H. J., Gallimore, J. R., Lovat, L. B., Bartfai, T., Alanine, A., Hertel, C., Hoffmann, T., Jakob-Roetne, R., Norcross, R. D., Kemp, J. A., Yamamura, K., Suzuki, M., Taylor, G. W., Murray, S., Thompson, D., Purvis, A., Kolstoe, S., Wood, S. P., and Hawkins, P. N. (2002). Targeted pharmacological depletion of serum amyloid P component for treatment of human amyloidosis. *Nature* 417:254–259.

Potter, M. A. and Luxton, G. (2002). Transthyretin measurement as a screening tool for protein calorie malnutrition in emergency hospital admissions. *Clin. Chem. Lab. Med.* 40:1349–1354.

Qiu, H., Shimada, K., and Cheng, Z. (1992). Chromosomal localization of the mouse prealbumin gene (*Ttr*) by *in situ* hybridization. *Cytogenet. Cell Genet.* 61:186–188.

Raz, A. (1969). The interaction of thyroxine with human plasma prealbumin and with the prealbumin-retinol-binding protein complex. *J. Biol. Chem.* 244:3230–3237.

Redondo, C., Damas, A. M., Olofsson, A., Lundgren, E., and Saraiva, M. J. (2000). Search for intermediate structures in transthyretin fibrillogenesis: soluble tetrameric Tyr78Phe TTR expresses a specific epitope present only in amyloid fibrils. *J. Mol. Biol.* 304:461–470.

Reilly, M. M., Adams, D., Davis, M., Said, G., and Harding, A. E. (1995a). Haplotype analysis of French, British and other European patients with familial amyloid polyneuropathy (MET30 and TYR77). *J. Neurol.* 242:664–668.

Reilly, M. M., Staunton, H., and Harding, A. E. (1995b). Familial amyloid polyneuropathy (TTR ala 60) in north west Ireland: a clinical, genetic, and epidemiological study. *J. Neurol. Neurosurg. Psych.* 59:45–49.

Reixach, N., Deechongkit, S., Jiang, X., Kelly, J. W., and Buxbaum, J. N. (2004). Tissue damage in the amyloidoses: Transthyretin monomers and non-native oligomers are the major cytotoxic species in tissue culture. *Proc. Natl. Acad. Sci. USA* 101:2817–2822.

Ritchie, R. F., Palomaki, G. E., Neveux, L. M., and Navolotskaia, O. (1999a). Reference distributions for the negative acute-phase proteins, albumin, transferrin, and transthyretin: a comparison of a large cohort to the world's literature. *J. Clin. Lab. Anal.* 13:280–286.

Ritchie, R. F., Palomaki, G. E., Neveux, L. M., Navolotskaia, O., Ledue, T. B., and Craig, W. Y. (1999b). Reference distributions for the negative acute-phase serum proteins, albumin, transferrin and transthyretin: a practical, simple and clinically relevant approach in a large cohort. *J. Clin. Lab. Anal.* 13:273–279.

Robbins, J. (2002). Transthyretin from discovery to now. *Clin. Chem. Lab. Med.* 40:1183–1190.

Rocken, C., Saeger, W., and Linke, R. P. (1994). Gastrointestinal amyloid deposits in old age. *Pathol. Res. Pract.* 190:641–649.

Rosales, F. J. (1998). A low molar ratio of retinol binding protein to transthyretin indicates vitamin A deficiency during inflammation: studies in rats and a posterior analysis of vitamin A-supplemented children with measles. *J. Nutr.* 128:1681–1687.

Rosen, H. N., Moses, A. C., Murrell, J. R., Liepnieks, J. J., and Benson, M. D. (1993). Thyroxine interactions with transthyretin: a comparison of 10 different naturally occurring human transthyretin variants. *J. Clin. Endocrinol. Metab.* 77:370–374.

Salvi, F., Salvi, G., Volpe, R., Mencucci, R., Plasmati, R., Michelucci, R., Gobbi, P., Santangelo, M., Ferlini, A., Forabosco, A., and Tassinari, C. A. (1993). Transthyretin-related TTR hereditary amyloidosis of the vitreous body: clinical and molecular characterization in two Italian families. *Ophthalmol. Paediatr. Genet.* 14:9–16.

Samadani, U. and Costa, R. H. (1996). The transcriptional activator hepatocyte nuclear factor 6 regulates liver gene expression. *Mol. Cell Biol.* 16:6273–6284.

Schenk, D., Barbour, R., Dunn, W., Gordon, G., Grajeda, H., Guido, T., Hu, K., Huang, J., Johnson-Wood, K., Khan, K., Kholodenko, D., Lee, M., Liao, Z., Lieberburg, I., Motter, R., Mutter, L., Soriano, F., Shopp, G., Vasquez, N., Vandevert, C., Walker, S., Wogulis, M., Yednock, T., Games, D., and Seubert, P. (1999). Immunization with amyloid-beta attenuates Alzheimer-disease-like pathology in the PDAPP mouse. *Nature* 400:173–177.

Schreiber, G., Tsykin, A., Aldred, A. R., Thomas, T., Fung, W. P., Dickson, P. W., Cole, T., Birch, H., De Jong, F. A., and Milland, J. (1989). The acute phase response in the rodent. *Ann. N. Y. Acad. Sci.* 557:61–85.

Schreiber, G. and Richardson, S. J. (1997). The evolution of gene expression, structure and function of transthyretin. *Comp. Biochem. Physiol.* 116B:137–160.

Schreiber, G. (2002). The evolutionary and integrative roles of transthyretin in thyroid hormone homeostasis. *J. Endocrinol.* 175:61–73.

Sekijima, Y., Tokuda, T., Kametani, F., Tanaka, K., Maruyama, K., and Ikeda, S. (2001). Serum transthyretin monomer in patients with familial amyloid polyneuropathy. *Amyloid* 8:257–262.

Sekijima, Y., Hammarstrom, P., Matsumura, M., Shimizu, Y., Iwata, M., Tokuda, T., Ikeda, S., and Kelly, J. W. (2002). The novel highly destabilized A25T TTR variant rapidly misfolds, resulting in CNS amyloidosis-degradation of a highly amyloidogenic variant as a protective mechanism? The 5th International Symposium on Familial Amyloidotic Polyneuropathy and Other Transthyretin Related Disorders, Matsumoto, Japan 42. 9-24-2002.

Sekijima, Y., Wiseman, R. L., Matteson, J., Hammarstrom, P., Miller, S. R., Sawkar, A. R., Balch, W. E., and Kelly, J. W. (2005). The biological and chemical basis for tissue-selective amyloid disease. *Cell* 121:73–85.

Skinner, M., Connors, L. H., Rubinow, A., Libbey, C., Sipe, J. D., and Cohen, A. S. (1985). Lowered prealbumin levels in patients with familial amyloid polyneuropathy (FAP) and their non-affected but at risk relatives. *Am. J. Med. Sci.* 289:17–21.

Skinner, M., Harding, J., Skare, I., Jones, L. A., Cohen, A. S., Milunsky, A., and Skare, J. (1992). A new transthyretin mutation associated with amyloidotic vitreous opacities. Asparagine for isoleucine at position 84. *Ophthalmology* 99:503–508.

Soares, M. L., Buxbaum, J., Sirugo, G., Coelho, T., Sousa, A., Kastner, D., and Saraiva, M. J. (1999). Genetic anticipation in Portuguese kindreds with familial amyloidotic polyneuropathy is unlikely to be caused by triplet repeat expansions. *Hum. Genet.* 104:480–485.

Soares, M. L., Centola, M., Chae, J., Saraiva, M. J., and Kastner, D. L. (2003). Human transthyretin intronic open reading frames are not independently expressed *in vivo* or part of functional transcripts. *Biochim. Biophys. Acta* 1626:65–74.

Soares, M. L., Coelho, T., Sousa, A., Holmgren, G., Saraiva, M. J., Kastner, D. L., and Buxbaum, J. N. (2004). Haplotypes and DNA sequence variation within and surrounding the transthyretin gene: genotype-phenotype correlations in familial amyloid polyneuropathy (V30M) in Portugal and Sweden. *Eur. J. Hum. Genet.* 12:225–237.

Soares, M. L. (2005). Susceptibility and modifier genes in Portuguese transthyretin V30M amyloid polyneuropathy: complexity in a single-gene disease. *Hum. Mol. Genet.* 14:543–553.

Sousa, A., Andersson, R., Drugge, U., Holmgren, G., and Sandgren, O. (1993). Familial amyloidotic polyneuropathy in Sweden: geographical distribution, age of onset, and prevalence. *Hum. Hered.* 43:288–294.

Sousa, J. L., Grandela, C., Fernandez-Ruiz, J., de Miguel, R., de Sousa, L., Magalhaes, A. I., Saraiva, M. J., Sousa, N., and Palha, J. A. (2004). Transthyretin is involved in depression-like behaviour and exploratory activity. *J. Neurochem.* 88(5), 1052–1058.

Sousa, M. M., Yan, S. D., Stern, D., and Saraiva, M. J. (2000). Interaction of the receptor for advanced glycation end products (RAGE) with transthyretin triggers nuclear transcription factor kB (NF-kB) activation. *Lab. Invest.* 80:1101–1110.

Sousa, M. M., Cardoso, I., Fernandes, R., Guimaraes, A., and Saraiva, M. J. (2001). Deposition of transthyretin in early stages of familial amyloidotic polyneuropathy: evidence for toxicity of nonfibrillar aggregates. *Am. J. Pathol.* 159:1993–2000.

Sousa, M. M. and Saraiva, M. J. (2001). Internalization of transthyretin. Evidence of a novel yet unidentified receptor-associated protein (RAP)-sensitive receptor. *J. Biol. Chem.* 276:14420–14425.

Sousa, M. M., Fernandes, R., Palha, J., Taboada, A., Vieira, P., and Saraiva, M. (2002). Evidence for early cytotoxicity aggregates in transgenic mice for human transthyretin Leu55Pro. *Am. J. Pathol.* 161:1935–1948.

Svendsen, I. H., Steensgaard-Hansen, F., and Nordvag, B. Y. (1998). A clinical, echocardiographic and genetic characterization of a Danish kindred with familial amyloid transthyretin methionine 111 linked cardiomyopathy. *Eur. Heart J.* 19:782–789.

Tagoe, C. E., Jacobson, D. R., Gallo, G., and Buxbaum, J. N. (2003). Mice transgenic for human TTR have the same frequency of renal TTR deposition whether maintained in conventional or specific pathogen-free environments. *Amyloid J. Protein Folding Disord.* 10:262–266.

Tagoe, C. E., French, D., Gallo, G., and Buxbaum, J. N. (2004). Amyloidogenesis is neither accelerated nor enhanced by injections of preformed fibrils in mice transgenic for wild-type human transthyretin: the question of infectivity. *Amyloid* 11:21–26.

Tajiri, T., Ando, Y., Hata, K., Kamide, K., Hashimoto, M., Nakamura, M., Terazaki, H., Yamashita, T., Kai, H., Haraoka, K., Imasato, A., Takechi, K., Nakagawa, K., Okabe, H., and Ishizaki, T. (2002). Amyloid formation in rat transthyretin: effect of oxidative stress. *Clin. Chim. Acta.* 323:129–137.

Takaoka, Y., Ohta, M., Miyakawa, K., Nakamura, O., Suzuki, M., Takahashi, K., Yamamura, K., and Sakaki, Y. (2004). Cysteine 10 is a key residue in amyloidogenesis of human transthyretin Val30Met. *Am. J. Pathol.* 164:337–345.

Teng, M. H., Yin, J. Y., Vidal, R., Ghiso, J., Kumar, A., Rabenou, R., Shah, A., Jacobson, D. R., Tagoe, C., Gallo, G., and Buxbaum, J. (2001). Amyloid and nonfibrillar deposits in mice transgenic for wild-type human transthyretin: a possible model for senile systemic amyloidosis. *Lab. Invest.* 81:385–396.

Tsuzuki, T., Mita, S., Maeda, S., Araki, S., and Shimada, K. (1985). Structure of the human prealbumin gene. *J. Biol. Chem.* 260:12224–12227.

Van Jaarsveld, P., Branch, W. T., and Robbins, J. (1973). Polymorphism of Rhesus monkey serum prealbumin. *J. Biol. Chem.* 248:7898–7903.

Vieira, A. V., Sanders, E. J., and Schneider, W. J. (1995). Transport of serum transthyretin into chicken oocytes. *J. Biol. Chem.* 270:2952–2956.

Ward, W. F. (2002). Protein degradation in the aging organism. *Prog. Mol. Subcell. Biol.* 29:35–42.

Wei, L. (2004). Deposition of transthyretin amyloid is not accelerated by the same amyloid *in vivo*. *Amyloid* 11:113–120.

Weizmann Institute of Science/GeneCards, Rehovot (December 31, 2004). Available at http://bioinfo.weizmann.ac.il/cards-bin/carddisp?TTR&search=TTR&suff=txt.

Westermark, P., Pitkänen, P., Benson, L., Vahlquist, A., Olofsson, B. O., and Cornwell, G. G., III (1985). Serum prealbumin and retinol-binding protein in the prealbumin-related senile and familial forms of systemic amyloidosis. *Lab. Invest.* 52:314–318.

Westermark, P., Sletten, K., Johansson, B., and Cornwell, G. G., III (1990). Fibril in senile systemic amyloidosis is derived from normal transthyretin. *Proc. Natl. Acad. Sci. USA* 87:2843–2845.

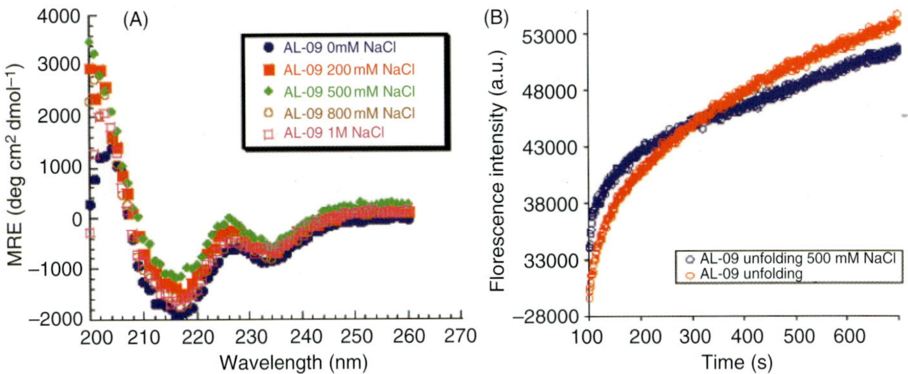

Figure 9.4. (A) The structure of AL-09 is not affected by NaCl. Far UV-CD spectra of AL-09 in the presence of different concentrations of NaCl. All reactions were done in the presence of 10 mM Tris-HCl at pH 7.4. (B) Unfolding kinetics of AL-09 in 10 mM Tris-HCl at pH 7.4 in the presence and the absence of 500 mM NaCl followed by the changes in fluorescence spectroscopy corresponding with Trp 35.

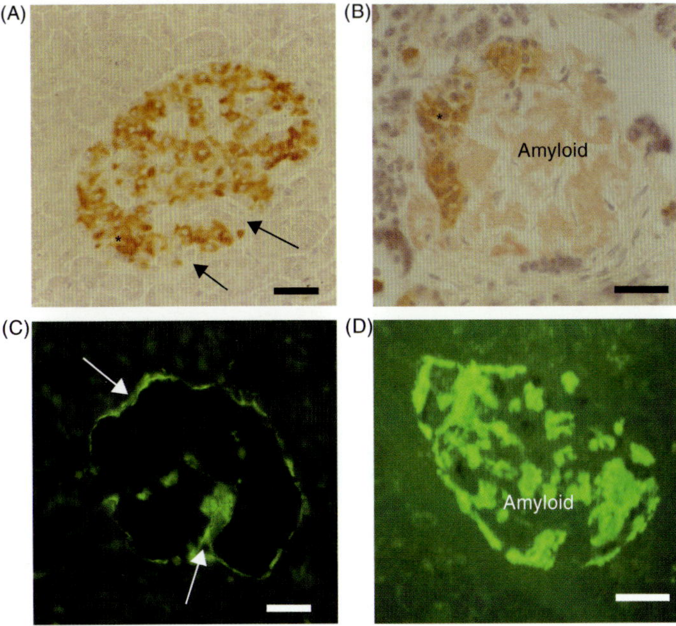

Figure 10.1. Distribution of islet amyloid and insulin-containing β-cells in the pancreas of nondiabetic (A) and type 2 diabetic patients (B, C, D). (A) Immunolabeling for insulin (brown) demonstrates that 60–80% of the islet mass contains β-cells. Capillaries (arrows) surround and penetrate into the islet centre. (B) Amyloid stained with Congo red (amorphous pink color) fills $>$ 60% of the islet space. The remaining β-cells (*) occupy $<$ 30% of the islet. (C) Islet from a subject diagnosed for $>$ 10 years with type 2 diabetes; amyloid labeled with thioflavine S (green fluorescence) is perivascular, adjacent to capillaries around and within the islet. (D) Amyloid in a type 2 diabetic subject (diabetes duration $>$ 10 years) (labeled with thioflavine S) occupies more than 80% of the islet. Scale bars = 20 μm.

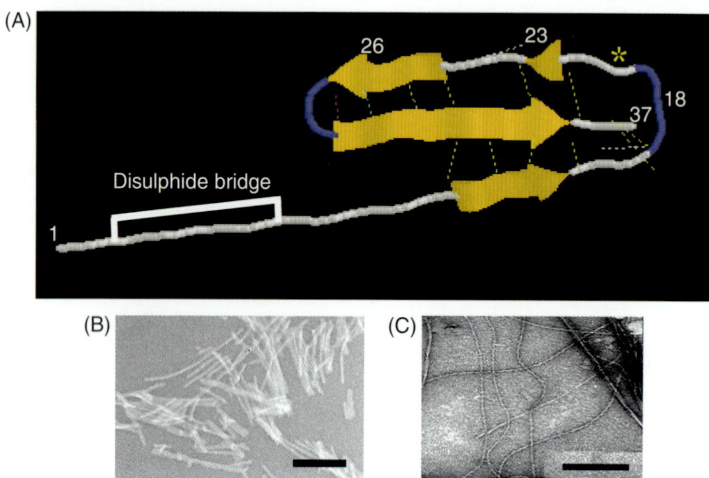

Figure 10.4. Molecular model of folded human IAPP and amyloid fibrils. (A) Hypothetical linear model of a folded β-sheet monomer of hIAPP generated using Insight II. Three major β-strands are predicted for hIAPP with hydrogen bonds (yellow dotted lines) between the strand regions; the region containing the cysteine bridge may not be incorporated into the fibril core. β-Turns in the region 17–19 and 28–29 are predicted. Serine 20 (asterisk) is replaced by glycine in a small number of Asian diabetic subjects. Replacement of residues 18, 23, 26 in rodent IAPP with the relevant human amino acids confers amyloidogenicity to rodent IAPP. (B) Amyloid fibrils extracted from the pancreas of a type 2 diabetic subject. Fibrils are short, straight, and do not show twisted fibril structure. (C) Fibrils formed *in vitro* from synthetic human IAPP demonstrating helical twisting of protofilaments. Scale bars = 200 nm.

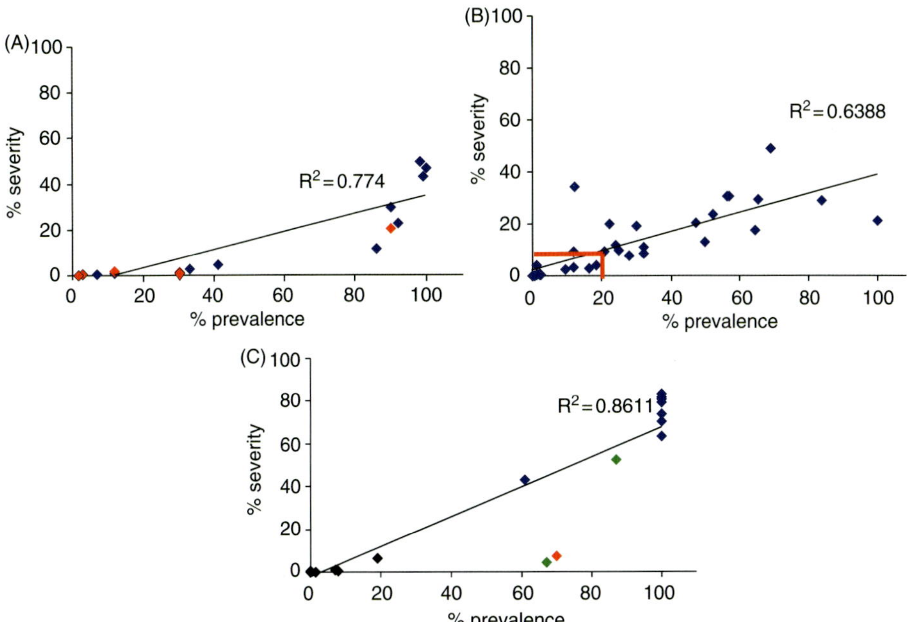

Figure 10.6. The relationship between islet amyloid prevalence (% islets affected) and amyloid severity (% islet area occupied) in human pancreas (A, B) and *Macaca mulatta* monkeys (C). (A) Data redrawn from Westermark *et al.* (1972) showing correlation between prevalence and severity in pancreatic tissue from 10 diabetic (blue) and 7 nondiabetic (red) patients. (B) Data was derived from studies on postmortem human diabetic pancreas (reported in Clark *et al.*, 1988); there was a significant relationship between prevalence and severity ($r^2 = 0.64$, $p < 0.05$); more than 50% of diabetic subjects had less than 10% islet area replaced in less than 20% of islets (red line). (C) Cross-sectional data from monkeys redrawn from Clark and Nilsson (2004) demonstrating a positive linear relationship of amyloid prevalence and severity in nondiabetic (red), glucose intolerant (green), and diabetic (blue) animals; in diabetes there was 100% prevalence with > 80% severity.

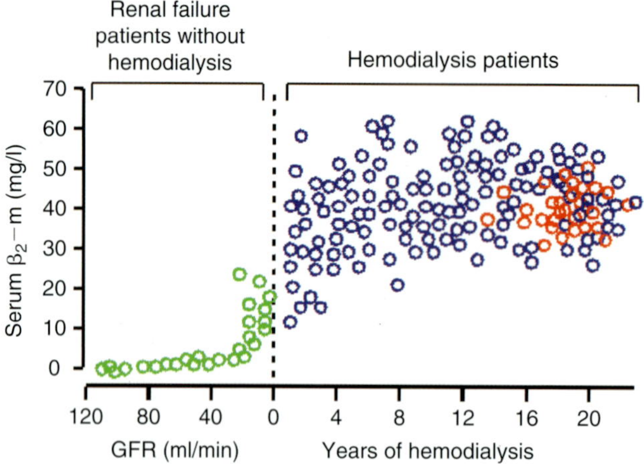

Figure 11.2. Serum β_2m concentrations in patients with chronic renal failure and those undergoing hemodialysis. The glomerular filtration rate (GFR) of patients in renal failure who are not undergoing hemodialysis is shown in green; patients undergoing hemodialysis who lack symptoms of DRA are shown in blue; and hemodialysis patients who have clinical symptoms of DRA are shown in red. The figure was kindly supplied by Professors Naiki and Gejyo (Department of Pathology, Fukui Medical University, Fukui, Japan, and Division of Clinical Nephrology and Rheumatology, Niigata University Graduate School of Medical and Dental Science, Niigata, Japan, respectively).

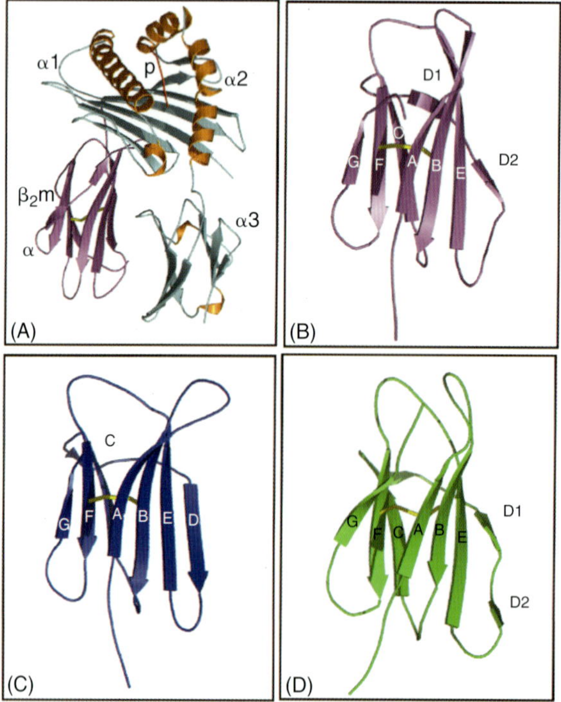

Figure 11.5. Structures of β_2m. (A) Ribbon diagram of the MHC class I molecule showing the extracellular domains: β_2m (purple), the heavy-chain subunits (α1–3), and the bound peptide (p) (red) are shown. (B) Enlargement of the structure of MHC bound β_2m. The disulfide bond that links strands B and F is shown in yellow. Structures were taken from PDB code 1UVQ (Siebold *et al.*, 2004). (C) Crystal structure of monomeric human β_2m. The single disulfide bond is shown in yellow. The structure was created using the coordinates 1LDS (Trinh *et al.*, 2002). Solution structure of monomeric human β_2m, determined by ^{1}H NMR spectroscopy, structure taken from PDB code 1JNJ (Verdone *et al.*, 2002). Each β-strand A–G is labeled. Note that the structure of the D strand differs significantly in (C) compared with (B) and (D). Diagrams were drawn using the program PyMol (DeLano, 2002).

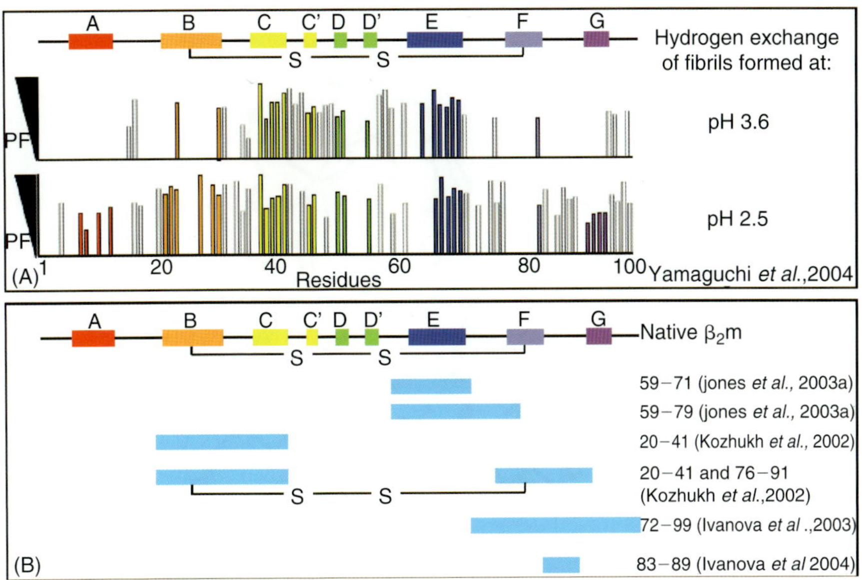

Figure 11.7. (A) Protection factors (PF) of residues in β_2m fibrils formed at pH 3.6 and pH 2.5 elucidated by hydrogen exchange. Hydrogen exchange data were kindly supplied by Professor Goto, Institute for Protein Research, Osaka University, Osaka, Japan (Yamaguchi *et al.*, 2004). (B) Secondary structural elements of human β_2m colored from the N-terminus (red) to the C-terminus (violet). Synthetic peptides shown to form fibrils *in vitro* are shown in blue with the residues involved in each peptide shown alongside (Kozhukh *et al.*, 2002; Ivanova *et al.*, 2003; Jones *et al.*, 2003a; Ivanova *et al.*, 2004).

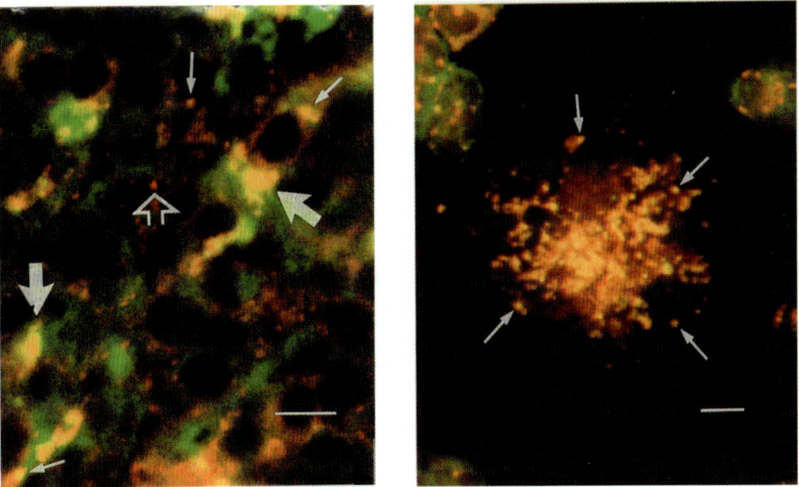

Figure 12.3. Confocal microscopic photomicrograph of serum amyloid A (SAA) and lysosome-associated membrane proteins (LAMP-1 and LAMP-2): a double-labeled spleen section (top), and a cytocentrifuged macrophage from amyloidotic mice. Fluorescein-labeled SAA and Texas red–labeled LAMP-1/LAMP-2 were colocalized, evident by the yellow color resulting from image overlays; small arrows indicate lysosomes and because of tight cell packing in the spleen section made the resolution of individual perikaryal (large arrows) lysosomes difficult. Bar = 10 μm. Reproduced from Chronopoulos S. *et al.*, 1994. *Journal of Pathology* 173: 361–369. © Pathological Society of Great Britain and Ireland. Permission is granted by John Wiley & Sons Ltd on behalf of PathSoc.

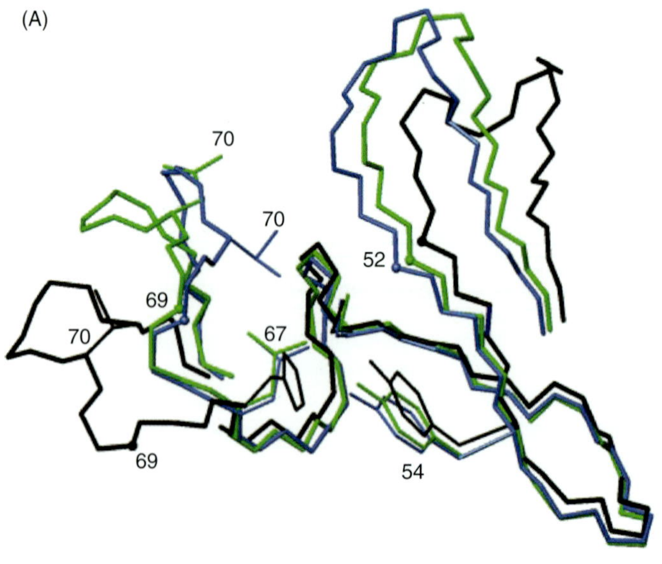

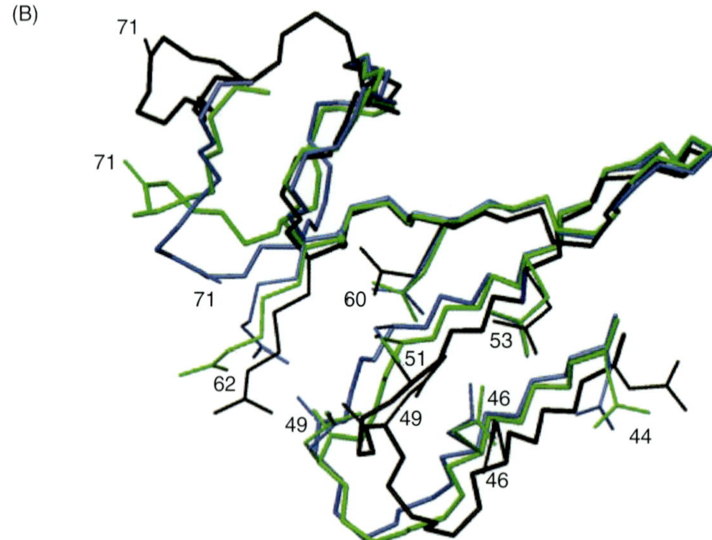

Figure 14.6. (A) Rearrangement of the β-domain of human lysozyme as the result of the mutations at residues 67 and 70. The figure shows residues 44–74, and the situation of the D67H (black), T70N (green), and wild-type (blue) proteins are overlaid. Side chains involved in hydrogen bonding are shown in bold, but all other side chains have been omitted for clarity. (B) As (A) but with the image rotated. Structures were drawn from PDB files as described in the caption of Figure 14.5.

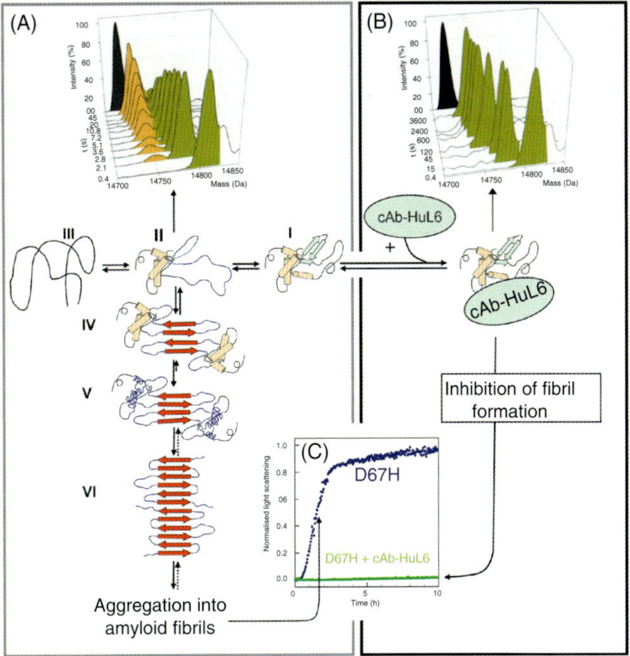

Figure 14.9. *Gray Box*: Schematic representation of the proposed mechanism for amyloid fibril formation by lysozyme. Under some physiologic conditions, the variant proteins (I) populate transiently an intermediate species (II) as revealed by H/D exchange experiments analyzed by ESI-MS (panel A). In these intermediate species, the β-domain and the C-helix are cooperatively unfolded whereas the remaining of the α-domain is in its native-like structure. These intermediate species then self-associate through the newly exposed aggregation-prone regions (IV) via the formation of intermolecular interactions to initiate fibril formation. Further rearrangement (V and VI) is likely to occur in the remainder of the structure, including the recruitment of additional regions of the polypeptide chain into the β-sheet structure prior to the formation of mature fibrils. Note also that the disulfide bridges are not represented in this scheme although they are present in the fibrils. *Black Box*: Proposed mechanism for the inhibition of fibril formation by a camelid antibody fragment. The electrospray mass spectra of D67H lysozyme in the presence of an equimolar concentration of the antibody fragment show a single peak (panel B), whose mass deceases with the length of time for which the exchange was allowed to proceed. The peaks of the species of lower mass species observed in the spectra of the free D67H variant (peaks colored yellow in panel A) and that result from a locally cooperative unfolding of the β-domain and the C-helix (Canet *et al.*, 2002) are therefore not observed in the spectra of the D67H protein in the presence of the antibody fragment. This result indicates that the binding of the antibody fragment to the D67H variant restores the stability and global cooperativity that is characteristic of wild-type lysozyme. (From Dumoulin *et al.*, 2003). Similar results have been obtained for the I56T variant (Dumoulin *et al.*, 2005a). Thus, in the presence of the antibody fragment, the variant proteins do not populate the partially folded intermediate (species II, box on the left-hand side) that otherwise can initiate the aggregation process. The result of antibody binding is therefore to prevent the ready conversion of the lysozyme variants into their aggregated states (panel C). Figure adapted from Dumoulin M., Bellotti V., and Dobson C. M. (2005b). Human lysozyme as an amyloidogenic protein. In: Sipe J. D. (eds), *The beta pleated sheet conformation and disease*, VCH Verlag GmbH & Co KgaA; pp. 635–656.

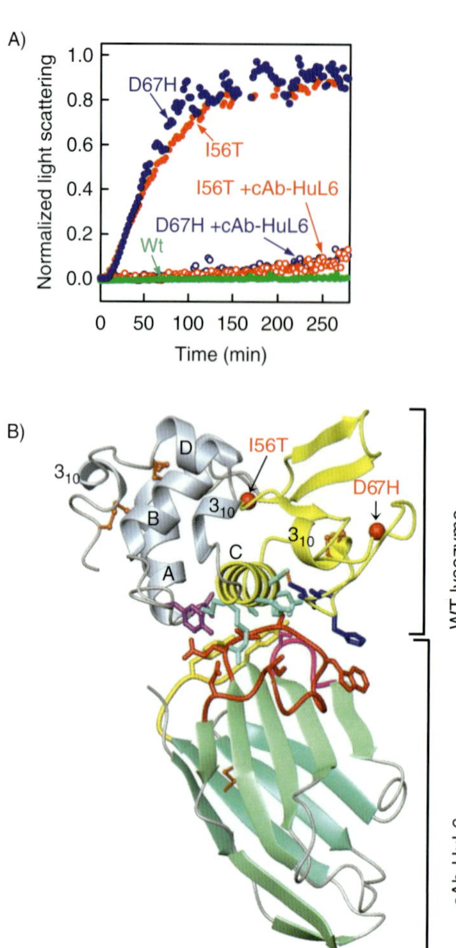

Figure 14.10. (A) Time course of the aggregation of the I56T and D67H variant lysozymes in the absence (● and ● for the I56T and D67H variants, respectively) and presence (○ and ○ for the I56T and D67H variants, respectively) of an equimolar quantity of cAb-HuL6 as monitored by light scattering. Data are also shown for wild-type lysozyme in the absence of cAb-HuL6 (●). The lysozyme concentration was 0.1 mg/mL in all cases, and the data were recorded at 65°C and pH 5.0 while the solutions were stirred. (B) Ribbon representation of the x-ray structure of wild-type lysozyme complexed with the cAb-HuL6. The β-strands in cAb-HuL6 are colored green. The lysozyme molecule is shown in light blue, the helices are labeled, and the disulfide bridges are colored orange. The sites of the I56T and D67H mutations are shown by a red ball, and the region of the molecule that has been found to unfold transiently in a locally cooperative manner in the amyloidogenic variants is colored yellow. The side chains of residues constituting the epitope are shown in violet, light blue, and dark blue for the α-domain, C-helix, and β-domain of lysozyme, respectively. Those residues constituting the paratope are shown in yellow, pink, and red and are from the CDR1, CDR2, and CDR3 loops of the cAb-HuL6 structure, respectively. (C) Ribbon diagram of lysozyme showing the Cα atoms of those residues significantly affected by the binding of cAb-HuL6 in space-filling representation as determined by NMR analysis. Those residues affected only in the I56T variant are shown in pink, those affected only in the D67H variant are shown in blue, and residues affected in both variants are shown in green. The region of the molecule that unfolds transiently in a locally cooperative manner is colored yellow. The lysozyme structure was generated from the x-ray coordinates (PDB entry 1LZ1) and the figure produced using MOLMOL (Koradi *et al.*, 1996). Perturbations to the NMR chemical shifts upon the binding of the antibody fragment include the amide resonance of residue 56 for the I56T variant (Dumoulin *et al.*, 2005a) and of residues 56 and 57 for the D67H variant (Dumoulin *et al.*, 2003). These two residues, which are themselves the locations of two of the well-defined pathogenic mutations of lysozyme that give rise to amyloid disease, are far from the binding interface with the antibody but are located in the interface between the α- and β-domain. This observation suggests that the antibody acts to restore the global cooperativity of the variant protein at least in part through the transmission of long-range conformational effects to the interface between the two structural domains of the protein.

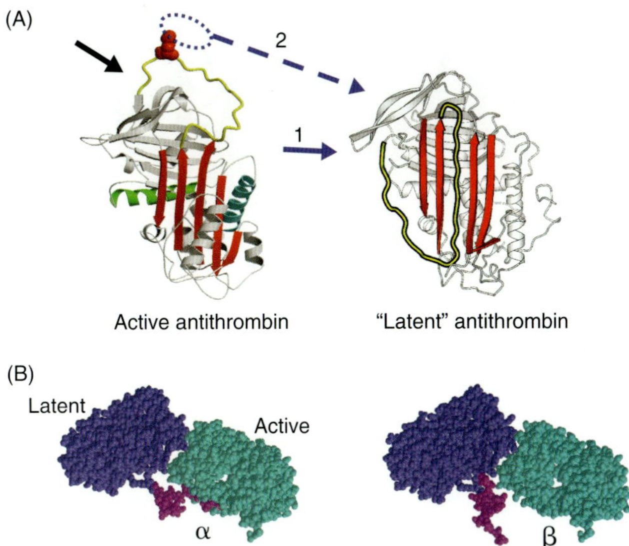

Figure 15.2. Monomeric transition, propagation, and glycoform selectivity. (A) Antithrombin is held in its active state by its distal hinge, s1C, the first strand of sheet C (arrowed, left), which prevents the reactive center peptide loop from entering fully into the middle of the A-sheet. Kinetic perturbations of s1C, however, allow some 3% of the total antithrombin per day to undergo this permanent change (arrow 1) to the inactive latent form. The full insertion of the reactive loop leaves the s1C position vacant, and this is rapidly filled by the reactive loop of a second active molecule (arrow 2) with an effective propagation of inactivity. The threat to health comes when mutations additionally destabilize s1C allowing massive transitions to the latent form, particularly when body temperatures rise, as with the fevers of infections. The loss of activity is magnified by the propagation of the inactivity of the mutant latent molecules to normal active molecules as shown in the space-filling representation in (B). The loss of active inhibitor due to dimer formation is compounded by the increased avidity of binding to the most active form of antithrombin, its glycoform β-antithrombin. The β-glycoform (on right) lacks an oligosaccharide that is encircled in the predominant but less active α_1-antithrombin (on left) impedes the linkage to the latent form.

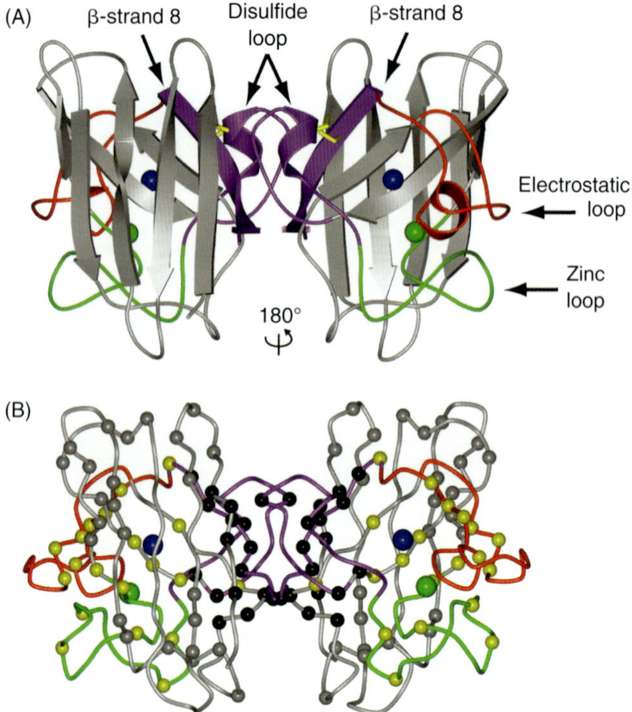

Figure 16.1. The human SOD1 dimeric structure and location of fALS mutations. (A) Each SOD1 monomer displays an eight-stranded Greek key β-barrel fold. The zinc and electrostatic loop elements are shown in green and red, respectively. The disulfide loop and β-strand 8 that form the bulk of the dimer interface are shown in purple. The disulfide bond between Cys 57 and Cys 146 is shown in yellow. The copper and zinc ions are represented as blue and green spheres, respectively. The molecular two-fold axis is indicated by the 180° symbol. (B) Human SOD1 showing the locations of fALS mutations. The color coding for the protein backbrone is the same as in (A). Those fALS mutations predicted to directly affect SOD1 dimerization are shown as black spheres, those that are predicted to affect dimerization indirectly through defects in metal binding are shown as yellow spheres, and those that fall in the β-barrel are shown as gray spheres. Reproduced from Doucette, P. A. *et al.* Dissociation of human copper-zinc superoxide dismutase dimers using chaotrope and reductant. Insights into the molecular basis for dimer stability. *J. Biol. Chem.* 279:54558–54566.

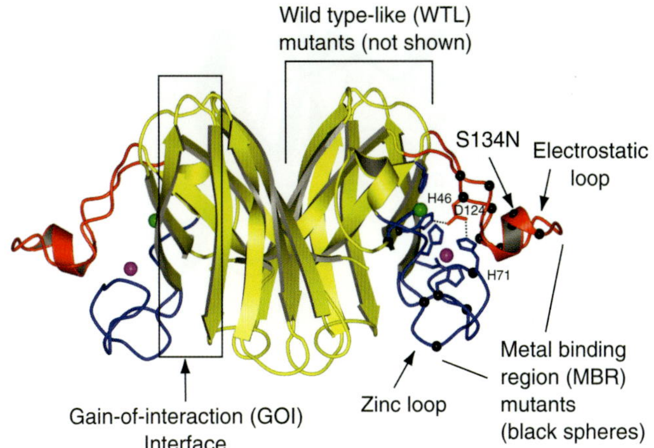

Figure 16.2. Negative design of human SOD1 and the locations of the wild type–like (WTL) and metal binding region (MBR) fALS mutations. In the left subunit, the gain-of-interaction (GOI) interface is boxed. In the wild-type protein, the zinc loop (blue) projects from the plane of the paper toward the viewer, preventing SOD1-SOD1 protein-protein interactions from occurring at the edge strands. The copper and zinc ions are shown as green and magenta spheres, respectively. In the right subunit, the positions of the MBR mutations are shown as black spheres and the WTL mutations (not shown for clarity) are scattered throughout the β-barrel depicted in yellow. Metal binding stabilizes the conformation of the zinc and electrostatic loop elements and therefore plays an intimate role in the negative design of the molecule (see text).

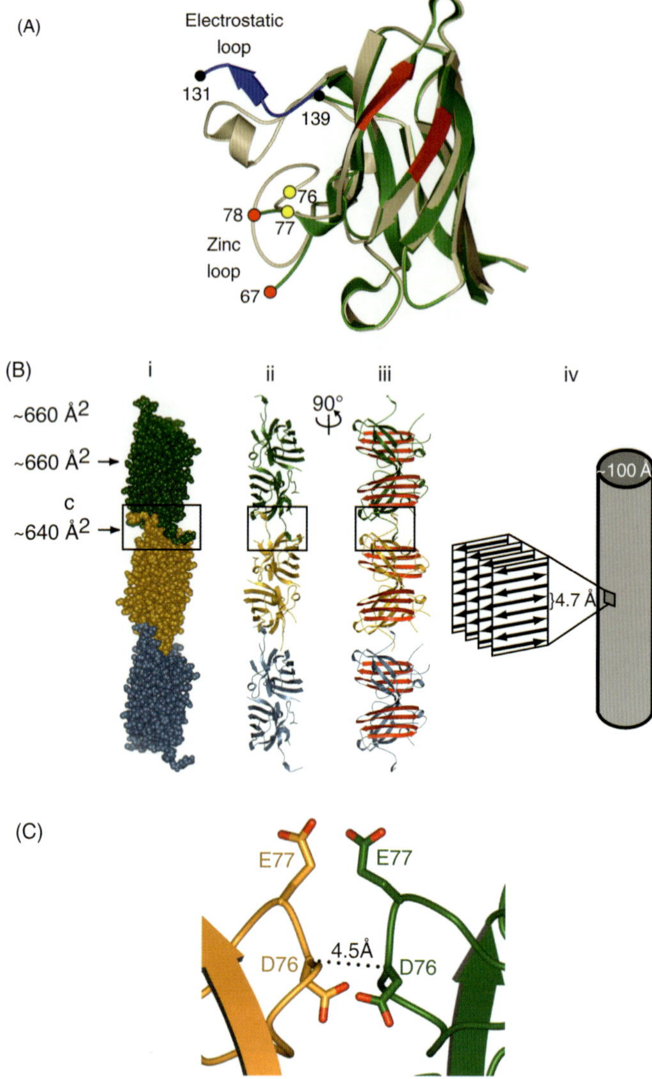

Figure 16.4. Linear, amyloid-like filaments formed by pathogenic SOD1 mutants H46R and S134N. (A) S134N and apo-H46R monomers (green) superimposed on human WT SOD1 (gray). In both structures, the zinc and electrostatic loops are disordered. Residues N- and C-terminal to these disordered regions are indicated by black and red spheres for the electrostatic and zinc loops, respectively. Residues of the electrostatic loop (blue) interact with the exposed cleft (red) of adjacent molecules in the crystal lattice (see boxed region in panel B). Asp 76 and Glu 77 of the zinc loop in the wild-type enzyme are represented by yellow spheres (see panel C). (B) Gain-of-interaction (GOI) in pathogenic SOD1 gives rise to cross-β fibrils in two different crystal systems. The SOD1 filaments are represented by three dimers from top to bottom in green, gold, and blue. The GOI interface is boxed. β strands 1, 2, 3, and 6 are shown in red. The long axes of the β strands run perpendicular to the long axis of the filament, an architecture similar to the "cross-β" structure observed in amyloid fibers as shown schematically in (iv). (C) Negative design of SOD1 prevents formation of the GOI interface. When metals are bound, the zinc loop elements are well-ordered and prevent self-association into the linear filaments through electrostatic and steric considerations (see text) Reproduced from Elam, J. S., *et al.* (2003a). An alternative mechanism of bicarbonate-mediated peroxidation by copper-zinc superoxide dismutase. *J. Biol. Chem.* 278:21032–21039.

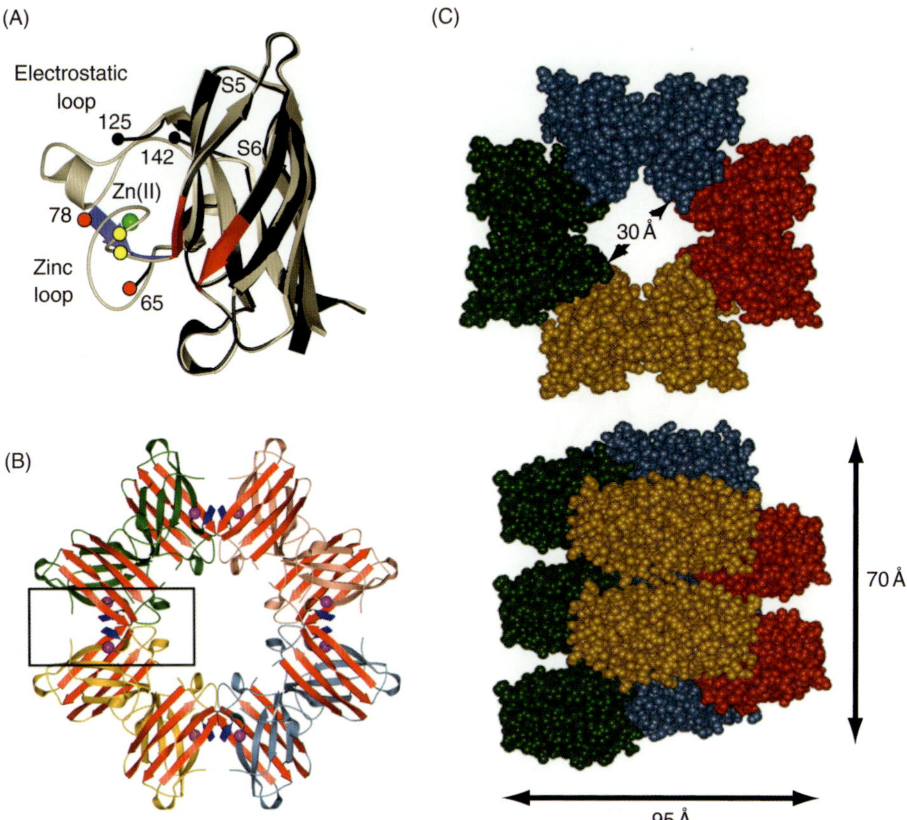

Figure 16.5. Helical, "pore-like" filaments formed between pathogenic SOD1 dimers. (A) Zn-H46R SOD1 monomer (black) superimposed on human wild-type SOD1 (gray). Portions of both the zinc and electrostatic loops are disordered. Residues N- and C-terminal to these disordered regions are indicated by black and red spheres for the electrostatic and zinc loop elements, respectively. The zinc ion is represented by a green sphere. This disorder exposes a cleft on the β-barrel (red). The GOI interface between Zn-H46R molecules is formed by reciprocal interactions between residues of the zinc loop (blue) with the exposed edges of β-strands 5 and 6 (red) of adjacent molecules in the crystal lattice. (B) The GOI in Zn-H46R forms a hollow helical pore-like structure. β-Strands 1, 2, 3, and 6 are shown in red. Residues of the zinc loop undergo a conformational change and form a short β strand (blue) that reciprocally add to this β-sheet in the neighboring dimers, stabilizing the GOI interface. Zn ions are shown in magenta spheres. (C) Annular pore-like structure observed for Zn-loaded H46R SOD1 similar to the pore-like structures observed in proteins that cause Parkinson's disease and Alzheimer's disease (see text). Reproduced from Elam, J. S., *et al.* (2003a). An alternative mechanism of bicarbonate-mediated peroxidation by copper-zinc superoxide dismutase. *J. Biol. Chem.* 278: 21032–21039.

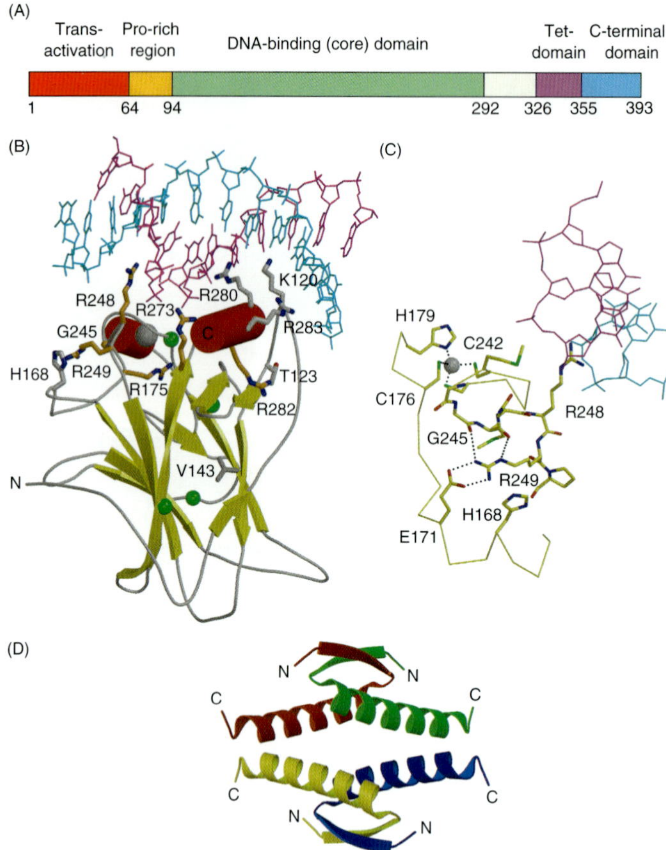

Figure 17.1. Structure of human p53. (A) Schematic view of the domain structure of human p53: The 393-residue protein contains an N-terminal transactivation domain, a proline-rich region, a DNA-binding (core) domain, a tetramerization domain (Tet-domain) and a C-terminal autoregulatory domain (see text for further details). (B) Structure of the DNA-binding (core) domain in complex with consensus DNA (PDB code 1TSR, molecule B). A β-sandwich provides the basic scaffold for a loop-sheet-helix motif and two large loops, L2 and L3, tethered by a zinc ion, which interact with the major and minor groove of the DNA, respectively. The zinc ion is shown as a gray sphere. The six hot-spot sites Arg175, Gly245, Arg248, Arg249, Arg273, and Arg282, which are most frequently mutated in human cancer, are highlighted in orange. The mutation sites in the superstable quadruple mutant M133L/V203A/N239Y/N268D are highlighted as small green spheres. (C) Close-up view of loops L2 and L3 in the DNA-binding surface, including the zinc coordination sphere, in the structure of wild type in complex with consensus DNA (PDB code 1TSR, molecule B). The orientation of the molecule is different to that shown in panel (B). The zinc ion is depicted as a gray sphere. Specific interactions mediated via the guanidinium group of Arg249 are highlighted with dotted lines. These include hydrogen bonds with backbone oxygens of residues Gly245 and Met246 on the same loop and a salt bridge with Glu171 on the L2 loop. DNA contacts are made via Arg248. Selected DNA residues in the proximity of Arg248 are shown in magenta and cyan. (D) Ribbon plot of the tetramerization domain structure (PDB code 1C26). Individual subunits within the tetramer are shown in different colors. Two monomers consisting of a β-strand and an α-helix are connected via an antiparallel β-sheet and an antiparallel helix-helix interface to form a dimer (e.g., the red and green chain). Two of these dimers are assembled into a tetramer via a parallel helix-helix interface. The figures were generated using MOLSCRIPT (Kraulis, 1991) and RASTER3D (Merritt and Bacon, 1997). Panels (B) and (C) were adapted from Joerger *et al.* (2005).

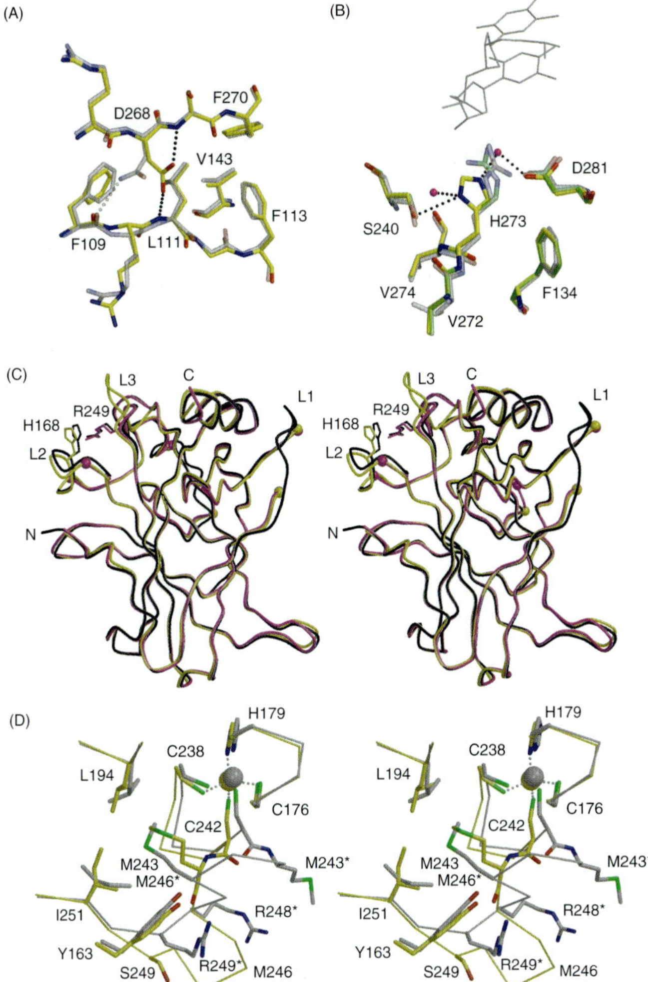

Figure 17.4. Crystal structures of p53 core domain mutants. (A) Close-up view of the N268D mutation site in the structure of the superstable quadruple mutant M133L/V203A/N239Y/N268D (*T*-p53C). The structure of *T*-p53C (PDB code 1UOL, chain A; yellow) is superimposed on the wild-type structure (PDB code 1TSR, chain A; transparent light gray). (B) View of the R273H mutation site in the structure of *T*-p53C-R273H (PDB code 2BIM, molecule A; yellow) superimposed on the structure of *T*-p53C (PDB code 1UOL, molecule A; transparent green) and DNA-bound wild type (PDB code 1TSR, molecule B; transparent light gray). Selected water molecules in the structure of *T*-p53C-R273H are represented by magenta spheres. Also shown are the two thymidylate moieties in the structure of DNA-bound wild type that are in close contact with Arg273 (thin gray lines). (C) Superposition of Cα atoms in the structures of *T*-p53C-H168R (PDB code 2BIN, magenta) and *T*-p53C-R249S (PDB code 2BIO, yellow) on those in the structure of *T*-p53C (PDB code 1UOL, molecule A; black). Cα atoms before and after chain breaks are marked with spheres in the color of the corresponding chain. (D) Structure of *T*-p53C-R249S (yellow) superimposed on the structure of p53 wild type (PDB code 1TSR, molecule A; light gray). The zinc ion in both structures is shown as a gray or yellow sphere. Wild-type residues are denoted by *. The figures were generated using MOLSCRIPT (Kraulis, 1991) and RASTER3D (Merritt and Bacon, 1997). Panel (A) was adapted from Joerger *et al.* (2004), panels (B) to (D) were adapted from Joerger *et al.* (2005).

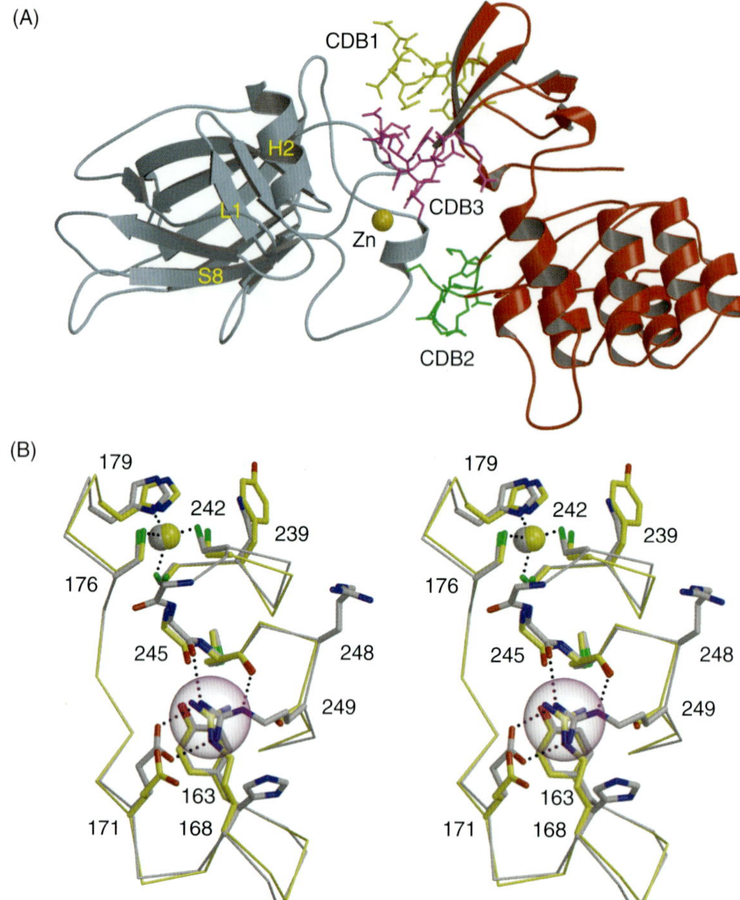

Figure 17.6. Reversing the structural effects of mutation in the p53 core domain. (A) Design of the chemical chaperone FL-CDB3. Ribbon diagram of the structure of the p53 core domain (gray)-53BP2 (red) complex (PDB code 1YCS; Gorina and Pavletich, 1996). The three p53-binding loops in 53BP2 are highlighted in different colors. CDB3 (residues 490–498 of 53BP2) is shown in purple. The binding site of the derived peptide CDB3 and its fluorescein-labeled variant FL-CDB3 was found to comprise mainly the L1 loop, β-strand S8, and helix H2 of p53 core domain, as indicated by HSQC NMR studies (Friedler *et al.*, 2002; Friedler *et al.*, 2004). The figure was generated using MOLSCRIPT (Kraulis, 1991) and RASTER3D (Merritt and Bacon, 1997). (B) Structural basis for the rescue of the cancer hot-spot mutant R249S by intragenic suppressor mutations. Stereo view of the structure of *T*-p53C-T123A/H168R/R249S (PDB code 2BIQ, yellow) superimposed on the structure of wild type (PDB code 1TSR, molecule A; light gray). Cα traces with selected side chains for parts of the L2 and L3 loops, including the zinc-binding site, are shown. In the case of Gly245 and Met246, all atoms of the residue are shown. The zinc ion in both structures is depicted as a gray or yellow sphere. Electron density for residues Met243 and Gly244 in *T*-p53C-T123A/H168R/R249S was very weak, and hence these residues were omitted from the model. Interactions mediated via the guanidinium group of Arg249 in wild type are shown by dotted lines. The common binding position of the guanidinium group of Arg168 and Arg249 in both structures is highlighted in magenta. The picture was adapted from Joerger *et al.* (2005).

White, J. T. and Kelly, J. W. (2001). Support for the multigenic hypothesis of amyloidosis: the binding stoichiometry of retinol-binding protein, vitamin A, and thyroid hormone influences transthyretin amyloidogenicity *in vitro*. *Proc. Natl. Acad. Sci. USA* 98:13019–13024.

Whitehead, A. S., Skinner, M., Bruns, G. A. P., Costello, W., Edge, M. D., Cohen, A. S., and Sipe, J. D. (1984). Cloning of human prealbumin complementary DNA. Localization of the gene to chromosome 18 and detection of a variant prealbumin allele in a family with familial amyloid polyneuropathy. *Mol. Biol. Med.* 2:411–423.

Wilce, J. A., Daly, N. L., and Craik, D. J. (2002). Synthesis and structural analysis of the N-terminal domain of the thyroid hormone-binding protein transthyretin. *Clin. Chem. Lab. Med.* 40:1221–1228.

Wojtczak, A. (1996). Structures of human transthyretin complexed with thyroxine at 2.0 A resolution and 3′,5′-dinitro-N-acetyl-L-thyronine at 2.2 A resolution. *Acta Crystallogr. D Biol. Crystallogr.* 52:758–765.

Yan, C., Costa, R. H., Darnell, J. E., Jr., Chen, J. D., and Van Dyke, T. A. (1990). Distinct positive and negative elements control the limited hepatocyte and choroid plexus expression of transthyretin in transgenic mice. *EMBO J.* 9:869–878.

Yazaki, M., Connors, L. H., Eagle, R. C., Jr., Leff, S. R., Skinner, M., and Benson, M. D. (2002). Transthyretin amyloidosis associated with a novel variant (Trp41Leu) presenting with vitreous opacities. *Amyloid* 9:263–267.

Yi, S., Takahashi, K., Naito, M., Tashiro, F., Wakasugi, S., Maeda, S., Shimada, K., Yamamura, K., and Araki, S. (1991). Systemic amyloidosis in transgenic mice carrying the human mutant transthyretin (Met30) gene: Pathologic similarity to human familial amyloidotic polyneuropathy, type I. *Am. J. Pathol.* 138:403–412.

Yoshioka, K., Furuya, H., Sasaki, H., Saraiva, M. J. M., Costa, P. P., and Sakaki, Y. (1989). Haplotype analysis of familial amyloidotic polyneuropathy. Evidence for multiple origins of the Val-Met mutation most common to the disease. *Hum. Genet.* 82:9–13.

Zeldenrust, S. R., Skinner, M., Skare, J., and Benson, M. D. (1994). A new transthyretin variant (His 69) associated with vitreous amyloid in an FAP family. *Amyloid Int. J. Exp. Clin. Invest.* 1:17–22.

Zhang, Q. and Kelly, J. W. (2003). Cys10 mixed disulfides make transthyretin more amyloidogenic under mildly acidic conditions. *Biochemistry* 42:8756–8761.

14

Human Lysozyme

Mireille Dumoulin, Russell J.K. Johnson, Vittorio Bellotti, and Christopher M. Dobson

Abstract

Lysozyme is perhaps the protein whose structure, stability, and folding behavior has been studied most widely over many years to define general principles underlying these complex phenomena. The relatively recent and unexpected observation that lysozyme is one of the 20 or so proteins whose conversion into amyloid deposits is associated with debilitating medical conditions has enabled these studies to be extended to probe the nature and origins of the misfolding events that underlie this type of disease. In this chapter, we summarize the results of these investigations and discuss both the specific mechanism through which lysozyme forms amyloid fibrils and the manner in which this process can be inhibited for potential therapeutic benefit. In addition, we discuss briefly how studies of lysozyme have provided new insights into links between normal and aberrant folding and into the way living systems avoid the consequences of the inherent tendency of polypeptides to convert into intractable and frequently toxic aggregates.

14.1. Introduction

Human lysozyme is a bacteriolytic enzyme (EC 3.2.1.17) that is widely distributed in a variety of tissues and body fluids, including the liver, articular cartilage, plasma, saliva, tears, and milk (Reitamo *et al.*, 1978). It hydrolyzes preferentially the β-1,4 glycosidic linkages between the *N*-acetylmuramic acid and *N*-acetylglucosamine groups that occur in the peptidoglycan cell wall structure of certain microorganisms, particularly of Gram-positive bacteria, and therefore appears to have a role in host defense (Jolles and Jolles, 1984). Lysozyme is highly expressed in hematopoietic cells and it is found in granulocytes, monocytes, and macrophages as well as in their bone marrow precursors (Reitamo *et al.*, 1978). The normal concentration of lysozyme in plasma ranges from 4 to 13 mg/L and only traces are detectable in the urine of healthy subjects. It has been estimated that at least 500 mg of lysozyme are produced per day by a normal subject, but the lifetime of the protein in plasma is very short; approximately 75% is eliminated within 1 h, mainly through clearance via the kidneys (Hansen *et al.*, 1972). Greatly increased concentrations of lysozyme in plasma and urine are, however, associated with a number of pathologic conditions and have been considered for many years to be a marker of monocytic leukemia (Osserman and Lawlor, 1966). In patients with myeloproliferative disorders but normal renal function, the production of lysozyme is increased by a factor up to 4.

Human lysozyme, along with hen egg white lysozyme, has been widely used over the past 25 years as a model system to study many aspects of protein structure and function, including the mechanism of protein folding and determinants of protein stability (Blake *et al.*, 1977; Redfield and Dobson, 1990; Herning *et al.*, 1991; Miranker *et al.*, 1991; Kuroki *et al.*, 1992; Radford *et al.*, 1992a,b; Harata *et al.*, 1993; Miranker *et al.*, 1993; Dobson *et al.*, 1994; Eyles *et al.*, 1994; Hooke *et al.*, 1994; Itzhaki *et al.*, 1994; Kidera *et al.*, 1994; Haezebrouck *et al.*, 1995; Takano *et al.*, 1995; Matagne *et al.*, 1997, 1998; Takano *et al.*, 1999; van den Berg *et al.*, 1999; Funahashi *et al.*, 2000; Matagne *et al.*, 2000; Takano *et al.*, 2000). One of the most interesting aspects of lysozyme folding is the existence of multiple pathways, involving a series of metastable intermediates that become structured in discrete domains that finally dock together to form the fully native protein (Miranker *et al.*, 1991; Radford *et al.*, 1992b; Miranker *et al.*, 1993; Hooke *et al.*, 1994; Matagne *et al.*, 1997). These and other observations concerning the folding of lysozyme have contributed significantly to the development of a "new view" of protein folding in terms of a stochastic search process on a highly evolved energy landscape determined by the protein sequence (Dobson *et al.*, 1994; Matagne and Dobson, 1998; Dinner *et al.*, 2000).

In addition to these insights into the nature of protein folding process, lysozyme has also been an extremely important system through which an understanding of protein misfolding and its potential consequences have developed. In the past 10 years or so, five natural mutations in the human lysozyme gene, which is located on chromosome 12 and organized in four exons and three introns (Peters *et al.*, 1989), have been reported (Pepys *et al.*, 1993; Booth *et al.*, 2000; Valleix *et al.*, 2002; Esposito *et al.*, 2003; Yazaki *et al.*, 2003). The amino acid substitutions are all located in the same region of the native structure of lysozyme, namely in the β-domain (see below), and give rise to six variant proteins (I56T, F57I, W64R, D67H, T70N, and F57I/T70N). All the mutations except T70N have been found to be associated with systemic non-neuropathic amyloidosis, a disease in which the variant proteins misfold and form amyloid fibrils in the extracellular space of multiple organs and tissues (Table 14.1).

The existence of a series of natural lysozyme variants, only some of which are associated with clinical pathologies, represents an excellent opportunity to investigate in detail how specific mutations can lead to amyloid disease. Wild-type lysozyme as well as the I56T, D67H, and T70N variants have been expressed in large quantities in a number of systems including *Saccharomyces cerevisiae* (Funahashi *et al.*, 1996), baculovirus (Booth *et al.*, 1997; Esposito

Table 14.1. Five kindreds with lysozyme amyloid disease (adapted from Yazaki *et al.*, 2003)

Family	1	2*	3	4	5
Mutations	I56T	F57I	W64R	D67H	D67H
Ethnic background	English	Italian	French	English	English
Residence	UK	Canada	France	UK	UK
Clinical features	Nephropathy Petechiae	Nephropathy	Nephropathy Intestinal obstruction	Nephropathy Gastrointestinal bleeding	Liver rupture
Organs affected	Kidney, liver spleen, skin, lymph nodule	Kidney	Kidney, gastrointestinal tract, salivary glands	Kidney, liver, spleen, gastrointestinal tract	Kidney, liver, spleen
Diagnosis	a, b, and c	b	a and b	a, b, and c	a and b
References	Pepys *et al.*, 1993	Yazaki *et al.*, 2003	Valleix *et al.*, 2002	Pepys *et al.*, 1993 Gillmore *et al.*, 1999	Harrison *et al.*, 1996

* One affected member of this family is heterozygous for both the mutations F57I and T70N.
a, immunochemical staining; b, sequence of the lysozyme gene; c, characterization of the protein from *ex vivo* fibrils.

et al., 2003), and *Aspergillus niger* (Spencer *et al.*, 1999). This has enabled detailed comparison of the properties of the variants, such as activity, stability, folding, dynamics, and aggregation, with those of the wild-type lysozyme. In addition, characterization of the more recently discovered variants, F57I and W64R, have recently been initiated (Kumita *et al.*, 2006). In this chapter, we summarize the results of all these studies, and discuss both the mechanism of lysozyme fibril formation and its significance for understanding the origin of the amyloid disorder with which this process is associated. In addition, we explore briefly the significance of the studies of lysozyme for understanding more general issues of protein aggregation and the origin of protein deposition diseases.

14.2. Lysozyme Amyloidosis

Since the first demonstration of the causative role of lysozyme in some familial forms of hereditary amyloidosis in 1993, five families have been found to be affected by this disease (Table 14.1) (Pepys *et al.*, 1993; Harrison *et al.*, 1996; Gillmore *et al.*, 1999; Valleix *et al.*, 2002; Yazaki *et al.*, 2003). In all the cases, the patients are heterozygous; the disease being transmitted through an autosomal dominant mechanism. In the initial families found to be carrying the mutations I56T or D67H, the molecular diagnosis was made on the basis of immunochemical staining, N-terminal amino acid sequencing of protein extracted from purified *ex vivo* fibrils, and by nucleotide sequencing of the lysozyme gene (Pepys *et al.*, 1993). For the more recently discovered mutations, F57I and W64R, the diagnosis was based on positive immunostaining of the tissues containing amyloid deposits with antibodies specific for lysozyme (for the W64R variant only) and the identification of the mutation in the lysozyme gene (for both variants) (Valleix *et al.*, 2002; Yazaki *et al.*, 2003). The age at which amyloid deposits appear, their distribution in tissue, and their clinical effects are very variable both within and between families (Pepys *et al.*, 1993; Gillmore *et al.*, 1999; Valleix *et al.*, 2002; Yazaki *et al.*, 2003). Such variability has been reported for other systemic amyloidoses, including the most common such disorder that is associated with transthyretin, but its causes remain largely unknown (Gillmore *et al.*, 1999). As for other amyloidoses, the mechanisms by which the amyloid deposits damage tissue and compromise organ function are very poorly understood. Clearly, the presence of massive deposits of fibrils (Fig. 14.1), which may involve kilograms of protein, will be structurally disruptive and incompatible with the normal biological functions of any organs in which they are located (Pepys, 2001). The systemic extracellular amyloid fibrils and in particular their smaller more diffusible precursor assemblies (oligomers and protofibrils) might also act in an amphipathic fashion to bind and pertub multiple cell-surface receptors and/or channels rather nonspecifically (Selkoe, 2003).

Analysis of amyloid fibrils in tissue from patients with the systemic amyloidosis associated with the I56T and D67H mutations has revealed that these aggregates are made up of the full-length variant proteins (Pepys *et al.*, 1993; Booth *et al.*, 1997). Thus, although wild-type lysozyme as well as the amyloidogenic variants are produced in these patients, the former does not convert into fibrillar structures. Moreover, the D67H variant extracted from *ex vivo* fibrils was shown to be able refold to the native state under appropriate conditions and, unlike the fibrils themselves, to exhibit enzyme activity (Booth *et al.*, 1997). In addition, experiments show that all four native disulfide bonds are present in the form of the variant protein found in the deposits (Booth *et al.*, 1997). These results suggest that neither cleavage of the polypeptide chain nor reduction of disulfide bonds is required for, or result from, fibril formation. The D67H variant must therefore fold correctly in the cell prior to deposition in tissue. This is probably also the case with the I56T variant although it has not been shown in

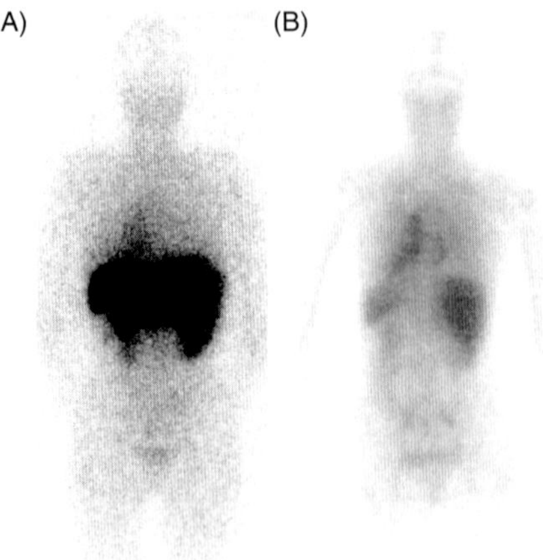

Figure 14.1. Posterior whole-body scintigraphic images after intraventous injection of ^{123}I-human SAP (serum amyloid protein that binds selectively to amyloid deposits). (A) From a patient with lysozyme amyloidosis. There is heavy amyloid deposition in the liver, spleen, and kidneys. (B) From a normal subject. The image shows the absence of protein deposits and that the tracer is distributed throughout the blood pool. Adapted from Gillmore, J. D. *et al.* (1999). Hereditary renal amyloidosis associated with variant lysozyme in a large English family. *Nephrol Dial Transplant* 14:2639–2644.

a completely definitive manner (Pepys *et al.*, 1993). Extraction and characterization of the protein in amyloid deposits from patients having the F57I or the W64R mutation has not yet been reported (Valleix *et al.*, 2002; Yazaki *et al.*, 2003). Attempts to identify the W64R variant in urine and plasma of affected patients by a combination of chromatographic separation and mass spectrometry have revealed no traces of the pathogenic variant protein although wild-type lysozyme is readily detected (Valleix *et al.*, 2002). This finding could result from selective and highly efficient incorporation of the amyloidogenic variant protein into the amyloid deposits or from higher levels of intracellular or extracellular degradation of the variants relative to the wild-type protein.

The *ex vivo* D67H lysozyme fibrils have been studied by a series of biophysical techniques. Electron microscopy after negative staining with uranyl acetate show that most of the fibrils are "wavy" in nature, with diameters ranging from 8 to 13 nm (Fig. 14.2A) (Jimenez *et al.*, 2001). Cryo-electron microscopy suggest that the fibrils are made of five or six protofilaments twisted around a hollow core (Serpell *et al.*, 2000; Jimenez *et al.*, 2001). X-ray diffraction studies reveal a meridional reflection at 4.6 to 4.8 Å and a broad equatorial reflection at 4 to 14 Å (Sunde *et al.*, 1997); these features are characteristic of the cross-β structure that is associated with all amyloid fibrils (Sunde *et al.*, 1997). This structure consists of a series of β-sheets that are orientated in such a way that the strands are perpendicular to the fibril axis.

The T70N mutation is the only known mutation in the lysozyme gene that does not appear to cause amyloidosis (Booth *et al.*, 2000; Esposito *et al.*, 2003; Yazaki *et al.*, 2003). It is much more common than the pathogenic mutations and has been reported to occur in 5–6% of human population, at least in Canada and the United Kingdom (Booth *et al.*, 2000; Yazaki *et al.*, 2003). The T70N mutation has also been found to occur in combination with the pathogenic mutation F57I (Yazaki *et al.*, 2003), but even in this case the mutation at position 70 has no detectable effect on the clinical phenotype of the disease (Yazaki *et al.*, 2003).

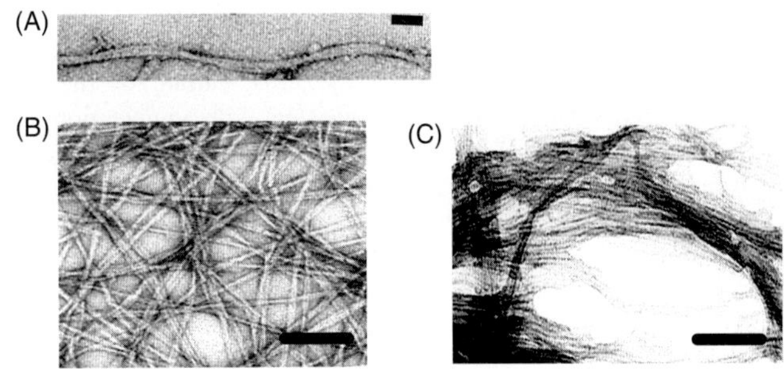

Figure 14.2. Electron microscopy images of fibrils resulting from the aggregation of the D67H variant of human lysozyme. (A) *Ex vivo* fibrils. Scale bar: 50 nm. (B) Fibrils formed *in vitro* by incubation at pH 2 and 37°C. The protein concentration is 10 mg/mL. Scale bar = 200 nm. In this latter case, the fibrils are composed of fragments of the protein as well as the full-length form (Dumoulin *et al.*, unpublished results). (C) Fibrils formed *in vitro* by incubation at pH 5.5, 48°C, 3 M urea while the solution is stirred. The protein concentration is 0.1 mg/mL. Scale bar = 200 nm. The fibrils have been found to contain the full-length protein (Dumoulin *et al.*, unpublished results).

14.3. *In Vitro* Studies of the Properties of the Variant Lysozymes

14.3.1. Wild-Type Lysozyme and Its Amyloidogenic Variants Can Form Amyloid Fibrils *In Vitro*

It is now widely accepted that the ability to form amyloid fibrils *in vitro* is a generic property of proteins resulting from stable interactions involving the main chain that is common to all polypeptides (Dobson, 1999, 2001, 2003b). In agreement with this idea, it has been found that the I56T, D67H, and T70N variants, and indeed wild-type human lysozyme, are able to form amyloid fibrils *in vitro* (Booth *et al.*, 1997; Morozova-Roche *et al.*, 2000; Dumoulin *et al.*, 2003; De Felice *et al.*, 2004; Johnson *et al.*, 2005; Dumoulin *et al.*, 2005a). All four proteins produce fibrils most readily under conditions such as low pH, high temperature, a moderate concentration of denaturant, or after application of high pressure, where a significant fraction of the protein molecules is at least partially unfolded. The two amyloidogenic variants form amyloid fibrils under similar conditions, but more extreme conditions are required for the T70N variant and especially for the wild-type protein (Dumoulin *et al.*, 2005b). This finding is in agreement with the fact that the two pathogenic variants are less stable—and therefore more easily denatured—than the T70N variant and wild-type lysozyme (see below).

Fibrils formed *in vitro* from the various lysozyme species have been analyzed by a wide variety of techniques including electron microscopy (Figs. 14.2B and 14.2C) (Booth *et al.*, 1997; Morozova-Roche *et al.*, 2000; Dumoulin *et al.*, 2003; De Felice *et al.*, 2004; Johnson *et al.*, 2005; Dumoulin *et al.*, 2005a), Congo red birefringence (Booth *et al.*, 1997; Morozova-Roche *et al.*, 2000), thioflavine-T fluorescence (Morozova-Roche *et al.*, 2000; Dumoulin *et al.*, 2003), FTIR (Booth *et al.*, 1997), and x-ray fiber diffraction (Morozova-Roche *et al.*, 2000; Dumoulin *et al.*, 2003), all of which show unequivocally their amyloid-like character. The kinetics of *in vitro* aggregation are sigmoidal, comprising a lag phase followed by an exponential growth phase

(Morozova-Roche *et al.*, 2000; Dumoulin *et al.*, 2003, 2005a). Moreover, it has been shown that fibril formation by I56T, D67H, and wild-type lysozyme can be greatly accelerated by seeding the solution with preformed fibrils from either the I56T variant or the wild-type protein (Morozova-Roche *et al.*, 2000). These results are consistent with a nucleation dependent growth process that has been suggested as a common mechanism of fibril formation (Harper and Lansbury, 1997). Amyloid fibrils containing wild-type human lysozyme can also seed the formation of fibrils from solutions of hen lysozyme, which has 60% sequence identity (Krebs *et al.*, 2004). The efficiency of the seeding in this case, however, is lower than that observed with preformed fibrils from proteins having a higher sequence similarity to hen lysozyme, such as turkey lysozyme or a single point mutant of hen lysozyme. This result is consistent with findings from other systems and indicates the importance of repetitive long-range interactions in stabilizing the core structure of amyloid fibrils. As mentioned above, amyloid fibrils are thought to result primarily as a consequence of stable interactions involving the main chain that is common to all polypeptides; the core structure will therefore be similar for all sequences. The side chains of the residues involved in the core of the fibrils, or even of the residues at the periphery of the core, will however influence both the propensity to form amyloid structures (Chiti *et al.*, 2003; Ventura *et al.*, 2004) and the specific details of the structures, for example the separation between β-sheets (Fandrich and Dobson, 2002) and the manner in which protofilaments assemble to form fibrils (Chamberlain *et al.*, 2000; Jimenez *et al.*, 2001).

14.3.2. The Variant Proteins Have a Native Fold Similar to that of the Wild-Type Protein

Human lysozyme is a protein of 130 residues (M_r = 14693) and belongs to the c-type class of lysozymes. The structure of the protein has been studied in detail by x-ray crystallography and NMR spectroscopy with several high-resolution structures under a range of different conditions available in the Protein Data Bank. The structure (Fig. 14.3) belongs to the $\alpha + \beta$ class and is very similar to that of other c-type lysozymes such as the much studied hen protein (Blake *et al.*, 1965). It is divided into two domains: an α-domain (residues 1–40 and 83–100) and a β-domain (residues 41–82). The α-domain comprises four helices (denoted A, B, C, and D) and two 3_{10} helices. The β-domain comprises a triple stranded β-sheet, a 3_{10} helix, and a long loop. The protein has four disulfide bonds of which two are located in the α-domain (C6–C128, C30–C116), and one in the long loop of the β-domain (C65–C81); the remaining disulfide bond links the two domains (C77–C95). The active site of the enzyme is formed in the cleft between the two domains with the acidic side-chains of E35 and D53 playing important roles in the catalytic mechanism (Muraki *et al.*, 1987). There are two side-chain hydrogen bonding networks in the wild-type protein (Fig. 14.4) (Artymiuk and Blake, 1981). Interestingly, both of these networks stabilize the β-domain, and cover the diverse regions of the primary structure that are brought into close proximity by the native fold.

As discussed earlier in this article, all the known natural mutations in the human lysozyme gene are located in the β-domain (Fig. 14.3). In the structure of the wild-type protein, I56, F57, and W64 are all at the α/β domain interface, whereas D67 and T70 occur in the long loop of the β-domain and are involved in one of the two hydrogen bonding networks mentioned above (Figs. 14.3 and 14.4). Although the I56T, D67H, and T70N variants have been found to be enzymatically active, the D67H and T70N lysozymes have significantly reduced activity relative to the wild-type protein (Table 14.2) (Booth *et al.*, 1997; Esposito *et al.*, 2003). This effect results from a combination of a lower substrate binding affinity and a less efficient turnover number (Booth *et al.*, 1997; Esposito *et al.*, 2003), perhaps resulting from structural distortions in the active site induced by the movement of the β-domain in these two variant proteins (see below).

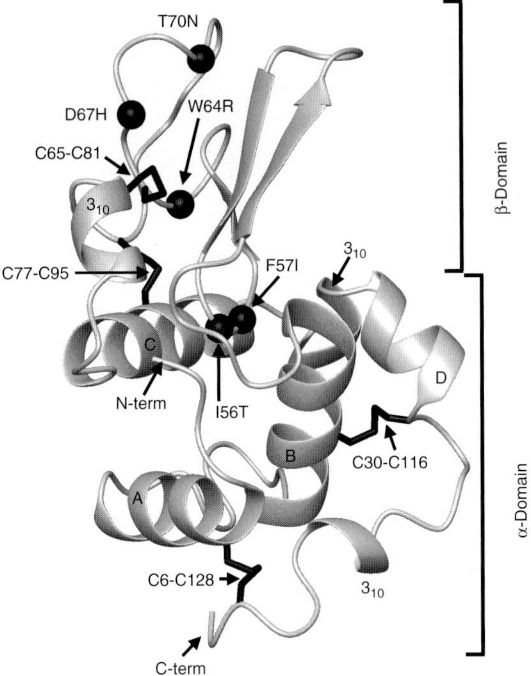

Figure 14.3. Structure of wild-type human lysozyme (PDB 1JSF) depicted in a ribbon representation showing the locations of the known natural mutations. The 3_{10} and α-helices are labelled and the disulfide bonds are shown in black. The structure was drawn with MOLMOL (Koradi R., Billeter M., and Wuthrich K., 1996. MOLMOL: a program for display and analysis of macromolecular structures. *J Mol Graph* 14:51–55).

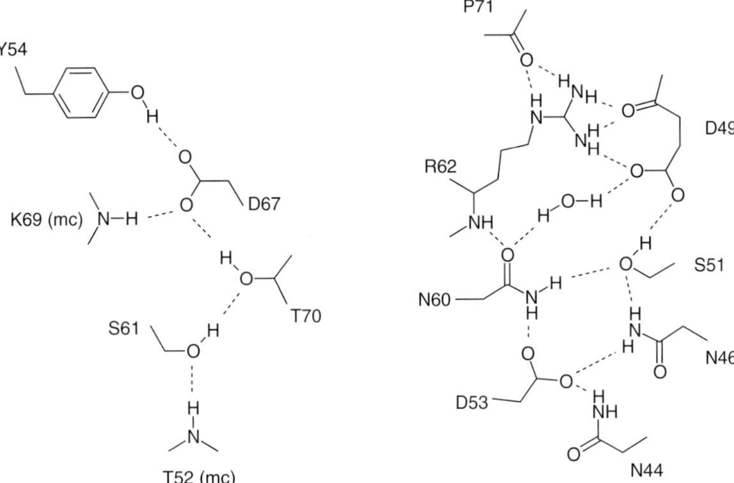

Figure 14.4. Diagram illustrating the two side-chain hydrogen bonding networks found in the β-domain of wild-type human lysozyme as discussed in the text. mc, main chain.

Table 14.2. Comparison of wild-type human lysozyme and three mutational variants (adapted from Dumoulin *et al.*, 2005b)

	WT	I56T	D67H	T70N
Enzymatic parameters[a]				
K_m (μM)	16.5	18.5	38.0	30.0
K_{cat} ($M\ s^{-1}$)	14.5	15.0	9.5	10.0
ΔG_{H20} (kJ mol^{-1})[b]	57.4	27.1	37.6	41.8
$\Delta\Delta G_{H20}$ (kJ mol^{-1})[c]	–	−30.3	−19.8	−15.6
T_m (°C)[d]	78 ± 1	68 ± 1	68 ± 1	74 ± 1
ANS fluorescence[e]	32	100	100	48

[a] Enzymatic activities determined with PNP-(GlcNAc)$_5$ at pH 5.0 and 37°C. Experimentral errors are ~20% for the K_m and ~5% for the K_{cat} (Esposito *et al.*, 2003).

[b] Difference in free energy between the native and unfolded conformations in the absence of denaturant. Determined at pH 6.5 and 20°C, the experimental error is ~10% (Esposito *et al.*, 2003).

[c] $\Delta\Delta G_{H20} = \Delta G_{H20}(\text{mutant}) - \Delta G_{H20}$ (wild type) (Esposito *et al.*, 2003).

[d] Determined by far UV CD measurements in 0.1 M sodium acetate buffer pH 5.0 (Dumoulin *et al.*, 2005a; Johnson *et al.*, 2005).

[e] Determined in 0.1 M sodium acetate buffer pH 5.0 and expressed as % relative to that of the amyloidogenic variants (Kumita *et al.*, unpublished results).

The structures of the I56T, the D67H, and the T70N variants have also been solved by x-ray diffraction methods (Funahashi *et al.*, 1996; Booth *et al.*, 1997; Johnson *et al.*, 2005). All three variants have global folds that are similar to that of the wild-type protein and in which all the four disulfide bonds are correctly formed. The structure of the I56T variant is almost identical to that of the wild-type protein (Funahashi *et al.*, 1996; Booth *et al.*, 1997). The mutation results in the replacement of a hydrophobic residue at the α/β domain interface with a more polar residue; however, it does not cause any detectable rearrangement of the polypeptide backbone or alter significantly the configuration of the neighboring side-chains. The greatest structural effects resulting from the I56T mutation are (i) the disruption, at the interface between the α- and β-domain, of the interactions formed in the wild-type protein by the isoleucine side-chain at position 56 and (ii) the introduction of a polar residue within a hydrophobic cluster (Fig. 14.5A). The hydrogen bonding potential of the polar side-chain in the variant appears to be at least partly satisfied though a hydrogen bond to an ordered water molecule near the surface (Booth *et al.*, 1997). Importantly, the native state solution structure has been shown by NMR spectroscopy to be very similar to that of the wild-type protein; the resonances of only four amide protons (A42, T56, F57, A92) differ by more than 0.1 ppm from the wild-type chemical shifts (Chamberlain *et al.*, 2001). Moreover, all of the residues involved are located within 5.5 Å of the mutation.

The D67H mutation introduces a relatively bulky histidine ring into a closely packed hydrogen bonding network, causing a rearrangement of part of the β-domain to prevent steric clashes with spatially close residues. Additionally, the histidine ring is unable to fulfill the hydrogen bonding role played by the aspartic acid side chain in the wild-type protein. The resulting rearrangement of the peptide backbone leads to the separation of two loops in the β-domain and causes a long range reordering of the native fold in this part of the protein structure (Fig. 14.6; **see color insert**). The largest deviations of C_α atoms from their wild-type positions occur for P71 (9.62 Å) and G48 (7.58 Å) (Booth *et al.*, 1997). The reorganization also results in the disruption of both side-chain hydrogen bonding networks in the β-domain (Booth *et al.*, 1997),

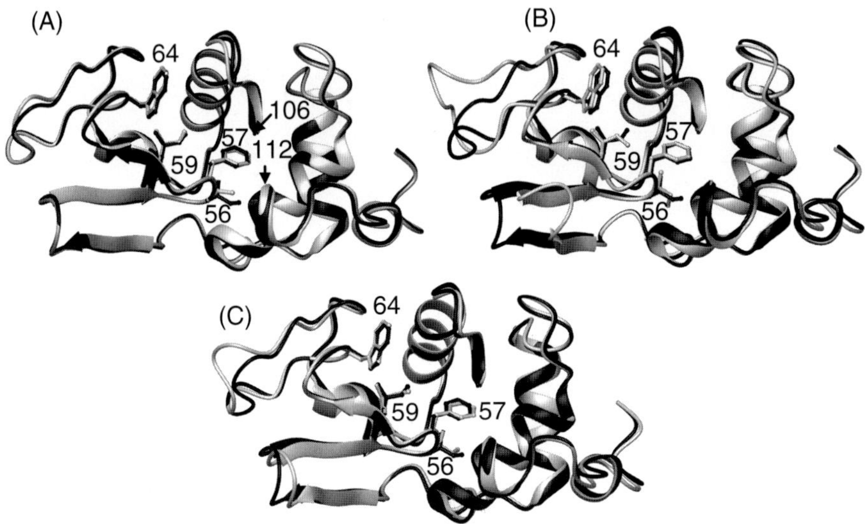

Figure 14.5. Structure of human lysozyme showing the residues at the domain interface region. The variant proteins (gray) are overlaid on a representation of the wild-type structure (black) showing in (A) the 156T variant; in (B) the D67H variant, and in (C) the T70N variant. The region encompassing residues 107–111 has been removed from all structures for clarity. The structures were drawn from PDB files for wild-type (1JSF), D67H (1LYY), 156T (1LOZ), and T70N (1W08) lysozymes using MOLMOL (Koradi, R., Billeter, M., and Wuthrich, K., 1996. MOLMOL: a program for display and analysis of macromolecular structures. *J Mol Graph* 14:51–55).

with effects that are transmitted through the protein structure to the domain interface. The result is that in the side chain of I56 adopts a different rotameric state, and the indole ring of W64 moves away from the C-helix (Fig. 14.5B). The chemical shifts of the backbone atoms of residues 48–53 and 60–82 are considerably perturbed relative to those of the wild-type protein showing that a substantial perturbation of the structure of the protein in solution results from the D67H mutation (Chamberlain *et al.*, 2001). A similar rearrangement is found in the x-ray structure of the T70N variant, although the magnitude of the atomic deviations is less than that found for the D67H variant and the rearrangement affects fewer residues (Fig. 14.6); the perturbations in NMR chemical shift values suggest that the situation in solution is similar (Johnson *et al.*, 2005). Crucially, however, most of the wild-type hydrogen bonds are formed in the native state of the T70N variant, and the α/β domain interface is only slightly perturbed by this mutation. Although the side chain of I59 adopts a new rotameric state in the T70N structure, the side chains of I56 and W64, the sites of two of the other amyloidogenic mutations, are both in their wild-type conformations (Fig. 14.5C).

These results strongly suggest that there is no direct correlation between the structural changes induced in the native state of lysozyme by the various mutations and the propensity of the variant proteins to form amyloid fibrils. They show, however, that both amyloidogenic mutations result in the loss of a number of long-range hydrophobic interactions that bridge the α/β domain interface, suggesting that the interface region of the I56T and D67H variants is significantly destabilized, either directly or indirectly, relative to the situation in the wild-type protein. This

(A)

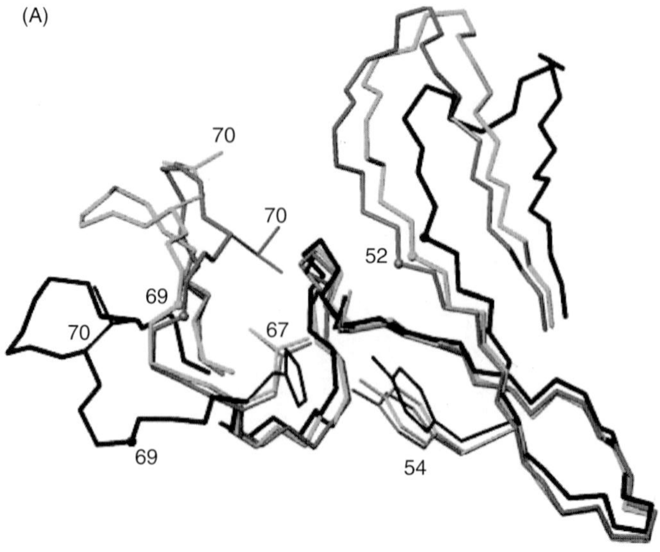

(B)

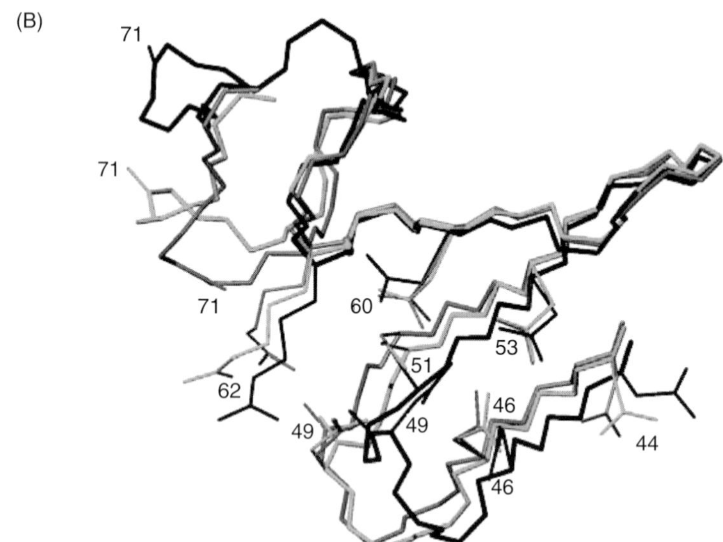

Figure 14.6. (A) Rearrangement of the β-domain of human lysozyme as the result of the mutations at residues 67 and 70. The figure shows residues 44–74, and the situation of the D67H (black), T70N (green), and wild-type (blue) proteins are overlaid. Side chains involved in hydrogen bonding are shown in bold, but all other side chains have been omitted for clarity. (B) As (A) but with the image rotated. Structures were drawn from PDB files as described in the caption of Figure 14.5.

common feature of these two amyloidogenic variants implies that it could be an important factor in their amyloidogenic properties (Booth *et al.*, 1997).

14.3.3. Effects of the Mutations on the Stability and Folding of Lysozyme

14.3.3.1. Equilibrium Unfolding

Chemically induced equilibrium unfolding studies carried out at pH 4.0 and 10°C (Funahashi *et al.*, 1996; Takano *et al.*, 2001) and at pH 6.5 and 20°C (Esposito *et al.*, 2003) have shown that the three variant proteins, I56T, D67H, and T70N, like the wild-type protein appear to unfold with a single cooperative and reversible transition characteristic of a two-state process. The native states of the three variants, however, are all destabilized relative to that of the wild-type protein (Table 14.2). The I56T protein is the most destabilized at pH 6.5 and 20°C ($\Delta\Delta G_{H20} = -30.3$ kJ mol^{-1}) with the D67H variant being destabilized just slightly more than the T70N protein under these conditions ($\Delta\Delta G_{H20} = -19.8$ and -15.6 kJ mol^{-1} for the D67H and T70N variant, respectively) (Esposito *et al.*, 2003).

Thermal unfolding experiments carried out at pH values between 2.5 and 7.0 show, however, that the I56T and D67H variants are decreased in thermostability relative to the wild-type protein to a similar extent; their unfolding transitions are about 10°C lower at all pH values studied (Funahashi *et al.*, 1996; Booth *et al.*, 1997; Morozova-Roche *et al.*, 2000; Takano *et al.*, 2001; Dumoulin *et al.*, 2003; Dumoulin *et al.*, 2005a). The T70N variant is also less thermostable than wild-type lysozyme by about 4°C (Table 14.2), but it is significantly more thermostable than the amyloidogenic variants. Most importantly, the thermal unfolding at pH 5.0 of the I56T and D67H variant lysozymes is substantially less cooperative than that of the T70N variant or wild-type lysozyme (Booth *et al.*, 1997; Esposito *et al.*, 2003; Dumoulin *et al.*, 2005a). Far and near ultraviolet circular dichroism (UV CD) measurements clearly show that the secondary and tertiary structures of the I56T and D67H variants do not unfold in a concomitant manner upon heating at pH 5.0, implying that a partially folded intermediate is populated. This conclusion is supported by a significant increase in the fluorescence of 8-anilino-1-naphthalene-sulfonic acid (ANS) observed near to the denaturation temperature of these proteins at pH 5.0 (Booth *et al.*, 1997; Dumoulin *et al.*, 2005a).

Although the UV CD studies suggest an apparent two-state process for the thermal unfolding of the T70N (Esposito *et al.*, 2003) and wild-type proteins at pH 5.0 (Booth *et al.*, 1997), these two proteins, also bind ANS in the vicinity of the midpoints of their unfolding transitions at this pH (Booth *et al.*, 1997; Dumoulin *et al.*, 2005a). This result indicates that the thermal unfolding deviate from a single two-state process. Importantly, the fluorescence intensity resulting from ANS binding to the T70N variant and the wild-type protein is only about 48% and 32%, respectively, of that observed for the amyloidogenic variants (Kumita et al., unpublished data), suggesting that the intermediate states are populated to a much smaller extent than for the latter species. Moreover, the thermal denaturation of the I56T variant and the wild-type protein at pH 2.0 and pH 7.0 has been monitored in detail by a range of techniques including fluorescence, far and near UV CD, and NMR spectroscopy (Haezebrouck *et al.*, 1995; Morozova-Roche *et al.*, 2000). These experiments allow the apparent fraction of molecules present in the native, intermediate, and unfolded states to be calculated during the denaturation process. This analysis reveal that the I56T variant populates an intermediate species to an extent of about 50% and 35–40% at its mid-denaturation temperatures at pH 2.0 and 7.0, respectively (Morozova-Roche *et al.*, 2000); whereas, for the wild-type protein the fraction of molecules in the intermediate state is only about 20–25% at pH 2.0 and negligible at pH 7.0 (Haezebrouck *et al.*, 1995; Morozova-Roche *et al.*, 2000).

Taken together, these results suggest that a partially unfolded intermediate species is formed upon thermal unfolding at the different pH values for all the proteins studied but is populated to much lower levels in the case of the T70N variant and particularly the wild-type protein. These results indicate that the ability to populate an equilibrium intermediate species is an intrinsic property of the lysozyme fold and is not restricted to the amyloidogenic variants. These latter proteins, however, populate such a species under much milder conditions and to a greater extent than do the T70N variant and wild-type lysozyme, implying that the I56T and D67H mutations stabilize the intermediate states relative to the native state. Very interestingly, the fact that the T70N variant is not associated with amyloidosis despite its destabilization relative to the wild-type protein suggests that a fine balance exists between a benign mutation and one that induces amyloid disease (Esposito *et al.*, 2003; Johnson *et al.*, 2005).

14.3.3.2. Kinetic Unfolding and Refolding

The kinetics of the unfolding of the I56T, D67H, and T70N variants and wild-type lysozyme in high concentrations of guanidinium chloride at pH 4.0 and 10°C (Funahashi *et al.*, 1996; Takano *et al.*, 2001) and at pH 5.0 and 20°C (Canet *et al.*, 1999; Esposito *et al.*, 2003) have all been found to fit single exponential functions. The unfolding of the three variants is, however, much faster than that of the wild-type protein. For example, the T70N, I56T, and D67H variants unfold, respectively, about 3, 30, and 160 times faster than wild-type lysozyme at pH 5 and 20°C in the presence of 5.4 M guanidinium chloride (Canet *et al.*, 1999; Esposito *et al.*, 2003).

A range of techniques, including stopped-flow circular dichroism, fluorescence, and pulsed hydrogen/deuterium (H/D) exchange analyzed by electrospray ionization mass spectrometry (ESI-MS) and NMR spectroscopy, has been used to monitor in detail the refolding of wild-type lysozyme and the I56T and D67H variants from their guanidinium chloride denatured states (Hooke *et al.*, 1994; Canet *et al.*, 1999). These studies have shown that wild-type human lysozyme refolds in the presence of its native disulfide bonds via multiple parallel tracks and through a series of well-defined intermediates. For the majority of the molecules that follow a slow track, a series of distinct steps is involved in the development of the fully native structure (Hooke *et al.*, 1994; Canet *et al.*, 1999). In the first step, the A and B helices and the C-terminal 3_{10} helix are stabilized in a locally cooperative manner. This step is followed by the cooperative folding of helices C and D, and finally by the folding of the β-domain. As for the homologous hen lysozyme (Itzhaki *et al.*, 1994), it is likely that a final step involving the docking of the two domains is required to generate the native close-packed structure with a functional active site. About 10% of the molecules fold along a fast track that arises from a population of molecules able to form the native state more efficiently than the remainder of the molecules. For these rapidly folding molecules, the β-domain becomes structured concomitantly with the formation of the α-domain.

No significant differences have been observed in the refolding behaviour of the T70N and the wild-type protein as monitored by fluorescence experiments at pH 5 and 20°C (Esposito *et al.*, 2003). The refolding of both the I56T and D67H variants *in vitro* has been studied in greater detail (Funahashi *et al.*, 1996; Canet *et al.*, 1999; Takano *et al.*, 2001) and has also been found to be closely similar to the wild-type protein (i.e., it occurs *via* multiple parallel pathways and through a series of well-defined intermediates). The rates of folding are, however, very different for the two variants (Canet *et al.*, 1999). The rate of the first refolding step (i.e., the development of structure in the A and B helices and the C-terminal 3_{10} helix) is not affected by either of the two the mutations. This result is fully consistent with the fact that neither of the mutated residues is in or near this region of the native fold. For the D67H variant, the subsequent steps in structure formation, that is, the coalescence first of the helices C and D on to helices A and B, and then of

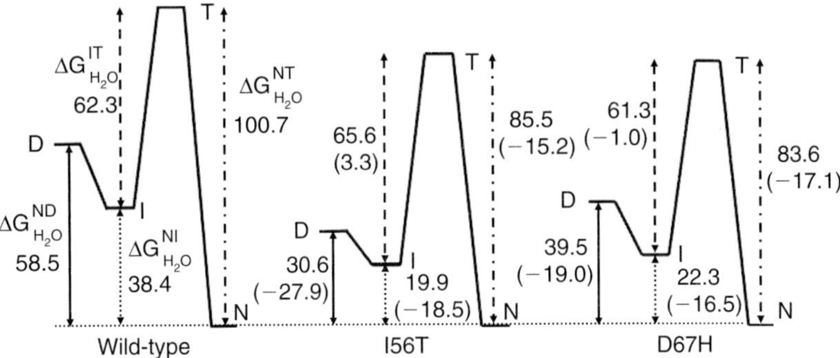

Figure 14.7. Free energy diagrams of the wild-type lysozyme and its amyloidogenic variant I56T and D67H. Values for $\Delta G^{ND}_{H_2O}$ were calculated from equilibrium experiments, whereas $\Delta G^{NT}_{H_2O}$ and $\Delta G^{IT}_{H_2O}$ were derived from kinetic studies; all these values are expressed in kJ/mol. The values in parentheses show the $\Delta\Delta G$ values between the wild-type and mutant proteins. Adapted from Takano *et al.* (2001). The stability and folding process of amyloidogenic mutant human lysozymes. *Eur J Biochem* 268:155–159.

the β-domain on to the now structured α-domain, also occur on similar time scales as for the wild-type protein. For the I56T protein, however, all these subsequent refolding steps take place more slowly (by about a factor of 10), reflecting the location of residue 56 at the interface between the α and β domains and the importance of the development of structure in this region of the protein for all these events.

To summarize, the lower stability of the D67H variant is almost entirely the result of an increase in the unfolding rate compared with the wild-type protein, whereas the reduction in stability of the I56T variant results from a combination of changes in both unfolding and refolding rates (Funahashi *et al.*, 1996; Canet *et al.*, 1999; Takano *et al.*, 2001). Free energy diagrams of the I56T and D67H variants can be derived from equilibrium and kinetic folding/unfolding data and compared with the equivalent diagram of the wild-type protein (Takano *et al.*, 2001) (Fig. 14.7). These diagrams show that, the I56T and D67H mutations decrease the free energy between the native and the unfolded states but also between the native and the intermediate states. Therefore, both variants have an increased equilibrium population of one or more species that are at least partially unfolded. These results suggest that the stabilization of the intermediate relative to the native state is a crucial factor in amyloidogenicity of human lysozyme (Takano *et al.*, 2001).

Preliminary studies on the oxidative folding of the I56T and D67H variants have shown that *in vitro* they can in fact fold faster than the wild-type protein, although there is evidence that aggregation occurs for all three proteins at the early stages of this process (Wain *et al.*, 2005). The efficient oxidative folding *in vitro* of the variant proteins is consistent with the observation that *ex vivo* fibrils are composed of full-length protein molecules containing all four native disulfide bonds (see above). The variant proteins are therefore likely to fold efficiently in the endoplasmic reticulum and to be secreted into extracellular space. In this environment, the variant proteins appear to function normally except that they have a higher tendency to aggregate under at least some conditions.

14.3.4. Effects of the Mutations on the Dynamic Properties of the Native State of Lysozyme

The internal dynamics of the native states of the I56T and D67H variants taking place on a fast timescale have been investigated by NMR spectroscopy and compared with those of the wild-type protein (Chamberlain *et al.*, 2001). The results of these experiments reveal that the behavior of the I56T variant is closely similar to that of the wild-type protein. By contrast, for the D67H variant there are large structural fluctuations in the region of the long loop (residues 55–90) of the β-domain that, as described above, has been found to adopt a substantially different conformation in the crystalline state (Booth *et al.*, 1997). Because the I56T variant shows no such dynamical behavior, the higher flexibility of a region of the β-domain in the native state of the D67H variant in solution does not appear to play a fundamental role in the process of conversion to amyloid fibrils (Booth *et al.*, 1997; Chamberlain *et al.*, 2001).

The conformational dynamics on a slower timescale for wild-type lysozyme and the I56T and D67H variant proteins in solution have been investigated by using a combination of ESI-MS and NMR spectroscopy to monitor the H/D exchange properties of the labile amide and side-chain hydrogens (Canet *et al.*, 2002; Dumoulin *et al.*, 2005a). These studies show that both mutations cause a cooperative destabilisation of a remarkably similar segment of the structure, comprising in both cases the β-domain and the adjacent C-helix. As a result, both variant proteins populate transiently a closely similar, partially unstructured intermediate in dynamic equilibrium with the native protein under physiologically relevant values of pH and temperature (Figs. 14.8A and 14.8B). In the partially structured species formed transiently in this way by the I56T and D67H variants, the β-domain and the adjacent C-helix have been found to be simultaneously unfolded while the remainder of the protein retains a largely native-like structure (Fig. 14.8C). Interestingly, this intermediate that is sampled occasionally by the variant proteins resembles that populated in the normal refolding of the protein from a highly denatured state (see above), emphasizing the close link between normal and aberrant folding behavior (Canet *et al.*, 2002).

The similarity in the behavior of the I56T and D67H variant proteins is remarkable given the different location of the mutations and their different effects on the native structure of the protein (Fig. 14.3). It can, however, be rationalized from the structural data described above. For both variants, the interface between the α- and β-domains is destabilized relative to the wild-type protein by the mutations, and the perturbation of this region of the structure is likely to be an important factor in the reduction in the global cooperativity of the native proteins. Indeed, the x-ray structure suggests that the reduction in the stability and global cooperativity of the I56T variant almost certainly arises from the loss of native contacts present in the structure of the wild-type protein and the associated disruption of the α/β domain interface. In case of the D67H variant, there could be in addition a more direct contribution to the loss of cooperativity resulting from the disruption of the hydrogen bonding networks and the associated reordering of the loop regions in the β-domain.

14.4. Mechanism of Amyloid Fibril Formation

As a result of the findings from the various biophysical studies described above, a mechanism for the formation of lysozyme amyloid fibrils can be proposed. The ability of the pathogenic variants of lysozyme to form amyloid fibrils *in vivo* appears to be primarily a result of the reduced stability of their native states relative to that of the wild-type protein, and especially to the reduction of their global cooperativity. The relatively low level of destabilization caused by the mutations then permits the variant proteins to fold efficiently enough to avoid degradation in the endoplasmic reticulum and hence to be secreted into extracellular space. By contrast, many possible mutational variants will be destabilized more substantially and therefore degraded rather than being secreted in significant quantities. It appears, however, that the amyloidogenic mutations reported so far destabilize the native state sufficiently that the proteins are able to populate

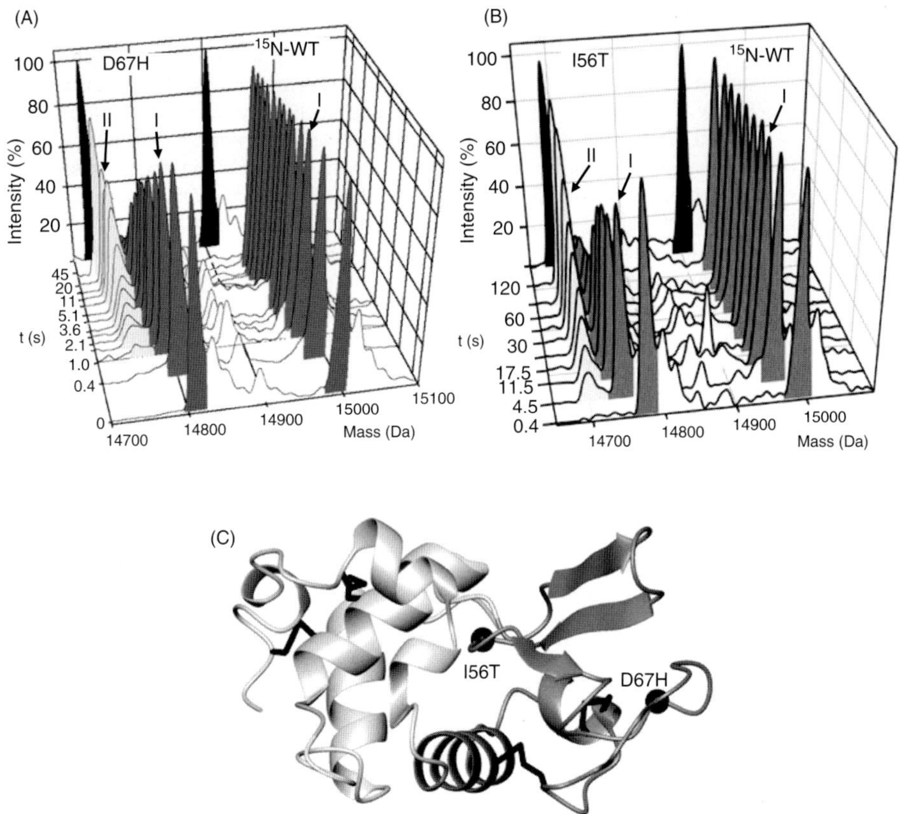

Figure 14.8. Electrospray mass spectra showing the effect of the amyloidogenic mutations on the dynamics of lysozyme. The spectra are of a mixture of (A) ^{14}N-D67H and ^{15}N-labeled wild-type lysozyme and (B) of ^{14}N-I56T and ^{15}N-labeled wild-type lysozyme. Mass spectra were recorded for equimolar mixtures of ^{15}N-labeled wild-type lysozyme and the D67H or I56T variant following incubation under hydrogen exchange conditions at pH 8.0–8.2 and 37°C. The three proteins were initially exposed to D_2O to replace all the labile hydrogens with deuterium atoms; the exchange process therefore involved the subsequent replacement of these deuterium atoms with hydrogen atoms from the solvent H_2O. The peaks observed in spectra of control samples recorded after complete H/D exchange are shown in black. The peaks colored in dark gray and labelled "I" arise from the gradual loss of deuterium during the course of exchange that occurs via an EX2 mechanism due to local structural fluctuations (Canet *et al.*, 2002). The peaks colored in light gray and labelled "II" were observed in the spectra of the I56T and D67H variants but not in that of the wild-type lysozyme. They result from a locally cooperative unfolding event giving rise to exchange by an EX1 mechanism. (C) Ribbon diagram of lysozyme showing in dark gray the regions involved in the transient cooperative unfolding observed for the I56T and D67H variants as determined by real time H/D experiments analyzed by NMR (Canet *et al.*, 2002; Dumoulin *et al.*, 2005a). These experiments reveal that, despite their different locations and the different effects of the two mutations on the structure and dynamics on fast timescale of the native state of lysozyme, both mutations cause the locally cooperative destabilization of a remarkably similar segment of the structure, in both cases the β-domain and the adjacent C-helix. The structure was drawn from the PDB file 1JSF as described in the caption of Figure 14.5. Disulfide bonds are shown in black. Adapted from Canet, D. P. *et al.* (2002). Local cooperativity in the unfolding of an amyloidogenic variant of human lysozyme. *Nat Struct Biol* 9:308–315; Dumoulin M. *et al.* (2005a). Reduced global cooperativity is a common feature underlying the amyloidogenicity of pathogenic lysozyme mutations. *J Mol Biol* 346:773–788.

a partially unfolded aggregation prone intermediate state under at least some physiologically relevant conditions to much greater extent than the wild-type protein.

It has been suggested that amyloid formation *in vivo* could be promoted by acidic environments (Colon and Kelly, 1992). In some intracellular compartments such as lysosome, pH values can be below 5 (Ohkuma and Poole, 1978), and such conditions could promote the formation of amyloid fibrils by human lysozyme (Morozova-Roche *et al.*, 2000). Calculations based on H/D exchange monitored by mass spectrometry suggest that the population of partially unfolded protein molecules at pH 5 and 37°C for the I56T and D67H variants is 1 in 1500 versus 1 in 10,000 for the wild-type protein (Canet *et al.*, 2002). As the tendency to aggregate is highly dependent on the concentration of aggregation-prone species (Lansbury, 1999), the amyloidogenic variants have a greater probability of forming amyloid fibrils than does the wild-type protein. Interestingly, T70N lysozyme appears not to be sufficiently destabilized relative to the wild-type protein to enable it to form detectable quantities of amyloid fibrils *in vivo*, showing that there is a narrow window of stability that results in amyloidogenic behavior.

In the partially unfolded intermediate species populated transiently by the I56T and D67H variants, the β-domain and the sequentially adjacent C-helix are substantially destabilized, whereas the remaining regions of the molecule largely retain their native like properties (Fig. 14.8C). As we have discussed, the transient formation of this intermediate (species II, gray box on the left-hand side on Fig. 14.9; **see color insert**) is thought to be the critical event that initiates the aggregation process that ultimately leads to the formation of fibrils. As a large part of this region of the protein forms β-sheet structure in the native protein, the initial steps in the aggregation process could simply involve the formation of intermolecular hydrogen bonds in the local region of the β-domain, rather than the intramolecular ones that characterize the native structure (species IV, gray box on the left-hand side on Fig. 14.9). This initial step could have similarities to domain swapping (Rousseau *et al.*, 2003) whereby oligomers with predominately native-like interactions form prior to the further conformational reorganization that is associated with the formation of extensive β-sheet structure (species V and VI, gray box on the left-hand side on Fig. 14.9). Indeed, species with native-like structural characteristic have been observed in other systems to form during the early stages of aggregation prior to reorganization to form the familiar amyloid structure (Bouchard *et al.*, 2000; Eakin *et al.*, 2004; Plakoutsi *et al.*, 2004).

The critical involvement of the partially folded species discussed above in the initiation of the aggregation process is supported by an important observation made with hen egg lysozyme is that fragments corresponding with parts of the β-domain and all the residues of the C-helix are readily cleaved from the protein by proteolysis at low pH and elevated temperatures where the native state is significantly destabilized (Frare *et al.*, 2004). These fragments correspond almost exactly with the regions of the human I56T and D67H variants that unfold to form the intermediates described earlier. In addition, they rapidly form amyloid fibrils in contrast to fragments corresponding with the remainder of the α-domain that remain largely soluble under the same conditions. This result indicates that the region corresponding with the β-domain and the C-helix not only can unfold with local cooperativity but also has a higher intrinsic propensity to aggregate than do other regions of the protein. Thus, it is likely that, after its exposure to the solvent as the result of partial unfolding, this region of the variant proteins could readily initiate the aggregation events that ultimately lead to the formation of fibrils. Indeed, the highly amyloidogenic character of the destabilized region is probably the reason why all the known amyloidogenic mutations are localized in the β-domain of the protein.

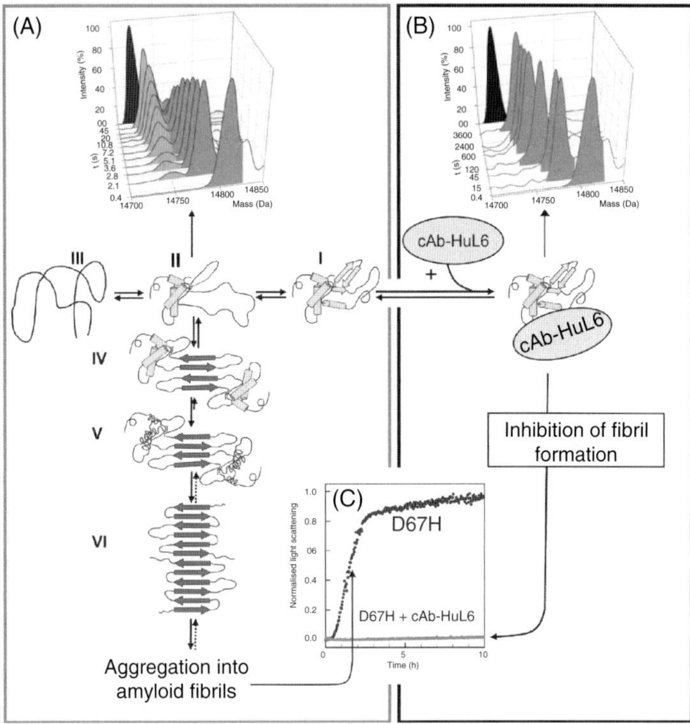

Figure 14.9. *Gray Box*: Schematic representation of the proposed mechanism for amyloid fibril formation by lysozyme. Under some physiologic conditions, the variant proteins (I) populate transiently an intermediate species (II) as revealed by H/D exchange experiments analyzed by ESI-MS (panel A). In these intermediate species, the β-domain and the C-helix are cooperatively unfolded whereas the remaining of the α-domain is in its native-like structure. These intermediate species then self-associate through the newly exposed aggregation-prone regions (IV) via the formation of intermolecular interactions to initiate fibril formation. Further rearrangement (V and VI) is likely to occur in the remainder of the structure, including the recruitment of additional regions of the polypeptide chain into the β-sheet structure prior to the formation of mature fibrils. Note also that the disulfide bridges are not represented in this scheme although they are present in the fibrils. *Black Box*: Proposed mechanism for the inhibition of fibril formation by a camelid antibody fragment. The electrospray mass spectra of D67H lysozyme in the presence of an equimolar concentration of the antibody fragment show a single peak (panel B), whose mass deceases with the length of time for which the exchange was allowed to proceed. The peaks of the species of lower mass species observed in the spectra of the free D67H variant (peaks colored yellow in panel A) and that result from a locally cooperative unfolding of the β-domain and the C-helix (Canet *et al.*, 2002) are therefore not observed in the spectra of the D67H protein in the presence of the antibody fragment. This result indicates that the binding of the antibody fragment to the D67H variant restores the stability and global cooperativity that is characteristic of wild-type lysozyme. (From Dumoulin *et al.*, 2003). Similar results have been obtained for the I56T variant (Dumoulin *et al.*, 2005a). Thus, in the presence of the antibody fragment, the variant proteins do not populate the partially folded intermediate (species II, box on the left-hand side) that otherwise can initiate the aggregation process. The result of antibody binding is therefore to prevent the ready conversion of the lysozyme variants into their aggregated states (panel C). Figure adapted from Dumoulin M., Bellotti V., and Dobson C. M. (2005b). Human lysozyme as an amyloidogenic protein. In: Sipe J. D. (eds), *The beta pleated sheet conformation and disease*, VCH Verlag GmbH & Co KgaA; pp. 635–656.

14.5. Inhibition of Amyloid Fibril Formation

As a result of the increasingly serious impact of amyloid disorders such as Alzheimer's disease and Parkinson's disease on the aging population of the developed world, a range of strategies for the prevention or the treatment of these diseases is being very actively explored (Dobson, 2003a). One of the most attractive therapeutic options is to prevent the accumulation of aggregation-prone species, thus inhibiting the earliest stages of the process that leads to their formation. Such an objective can be achieved by stabilizing the native state of an amyloidogenic protein by binding small synthetic ligands, and strategy of this type has been demonstrated in the case of transthyretin (McCammon *et al.*, 2002; Hammarstrom *et al.*, 2003). We have investigated an alternative approach for lysozyme, involving the use of an antibody to perturb the stability and cooperativity, and hence the aggregation propensities, of the lysozyme variants. Even the amyloidogenic variants of lysozyme must be incubated under somewhat denaturing conditions (e.g., at low pH and elevated temperature or in the presence of chemical denaturants; (Fig. 14.10A; **see color insert**)) to form amyloid fibrils *in vitro* in a reasonable timescale. Under such conditions, however, most antibodies or antibody fragments will not retain their functional properties (van der Linden *et al.*, 1999). Fortuitously, however, human lysozyme was one of a series of protein antigens injected into dromedaries in order to produce and study the properties of recombinant heavy-chain antibody fragments (Dumoulin *et al.*, 2002). Heavy-chain antibodies are unique functional antibodies that are produced by camelids (camels, dromedaries, and llamas) (Hamers-Casterman *et al.*, 1993). As their name suggests, these antibodies are devoid of the light chains of conventional antibodies: their antigen binding site is therefore limited to a single domain, referred to as the V_HH domain. After immunization, recombinant V_HHs can be isolated and produced as soluble monomers in *E. coli* (Muyldermans, 2001). The V_HHs are typically highly stable and soluble (van der Linden *et al.*, 1999; Perez *et al.*, 2001; Dumoulin *et al.*, 2002).

After the immunization of dromedaries with wild-type human lysozyme, a recombinant V_HH referred to as cAb-HuL6, was selected and characterized (Dumoulin *et al.*, 2002). The dissociation constant (K_D) for binding of this antibody fragment to wild-type lysozyme at pH 7.5 and 25°C is 0.7 nM (Dumoulin *et al.*, 2002), and the antibody fragment also binds both the I56T and D67H variants with high affinity: K_D = 1.2 nM and 0.6 nM for the I56T and D67H variants, respectively (Dumoulin *et al.*, 2003; Dumoulin *et al.*, 2005a). This fragment is indeed very stable [Difference in free energy between the native and unfolded conformations in the absence of denaturant ΔG_{H2O}, at pH 7.0 and 25°C is 40 kJ/mol (Dumoulin *et al.*, 2002)], and the temperature of the midpoint of denaturation at pH 5.0 is 78°C (Dumoulin et al., unpublished results). This V_HH is therefore very suitable for investigating the possibility of using an antibody fragment to stabilize the native state of the lysozyme variants as a means of reducing their aggregation propensities.

Binding the camelid antibody fragment to both the I56T and D67H variant proteins was found in fact to inhibit dramatically their ability to aggregate *in vitro* and hence to form amyloid fibrils (Dumoulin *et al.*, 2003; Dumoulin *et al.*, 2005a) (Fig. 14.10A). The effect of the binding of the antibody fragment on the global cooperativity of the variant proteins was investigated by pulse-labeling H/D exchange monitored by ESI-MS (Fig. 14.9B). In the presence of the antibody fragment, virtually none of the molecules of the I56T (Dumoulin *et al.*, 2005a) and D67H (Dumoulin *et al.*, 2003) variant proteins undergoes, on the timescale of the experiment (up to 1 h), even a single locally cooperative unfolding event of the type observed for the free lysozyme variants (see Figs. 14.8A and 14.8B). These results indicate that the frequency of such fluctuations in both variants is drastically reduced as a result of binding to the antibody fragment. In addition, real-time ESI-MS and NMR experiments have confirmed that the binding of the antibody restores the high degree of global cooperativity that is characteristic of the wild-type protein (Dumoulin *et al.*, 2005a). These results, therefore, provide further evidence that the

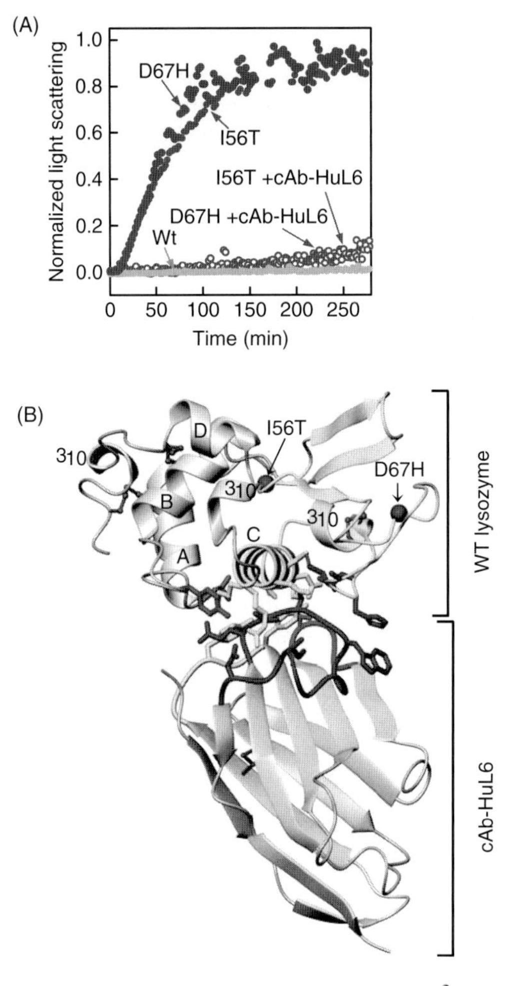

Figure 14.10. (A) Time course of the aggregation of the I56T and D67H variant lysozymes in the absence (● and ● for the I56T and D67H variants, respectively) and presence (○ and ○ for the I56T and D67H variants, respectively) of an equimolar quantity of cAb-HuL6 as monitored by light scattering. Data are also shown for wild-type lysozyme in the absence of cAb-HuL6 (●). The lysozyme concentration was 0.1 mg/mL in all cases, and the data were recorded at 65°C and pH 5.0 while the solutions were stirred. (B) Ribbon representation of the x-ray structure of wild-type lysozyme complexed with the cAb-HuL6. The β-strands in cAb-HuL6 are colored green. The lysozyme molecule is shown in light blue, the helices are labeled, and the disulfide bridges are colored orange. The sites of the I56T and D67H mutations are shown by a red ball, and the region of the molecule that has been found to unfold transiently in a locally cooperative manner in the amyloidogenic variants is colored yellow. The side chains of residues constituting the epitope are shown in violet, light blue, and dark blue for the α-domain, C-helix, and β-domain of lysozyme, respectively. Those residues constituting the paratope are shown in yellow, pink, and red and are from the CDR1, CDR2, and CDR3 loops of the cAb-HuL6 structure, respectively. (C) Ribbon diagram of lysozyme showing the Cα atoms of those residues significantly affected by the binding of cAb-HuL6 in space-filling representation as determined by NMR analysis. Those residues affected only in the I56T variant are shown in pink, those affected only in the D67H variant are shown in blue, and residues affected in both variants are shown in green. The region of the molecule that unfolds transiently in a locally cooperative manner is colored yellow. The lysozyme structure was generated from the x-ray coordinates (PDB entry 1LZ1) and the figure produced using MOLMOL (Koradi *et al.*, 1996). Perturbations to the NMR chemical shifts upon the binding of the antibody fragment include the amide resonance of residue 56 for the I56T variant (Dumoulin *et al.*, 2005a) and of residues 56 and 57 for the D67H variant (Dumoulin *et al.*, 2003). These two residues, which are themselves the locations of two of the well-defined pathogenic mutations of lysozyme that give rise to amyloid disease, are far from the binding interface with the antibody but are located in the interface between the α- and β-domain. This observation suggests that the antibody acts to restore the global cooperativity of the variant protein at least in part through the transmission of long-range conformational effects to the interface between the two structural domains of the protein.

formation of a partially unfolded species with a high propensity to aggregate, resulting from the locally cooperative unfolding of the β-domain and the C-helix, is the critical event that triggers the aggregation process in the absence of the antibody fragment.

Structural studies of the complex between the wild-type protein and the camelid antibody fragment reveal, however, that the epitope includes neither the site of the mutations nor most of the residues in the region of the protein structure that is destabilized by the mutations (Fig. 14.10B; **see color insert**) (Dumoulin *et al.*, 2003). Thus, the effects of binding are not simply to mask the entire region of the protein destabilized by the mutation and hence to prevent its unfolding from the remainder of the structure. Rather, it appears that the binding of the antibody fragment restores by a more subtle mechanism the global cooperativity of the lysozyme structure that is disrupted by these two amyloidogenic mutations. Analysis of the NMR chemical shifts changes resulting from binding of the antibody fragment (Fig. 14.10C; **see color insert**) suggest that restoration of their cooperativity occurs, at least in part, through the transmission of long-range conformational effects to the interface between the two structural domains (Dumoulin *et al.*, 2003; Dumoulin *et al.*, 2005a).

These results clearly demonstrate that there is a close link between the ability of an amyloidogenic variant to populate an intermediate state and its propensity to convert into amyloid fibrils. This conclusion reinforces the view that the remarkable cooperativity of native protein structures is an essential evolutionary development to enable otherwise marginally stable structures to resist aggregation under conditions in which they exert their biological function (Dobson, 1999; 2003b). Evolution through natural selection has resulted in proteins with native states that are not highly prone to aggregation (Richardson and Richardson, 2002). It is clear from the native structure of human lysozyme that the region of the protein implicated in the initial aggregation events (the β-domain and the C-helix) does not self-associate significantly in the native state; significant reorganization is therefore required to allow the formation of the intermolecular interactions necessary for the cross-β structure of amyloid fibrils.

Our results also suggest that exploration of a series of antibodies raised against a given protein antigen could well result in the discovery of species able to overcome the effects of a wide range of protein misfolding diseases. Recent studies indeed report the success of both active and passive vaccination approaches in slowing and/or reversing the aggregation process and its pathologic consequences in mouse models of light-chain amyloidosis, Alzheimer's disease, and mammalian prion diseases (Dumoulin and Dobson, 2004).

14.6. Conclusion

Human lysozyme is the best characterised, in terms of stability, dynamics, folding, and aggregation, of any of the proteins associated with amyloid disease. Our knowledge, at a molecular level, of the effects of natural amyloidogenic mutations on the properties of the protein has enabled the reduction of global stability and cooperativity to be identified as a major factor underlying the amyloidogenicity of pathogenic lysozyme mutations. The formation of fibrils *in vivo*, from the I56T and D67H variants at least, results from a series of events that by themselves might not be seriously harmful but together can be catastrophic. In particular: (i) The degree of destabilization of the native state of the mutational variants is small enough that they are able to fold correctly and efficiently and hence to be secreted into the extracellular space where they normally function; (ii) The mutations decrease sufficiently the stability and the global cooperativity of the proteins to enable them to populate under some conditions, perhaps associated with some stress conditions, partially unfolded states in which at least a part of the main chain is exposed; and (iii) The region exposed in this manner is highly aggregation prone and readily initiates events that ultimately lead to the formation of amyloid fibrils.

In summary, therefore, the fact that lysozyme, one of the best characterized of all proteins, is associated with amyloid disease has enabled the nature and origins of protein misfolding, and of the debilitating disorders with which it can be associated, to be probed in very great details. In addition, it has proved possible to use the lysozyme system to explore the general principles behind potential therapeutic strategies. These studies have contributed substantially to our current understanding of the nature of protein aggregation and the way that its consequences are prevented in normally functioning organisms. As these aspects of protein behavior appear to be to a large degree generic in character, these studies have also contributed to new ideas about the nature of proteins, the ubiquitous biological molecules that regulate virtually all the events taking place in living systems, and the manner in which they and the environments in which they function have evolved.

Acknowledgments

The authors wish to thank Janet Kumita for critical reading of this manuscript. M.D. and C.M.D. are grateful to the European Commission, the BBRSC, the Wellcome Trust, and the Belgium Government, under the framework of the Interuniversity Attraction Poles, for their support of those parts of our own research that are described in this chapter. R.J.K.J. was supported by a BBSRC studentship. This article is a substantially revised and extended version of an earlier review (Dumoulin *et al.*, 2005b).

References

Artymiuk P. J., and Blake C. C. F. (1981). Refinement of human lysozyme at 1.5 A resolution analysis of non-bonded and hydrogen-bond interactions. *J Mol Biol* 152:737–762.

Blake C. C. F., Koenig D. F., Mair G. A., North A. C., Phillips D. C., and Sarma V. R. (1965). Structure of hen egg-white lysozyme. A three-dimensional Fourier synthesis at 2 Angstrom resolution. *Nature* 206:757–761.

Blake C. C. F., Grace D. E., Johnson L. N., Perkins S. J., Phillips D. C., Cassels R., Dobson C. M., Poulsen F. M., and Williams R. J. (1977). Physical and chemical properties of lysozyme. *Ciba Found Symp* 137–185.

Booth D. R., Sunde M., Bellotti V., Robinson C. V., Hutchinson W. L., Fraser P. E., Hawkins P. N., Dobson C. M., Radford S. E., Blake C. C. F., and Pepys M. B. (1997). Instability, unfolding and aggregation of human lysozyme variants underlying amyloid fibrillogenesis. *Nature* 385:787–793.

Booth D. R., Pepys M. B., and Hawkins P. N. (2000). A novel variant of human lysozyme (T70N) is common in the normal population. *Hum Mutat* 16:180.

Bouchard M., Zurdo J., Nettleton E. J., Dobson C. M., and Robinson C. V. (2000). Formation of insulin amyloid fibrils followed by FTIR simultaneously with CD and electron microscopy. *Protein Sci* 9:1960–1967.

Canet D., Sunde M., Last A. M., Miranker A., Spencer A., Robinson C. V., and Dobson C. M. (1999). Mechanistic studies of the folding of human lysozyme and the origin of amyloidogenic behavior in its disease-related variants. *Biochemistry* 38:6419–6427.

Canet D., Last A. M., Tito P., Sunde M., Spencer A., Archer D. B., Redfield C., Robinson C. V., and Dobson C. M. (2002). Local cooperativity in the unfolding of an amyloidogenic variant of human lysozyme. *Nat Struct Biol* 9:308–315.

Chamberlain A. K., MacPhee C. E., Zurdo J., Morozova-Roche L. A., Hill H. A., Dobson C. M., and Davis J. J. (2000). Ultrastructural organization of amyloid fibrils by atomic force microscopy. *Biophys J* 79:3282–3293.

Chamberlain A. K., Receveur V., Spencer A., Redfield C., and Dobson C. M. (2001). Characterization of the structure and dynamics of amyloidogenic variants of human lysozyme by NMR spectroscopy. *Protein Sci* 10:2525–2530.

Chiti F., Stefani M., Taddei N., Ramponi G., and Dobson C. M. (2003). Rationalization of the effects of mutations on peptide and protein aggregation rates. *Nature* 424:805–808.

Colon W., and Kelly J. W. (1992). Partial denaturation of transthyretin is sufficient for amyloid fibril formation *in vitro. Biochemistry* 31:8654–8660.

De Felice F. G., Vieira M. N., Meirelles M. N., Morozova-Roche L. A., Dobson C. M., and Ferreira S. T. (2004). Formation of amyloid aggregates from human lysozyme and its disease-associated variants using hydrostatic pressure. *FASEB J* 18:1099–1101.

Dinner A. R., Sali A., Smith L. J., Dobson C. M., and Karplus M. (2000). Understanding protein folding via free-energy surfaces from theory and experiment. *Trends Biochem Sci* 25:331–339.

Dobson C. M. (1999). Protein misfolding, evolution and disease. *Trends Biochem Sci* 24:329–332.

Dobson C. M. (2001). The structural basis of protein folding and its links with human disease. *Philos Trans R Soc Lond B Biol Sci* 356:133–145.

Dobson C. M. (2003a). Protein folding and disease: a view from the first Horizon Symposium. *Nat Rev Drug Discov* 2:154–160.

Dobson C. M. (2003b). Protein folding and misfolding. *Nature* 426:884–890.

Dobson C. M., Evans P. A., and Radford S. E. (1994). Understanding how proteins fold: the lysozyme story so far. *Trends Biochem Sci* 19:31–37.

Dumoulin M., and Dobson C. M. (2004). Probing the origins, diagnosis and treatment of amyloid diseases using antibodies. *Biochimie* 86:589–600.

Dumoulin M., Conrath K., Van Meirhaeghe A., Meersman F., Heremans K., Frenken L. G., Muyldermans S., Wyns L., and Matagne A. (2002). Single-domain antibody fragments with high conformational stability. *Protein Sci* 11:500–515.

Dumoulin M., Last A. M., Desmyter A., Decanniere K., Canet D., Larsson G., Spencer A., Archer D. B., Sasse J., Muyldermans S., Wyns L., Redfield C., Matagne A., Robinson C. V., and Dobson C. M. (2003). A camelid antibody fragment inhibits the formation of amyloid fibrils by human lysozyme. *Nature* 424:783–788.

Dumoulin M., Canet D., Last A. M., Pardon E., Archer D. B., Muyldermans S., Wyns L., Matagne A., Robinson C. V., Redfield C., and Dobson C. M. (2005a). Reduced global cooperativity is a common feature underlying the amyloidogenicity of pathogenic lysozyme mutations. *J Mol Biol* 346:773–788.

Dumoulin M., Bellotti V., and Dobson C. M. (2005b). Human lysozyme as an amyloidogenic protein. In: Sipe J. D. (eds), *The beta pleated sheet conformation and disease*, VCH Verlag GmbH & Co KgaA; pp. 635–656.

Eakin C. M., Attenello F. J., Morgan C. J., and Miranker A. D. (2004). Oligomeric assembly of native-like precursors precedes amyloid formation by beta-2 microglobulin. *Biochemistry* 43:7808–7815.

Esposito G., Garcia J., Mangione P., Giorgetti S., Corazza A., Viglino P., Chiti F., Andreola A., Dumy P., Booth D., Hawkins P. N., and Bellotti V. (2003). Structural and folding dynamics properties of T70N variant of human lysozyme. *J Biol Chem* 278:25910–25918.

Eyles S. J., Radford S. E., Robinson C. V., and Dobson C. M. (1994). Kinetic consequences of the removal of a disulfide bridge on the folding of hen lysozyme. *Biochemistry* 33:13038–13048.

Fandrich M., and Dobson C. M. (2002). The behavior of polyamino acids reveals an inverse side chain effect in amyloid structure formation. *EMBO J* 21:5682–5690.

Frare E., Polverino de Laureto P., Zurdo J., Dobson C. M., and Fontana A. (2004). A highly amyloidogenic region of hen lysozyme. *J Mol Biol* 340:1153–1165.

Funahashi J., Takano K., Ogasahara K., Yamagata Y., and Yutani K. (1996). The structure, stability, and folding process of amyloidogenic mutant human lysozyme. *J Biochem (Tokyo)* 120:1216–1223.

Funahashi J., Takano K., Yamagata Y., and Yutani K. (2000). Role of surface hydrophobic residues in the conformational stability of human lysozyme at three different positions. *Biochemistry* 39:14448–14456.

Gillmore J. D., Booth D. R., Madhoo S., Pepys M. B., and Hawkins P. N. (1999). Hereditary renal amyloidosis associated with variant lysozyme in a large English family. *Nephrol Dial Transplant* 14:2639–2644.

Haezebrouck P., Joniau M., Van Dael H., Hooke S. D., Woodruff N. D., and Dobson C. M. (1995). An equilibrium partially folded state of human lysozyme at low pH. *J Mol Biol* 246:382–387.

Hamers-Casterman C., Atarhouch T., Muyldermans S., Robinson G., Hamers C., Songa E. B., Bendahman N., and Hamers R. (1993). Naturally occurring antibodies devoid of light chains. *Nature* 363:446–448.

Hammarstrom P., Wiseman R. L., Powers E. T., and Kelly J. W. (2003). Prevention of transthyretin amyloid disease by changing protein misfolding energetics. *Science* 299:713–716.

Hansen N. E., Karle H., Andersen V., and Olgaard K. (1972). Lysozyme turnover in man. *J Clin Invest* 51:1146–1155.

Harata K., Muraki M., and Jigami Y. (1993). Role of Arg115 in the catalytic action of human lysozyme. X-ray structure of His115 and Glu115 mutants. *J Mol Biol* 233:524–535.

Harper J. D., and Lansbury P. T., Jr. (1997). Models of amyloid seeding in Alzheimer's disease and scrapie: mechanistic truths and physiological consequences of the time-dependent solubility of amyloid proteins. *Annu Rev Biochem* 66:385–407.

Harrison R. F., Hawkins P. N., Roche W. R., MacMahon R. F., Hubscher S. G., and Buckels J. A. (1996). 'Fragile' liver and massive hepatic haemorrhage due to hereditary amyloidosis. *Gut* 38:151–152.

Herning T., Yutani K., Taniyama Y., and Kikuchi M. (1991). Effects of proline mutations on the unfolding and refolding of human lysozyme: the slow refolding kinetic phase does not result from proline cis-trans isomerization. *Biochemistry* 30:9882–9891.

Hooke S. D., Radford S. E., and Dobson C. M. (1994). The refolding of human lysozyme: a comparison with the structurally homologous hen lysozyme. *Biochemistry* 33:5867–5876.

Itzhaki L. S., Evans P. A., Dobson C. M., and Radford S. E. (1994). Tertiary interactions in the folding pathway of hen lysozyme: kinetic studies using fluorescent probes. *Biochemistry* 33:5212–5220.

Jimenez J. L., Tennent G. A., Pepys M. B., and Saibil H. R. (2001). Structural diversity of ex vivo amyloid fibrils studied by cryo-electron microscopy. *J Mol Biol* 311:241–247.

Johnson J. R. K., Christodoulou J., Dumoulin M., Caddy G., Alcocer M. J., Murtagh G. J., Kumita J. R., Larsson G., Robinson C. V., Archer D. B., Luisi B., and Dobson C. M. (2005). Rationalising lysozyme amyloidosis: Insight from the strucutre and solution dynamics of T70N lysozyme. *J Mol Biol* 352:823–836.

Jolles P., and Jolles J. (1984). What's new in lysozyme research? *Mol Cell Biochem* 63:165–189.

Kidera A., Inaka K., Matsushima M., and Go N. (1994). Response of dynamic structure to removal of a disulfide bond: normal mode refinement of C77A/C95A mutant of human lysozyme. *Protein Sci* 3:92–102.

Koradi R., Billeter M., and Wuthrich K. (1996). MOLMOL: a program for display and analysis of macromolecular structures. *J Mol Graph* 14:51–55.

Krebs M. R., Morozova-Roche L. A., Daniel K., Robinson C. V., and Dobson C. M. (2004). Observation of sequence specificity in the seeding of protein amyloid fibrils. *Protein Sci* 13:1933–1938.

Kumita J. R., Johnson R. J., Alcocer M. J., Dumoulin M., Holmqvist F., McCammon M. G., Robinson C. V., Archer D. B., and Dobson C. M. (2006). Impact of the native-state stability of human lysozyme variants on protein secretion by *Pichia pastoris*. *Febs J* 273:711–720.

Kuroki R., Inaka K., Taniyama Y., Kidokoro S., Matsushima M., Kikuchi M., and Yutani K. (1992). Enthalpic destabilization of a mutant human lysozyme lacking a disulfide bridge between cysteine-77 and cysteine-95. *Biochemistry* 31:8323–8328.

Lansbury P. T., Jr. (1999). Evolution of amyloid: what normal protein folding may tell us about fibrillogenesis and disease. *Proc Natl Acad Sci USA* 96:3342–3344.

Matagne A., Radford S. E., and Dobson C. M. (1997). Fast and slow tracks in lysozyme folding: insight into the role of domains in the folding process. *J Mol Biol* 267:1068–1074.

Matagne A., Chung E. W., Ball L. J., Radford S. E., Robinson C. V., and Dobson C. M. (1998). The origin of the alpha-domain intermediate in the folding of hen lysozyme. *J Mol Biol* 277:997–1005.

Matagne A., and Dobson C. M. (1998). The folding process of hen lysozyme: a perspective from the "new view." *Cell Mol Life Sci* 54:363–371.

Matagne A., Jamin M., Chung E. W., Robinson C. V., Radford S. E., and Dobson C. M. (2000). Thermal unfolding of an intermediate is associated with non-Arrhenius kinetics in the folding of hen lysozyme. *J Mol Biol* 297:193–210.

McCammon M. G., Scott D. J., Keetch C. A., Greene L. H., Purkey H. E., Petrassi H. M., Kelly J. W., and Robinson C. V. (2002). Screening transthyretin amyloid fibril inhibitors. Characterization of novel multiprotein, multiligand complexes by mass spectrometry. *Structure (Camb)* 10:851–863.

Miranker A., Radford S. E., Karplus M., and Dobson C. M. (1991). Demonstration by NMR of folding domains in lysozyme. *Nature* 349:633–636.

Miranker A., Robinson C. V., Radford S. E., Aplin R. T., and Dobson C. M. (1993). Detection of transient protein folding populations by mass spectrometry. *Science* 262:896–900.

Morozova-Roche L. A., Zurdo J., Spencer A., Noppe W., Receveur V., Archer D. B., Joniau M., and Dobson C. M. (2000). Amyloid fibril formation and seeding by wild-type human lysozyme and its disease-related mutational variants. *J Struct Biol* 130:339–351.

Muraki M., Morikawa M., Jigami Y., and Tanaka H. (1987). The roles of conserved aromatic amino-acid residues in the active site of human lysozyme: a site-specific mutagenesis study. *Biochim Biophys Acta* 916:66–75.

Muyldermans S. (2001). Single domain camel antibodies: current status. *J Biotechnol* 74:277–302.

Ohkuma S., and Poole B. (1978). Fluorescence probe measurement of the intralysosomal pH in living cells and the perturbation of pH by various agents. *Proc Natl Acad Sci USA* 75:3327–3331.

Osserman E. F., and Lawlor D. P. (1966). Serum and urinary lysozyme (muramidase) in monocytic and monomyelocytic leukemia. *J Exp Med* 124:921–952.

Pepys M. B., Hawkins P. N., Booth D. R., Vigushin D. M., Tennent G. A., Soutar A. K., Totty N., Nguyen O., Blake C. C. F., Terry C. J., Feest T. G., Zalin A. M., and Hsuan J. J. (1993). Human lysozyme gene mutations cause hereditary systemic amyloidosis. *Nature* 362:553–557.

Pepys M. B. (2001). Pathogenesis, diagnosis and treatment of systemic amyloidosis. *Philos Trans R Soc Lond B Biol Sci* 356:203–210; discussion 210–201.

Perez J. M., Renisio J. G., Prompers J. J., van Platerink C. J., Cambillau C., Darbon H., and Frenken L. G. (2001). Thermal unfolding of a llama antibody fragment: a two-state reversible process. *Biochemistry* 40:74–83.

Peters C. W., Kruse U., Pollwein R., Grzeschik K. H., and Sippel A. E. (1989). The human lysozyme gene. Sequence organization and chromosomal localization. *Eur J Biochem* 182:507–516.

Plakoutsi G., Taddei N., Stefani M., and Chiti F. (2004). Aggregation of the acylphosphatase from S. solfataricus. The folded and partially unfolded states can be both precursors for amyloid formation. *J Biol Chem* 270:14111–14119.

Radford S. E., Buck M., Topping K. D., Dobson C. M., and Evans P. A. (1992a). Hydrogen exchange in native and denatured states of hen egg-white lysozyme. *Proteins* 14:237–248.

Radford S. E., Dobson C. M., and Evans P. A. (1992b). The folding of hen lysozyme involves partially structured intermediates and multiple pathways. *Nature* 358:302–307.

Redfield C., and Dobson C. M. (1990). 1H NMR studies of human lysozyme: spectral assignment and comparison with hen lysozyme. *Biochemistry* 29:7201–7214.

Reitamo S., Klockars M., Adinolfi M., and Osserman E. F. (1978). Human lysozyme (origin and distribution in health and disease). *Ric Clin Lab* 8:211–231.

Richardson J. S., and Richardson D. C. (2002). Natural beta-sheet proteins use negative design to avoid edge-to-edge aggregation. *Proc Natl Acad Sci USA* 99:2754–2759.

Rousseau F., Schymkowitz J. W., and Itzhaki L. S. (2003). The unfolding story of three-dimensional domain swapping. *Structure (Camb)* 11:243–251.

Selkoe D. J. (2003). Folding proteins in fatal ways. *Nature* 426:900–904.

Serpell L. C., Sunde M., Benson M. D., Tennent G. A., Pepys M. B., and Fraser P. E. (2000). The protofilament substructure of amyloid fibrils. *J Mol Biol* 300:1033–1039.

Spencer A., Morozova-Roche L. A., Noppe W., MacKenzie D. A., Jeenes D. J., Joniau M., Dobson C. M., and Archer D. B. (1999). Expression, purification, and characterization of the recombinant calcium-binding equine lysozyme secreted by the filamentous fungus *Aspergillus niger*: comparisons with the production of hen and human lysozymes. *Protein Expr Purif* 16:171–180.

Sunde M., Serpell L. C., Bartlam M., Fraser P. E., Pepys M. B., and Blake C. C. F. (1997). Common core structure of amyloid fibrils by synchrotron X-ray diffraction. *J Mol Biol* 273:729–739.

Takano K., Ogasahara K., Kaneda H., Yamagata Y., Fujii S., Kanaya E., Kikuchi M., Oobatake M., and Yutani K. (1995). Contribution of hydrophobic residues to the stability of human lysozyme: calorimetric studies and X-ray structural analysis of the five isoleucine to valine mutants. *J Mol Biol* 254:62–76.

Takano K., Tsuchimori K., Yamagata Y., and Yutani K. (1999). Effect of foreign N-terminal residues on the conformational stability of human lysozyme. *Eur J Biochem* 266:675–682.

Takano K., Yamagata Y., and Yutani K. (2000). Role of amino acid residues at turns in the conformational stability and folding of human lysozyme. *Biochemistry* 39:8655–8665.

Takano K., Funahashi J., and Yutani K. (2001). The stability and folding process of amyloidogenic mutant human lysozymes. *Eur J Biochem* 268:155–159.

Valleix S., Drunat S., Philit J. B., Adoue D., Piette J. C., Droz D., MacGregor B., Canet D., Delpech M., and Grateau G. (2002). Hereditary renal amyloidosis caused by a new variant lysozyme W64R in a French family. *Kidney Int* 61:907–912.

van den Berg B., Chung E. W., Robinson C. V., and Dobson C. M. (1999). Characterisation of the dominant oxidative folding intermediate of hen lysozyme. *J Mol Biol* 290:781–796.

van der Linden R. H., Frenken L. G., de Geus B., Harmsen M. M., Ruuls R. C., Stok W., de Ron L., Wilson S., Davis P., and Verrips C. T. (1999). Comparison of physical chemical properties of llama VHH antibody fragments and mouse monoclonal antibodies. *Biochim Biophys Acta* 1431:37–46.

Ventura S., Zurdo J., Narayanan S., Parreno M., Mangues R., Reif B., Chiti F., Giannoni E., Dobson C. M., Aviles F. X., and Serrano L. (2004). Short amino acid stretches can mediate amyloid formation in globular proteins: the Src homology 3 (SH3) case. *Proc Natl Acad Sci USA* 101: 7258–7263.

Wain R., Smith L. J., and Dobson C. M. (2005). Oxidative refolding of amyloidogenic variants of human lysozyme. *J Mol Biol* 351:662–671.

Yazaki M., Farrell S. A., and Benson M. D. (2003). A novel lysozyme mutation Phe57Ile associated with hereditary renal amyloidosis. *Kidney Int* 63:1652–1657.

15

Serpins and the Diversity of Conformational Diseases

Robin W. Carrell

Abstract

The diverse members of the serpin family of protease inhibitors can each typically undergo a profound change in conformation. Aberrations of this conformational transition frequently occur and lead to a range of diseases, including emphysema, cirrhosis, thrombosis, and dementia, each of the diseases reflecting the functions and site of synthesis of the individual serpins, as illustrated here with α_1-antitrypsin, antithrombin, and neuroserpin. The molecular pathology of these aberrations has been studied in crystallographic detail. So the serpins now provide an archetypal example of the diversity of changes that underlie the conformational diseases. The studies show, in structural detail, how a range of consequences can arise from even a single conformational aberration. The mutant serpins can exceptionally align to give highly ordered amyloid-like fibrils, but the most studied manifestation of their structural instability is the formation of beta-linked polymers that aggregate as endoplasmic inclusion bodies. However, the same mutations that cause the intracellular polymerization of a serpin can also give rise, in the same serpin, to monomeric transitions and to the formation of soluble dimers and oligomers. Each of these forms has potentially toxic effects, and the different proportions and consequences of each account for the variability of the associated diseases.

15.1. Introduction

The serpins are the predominant family of serine protease inhibitors in higher organisms and are ubiquitously present as the inhibitors that control both intra- and extra-cellular proteolytic cascades (Carrell and Boswell, 1986; Silverman *et al.*, 2001). Their primacy is due to their unique mobile mechanism, which allows the destruction as well as the inhibition of their target proteases. This mechanism, involving a radical change in fold, has been well characterized (Loebermann *et al.*, 1985; Stein *et al.*, 1990; Huntington *et al.*, 2000). There are now more than 50 high-resolution crystallographic structures of a range of conformations, which when placed in sequence, provide accurate video depictions of the various ways in which serpins can change their fold. The complexity of the serpin mechanism has two special implications: all of the members of the family share the same tightly conserved template structure (Huber and Carrell, 1989), and all are similarly vulnerable to mutations affecting their key mobile regions (Stein and Carrell, 1995). The consequences of such mutations can now be seen in molecular detail, such that the serpins currently provide the best structural model of the way that protein conformational changes result in dysfunction and disease (Carrell and Lomas, 2002).

15.1.1. What is a Conformational Disease?

There are striking similarities between the diseases arising from mutations in the critical mobile regions of the serpins and numbers of other unrelated late-onset disorders. In antithrombin, the serpin that controls anticoagulation, such mutations result in changes reminiscent of those seen in the prion encephalopathies, with a transition of the mutant antithrombin to a protease resistant and hyperstable form that can then propagate conformational changes to normal antithrombin molecules (Carrell and Gooptu, 1998). The same mutations in the main elastase inhibitor in blood, called α_1-antitrypsin, result in its polymerization immediately after its synthesis in the hepatocyte (Lomas *et al.*, 1992). The consequent intracellular aggregation of the polymerized α_1-antitrypsin causes a slow loss of hepatocytes with progressive liver degeneration and eventual cirrhosis (Sharp *et al.*, 1969; Lomas and Carrell, 2002). The similarity of this association of protein aggregation and progressive degeneration with the common neurodegenerative disorders, including Alzheimer's disease and Parkinson's disease, led to the proposal by Carrell and Lomas (1997) of the new entity of the conformational diseases. Conformational diseases were defined as arising when "a constituent protein undergoes a change in size or conformation, with resultant self-association and tissue deposition." The proposal that the findings with the serpins were relevant to these diseases as a whole was a radical one, and in particular there was scepticism that the cirrhosis arising from abnormalities of α_1-antitrypsin provided a valid model for the changes observed in the common dementias. Vindication came however, when a family originally labeled as having "atypical Alzheimer's disease" was found to have intraneuronal aggregates of the brain-specific serpin, neuroserpin. The clinching finding was that the causative mutation in neuroserpin was at an identical site to a mutation in α_1-antitrypsin that results in late-onset liver degeneration (Davis *et al.*, 1999).

15.1.2. Aggregation Rather Than Amyloid

The shared feature of all the conformational diseases is the formation of aberrant intermolecular linkages. In almost all instances, these result from β-bonding, usually as strand to strand, or strand to sheet, β-linkages. The attention of the field, however, has long been focused on just one consequence of such linkages, the formation of the highly ordered end-product, amyloid. Amyloid is readily recognizable in tissue sections because of its characteristic staining properties and birefringence. Moreover, the large-scale deposition of amyloid is known to lead to organ failure, as in the group of diseases classified by Pepys (2001) as the amyloidoses. The difficulty arises in the extrapolation of these observations with massive deposition of amyloid, to diseases such as Alzheimer's or the spongiform encephalopathies, where amyloid formation is a minor, and often late or inconsistent, tissue finding. The danger is the too ready conclusion that amyloid is the toxic species causing the disease or is necessarily a component of the pathogenic process.

The serpins provide a clear example of the way that aberrant protein aggregation can lead to disease without progression to amyloid formation. It is possible to induce amyloid formation by serpins *in vitro* (Janciauskiene *et al.*, 1995), but the interlinkage of mutant serpins causing liver degeneration with α_1-antitrypsin or neurodegeneration with neuroserpin results from the formation of disordered and entangled polymers with no associated amyloid formation. Furthermore, other serpin conformational diseases, including familial thrombosis with mutants of antithrombin or angioedema with mutants of C1-inhibitor, can be shown to result from dimer or small oligomer formation (Aulak *et al.*, 1993; Eldering *et al.*, 1995; Lindo *et al.*, 1995; Tanaka *et al.* 2002, Corral, Huntington *et al.* 2003). Similarly, there is now increasing evidence that the cell pathology of Alzheimer's and other conformational dementias arise from early oligomer or filament formation, well before ordered amyloid formation and deposition (Takahashi *et al.* 2004;

Crowther *et al.* 2005; Demuro *et al.* 2005). It is an important distinction to make, as oligomer formation is primarily an intra- or pericellular event whereas the deposition of amyloid is most evident as an extracellular deposit. Also, the focus on amyloid has been accompanied by the belief that its formation, as well as the associated diseases, arises from a complete "molten globule" protein unfolding. However, the finding with the serpins that diseases result from only minor changes in fold has been borne out with other proteins, including some resulting in amyloid formation. In several of these, the intermolecular linkages occur due to domain swapping, with only minor perturbations of the overall tertiary structure (Bennett and Eisenberg, 2004).

15.1.3. Gain-of-Function Disadvantage

A feature of the conformational diseases, differentiating them from other genetic disorders, is that their pathology arises from a gain-of-function disadvantage rather than a straightforward loss of function. Most severely destabilizing mutations in proteins, such as a major deletion of sequence or the insertion of a proline in the middle of a helix, result in inefficient folding and hence in rapid proteolysis of the newly formed polypeptide. However, the mutations causing conformational disease are generally subtler in nature and may cause only borderline changes in stability. To give an example; the mutation of a threonine to a methionine in an internal helix in antithrombin does not affect its inhibitory function but does cause a slight (1°C) decrease in the thermal stability of the antithrombin (Beauchamp *et al.*, 1998). The mutant antithrombin is synthesized and secreted and circulates as an effective inhibitor of thrombin until an unrelated infection causes an increase in body temperature. This rise in ambient temperature instigates the gain-of-function disadvantage by triggering a conformational perturbation that results in a propagating inactivation and severe thrombosis. The two different effects can also be seen with the common Z mutation of α_1-antitrypsin (Laurell and Eriksson, 1963; Carrell and Lomas, 2002). This causes a slight change in stability, which is sufficient to allow the intracellular polymerization of the mutant protein. There is a consequent loss-of-function disadvantage, as the failure in the secretion of this elastase inhibitor leaves the lungs vulnerable to attack by proteases, with eventual development of the lung disease, emphysema. At the same time, the aggregation of the mutant α_1-antitrypsin in the hepatocytes is a gain-of-function disadvantage, a positive effect, resulting in progressive cell loss with eventual liver cirrhosis.

15.2. The Serpins

15.2.1. Mobility and Vulnerability

The inhibitory serpins are metastable proteins that can undergo a remarkable change in conformation, from an active native form with a T_m near 60°C, to a hyperstable inert form with a T_m above 100°C. The driving force in this change is the transition of the main (A) β-sheet of the molecule from a five-stranded parallel form to a six-stranded antiparallel sheet (Loebermann *et al.* 1984; Huber and Carrell, 1989; Huntington *et al.*, 2000). As shown in Figure 15.1A, the transition occurs because of entry into the middle (s4A) position of the sheet by the peptide loop released through cleavage of the reactive center by a target protease. It is this spring-like entry of the reactive center loop into the A-sheet that allows the destruction as well as the inhibition of the protease. The protease is forcibly displaced together with the cleaved reactive loop to the other pole of the molecule. The displacement causes a plucking disruption of the active site of the protease, with an accompanying distortion of 40% of its ordered structure, and hence its subsequent destruction. This precisely coordinated inhibitory mechanism involves a sequence of movements, triggered by

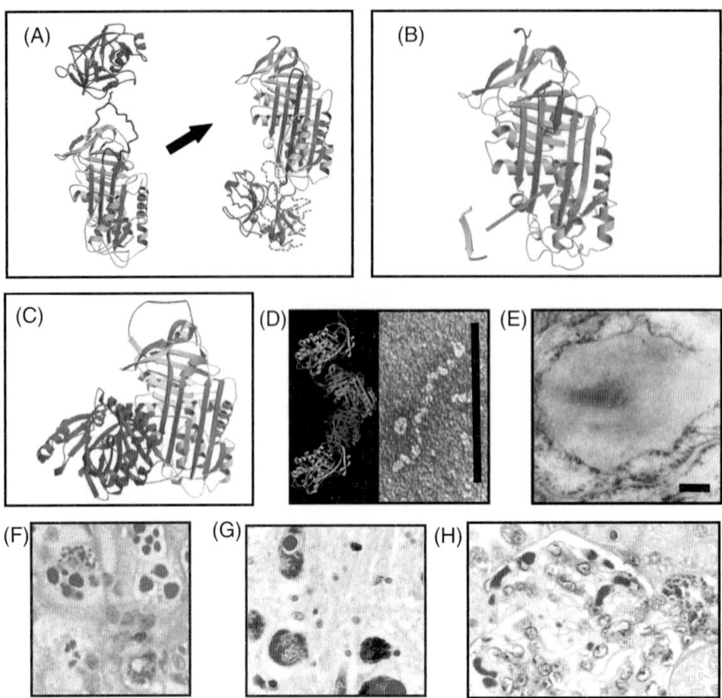

Figure 15.1. Serpins: minor conformational change, aggregation, and tissue-specific disease. The serpin family of protease inhibitors illustrates how minor changes in conformation can lead to intracellular aggregation and disease. The inhibitory mechanism of the serpins (A) depends on the entrapment of the protease when it cleaves the exposed reactive center peptide loop of the serpin. This triggers the opening of the main β-sheet of the serpin to allow entry of the reactive center peptide loop as its central fourth strand. This action forcibly displaces the protease to the opposite pole of the serpin, with the consequent loss of ordered structure (dotted) resulting in the destruction of the protease. If the entry of the reactive loop is aberrantly impeded, then the reactive loop of a second molecule can insert in its place (B), to give formation of a dimer as in (C). The dimer can then catalyze serial domain exchanges with the reactive loop of each molecule sequentially inserting into the β-sheet of the next to give polymer formation. The polymers are disordered in appearance on electron microscopy (D, right) but the molecular units (D, left) essentially retain their normal conformation. The polymerized serpin accumulates (E) in the rough endoplasmic reticulum (RER) and is characteristically evident on light microscopy as punctate periodic acid–Schiff (PAS)-staining inclusions (F–H). The inclusions occur in the cells of primary synthesis of each serpin and hence result in a diversity of pathologies. Thus the aggregation of α_1-antitrypsin subsequent to its synthesis in hepatocytes (F) results in liver cirrhosis, the aggregation of neuroserpin in neurons (G) results in neurodegeneration, and of megsin in glomerular epithelial cells (H) of the transgenic rat results in renal failure Miyata *et al.* (2005). Scale bar in (D) and (E) is 100 nm. Figure adapted in black and white from Carrrell, R. W. (2005). Cell toxicity and conformational disease. *Trends in Cell Biology* 15:574–580, with permission of Elsevier.

cleavage of the reactive center peptide loop. The A-sheet opens between strands 3 and 5 with a sliding shutter-like movement and at the same time the cleaved loop, which is hinged at the top of strand 5, inserts into the open strand 4 position. The elegance of the mechanism has been enhanced by the evolution in individual serpins of adaptations that allow the modulation of inhibitory activity. This is seen with the activation of antithrombin by heparin oligosaccharides (Jin *et al.*, 1997) or of plasminogen activator-1 (PAI-1) by the protein vitronectin (Zhou *et al.*, 2003), both of which tighten the 5-stranded A-sheet to give an optimal inhibitory conformation of the reactive loop.

Elegant mechanisms bring with them a vulnerability to breakdown. The central transition of the serpin mechanism, from a 5- to a 6-stranded A-sheet, is readily subverted. Such subversion occurs physiologically when the reactive loop is attacked by non-target proteases to give the hyperstable and inactive 6-stranded cleaved conformation. But the change in fold can also occur spontaneously when random thermal movements perturb the anchoring of the distal hinge of the reactive center loop (Fig. 15.2A; **see color insert**). This allows the release of the hinge and hence the movement of the freed loop into the strand 4 position. Evolution has adapted this change to advantage in PAI-1, with the ready transition of the intact molecule to what is known as the latent form (Mottonen *et al.*, 1992). This ability to transform to an inactive conformation provides PAI-1 with the potential to reversibly switch-off its inhibitory activity. The same transition, to an inert 6-stranded form, also occurs spontaneously, although to a lesser extent, in antithrombin and neuroserpin (Mushunje *et al.*, 2004; Onda *et al.*, 2004). Although the transition in these and most other serpins is effectively irreversible, the convention is to refer to the change in each of them as being a transition to the latent form. Normally, however, the distal hinge firmly pinions the reactive loop and prevents its entry into the A-sheet. Thus if aberrant opening of the A-sheet occurs (Figs. 15.1B and 15.1C), the molecule's own reactive loop is usually unable to enter the

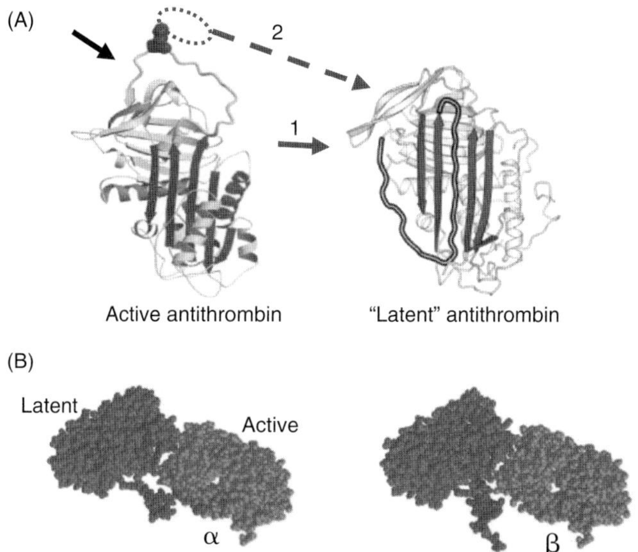

Figure 15.2. Monomeric transition, propagation, and glycoform selectivity. (A) Antithrombin is held in its active state by its distal hinge, s1C, the first strand of sheet C (arrowed, left), which prevents the reactive center peptide loop from entering fully into the middle of the A-sheet. Kinetic perturbations of s1C, however, allow some 3% of the total antithrombin per day to undergo this permanent change (arrow 1) to the inactive latent form. The full insertion of the reactive loop leaves the s1C position vacant, and this is rapidly filled by the reactive loop of a second active molecule (arrow 2) with an effective propagation of inactivity. The threat to health comes when mutations additionally destabilize s1C allowing massive transitions to the latent form, particularly when body temperatures rise, as with the fevers of infections. The loss of activity is magnified by the propagation of the inactivity of the mutant latent molecules to normal active molecules as shown in the space-filling representation in (B). The loss of active inhibitor due to dimer formation is compounded by the increased avidity of binding to the most active form of antithrombin, its glycoform β-antithrombin. The β-glycoform (on right) lacks an oligosaccharide that is encircled in the predominant but less active α_1-antithrombin (on left) impedes the linkage to the latent form.

sheet but the loops of other molecules can readily fill the strand 4 vacancy to give loop-sheet intermolecular linkages (Lomas *et al.*, 1992; Huntington *et al.*, 1999; Dunstone *et al.*, 2000). The formation of loop-sheet oligomers and polymers characteristically occurs when inhibitory serpins are subjected to thermal or other denaturant strain.

15.2.2. Mutations and Disease

More than 200 different mutations have been identified in serpins as the cause of a variety of diseases. An unexpected finding is the high proportion of mutations that directly affect the structural mechanisms of the serpin and consequently result in a gain-of-function conformational disease. Even more striking is the concentration of these mutations in the critical mobile regions of the molecule, particularly the proximal and distal hinges of the reactive loop and the structures involved in the sliding movement of the strands of the A-sheet (Stein and Carrell, 1995). This is especially so with the circumscribed "shutter" region that underlies the point of bifurcation of strands 3 and 5.

Mutations in the distal hinge of the reactive loop disrupt its linkage to the body of the serpin molecule and hence favor the insertion of the loop into the A-sheet to give the latent conformation. In doing so, however, the released distal hinge leaves a vacant strand position in the C-sheet, which becomes available for aberrant intermolecular linkages. Mutations in the other mobile regions including the proximal hinge of the loop and the shutter region seem primarily to result in a premature or inappropriate opening of the A-sheet, which allows the insertion of a reactive center loop from another molecule to give intermolecular loop-sheet linkages. Thus two of the notable mutations in α_1-antitrypsin have precisely similar consequences even though they occur in well separated regions of the molecule. The common Z mutation in Europeans, which occurs in the proximal hinge, and the less common Japanese Siiyama mutation, which occurs in the shutter region, both result in the intracellular loop-sheet polymerization of α_1-antitrypsin and hence its aggregation, with consequent liver as well as lung degeneration (Figs. 15.1C–15.1E) (Lomas and Parfrey, 2004).

15.2.3. Monomers, Oligomers, and β-Promiscuity

A clear lesson from the serpins is the unexpected diversity of changes that result from aberrant conformational transitions and the variety, indeed promiscuity, of the β-linkages that underlie these changes (Chow *et al.*, 2004). To date, seven different β-strand intermolecular linkages have been crystallographically identified, only some of which have been published. No doubt more still will be identified. The linkages involve strand insertions in various sheets, or edge-strand sheet additions, or sheet-to-sheet linkages. The most common underlying aberration is an uncoordinated opening of the A-sheet, which can lead either to the monomeric latent transition or to loop-sheet oligomer formation. Whichever of these alternatives happens seems to depend not so much on differences between individual serpins as on their concentration in the period after synthesis. Thus mutants of serpins that are synthesized in large amounts, such as α_1-antitrypsin, readily polymerize intracellularly with only a small proportion of the protein being secreted in monomeric form, whereas homologous mutations in a serpin such as antithrombin, with a synthesis rate 10-fold less than α_1-antitrypsin, result extracellularly in the monomeric latent transformation or in loop-sheet dimer or trimer formation. Although the most frequent intermolecular linkages occur through strand insertion into the A-sheet, other linkages also occur, as with strand addition to the C-sheet in the antithrombin dimer or A-sheet-to-A-sheet interactions with PAI-1 (Sharp, 1969) or side-by-side loop-sheet linkages in the heterochromatin associated serpin, MENT (Springhetti, 2003). The lesson is that conformational perturbations readily, and often unpredictably, allow a range of β-sheet and β-strand linkages that can lead to progressive intermolecular linkages. The requirements for intermolecular β-strand linkage and sheet formation are undemanding and once under way, such linkages in other proteins frequently progress to the highly ordered association seen in amyloid. But the clear message from the serpins is that amyloid is only one consequence of aberrant β-linkage and that pathology can occur well prior to, and independent of, amyloid formation (Carrell, 2005).

15.2.4. Latent Antithrombin, Propagation, and Prions

Although it was early proposed that the serpins in general had the potential to undergo a transformation to the inert (6-stranded A-sheet) latent form, the results with the candidate serpin, antithrombin, were initially confusing (Carrell *et al.*, 1991). The reason for this became apparent with later findings showing that latent antithrombin rapidly links with a molecule of active antithrombin to give a dimer, which has an electrophoretic mobility close to that of the normal monomer (Fig. 15.2). It has now been shown that antithrombin does indeed spontaneously undergo this transition, with some 3% of the antithrombin circulating in the blood being converted to the latent form each day (Mushunje *et al.*, 2004). The body copes with this as part of the normal turnover mechanism but danger arises if mutations occur at sites, such as the distal hinge of the reactive loop, which facilitate the release of the intact peptide loop. In affected individuals, this results in a doubling of the level of circulating latent antithrombin, but the risk to health occurs when the transition to the latent form is further increased by other stresses, as with increases in body temperature. So typically, families that inherit such mutations give a history of episodes of thrombosis triggered by the rises in body temperature that occur with otherwise minor respiratory or urinary infections. The puzzling feature has been the disproportionate seriousness of these episodes in affected heterozygotes, with often life-threatening massive thrombosis over and above that experienced with a comparable complete allelic loss of expression.

The reason for this gain-of-disadvantage becomes clear from the study of the structural consequences of the latent transition (Carrell *et al.*, 2001). The latent transition in antithrombin initiates a propagation of inactivation that in all but one respect is as arcane as the propagation observed with the prion encephalopathies—the difference being that the propagation of antithrombin inactivation is reversible and self-limiting. But the sequence illustrated in Figure 15.2 shows how the changes that have seemed so bewildering with the prion diseases can take place. The latent transition involves an initiating change in fold from a protease-sensitive to a protease-resistant form, with an accompanying profound increase in thermal stability. The conformational change of the latent transition releases a strand from β-sheet C of antithrombin, with the vacated position being filled by the reactive loop of an active (nonmutant) molecule of antithrombin. This propagation of inactivation is glycoform selective, with preferential linkage occurring to the most active form of antithrombin, β-antithrombin, which lacks an oligosaccharide chain in the region near the interface of the two molecules. Thus the transition of a mutant molecule of antithrombin to the latent form also results in the selective inactivation of the antithrombin glycoform that forms the first line of defense against intravascular thrombosis.

15.2.5. α1-Antitrypsin and Hepatocellular Toxicity

The most common pathologies arising from mutations in the serpins are due to the formation of longer oligomers and polymers by those members of the serpin family that are expressed in high concentration; with the polymers becoming entangled in the endoplasmic reticulum (ER) just prior to transport in the Golgi. The prime example of such polymerization and ER aggregation occurs due to mutations in the predominant protease inhibitor in human plasma, α_1-antitrypsin

(Carrell and Lomas, 2002). Some 4% of people of Northern European descent are heterozygotes for the Z mutation, which results in a replacement of the conserved glutamate at the hinge-point of the reactive center loop. This disruption of the hinge more readily allows an untriggered opening of the A-sheet with the consequent formation of loop-sheet intermolecular linkages. The resultant long-chain polymers of Z α_1-antitrypsin entangle and aggregate as large inclusions in the ER of the hepatocyte (Figs. 15.1E and 15.1F). The aggregation of α_1-antitrypsin immediately subsequent to its synthesis has two consequences. The first is a direct loss-of-function effect, due to the blockage of secretion of α_1-antitrypsin into the plasma. The plasma deficiency of α_1-antitrypsin leaves the lung unprotected from attack by the elastases released by inflammatory white cells and predisposes to the degenerative lung disease, emphysema. But it is the second, gain-of-function, disadvantage that is a characteristic feature of the conformational diseases. The aggregation of α_1-antitrypsin in the hepatocytes, as well as causing a loss of inhibitory protection, also has a positive disadvantage, which leads in homozygotes to a slowly progressive loss of hepatocytes and eventual liver cirrhosis.

There is persuasive evidence that the liver damage in homozygotes for the mutant α_1-antitrypsin is an inherent consequence of the intracellular protein aggregation. It only occurs with mutations that cause intracellular aggregation of the newly synthesized protein and not from stop-codon or other null mutations that result in a complete loss of synthesis. The liver pathology is clearly related to the magnitude of the aggregation. Heterozygotes also develop intracellular inclusions of the mutant Z α_1-antitrypsin, identical to those shown in Figure 15.2F, but the development of progressive liver damage only occurs in homozygotes for the Z mutant who have a double aggregative load.

The loss of hepatocytes is due to the aggregation of the mutant protein and not to its deficiency. The presence of a grossly destabilizing mutation will result in the newly synthesized protein being misfolded and hence recognized and degraded by the proteosomal pathway (Wu *et al.*, 2003). But cellular toxicity, however, arises from lesser destabilizing mutations, when a proportion of the protein does fold correctly and then moves into the endoplasmic pathway. Here the correctly folded but perturbed molecules are present at concentrations that allow self-extending intermolecular linkages, with resultant polymer formation and aggregation in the ER. The individual α_1-antitrypsin molecules within these polymers essentially retain their normal conformation and hence are not recognized by the ubiquitin-proteosomal pathway. The cellular response to this protein aggregation is referred to as ER stress. This stress occurs due to an imbalance between the load of aggregated protein within the ER and the ability of the organelle to process that load. ER stress includes both an ER overload response, with the activation of transcription factors, and an unfolded protein response, with an attenuation of protein biosynthesis and an upregulation of chaperones and other ER-resident proteins (Ron, 2002). The aggregation of α_1-antitrypsin in the hepatocyte, however, results in ER stress (Rutowski and Kaufman, 2004) with an ER overload response (Lawless *et al.*, 2004) but not in an unfolded protein response (Graham *et al.*, 1990). Hepatocellular proliferation takes place accompanied by caspase activation (Teckman *et al.*, 2003; Rudnick *et al.* 2004) and the assumed eventual consequence is apoptosis and cumulative hepatocyte loss.

15.2.6. ER Aggregation and the Neuroserpin Dementias

The findings with α_1-antitrypsin liver disease, that aberrant protein aggregation can inherently affect cell viability, led us to propose that the same process could also underlie the common inclusion-body dementias (Carrell and Lomas, 1997). But it seemed a long step to relate findings in liver disease to the brain damage that results from a loss of neurons. Justification, however, came in a striking way, with the subsequent finding of a family with a novel "Alzheimer-like" history of late-onset dementia (Davis *et al.*, 1999). A distinctive feature of this familial neurodegenerative disease is the widespread presence in neurons of inclusion bodies formed of neuroserpin, a brain-specific homologue of α_1-antitrypsin. The neuronal inclusions are identical in appearance and position in the ER to the hepatocyte inclusions of α_1-antitrypsin (Fig. 15.1G). Similarly, the neuroserpin in the neuronal inclusion bodies shows the same monomolecular change to the latent conformation (Onda *et al.*, 2005) observed in mutant antithrombins, together with polymerization and entanglement identical to that seen with mutant α_1-antitrypsin (Davis *et al.*, 1999). Compelling evidence for the close identity of molecular mechanisms in both the hepatocyte and the neuronal pathology has come with the finding (Davis *et al.*, 2002) of three further families, each with a different mutation in neuroserpin and with varying manifestations of neurodegenerative disease. All four of the neuroserpin mutations are at sites crucial to the maintenance of the serpin conformation, where mutations in other members of the serpin family are known to cause intermolecular linkage and polymerization.

The degeneration associated with the mutants of neuroserpin is clearly due to a toxic gain-of-function. It is not the loss of function of neuroserpin that results in the disease, as in each case the neurodegeneration results from heterozygous mutations, which will give a reduction rather than cessation of neuroserpin inhibitory activity. Moreover, neuroserpin-deficient mice show behavioral rather than neurodegenerative changes (Madani *et al.*, 2003). The inherent disadvantage of the intracellular aggregation of neuroserpin is also convincingly apparent in the close correlation of the predicted molecular instability of each of the four mutant neuroserpins with the observed severity of the disease and magnitude of inclusion body formation. These correlations are confirmed by *in vitro* studies of the instability and of the rates of polymerization of each of the mutants (Belorgey *et al.*, 2002; Belorgey *et al.*, 2004; Onda *et al.*, 2005) and by cell culture studies of inclusion formation (Miranda *et al.*, 2004).

15.3. What Causes Cell Toxicity?

The consistent genotype-phenotype correlation of mutations and disease, with both α_1-antitrypsin and neuroserpin, is compelling evidence that protein aggregation in the ER is, in itself, a sufficient cause of cell toxicity and attrition. This conclusion fits with findings from other proteins (Brennan *et al.*, 2002; Ron, 2002; Rutishauser and Spiess, 2002), as well as the serpins. The serpins, however, provide the clearest evidence of the way structural changes lead to protein aggregation in the ER and hence to the accompanying cellular and disease consequences. The aggregation of α_1-antitrypsin and of neuroserpin result in ER stress, but the details of the subsequent events leading to the attrition of hepatocytes and neurons are still sketchy, and the occurrence of apoptosis in the human diseases is still largely based on deductive evidence. The finding of a remarkable new serpinopathy in a laboratory animal is now filling these missing gaps in the story. Miyata and colleagues in the course of a study of the expression of the mesangial serpin, megsin, noted that transgenic rats with over expression of megsin developed renal and pancreatic disease (Inagi *et al.*, 2005a). The changes in the kidney and in the pancreas were typical of a serpinopathy, with characteristically staining inclusions of megsin in the RER (Fig. 15.1H). Other biochemical and clinical findings in the transgenic rats provide consistent support for the conclusion from the human disorders that the pathology inherently arises from the cumulative aggregation of the serpin within the ER. The clinical severity and time of onset of the disease in the rat, as with the human, is dose-dependent in terms of the magnitude of the protein aggregation. Furthermore, the transgenic rats convincingly show the accompanying occurrence of apoptosis in the affected cells along with a marked upregulation of ER stress chaperones (Inagi *et al.*, 2005b).

The pancreatic pathology in the megsin transgenic rats resembles that seen in type 2 diabetes in man and supports the proposal that type 2 diabetes is a conformational disease (Hayden *et al.*, 2005). Further support, along with detailed demonstration of ER stress–induced changes, comes from another animal model of an ER retention disease. The Akita mouse (Oyadamori *et al.*, 2002) develops diabetes due to a loss of pancreatic β-cells analogous to the similar loss of islet β-cells in the pancreas of the megsin-transgenic rat. In the Akita mouse, an aberrant proinsulin forms polymers that accumulate as dense inclusions in the ER of the β-cells. The formation of these inclusions is accompanied by a demonstrable ER stress response with increased expression of chaperones, a strong upregulation of the transcription factor CHOP, and eventual β-cell apoptosis.

15.4. Are Oligomers Toxic?

There is now increasing evidence that the toxic species in numbers of the conformational diseases, including the common dementias, are the early oligomers that well precede polymer and fibril formation. The innate toxicity of small polymers is evident with the serpins. Whereas mutants of α_1-antitrypsin and neuroserpin rapidly form long polymers, similar mutations in other plasma serpins, which are synthesized in lesser amounts, predominately result in short oligomers. This is seen with reactive loop hinge mutations in C1-inhibitor, which result in the formation in the plasma of small oligomers, principally dimers and trimers (Aulak *et al.*, 1993; Eldering *et al.*, 1995). The toxicity of such small oligomers is most directly seen with hinge and shutter mutants of antithrombin. These, too, result in dimer and trimer formation in the plasma but are also accompanied by unexpectedly severe and episodic deep vein thrombosis (Lindo *et al.*, 1995; Corral *et al.*, 2004). In addition to this there is now increasing evidence that the pathology associated with homozygosity for the Z variant of α_1-antitrypsin is due not only to its intracellular aggregation and inadequate secretion but also to the subsequent formation within the plasma of short-chain polymers. The complexes of α_1-antitrypsin with its target proteases are known to act as both an attractant and an activator of inflammatory white cells (Banda *et al.*, 1988; Kurdowska and Travis, 1990) so it is not surprising that polymers similarly have striking proinflammatory effects on human neutrophils *in vitro* (Parmar *et al.*, 2002; Mahadeva *et al.*, 2005). The *in vivo* relevance of this is underlined by the finding of polymers of Z α_1-antitrypsin in lung washings from patients with emphysema arising from α_1-antitrypsin deficiency (Elliott *et al.*, 1998; Aldonyte *et al.*, 2004).

15.4.1. Oligomer Infectivity

The specific proinflammatory effects of serpin oligomers explain some but not all of their observed gain-of-disadvantage consequences. Other interactions must also be responsible for the observed exacerbation of thromboembolic disease by circulating antithrombin oligomers. In particular, a more general explanation is needed for the intracellular toxicity of small oligomers, which has recently been implicated as the cause of neuronal degeneration in a several of the conformational dementias (Reixach *et al.*, 2004; Takahashi *et al.*, 2004; Crowther *et al.*, 2005; Cleary *et al.*, 2005; Tomidokoro *et al.*, 2005). This is likely to be due, at least in part, to the inherent infectivity of the initial oligomers. By definition, these oligomers must have the capability, not seen in the native proteins, of forming progressive linkages. This capability for intermolecular linkage will be most apparent when expressed as autolinkages to give fibril and eventually amyloid formation. But the β-bonding that is predominately involved is promiscuous in nature. Thus the ability for autolinkages must also be accompanied by the potential to form linkages to other cellular proteins and peptides, with likely toxic consequences. This is well illustrated with the serpins, not only with the propagation of inactivation of latent antithrombin but also in the loop-sheet linkages that more frequently result in disease, as with the intracellular polymerization and aggregation of the common Z variant of α_1-antitrypsin. The problem is not so much in the instability of the monomer but in the infectivity of the oligomers (Zhou and Carrell, 2005). This concept, of an infective end-unit, with an induced conformational change favoring β-strand linkages, provides an explanation for the diverse pathologies seen in the serpinopathies.

Structural evidence for such a general mechanism comes from the simplest infective serpin unit—monomeric latent antithrombin (Fig. 15.2). The conformational infectivity of latent antithrombin is due to the presence of a vacant strand in the C-sheet of the molecule left after entry of the reactive loop into the A-sheet. As shown in the figure, the vacated strand can be rapidly replaced by the reactive loop of another antithrombin molecule to give an antithrombin dimer. Surprisingly, however, this is not a specific interaction as the vacated strand can alternatively be replaced by other peptides of varied sequences: the requirement primarily being for sequences that have alternating hydrophobic amino acid side chains (Mushunje *et al.*, 2005). Thus there is a clear message that is relevant to protein aggregation in general. The conformational infectivity responsible for protein aggregation is not limited to protein auto-linkages but may also result in β-strand and β-sheet linkages to other unrelated proteins. This potential of small oligomers to promiscuously form β-linkages helps explain their toxic threat within the complex pathways and structures of the cell.

15.4.2. Implications for Treatment

Knowledge of the structural changes underlying the conformational diseases opens the prospect of designed therapies. As discussed, these diseases arise from protein instability that results in intracellular aggregation, which in the ER can directly result in cell toxicity and death. A first approach to therapy is to remedy the underlying protein instability. This is not as daunting as may initially be thought. Almost all proteins have natural ligands, which on binding have the potential to convert an unstable protein to a hyperstable form. This is seen physiologically with PAI-1, the serpin inhibitor of fibrinolysis (Preissner *et al.*, 1990; Seiffert and Loskutoff, 1991). PAI-1 rapidly undergoes a transition to the inert latent conformation at body temperature. But the binding of a domain of the plasma protein vitronectin preserves PAI-1 in its active form even at temperatures 10°C higher than body temperature (Zhou *et al.*, 2003). An example of how the same principle has

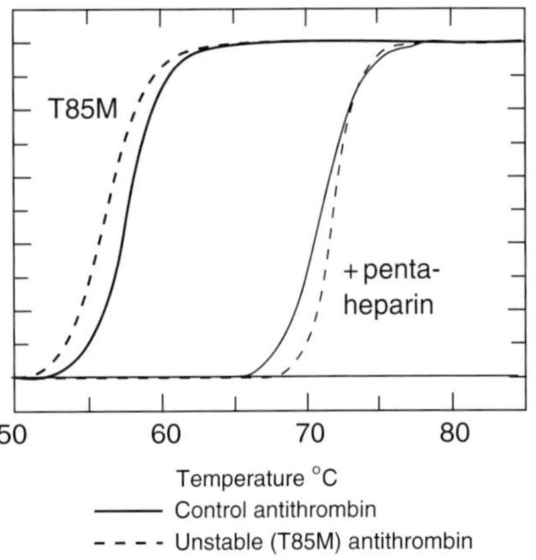

Figure 15.3. Effective therapy for a conformational disease. A plot of changes in ellipticity versus temperature, with an unstable mutant of antithrombin (on left) that results in episodic severe thrombosis, shows a 1°C loss of thermal stability (T_m). The administration, however, of the synthetic pentapeptide of heparin (on right) causes a 10°C increase in T_m with consequent stabilization of the mutant antithrombin (Beauchamp *et al.*, 1998).

been used therapeutically is shown in Figure 15.3 with a conformationally unstable variant of antithrombin associated with severe episodes of thrombosis (Beauchamp *et al.*, 1998). The variant has a mutation in its shutter region (Thr85Met) that results a small decrease in its thermal stability. This is sufficient, however, to cause the triggering of a conformational change to the latent form with the slight increases in body temperature that occur in all individuals from time to time with the fevers of infections. The change to the latent form results in a propagation of inactivation with the consequent onset of massive thrombosis. These changes are prevented however, by the administration of a therapeutic pentasaccharide fragment of heparin, which converts both the variant and the normal antithrombin to an active hyperstable form.

Another approach to the therapy of conformational diseases is to increase the turnover of the aggregated protein. In some conformational diseases, antibodies are being utilized to mobilize the deposited protein or its oligomeric precursors. With the serpins, however, the aggregation occurs within the endoplasmic compartment and thus beyond the access of antibodies. For this reason, attempts to prevent or to mobilize serpin aggregates have centered on attempts to increase chaperone activity with, as yet, little success. Where the serpins are leading the way is in a third approach. This is the use of structure-based design of peptides or their mimetics to specifically block the sites involved in intermolecular β-linkages (Mahadeva *et al.*, 2002; Zhou *et al.*, 2004). This has now been convincingly achieved *in vitro* with the preferential blockage by small peptides of linkage sites in the abnormal as compared with the functionally normal forms of the serpin. The challenge now is to convert these peptides into forms that will be pharmaceutically effective, with particular reference to their accessibility to the ER.

15.5. Conclusion: Why "Conformational Diseases"?

The conformational instability of a protein, whether it is acquired or inherited, may result in a variety of endpoints. The resultant diseases have previously been classified and named on the basis of these individual endpoints—notably as amyloidoses or as inclusion-body diseases or as ER storage disorders. But the arbitrary basis of these endpoint classifications is highlighted by studies of the serpins. These show, in structural detail, how a range of consequences can arise from even a single conformational aberration. The serpins can aberrantly align to give highly ordered amyloid-like fibrils, but the most studied manifestation of their structural instability is the formation of beta-linked polymers that aggregate as endoplasmic inclusion bodies. However, the same mutations that cause the intracellular polymerization of a serpin can also give rise, in the same serpin, to monomeric transitions and to the formation of soluble dimers and oligomers. Each of these forms has potentially toxic effects, and the different proportions and consequences of each account for the variability of the associated diseases. All of these disorders, despite their different clinical presentations, arise from aberrant protein conformational transitions and interactions and hence are most appropriately described as conformational diseases.

The direct relevance of the serpinopathies to conformational diseases associated with other proteins became apparent with the realization that the toxic species in numbers of these diseases, including the common dementias, are primarily the oligomers that precede amyloid fibril formation. This recognition of the toxicity of early oligomeric forms has shifted the focus of research to the cell and to the cytotoxicity that underlies most of the conformational diseases. It is the resultant cumulative cell loss, with eventual organ failure, that accounts for the characteristic clinical feature of these diseases—their late and slow onset. The other special feature of the conformational diseases, the way external factors can accelerate or precipitate their clinical onset, is again well illustrated by serpins, as with the temperature-induced transition of antithrombin to give the severe and sudden onset of thrombosis.

References

Aldonyte, R., Jansson, L., Ljungberg, O., Larsson, S., and Janciauskiene, S. (2004). Polymerized alpha-antitrypsin is present on lung vascular endothelium. New insights into the biological significance of alpha-antitrypsin polymerization. *Histopathology* 45:587–592.

Aulak, K. S., Eldering, E., Hack, C. E., Lubbers, Y. P., Harrison, R. A., Mast, A., Cicardi, M., and Davis, A. E. (1993). A hinge region mutation in C1-inhibitor (Ala436→Thr) results in nonsubstrate-like behavior and in polymerization of the molecule. *J Biol Chem* 268:18088–18094.

Banda, M. J., Rice, A. G., Griffin, G. L., and Senior, R. M. (1988). The inhibitory complex of human alpha 1-proteinase inhibitor and human leukocyte elastase is a neutrophil chemoattractant. *J Exp Med* 167:1608–1615.

Beauchamp, N. J., Pike, R. N., Daly, M., Butler, L., Makris, M., Dafforn, T. R., Zhou, A., Fitton, H. L., Preston, F. E., Peake, I. R., and Carrell, R. W. (1998). Antithrombins Wibble and Wobble (T85M/K): Archetypal conformational diseases with *in vivo* latent-transition, thrombosis and heparin activation. *Blood* 92:2696–2706.

Belorgey, D., Crowther, D. C., Mahadeva, R., Lomas, D. A. (2002). Mutant Neuroserpin (S49P) that causes familial encephalopathy with neuroserpin inclusion bodies is a poor proteinase inhibitor and readily forms polymers *in vitro*. *J Biol Chem* 277:17367–17373.

Belorgey, D., Sharp, L. K., Crowther, D. C., Onda, M., Johansson, J., and Lomas, D. A. (2004). Neuroserpin Portland (Ser52Arg) is trapped as an inactive intermediate that rapidly forms polymers: implications for the epilepsy seen in the dementia FENIB. *Eur J Biochem* 271:3360–3367.

Bennett, M. J., and Eisenberg, D. (2004). The evolving role of 3D domain swapping in proteins. *Structure* 12:1339–1341.

Brennan, S. O., Maghzal, G., Shneider, B. L., Gordon, R., Magid, M. S., and George, P. M. (2002). Novel fibrinogen gamma375 Arg→Trp mutation (fibrinogen aguadilla) causes hepatic endoplasmic reticulum storage and hypofibrinogenemia. *Hepatology* 36:652–658.

Carrell, R. W., and Boswell, D. R. (1986). Serpins: the superfamily of plasma serine proteinase inhibitors. In: *Proteinase Inhibitors*. (Eds Barrett A and Salvesen G), Chapter 12, pp 403–419, Elsevier Biomedical Press, Amsterdam.

Carrell, R. W., Evans, D. L., and Stein, P. E. (1991). Mobile reactive center of serpins and the control of thrombosis. *Nature* 353:576–579.

Carrell, R. W., and Lomas, D. A. (1997). Conformational disease: A new clinical entity. *Lancet* 350:134–138.

Carrell, R. W., and Gooptu, B. (1998). Conformational changes and disease—including serpins, prions and Alzheimer's. *Curr Opin Struct Biol* 8:799–809.

Carrell, R. W., Huntington, J. A., Mushunje, A., and Zhou, A. (2001). The conformational basis of thrombosis. *Thromb Haemost* 86:14–22.

Carrell, R. W., and Lomas, D. A. (2002). Alpha1-antitrypsin deficiency—a model for conformational diseases. *N Engl J Med* 346:45–53.

Carrrell, R W. (2005). Cell toxicity and conformational disease. *Trends Cell Biol.* 15:574–580.

Chow, M. K., Lomas, D. A., and Bottomley, S. P. (2004). Promiscuous beta-strand interactions and the conformational diseases. *Curr Med Chem* 11:491–499.

Cleary, J. P., Walsh, D. M., Hofmeister, J. J., Shankar, G. M., Kuskowski, M. A., Selkoe, D. J., and Ashe, K. H. (2005). Natural oligomers of the amyloid-beta protein specifically disrupt cognitive function. *Nat Neurosci* 8:79–84.

Corral, J., Huntington, J. A., Gonzalez-Conejero, R., Mushunje, A., Navarro, M., Marco, P., Vicente, V., and Carrell, R. W. (2004). Mutations in the shutter region of antithrombin result in formation of disulfide-linked dimers and severe venous thrombosis. *J Thromb Haemost* 2:931–939.

Crowther, D. C., Kinghorn, K. J., Miranda, E., Page, R., Curry, J. A., Duthie, F. A., Gubb, D. C., and Lomas, D. A. (2005). Intraneuronal Aβ, non-amyloid aggregates and neurodegeneration in a Drosophila model of Alzheimer's disease. *Neuroscience* 132:123–135.

Davis, R. L., Shrimpton, A. E., Holohan, P. D., Bradshaw, C., Feiglin, D., Sonderegger, P., Kinter, J., Becker, L. M., Lacbawan, F., Krasnewich, D., Muenke, M., Lawrence, D. A., Yerby, M. S., Shaw, C.-M., Gooptu, B., Finch, J. T., Carrell, R. W., and Lomas, D. A. (1999). Familial dementia caused by polymerisation of mutant neuroserpin. *Nature* 401:376–379.

Davis, R. L., Shrimpton, A. E., Carrell, R. W., Lomas, D. A., Gerhard, L., Baumann, B., Lawrence, D. A., Yepes, M., Kim, T. S., Ghetti, B., Piccardo, P., Takao, M., Lacbawan, F., Muenke, M., Sifers, R. N., Bradshaw, C. B., Kent, P. F., Collins, G. H., Larocca, D., and Holohan, P. D. (2002). Association between conformational mutations in neuroserpin and onset and severity of dementia. *Lancet* 359:2242–2247.

Demuro, A., Mina, E., Kayed, R., Milton, S. C., Parker, I., and Glabe, C. G. (2005). Calcium dysregulation and membrane disruption as a ubiquitous neurotoxic mechanism of soluble amyloid oligomers. *J Biol Chem* 280:17294–17300.

Dunstone, M. A., Dai, W., Whisstock, J. C., Rossjohn, J., Pike, R. N., Feil, S. C., Le Bonniec, B. F., Parker, M. W., and Bottomley, S. P. (2000). Cleaved antitrypsin polymers at atomic resolution. *Protein Sci* 9:417–20.

Eldering, E., Verpy, E., Roem, D., Meo, T., and Tosi, M. (1995). COOH-terminal substitutions in the serpin C1 inhibitor that cause loop overinsertion and subsequent multimerization. *J Biol Chem* 270:2579–87.

Elliott, P. R., Bilton, D., and Lomas, D. A. (1998). Lung polymers in Z alpha1-antitrypsin deficiency-related emphysema. *Am J Respir Cell Mol Biol* 18:670–4.

Graham, K. S., Le, A., and Sifers, R. N. (1990) .Accumulation of the insoluble PiZ variant of human a_1-antitrypsin within the hepatic endoplasmic reticulum does not elevate the steady-state level of grp78/BiP. *J Biol Chem* 265:20463–20468.

Hayden, M. R., Tyagi, S. C., Kerklo, M. M., and Nicolls, M. R. (2005). Type 2 diabetes mellitus as a conformational disease. *JOP* 6:287–302.

Huber, R., and Carrell, R. W. (1989). Implications of the three-dimensional structure of α_1-antitrypsin for structure and function of serpins. *Biochemistry* 28:8951–8966.

Huntington, J. A., Pannu, N. S., Hazes, B., Read, R., Lomas, D. A., and Carrell, R. W. (1999). A 2.6Å Structure of a serpin polymer and implications for conformational disease. *J Mol Biol* 293:449–455.

Huntington, J. A., Read, R. J., and Carrell, R. W. (2000). Structure of a serpin-protease complex shows inhibition by deformation. *Nature* 407:923–926.

Inagi, R., Nangaku, M., Usuda, N., Shimizu, A., Onogi, H., Izuhara, Y., Nakazato, K., Ueda, Y., Oishi, H., Takahashi, S., Yamamoto, M., Suzuki, D., Kurokawa, K., van Ypersele de Strihou, C., and Miyata, T. (2005a) Novel serpinopathy in rat kidney and pancreas induced by overexpression of megsin. *J Am Soc Nephrol* 16:1339–1349.

Inagi, R., Nangaku, M., Onogi, H., Ueyama, H., Kitao, Y., Nakazato, K., Ogawa, S., Kurokawa, K., Couser, W. G., and Miyata, T. (2005b) Involvement of endoplasmic reticulum (ER) stress in podocyte injury induced by excessive protein accumulation. *Kidney Int* 68:2639–2650.

Janciauskiene, S., Carlemalm, E., and Eriksson, S. (1995). *In vitro* amyloid fibril formation from alpha 1-antitrypsin. *Biol Chem Hoppe Seyler* 376:103–109.

Jin, L., Abrahams, J. P., Skinner, R., Petitou, M., Pike, R. N., Carrell, R. W. (1997). The anticoagulant activation of antithrombin by heparin. *Proc Natl Acad Sci USA* 94:14683–14688.

Kurdowska, A., and Travis, J. (1990). Acute phase protein stimulation by alpha 1-antichymotrypsin-cathepsin G complexes. Evidence for the involvement of interleukin-6. *J Biol Chem* 265:21023–21026.

Laurell, C.-B., and Eriksson, S. (1963). The electrophoretic α_1-globulin pattern of serum in α_1-antitrypsin deficiency. *Scand J Clin Lab Invest* 15:132–140.

Lawless, M. W., Greene, C. M., Mulgrew, A., Taggart, C. C., O'Neill, S. J., and McElvaney, N. G. (2004). Activation of endoplasmic reticulum-specific stress responses associated with the conformational disease Z alpha 1-antitrypsin deficiency. *J Immunol* 172, 5722–5726.

Lindo, V. S., Kakkar, V. V., Learmonth, M., Melissari, E., Zappacosta, F., Panico, M., and Morris, H. R. (1995). Antithrombin-TRI (Ala382 to Thr) causing severe thromboembolic tendency undergoes the S-to-R transition and is associated with a plasma-inactive high-molecular-weight complex of aggregated antithrombin. *Br J Haematol* 89:589–601.

Loebermann, H., Tokuoka, R., Deisenhofer, J., and Huber, R. (1984). Human a1-proteinase inhibitor. Crystal structure analysis of two crystal modifications, molecular model and preliminary analysis of the implications for function. *J Mol Biol* 177:531–556.

Lomas, D. A., and Carrell, R. W. (2002). Serpinopathies and the conformational dementias, *Nat Rev Genet* 3:759–768.

Lomas, D. A., and Parfrey, H. (2004). Alpha1-antitrypsin deficiency. 4: Molecular pathophysiology. *Thorax* 59:529–35.

Lomas, D. A., Evans, D. L., Finch, J. T., and Carrell, R. W. (1992). Molecular mechanism of Z α_1-antitrypsin accumulation in the liver. *Nature* 357:605–607.

Madani, R., Kozlov, S., Akhmedov, A., Cinelli, P., Kinter, J., Lipp, H. P., Sonderegger, P., and Wolfer, D. P. (2003). Impaired explorative behavior and neophobia in genetically modified mice lacking or overexpressing the extracellular serine protease inhibitor neuroserpin. *Mol Cell Neurosci* 23:473–494.

Mahadeva, R., Dafforn, T. R., Carrell, R. W., and Lomas, D. A. (2002). 6-mer peptide selectively anneals to a pathogenic serpin and blocks polymerization. *J Biol Chem* 277:6771–6774.

Mahadeva, R., Atkinson, C., Li, Z., Stewart, S., Janciauskiene, S., Kelley, D. G., Parmar, J., Pitman, R., Shapiro, S. D., and Lomas, D. A. (2005). Polymers of Z alpha1-antitrypsin co-localize with neutrophils in emphysematous alveoli and are chemotactic *in vivo*. *Am J Pathol* 166:377–86.

Miranda, E., Romisch, K., and Lomas, D. A. (2004). Mutants of neuroserpin that cause dementia accumulate as polymers within the endoplasmic reticulum. *J Biol Chem* 279:28283–28291.

Mottonen, J., Strand, A., Symersky, J., Sweet, R. M., Danley, D. E., Geoghegan, K. F., Gerard, R. D., and Goldsmith, E. J. (1992). Structural basis of latency in plasminogen activator inhibitor-1. *Nature* 355:270–273.

Mushunje, A., Evans, G., Bennan, S. O., Carrell, R. W., and Zhou, A. (2004). Latent antithrombin and its detection, formation and turnover in the circulation *J Thromb Haemost* 2:2170–2177.

Mushunje, A., Zhou, A., and Carrell, R. W. (2005). Defining the specificity of latent antithrombin heterodimer formation and their *in vivo* pathological significance. Abstract 125, Proceedings 4th International Symposium on Serpin Structure, Function and Biology, Cairns, Australia.

Onda, M., Belorgey, D., Sharp, L. K., and Lomas, D. A. (2005). Latent Ser49Pro neuroserpin forms polymers in the dementia familial encephalopathy with neuroserpin inclusion bodies. *J Biol Chem* 280:13735–13741.

Oyadomari, S., Koizumi, A., Takeda, K., Gotoh, T., Akira, S., Araki, E., Mori, M. (2002). Targeted disruption of the Chop gene delays endoplasmic reticulum stress-mediated diabetes. *J Clin Invest* 109:525–532.

Parmar, J. S., Mahadeva, R., Reed, B. J., Farahi, N., Cadwallader, K. A., Keogan, M. T., Bilton, D., Chilvers, E. R., and Lomas, D. A. (2002). Polymers of alpha(1)-antitrypsin are chemotactic for human neutrophils: a new paradigm for the pathogenesis of emphysema. *Am J Respir Cell Mol Biol* 26:723–730.

Pepys, M. B. (2001). Pathogenesis, diagnosis and treatment of systemic amyloidosis. *Philos Trans R Soc Lond B Biol Sci.* 3569:203–210.

Preissner, K. T., Grulich-Henn, J., Ehrlich, H. J., Declerck, P., Justus, C., Collen, D., Pannekoek, H., and Muller-Berghaus, G. (1990). Structural requirements for the extracellular interaction of plasminogen activator inhibitor 1 with endothelial cell matrix-associated vitronectin. *J Biol Chem* 265:18490–18498.

Reixach, N., Deechongkit, S., Jiang, X., Kelly, J. W., and Buxbaum, J. N. (2004). Tissue damage in the amyloidoses: Transthyretin monomers and nonnative oligomers are the major cytotoxic species in tissue culture. *Proc Natl Acad Sci USA* 101:2817–2822.

Ron, D. (2002) Translational control in the endoplasmic reticulum stress response. *J Clin Invest* 110:1383–1388.

Rudnick, D. A., Liao, Y., An, J. K., Muglia, L. J., Perlmutter, D. H., and Teckman, J. H. (2004). Analyses of hepatocellular proliferation in a mouse model of alpha-1-antitrypsin deficiency. *Hepatology* 39:1048–1055.

Rutishauser, J., and Spiess, M. (2002). Endoplasmic reticulum storage diseases. *Swiss Med Wkly* 132:211–222.

Rutkowski, D. T., and Kaufman, R. J. (2004). A trip to the ER: coping with stress. *Trends Cell Biol* 14:20–28.

Seiffert, D., and Loskutoff, D. J. (1991). Kinetic analysis of the interaction between type 1 plasminogen activator inhibitor and vitronectin and evidence that the bovine inhibitor binds to a thrombin-derived amino-terminal fragment of bovine vitronectin. *Biochim Biophys Acta* 1078:23–30.

Sharp, H. L., Bridges, R. A., Krivit, W., and Freier, E. F. (1969). Cirrhosis associated with alpha-1-antitrypsin deficiency: a previously unrecognised inherited disorder. *J Lab Clin Med* 73:934–939.

Silverman, G. A., Bird, P. I., Carrell, R. W., Church, F. C., Coughlin, P. B., Gettins, G. W., Irving, J. A., Lomas, D. A., Luke, C. J., Moyer, R. W., Pemberton, P. A., Remold-O'Donnell, E., Salvesen, G. S., Travis, J., and Whisstock, J. C. (2001). The serpins are an expanding superfamily of structurally similar but functionally diverse proteins. *J Biol Chem* 276:33293–33296.

Springhetti, E. M., Istomina, N. E., Whisstock, J. C., Nikitina, T., Woodcock, C. L., and Grigoryev, S. A. (2003). Role of the M-loop and reactive center loop domains in the folding and bridging of nucleosome arrays by MENT. *J Biol Chem* 278:43384–43393.

Stein, P. E., Leslie, A. G. W., Finch, J. T., Turnell, W. G., McLaughlin, P. J., and Carrell, R. W. (1990). Crystal structure of ovalbumin as a model for the reactive center of serpins. *Nature* 347:99–102.

Stein, P. E., and Carrell, R. W. (1995). What do dysfunctional serpins tell us about molecular mobility and disease? *Nat Struct Biol* 2:96–104.

Takahashi, R. H., Almeida, C. G., Kearney, P. F., Yu, F., Lin, M. T., Milner, T. A., and Gouras, G. K. (2004). Oligomerization of Alzheimer's beta-amyloid within processes and synapses of cultured neurons and brain. *J Neurosci* 24:3592–3599.

Tanaka, Y., Ueda, K., Ozawa, T., Sakuragawa, N., Yokota, S., Sato, R., Okamura, S., Morita, M., and Imanaka, T. (2002). Intracellular accumulation of antithrombin Morioka (C95R), a novel mutation causing type I antithrombin deficiency. *J Biol Chem* 277:51058–51067.

Teckman, J. H., An, J. K., Blomenkamp, K., Schmidt, B., and Perlmutter, D. (2003). Mitochondrial autophagy and injury in the liver in alpha 1-antitrypsin deficiency. *Am J Physiol Gastrointest Liver Physiol* 286:G851–862.

Tomidokoro, Y., Lashley, T., Rostagno, A., Neubert, T. A., Bojsen-Moller, M., Braendgaard, H., Plant, G., Holton, J., Frangione, B., Revesz, T., and Ghiso, J. (2005). Familial Danish dementia: co-existence of Danish and Alzheimer amyloid subunits (ADan AND A{beta}) in the absence of compact plaques. *J Biol Chem* 280:36883–36894.

Wu, Y., Swulius, M. T., Moremen, K. W., Sifers, R. N. (2003). Elucidation of the molecular logic by which misfolded alpha 1-antitrypsin is preferentially selected for degradation. *Proc Natl Acad Sci USA* 100:8229–8234.

Zhou, A., Huntington, J. A., Pannu, N. S., Carrell, R. W., and Read, R. J. (2003). How vitronectin binds PAI-1 to modulate fibrinolysis and cell migration. *Nat Struct Biol* 10:541–544.

Zhou, A., Stein, P. E., Huntington, J. A., Sivasothy, P., Lomas, D. A., and Carrell, R. W. (2004). How small peptides block and reverse serpin polymerisation. *J Mol Biol* 342:931–941.

Zhou, A., and Carrell, R. W. (2005). Serpin polymers are infectivethe propagation of polymerisation. Abstracts 31 and 155, Proceedings 4th International Symposium on Serpin Structure, Function and Biology, Cairns, Australia.

Altered Protein Structure and Impaired Function

16

Human Copper-Zinc Superoxide Dismutase and Familial Amyotrophic Lateral Sclerosis

Ahmad Galaleldeen and P. John Hart

Abstract

The structural and biochemical properties of human copper-zinc superoxide dismutase (SOD1) have been studied for more than 30 years. Although most enzymes give up their secrets over such a long period, SOD1 has continued to be the subject of intense investigation, most recently due to the discovery of its linkage to an inherited form of the fatal neurodegenerative disorder amyotrophic lateral sclerosis (ALS; motor neuron disease, Lou Gehrig's disease). Approximately 100 mutations in SOD1 are now known to give rise to ALS, most of which are single amino acid substitutions. Accumulating evidence strongly suggests that the majority of these lesions lead to SOD1 aggregation, and it appears that this nonnative SOD1 self-association is somehow injurious to motor neurons. However, the detailed molecular mechanism(s) through which these SOD1 mutants exert their toxic effects remain largely undefined. The information contained in this chapter is intended to provide a synopsis of what is currently known and/or hypothesized about the molecular basis for SOD1-linked ALS, with a strong emphasis on the biophysical and structural properties of the enzyme. Enhanced understanding of the structural basis for pathogenic SOD1 misfolding and self-association is likely to be a prerequisite for the development of therapeutic avenues of this progressive neurodegenerative disease.

16.1. Amyotrophic Lateral Sclerosis

Amyotrophic lateral sclerosis (ALS) is the most common adult motor neuron disease, characterized by the progressive destruction of both upper and lower motor neurons. The disorder typically manifests as muscle weakness in the extremities that expands to affect all muscles of the body, resulting in paralysis and death generally within 5 years postdiagnosis (Haverkamp *et al.*, 1995). The disorder was originally known as Charcot's sclerosis after the French physician Jean-Martin Charcot, who was the first to describe ALS in 1869 (Charcot and Joffroy, 1869). In the United States the condition became known as Lou Gehrig's disease after that New York Yankee (and one of the greatest baseball players of all time) was afflicted in the 1930s. According to the ALS CARE database, ~30,000 Americans have ALS at any given time and each year ~5600 people in the United States are diagnosed (see www.alsa.org). The average age of diagnosis is 55 years and the condition is somewhat more common in men than in women. Despite the fact that ALS has been known for well over a century, there remains no effective therapy.

Most cases (90–95%) of ALS are termed *sporadic* (sALS) with no established genetic link. The remaining cases have an inherited component and are termed *familial* ALS (fALS). Approximately 20% of the familial cases are linked to chromosome 21q22.1 and autosomal dominant mutations in the gene encoding the ubiquitously expressed cytosolic antioxidant protein copper-zinc superoxide dismutase (SOD1) (Deng *et al.*, 1993; Rosen *et al.*, 1993). Because sALS and fALS cases share clinical and pathologic features, the discovery of a role of SOD1 in ALS generated enormous excitement in the research community because the structure and function of SOD1 is known (Tainer *et al.*, 1982; Strange *et al.*, 2003). The linkage of cases of ALS to lesions in *SOD1* provided one of the first focused research efforts aimed at understanding the etiology of the disease, with the hope that lessons learned from SOD1-linked ALS may be applicable to the sporadic forms.

16.2. SOD1 Structure and Enzymatic Activity

Eukaryotic SOD1 is a 32-kDa homodimeric protein found in the cytosol, nucleus, and intermembrane space of mitochondria (Lyons *et al.*, 1999; Okado-Matsumoto and Fridovich 2001; Sturtz *et al.*, 2001). The enzyme catalyzes the disproportionation of superoxide, a normal by-product of cellular respiration, into dioxygen and hydrogen peroxide ($2O_2^- + 2H^+ \rightarrow O_2 + H_2O_2$) at nearly diffusion-controlled rates (Getzoff *et al.*,1983). The reaction is mediated by protein-bound copper ions alternating between reduced and oxidized forms (reactions 1 and 2) (Rotilio *et al.*,1972; Fielden *et al.*, 1974).

$$O_2^- + \text{SOD1} - \text{Cu(II)} \rightarrow O_2 + \text{SOD1} - \text{Cu(I)} \tag{16.1}$$

$$O_2^- + \text{SOD1} - \text{Cu(I)} + 2H^+ \rightarrow H_2O_2 + \text{SOD1} - \text{Cu(II)} \tag{16.2}$$

Each SOD1 monomer displays an eight-stranded Greek key β-barrel fold, binds one copper and one zinc ion, and possesses a conserved intrasubunit disulfide bond (Fig. 16.1A; **see color insert**). The presence of the latter is somewhat uncommon in cytosolic proteins. The copper cycles between a distorted square planer geometry in its oxidized form and a nearly trigonal planer coordination geometry in its reduced form (Ogihara *et al.*, 1996; Hart *et al.*, 1999). Two lengthy loop elements that protrude from the β-barrel, termed the *zinc* and *electrostatic* loops, form the walls of the active site and are intimately involved in metal binding. The zinc loop (green in Fig. 16.1A) consists of residues 63–84 and contains most of the metal binding ligands. The electrostatic loop (red in Fig. 16.1A) consists of residues 121–144 and contains charged residues that help to guide the negatively charged substrate to the active site entrance while sterically excluding larger nonsubstrate ions (Getzoff *et al.*, 1983; Bertini *et al.*, 1998). The "disulfide loop" (purple in Fig. 16.1A) is covalently linked to the β-barrel through a bond between Cys 57 and Cys 146. This disulfide loop, along with β-strand 8 of the β-barrel (also purple) make up the bulk of the dimer interface through reciprocal interactions across the molecular two-fold.

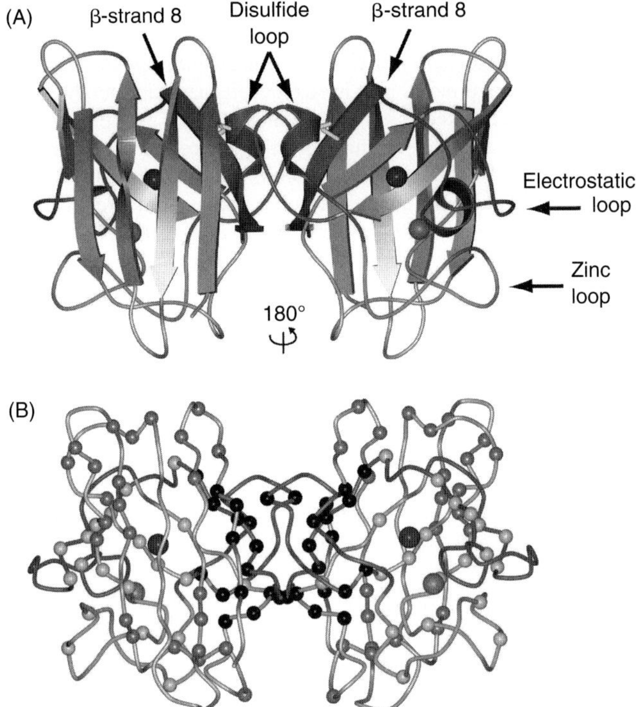

Figure 16.1. The human SOD1 dimeric structure and location of fALS mutations. (A) Each SOD1 monomer displays an eight-stranded Greek key β-barrel fold. The zinc and electrostatic loop elements are shown in green and red, respectively. The disulfide loop and β-strand 8 that form the bulk of the dimer interface are shown in purple. The disulfide bond between Cys 57 and Cys 146 is shown in yellow. The copper and zinc ions are represented as blue and green spheres, respectively. The molecular two-fold axis is indicated by the 180° symbol. (B) Human SOD1 showing the locations of fALS mutations. The color coding for the protein backbrone is the same as in (A). Those fALS mutations predicted to directly affect SOD1 dimerization are shown as black spheres, those that are predicted to affect dimerization indirectly through defects in metal binding are shown as yellow spheres, and those that fall in the β-barrel are shown as gray spheres. Reproduced from Doucette, P. A. *et al.* (2004) Dissociation of human copper-zinc superoxide dismutase dimers using chaotrope and reductant. Insights into the molecular basis for dimer stability. *J. Biol. Chem.* 279:54558–54566.

16.3. SOD1 and Toxic Gain-of-Function

In seminal papers published in 1993, 11 distinct single amino acid substitutions were identified in different families with a history of ALS (Deng *et al.*, 1993; Rosen *et al.*, 1993). Since that time, the number of fALS mutations has risen to ~114, occurring at over a third of the 153 residues in the primary sequence (Andersen, 2000). The vast majority of the alterations are point mutations, although there are a few changes that result in amino acid deletions or truncations of the polypeptide. These disease-causing mutations are spread throughout the three-dimensional structure of the enzyme, including the dimer interface, the β-strands, the zinc and electrostatic loop elements, and the copper- and zinc-binding sites (Fig. 16.1B; **see color insert**). An intriguing question is how such a structurally diverse collection of mutations in SOD1 can all elicit the disease (see Section 16.7 below).

Indeed, SOD1-mediated toxicity in fALS has been a controversial issue. Early studies suggested that some SOD1 mutant proteins may have reduced enzymatic activity (Deng *et al.*, 1993; Rosen *et al.*, 1994), which led to the assumption that SOD1-linked ALS was a loss of function disorder because SOD1 plays a crucial role in protecting the cells from free radicals. It was later determined, however, that many of the pathogenic amino acid substitutions do not result in a loss of activity and that certain pathogenic mutants devoid of copper ion also elicit paralysis in transgenic mouse models (Borchelt *et al.*, 1994; Subramaniam *et al.*, 2002; Wang *et al.*, 2002a; Wang *et al.*, 2003). Moreover, SOD1 knockout mice do not develop motor neuron disease (Reaume *et al.*, 1996).

A gain of toxic function rather than loss of function was proposed when transgenic mice overexpressing fALS SOD1 mutant proteins in addition to their own wild-type SOD1 (and therefore have normal SOD1 activity) became paralyzed (Bruijn *et al.*, 1997). This observation nicely parallels the autosomal dominant nature of the disorder in humans. Explanatory hypotheses as to the nature of this toxic property ranged from aberrant copper-mediated chemistry (Yim *et al.*, 1993; Wiedau-Pazos *et al.*, 1996; Yim *et al.*, 1996) to aggregation of the mutant proteins (Bruijn *et al.*, 1997, 1998). In recent years, a consensus has emerged that the propensity of pathogenic SOD1 proteins to self-associate very likely underlies disease etiology. Thus, the molecular determinants of mutant SOD1 self-association and pathogenesis has become a subject of keen interest.

16.3.1. Pathogenic SOD1 Aggregation

Proteinaceous inclusions are observed in patients with both sporadic and familial ALS (Shibata *et al.*, 1996). In addition, insoluble protein complexes (aggregates) that contain pathogenic SOD1 are routinely observed in transgenic mice overexpressing various mutant SOD1 proteins (Bruijn *et al.*, 1998; Wang *et al.*, 2002b). These findings are strongly suggestive that SOD1 aggregates mediate toxicity in these models. Other molecules, such as heat shock proteins or other essential signaling molecules, may also become sequestered in these aggregates, which may indirectly impair motor neuron function (Bruening *et al.*, 1999; Okado-Matsumoto and Fridovich, 2002).

Although it is certainly possible that the insoluble aggregates are somehow toxic, an alternative idea is that they are actually somewhat benign, resulting from a cellular defense mechanism that occurs when the levels of misfolded or damaged proteins exceed the capacity of the protein degradation machinery to eliminate them (Sherman and Goldberg, 2001). Mounting evidence suggests that instead of these insoluble aggregates, it may be their soluble oligomeric precursors termed *protofibrils* that are toxic, possibly by forming pore-like structures that can damage cell or mitochondrial membranes or adversely affect the proteasome (Lashuel *et al.*, 2002). Such soluble oligomers are found to be intermediates in the fibrillization of β-amyloid in Alzheimer's disease and α-synuclein in Parkinson's disease *in vitro* (Lansbury, 1999; Goldberg and Lansbury, 2000). The concept that the soluble protofibrils are toxic in neurodegenerative diseases is supported by the observation that there is no clear correlation between the quantity of fibrillar, insoluble deposits at autopsy and the clinical severity of Alzheimer's disease or Parkinson's disease. In addition, transgenic mouse models of these disorders have disease-like phenotypes well before fibrillar deposits can be detected (Wang *et al.*, 2002a, b). Defects in axonal transport are observed in mouse models of ALS at around 30 days after birth, before symptoms or insoluble proteinaceous deposits are visible, suggesting that soluble oligomers of pathogenic SOD1 could be the interfering entity (Johnston *et al.*, 2000; Shah *et al.*, 2002). Whether or not the toxic oligomeric SOD1 species is soluble or insoluble, it is clear that these pathogenic SOD1 proteins must have *some* feature distinct from the wild-type protein that facilitates their self-association (see below), and it may be this feature that underlies the toxic gain-of-function in these proteins.

16.3.2. Impairment of Proteasome Activity in ALS

A hallmark of fALS is the presence of aggregates that are enriched for pathogenic SOD1. Their presence suggests that the burden of misfolded pathogenic SOD1 proteins eventually exceeds the capacity of the protein degradation machinery in the motor neuron. Mild oxidation to introduce protein carbonyls (and hence partial unfolding) may be part of a normal process where mistranslated or misfolded proteins are targeted for proteolysis (Dunlop *et al.*, 2002; Grune *et al.*, 2003; Shringarpure *et al.*, 2003). As a result of mutation, some fALS SOD1 proteins may be partially unfolded or mimic oxidatively modified SOD1 and thus adversely affect proteasome function *in vivo*. Oxidatively modified wild-type SOD1 proteins containing abnormal protein carbonyls have been detected in cell culture after addition of proteasome inhibitors before or after addition of a source of iron, which causes levels of oxidized proteins to increase (Drake *et al.*, 2002). Moreover, Davies and co-workers used SOD1 as a model system and demonstrated that mildly oxidized SOD1 is one of the proteins that is specifically targeted for proteolysis by the 20 S proteasome as a consequence of the damage. The pathway for degradation of oxidized SOD1 requires no ubiquitination or ATP and they concluded that recognition of the mildly oxidized SOD1 protein is due to partial unfolding leading to exposure of hydrophobic portions of the protein (Grune *et al.*, 2003).

Recent studies have strongly suggested that proteasome levels and/or activities are adversely affected by the presence of the fALS-mutant SOD1 proteins (Urushitani *et al.*, 2002; Allen *et al.*, 2003; Puttaparthi *et al.*, 2003; Kabashi *et al.*, 2004). The level of 20 S proteasome activity was substantially reduced in lumbar spinal neurons relative to the surrounding neuropil in the G93A SOD1 transgenic mouse; it seems that impairment of the proteasome is an early event and contributes to ALS pathogenesis (Kabashi *et al.*, 2004). We hypothesize, on the basis of our biophysical studies (see below), that some of the fALS-mutant SOD1 proteins are likely to monomerize or be partially unfolded *in vivo* either as a consequence of their loss of metal ion binding ability, destabilization of their apo-proteins, or because of enhanced oxidative damage to the mutant SOD1 that may or may not be mediated by bound copper ions. In any of these cases, the partially unfolded fALS-mutant SOD1 proteins are likely to expose hydrophobic portions of their structures, and they may be abnormally targeted to the proteasome even without oxidation. If this is true, it could help explain the inhibition of proteasome function observed in the disease models.

16.4. Classification of SOD1 Mutants

Based on their positions in the 3-D structure and on metal content as determined by inductively coupled plasma mass spectrometry, the fALS mutant SOD1 proteins are divided into two categories (Valentine and Hart, 2003). The first we term *wild type–like* (WTL) mutants because they are isolated from their expression systems with copper and zinc levels nearly identical to those found for the wild-type protein expressed in the same system. The SOD1 activities and the spectroscopic characteristics of the WTL SOD1 proteins are the same as those of wild-type SOD1 (Goto *et al.*, 2000; Hayward *et al.*, 2002). The second we term *metal-binding region* (MBR) mutants. These include mutations in the metal-binding ligands themselves and/or in the electrostatic and zinc loop elements that are intimately involved with metal binding. The right subunit in Fig. 16.2 (**see color insert**) shows the positions of the MBR mutations represented as black spheres. The WTL mutations are not explicitly shown but are scattered throughout the β-barrel of the protein shown in yellow. The electrostatic (red) and zinc (blue) loop elements are in contact, and the conformation of

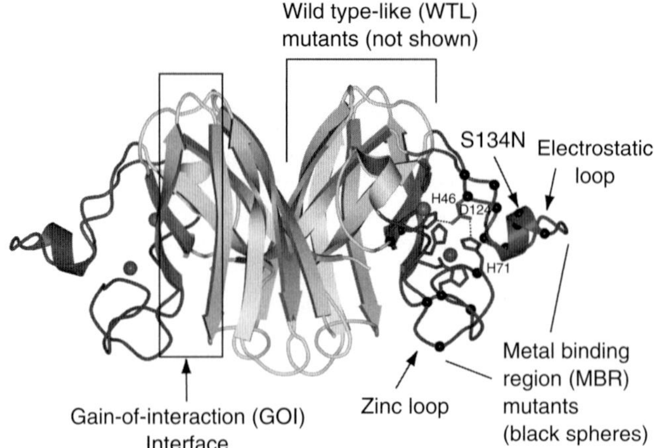

Figure 16.2. Negative design of human SOD1 and the locations of the wild type–like (WTL) and metal binding region (MBR) fALS mutations. In the left subunit, the gain-of-interaction (GOI) interface is boxed. In the wild-type protein, the zinc loop (blue) projects from the plane of the paper toward the viewer, preventing SOD1-SOD1 protein-protein interactions from occurring at the edge strands. The copper and zinc ions are shown as green and magenta spheres, respectively. In the right subunit, the positions of the MBR mutations are shown as black spheres and the WTL mutations (not shown for clarity) are scattered throughout the β-barrel depicted in yellow. Metal binding stabilizes the conformation of the zinc and electrostatic loop elements and therefore plays an intimate role in the negative design of the molecule (see text).

one of these loops directly affects the conformation of the other. The side chain of Asp 124 links the electrostatic and zinc loops and contributes directly to the stabilization of both the copper and zinc binding sites by forming hydrogen bonds simultaneously to the non-liganding imidazole nitrogen atoms of copper ligand His46 and zinc ligand His71. Due to this intimate cross-talk between these loop elements, it is not surprising that the MBR mutants are isolated from their expression systems with very low zinc and copper (Hayward *et al.*, 2002; Rodriguez *et al.*, 2002).

16.5. Negative Design of Wild-Type SOD1

The left subunit in Figure 16.2 shows that although each subunit of SOD1 possesses an eight-stranded Greek key β-barrel fold, not all of the strands forming the β-barrel are continuously hydrogen-bonded to each other. β-Strands 5 and 6 (boxed) are considered to be "edge" strands. Edge strands are potentially deleterious because they provide a hydrogen-bonding surface that can accommodate interaction with other edge strands, which in turn can lead to the formation of higher order structures. As a consequence, proteins containing β-sheets have evolved a variety of strategies to protect their edge strands from nonproductive interactions by incorporating "negative design" into their architecture [an excellent review on this topic can be found in Richardson and Richardson (2002)]. As shown in Figure 16.2, the negative design of SOD1 is derived from the well-ordered electrostatic and zinc loop elements (particularly the latter) that protrude from the β-barrel and prevent SOD1-SOD1 protein-protein interactions from occurring at these edge strands. Because the ligands for the copper and zinc ions are located in these loops, their conformations are strongly influenced by the presence of metal ions (see below).

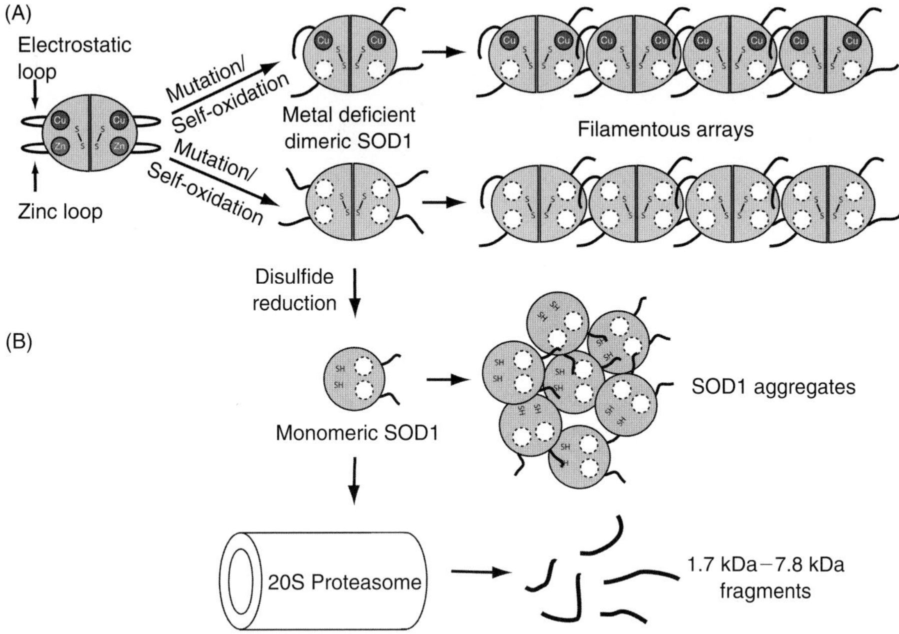

Figure 16.3. The monomer and dimer models of pathogenic SOD1 aggregation. (A) The top two rows show schematically how the dimeric form of SOD1 can self-associate. Metal deficiency (particularly in the zinc binding site) allows conformational flexibility of the zinc and electrostatic loop elements, permitting non-native SOD1-SOD1 interactions (see Fig. 16.4). (B) Reduction of the disulfide bond in apo-SOD1 facilitates monomerization. SOD1 monomers show an increased propensity to aggregate in biophysical studies [see Rakhit *et al.* (2004)]. Monomeric SOD1 also acts as a substrate for the 20 S proteasome. The products of proteasomal digestion are discrete fragments that may themselves aggregate or exert toxic effects (see text).

16.6. Models of SOD1 Self-Association

Currently, there exist two models for pathogenic SOD1 aggregation. As shown in Figure 16.3, these are known as the "dimer" and the "monomer" models, respectively. In the sections below, we outline structural and biophysical studies that address each of these possibilities.

16.6.1. Dimer Models of Aggregation

16.6.1.1. Linear Amyloid-like Filaments of Pathogenic SOD1

X-ray crystal structures of human MBR SOD1 mutants S134N and apo-H46R have been determined (Elam *et al.*, 2003b; Antonyuk *et al.*, 2005). As shown in the right subunit of Figure 16.2, the S134N substitution is located in the small helix located in the electrostatic loop and the H46R substitution affects a copper ligand. The S134N and H46R proteins are both metal deficient when isolated, and their low levels of metallation *in vitro* are quite likely to represent their states *in vivo*. Figure 16.4A (**see color insert**) shows that the β-barrel elements of both of these mutants (green) are preserved relative to the wild-type protein (gray). In contrast, these mutants demonstrate substantial disorder and conformational changes in their zinc and electrostatic

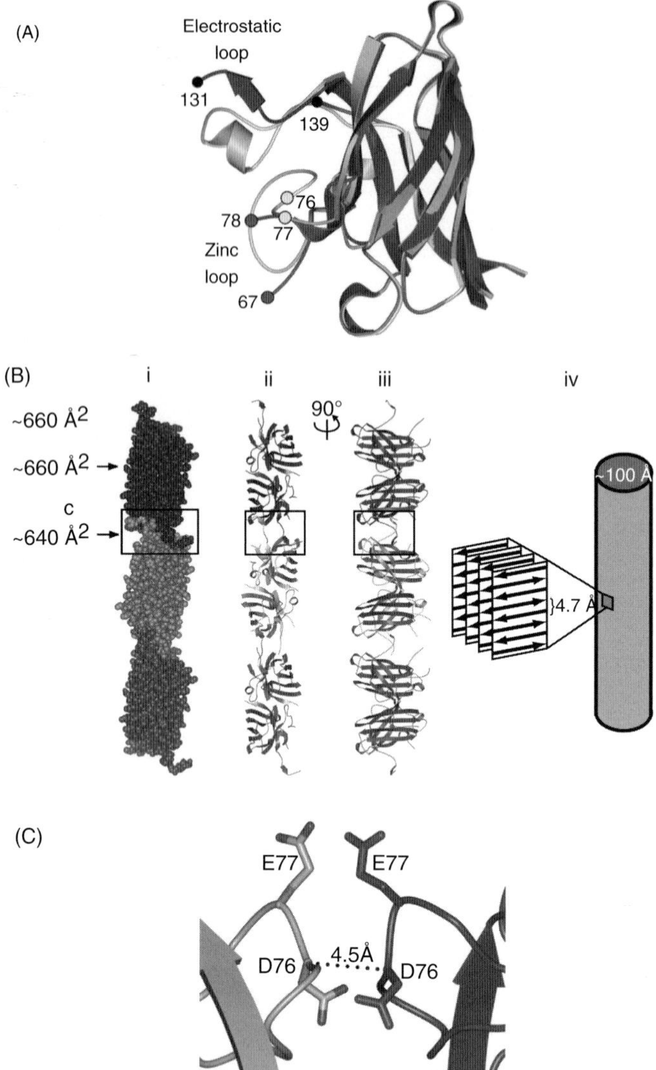

Figure 16.4. Linear, amyloid-like filaments formed by pathogenic SOD1 mutants H46R and S134N. (A) S134N and apo-H46R monomers (green) superimposed on human WT SOD1 (gray). In both structures, the zinc and electrostatic loops are disordered. Residues N- and C-terminal to these disordered regions are indicated by black and red spheres for the electrostatic and zinc loops, respectively. Residues of the electrostatic loop (blue) interact with the exposed cleft (red) of adjacent molecules in the crystal lattice (see boxed region in panel B). Asp 76 and Glu 77 of the zinc loop in the wild-type enzyme are represented by yellow spheres (see panel C). (B) Gain-of-interaction (GOI) in pathogenic SOD1 gives rise to cross-β fibrils in two different crystal systems. The SOD1 filaments are represented by three dimers from top to bottom in green, gold, and blue. The GOI interface is boxed. β strands 1, 2, 3, and 6 are shown in red. The long axes of the β strands run perpendicular to the long axis of the filament, an architecture similar to the "cross-β" structure observed in amyloid fibers as shown schematically in (iv). (C) Negative design of SOD1 prevents formation of the GOI interface. When metals are bound, the zinc loop elements are well-ordered and prevent self-association into the linear filaments through electrostatic and steric considerations (see text) Reproduced from Elam, J. S., *et al.* (2003a). An alternative mechanism of bicarbonate-mediated peroxidation by copper-zinc superoxide dismutase. *J. Biol. Chem.* 278:21032–21039.

loop elements relative to the wild-type enzyme. In both structures, the conformational mobility of these loop elements due to metal deficiency exposes the edges of β-strands 5 and 6, which together form a cleft (red in Fig. 16.4A) between the two β-sheets of the SOD1 β-barrel. This solvent-accessible depression serves as a "hot spot" for non-native SOD1-SOD1 protein-protein interactions, where it contacts a portion of the electrostatic loop (residues 125–131) of a neighboring molecule in the crystal lattice (blue in Fig. 16.4A). These new contacts are termed *gain-of-interaction* (GOI) contacts to differentiate them from the naturally occurring interfaces between subunits of the native SOD1 homodimer. When combined, the GOI contacts and the native dimeric interactions result in the formation of the linear filamentous arrays shown in Figure 16.4B (**see color insert**).

The S134N and apo-H46R linear filaments are practically identical. Their GOI interfaces (boxed) are symmetrical and run parallel to the symmetry dyad of the natural dimers. Figure 16.4B(i) reveals that GOI interfaces bury nearly the same amount of solvent accessible surface area per polypeptide as the highly stable native SOD1 homodimeric interface. The β-sheets of the pathogenic SOD1 molecules run parallel to the long axis of the fiber, while their constituent β-strands run perpendicular to this axis [Fig. 16.4B(iii)], an arrangement similar to what is observed in amyloid fibers. Superposition of metal-loaded, native SOD1 dimers onto mutant dimers that comprise these linear filaments results in steric and electrostatic repulsion through the zinc loops at the mutant GOI interface (Fig. 16.4C; **see color insert**). Well-ordered SOD1 electrostatic and zinc loops are therefore critical to the negative design of SOD1 and to prevent filamentous assembly. Analyses of newly obtained crystals of metal-deficient pathogenic SOD1 mutants H80R (zinc ligand) and E133Δ (electrostatic loop residue) reveal that they also form filamentous arrays analogous to those formed by the S134N and H46R proteins.

16.6.1.2. Helical Filamentous Arrays of Pathogenic SOD1

As mentioned previously in the chapter, biophysical studies support the emerging notion that a prefibrillar soluble oligomer (protofibril) could be responsible for cell death and that the insoluble forms that are typically observed after patient death may be neuroprotective. Peter Lansbury and colleagues have suggested that a subpopulation of the soluble protofibrils may function as pores with the ability to permeabilize cell or mitochondrial membranes (Caughey and Lansbury, 2003). In support of this, annular structures are observed in familial mutants of α-synuclein (Parkinson's disease) and Alzheimer's precursor protein (Alzheimer's disease) (Lashuel *et al.*, 2002).

Intriguingly, pore-like entities are also observed in the x-ray crystal structure of zinc-loaded MBR mutant H46R (Elam *et al.*, 2003b; Antonyuk *et al.*, 2005). Figure 16.5 (**see color insert**) shows a variation on the theme observed for the amyloid-like filaments, where GOI interface between Zn-H46R dimers results in the assembly of helical filaments comprised of four Zn-H46R SOD1 dimers per turn. A helical filament is formed because the GOI intermolecular contact surface is shifted several amino acids relative to that of the linear filaments and zinc loop residues rather than residues of the electrostatic loop participate in non-native inter-dimer interactions. Zinc loop residues 78–81 (blue in Figs. 16.5A and 16.5B) hydrogen bond to the solvent exposed edge of β-strand 6, extending the β-sheet formed by β-strands 1, 2, 3, and 6 by one strand. Figures 16.5B and 16.5C show that repetition of this GOI leads to the formation of hollow "tubes" with an overall diameter of ~95 Å and an inner, water-filled cavity approximately 30 Å across. These dimensions are intriguingly similar to those observed for the "pores" formed by α-synuclein and Alzheimer's precursor protein (Lashuel *et al.*, 2002).

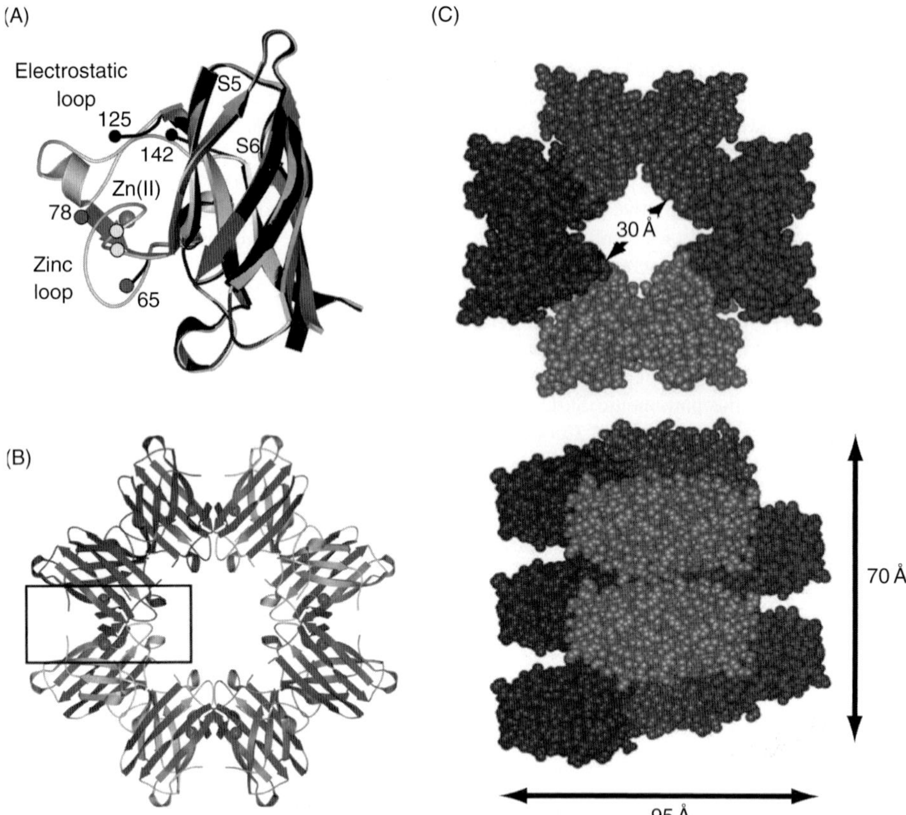

Figure 16.5. Helical, "pore-like" filaments formed between pathogenic SOD1 dimers. (A) Zn-H46R SOD1 monomer (black) superimposed on human wild-type SOD1 (gray). Portions of both the zinc and electrostatic loops are disordered. Residues N- and C-terminal to these disordered regions are indicated by black and red spheres for the electrostatic and zinc loop elements, respectively. The zinc ion is represented by a green sphere. This disorder exposes a cleft on the β-barrel (red). The GOI interface between Zn-H46R molecules is formed by reciprocal interactions between residues of the zinc loop (blue) with the exposed edges of β-strands 5 and 6 (red) of adjacent molecules in the crystal lattice. (B) The GOI in Zn-H46R forms a hollow helical pore-like structure. β-Strands 1, 2, 3, and 6 are shown in red. Residues of the zinc loop undergo a conformational change and form a short β strand (blue) that reciprocally add to this β-sheet in the neighboring dimers, stabilizing the GOI interface. Zn ions are shown in magenta spheres. (C) Annular pore-like structure observed for Zn-loaded H46R SOD1 similar to the pore-like structures observed in proteins that cause Parkinson's disease and Alzheimer's disease (see text). Reproduced from Elam, J. S., *et al.* (2003a). An alternative mechanism of bicarbonate-mediated peroxidation by copper-zinc superoxide dismutase. *J. Biol. Chem.* 278: 21032–21039.

16.6.2. Monomer Model of Pathogenic SOD1 Aggregation

The x-ray crystal structures described above suggest a model for pathogenic SOD1 self-association where the metal-deficient SOD1 dimer undergoe conformational changes in loop elements that lead to non-native SOD1-SOD1 protein-protein interactions with the dimer serving as the building block in the formation of filamentous arrays (Elam *et al.*, 2003b).

However, as shown in Figure 16.1B, a wide range of pathogenic SOD1 mutations fall at the dimer interface, in regions critical for metal binding, and at or near the intrasubunit disulfide bond. It has long been known that metal-binding and the intrasubunit disulfide bond both stabilize the dimeric form of the enzyme (Forman and Fridovich, 1973; Abernethy *et al.*, 1974). Thus, many pathogenic mutations are predicted to lead to an increased propensity for monomerization. For example, mutations falling at the positions of the black spheres in Figure 16.1B directly affect the dimer interface and/or the intrasubunit disulfide bond while mutations falling at the positions of the yellow spheres affect dimer stability indirectly through the impairment of metal binding. In support of these concepts, a recent study by Chakrabartty and colleagues suggests that oxidized, monomeric pathogenic human SOD1 may be on the pathway to aggregation (Rakhit *et al.*, 2004).

As a benchmark against which to compare the dissociation of fALS mutant SOD1 proteins, we recently examined the apo-wild type protein rigorously using analytical ultracentrifugation sedimentation velocity experiments and found that reduction of its disulfide bond with TCEP [tris-(2-carboxyethyl) phosphine], an effective reducing agent that does not absorb at 280 nm, results in the dissociation of the dimer into its monomeric subunits (Doucette *et al.*, 2004) (Fig. 16.6A). If the intrasubunit disulfide bond is allowed to remain intact in each subunit, however, we found that guanidine HCl must be added to a concentration of approximately 1 M before dissociation is observed (not shown). In contrast, in the human wild-type metal-bound SOD1 protein, the concentration of guanidine HCl must be increased to ~3 M before dissociation is

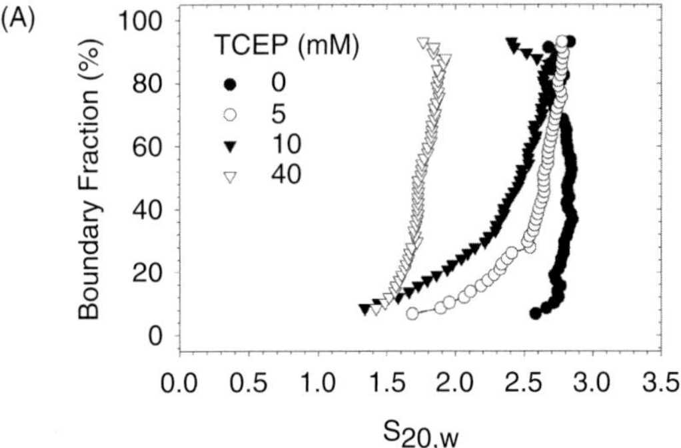

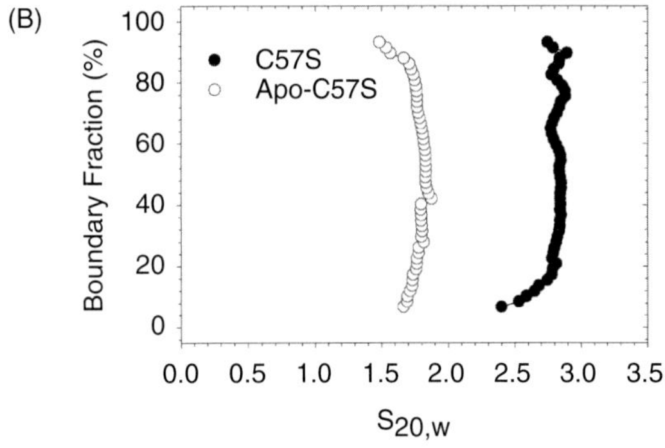

Figure 16.6. Sedimentation velocity analyses of apo-human SOD1 and apo- and metal-bound forms of the disulfide mutant C57S. (A) In the absence of reducing agent, apo-SOD1 sediments as a single ~2.8 S dimeric species (filled circles). At higher TCEP concentration where the disulfide bond is reduced, apo-SOD1 sediments as a ~1.7 S monomeric species (open triangles). (B) Sedimentation velocity analyses of C57S human SOD1 mutant in regular buffer (no TCEP). Metal bound SOD1 is represented by filled circles, sedimenting as a single ~2.8 S dimeric species while apo-C57S is represented by open circles, sedimenting as a single ~1.8 S monomeric species. Reproduced from Doucette P. A. *et al.* (2004). Dissociation of human copper-zinc superoxide dismutase dimers using chaotrope and reductant. Insights into the molecular basis for dimer stability. *J. Biol. Chem.* 279: 54558–54566.

observed. The importance of metal binding and the integrity of the intrasubunit disulfide bond to SOD1 quaternary structure is demonstrated further in the sedimentation behavior of the non-fALS mutant C57S, where the conversion of Cys57 to serine abrogates the intrasubunit disulfide bond. As shown in Figure 16.6B, apo-C57S sediments as a homogeneous ~1.7 S monomeric species in regular buffer (no TCEP) while the metal-bound form sediments as a nearly homogeneous ~2.8 S dimeric species. Thus, even in the absence of the intrasubunit disulfide bond, metal binding strongly favors dimerization of SOD1 (Doucette *et al.*, 2004). Finally, some pathogenic SOD1 mutants are more prone to reduction or loss of the intrasubunit disulfide bond. Hayward and colleagues demonstrated using partially denaturing gels that the disulfide bond in many fALS mutants appear more susceptible to reduction and the mutants more susceptible to monomerization than the wild-type protein (Tiwari and Hayward, 2003). The C146R pathogenic mutant of human SOD1 is predicted to behave similarly to C57S, as both mutations compromise the intrasubunit disulfide bond.

16.6.2.1. Degradation of Monomeric Pathogenic SOD1 by the 20 S Proteasome

An intriguing hint that monomeric SOD1 may be important in ALS pathogenesis comes from *in vitro* assays that look at the digestion of SOD1 by the 20 S proteasome. Neither pathogenic SOD1 proteins from the WTL or MBR mutant classes act as substrates for the 20 S proteasome when they have metals bound and are thus in their dimeric forms (data not shown). Figure 16.7 shows that the apo-wild type and the apo-MBR mutant E133Δ are poor substrates for the 20 S proteasome at 30°C and 37°C by noting the fraction of intact starting material that remains after 90 min incubation in the digestion assay (Di Noto *et al.*, 2005). In contrast, however, Figure 16.7 shows that apo-A4V, apo-G37R, and apo-G93A are all reasonable substrates with approximately 65% of apo-A4V, 75% of apo-G37R, and 80% of apo-G93A remaining intact at 37°C after this interval, so their order of susceptibility to proteasomal attack is apo-A4V > apo-G37R > apo-G93A. This is an intriguing result given that the order of susceptibility to proteasomal digestion correlates with the melting temperature of these apo-proteins (although apo-G93A has not yet been tested). apo-A4V melts at 40.5°C, apo-G37R melts at 44.3°C, and their half-lives *in vivo* are 7.5 and 13 h, respectively (Borchelt *et al.*, 1994).

The apo-C57S protein is the best substrate for digestion with only ~60% of the protein remaining after the digestion assay at 37°C. As shown above in Figure 16.6B, apo-C57S sediments exclusively as a monomer. The implication is that the apo-proteins are dissociating in this assay to become substrates for the 20 S proteasome. This is intriguing because the 20 S proteasome in the absence of its 19 S activator complex is thought to recognize hydrophobic patches on partially unfolded substrates. If the burden of misfolded SOD1 proteins exceed the capacity of the protein degradation system *in vivo*, these (monomeric?) proteins may accumulate and self-associate. All of the apo-SOD1 proteins studied, including wild type at higher temperatures, are proteolyzed by the 20 S proteasome into four distinct fragments ranging from 1.7 kDa to 7.8 kDa in size (Fig. 16.3B). It is possible that the products of this proteolytic digestion may be toxic to motor neurons, either through direct aggregation of the fragments themselves or through their potential ability to inhibit the proteasome.

A final indication that monomeric pathogenic SOD1 could be important in fALS comes from a recent study that identified a pathogenic C-terminally truncated SOD1 protein from a Danish family (G127X). This SOD1 mutant is incapable of forming a dimer because the C-terminal strand (strand 8) that makes up a large portion of the dimer interface is missing (purple strand in Fig. 16.1A). However, the G127X protein was shown to cause disease when expressed in transgenic mice (Jonsson *et al.*, 2004).

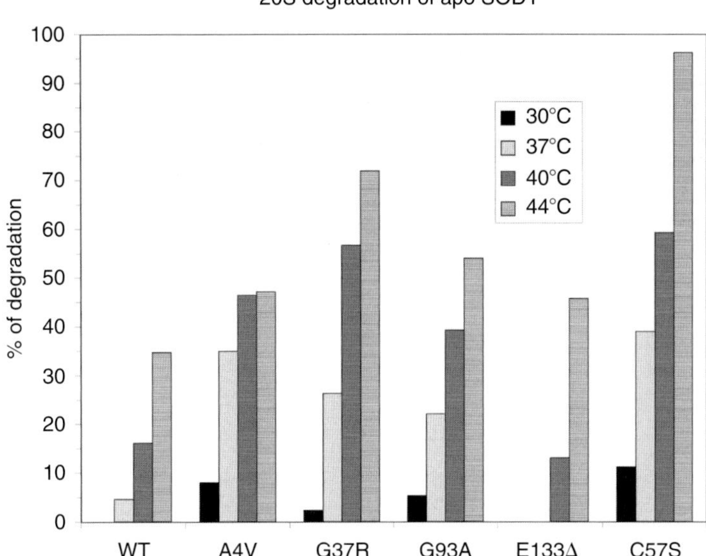

Figure 16.7. 20 S proteasomal digestion of apo-SOD1 proteins as a function of temperature over 90 min. Overall proteolytic activity is unaffected by the temperature differences. Apo-C57S, a monomeric form of SOD1 (see Fig. 16.5B), is the best substrate. Metal-bound (dimeric) forms of SOD1 proteins are not digested (data not shown). We interpret the data to mean that at higher temperatures, wild type, A4V, G37R, G93A, and E133Δ are dissociating to monomers prior to digestion.

16.7. Oxidative Damage and Models of Aggregation

As described above, the MBR mutant class of pathogenic SOD1 proteins are metal-deficient and appear to be primed and ready to self-associate as a result of conformational changes in loop elements involved in metal binding. How then, might the WTL class of pathogenic SOD1 proteins with well-ordered loop elements be toxic? The answer may lie in an alternate reactivity of SOD1 known as the "peroxidase" activity. The following reaction scheme has been proposed in reactions (16.3) and (16.4):

$$\text{SOD} - \text{Cu(I)} + \text{H}_2\text{O}_2 \rightarrow \text{SOD} - \text{Cu(II)(*OH)} + \text{OH} \quad (16.3)$$

$$\text{SOD} - \text{Cu(II)(*OH)} + \text{XH} \rightarrow \text{SOD} - \text{Cu(II)} + \text{H}_2\text{O} + \text{X*} \quad (16.4)$$

where XH represents amino acids at the active site (Hodgson and Fridovich, 1975; Liochev and Fridovich, 1999). This requires that copper (or some other redox active metal ion) be bound to the mutant SOD1 protein to promote the oxidative reaction. Indeed, fALS mutant SOD1 proteins have been reported to function as nonspecific peroxidases, using hydrogen peroxide as a substrate to generate hydroxyl radical (*OH). Studies of two human WTL fALS mutants, A4V and G93A, have demonstrated that this peroxidative reaction occurs at significantly higher rates for these fALS mutant SOD1 proteins than for wild-type human SOD1 (Wiedau-Pazos *et al.*, 1996; Yim *et al.*, 1996, 1997). Figure 16.8A shows that in the presence of H_2O_2, wild-type human SOD1 self-inactivates as a function of time. Intriguingly, the presence of bicarbonate anion enhances the rates of self-inactivation two- to three-fold (Elam *et al.*, 2003a). The potential relevance of this activity is underscored by the significant concentration of HCO_3^- found *in vivo* (~25 mM). Figure 16.8B represents a schematic version of reaction (16.4), where the copper-bound oxidant (hydroxyl radical) can attack one of the histidine copper ligands to form 2-oxo-histidine. Over time, oxidative modification of metal ligands could compromise the ability of these mutants to bind metal ions and may convert them to forms that are much more prone to aggregation. These forms would be relevant under both the dimer and monomer models of pathogenic SOD1 aggregation, as metal deficiency can lead to filamentous assembly of SOD1 dimers as described previously in the chapter and can promote monomerization as also described previously. In this way, the WTL and the MBR classes of pathogenic SOD1 mutants can become fused into a single class of molecules that can oligomerize and lead to motor neuron disfunction (Valentine and Hart, 2003).

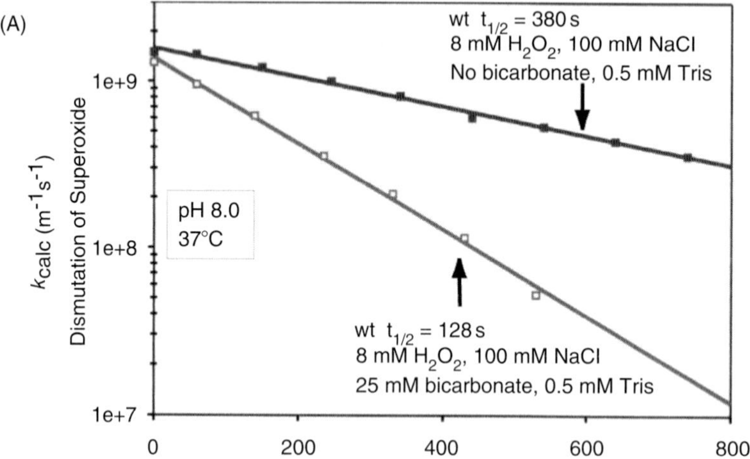

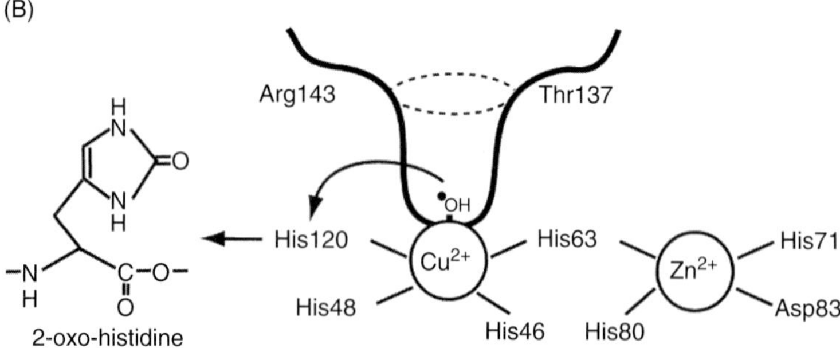

Figure 16.8. Self-oxidation of human SOD1 results in inactivation and metal ion loss. (A) Effect of bicarbonate on hydrogen peroxide–mediated SOD1 self-inactivation. (B) The attack of an oxygen radical species on one of the histidine coper ligands in SOD1 leads for the formation of 2-oxo histidine, leading to cofactor loss and enzyme inactivation (see text). Reproduced from Elam, J. S., *et al.* (2003a). An alternative mechanism of bicarbonate-mediated peroxidation by copper-zinc superoxide dismutase. *J. Biol. Chem.* 278:21032–21039.

16.8. Therapeutic Approaches for ALS

Most neurodegenerative diseases are difficult to diagnose. A lack of relevant biomarkers often allows progression of these diseases for an extended time before medications are administered. Through the long preclinical phase, neuronal loss occurs rapidly and by the time of the diagnosis some patients have lost approximately 50% of their motor neurons (Lansbury, 2004). Currently, the only U.S. FDA–approved drug for the treatment of ALS is riluzole. This drug slows the disease progression and extends survival, but not to any great extent.

Given the models of pathogenic SOD1 aggregation described above, one potential therapeutic avenue could come from the design of small molecules that inhibit the non-native SOD1-SOD1 protein-protein interactions under the dimer model of aggregation or alternatively stabilize the SOD1 homodimer under the monomer model of aggregation. With regard to the latter, Peter Lansbury and colleagues have identified a series of FDA-approved compounds that appear to bind at the dimer interface, stabilizing the pathogenic SOD1 proteins against dissociation and aggregation (Ray *et al.*, 2005). The research community eagerly awaits the testing of these compounds in transgenic mice overexpressing pathogenic SOD1 to determine their efficacy.

Acknowledgments

This work was supported by grants from the NIH NS39112, the Robert A. Welch Foundation AQ-1399, and the ALS Association. Collaborative efforts on the Zn-H46R structure by R. Strange, M. Hough, S. Antonyuk, and S.S. Hasnain (Daresbury Lab, UK) and on the 20 S proteasome digestion assays by R. Levine and L. Di Noto (NIH) are also gratefully acknowledged. We thank Alex Taylor and Stephen Holloway for help with the figures.

References

Abernethy, J. L., Steinman, H. M., and Hill, R. L. (1974). Bovine erythrocyte superoxide dismutase. *J. Biol. Chem.* 249: 7339–7347.

Allen, S., Heath, P. R., Kirby, J., Wharton, S. B., Cookson, M. R., Menzies, F. M., Banks, R. E., and Shaw, P. J. (2003). Analysis of the cytosolic proteome in a cell culture model of familial amyotrophic lateral sclerosis reveals alterations to the proteasome, antioxidant defenses, and nitric oxide synthetic pathways. *J. Biol. Chem.* 278: 6371–6383.

Andersen, P. (2000). Genetic factors in the early diagnosis of ALS. *Amyotroph. Lateral Scler. Other Motor Neuron Disord.* Suppl. 1: S31–S42.

Antonyuk, S., Elam, J. S., Hough, M. A., Strange, R. W., Doucette, P. A., Rodriguez, J. A., Hayward, L. J., Valentine, J. S., Hart, P. J., and Hasnain, S. S. (2005). Structural consequences of the familial amyotrophic lateral sclerosis SOD1 mutant His46Arg. *Protein Sci.* 14:1201–1203.

Bertini, I., Mangani, S., and Viezzoli, M. S. (1998). Structure and properties of copper-zinc superoxide dismutase. *Adv. Inorg. Chem.* 45: 127–251.

Borchelt, D. R., Lee, M. K., Slunt, H. S., Guarnieri, M., Xu, Z. S., Wong, P. C., Brown, R. H., Jr., Price, D. L., Sisodia, S. S., and Cleveland, D. W. (1994). Superoxide dismutase 1 with mutations linked to familial amyotrophic lateral sclerosis possesses significant activity. *Proc. Natl. Acad. Sci. USA* 91: 8292–8296.

Bruening, W., Roy, J., Giasson, B., Figlewicz, D. A., Mushynski, W. E., and Durham, H. D. (1999). Up-regulation of protein chaperones preserves viability of cells expressing toxic Cu/Zn-superoxide dismutase mutants associated with amyotrophic lateral sclerosis. *J. Neurochem.* 72: 693–699.

Bruijn, L. I., Becher, M. W., Lee, M. K., Anderson, K. L., Jenkins, N. A., Copeland, N. G., Sisodia, S. S., Rothstein, J. D., Borchelt, D. R., Price, D. L., and Cleveland, D. W. (1997). ALS-linked SOD1 mutant G85R mediates damage to astrocytes and promotes rapidly progressive disease with SOD1-containing inclusions. *Neuron* 18: 327–338.

Bruijn, L. I., Houseweart, M. K., Kato, S., Anderson, K. L., Anderson, S. D., Ohama, E., Reaume, A. G., Scott, R. W., and Cleveland, D. W. (1998). Aggregation and motor neuron toxicity of an ALS-linked SOD1 mutant independent from wild-type SOD1. *Science* 281: 1851–1854.

Caughey, B., and Lansbury, P. T. (2003). Protofibrils, pores, fibrils, and neurodegeneration: Separating the responsible protein aggregates from the innocent bystanders. *Annu. Rev. Neurosci.* 26: 267–298.

Charcot, J. M., and Joffroy, A. (1869). Deux cas d'atrophie musculaire progressive avec lesions de la substance grise et de faisceaux anterolateraux de la moelle epiniere. *Arch. Physiol. Norm. Pathol.* 1: 354–367.

Deng, H. X., Hentati, A., Tainer, J. A., Iqbal, Z., Cayabyab, A., Hung, W. Y., Getzoff, E. D., Hu, P., Herzfeldt, B., Roos, R. P., Warner, C., Deng, G., Soriano, E., Smyth, C., Parge, H. E., Ahmed, A., Roses, A. D., Hallewell, R. A., Pericak-Vance, M. A., and Siddique, T. (1993). Amyotrophic lateral sclerosis and structural defects in Cu-Zn superoxide dismutase. *Science* 261: 1047–1051.

Di Noto, L., Whitson, L. J., Hart, P. J., and Levine, R. L. (2005). Proteasomal degradation of mutant superoxide dismutases linked to amyotrophic lateral sclerosis. *J. Biol. Chem.* 280:39907–39913.

Doucette, P. A., Whitson, L. J., Cao, X., Schirf, V., Demeler, B., Valentine, J. S., Hansen, J. C., and Hart, P. J. (2004). Dissociation of human copper-zinc superoxide dismutase dimers using chaotrope and reductant. Insights into the molecular basis for dimer stability. *J. Biol. Chem.* 279: 54558–54566.

Drake, S. K., Bourdon, E., Wehr, N. B., Levine, R. L., Backlund, P. S., Yergey, A. L., and Rouault, T. A. (2002). Numerous proteins in Mammalian cells are prone to iron-dependent oxidation and proteasomal degradation. *Dev. Neurosci.* 24: 114–124.

Dunlop, R. A., Rodgers, K. J., and Dean, R. T. (2002). Recent developments in the intracellular degradation of oxidized proteins. *Free Radic. Biol. Med.* 33: 894–906.

Elam, J. S., Malek, K., Rodriguez, J. A., Doucette, P. A., Taylor, A. B., Hayward, L. J., Cabelli, D. E., Valentine, J. S., and Hart, P. J. (2003a). An alternative mechanism of bicarbonate-mediated peroxidation by copper-zinc superoxide dismutase. *J. Biol. Chem.* 278: 21032–21039.

Elam, J. S., Taylor, A. B., Strange, R., Antonyuk, A., Doucette, P. A., Rodriguez, J. A., Hasnain, S. S., Hayward, L. J., Valentine, J. S., Yeates, T. O., and Hart, P. J. (2003b). Amyloid-like filaments and water-filled nanotubes formed by SOD1 mutants linked to familial ALS. *Nat. Struct. Biol.* 10: 461–467.

Fielden, E. M., Roberts, P. B., Bray, R. C., Lowe, D. J., Mautner, G. N., Rotilio, G., and Calabrese, L. (1974). Mechanism of action of superoxide dismutase from pulse radiolysis and electron paramagnetic resonance. Evidence that only half the active sites function in catalysis. *Biochem. J.* 139: 49–60.

Forman, H. J., and Fridovich, I. (1973). On the stability of bovine superoxide dismutase: The effects of metals. *J. Biol. Chem.* 248: 2645–2649.

Getzoff, E. D., Tainer, J. A., Weiner, P. K., Kollman, P. A., Richardson, J. S., and Richardson, D. C. (1983). Electrostatic recognition between superoxide and copper, zinc superoxide dismutase. *Nature* 306: 287–290.

Goldberg, M. S., and Lansbury, P. T., Jr. (2000). Is there a cause-and-effect relationship between alpha-synuclein fibrillization and Parkinson's disease? *Nat. Cell Biol.* 2: E115–119.

Goto, J. J., Zhu, H., Sanchez, R. J., Nersissian, A., Gralla, E. B., Valentine, J. S., and Cabelli, D. E. (2000). Loss of *in vitro* metal ion binding specificity in mutant copper-zinc superoxide dismutases associated with familial amyotrophic lateral sclerosis. *J. Biol. Chem.* 275: 1007–1014.

Grune, T., Merker, K., Sandig, G., and Davies, K. J. (2003). Selective degradation of oxidatively modified protein substrates by the proteasome. *Biochem. Biophys. Res. Commun.* 305: 709–718.

Hart, P. J., Balbirnie, M. M., Ogihara, N. L., Nersissian, A. M., Weiss, M. S., Valentine, J. S., and Eisenberg, D. (1999). A structure-based mechanism for copper-zinc superoxide dismutase. *Biochemistry* 38: 2167–2178.

Haverkamp, L. J., Appel, V., and Appel, S. H. (1995). Natural history of amyotrophic lateral sclerosis in a database population. Validation of a scoring system and a model for survival prediction. *Brain* 118: 707–719.

Hayward, L. J., Rodriguez, J. A., Kim, J. W., Tiwari, A., Goto, J. J., Cabelli, D. E., Valentine, J. S., and Brown, R. H., Jr. (2002). Decreased metallation and activity in subsets of mutant superoxide dismutases associated with familial amyotrophic lateral sclerosis. *J. Biol. Chem.* 277: 15923–15931.

Hodgson, E. K., and Fridovich, I. (1975). The interaction of bovine erythrocyte superoxide dismutase with hydrogen peroxide: Chemiluminescence and peroxidation. *Biochemistry* 14: 5299–5303.

Johnston, J. A., Dalton, M. J., Gurney, M. E., and Kopito, R. R. (2000). Formation of high molecular weight complexes of mutant Cu, Zn-superoxide dismutase in a mouse model for familial amyotrophic lateral sclerosis. *Proc. Natl. Acad. Sci. USA* 97: 12571–12576.

Jonsson, P. A., Ernhill, K., Andersen, P. M., Bergemalm, D., Brannstrom, T., Gredal, O., Nilsson, P., and Marklund, S. L. (2004). Minute quantities of misfolded mutant superoxide dismutase-1 cause amyotrophic lateral sclerosis. *Brain* 127: 73–88.

Kabashi, E., Agar, J. N., Taylor, D. M., Minotti, S., and Durham, H. D. (2004). Focal dysfunction of the proteasome: A pathogenic factor in a mouse model of amyotrophic lateral sclerosis. *J. Neurochem.* 89: 1325–1335.

Lansbury, P. T., Jr. (1999). Evolution of amyloid: What normal protein folding may tell us about fibrillogenesis and disease. *Proc. Natl. Acad. Sci. USA* 96: 3342–3344.

Lansbury, P. T., Jr. (2004). Back to the future: the "old-fashioned" way to new medications for neurodegeneration. *Nat. Med.* 10: Suppl:S51–S57.

Lashuel, H. A., Hartley, D., Petre, B. M., Walz, T., and Lansbury, P. T., Jr. (2002). Neurodegenerative disease: amyloid pores from pathogenic mutations. *Nature* 418: 291.

Liochev, S. I., and Fridovich, I. (1999). On the role of bicarbonate in peroxidations catalyzed by Cu,Zn superoxide dismutase. *Free. Radic. Biol. Med.* 27: 1444–1447.

Lyons, T., Gralla E. B., and Valentine, J. S. (1999). Biological chemistry of copper-zinc superoxide dismutase and its link to amyotrophic lateral sclerosis. In: Sigel, A., and Sigel, H. (eds.), *Metal Ions in Biological Systems*, Vol. 36. Marcel Dekker, Inc., New York, Basel, pp. 125–177.

Ogihara, N. L., Parge, H. E., Hart, P. J., Weiss, M. S., Goto, J. J., Crane, B. R., Tsang, J., Slater, K., Roe, J. A., Valentine, J. S., Eisenberg, D., and Tainer, J. A. (1996). Unusual trigonal-planar copper configuration revealed in the atomic structure of yeast copper-zinc superoxide dismutase. *Biochemistry* 35: 2316–2321.

Okado-Matsumoto, A., and Fridovich, I. (2001). Subcellular distribution of superoxide dismutases (SOD) in rat liver: Cu,Zn-SOD in mitochondria. *J. Biol. Chem.* 276: 38388–38393.

Okado-Matsumoto, A., and Fridovich, I. (2002). Amyotrophic lateral sclerosis: A proposed mechanism. *Proc. Natl. Acad. Sci. USA* 99: 9010–9014.

Puttaparthi, K., Wojcik, C., Rajendran, B., DeMartino, G. N., and Elliott, J. L. (2003). Aggregate formation in the spinal cord of mutant SOD1 transgenic mice is reversible and mediated by proteasomes. *J. Neurochem.* 87: 851–860.

Rakhit, R., Crow, J. P., Lepock, J. R., Kondejewski, L. H., Cashman, N. R., and Chakrabartty, A. (2004). Monomeric Cu,Zn-superoxide dismutase is a common misfolding intermediate in the oxidation models of sporadic and familial amyotrophic lateral sclerosis. *J. Biol. Chem.* 279: 15499–15504.

Ray, S. S., Nowak, R. J., Brown, R. H., Jr., and Lansbury, P. T., Jr. (2005). Small-molecule-mediated stablization of familial amyotrophic lateral sclerosis-linked superoxide dismutase mutants against unfolding and aggregation. *Proc. Natl. Acad. Sci. USA.* 102: 3639–3644.

Reaume, A. G., Elliott, J. L., Hoffman, E. K., Kowall, N. W., Ferrante, R. J., Siwek, D. F., Wilcox, H. M., Flood, D. G., Beal, M. F., Brown, R. H., Jr., Scott, R. W., and Snider, W. D. (1996). Motor neurons in Cu/Zn superoxide dismutase-deficient mice develop normally but exhibit enhanced cell death after axonal injury. *Nat. Genet.* 13: 43–47.

Richardson, J. S., and Richardson, D. C. (2002). Natural beta-sheet proteins use negative design to avoid edge-to-edge aggregation. *Proc. Natl. Acad. Sci. USA.* 99: 2754–2759.

Rodriguez, J. A., Valentine, J. S., Eggers, D. K., Roe, J. A., Tiwari, A., Brown, R. H., Jr., and Hayward, L. J. (2002). Familial ALS-associated mutations decrease the thermal stability of distinctly metallated species of human copper-zinc superoxide dismutase. *J. Biol. Chem.* 277: 15932–15937.

Rosen, D. R., Siddique, T., Patterson, D., Figlewicz, D. A., Sapp, P., Hentati, A., Donaldson, D., Goto, J., O'Regan, J. P., Deng, H. X., Rahmani, Z., Krizus, A., Mckenna-Yasek, D., Cayabyab, A., Gatson, S. M., Berger, R., Tanzi, R. E., Halperin, J. J., Herzfeldt, B., Van den Bergh, R., Hung, W. Y., Bird, T., Deng, G., Mulder, D. W., Smyth, C., Laing, N. G., Soriano, E., Pericak-Vance, M. A., Haines, J., Rouleau, G. A., Gusella, J. S., Horvitz, H. R., and Brown, R. H., Jr. (1993). Mutations in Cu/Zn superoxide dismutase gene are associated with familial amyotrophic lateral sclerosis. *Nature* 362: 59–62.

Rosen, D., Sapp, P., O'Regan, J., McKenna-Yasek, D., Schlumpf, K. S., Gusella, J. F., Horvitz, H. R., and Brown, R. H., Jr. (1994). Genetic linkage analysis of familial amyotrophic lateral sclerosis using human chromosome 21 microsatellite DNA markers. *Am. J. Med. Genet.* 51: 61–69.

Rotilio, G., Bray, R. C., and Fielden, E. M. (1972). A pulse radiolysis study of superoxide dismutase. *Biochim. Biophys. Acta.* 268: 605–609.

Shah, J. V., Cleveland, D. W., and Williamson, T. L. (2002). Slow axonal transport: Fast motors in the slow lane. *Curr. Opin. Cell Biol.* 14: 58–62.

Sherman, M. Y., and Goldberg, A. L.. (2001). Cellular defenses against unfolded proteins: A cell biologist thinks about neurodegenerative diseases. *Neuron* 29: 15–32.

Shibata, N., Hirano, A., Kobayashi, M., Siddique, T., Deng, H. X., Hung, W. Y., Kato, T., and Asayama, K. (1996). Intense superoxide dismutase-1 immunoreactivity in intracytoplasmic hyaline inclusions of familial amyotrophic lateral sclerosis with posterior column involvement. *J. Neuropathol. Exp. Neurol.* 55: 481–490.

Shringarpure, R., Grune, T., Mehlhase, J., and Davies, K. J. (2003). Ubiquitin conjugation is not required for the degradation of oxidized proteins by proteasome. *J. Biol. Chem.* 278: 311–318.

Strange, R. W., Antonyuk, S., Hough, M. A., Doucette, P. A., Rodriguez, J. A., Hart, P. J., Hayward, L. J., Valentine, J. S., and Hasnain, S. S. (2003). The structure of holo- and metal-deficient wild type human Cu,Zn superoxide dismutase and its relevance to familial amyotrophic lateral sclerosis. *J. Mol. Biol.* 328: 877–891.

Sturtz, L. A., Diekert, K., Jensen, L. T., Lill, R., and Culotta, V. C. (2001). A fraction of yeast Cu,Zn-superoxide dismutase and its metallochaperone, CCS, localize to the intermembrane space of mitochondria: A physiological role for SOD1 in guarding against mitochondrial oxidative damage. *J. Biol. Chem.* 276: 38084–38089.

Subramaniam, J. R., Lyons, W. E., Liu, J., Bartnikas, T. B., Rothstein, J., Price, D. L., Cleveland, D. W., Gitlin, J. D., and Wong, P. C. (2002). Mutant SOD1 causes motor neuron disease independent of copper chaperone- mediated copper loading. *Nat. Neurosci.* 5: 301–307.

Tainer, J. A., Getzoff, E. D., Beem, K. M., Richardson, J. S., and Richardson, D. C. (1982). Determination and analysis of the 2 A-structure of copper, zinc superoxide dismutase. *J. Mol. Biol.* 160: 181–217.

Tiwari, A., and Hayward, L. J. (2003). Familial amyotrophic lateral sclerosis mutants of copper/zinc superoxide dismutase are susceptible to disulfide reduction. *J. Biol. Chem.* 278: 5984–5992.

Urushitani, M., Kurisu, J., Tsukita, K., and Takahashi, R. (2002). Proteasomal inhibition by misfolded mutant superoxide dismutase 1 induces selective motor neuron death in familial amyotrophic lateral sclerosis. *J. Neurochem.* 83: 1030–1042.

Valentine, J. S., and Hart, P. J. (2003). Misfolded CuZnSOD and ALS. *Proc. Natl. Acad. Sci. USA.* 100: 3617–3622.

Wang, J., Xu, G., and Borchelt, D. R. (2002a). High molecular weight complexes of mutant superoxide dismutase 1: Age-dependent and tissue-specific accumulation. *Neurobiol. Dis.* 9: 139–148.

Wang, J., Xu, G., Gonzales, V., Coonfield, M., Fromholt, D., Copeland, N. G., Jenkins, N. A., and Borchelt, D. R. (2002b). Fibrillar inclusions and motor neuron degeneration in transgenic mice expressing superoxide dismutase 1 with a disrupted copper-binding site. *Neurobiol. Dis.* 10: 128–138.

Wang, J., Slunt, H., Gonzales, V., Fromholt, D., Coonfield, M., Copeland, N. G., Jenkins, N. A., and Borchelt, D. R. (2003). Copper-binding-site-null SOD1 causes ALS in transgenic mice: aggregates of non-native SOD1 delineate a common feature. *Hum. Mol. Genet.* 12: 2753–2764.

Wiedau-Pazos, M., Goto, J. J., Rabizadeh, S., Gralla, E. B., Roe, J. A., Lee, M. K., Valentine, J. S., and Bredesen, D. E. (1996). Altered reactivity of superoxide dismutase in familial amyotrophic lateral sclerosis. *Science* 271: 515–518.

Yim, M. B., Chock, P. B., and Stadtman, E. R. (1993). Enzyme function of copper, zinc superoxide dismutase as a free radical generator. *J. Biol. Chem.* 268: 4099–4105.

Yim, M. B., Kang, J. H., Yim, H. S., Kwak, H. S., Chock, P. B., and Stadtman, E. R. (1996). A gain-of-function of an amyotrophic lateral sclerosis-associated Cu-Zn superoxide dismutase mutant: An enhancement of free radical formation due to a decrease in Km for hydrogen peroxide. *Proc. Natl. Acad. Sci. USA.* 93: 5709–5714.

Yim, H. S., Kang, J. H., Chock, P. B., Stadtman, E. R., and Yim, M. B. (1997). A familial amyotrophic lateral sclerosis-associated A4V Cu-Zn superoxide dismutase mutant has a lower Km for hydrogen peroxide: Correlation between clinical severity and the Km value. *J. Biol. Chem.* 272: 8861–8863.

17

Understanding the Effects of Cancer-Associated Mutations in the Tumor Suppressor Protein p53: Structural Consequences of Mutations and Possible Ways of Rescuing Oncogenic Mutants

Andreas C. Joerger, Assaf Friedler, and Alan R. Fersht

Abstract

The tumor suppressor protein p53 is a key control in the cell cycle and plays a crucial role in the prevention of cancer development. It is mutated in approximately half of all human cancers and has, therefore, become an important target for the development of novel cancer therapies. Here, we review the structure of the protein, the effects of mutation and how they may be reversed. p53 has a highly complex domain organization consisting of structured regions combined with largely unstructured domains. Most cancer-associated mutations are located in the DNA-binding core domain of the protein. The molecular basis for the detrimental effect of these mutations has been elucidated by structural and biophysical studies. Whereas some mutations affect residues that make direct contact with target DNA, others induce structural perturbations that reduce the thermodynamic stability of the protein. Because p53 core domain is only marginally stable above body temperature, many cancer mutations not only induce local conformational changes but also cause global unfolding of the core domain under physiologic conditions. Novel therapeutic strategies aim, therefore, to develop chemical chaperones that help refold p53 mutants to their correct native structure. Valuable lessons can be learned from studies on so-called second-site suppressor mutations that reverse the effects of cancer mutations. These may provide a basis for the rational design of novel therapeutics.

17.1. Introduction

The tumor suppressor p53 is a transcription factor that is at the center of a network of interactions that affect the cell cycle and apoptosis (Vogelstein *et al.*, 2000; Ryan *et al.*, 2001). The protein is induced by a variety of stresses that include oncogene activation and DNA damage caused by chemotherapy and radiotherapy. On induction, it activates a variety of genes whose products lead to G1 and G2 cell cycle arrest and apoptosis (Vogelstein *et al.*, 2000; Ryan *et al.*,

2001). It is such an effective tumor suppressor that it is inactivated in virtually all cancers; in about 50% of cancers p53 is directly inactivated by mutation, and in the remainder its activity is lost by perturbations of its associated pathways and interactions (Hainaut and Hollstein, 2000). Reactivating mutant p53 is an important target in the development of novel therapies for cancer (Lane and Lain, 2002; Lane and Hupp, 2003). To understand how p53 is inactivated, it is necessary to understand its structure and how it responds to mutation. Such knowledge will provide a basis for the rational design of novel therapeutics that may reverse the effects of mutation. In this chapter, we survey the structure of the protein, the effects of mutation, and how they may be reversed.

17.2. The Domain Structure of Human p53

Human p53 is a 393-residue protein, which is active as a homotetramer. Each chain has a complex domain structure that is organized into several functional and structural entities (Fig. 17.1A; **see color insert**) (Vousden and Lu, 2002; Courtois *et al.*, 2004): The N-terminal part of the protein comprises the transactivation domain (residues 1–63). It contains regions forming protein-protein interactions with regulatory proteins such as MDM2 (residues 19–26 of p53), which triggers ubiquitination and ultimately degradation of the protein in the proteasome (Momand *et al.*, 2000; Michael and Oren, 2002, 2003), and p300/CBP, which is responsible for acetylation of residues in the C-terminal part of p53 (Grossman, 2001). This domain is followed by a proline-rich region (residues 64–92), which binds SH3 domains and was suggested to have a regulatory role (Walker and Levine, 1996; Müller-Tiemann *et al.*, 1998). The central (core) domain is responsible for specific DNA-binding (residues 94–292) and is the most important independent domain of p53 (Cho *et al.*, 1994). The C-terminal part of the protein includes a tetramerization domain (residues 326–355) (Clore *et al.*, 1994) and the negative autoregulatory domain at the extreme C-terminus (363–393), which contains acetylation and phosphorylation sites and regulates the DNA-binding activity of p53 (Prives and Manley, 2001; Weinberg *et al.*, 2004; Friedler *et al.*, 2005a). Further elements comprise a highly conserved leucine-rich nuclear export signal within the tetramerization domain (residues 340–350) (Stommel *et al.*, 1999) and a nuclear localization signal, which is located between the core domain and tetramerization domain (residues 302–322) (Shaulsky *et al.*, 1990). Most cancer-associated mutations map to the DNA-binding core domain (Hainaut and Hollstein, 2000), and six hot-spot mutation sites stand out as most frequently associated with human cancer (Arg175, Gly245, Arg248, Arg249, Arg273, and Arg282).

The crystal structures have been solved of both the DNA-binding (Cho *et al.*, 1994) and the tetramerization domains (Jeffrey *et al.*, 1995). NMR studies have revealed a very similar structure and tetrameric assembly of the tetramerization domain in solution (Clore *et al.*, 1994, 1995). The tetramer exhibits D2-symmetry and can be best described as a dimer of dimers (Fig. 17.1D; **see color insert**). Other domains such as the N-terminal transactivation domain are thought to be mainly unstructured in the free protein but at least partially structured when interacting with regulatory proteins. Crystallographic studies have shown that a 15-residue transactivation domain peptide of p53 binds to a hydrophobic cleft of the N-terminal domain of MDM2 as an amphipathic α-helix (Kussie *et al.*, 1996). A similar result was found for a peptide derived from the C-terminal negative regulatory domain of p53 (residues 367–388). NMR spectroscopy revealed that this peptide, which has no regular structure in its native form, becomes helical upon binding to S100B (Rustandi *et al.*, 2000). How the various domains of p53 are assembled in the full-length tetramer is, however, still not known.

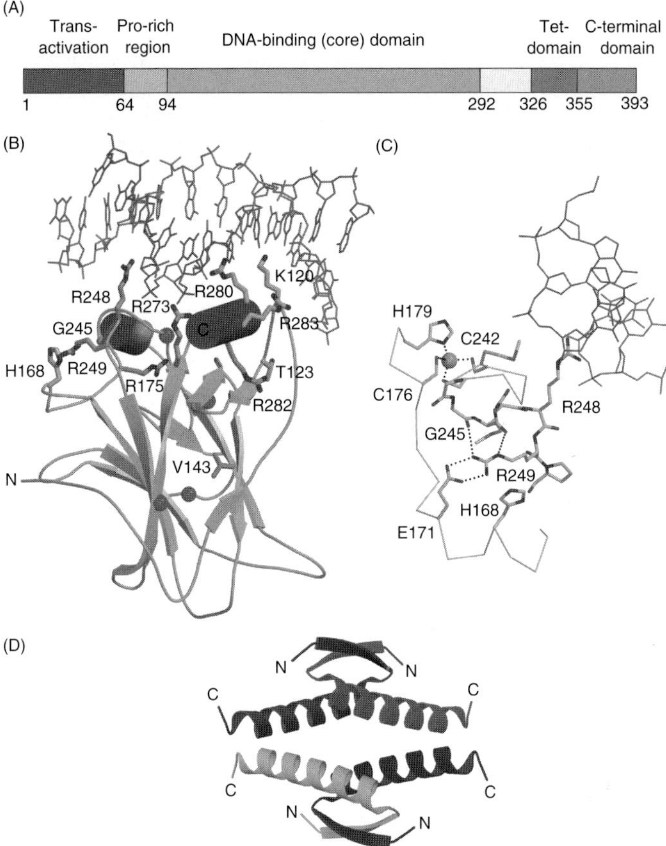

Figure 17.1. Structure of human p53. (A) Schematic view of the domain structure of human p53: The 393-residue protein contains an N-terminal transactivation domain, a proline-rich region, a DNA-binding (core) domain, a tetramerization domain (Tet-domain) and a C-terminal autoregulatory domain (see text for further details). (B) Structure of the DNA-binding (core) domain in complex with consensus DNA (PDB code 1TSR, molecule B). A β-sandwich provides the basic scaffold for a loop-sheet-helix motif and two large loops, L2 and L3, tethered by a zinc ion, which interact with the major and minor groove of the DNA, respectively. The zinc ion is shown as a gray sphere. The six hot-spot sites Arg175, Gly245, Arg248, Arg249, Arg273, and Arg282, which are most frequently mutated in human cancer, are highlighted in orange. The mutation sites in the superstable quadruple mutant M133L/V203A/N239Y/N268D are highlighted as small green spheres. (C) Close-up view of loops L2 and L3 in the DNA-binding surface, including the zinc coordination sphere, in the structure of wild type in complex with consensus DNA (PDB code 1TSR, molecule B). The orientation of the molecule is different to that shown in panel (B). The zinc ion is depicted as a gray sphere. Specific interactions mediated via the guanidinium group of Arg249 are highlighted with dotted lines. These include hydrogen bonds with backbone oxygens of residues Gly245 and Met246 on the same loop and a salt bridge with Glu171 on the L2 loop. DNA contacts are made via Arg248. Selected DNA residues in the proximity of Arg248 are shown in magenta and cyan. (D) Ribbon plot of the tetramerization domain structure (PDB code 1C26). Individual subunits within the tetramer are shown in different colors. Two monomers consisting of a β-strand and an α-helix are connected via an antiparallel β-sheet and an antiparallel helix-helix interface to form a dimer (e.g., the red and green chain). Two of these dimers are assembled into a tetramer via a parallel helix-helix interface. The figures were generated using MOLSCRIPT (Kraulis, 1991) and RASTER3D (Merritt and Bacon, 1997). Panels (B) and (C) were adapted from Joerger *et al.* (2005).

17.3. The Crystal Structure of the DNA-Binding Domain

The crystal structure of the human p53 core domain has been solved at 2.2 Å resolution in complex with a consensus DNA sequence (Cho *et al.*, 1994). Two of the three molecules in the asymmetric unit of the crystal bind to the same face of the DNA duplex, forming a head-to-tail dimer with weak protein-protein interactions between the two molecules, whereas the third molecule makes no significant contact with the DNA. Overall, the structures of the three molecules are very similar, which suggests that DNA binding induces only minor structural changes within the core domain. Additional confirmation comes from the structure of p53 core domain from mouse, which has been solved in the absence of DNA (Zhao *et al.*, 2001). The core domain structure consists of two antiparallel β-sheets of four and five twisted strands, forming a β-sandwich with a "Greek key" topology (Fig. 17.1B; **see color insert**). This compact, barrel-like structure provides the basic scaffold for two large loops, L2 (residues 163–195) and L3 (residues 236–251), and a loop-sheet-helix motif at the same end of the β-sandwich. The conformation of the two large loops L2 and L3 is stabilized by a zinc ion, which is tetrahedrally coordinated by three cysteines and a histidine (Cys176, His179, Cys238, and Cys242; cf. Fig. 17.1C; **see color insert**). The loop-sheet-helix motif and the two large loops form an extended surface, rich in basic amino acids, which binds DNA. In the complex with consensus DNA, these structural elements make an extensive number of contacts with the major groove and minor groove of the DNA. Parts of the helix and the L1 loop of the loop-sheet-helix motif bind to the major groove. This involves specific interactions of Lys120 and Arg280 with bases of the pentameric consensus sequence. Arg248 from loop L3 protrudes into the adjacent minor groove where its positively charged guanidinium group is in close contact with the negatively charged DNA backbone. Interestingly, parts of this DNA-binding surface were found to overlap with the binding sites for the 53BP1 and 53BP2 proteins. 53BP1 functions as a DNA damage checkpoint protein (DiTullio *et al.*, 2002), whereas 53BP2 forms the C-terminal half of a larger protein (ASPP2), which specifically enhances p53-induced apoptosis (Samuels-Lev *et al.*, 2001). Crystal structures of complexes with the p53 core domain reveal that in both cases the contacting residues are concentrated on the L3 loop (Gorina and Pavletich, 1996; Derbyshire *et al.*, 2002; Joo *et al.*, 2002). The DNA binding-surface in the core domain is a highly promiscuous binding-site and was found to interact with a number of other proteins such as Rad51, which is involved in homologous recombination (Buchhop *et al.*, 1997; Linke *et al.*, 2003; Friedler *et al.*, 2005b).

The crystal structure of the core domain provided a framework for understanding how the common cancer-associated mutations inactivate p53 (Fig. 17.1B). It rationalized observations that these mutations can be subdivided into two classes. The first class of mutations are "contact mutations," for example, the cancer hot-spot mutations R248Q and R273H. These mutations directly affect DNA-binding, because an essential DNA-contacting residue is lost. In contrast, the second class of mutations, the "structural mutations," are thought to cause structural perturbations within the core domain and thus indirectly affect DNA binding. The hot-spot mutations R175H, G245S, R249S, and R282W belong to this second group.

17.4. The Consequences and Effects of p53 Mutations at the Molecular Level

Several experimental techniques are used to study the effect of mutation on p53. The effects on the thermodynamic stability of the protein are analyzed using fluorescence spectroscopy and calorimetry, while the structural effects can be studied mainly by NMR and x-ray crystallography.

NMR is especially suitable for studying the dynamic behavior of p53 mutants (Bullock *et al.*, 1997; Wong *et al.*, 1999; Bullock *et al.*, 2000; Bullock and Fersht, 2001).

17.4.1. Effect of Mutations on Stability

First we must define what is meant by stability, as "stability" is often used in a confusing way in the p53-related literature. An increase in the cellular level of p53 is often qualitatively referred to as "stabilization of p53," regardless of whether the protein is denatured or not. This "stability" is often just a consequence of whether or not the p53 protein is degraded via MDM2, and so inactive, unstable, mutants of p53 accumulate because the mdm2 gene is not activated by p53. The correct quantitative definition of p53 stabilization, however, refers to its thermodynamic properties: stabilization of p53 means an increase in the thermodynamic stability of the protein, derived from its folding-unfolding equilibrium (i.e., in the ΔG for the denaturation of a mutant compared with that of wild type). Here, when we refer to p53 stabilization, it is always the thermodynamic stabilization of the protein.

The core domain of the p53 protein has a low intrinsic stability and its melting temperature, 42–44°C, is only slightly above body temperature (Fig. 17.2) (Bullock *et al.*, 1997; Bullock *et al.*, 2000). The full-length protein melts at a similar temperature. Its low stability may be a consequence of the need to regulate the level of p53 by degradation by the proteasome. The consequence of the intrinsic low stability is that destabilizing mutations in the protein have a detrimental effect on protein function, and slightly destabilized proteins melt at below body temperature (Fig. 17.2) (Bullock *et al.*, 1997, 2000). The mutations can be divided into three classes (Fig. 17.3): (I) DNA contact mutations, which have little or no effect on protein stability, but impair function due to the loss of key residues that mediate specific DNA binding. (II) Mutations that cause local distortion of the structure and are moderately destabilizing (by < 2 kcal/mol compared with the wild-type protein). Most of the non-contact mutations

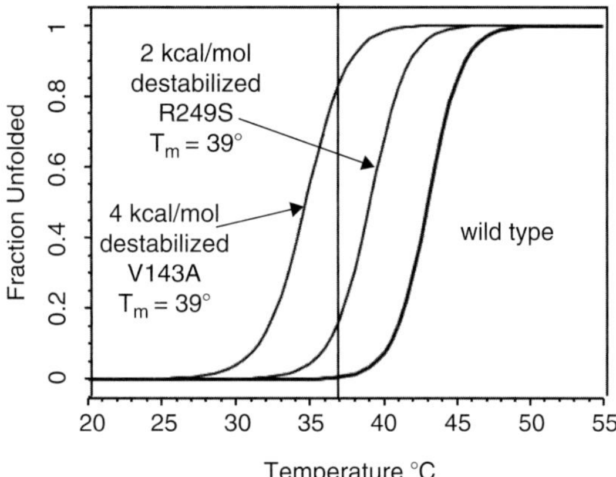

Figure 17.2. Extrapolated temperature denaturation curves of wild-type p53 core domain and mutants that are destabilized by 2 and 4 kcal/mol, respectively. The temperature-dependence was calculated from urea denaturation studies at pH = 7.2 monitored by normalized fluorescence emission at 356 nm using the Gibbs-Helmholtz equation. See Bullock *et al.* (2000) for details. At a physiologic temperature of 37°C (shown by a vertical line), only a small change in the fraction of unfolded protein occurs for mutants that are destabilized by 2 kcal/mol or less (e.g., mutant R249S). For mutants that are destabilized between 2 and 4 kcal/mol, large changes in the fraction of unfolded protein occur. Accordingly, a mutant such as V143A, which is destabilized by 3.5 kcal/mol, is mostly unfolded at 37°C.

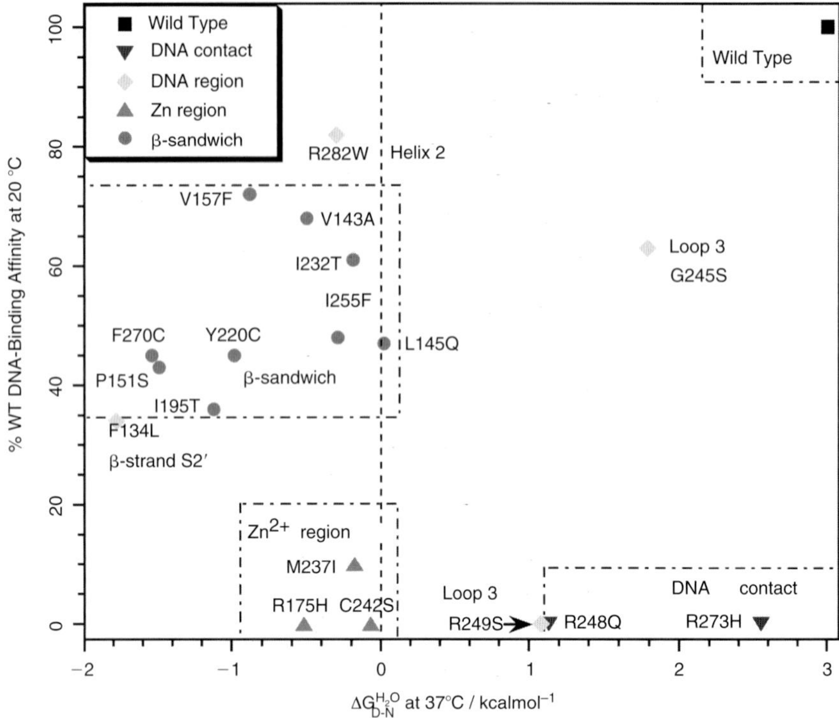

Figure 17.3. The different classes of mutations in p53, grouped according to stability (x-axis) and DNA binding properties (y-axis). Mutant phenotypes correlate with the site of mutation as shown by a plot of stability (estimated at 37°C) against DNA-binding affinity at 20°C (a temperature at which all mutants are folded). A free energy of unfolding in water, ΔG, of 0 kcal/mol (shown by a dashed line) corresponds with 50% denatured protein. Data taken from Bullock *et al.* (2000).

in the DNA-binding interface fall into this category (e.g., the loop L3 mutations G245S and R249S). (III) Mutations that result in global unfolding of the protein and are highly destabilizing (by >3 kcal/mol compared with the wild type). All the β-sandwich mutants fall into this category. The destabilizing mutations inactivate p53 by lowering the melting temperature of its core domain to below body temperature, making it denature under physiologic conditions (Bullock and Fersht, 2001).

17.4.2. Design of a Superstable p53 Core Domain Mutant

Not only is the inherently low thermodynamic stability of the wild-type p53 protein the Achilles' heel of human p53, but it also makes the protein difficult to study by biophysical and structural methods. As we have seen above, even weakly destabilizing mutations can trigger the unfolding and hence the loss of activity of human p53 core domain at physiologic temperature. Conversely, it should be possible to stabilize both wild-type and mutant proteins by single amino acid substitutions that contribute to the overall stability of the structural scaffold without compromising its DNA-binding activity and hence its function *in vivo*. Such stabilized p53 variants could be better suited for many experimental purposes and could, furthermore, show possible ways to reactivate tumorigenic p53 mutants.

By adopting a semirational approach on the basis of the molecular evolution of p53, a superstable mutant of the human p53 core domain was designed (Nikolova *et al.*, 1998). A comparison of sequences of p53 homologues from more than 20 species revealed many naturally occurring mutations. A number of the identified mutations were introduced into human p53 core domain, and the effect on the thermodynamic stability of the protein was measured. Such a strategy should minimize the selection of mutations that impair activity. In addition, the N239Y substitution was included, which is not naturally occurring, but had been reported to restore activity in some of the common cancer mutants (Brachmann *et al.*, 1998) (see below). The most stable substitutions were finally combined in a quadruple mutant (*T*-p53C). Four point mutations, M133L, V203A, N239Y, and N268D, stabilize the core domain by 2.65 kcal/mol without impairing binding to the *gadd45* recognition element (Nikolova *et al.*, 1998). The effects of the point mutations, which are located in different regions of the core domain (cf. Fig. 17.1B), are nearly additive. The main contribution to stability increase comes from the N239Y and N268D substitutions, which are of particular interest, because they are both known to act as second-site suppressors for various cancer-associated mutations (Brachmann *et al.*, 1998). Interestingly, these mutations were also found in a study on the *in vitro* evolution of thermostable p53 variants (Matsumura and Ellington, 1999).

17.4.3. Crystal Structure of the Superstable p53 Core Domain Mutant (*T*-p53C)

In order to understand the molecular basis for its increased stability, the high-resolution crystal structure (1.9 Å) of mutant *T*-p53C in the absence of DNA has been solved (Joerger *et al.*, 2004). This structure reveals that the four point mutations cause only local structural changes, whereas the overall structure of the β-sandwich and the DNA-binding surface is conserved. As is to be expected, the largest deviations from the wild-type structure can be found in the loop regions. This is either due to an induced fit movement upon DNA binding (loop L1) or is a direct consequence of a different crystallographic environment of the molecules in both structures and thus reflects the inherent flexibility of some of these regions (e.g., S7/S8 region).

As shown by urea-induced unfolding studies, the mutations M133L and V203A make the smallest contribution to the stability increase in *T*-p53C (Nikolova *et al.*, 1998). The crystal structure reveals that this is achieved by only small structural changes, mainly by local repacking of the side chains around the mutation sites in the loop-sheet-helix motif and the S5/S6 turn, respectively. The structure provides an explanation for the large effect of the N239Y and N268D mutations on stability increase. The N239Y substitution is located in the L3 loop, which forms part of the DNA-binding surface. The mutation induces minor changes in neighboring side chains, such as Leu137, to accommodate the tyrosyl moiety. Tyr239 is within a 5 Å distance of the zinc ligands Cys238, Cys242, and His179 and thus directly connects with zinc binding. Tyr239 seems to stabilize loop L3 as reflected by the reduced crystallographic temperature factors in the structure of *T*-p53C in this region.

The N268D mutation alters the hydrogen-bonding network (Fig. 17.4A; **see color insert**). In wild type, the side-chain amide of Asn268 on β-strand S10 forms a hydrogen bond with the backbone oxygen of Phe109 on β-strand S1. In the quadruple mutant, the side chain of Asp268 has flipped relative to the position of the asparagine in wild type. It uses both of its carboxylate oxygens to form two new hydrogen bonds with backbone nitrogen atoms of Ser269 on the same

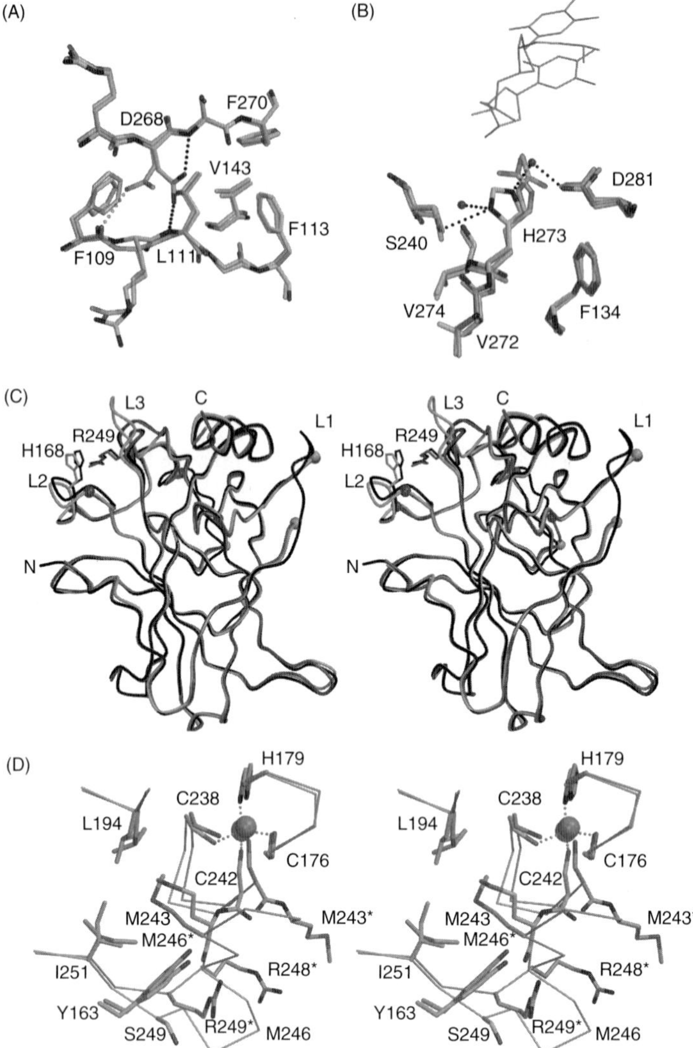

Figure 17.4. Crystal structures of p53 core domain mutants. (A) Close-up view of the N268D mutation site in the structure of the superstable quadruple mutant M133L/V203A/N239Y/N268D (*T*-p53C). The structure of *T*-p53C (PDB code 1UOL, chain A; yellow) is superimposed on the wild-type structure (PDB code 1TSR, chain A; transparent light gray). (B) View of the R273H mutation site in the structure of *T*-p53C-R273H (PDB code 2BIM, molecule A; yellow) superimposed on the structure of *T*-p53C (PDB code 1UOL, molecule A; transparent green) and DNA-bound wild type (PDB code 1TSR, molecule B; transparent light gray). Selected water molecules in the structure of *T*-p53C-R273H are represented by magenta spheres. Also shown are the two thymidylate moieties in the structure of DNA-bound wild type that are in close contact with Arg273 (thin gray lines). (C) Superposition of Cα atoms in the structures of *T*-p53C-H168R (PDB code 2BIN, magenta) and *T*-p53C-R249S (PDB code 2BIO, yellow) on those in the structure of *T*-p53C (PDB code 1UOL, molecule A; black). Cα atoms before and after chain breaks are marked with spheres in the color of the corresponding chain. (D) Structure of *T*-p53C-R249S (yellow) superimposed on the structure of p53 wild type (PDB code 1TSR, molecule A; light gray). The zinc ion in both structures is shown as a gray or yellow sphere. Wild-type residues are denoted by *. The figures were generated using MOLSCRIPT (Kraulis, 1991) and RASTER3D (Merritt and Bacon, 1997). Panel (A) was adapted from Joerger *et al.* (2004), panels (B) to (D) were adapted from Joerger *et al.* (2005).

strand and Leu111 on strand S10. This links the two sheets of the β-sandwich in an energetically much more favorable way than is achieved in the wild type. Although the N239Y and N268D substitutions act on different regions of the core domain, and do this in a very different way, they contribute to the thermodynamic stability of the protein to a similar extent. Overall, they seem to reduce the structural plasticity of the protein. This is evident from both the reduced relative temperature factors in certain regions of the quadruple mutant crystal structure and from NMR studies on the dynamic properties of the protein in solution (Ang H.C., Freund S.M.V. and Fersht A.R. unpublished results).

The structure of the stabilized variant *T*-p53C not only provides us with insights into the molecular basis for stabilizing mutations in p53 but also shows the potential practical use of this mutant for experimental purposes. So far, many biophysical studies on p53 have been hampered by the low thermodynamic stability of the destabilized mutants and even the wild type itself. The superstable quadruple mutant *T*-p53C provides a more rigid and stable structural framework, while maintaining the overall structural characteristics of the wild-type protein.

17.4.4. Structural Effects of Destabilizing Mutations in p53 Core Domain

Initial information about the structural changes that occur in the p53 core domain upon mutation was obtained using heteronuclear single quantum correlation (HSQC) NMR spectroscopy. The chemical shift changes between the wild type and the mutant protein for each residue were taken as an indication of a structural change around that residue upon mutation, though the magnitude of the changes was not defined (Wong *et al.*, 1999; Friedler *et al.*, 2004). Different hot-spot mutants show different structural changes. Mutations in loop L3 (G245S, R248Q, and R249S), situated in the DNA-contact region, result in structural changes in the loop L3 and L2 regions, which affect the whole DNA-binding interface but not the β-sandwich. The contact mutation R273H results mainly in changes around the mutation site. In contrast, the β-sandwich mutation V143A (cf. Fig. 17.1B) results in chemical shift changes across all of the β-sandwich and the DNA-binding surface (Wong *et al.*, 1999; Friedler *et al.*, 2004).

Recently, oncogenic mutations were introduced into the stabilized variant *T*-p53C, and the crystal structures of the modified proteins were successfully solved at high resolution (Joerger *et al.*, 2005). This allowed a more detailed look at the effects of two of the most important cancer mutations and revealed the fundamentally different mechanisms by which "contact" and "structural" mutations inactivate p53. The structure of *T*-p53C-R273H confirms that this mutant has all the trademarks of a classical contact mutant. The mutation replaces a DNA-contact residue without significantly affecting the conformation of neighboring residues in the DNA-binding surface (Fig. 17.4B; **see color insert**). Hence, the properties of mutant R273H can be directly attributed to the loss of one particular DNA contact and are not due to secondary effects caused by structural distortions in neighboring regions. This is consistent with the observation that binding of specific DNA is not completely abolished in this mutant.

In contrast, R249S is a structural mutation. The mutation induces major perturbations in the DNA-binding surface, whereas the structure of the β-sandwich is conserved (Figs. 17.4C and 17.4D; **see color insert**). It disrupts a network of interactions that stabilize the L3 loop in a conformation enabling Arg248 to make DNA contacts. This loop is highly flexible in *T*-p53C-R249S and adopts a conformation very different from that found in wild type. It is not only more flexible, but the substantial displacement of Met243 and Met246 in *T*-p53C-R249S indicates that the R249S mutation induces partial misfolding and thus favors non-native conformations of the L3 loop. This accounts for the reduced DNA-binding affinity. While the nature of the structural distortions caused by the R249S mutation is in essence the same in wild type as indicated by NMR studies, they may be more pronounced and communicated to a larger extent to more distant regions of the protein because of the lower stability of the underlying structural scaffold.

At only slightly more than 0.1% of all reported cancer mutations, the H168R mutation is much less frequent than the hot-spot mutations R249S and R273H (www-p53.iarc.fr; Olivier *et al.* 2002). This mutation destabilizes the core domain by ≈3 kcal/mol and also belongs to the group of structural mutations. The crystal structure of *T*-p53C-H168R has major distortions around the mutation site in the L2 loop, that is, there was no interpretable electron density for residues 166 to 170, indicative of an increased mobility of these residues (Joerger *et al.*, 2005). As in the case of *T*-p53C-R249S, the structure of the central β-sandwich was not affected by the mutation (Fig. 17.4C).

17.5. Reversing the Structural Effects of p53 Mutations with Small Molecules

Many mutations in p53 lead to its misfolding and aggregation. This is an important example of a more general phenomenon, where misfolding of proteins leads to disease. Many other human diseases, such as amyloidoses, cystic fibrosis, and lysosomal storage diseases, are caused by protein misfolding, which occurs mainly due to mutations that disrupt the three-dimensional structure of the protein (Dobson, 1999; Fan *et al.*, 1999; Morello *et al.*, 2000; Bullock and Fersht, 2001; Fan, 2003). A novel therapeutic strategy for such diseases is to develop small molecules that will assist in refolding of the proteins, which will result in their reactivation. Such molecules are termed *chemical chaperones* (Morello *et al.*, 2000; Fan, 2003).

Refolding of p53 mutants to their correct native structure should lead to their reactivation. Destabilized p53 mutants undergo a folding–unfolding equilibrium, which involves numerous states ranging from native and native-like structures, through distorted structures, to globally unfolded states (Bullock *et al.*, 1997, Bullock and Fersht, 2001). A small-molecule chemical chaperone that binds preferentially to the native state would result in a shift of the equilibrium toward the native state, leading to thermodynamic stabilization and to restoration of activity (Bullock and Fersht, 2001; Friedler *et al.*, 2002) (Fig. 17.5A).

17.5.1. Kinetic Instability of p53 Core Domain Mutants

The thermodynamically unstable p53 core domain mutants are also kinetically unstable and denature within minutes at 37°C (Friedler *et al.*, 2003). The half-life ($t_{1/2}$) of unfolding of wild-type p53 core domain is 9 min. The hot-spot mutant G245S, which is moderately destabilized, has $t_{1/2}$ = 4.6 min. The highly destabilized mutant I195T has a $t_{1/2}$ of less than 1 min. After unfolding, the denatured proteins aggregate, the rate increasing with higher concentration of protein (Friedler *et al.*, 2003).

The kinetic instability and rapid unfolding rates of oncogenic p53 mutants define the suitable rescue strategy for these mutants. In principle, the binding of a genuine ligand of p53, (e.g., DNA or a p53-binding protein) will also shift the conformational equilibrium in favor of a wild-type-like structure. However, because of their short half-lives at body temperature, many p53 mutants denature very rapidly after biosynthesis and never get the chance to bind their natural ligand. This kinetic instability may thus be the underlying reason for their inactivation. p53-stabilizing chemical chaperones have to act immediately after biosynthesis, before the protein unfolds, and keep it folded until it enters the nucleus and binds its sequence-specific target DNA or other natural ligands, which then take over (Fig. 17.5B) (Friedler *et al.*, 2003).

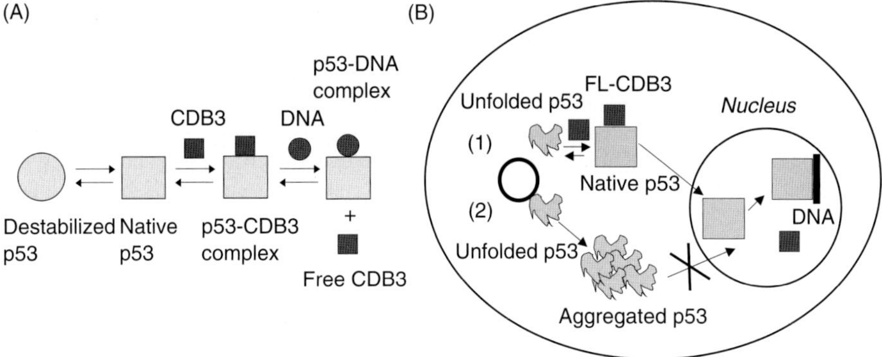

Figure 17.5. Mechanism of action of FL-CDB3, a chemical chaperone that refolds destabilized p53 core domain: (A) Chaperone strategy for rescue of p53: A schematic model of the proposed mechanism of action for FL-CDB3 (see text for details), adapted from Friedler *et al.* (2002). (B) A model for the action of p53 stabilizing drugs. Newly synthesized mutant p53 is released from the ribosome (circle) in its unfolded form. Then, two pathways are possible: (1) In the presence of a stabilizing drug such as FL-CDB3 (small square), the protein refolds to its native conformation (large square), is transported into the nucleus as a complex with the drug, and then binds its natural ligand (e.g., its target DNA). (2) In the absence of the drug, the unfolded protein aggregates and cannot be transported into the nucleus and bind DNA. Adapted from Friedler *et al.* (2003).

17.5.2. Search for Small-Molecule Chemical Chaperones that Stabilize p53 Core Domain

Potential p53-stabilizing small-molecule drugs may be found by random screening or by rational design. Various p53-activating compounds have been discovered using cell-based assay screening (Selivanova *et al.*, 1997, 1998, 1999; Bykov *et al.*, 2002, 2003; Peng *et al.*, 2003; Issaeva *et al.*, 2004). Biophysical studies show that most of them do not bind p53 core domain and so function by different mechanisms (Friedler *et al.*, 2002; Rippin *et al.*, 2002; Issaeva *et al.*, 2003; Wang *et al.*, 2003). The compound CP-31398 (Foster *et al.*, 1999), found by random screening of the Pfizer drug library for compounds that stabilize p53, also does not bind to the core domain *in vitro* (Rippin *et al.*, 2002) but functions *in vivo* by inhibiting ubiquitination of p53 and hence its degradative pathway (Wang *et al.*, 2003). A more recent study did not detect any stimulation of mutant p53 activity *in vivo* by CP-31398, but the compound was found to be highly toxic to human cells at concentrations where the original experiments had been carried out (Tanner and Barberis, 2004).

The surface of p53 lacks well-defined clefts for binding small molecules. Rational design of p53-stabilizing compounds has focused so far on peptides that mimic natural proteins that bind p53 core domain (Friedler *et al.*, 2002). These are excellent candidates to serve as chemical chaperones for p53, because they have already been optimized by nature to bind the native state of the p53 core domain. Further, being larger and more flexible than the typical small-molecule drugs, peptides could target the DNA-binding interface of p53 more efficiently even in the absence of a well-defined binding pocket.

17.5.3. CDB3: A Chemical Chaperone that Binds and Stabilizes Mutant p53 *In Vitro* and *In Vivo*

The structure of the complex between p53 core domain and the p53 binding protein 2 (53BP2 or ASPP2) (Gorina and Pavletich, 1996) was used as a basis for the design of small binding peptides (Fig. 17.6A; **see color insert**). Initial screening for binding used HSQC NMR and surface plasmon resonance (BIAcore) to identify a peptide, CDB3 (Core Domain Binding peptide number 3), which binds to p53 core domain. CDB3 is derived from the p53-binding loop of the protein 53BP2 (residues 490–498) and its sequence is REDEDEIEW (Friedler *et al.*, 2002).

NMR HSQC chemical shift analysis shows that CDB3 binds p53 in loop L1, helix H2, and β-strand S8 at the edge of the DNA binding site, partly overlapping it (cf. Fig. 17.6A) (Friedler *et al.*, 2002). The fluorescein-labeled peptide (FL-CDB3) binds wild-type p53 core domain with a dissociation constant of 600 nM, as measured by fluorescence anisotropy. It stabilizes mutant p53 core domain against chemical and thermal denaturation (Friedler *et al.*, 2002) and restores DNA binding of the highly destabilized mutant I195T back to wild-type levels (Friedler *et al.*, 2002). CDB3 also slows down the unfolding rate of p53 core domain (Friedler *et al.*, 2003). Further, FL-CDB3 was found to reverse structural changes caused by the cancer hot-spot mutation R249S. This mutation is very frequent in hepatocellular carcinoma (Aguilar *et al.*, 1993). Rescuing mutant R249S is, therefore, an important therapeutic target. HSQC and relaxation NMR techniques have shown that the R249S mutation results in an ensemble of native and native-like conformations in a dynamic equilibrium (Friedler *et al.*, 2004). As FL-CDB3 was designed to rescue mutants of p53 by binding specifically to their native structure, and because R249S exists in a dynamic equilibrium, R249S is a natural target for rescue by chemical chaperones such as FL-CDB3. NMR studies show that FL-CDB3 binds to the same site of the core domain as in wild-type and changes the chemical shifts of R249S back toward the wild-type values and thus reverses the structural effects of mutation (Friedler *et al.*, 2004).

FL-CDB3 is 50 times more active than the nonlabeled CDB3 and was used as a lead compound for *in vivo* studies. *In vivo*, FL-CDB3 acts as a chemical chaperone and rescues the tumor-suppressing function of oncogenic mutants of p53 in living cells (Issaeva *et al.*, 2003).

17.6. Reversing the Structural Effects of Tumorigenic Mutations by Second-Site Suppressor Mutations

The deleterious effect of many common cancer mutations can be reversed by intragenic suppressor mutations. Using a yeast selection system in combination with mammalian reporter gene and apoptosis assays, Brachmann *et al.* (1998) have identified several second-site suppressor mutations for the cancer hot-spot mutations G245S and R249S, as well as V143A, the classic example of a temperature-sensitive cancer mutant (Zhang *et al.*, 1994). Most of these suppressor mutations are not specific for only one single cancer mutation but restore activity in a whole subset of different mutants albeit at different levels. The suppressor mutations N239Y and S240N, which are not found in any sequence of the known p53 homologues, show a very similar rescue profile. They both suppress the cancer hot-spot mutant G245S and to a lesser extent restore activity in V143A. N268D, another of these second-site suppressor mutations, is naturally occurring in p53 of various rodents and was found to restore activity in V143A. Remarkably, H168R, which in conjunction with T123A is reported to rescue the cancer hot-spot mutant R249S, causes cancer when on its own. A more recent study reported that 16 out of 30 of the most frequently occurring cancer mutants can be rescued by changes in residues 235, 239, and 240, either alone or in combination (Baroni *et al.*, 2004).

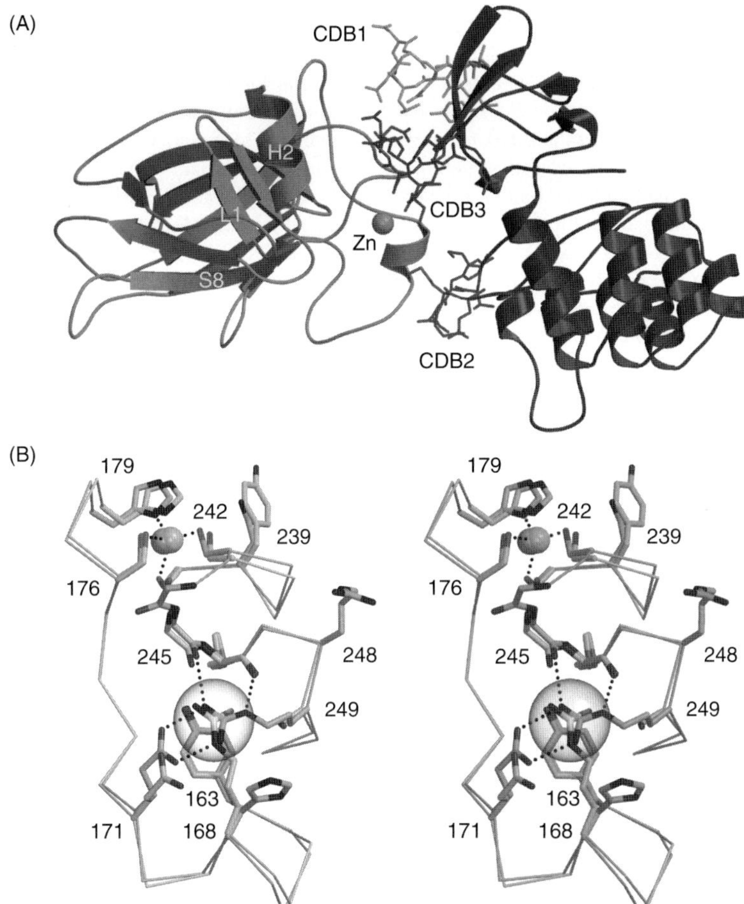

Figure 17.6. Reversing the structural effects of mutation in the p53 core domain. (A) Design of the chemical chaperone FL-CDB3. Ribbon diagram of the structure of the p53 core domain (gray)-53BP2 (red) complex (PDB code 1YCS; Gorina and Pavletich, 1996). The three p53-binding loops in 53BP2 are highlighted in different colors. CDB3 (residues 490–498 of 53BP2) is shown in purple. The binding site of the derived peptide CDB3 and its fluorescein-labeled variant FL-CDB3 was found to comprise mainly the L1 loop, β-strand S8, and helix H2 of p53 core domain, as indicated by HSQC NMR studies (Friedler *et al.*, 2002; Friedler *et al.*, 2004). The figure was generated using MOLSCRIPT (Kraulis, 1991) and RASTER3D (Merritt and Bacon, 1997). (B) Structural basis for the rescue of the cancer hot-spot mutant R249S by intragenic suppressor mutations. Stereo view of the structure of *T*-p53C-T123A/H168R/R249S (PDB code 2BIQ, yellow) superimposed on the structure of wild type (PDB code 1TSR, molecule A; light gray). Cα traces with selected side chains for parts of the L2 and L3 loops, including the zinc-binding site, are shown. In the case of Gly245 and Met246, all atoms of the residue are shown. The zinc ion in both structures is depicted as a gray or yellow sphere. Electron density for residues Met243 and Gly244 in *T*-p53C-T123A/H168R/R249S was very weak, and hence these residues were omitted from the model. Interactions mediated via the guanidinium group of Arg249 in wild type are shown by dotted lines. The common binding position of the guanidinium group of Arg168 and Arg249 in both structures is highlighted in magenta. The picture was adapted from Joerger *et al.* (2005).

17.6.1. Insights into the Mechanism of Rescue

The effect of the above mutants on the stability and DNA-binding activity of p53 core domain was studied to further elucidate how they rescue tumorigenic p53 mutants (Nikolova *et al.*, 2000). Double mutant cycles show that N268D and N239Y act as "global stability" suppressors by increasing the stability of the cancer mutants G245S and V143A. The free-energy changes are additive. The crystal structure of the stabilized variant *T*-p53C provides further insight into how the two second-site suppressor mutations N239Y and N268D rescue some of the common structurally destabilized cancer mutants (Joerger *et al.*, 2004). Tyr239 stabilizes the conformation of the L2/L3 loop region, in particular the L3 region harboring the DNA contact residue Arg248 (cf. Fig. 17.1C). This stabilizing effect seems to compensate for the structural perturbation caused by the G245S mutation, resulting in a rescue of this cancer mutant. The rescue of V143A by N268D can also be readily explained. NMR studies indicate that the V143A mutation results in changes in almost all the residues of the β-sandwich (Wong *et al.*, 1999). In wild type and the superstable quadruple mutant (cf. Figs. 17.1B and 17.4A), the side chain of Val143, which is located on β-strand S3, points toward the hydrophobic core of the β-sandwich, where it is completely buried (Cho *et al.*, 1994; Joerger *et al.*, 2004). Val143 lies at the very heart of a cluster of hydrophobic residues contributed by different regions of the protein. The truncation of the methyl groups in mutant V143A is detrimental to the packing of these hydrophobic core residues. The mechanism of rescue by an N268D substitution can be pinpointed to the formation of new hydrogen bonds. The energy loss of more than 3 kcal/mol arising from the creation of a hydrophobic cavity in mutant V143A can be partly compensated for by Asp268, which bridges the two sheets of the β-sandwich via hydrogen bonds in an energetically more favorable way than was achieved by the asparagine in wild type (Fig. 17.4A). Asp268 may directly counteract the loss of interactions by the V143A mutation by providing a more rigid structural framework.

In contrast with the global suppressor mutations N239Y and N268D, the suppressor mutation H168R is specific for the R249S mutation. Despite destabilizing wild type and *T*-p53C by 3 kcal/mol, the H168R mutation has virtually no effect on the stability of R249S but restores its binding affinity for the *gadd45* promoter (Nikolova *et al.*, 2000; Joerger *et al.*, 2005). The free-energy changes are not additive, which shows that there are interacting structural changes upon mutation. Recently solved crystal structures of *T*-p53C mutants revealed the molecular basis of this rescue mechanism (Joerger *et al.*, 2005). As we have seen above, the cancer hot-spot mutation R249S induces structural perturbations in the DNA-binding surface and favors non-native conformations of the L3 loop harboring the DNA-contact residue Arg248. This results in impaired DNA-binding. The H168R mutation, on the other hand, was found to cause substantial structural perturbations in the L2 loop of *T*-p53C-H168R.

When both cancer mutations are present in the same molecule, however, the effects induced by the single oncogenic mutations are reversed. The guanidinium group of Arg168 in the structures of *T*-p53C-H168R/R249S and *T*-p53C-T123A/H168R/R249S superimposes strikingly well with the position of the guanidinium group of Arg249 in the wild type and *T*-p53C (Fig. 17.6B; **see color insert**). Thus, Arg168 rescues R249S by mimicking the structural role of Arg249 in wild type. As a consequence, wild-type conformation is largely restored in both the L2 and L3 loop, concomitant with a restoration of DNA-binding activity to wild-type levels. How the T123A mutation, which is located on the opposite site of the DNA-binding surface, contributes to the rescue of R249S was not seen (cf. Fig. 17.1B). The structures of *T*-p53C-H168R/R249S and *T*-p53C-T123A/H168R/R249S were virtually identical except for the truncation of the side chain of Thr123. The T123A mutation, however, is not crucial for the rescue of R249S, because H168R was found to rescue R249S also in combination with mutations other than T123A, for example, in combination with K139R and N239Y (Baroni *et al.*, 2004). It has been reported that the T123A mutation and several other mutations in the immediate environment of Thr123 show an increased transactivation activity for a number of response elements in yeast-based assays using the full-length protein (Resnick and Inga, 2003). This mutation may possibly be effective to its full extent only in the context of the tetrameric full-length protein.

The mechanism of specific rescue of mutant R249S by the H168R mutation is fundamentally different from that by which global suppressors rescue mutant p53. Prime examples of such global suppressor mutations in p53 are the N239Y and N268D mutations. They increase the thermodynamic stability of the protein without compromising the structure and hence the function of the core domain. Accordingly, these mutations can fully or partially rescue a whole subset of structurally destabilized cancer mutants.

17.7. Conclusion

For more than 25 years, studies carried out using molecular biology in combination with protein engineering experiments and structural studies by both NMR and protein crystallography have resulted in a wealth of information about the p53 protein and the complex p53 network as a whole. This has vastly improved our understanding of structure-function and structure-mutation relationships in p53. We now have a much clearer picture of how human p53 is inactivated by cancer mutations. In particular, the effects of the cancer hot-spot mutations in the DNA-binding core domain of human p53 are well understood. They either directly remove essential DNA contacts or destabilize the protein, both locally and globally, causing it to unfold at body temperature. Studies on second-site suppressor mutations have shown possible ways of rescuing mutant p53 and have revealed general principles by which proteins can adapt to otherwise deleterious mutations. Many studies are now focusing on mimicking these effects by using generic molecules. Both semirational design and screening projects aim to find chemical chaperones that restore activity in a wide range of oncogenic mutants, in the ultimate pursuit of developing an effective anticancer drug.

References

Aguilar, F., Hussain, S. P. and Cerutti, P. (1993). Aflatoxin B1 induces the transversion of G → T in codon 249 of the p53 tumor suppressor gene in human hepatocytes. *Proc. Natl. Acad. Sci. USA.* 90: 8586–8590.

Baroni, T. E., Wang, T., Qian, H., Dearth, L. R., Truong, L. N., Zeng, J., Denes, A. E., Chen, S. W. and Brachmann, R. K. (2004). A global suppressor motif for p53 cancer mutants. *Proc. Natl. Acad. Sci. USA* 101: 4930–4935.

Brachmann, R. K., Yu, K., Eby, Y., Pavletich, N. P. and Boeke, J. D. (1998). Genetic selection of intragenic suppressor mutations that reverse the effect of common p53 cancer mutations. *EMBO J.* 17: 1847–1859.

Buchhop, S., Gibson, M. K., Wang, X. W., Wagner, P., Sturzbecher, H. W. and Harris, C. C. (1997). Interaction of p53 with the human Rad51 protein. *Nucleic Acids Res.* 25: 3868–74.

Bullock, A. N., Henckel, J., DeDecker, B. S., Johnson, C. M., Nikolova, P. V., Proctor, M. R., Lane, D. P. and Fersht, A. R. (1997). Thermodynamic stability of wild-type and mutant p53 core domain. *Proc. Natl. Acad. Sci. USA* 94: 14338–14342.

Bullock, A. N., Henckel, J. and Fersht, A. R. (2000). Quantitative analysis of residual folding and DNA binding in mutant p53 core domain: definition of mutant states for rescue in cancer therapy. *Oncogene* 19: 1245–1256.

Bullock, A. N. and Fersht, A. R. (2001). Rescuing the function of mutant p53. *Nat. Cancer Rev.* 1: 68–76.

Bykov, V. J., Issaeva, N., Shilov, A., Hultcrantz, M., Pugacheva, E., Chumakov, P., Bergman, J., Wiman, K. G. and Selivanova, G. (2002). Restoration of the tumor suppressor function to mutant p53 by a low- molecular-weight compound. *Nat. Med.* 8: 282–288.

Bykov, V. J., Selivanova, G. and Wiman, K. G. (2003). Small molecules that reactivate mutant p53. *Eur. J. Cancer* 39: 1828–1834.

Cho, Y., Gorina, S., Jeffrey, P. D. and Pavletich, N. P. (1994). Crystal structure of a p53 tumor suppressor-DNA complex: understanding tumorigenic mutations. *Science* 265: 346–355.

Clore, G. M., Ernst, J., Clubb, R., Omichinski, J. G., Kennedy, W. M., Sakaguchi, K., Appella, E. and Gronenborn, A. M. (1995). Refined solution structure of the oligomerization domain of the tumour suppressor p53. *Nat. Struct. Biol.* 2: 321–333.

Clore, G. M., Omichinski, J. G., Sakaguchi, K., Zambrano, N., Sakamoto, H., Appella, E. and Gronenborn, A. M. (1994). High-resolution structure of the oligomerization domain of p53 by multidimensional NMR. *Science* 265: 386–391.

Courtois, S., de Fromentel, C. C. and Hainaut, P. (2004). p53 protein variants: structural and functional similarities with p63 and p73 isoforms. *Oncogene* 23: 631–638.

Derbyshire, D. J., Basu, B. P., Serpell, L. C., Joo, W. S., Date, T., Iwabuchi, K. and Doherty, A. J. (2002). Crystal structure of human 53BP1 BRCT domains bound to p53 tumour suppressor. *EMBO J.* 21: 3863–3872.

DiTullio, R. A., Jr., Mochan, T. A., Venere, M., Bartkova, J., Sehested, M., Bartek, J. and Halazonetis, T. D. (2002). 53BP1 functions in an ATM-dependent checkpoint pathway that is constitutively activated in human cancer. *Nat. Cell Biol.* 4:998–1002.

Dobson, C. M. (1999). Protein misfolding, evolution and disease. *Trends Biochem. Sci.* 24: 329–332.

Fan, J. Q., Ishii, S., Asano, N. and Suzuki, Y. (1999). Accelerated transport and maturation of lysosomal alpha-galactosidase A in Fabry lymphoblasts by an enzyme inhibitor. *Nat. Med.* 5: 112–115.

Fan, J. Q. (2003). A contradictory treatment for lysosomal storage disorders: inhibitors enhance mutant enzyme activity. *Trends Pharmacol. Sci.* 24: 355–360.

Foster, B. A., Coffey, H. A., Morin, M. J. and Rastinejad, F. (1999). Pharmacological rescue of mutant p53 conformation and function. *Science* 286: 2507–2510.

Friedler, A., Hansson, L. O., Veprintsev, D. B., Freund, S. M., Rippin, T. M., Nikolova, P. V., Proctor, M. R., Rüdiger, S. and Fersht, A. R. (2002). A peptide that binds and stabilizes p53 core domain: chaperone strategy for rescue of oncogenic mutants. *Proc. Natl. Acad. Sci. USA* 99: 937–942.

Friedler, A., Veprintsev, D. B., Hansson, L. O. and Fersht, A. R. (2003). Kinetic instability of p53 core domain mutants: implications for rescue by small molecules. *J. Biol. Chem.* 278: 24108–24112.

Friedler, A., DeDecker, B. S., Freund, S. M., Blair, C., Rüdiger, S. and Fersht, A. R. (2004). Structural distortion of p53 by the mutation R249S and its rescue by a designed peptide: implications for "mutant conformation." *J. Mol. Biol.* 336: 187–196.

Friedler, A., Veprintsev, D. B., Freund, S. M., von Glos, K. I. and Fersht, A. R. (2005a). Modulation of binding of DNA to the C-terminal domain of p53 by acetylation. *Structure* 13: 629–36.

Friedler, A., Veprintsev, D. B., Rutherford, T., von Glos, K. I. and Fersht, A. R. (2005b). Binding of Rad51 and other peptide sequences to a promiscuous, highly electrostatic, binding site in p53. *J. Biol. Chem.* 280: 8051–8059.

Gorina, S. and Pavletich, N. P. (1996). Structure of the p53 tumor suppressor bound to the ankyrin and SH3 domains of 53BP2. *Science* 274: 1001–1005.

Grossman, S. R. (2001). p300/CBP/p53 interaction and regulation of the p53 response. *Eur. J. Biochem.* 268: 2773–2778.

Hainaut, P. and Hollstein, M. (2000). p53 and human cancer: the first ten thousand mutations. *Adv. Cancer Res.* 77: 81–137.

Issaeva, N., Friedler, A., Bozko, P., Wiman, K. G., Fersht, A. R. and Selivanova, G. (2003). Rescue of mutants of the tumor suppressor p53 in cancer cells by a designed peptide. *Proc. Natl. Acad. Sci. USA* 100: 13303–13307.

Issaeva, N., Bozko, P., Enge, M., Protopopova, M., Verhoef, L. G., Masucci, M., Pramanik, A. and Selivanova, G. (2004). Small molecule RITA binds to p53, blocks p53-HDM-2 interaction and activates p53 function in tumors. *Nat. Med.* 10:1321–1328.

Jeffrey, P. D., Gorina, S. and Pavletich, N. P. (1995). Crystal structure of the tetramerization domain of the p53 tumor suppressor at 1.7 angstroms. *Science* 267: 1498–1502.

Joerger, A. C., Allen, M. D. and Fersht, A. R. (2004). Crystal structure of a superstable mutant of human p53 core domain. Insights into the mechanism of rescuing oncogenic mutations. *J. Biol. Chem.* 279: 1291–1296.

Joerger, A. C., Ang, H. C., Veprintsev, D. B., Blair, C. M. and Fersht, A. R. (2005). Structures of p53 cancer mutants and mechanism of rescue by second-site suppressor mutations. *J. Biol. Chem.* 280: 16030–16037.

Joo, W. S., Jeffrey, P. D., Cantor, S. B., Finnin, M. S., Livingston, D. M. and Pavletich, N. P. (2002). Structure of the 53BP1 BRCT region bound to p53 and its comparison to the Brca1 BRCT structure. *Genes Dev.* 16: 583–593.

Kraulis, P. J. (1991). MOLSCRIPT: a program to produce both detailed and schematic plots of protein structures. *J. Appl. Crystallogr.* 24: 946–950.

Kussie, P. H., Gorina, S., Marechal, V., Elenbaas, B., Moreau, J., Levine, A. J. and Pavletich, N. P. (1996). Structure of the MDM2 oncoprotein bound to the p53 tumor suppressor transactivation domain. *Science* 274: 948–953.

Lane, D. P. and Lain, S. (2002). Therapeutic exploitation of the p53 pathway. *Trends Mol. Med.* 8: S38-S42.

Lane, D. P. and Hupp, T. R. (2003). Drug discovery and p53. *Drug Discov. Today* 8: 347–355.

Linke, S. P., Sengupta, S., Khabie, N., Jeffries, B. A., Buchhop, S., Miska, S., Henning, W., Pedeux, R., Wang, X. W., Hofseth, L. J., Yang, Q., Garfield, S. H., Sturzbecher, H. W. and Harris, C. C. (2003). p53 interacts with hRAD51 and hRAD54, and directly modulates homologous recombination. *Cancer Res.* 63: 2596–605.

Matsumura, I. and Ellington, A. D. (1999). *In vitro* evolution of thermostable p53 variants. *Protein Sci.* 8: 731–740.

Merritt, E. A. and Bacon, D. J. (1997). Raster3D: Photorealistic molecular graphics. *Methods Enzymol.* 277: 505–524.

Michael, D. and Oren, M. (2002). The p53 and Mdm2 families in cancer. *Curr. Opin. Genet. Dev.* 12: 53–59.

Michael, D. and Oren, M. (2003). The p53-Mdm2 module and the ubiquitin system. *Semin. Cancer Biol.* 13: 49–58.

Momand, J., Wu, H. H. and Dasgupta, G. (2000). MDM2—master regulator of the p53 tumor suppressor protein. *Gene* 242: 15–29.

Morello, J. P., Petaja-Repo, U. E., Bichet, D. G. and Bouvier, M. (2000). Pharmacological chaperones: a new twist on receptor folding. *Trends Pharmacol. Sci.* 21: 466–469.

Müller-Tiemann, B. F., Halazonetis, T. D. and Elting, J. J. (1998). Identification of an additional negative regulatory region for p53 sequence-specific DNA binding. *Proc. Natl. Acad. Sci. USA* 95: 6079–6084.

Nikolova, P. V., Henckel, J., Lane, D. P. and Fersht, A. R. (1998). Semirational design of active tumor suppressor p53 DNA binding domain with enhanced stability. *Proc. Natl. Acad. Sci. USA* 95: 14675–14680.

Nikolova, P. V., Wong, K. B., DeDecker, B., Henckel, J. and Fersht, A. R. (2000). Mechanism of rescue of common p53 cancer mutations by second-site suppressor mutations. *EMBO J.* 19: 370–378.

Olivier, M., Eeles, R., Hollstein, M., Khan, M. A., Harris, C. C. and Hainaut, P. (2002). The IARC TP53 database: new online mutation analysis and recommendations to users. *Hum. Mutat.* 19: 607–614.

Peng, Y., Li, C., Chen, L., Sebti, S. and Chen, J. (2003). Rescue of mutant p53 transcription function by ellipticine. *Oncogene* 22: 4478–4487.

Prives, C. and Manley, J. L. (2001). Why is p53 acetylated? *Cell* 107: 815–818.

Resnick, M. A. and Inga, A. (2003). Functional mutants of the sequence-specific transcription factor p53 and implications for master genes of diversity. *Proc. Natl. Acad. Sci. USA* 100: 9934–9939.

Rippin, T. M., Bykov, V. J., Freund, S. M., Selivanova, G., Wiman, K. G. and Fersht, A. R. (2002). Characterization of the p53-rescue drug CP-31398 *in vitro* and in living cells. *Oncogene* 21: 2119–2129.

Rustandi, R. R., Baldisseri, D. M. and Weber, D. J. (2000). Structure of the negative regulatory domain of p53 bound to S100B(betabeta). *Nat. Struct. Biol.* 7: 570–574.

Ryan, K. M., Phillips, A. C. and Vousden, K. H. (2001). Regulation and function of the p53 tumor suppressor protein. *Curr. Opin. Cell Biol.* 13: 332–337.

Samuels-Lev, Y., O'Connor, D. J., Bergamaschi, D., Trigiante, G., Hsieh, J. K., Zhong, S., Campargue, I., Naumovski, L., Crook, T. and Lu, X. (2001). ASPP proteins specifically stimulate the apoptotic function of p53. *Mol. Cell* 8: 781–794.

Selivanova, G., Iotsova, V., Okan, I., Fritsche, M., Strom, M., Groner, B., Grafstrom, R. C. and Wiman, K. G. (1997). Restoration of the growth suppression function of mutant p53 by a synthetic peptide derived from the p53 C-terminal domain. *Nat. Med.* 3: 632–638.

Selivanova, G., Kawasaki, T., Ryabchenko, L. and Wiman, K. G. (1998). Reactivation of mutant p53: a new strategy for cancer therapy. *Semin. Cancer Biol.* 8: 369–378.

Selivanova, G., Ryabchenko, L., Jansson, E., Iotsova, V. and Wiman, K. G. (1999). Reactivation of mutant p53 through interaction of a C-terminal peptide with the core domain. *Mol. Cell Biol.* 19: 3395–3402.

Shaulsky, G., Goldfinger, N., Ben-Zeev, A. and Rotter, V. (1990). Nuclear accumulation of p53 protein is mediated by several nuclear localization signals and plays a role in tumorigenesis. *Mol. Cell. Biol.* 10: 6565–6577.

Stommel, J. M., Marchenko, N. D., Jimenez, G. S., Moll, U. M., Hope, T. J. and Wahl, G. M. (1999). A leucine-rich nuclear export signal in the p53 tetramerization domain: regulation of subcellular localization and p53 activity by NES masking. *EMBO J.* 18: 1660–1672.

Tanner, S. and Barberis, A. (2004). CP-31398, a putative p53-stabilizing molecule tested in mammalian cells and in yeast for its effects on p53 transcriptional activity. *J. Negat. Results Biomed.* 3: 5.

Vogelstein, B., Lane, D. and Levine, A. J. (2000). Surfing the p53 network. *Nature* 408: 307–310.

Vousden, K. H. and Lu, X. (2002). Live or let die: the cell's response to p53. *Nat. Rev. Cancer* 2: 594–604.

Walker, K. K. and Levine, A. J. (1996). Identification of a novel p53 functional domain that is necessary for efficient growth suppression. *Proc. Natl. Acad. Sci. USA* 93: 15335–15340.

Wang, W., Takimoto, R., Rastinejad, F. and El-Deiry, W. S. (2003). Stabilization of p53 by CP-31398 inhibits ubiquitination without altering phosphorylation at serine 15 or 20 or MDM2 binding. *Mol. Cell. Biol.* 23: 2171–81.

Weinberg, R. L., Freund, S. M., Veprintsev, D. B., Bycroft, M. and Fersht, A. R. (2004). Regulation of DNA binding of p53 by its C-terminal domain. *J. Mol. Biol.* 342: 801–11.

Wong, K. B., DeDecker, B. S., Freund, S. M., Proctor, M. R., Bycroft, M. and Fersht, A. R. (1999). Hot-spot mutants of p53 core domain evince characteristic local structural changes. *Proc. Natl. Acad. Sci. USA* 96: 8438–8442.

Zhang, W., Guo, X. Y., Hu, G. Y., Liu, W. B., Shay, J. W. and Deisseroth, A. B. (1994). A temperature-sensitive mutant of human p53. *EMBO J.* 13: 2535–2544.

Zhao, K., Chai, X., Johnston, K., Clements, A. and Marmorstein, R. (2001). Crystal structure of the mouse p53 core DNA-binding domain at 2.7 A resolution. *J. Biol. Chem.* 276: 12120–12127.

IV

Changes in Supramolecular Structure

18

Protein Aggregation in Muscle Fibers and Respective Neuromuscular Disorders

Alexandra Vrabie and Hans H. Goebel

Abstract

Protein aggregation in muscle fibers may be a nonspecific phenomenon such as occurring in cores or ragged red fibers. However, it may also be a disease-specific and disease-significant phenomenon constituting protein aggregate myopathies (PAMs). These may be divided into two classes: The first one is marked by impaired extralysosomal degradation of proteins, catabolic PAM, encompassing desmin-related myopathies. Mutant proteins, that is, desmin, myotilin, or α-B crystallin, defy protein degradation, aggregate and associate with other proteins within muscle fibers, hence marking desminopathies, myotilinopathies, and α-B crystallinopathies. A second class of PAM encompasses those apparently associated with developmental errors though, again, based on mutations in genes for myofibrillar proteins, foremost sarcomeric actin and myosin. Here, actinopathies and myosinopathies often occur early in childhood while catabolic PAMs are largely of adult or even late onset. The common principle of these PAMs is that immunohistochemical identification of certain proteins resulted in subsequent molecular analysis of respective genes, identification of mutations, and demonstration of mutant proteins as important components of these protein aggregates.

18.1. Introduction

Immunohistochemistry has enabled recognition of extralysosomal protein aggregation within muscle fibers and other cells, foremost nerve and glial cells. Formerly, these protein aggregates had received various names according to their light and electron microscopic structures, generally termed inclusions, specifically designated as a motley gamut of structures. Roughly, these inclusions or inclusion bodies can be divided into two groups, those with, albeit not necessarily completely, identified proteins such as rods or nemaline bodies, and cytoplasmic bodies, hyaline bodies, or core-like lesions and those without known protein contents such as cylindrical spirals, reducing bodies, fingerprint bodies, Zebra bodies, and tubular aggregates. Another principle of protein aggregate description may be a pathogenetic one, related to impaired formation of normal structures and to integration or anabolic, for example, actin aggregates and myosin aggregates, or related to extralysosomal degradation or catabolic, for example, desmin-related inclusions, core-like lesions. Identification of proteins within inclusions and inclusion bodies, hence, in several instances, resulted in analysis of respective genes and their mutations, for instance of actin, myosin, and

Table 18.1. Protein aggregate myopathies

• Actinopathy
• Myosinopathy [hyaline body myopathy (HBM)]
• α-B crystallinopathy
• Myotilinopathy
• Selenoproteinopathy
• Desminopathy
• Desmin-related myopathies
2q21 myopathy
10q23 myopathy
15q22 CRD myopathy
Spheroid body myopathy
• Hereditary inclusion body myopathy (h-IBM)
• Core diseases
• *Other:* Myopathies marked by inclusions (putative)
• Oculopharyngeal muscular dystrophy

desmin. Myopathies marked by such protein aggregates associated with respective gene mutations are, therefore, named actinopathies, myosinopathies, and desminopathies, similar to the term dystrophinopathies when the protein dystrophin is affected and the dystrophin gene contains mutations. This and other mutation-related proteinopathies are, together with sarcoglycanopathies, dysferlinopathies, caveolinopathies, laminopathies, and emerinopathy, largely recognized as muscular dystrophies. However, contrary to protein aggregation, these muscular dystrophies are marked by reduction in or complete absence of respective proteins, evidenced by immunohistochemistry and/or immunoblotting. Myopathies marked by aggregation of proteins within muscle fibers are, therefore, also termed protein aggregate myopathies (PAMs) (Table 18.1). Many of these PAMs are hereditary in nature, owing to mutations in the already mentioned genes: actin, myosin, desmin, myotilin, selenoprotein, and α-B crystallin. Others may be sporadic although still originating from gene mutations, such as actinopathies marked by a large number of *de novo* mutations in the ACTA1 gene or may be of nongenetic nature, for example, certain myofibrillar myopathies/desmin-related myopathies or inclusion body myositis.

Morphologic investigation, almost exclusively by muscle biopsy, still remains the most crucial diagnostic technique to identify PAM. On the other hand, the morphology of PAM, the immunohistochemistry and subsequent molecular analysis, has also led (from morphologic lesion to gene mutation) to genetic identification of individual PAMs.

18.2. Actinopathies

Following the definition of individual proteinopathies, actinopathies are defined as neuromuscular disorders owing to mutations in the (sarcomeric) actin gene (ACTA1). Such ACTA1 mutations are seen in roughly one third of patients with nemaline myopathy (Agrawal *et al.*, 2004; Wallgren-Pettersson *et al.*, 2004). Mutations in the remaining two thirds of patients encompass those of nebulin, tropomyosin 2 and 3, troponin genes, and still unidentified genes. Belonging to the group of nemaline myopathies, ACTA1 mutations are morphologically characterized by nemaline bodies or rods, derivatives of the sarcomeric Z-disk. As nemaline myopathies have molecularly been subdivided according to the mutated genes (Wallgren-Pettersson *et al.*, 2004), the term actinopathy used here denotes not only a disorder marked by mutations in the sarcomeric

ACTA1 gene but also by accumulation of actin filaments, with or without rods, either in the sarcoplasm or in the muscle fiber nuclei or both. Hence, actinopathy is an actin filament–accumulating neuromuscular disease owing to mutations in the *ACTA1* gene, a true PAM.

Aggregation of actin filaments had earlier been observed though in a somewhat passing and, thus, unappreciated fashion (Dubowitz, 1985) and in conjunction with nemaline bodies/rods in nemaline myopathies (Karpati and Carpenter, 1992). However, the first thorough description of actin filament aggregation as a hallmark of a congenital myopathy was given by Bornemann et al. (Bornemann *et al.*, 1996) in an infant who later was found to have a mutational duplication in the *ACTA1* gene (Sparrow *et al.*, 2003). This publication was followed by similar observations in three young children with actin filament aggregates in their biopsied muscle specimens (Goebel *et al.*, 1997), two of them with intranuclear rods only (Goebel *et al.*, 1997), who were later found to have different missense mutations in the *ACTA1* gene (Nowak *et al.*, 1999). These reports established actinopathy as a new congenital myopathy, partly within the group of nemaline myopathies.

18.2.1. Actin

Of the numerous actins belonging to the microfilament family, it is the α-skeletal muscle or sarcomeric actin having a MW of 42 kDa that is of importance in muscle fibers (Kreis and Vale, 1998) while a separately coded type of actin, cardiac actin, predominates within cardiac myocytes. Two types of sarcomeric actins occur, the monomere globular type or G-actin from which the more stable filament type, F-actin, originates. To which extent globular actin is a component of the actin filament aggregates is not yet known. Similarly, it is not known whether the actin filament aggregates consist of mutant actin only or of both mutant and wild-type forms of actin.

Other proteins than actin making up the actin filament are nebulin, tropomyosin, and troponin. Mutations in these proteins can be identified in other patients with nemaline myopathy, thus, rendering most genetically recognized nemaline myopathies so far actin filament–related conditions. However, separate accumulation of any of these actin-related proteins in patients with these mutations in the respective genes, to our knowledge, has not been observed, whereas some nebulin has been identified in actin filament aggregates (unpublished observation).

18.2.2. Genetics

The mode of inheritance in this rather newly identified hereditary myopathy is determined by molecular data rather than by family studies.

Mutations in the *ACTA1* gene, which is located at 1q42.1 and consists of seven exons, are very numerous, most of them representing missense mutations (Sparrow *et al.*, 2003), whereas frameshift, nonsense, duplication, and splice-site mutations have only been observed as single examples. The majority of these ACTA1 mutations are dominant and most of them are associated with nemaline myopathy, rendering ACTA1 mutations a major component of the genetic spectrum of nemaline myopathies (Wallgren-Pettersson *et al.*, 2004). Most of these dominant mutations are *de novo* ones. Among the multitude of ACTA1 mutations only a small minority of six have been encountered together with aggregation of actin filaments (Sparrow *et al.*, 2003), comprising five missense mutations, G15R (Goebel *et al.*, 1997), G146D (Agrawal *et al.*, 2004), V163L (Goebel *et al.*, 1997), D154N (Schröder *et al.*, 2004), S348L (apparently not yet separately published), and one dupliation, AS144/5 (Bornemann *et al.*, 1996). Although the ACTA1 mutations associated with actin filament aggregation spread across the entire *ACTA1* gene, they share certain features such as being situated in "highly conserved residues," that is, subdomains 1 and 3 (Sparrow *et al.*, 2003). As regards genotyping-phenotyping, most of these mutations are associated with rapid progression of the disease and early death. However, it is curious to note that the Val163Leu mutation, so far found in only two unrelated children having intranuclear rod-only myopathy, was associated with different degrees of clinical severity. One of the two children died at the age of 4 months while the other one is alive in his teens not severely incapacitated (Goebel *et al.*, 1997; Sparrow *et al.*, 2003).

18.2.3. Clinical Features

The patient who has an actinopathy is usually a floppy infant, weak since birth and markedly hypotonic. Patients may already show contractures at birth, thus making actinopathy a member of the arthrogryposis multiplex congenita group. Limb movements may be decreased, which sometimes may already be noticeable during the fetal stage. The children have feeding difficulties and often require ventilatory support because of respiratory insufficiency. Creatine kinase may be normal or mildly elevated while the electromyogram appears "myopathic." Although cardiac abnormalities such as P-Q enlarged in QRS amplitudes as well as cardiomegaly have been noted (Goebel *et al.*, 1997), cardiomyopathy does not seem to be a consistent clinical feature. Largely due to respiratory failure and pneumonia, the life of the patients is often shortened to less than one year, rendering actinopathy a rather severe congenital myopathy (Bornemann *et al.*, 1996; Goebel *et al.*, 1997; Agrawal *et al.*, 2004; Schröder *et al.*, 2004).

However, occasionally actinopathy may not be an early fatal disorder but may rather run a milder course. Although starting with floppiness and even reduced fetal movements with some bouts of pulmonary infections, respiratory insufficiency may not be too severe, allowing continuous development of the patient with actinopathy and resulting in a milder course (Goebel *et al.*, 1997). However, to our knowledge, adults with actinopathy have not yet been identified.

18.2.4. Morphologic Features

There is a nonspecific, usually solitary or oligofocal phenomenon in muscle fibers that is known as the filamentous body (Fig. 18.1). A particular nosological association of filamentous bodies is not known. Aggregation of actin filaments in actinopathies is much more intense and widespread. It is the morphologic hallmark of actinopathies. Large plaques (Fig. 18.2) may be encountered close or distant to peripheral nuclei in the subsarcolemmal region. In regular histologic or semithin resin-embedded sections, these actin filament aggregates appear rather pale and opaque when compared with the sarcomere-containing sarcoplasm. They may be sharply demarcated from surrounding intact or disarrayed sarcomeres. They may also be seen among sarcomeres deeper within muscle fibers. Actin aggregates may consist exclusively of actin filaments (Fig. 18.3) or may contain thicker filaments, even fragments of sarcomeres. Activities of oxidative enzymes and ATPases may be absent from actin filament aggregates giving the lesions a "rubbed-out" appearance, similar to protein aggregates in desmin-related myopathies. The nature of the filaments as being composed of actin is recognized by thickness of the filaments, which are in the actin range of 5 to 8 nm, labeling at the ultrastructural level with an antibody against actin (Fig. 18.3) and as actin-related light microscopic plaques (Fig. 18.2). Only few other proteins have been identified in these actin filament aggregates, largely α-actinin and nebulin as well as α-B crystallin. Although, occasionally, actin filaments may be found invaginated into the muscle fiber nucleus (Goebel *et al.*, 1997), true intranuclear formation of actin filament aggregates has not been documented.

As mutations in the cardiac actin gene may result in cardiomyopathies [gene table; Kaplan and Fontaine (2002)], similar aggregates of cardiac actin filaments may possibly form within

Figure 18.1. A filamentous body is situated close to a nucleus beneath the sarcolemma.

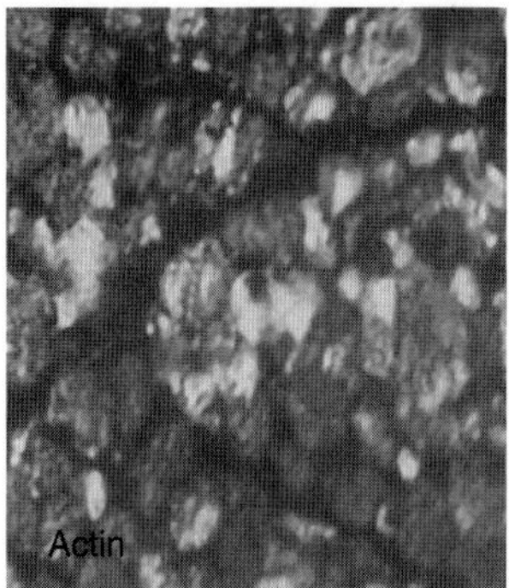

Figure 18.2. Actin filament aggregates appear as large, light areas within muscle fibers; immunofluorescence microscopy.

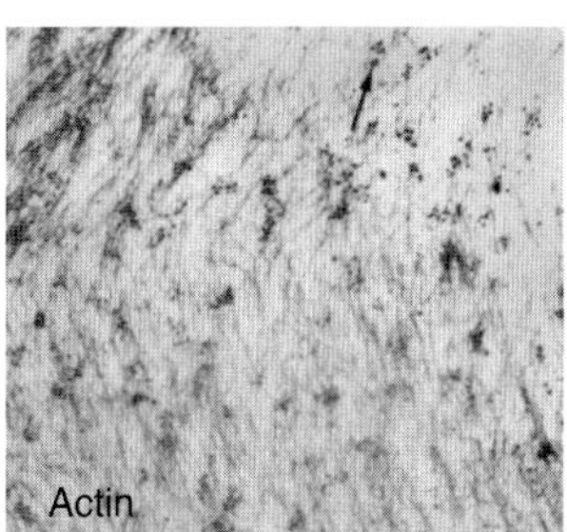

Figure 18.3. Gold grains (arrow) of actin filaments indicate immunoelectron microscopic presence of actin.

cardiac myocytes but, to our knowledge, they have not been convincingly documented (Arbustini *et al.*, 1998).

Nemaline bodies have been found associated with actin filament aggregates. Surprisingly frequent, these rods have been located within muscle fiber nuclei, so-called intranuclear rods, not associated with sarcoplasmic rods (Goebel *et al.*, 1997) or associated with sarcoplasmic minirods (Agrawal *et al.*, 2004) only. Apparently, one patient only has been encountered to have all three features: sarcoplasmic rods, intranuclear rods, and actin filament aggregates (Schröder *et al.*, 2004). However, there is a number of patients with actin filament aggregates who do not have sarcoplasmic or intranuclear rods (Bornemann *et al.*, 1996) (patient 2) (Goebel *et al.*, 1997; Goebel *et al.*, 2004). This absence of rods and presence of actin filament aggregates, related to ACTA1 mutations, poses the interesting notion that some of the actinopathies cannot necessarily be considered nemaline myopathies (Bornemann *et al.*, 1996; Goebel *et al.*, 2004) and that aggregation of actin filaments may be unrelated to formation of rods.

Morphogenesis and pathogenesis of the actin filament aggregation are unknown, as is any genetic background of filamentous bodies. However, the dearth of other proteins in actin filament aggregates, when compared with the wealth of proteins accumulating in desmin-related aggregates, as well as the early clinical onset of actinopathies and their often fatal outcome suggest defects in or incomplete development and maturation of actin filaments and their failure to be integrated into sarcomeres. As, at this time, it is not known whether these actin filament aggregates contain both mutant and wild-type actins or only mutant actin, the pathogenetic significance of actin filament aggregates is unclear. Perhaps, mutant actin, able to form filaments but not of their arrangement with thick or myosin filaments to form normal sarcomeres, is displaced within the muscle fiber to impair sarcomere contraction and sarcomere function as little as possible as may be the case with filamentous bodies. Actually, filamentous bodies are not known to contain mutant actin and being a regular feature of many diverse neuromuscular disorders may be evidence of occasional mismananged integration of normal actin filaments into sarcomeres and functional isolation by subsarcolemmal displacement.

18.3. Myosinopathies

Myosinopathies, by our definition, are neuromuscular conditions marked by the accumulation of myosin as protein aggregates and by mutations in the myosin gene. This group of disorders is the most recent one among PAMs. Myosinopathies have changed their name over the years, originating from myofibrillar lysis myopathy (Cancilla *et al.*, 1971), then later reported as hyaline body myopathies (Sahgal and Sahgal, 1977; Ceuterick *et al.*, 1993; Barohn *et al.*, 1994; Masuzugawa *et al.*, 1997). It is more than 30 years after the original description (Cancilla *et al.*, 1971) that mutations in the β-myosin chain *MYH7* gene have been identified as one cause of hyaline body myopathy, then newly designated as myosin myopathy or myosinopathy (Tajsharghi *et al.*, 2003). Contrary to aggregates of desmin and actin, aggregates of myosin give a granular rather than filamentous appearance by electron microscopy.

Based on mutations in the myosin heavy chain IIa gene, another type of myosin myopathy has been recognized (Tajsharghi *et al.*, 2002), formerly designated as autosomal-dominant inclusion body myopathy type 3 (Martinsson *et al.*, 2000). In this condition, inclusion bodies of the inclusion body myopathy (IBM) type define protein aggregation, but myosin has not unequivocally been found as a component of these inclusion bodies (discussed further later in the chapter).

18.3.1. Myosin

Among the numerous types of myosin (Kreis and Vale, 1998), there is a particular type of muscle myosin present in the thick filaments of the A-bands in sarcomeres of striated muscle cells (i.e., skeletal muscle fibers and cardiac myocytes). This myosin consists of a pair of myosin-heavy chains (MyHC) and two pairs of light chains. In different muscle fiber types, different types of MyHC isoforms are present, MyHC2a in 2A muscle fibers, MyHC2x in 2B muscle fibers, and MyHC1 in slow muscle fibers, or in cardiomyocytes as slow/β-cardiac MyHC. The rod part of the myosin protein connects to a globular head with actin and ATP-binding regions. It hydrolyzes ATP actin as an enzyme, adenosine triphosphatase (ATPase), which, therefore, is active in hyaline bodies.

18.3.2. Genetics

Of the different MyHC isoforms related to hyaline bodies, it is the slow/β-cardiac MyHC that is encoded by the *MYH7* gene located on chromosome 14q11.2. On the one hand, numerous mutations in this gene cause familial hypertrophic cardiomyopathy though not a skeletal myopathy; on the other hand, two separate missense mutations, Arg1845Trp (Tajsharghi *et al.*, 2003) and H1904L (Bohlega *et al.*, 2004) in the rod domain have been described in separate families causing skeletal myopathy, apparently not cardiomyopathy, and myosin storage in muscle fibers. So far, only these two missense mutations have been reported.

Different from patients with this *MYH7* gene and its disease hyaline body myopathy are mutations in the fast skeletal myosin encoded by the *MYH2* gene leading to a particular type of inclusion body myopathy (Martinsson *et al.*, 2000; Tajsharghi *et al.*, 2002). However, the protein composition of these inclusions has, to our knowledge, not completely been identified. Thus, whether myosin, even in a mutant form, may be a component of these inclusions and, perhaps, even constitute the seed for protein aggregation and inclusion body formation in this condition also remains to be elucidated.

18.3.3. Clinical Features

While myosinopathy has been described in only very few patients though in a familial fashion and once as a sporadic condition, its nomenclatorial predecessors, myofibrillar lysis myopathy and hyaline body myopathy, encompass a broad clinical spectrum, having been observed in siblings at a very young age (Cancilla *et al.*, 1971), that is, in the original report, as well as in adults with late onset (Sahgal and Sahgal, 1977; Ceuterick *et al.*, 1993).

18.3.4. Morphologic Features

Plaques of granular material, the hyaline bodies (Fig. 18.4), immunohistochemically and enzyme histochemically showing myosin and activity of ATPase (Fig. 18.5), respectively, contrary to the absence of ATPase activity in aggregates of desmin and actin, are the hallmark of myosinopathies, the hyaline bodies. By light microscopy, they may appear as pale plaques (Fig. 18.4), usually beneath the sarcolemma, but also deeper inside the muscle fiber. Sarcomeres among the plaques are intact, often sharply demarcated from granular aggregates. Granular aggregates appear homogeneous, without any other organelles or structures inside. They contain myosin (Ceuterick *et al.*, 1993; Barohn *et al.*, 1994; Masuzugawa *et al.*, 1997), apparently lacking other proteins (Ceuterick *et al.*, 1993). However, whether these granular aggregates within muscle fibers consist exclusively of mutant myosin or of both mutant and wild-type myosins has not been determined. Likewise, it is unclear whether similar granular aggregates are present in cardiac myocytes of patients with myosinopathy or in those cardiac myocytes of other patients who have mutations in the *MYH7* gene and cardiomyopathy, but not skeletal myopathy (Arbustini *et al.*, 1998).

Morphogenesis and pathogenesis of granular myosin aggregation are still unknown. Lack of other proteins accumulating in these aggregates suggests not failure in protein degradation but

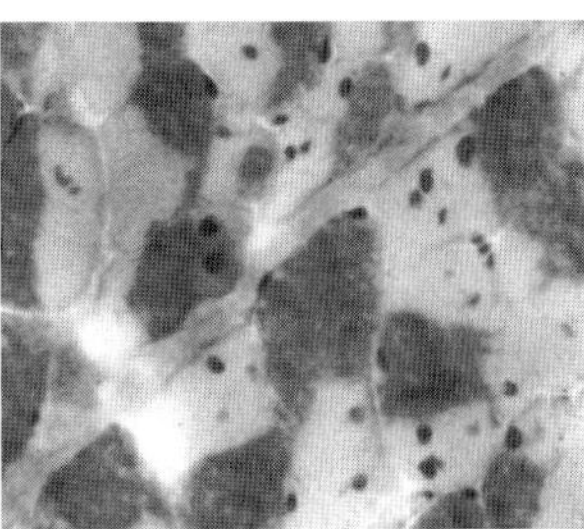

Figure 18.4. Hyaline bodies appear as light or opaque plaques, sharply demarcated from the remainder of the muscle fibers; modified Gömöri trichrome stain.

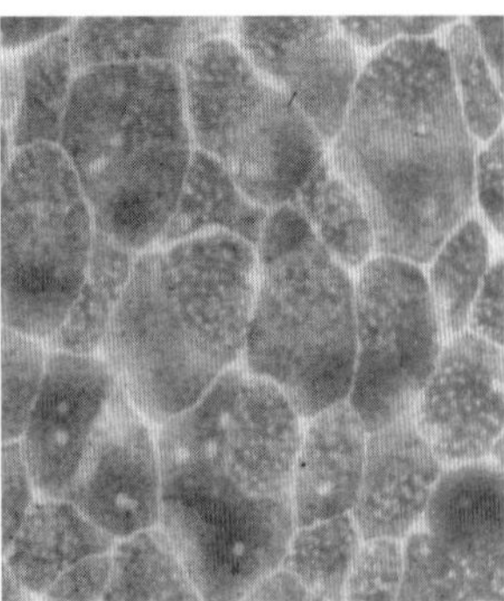

Figure 18.5. Hyaline plaques demonstrate enzyme histochemical activity of adenosine triphosphatase.

in maturation and formation of myosin filaments, while a broad clinical spectrum, including patients with late onset, renders this interpretation more tenuous than that in actinopathies.

18.4. Desminopathies

The desminopathies are defined as a new emerging group of diseases caused by mutations in the desmin gene and marked by accumulation of desmin as deposits, seen as filamentous or granulofilamentous material on electron microscopic examination and strongly immunopositive for desmin in each abnormal region of the muscle fiber. The cases showing mutations in the α-B crystallin gene (Vicart *et al.*, 1998; Fardeau *et al.*, 2000; Selcen and Engel, 2003) are designated as "α-B crystallinopathy" (Goebel and Warlo, 2000). Reduction or even lack of desmin has never been documented in humans. The term myofibrillar myopathy was proposed as a noncommittal term for a broad spectrum of morphologic changes in muscle biopsy specimens characterized by myofibrillar dissolution and accumulation of degradation protein products (Nakano *et al.*, 1996; Engel, 1999) including desmin and, therefore, also generally termed desmin-related myopathies (DRMs). Other such mutation-related DRMs are selenoproteinopathy (Ferreiro *et al.*, 2004), due to mutations in the selenoprotein N1 gene, and myotilinopathy owing to mutations in the myotilin gene (Selcen and Engel, 2004). Thus, desminopathy is a subgroup of myofibrillar myopathy (Dalakas *et al.*, 2000).

18.4.1. Desmin

Desmin represents the most prominent member of the superfamily of intermediate filament proteins in skeletal, cardiac, and certain smooth muscle cells. The cytoskeleton of all eukaryotic cells is composed of microtubules (tubulins), microfilaments (actin), and intermediate filaments (designated on the basis of their filament diameter, 8–10 nm, an integrated network including more than 60 members, the main role of which being to withstand increased mechanical stress (Fuchs, 1996)). In close relation with them, designed to interconnect different components of the cytoskeleton and to regulate its assembly, the so-called intermediate-filament-associated proteins maintain the integrity of the cellular network of intermediate filaments.

Desmin, formerly also named skeletin, is a 53-kDa protein, located in the mature skeletal muscle between the subsarcolemmal region and the nuclear membrane, associated with lamin B and around the myofibrillar Z disks, encircling and interconnecting the myofibrils at this level, thus aligning the myofibrils and linking them to the nuclei, to the plasma membrane, especially in the region of the costameres and to cytoplasmic organelles such as mitochondria [see the now widely reproduced drawing by Dalakas *et al.* (2000)]. In the heart, desmin is increased in the Purkinje fibers, as a major component, and at the level of the intercalated disks.

Expression of desmin can easily be detected during early embryogenesis and appears to participate in myogenesis, being more strongly expressed during fetal life, when desmin is longitudinally arranged, as well as in regenerating muscle fibers. Normally, increased levels of desmin are present at the neuromuscular and myotendinous junctions. Desmin is also increased in certain neoplastic cells of the rhabdomyoma-rhabdomyosarcoma types and is used as a diagnostic marker for these types of tumors. Nonspecific increase in desmin is a feature of many pathologic states involving muscles, and increased desmin is found in atrophic fibers in spinal muscular atrophy, in congenital myotonic dystrophy, as well as in target fibers in denervated muscle fibers, in nemaline myopathy, and myotubular myopathy. In all these circumstances, the diffuse or focal increase in desmin is nonspecific, and these diseases are not considered desmin-related myopathies.

18.4.2. Genetics

Desmin is encoded by a single gene (DES) located on chromosome 2q35 (Viegas-Péquignot *et al.*, 1989). Mutations in the desmin gene, as well as in α-B crystallin, myotilin, and selenoprotein N genes cause familial or sporadic forms of skeletal myopathy, morphologically characterized by myofibrillar abnormalities and accretion of proteins, among them desmin, thus belonging to the group of desmin-related myopathies or myofibrillar myopathies. These disorders are clinically and genetically heterogenous. Only a minority of them were proved to be caused by mutations in respective genes (Selcen *et al.*, 2004), leading to the conclusion that the majority of them are due to yet unidentified gene defects or nongenetic at all. Mutational analyses of other genes such as those of paranemin, synemin, and syncoilin will be essential for establishing the complete and accurate genetic background of myofibrillar myopathies.

The pattern of inheritance in familial desminopathy is autosomal dominant, autosomal recessive, or without a family history, nonfamilial cases being caused by *de novo* mutations in the desmin gene. Multiple types of mutations have been described in the desmin gene: insertion, point substitution, small in-frame deletions, and larger exon-skipping deletions. Mutant desmin is assembly-unable, accumulates as desmin-immunoreactive deposits, and may disrupt the regular desmin network within the muscle fiber, a feature that has been shown in transfected cells *in vitro* (Goldfarb *et al.*, 2004). The gene, highly conserved among vertebrate species, is composed of nine exons within an 8.4-kb region and codes for 476 amino acids. In the desmin protein molecule, an alpha-helical coiled-coil rod domain made up of 303 amino acid residues maintaining a seven residue (heptad) repeat pattern is flanked by nonhelical head and tail structures (N and C terminals). There are no mutations identified in the 1A desmin helix, and only two mutations were recently described in the head domain as well as in the tail area (Selcen *et al.*, 2004). A hot spot for mutations is exon 6, encoding for the C-terminal part of segment 2 B of the helix, most of the identified mutations being located in this area. Six missense mutations have introduced proline, a well-known helix breaker and normally absent in the desmin rod domain.

18.4.3. Clinical Features

The clinical picture of desmin-related myopathy is very heterogeneous, largely affecting adults, but also children (Goebel and Borchert, 2002), the distinct clinical phenotypes apparently being the result of different mutations in the gene and of the type of inheritance, dominant, recessive, and *de novo* mutations causing distinct clinical syndromes (Dalakas *et al.*, 2000). The usual clinical picture is that of progressive muscle weakness and atrophy first distal in the lower limbs, mostly in the anterior compartment and then in arms and hands. Later, weakness spreads to truncal, neck-flexor, and even facial muscles. Possible bulbar signs encompass swallowing difficulties and impairment of respiratory function. Cardiac disease manifests with arrhythmias, conduction blocks, and restrictive dysfunction with diminished ventricular ejection fraction. Patients with autosomal-recessive inheritance show the most severe form of disease with early age of onset usually during the childhood with signs of cardiac disease followed by generalized weakness first in the lower limbs, then progressing to the upper limbs and sudden death due to cardiac complications. The autosomal-dominant pattern of inheritance associates with later onset of disease most often in early to middle adulthood, rather slower progression, and five distinct clinical pictures: (1) skeletal myopathy and subsequently cardiac disease with arrhythmias and conduction blocks (Horowitz and Schmalbruch, 1994; Sjöberg *et al.*, 1999); (2) skeletal myopathy with slow progression and without signs of cardiomyopathy (Dalakas *et al.*, 2000, 2003); (3) cardiomyopathy followed by skeletal myopathy (Goldfarb *et al.*, 1998); (4) skeletal myopathy and subsequently involvement of the respiratory muscles, with no cardiomyopathy (Dagvadorj *et al.*, 2003); (5) cardiomyopathy with no signs of skeletal myopathy (Li *et al.*, 1999).

De novo mutations in the desmin gene represent a complex group. The four Western European patients recently described by (Dagvadorj *et al.*, 2004) had a *de novo* Arg406Trp mutation presenting with early onset of cardiomyopathy, quickly followed by muscle involvement and rapid progression within 5 years to severe incapacity, pointing to the desmin position 406 as a hot spot for spontaneous mutations, at the C-terminal end of the 2B helix, with a critical role in the normal assembly of desmin filaments.

The considerable heterogeneity of clinical presentations is presumed to be the main explanation for the fact that many patients are still misdiagnosed, though the disease seems to be a relatively frequent one. The precise prevalence of desminopathy can be assessed only after extensive genetic tests. The electromyogram usually shows a myopathic pattern or a mixed myopathic/neuropathic one. The creatine kinase (CK) level may be normal or mildly elevated; echocardiography is useful for correct evaluation of the cardiomyopathy.

18.4.4. Morphologic Features

The morphologic hallmark is the presence, in the biopsied muscle, of multifocal or disseminated excess of desmin and other proteins, forming inclusion bodies or patches. Cytoplasmic bodies were observed and demonstrated on electron microscopy even before the introduction of immunohistochemistry and were interpreted as a myopathy-characteristic feature (Goebel *et al.*, 1981) in a congenital myopathy. After antidesmin antibodies became available, these cytoplasmic bodies showed an increased amount of desmin (Osborn and Goebel, 1983).

The pathologic changes in the muscle fibers are of various intensity even in the same patient in individual muscle groups, and the diagnosis can be missed at biopsy if a clinically less affected or strong muscle is chosen. Usual features, revealed by light microscopy of the muscle tissue, include increased variation in the muscle fiber size and shape, internalization of nuclei, rare necrotic and regenerating fibers and, occasionally, scattered rimmed vacuoles. The pathognomonic changes are areas of myofibrillar lesions that are blue or blue-red in sections stained with the Gömöri modified trichrome stain and the presence of dense hyaline structures intensely eosinophilic in hematoxylin and eosin sections and stained blue to blue-purple in trichrome-stained sections, located mainly under the sarcolemma, where they often appear as a crescent or semicircular area, as well as in the center of the muscle fiber. Morphologically, two types of abnormalities are encountered: inclusion bodies termed cytoplasmic or spheroid bodies and granulofilamentous material (Fig. 18.6), occasionally mixed in the same biopsy specimen. The granular and filamentous components may sometimes occur in an orderly fashion, the granular component forming a core surrounded by a halo of radiating filaments or the cytoplasmic bodies

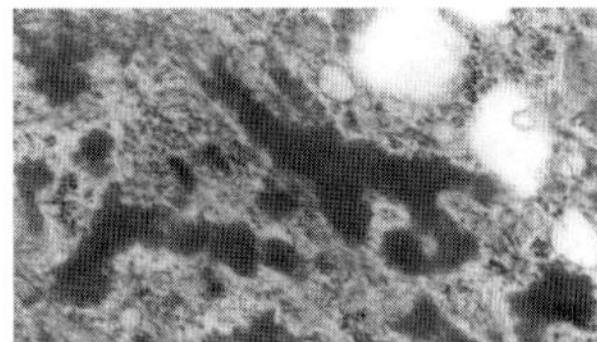

Figure 18.6. Granulofilamentous material identifies desmin-related myopathy.

may be mixed in a rather disorderly fashion, the granular component of these structures being more consistently present than the filaments (Goebel, 1995, 1997). Correlating the clinical and pathologic data, it appears that granulofilamentous material is more often associated with cardiomyopathy while the inclusion bodies better correlate with respiratory distress. Desmin, at least in part, seems to be abnormally phosphorylated (Rappaport *et al.*, 1988; Bertini *et al.*, 1991; Sabatelli *et al.*, 1992). Additional features that may be encountered are honeycomb structures (Goebel *et al.*, 1978; Nakano *et al.*, 1996) and autophagic or rimmed vacuoles close to which or in which tubulofilamentous aggregates (Reichmann *et al.*, 1997; Fidzianska *et al.*, 1999), usually seen in sporadic and hereditary inclusion body myopathy/myositis, are localized. These observations suggest a possible pathogenic overlap between these disorders. Another ultrastructural feature is an increase in the number of mitochondria with increased oxidative enzyme histochemical activities (Goebel *et al.*, 1978; Nakano *et al.*, 1996). Transfection experiments provided further evidence that mitochondrial abnormalities are related to the expression of truncated desmin (Schröder *et al.*, 2003). The concomitant presence of reducing bodies (Hübner and Pongratz, 1982; Goebel and Lenard, 1992; Bertini *et al.*, 1994) is of yet unknown significance.

Immunohistochemical and immunoelectron microscopic studies (Figs. 18.7A–18.7C) demonstrate in each specimen a markedly increased expression of desmin (Fig. 18.7A) in the intrasarcoplasmic deposits, as well as many additional proteins (Tables 18.2 and 18.3) of cytoskeletal type including nestin, vimentin (normally expressed only in fetal predesmin stages and in regenerating muscle fibers), and synemin. Additional aggregation includes sarcomeric proteins like actin, α-actinin, titin, nebulin, fast and slow myosins; transsarcolemmal proteins like dystrophin (Dys 1, 2, 3) (Fig. 18.7B), utrophin [dystrophin-related protein (DRP) 2; present in immature stages, at motor end-plate areas, and in regenerating fibers], dysferlin, merosin, sarcoglycans, caveolin, α- and β-dystroglycans; nuclear proteins emerin, lamin B; chaperone types: ubiquitin and α-B crystallin (small heat shock protein co-localizing with desmin at the Z-disk level that stabilizes and protects desmin by preventing an irreversible aggregation) (Fig. 18.7C); Alzheimer's disease type proteins: amyloid-β (residues 8–17, 17–24), β-amyloid precursor protein (β-APP, N and C terminal epitopes KPI domain), cyclin-dependent kinases including CDK1, CDK2, CDK4, CDK5, CDK7, CDK's kinase p 21 (CDK inhibitor) and others: cathepsin B, calpain, gelsolin, α_1-antichymotrypsin, neural cell adhesion molecule (N-CAM), and prion protein. Because of the diversity of these proteins accruing in the muscle fiber, only immunohistochemical studies of muscle proteins are insufficient for the diagnosis, and genetic testing is essential for establishing a precise nosologic diagnosis and enables reliable genetic counselling.

Accumulation of desmin is not only confined to skeletal muscle fibers but is a characteristic feature in desminopathies with cardiomyopathy and in patients with cardiac signs preceding any sign of skeletal muscle involvement. Identification of desmin aggregates in myocytes from myocardial biopsy is a useful diagnostic tool, showing the same morphologic changes as described in skeletal muscle (Bertini *et al.*, 1991; Lobrinus *et al.*, 1998) (Fig. 18.8).

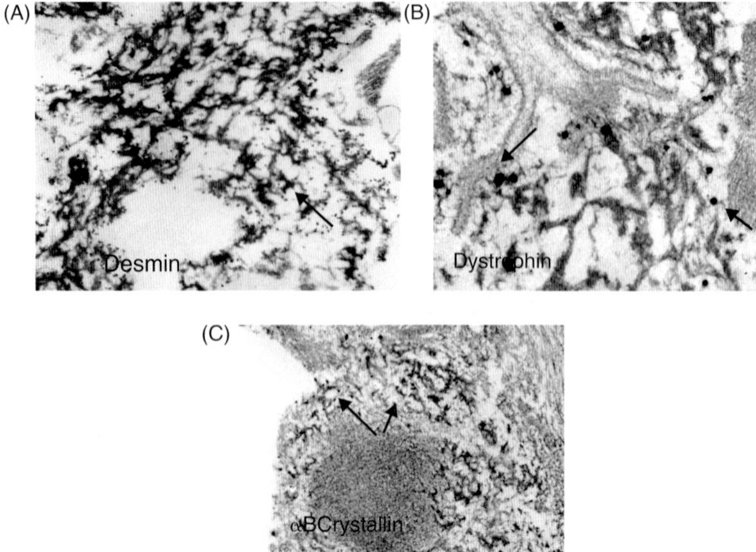

Figure 18.7. Immunoelectron microscopy: The fine filaments, largely outside of the electron-dense granular material, contain (A) desmin (arrow), (B) dystrophin (arrows), (C) α-B crystallin (arrows).

Table 18.2. Proteins found in relation with desmin-related deposits (collected from the literature)

Cytoskeletal proteins	Transsarcolemmal proteins	Cyclin-dependent kinases
Desmin	β-Spectrin	CDK1
Vimentin	Dystrophins (Dys 1, 2, 3)	CDK2
Nestin	Utrophin	CDK4
Synemin	β-Sarcoglycan	CDK5
	Dysferlin	CDK7
Sarcomeric proteins	Merosin 300	CDK s kinase
Nebulin	Caveolin	p21 (CDK inhibitor)
Titin	α- and β-Dystroglycan	
Actin		*Other proteins*
α-Actinin	*Nuclear proteins*	Cathepsin B
Myosin fast	Emerin	Calpain
Myosin slow	Lamin B	Gelsolin
Myotilin	Nuclear matrix-associated protein	α_1-Antichymotrypsin
		Neural cell adhesion molecule (N-CAM)
Chaperone proteins		Prion protein
Ubiquitin		
α-B Crystallin		

Table 18.3. Comparative immunocytochemical results in myofibrillar myopathy (MM)/inflammatory inclusion body myositis (i-IBM)

Proteins	MM	i-IBM	Source
Chaperone proteins			
Ubiquitin	++	+	Dako
α-B Crystallin	++	+	Serotec
Cytoskeletal proteins			
Desmin	++	++	Dako
Vimentin	+	+	Dako
Plectin	+	+	BD Transduction Lab.
Sarcomeric proteins			
Actin	+	+	Sigma
α-Actinin	+	+	Sigma
Myotilin	+	+	Novocastra
Transsarcolemmal proteins			
Dystrophin 1	+	−	Novocastra
Dystrophin 2	+	−	Novocastra
Dystrophin 3	+	−	Novocastra
Utrophin (DRP2)	+/−	−/+	Novocastra
α-Sarcoglycan	−/+	−	Novocastra
γ-Sarcoglycan	−/+	−	Novocastra
δ-Sarcoglycan	+	+/−	Novocastra
Dysferlin	+	+	Novocastra
Merosin 80	+/−	−	Chemicon/Novocastra
Merosin 300	+/−	−	Chemicon/Novocastra
Neural nitric oxide synthase (nNOS)	+	+	Upstate/Biosol
β-Laminin	−/+	−	Chemicon
γ-Laminin	−/+	−	Chemicon
Collagen 6	−/+	−	JCN
Caveolin	+	+	Dianova
α-Dystroglycan	+	−	Dr. Kröger—Uni Mainz
β-Dystroglycan	+/−	−	Novocastra
Nuclear proteins			
Emerin	−	−	Novocastra
Lamin A/C	−	−	Novocastra
Others			
Heat shock protein 72/73	+	+	Calbiochem
Prion protein	−	−	Dako

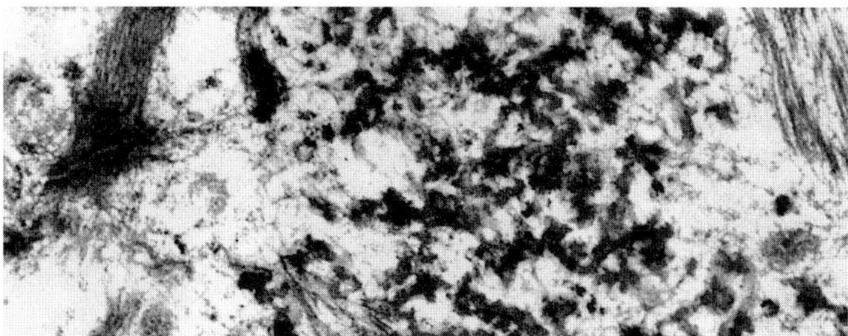

Figure 18.8. Granulofilamentous material within cardiac myocytes of desminopathy is identical to the granulofilamentous material in skeletal muscle fibers.

A similar though not identical desmin-related myopathy of late-onset and distal muscle weakness has recently been described in two siblings, affecting skeletal muscle, peripheral nervous system, and heart. The respective protein-aggregating lesions, called hyaline masses, showed accumulation of a diverse number of proteins, among them desmin, α-B crystallin, congophilic material, ubiquitin, dystrophin, and α_1-antichymotrypsin but lacked actin, α-actinin, nebulin, titin, N-CAM, and amyloid-β proteins (Selcen *et al.*, 2002) earlier found in hyaline structures of myofibrillar myopathy (MM) (De Bleecker *et al.*, 1996*b*).

18.5. Inflammatory and Hereditary Inclusion Body Myopathies

Sporadic inclusion body myositis is an *inflammatory*, apparently nonhereditary myopathy (i-IBM) mainly affecting elderly people (after 50 years of age in 80% of the patients), with a male preponderance (male to female ratio being 3:1), a duration of disease more than 6 months, and a progressive course leading to severe disability within 10–15 years after onset of first symptoms. Since its recognition in the early 1970s, it has become clear that i-IBM is the most commonly acquired myopathy in middle-aged and older people (with an overall incidence of 1:100,000 p.a.) and has to be an important diagnostic consideration in the evaluation of progressive weakness in older Caucasian males, as i-IBM is reported to account for 15–28% of cases in series of inflammatory myopathies. Still classified as one of the inflammatory myopathies, together with polymyositis and dermatomyositis, some features, including its resistance to immunosuppressant therapy, set it apart.

Hereditary inclusion body myopathies (h-IBM) are *noninflammatory* diseases with early age of onset (second or third decade of life), inherited in autosomal-dominant or autosomal-recessive forms. In autosomal-recessive h-IBM, linkage to chromosome 9p1-q1 has been established in different ethnic groups (Argov and Soffer, 2002) but the abnormal gene has not yet been identified. Welander distal myopathy and tibial muscular dystrophy are autosomal-dominant distal myopathies, which have been linked to chromosomes 2p13 and 2q31, respectively. Titin is the candidate gene in tibial muscular dystrophy. A Swedish autosomal-dominant h-IBM with proximal muscle weakness was recently described and linked to chromosome 17p13, the genetic defect associated being a missense mutation in the myosin heavy chain (MyHC) IIa gene (Martinsson *et al.*, 2000).

18.5.1. Inclusion Bodies

The inclusion bodies observed in IBM, both the sporadic/inflammatory and hereditary forms, consist of tubulofilaments, measuring 16–21 nm in diameter. These inclusion bodies are located within the sarcoplasm, often close to autophagic vacuoles and, albeit rarer, in muscle fiber nuclei. They appear, by electron microscopy, as paired helical filaments similar to those making up neurofibrillary tangles of Alzheimer's disease. Their main protein is tau, but other proteins have also been identified such as prion protein, α-synuclein, and β-amyloid precursor protein. Tubulofilamentous aggregates do not seem to differ in their fine structure between sporadic/inflammatory IBM and various genetic forms of hereditary IBM, but differences in protein composition have been claimed, using the SMI 130 antibody (Mirabella *et al.*, 1996). However, tubulofilamentous aggregates are not unique to IBM. They have been seen in a variety of other conditions, among them desmin-related myopathies (Reichmann *et al.*, 1997; Fidzianska *et al.*, 1999; Goebel and Warlo, 2000). Although several h-IBMs have been associated with gene loci, only the IIa myosin of h-IBM III has been identified as a mutant protein. Whether it is a component of the respective tubulofilamentous aggregates remains to be elucidated.

18.5.2. Genetics

A large number of both dominant and recessive forms of h-IBM (Argov and Soffer, 2002) have now been identified, some in single families only (Askanas and Engel, 2003), of which only few gene loci are known, that is, on chromosome 9p1-q1 (IBM2), which is allelic to the Nonaka type of distal myopathy. The gene MYH2 on chromosome 17p13.1 has been shown to harbor a Glu706Lys missense mutation in exon 17 (Oldfors *et al.*, 2002).

18.5.3. Clinical Features

Both proximal and distal muscle weakness of upper and lower extremities can be present, distal weakness being predominant in 20% of cases and usually asymmetric, unlike in polymyositis and dermatomyositis (Amato *et al.*, 1996). More severe weakness is commonly noted in arms involving wrist and especially finger flexors, so that loss of finger dexterity and grip strength may be a prominent presenting symptom. In legs, weakness is mostly present in foot dorsiflexors and quadriceps muscles. Symptoms relating to quadriceps muscle weakness like "knees giving way," difficulty in rising from a chair and squatting are typical symptoms. Associated quadriceps atrophy in i-IBM patients is the most common cause of "isolated quadriceps myopathy." Commonly spared are pectoralis, deltoid, hand interossei, and facial muscles. Other symptoms include isolated erector spinae weakness or "droopy neck" syndrome. Dysphagia is an early or late symptom, secondary to direct involvement of the cricopharyngeal musculature and affecting the upper third of the esophagus with difficulties in chewing/swallowing in 30% of the patients, especially in females. Fatigue, reduced tolerance of exertion, and diminished tendon reflexes, especially in the knee, are additional features. Sensory and autonomic dysfunctions in some of the patients are due to a concurrent polyneuropathy. Some patients may have signs of an autoimmune disorder such as Sjögren's syndrome, psoriasis, systemic lupus erythematosus, idiopathic interstitial pneumonitis, scleroderma, idiopathic thrombocytopenic purpura, thyroid dysfunction, and sarcoidosis. Association with malignancy is not certain. Clear diagnostic criteria based on clinical and laboratory features as well as on family history for "possible," "probable," or "definite" IBM have been established in 1995 by Griggs et al. (1995).

The serum CK level is normal in 10–20% of the patients or mildly elevated, especially in an advanced stage, most often 2 to 5 times normal and, in a minority of patients, up to 12 times increased. Electromyography shows myopathic changes and irritative phenomena (fibrillation potentials and positive sharp waves that, in the chronic stage, may be of low amplitude and infrequent or absent; however, long-duration potentials are commonly observed and do not exclude a diagnosis of i-IBM), but none of these findings are specific. The confirmation of the diagnosis requires a muscle biopsy, which can be preceded by a needle electrode examination to determine which muscle is optimal for biopsy, but not directly from the site of the needle insertion. Because morphologic changes in muscle may be patchy, examination of sufficient tissue is required to avoid sampling error.

18.5.4. Morphologic Features

A common pathologic finding in both i-IBM and h-IBM are "rimmed vacuoles" with or without "inclusion bodies" in muscle fibers, the term "inclusion body myositis" having been used for the first time by Yunis and Samaha (1971) to describe a patient with features of chronic polymyositis showing sarcoplasmic inclusions. Variation in myofiber diameters, endomysial fibrosis, occasional necrotic and regenerating fibers, angulated atrophic, rarely grouped muscle fibers, and ragged red and cytochrome C oxidase–negative fibers are further myopathic features.

The main difference between i-IBM and h-IBM is that in i-IBM, varying degrees of inflammatory infiltrates consisting of large numbers of autoaggressive CD8+ T-lymphocytes and $CD68^+$ macrophages are present, the latter reacting histochemically for acid phosphatase and surrounding and partially invading non-necrotic muscle fibers, while other regions of the muscle fiber may appear intact. Usually, muscle fibers show upregulation of MHC I (major histocompatibility complex). Endomysial inflammatory infiltrates, in the form of linear distribution of cells among muscle fibers or as larger collections around muscle fibers, are mostly T lymphocytes around vessels and mainly macrophages near and within muscle fibers. Recent studies have shown that autoinvasive $CD8^+$ T cells are specifically selected and clonally expanded *in situ* (Oldfors and Lindberg, 1999). Clonal restriction of T-cell-receptor expression was shown to persist over time (Amemiya and Dalakas, 1998), suggesting a response to persistent antigenic stimulation in muscle though the type of antigen has not yet been identified. Cytokines are important molecules in inflammatory response and immune regulation. The expression of proinflammatory cytokines, such as IL-1α, IL-1β and TNF-α, as well as the inhibitory cytokine TGF-β has been demonstrated in muscle tissue of i-IBM as well as in other inflammatory myopathies. Chemokines, important proteins that induce directional migration of inflammatory cells to immunologically active sites, were investigated and the expression of monocyte chemoattractant protein 1 and macrophage inflammatory protein 1α has been demonstrated both in infiltrating inflammatory cells and in endothelial cells.

The presence of vacuoles (at least 1:1000 muscle fibers) containing or being rimmed by basophilic granular material is characteristic of both i-IBM and h-IBM, but these may also be encountered in other diseases, such as inherited distal myopathies or oculopharyngeal muscular dystrophy. Vacuoles are present in small or large muscle fibers, often associated with a nucleus and are better seen by modified Gömöri trichrome stain. These vacuolated muscle fibers are often ultramorphologically associated with intracytoplasmic and, though less frequent, also intranuclear, tubulofilaments with diameters of around 16–21 nm that appear as paired helical filaments, a morphologic hallmark of IBM.Other common findings at the electron microscopic level are myelin-like bodies (membranous whorls) of varying size and shape disrupting the myofibrils and aggregates of mitochondria.

Mendell, Askanas, and Engel (Mendell *et al.*, 1991; Askanas *et al.*, 1992; Askanas and Engel, 2001) described the presence of intracellular deposits staining positively for amyloid protein, as well as abnormal accumulation of various Alzheimer's disease-like proteins, including α_1-antichymotrypsin, β-amyloid precursor protein (βAPP) epitopes, hyperphosphorylated τau, ubiquitin, apolipoprotein E, and prion protein, as well as presenilin-1, neuronal (n) and inducible (i) forms of nitric oxide synthase (NOS) raising the hypothesis of degeneration as pathogenesis in i-IBM. Another protein, recently found to be pathologically accumulated in muscle in i-IBM as well as in the brain in neurodegenerative diseases, is α-synuclein, which colocalizes with possibly secondarily deposited Aβ (Askanas and Engel, 2001). In addition, several kinases are strongly expressed in IBM (Nakano *et al.*, 2001). The new finding of accumulation of RNA at the same site as phosphorylated tau epitopes adds to the list of deposits accumulating in the vacuolated fibers, as well as the survival motor neuron protein demonstrated in association with paired helical filaments (Broccolini *et al.*, 2000). Thus, a multitude of aggregating proteins has been recognized in IBM, similar to those in MM (Fig. 18.9; Tables 18.3 and 18.4). A possible clue to the mysterious etiology of i-IBM came from the study of α-B crystallin in i-IBM that showed upregulation not only in vacuolated fibers and in fibers invaded by mononuclear cells but also in normal muscle fibers. Because α-B crystallin is a small heat shock protein upregulated in cellular stress, it was suggested that biologic stress of muscle fibers precedes the structural alterations and that α-B crystallin can be the autoantigen triggering the inflammatory response (Banwell and Engel, 2000). More recently (Ferrer *et al.*, 2004) (Table 18.4), the involvement of the ubiquitin-proteasome system and of immunoproteasome subunits was shown in i-IBM as well as in another group of muscular diseases marked by abnormal accumulation of proteins in muscle fibers, the genetically and clinically heterogeneous group of so-called myofibrillar myopathies (MMs) (Ferrer *et al.*, 2004). This study showed increased immunoreactivity of the 20S proteasome, 19S complex, and PA28α/β colocalizing abnormal deposits of different proteins in both types of diseases. In all the specimens studied, subunits of the immunoproteasome lysosomal membrane protein (LMP)-2, LMP7, and Multicatalytic endopeptidase complex-like 1 (MECL1) colocalized with proteasomal immunoreactivity and abnormal protein accretion. Thus, this study provides a link between abnormal protein accumulation, altered proteasomal expression and MHC class I antigen production in IBM and MM (Tables 18.3 and 18.4). Further research may generate clues to the enigmatic pathogenesis of i-IBM in which, up to now, no treatment has been proved effective and will identify primary genetic defects in h-IBM families.

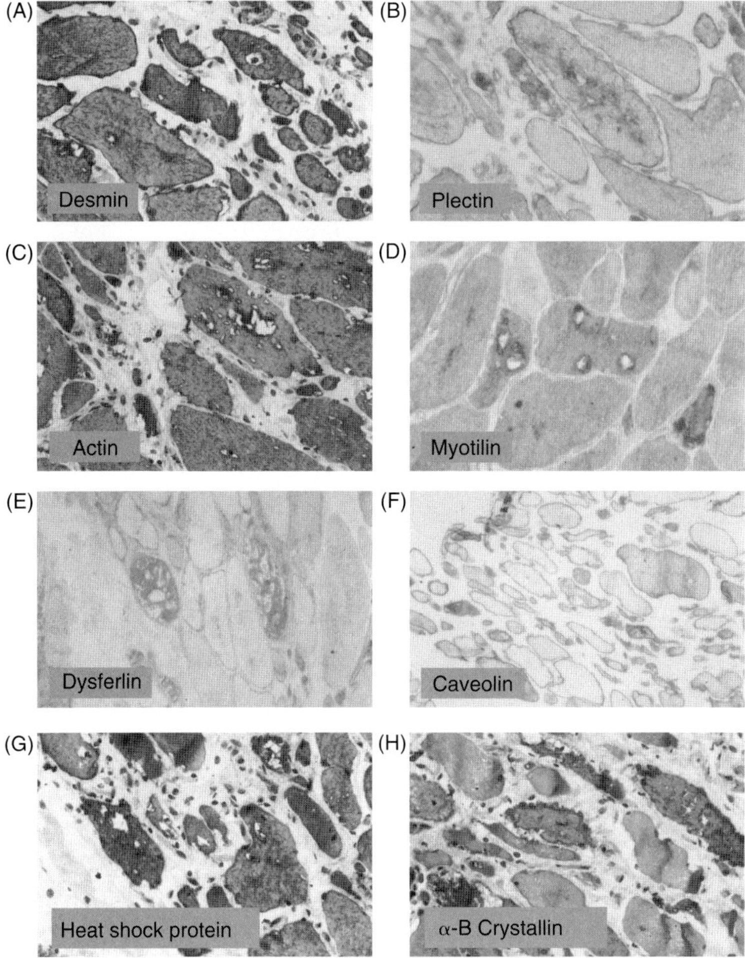

Figure 18.9. Comparative immunohistochemical aggregation of several proteins similar in i-IBM (A–H) and DRM/MM (I–R).

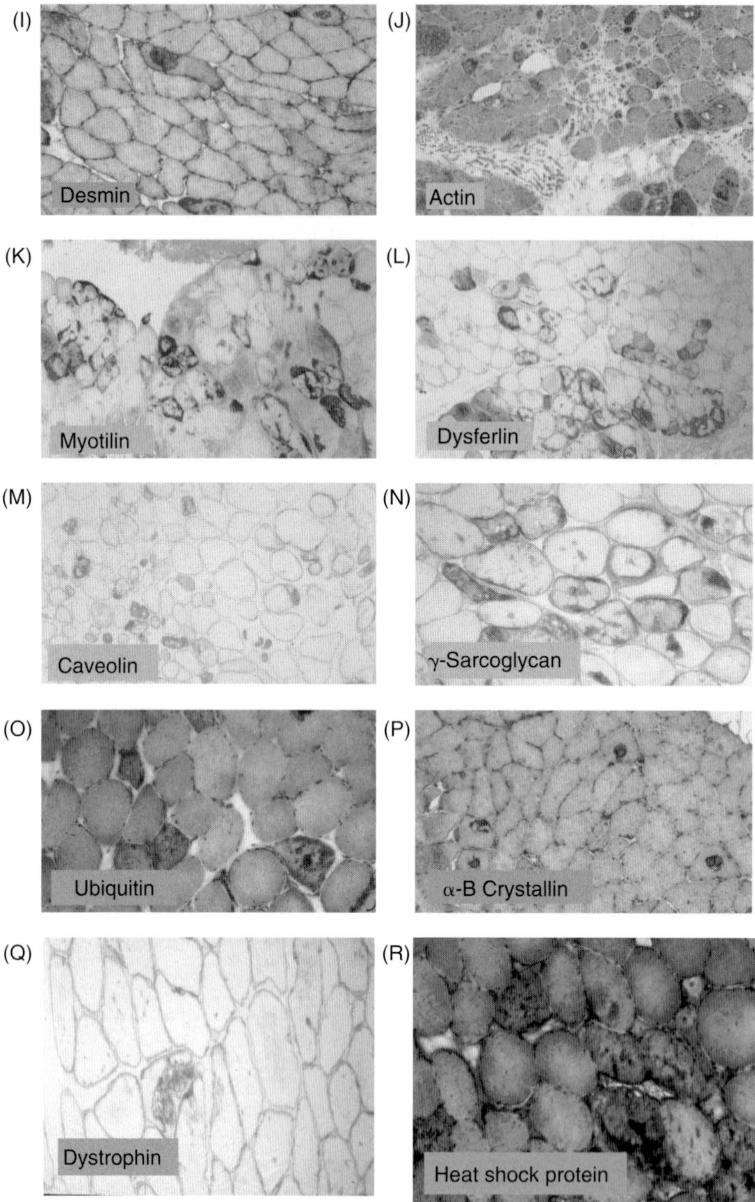

Figure 18.9. Continued

Table 18.4. Aggregation of ubiquitin-proteasome pathway-related proteins in Desmin-related myopathy (DRM)/myofibrillar myopathy (MM) and inflammatory inclusion body myopathy (i-IBM) (Ferrer *et al.*, 2004)

Proteins	DRM/MM	i-IBM
20S-Proteasome	+	+
20S-Z-proteasome	+	+
19S-Subunit S1-proteasome	+	+
PA28α/β Activator	+	+
Proteasome subunits		
LMP2	+	+
LMP7	+	+
MECLI1	+	+
MHC-I	+	+
SAPK/JNK-P	+	+
Phospho-ATF-2	+	+
Phospho-c-Jun^{Ser63}	+	+
Phospho-CREB	+	+
Phospho-Elk-1	+	+
Phospho-*tau*^{Ser262}	+	+
Phospho-*tau*^{Ser422}	+	+
Phospho-38-P kinase	+	+
Desmin	+	+
Actin	+	+
Gelsolin	+	+
Ubiquitin	+	+
α-B Crystallin	+	+

The presence of ragged red fibers in muscle tissue, as well as of cytochrome C oxidase–negative muscle fiber segments and of multiple mitochondrial deletions in 50% of the cases (Desnuelle *et al.*, 1998) may indicate the possible contribution of a mitochondrial component in the etiology of sporadic IBM (s-IBM). Abnormal accumulation of components related to lipid metabolism, for example cholesterol, is possibly due to its abnormal trafficking (Askanas and Engel, 2003). Other debates in the literature concern the role of a neurogenic component in i-IBM, as almost 20% of i-IBM cases have a concomitant morphologically and electrophysiologically proven polyneuropathy, predominately of sensory type and usually subclinical, with loss of myelinated axons in sural nerve specimens (Lindberg *et al.*, 1990).

18.6. Other Diseases Associated with Protein Aggregation

18.6.1. Core Diseases

Similar to the bi-etiological aggregation of proteins in desmin-related myopathies/myofibillar myopathies or h-IBM and i-IBM (i.e., genetic or acquired), aggregation of proteins within cores of central core disease (CCD) and multi-minicore disease (MmD) may fall into the same two categories. Earlier, accumulation of diverse proteins was recorded in cores of CCD (Vita *et al.*, 1994; De Bleecker *et al.*, 1996a) and in MmD (Bönnemann *et al.*, 2003). Although the presence of mutant proteins, based on mutations in respective protein-coding genes, is not clearly identified, association of CCD with mutations in the ryanodine receptor 1 (*RYR1*) gene and of MmD with mutations in the selenoprotein N gene (Ferreiro *et al.*, 2002) suggests that respective mutant proteins may accumulate in cores of these disorders. The fact that these putative mutant proteins are not the only proteins accumulating in the cores, but also many others (Fig. 18.10; Table 18.5), certainly of wild type, supports the interpretation that this is catabolic extralysosomal protein aggregation due to impaired degradation of proteins. This assumption is further supported by the fact that similar proteins also accrue in target lesions of the muscle fibers in denervating processes (De Bleecker *et al.*, 1996a) in which both genetic and nongenetic causes are not primarily located within muscle fibers but in the second motor neuron between the neuronal perikaryon and the presynaptic part of the motor end-plate. A recent observation, which requires independent confirmation, concerns absence of the RYR1 protein from cores in RYR1 mutation–associated CCD specimens (Parain *et al.*, 2004). If this observation is confirmed, it may also serve as a diagnostic clue to expect mutations in the *RYR1* gene in CCD in case the RYR1 protein is absent from cores and to expect a non-RYR1 mutation if the RYR1 protein is present within the cores of the CCD specimen. Recently, cores only have been identified in two families, based on missense mutations in the *ACTA1* gene (Kaindl *et al.*, 2004).

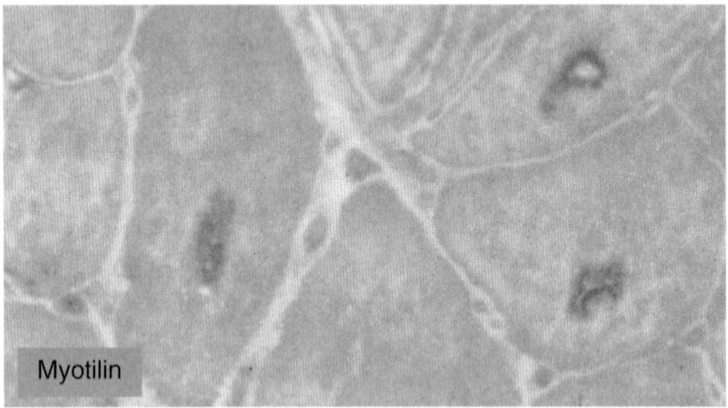

Figure 18.10. There is accumulation of myotilin within central cores in central core disease.

Table 18.5. Aggregation of proteins in cores

Central cores (Vita *et al.*, 1994; De Bleecker *et al.*, 1996a)		Minicores (Bönnemann *et al.*, 2003)
Actin	γ-Filamin/filamin 2C	α-B Crystallin
α-Actinin	Gelsolin	Desmin
α_1-Antichymotrypsin	β- and γ-1 Laminins	γ-Filamin/filamin 2C
β-APP C-terminal	Merosins 80 + 300	FATZ/myozenin/calsarcin
β-APP N-terminal, residues 8–17	β_2-Microglobulin	
β-AP, residues 17–24	Myosin slow	
Caveolin	Nebulin	
Collagen 6	N-CAM	
Dysferlin	β-, γ-, δ-Sarcoglycans	
α- and β-Dystroglycans	Ubiquitin	
Dystrophin	Utrophin	

FATZ, filamin-, actinin-, and telethonin-binding protein of the Z-disk of skeletal muscle; N-CAM, neural cell adhesion molecule.

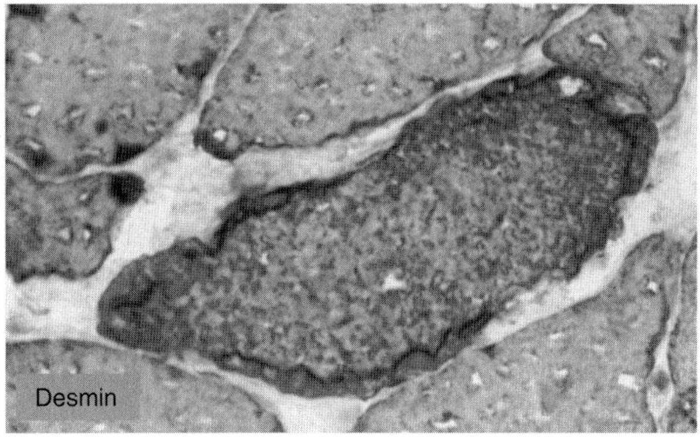

Figure 18.11. Immunohistochemical demonstration of aggregated desmin in the subsarcolemmal "ragged" area of a ragged red muscle fiber.

18.6.2. Ragged Red Fiber Disease

Ragged red fibers are marked by accumulation of abnormal mitochondria within muscle fibers, amassing beneath the sarcolemma as well as across the muscle fiber, often in a segmental fashion. At this area of red raggedness, within the muscle fiber, which is usually devoid of sarcomeres containing sarcoplasm and other nonspecific structures, aggregation of proteins may also occur. Indeed, aggregating proteins such as desmin (Fig. 18.11), α-B crystallin, as well as dysferlin could be identified. Again, the multitude of accruing proteins suggests impaired catabolism of these proteins rendering ragged red fiber disease another though not specific PAM. Abnormalities in mitochondria, also suggesting dysfunction, have earlier been reported in a muscle biopsy specimen associated with a mutation in the desmin gene (Schröder *et al.*, 2003). The association of protein aggregation and mitochondrial abnormalities suggests a disturbed mutual interplay of these intracellular components but, at this time, does not allow distinction of primary events and influences from secondary ones.

18.6.3. Other Possible Lesions Associated with Protein Aggregation

If areas of red raggedness and mitochondrial aggregation devoid of sarcomeres or structural disturbance in sarcomeres such as cores and myofibrillar lesions form the basis of protein aggregation, other structural abnormalities may also be related to protein aggregation such as sarcoplasmic masses in myotonic dystrophy (DM1), caps in cap disease, or ringbinden, and areas of lobulation in lobulated fibers, apart from inclusion bodies of various kinds.

Protein aggregation seen in denervated atrophic fibers, though not necessarily in those fibers of a neurogenic process that contains core-like lesions (De Bleecker *et al.*, 1996*a*) also suggests impaired degradation of proteins in structurally disturbed sarcomeres as they develop during denervation and atrophy. Thus, protein aggregation in denervated muscle fibers may be an association of impaired metabolism and immaturity, indicating upregulation of proteins, which is, for instance, suggested in denervated muscle fibers by the reappearance of extrajunctional acetylcholine receptors.

18.7. Conclusion

Protein aggregation within muscle fibers and respective protein aggregate myopathies represent a newly emerging scientific field of research. Compared with the wealth of data obtained in protein aggregate disorders of the nervous system, foremost Alzheimer's and Parkinson's diseases, the scientific yield in PAM appears still rather small. Biological similarities between nerve cells and muscle fibers may allow conclusions for PAM, drawn from the protein aggregate neurodegeneration. One such gain of knowledge and insight is delineating similarities and dissimilarities between protein aggregate encephalopathies (PAEs) and PAM. For instance, while the genetic aspect of PAM dominates the discussion, the overwhelmingly sporadic appearance of Alzheimer's and Parkinson's diseases constitutes major research aspects in PAE. Sporadic PAM may outnumber genetic PAM (Selcen *et al.*, 2004). Little is known about the pathogenesis and morphogenesis or the cause(s) of sporadic PAM, especially myofibrillar myopathy. Hence, to recognize, clearly delineate, and investigate sporadic PAM will be one of the major future issues in human protein aggregation research. Another dissimilarity between PAE and PAM is the appearance of actin and myosin aggregates in certain PAMs, which hardly seem to have an equivalent among the PAE whose pathogenetic background is even more enigmatic at this time than that of the catabolic PAM caused by impaired extralysosomal protein degradation. Progress in understanding PAM will certainly be promoted by increasing awareness of this still rare and, perhaps, underdiagnosed group of neuromuscular conditions.

Acknowledgments

We are gateful to Ms. M. Schlie for immunohistochemical preparations, to Ms. I. Warlo for immunoelectron microscopic studies, to Mr. W. Wagner for photography, and to Ms. A. Wöber for editorial assistance. A.V. has been supported by a fellowship from the European Society of Neurology. Further support has come from the European Neuromuscular Centre (ENMC) Consortium on "Desmin and Associated Disorders" as well as from the Deutsche Gesellschaft für Muskelkranke e.V.

References

Agrawal, P. B., Strickland, C. D., Midgett, C., Morales, A., Newburger, D. E., Poulos, M. A., Tomczak, K. K., Ryan, M. M., Iannaccone, S. T., Crawford, T. O., Laing, N. G., and Beggs, A. H. (2004). Heterogeneity of nemaline myopathy cases with skeletal muscle α-actin gene mutations. *Ann. Neurol.* 56:86–96.

Amato, A. A., Gronseth, G. S., Jackson, C.-E., Wolfe, G. L., Katz, J. S., Bryan, W. W., and Barohn, R. J. (1996). Inclusion body myositis: clinical and pathological boundaries. *Ann. Neurol.* 40:581–586.

Amemiya, K., and Dalakas, M. C. (1998). The T-cell receptor repertoire of endomysial lymphocytes in patients with inclusion body myositis (IBM) remains restricted over time: studies in repeated muscle biopsies (P04.015). *Neurology* 50:A204.

Arbustini, E., Morbini, P., Grasso, M., Fasani, R., Verga, L., Bellini, O., Dal Bello, B., Campana, C., Piccolo, G., Febo, O., Opasich, C., Gavazzi, A., and Ferrans, V. J. (1998). Restrictive cardiomyopathy, atrioventricular block and mild to subclinical myopathy in patients with desmin-immunoreactive material deposits. *J. Am. Coll. Cardiol.* 32:645–653.

Argov, Z., and Soffer, D. (2002). Hereditary inclusion body myopathies. In: Karpati, G. (ed), *Structural and Molecular Basis of Skeletal Muscle Diseases*. ISN Neuropath Press, Basel, pp. 274–276.

Askanas, V., and Engel, W. K. (2001). Inclusion-body myositis. Newest concepts of pathogenesis and relation to aging and Alzheimer disease. *J. Neuropathol. Exp. Neurol.* 60:1–14.

Askanas, V., and Engel, A. G. (2003). Unfolding story of inclusion-body myositis and myopathies: role of misfolded proteins, amyloid-beta, cholesterol, and aging. *J. Child Neurol.* 18:185–190.

Askanas, V., Engel, W. K., and Alvarez, R. B. (1992). Light and electron microscopic localization of ß-amyloid protein in muscle biopsies of patients with inclusion-body myositis. *Am. J. Pathol.* 141:31–36.

Banwell, B. L., and Engel, A. G. (2000). α–B–crystallin immunolocalization yields new insights into inclusion body myositis. *Neurology* 54:1033–1041.

Barohn, R. J., Brumback, R. A., and Mendell, J. R. (1994). Hyaline body myopathy. *Neuromuscul. Disord.* 4:257–262.

Bertini, E., Bosman, C., Ricci, E., Servidei, S., Boldrini, R., Sabatelli, M., and Salviati, G. (1991). Neuromyopathy and restrictive cardiomyopathy with accumulation of intermediate filaments: a clinical, morphological and biochemical study. *Acta Neuropathol. (Berl.)* 81:632–640.

Bertini, E., Salviati, G., Apollo, F., Ricci, E., Servidei, S., Broccolini, A., Papacci, M., and Tonali, P. (1994). Reducing body myopathy and desmin storage in skeletal muscle: morphological and biochemical findings. *Acta Neuropathol. (Berl.)* 87:106–112.

Bohlega, S., Abu-Amero, S. N., Wakil, S. M., Carroll, P., Al-Amr, R., Lach, B., Al-Sayed, Y., Cupler, J., and Meyer, B. F. (2004). Mutation of the slow myosin heavy chain rod domain underlies hyaline body myopathy. *Neurology* 62:1518–1521.

Bornemann, A., Bloch, P., Petersen, M., and Schmalbruch, H. (1996). Fatal congenital myopathy with actin filament deposits. *Acta Neuropathol. (Berl.)* 92:104–108.

Bönnemann, C., Thompson, T. G., van der Ven, P. F. M., Goebel, H. H., Warlo, I., Vollmers, B., Reimann, J., Herms, J., Gautel, M., Takada, F., Beggs, A. H., Fürst, D. O., Kunkel, L. M., Hanefeld, F., and Schröder, R. (2003). Filamin C accumulation is a strong but nonspecific immunohistochemical marker of core formation in muscle. *J. Neurol. Sci.* 206:71–78.

Broccolini, A., Engel, K. W., Alvarez, R. B., and Askanas, V. (2000). Paired helical filaments of inclusion-body myositis muscle contain RNA and survival motor neuron protein. *Am. J. Pathol.* 156:1151–1155.

Cancilla, P. A., Kalyanaraman, K., Verity, M. A., Munsat, T., and Pearson, C. M. (1971). Familial myopathy with probable lysis of myofibrils in type I fibres. *Neurology* 21:579–585.

Ceuterick, C., Martin, J.-J., and Martens, C. (1993). Hyaline bodies in skeletal muscle of a patient with a mild chronic non-progressive congenital myopathy. *Clin. Neuropathol.* 12:79–83.

Dagvadorj, A., Goudeau, B., Hilton-Jones, D., Blancato, J. K., Shatunov, A., Simon-Casteras, M., Squier, W., Nagle, J. W., Goldfarb, L. G., and Vicart, P. (2003). Respiratory insufficiency in desminopathy patients caused by introduction of proline residues in desmin c-terminal alpha-helical segment. *Muscle Nerve* 27:643–645.

Dagvadorj, A., Olivé, M., Urtizberea, J.-A., Halle, M., Bönnemann, C., Park, K.-Y., Goebel, H. H., Ferrer, I., Vicart, P., Dalakas, M. C., and Goldfarb, L. (2004). A series of West European patients with severe cardiac and skeletal myopathy associated with a de novo R406W mutation in desmin. *J. Neurol.* 251:143–149.

Dalakas, M. C., Park, K.-Y., Semino-Mora, C., Lee, H.-S., Sivakumar, K., and Goldfarb, L. (2000). Desmin myopathy, a skeletal myopathy with cardiomyopathy caused by mutations in the desmin gene. *N. Engl. J. Med.* 342:770–780.

Dalakas, M. C., Dagvadorj, A., Goudeau, B., Park, K.-Y., Takeda, K., Simon-Casteras, M., Vasconcelos, O., Sambuughin, N., Shatunov, A., Nagle, J. W., Sivakumar, K., Vicart, P., and Goldfarb, L. G. (2003). Progressive skeletal myopathy, a phenotypic variant of desmin myopathy associated with desmin mutations. *Neuromuscul. Disord.* 55:563–577.

De Bleecker, J. L., Ertl, B. B., and Engel, A. G. (1996a). Patterns of abnormal protein expression in target formations and unstructured cores. *Neuromuscul. Disord.* 6:339–349.

De Bleecker, J. L., Engel, A. G., and Ertl, B. B. (1996b). Myofibrillar myopathy with abnormal foci of desmin positivity. II. Immunocytochemical analysis reveals accumulation of multiple other proteins. *J. Neuropathol. Exp. Neurol.* 55:563–577.

Desnuelle, C., Paquis, V., Paul, R., Saunières, A., Alvarez, R.B., and Askanas, V. (1998). mtDNA analysis in muscle of patients with sporadic inclusion-body myositis and hereditary inclusion-body myopathy. In: Askanas, V., Serratrice, G., and Engel, K. W. (eds.), *Inclusion-Body Myositis and Myopathies*.Cambridge University Press, Cambridge, pp. 318–328.

Dubowitz, V. (1985). *Muscle Biopsy: A Practical Approach*. Baillière Tindall, London, pp. 664–670.

Engel, A. G. (1999). Myofibrillar myopathy (editorial). *Ann. Neurol.* 46:681–683.

Fardeau, M., Vicart, P., Caron, A., Chateau, D., Chevalay, M., Collin, H., Chapon, F., Duboc, D., Eymard, B., Tomé, F. M. S., Dupret, J. M., Paulin, D., and Guicheney, P. (2000). Myopathie familiale avec surcharge en desmine, sous forme de matériel granulo-filamentaire dense en microscopie électronique, avec mutation dans le gène de l'alpha-B-cristalline. *Rev. Neurol. (Paris)* 156:497–504.

Ferreiro, A., Quijano-Roy, S., Pichereau, C., et al. (2002). Mutations of the selenoprotein N gene, implicated in rigid spine muscular dystrophy, cause the classical phenotype of multi-minicore disease. *Am. J. Hum. Genet.* 71:739–749.

Ferreiro, A., Ceuterick-de Groote, C., Marks, J. J., Goesmans, N., Schreiber, G., Hanefeld, F., Fardeau, M., Martin, J.-J., Goebel, H. H., Richard, P., Guicheney, P., and Bönnemann, C. (2004). Desmin-related myopathy with Mallory body-like inclusions is caused by mutations of the selenoprotein N gene. *Ann. Neurol.* 55: 676–686.

Ferrer, I., Martín, B., Castaño, J. G., Lucas, J. J., Moreno, D., and Olivé, M. (2004). Proteasomal expression, induction of immunoproteasome subunits, and local MHC class I presentation in myofibrillar myopathy and inclusion body myositis. *J. Neuropathol. Exp. Neurol.* 63:484–498.

Fidzianska, A., Drac, H., and Kaminska, A. (1999). Familial inclusion body myopathy with desmin storage. *Acta Neuropathol. (Berl.)* 97:509–514.

Fuchs, E. (1996). The cytoskeleton and disease: genetic disorders of intermediate filaments. *Annu. Rev. Genet.* 30:197–231.

Goebel, H. H., Muller, J., Gillen, H. W., and Merritt, A. D. (1978). Autosomal dominant "spheroid body myopathy." *Muscle Nerve* 1:14–26.

Goebel, H. H., Schloon, H., and Lenard, H. G. (1981). Congenital myopathy with cytoplasmic bodies. *Neuropediatrics* 12:166–180.

Goebel, H. H., and Lenard, H. G. (1992). Congenital myopathies. In: Rowland, L. P., and DiMauro, S. (eds.), *Handbook of Clinical Neurology*, Volume 18/62. Elsevier Science Publishers B.V., Amsterdam, pp. 331–367.

Goebel, H. H. (1995). Desmin-related neuromuscular disorders. *Muscle Nerve* 18:1306–1320.

Goebel, H. H. (1997). Desmin-related myopathies. *Curr. Opin. Neurol.* 10:426–429.

Goebel, H. H., Anderson, J. R., Hübner, C., Oexle, K., and Warlo, I. (1997). Congenital myopathy with excess of thin myofilaments. *Neuromuscul. Disord.* 7:160–168.

Goebel, H. H., and Warlo, I. A. P. (2000). Progress in desmin-related myopathies. *J. Child Neurol.* 15:565–572.

Goebel, H. H., and Borchert, A. (2002). Protein surplus myopathies and other rare congenital myopathies. *Semin. Pediatr. Neurol.* 9:160–170.

Goebel, H. H., Brockmann, K., Bönnemann, C., Warlo, I. A. P., Hanefeld, F., Labeit, S., and Durling, H. J. (2004). Actin-related myopathy without any missense mutation in the *ACTA1* gene. *J. Child Neurol.* 19:149–153.

Goldfarb, L. G., Park, K.-Y., Cervenáková, S., Lee, H.-S., Vasconcelos, O., Nagle, J. W., Semino-Mora, C., Sivakumar, K., and Dalakas, M. C. (1998). Missense mutations in desmin associated with familial cardiac and skeletal myopathy (letter). *Nat. Genet.* 19:402–403.

Goldfarb, L. G., Vicart, P., Goebel, H. H., and Dalakas, M. C. (2004). Desmin myopathy. *Brain* 127:723–734.

Griggs, R. C., Askanas, V., DiMauro, S., Engel, A., Karpati, G., Mendell, J. R., and Rowland, L. P. (1995). Inclusion body myositis and myopathies. *Ann. Neurol.* 38:705–713.

Horowitz, S. H., and Schmalbruch, H. (1994). Autosomal dominant distal myopathy with desmin storage: a clinicopathologic and electrophysiologic study of a large kinship. *Muscle Nerve* 17:151–160.

Hübner, G., and Pongratz, D. (1982). Granularkörpermyopathie (sog. reducing body myopathy). *Pathologe* 3:111–113.

Kaindl, A. M., Rüschendorf, F., Krause, S., Goebel, H. H., Koehler, K., Becker, C., Pongratz, D., Müller-Höcker, J., Nürnberg, P., Stoltenburg-Didinger, G., Lochmüller, H., and Huebner, A. (2004). Missense mutations of *ACTA1* cause dominant congenital myopathy with cores. *J. Med. Genet.* 41:842–848.

Karpati, G., and Carpenter, S. (1992). Skeletal muscle in neuromuscular diseases. In: Rowland, L. P., and DiMauro, S. (eds.), *Myopathies—Handbook of Clinical Neurology*, Volume 18/62. Elsevier Science Publishers B.V., Amsterdam, pp. 1–48.

Kreis, T., and Vale, R., Eds. (1998). *Guidebook to the Cytoskeletal and Motor Proteins*, 2nd ed. Oxford University Press, Oxford.

Li, D., Tapscoft, T., Gonzales, O., Burch, P. E., Quinones, M. A., Zoghbi, W. A., Hill, R., Bachinski, L. L., Mann, D. L., and Roberts, R. (1999). Desmin mutation responsible for idiopathic dilated cardiomyopathy. *Circulation* 100:461–464.

Lindberg, C., Oldfors, A., and Hedström, A. (1990). Inclusion body myositis: peripheral nerve involvement: combined morphological and electrophysiological studies on peripheral nerves. *J. Neurol. Sci.* 99:327–338.

Lobrinus, J. A., Janzer, R. C., Kuntzer, T., Matthieu, J.-M., Pfend, G., Goy, J.-J., and Bogousslavsky, J. (1998). Familial cardiomyopathy and distal myopathy with abnormal desmin accumulation and migration. *Neuromuscul. Disord.* 8:77–86.

Martinsson, T., Oldfors, A., Darin, N., Berg, K., Tajsharghi, H., Kyllerman, M., and Wahlström, J. (2000). Autosomal dominant myopathy: missense mutation (Glu-706 to Lys) in the myosin heavy chain IIa gene. *Proc. Natl. Acad. Sci. USA* 96:14614–14619.

Masuzugawa, S., Kuzuhara, S., Narita, Y., Naito, Y., Taniguchi, A., and Ibi, T. (1997). Autosomal dominant hyaline body myopathy presenting as scapuloperoneal syndrome: clinical features and muscle pathology. *Neurology* 48:253–257.

Mendell, J. R., Sahenk, Z., Gales, T., and Paul, L. (1991). Amyloid filaments in inclusion body myositis. Novel findings provide insight into nature of filaments. *Arch. Neurol.* 48:1229–1234.

Mirabella, M., Alvarez, R. B., Bilak, M., Engel, W. K., and Askanas, V. (1996). Difference in expression of phosphorylated tau epitopes between sporadic inclusion-body myositis and hereditary inclusion-body myopathy. *J Neuropathol. Exp. Neurol.* 55:774–786.

Nakano, S., Engel, A. G., Waclawik, A. J., Emslie-Smith, A. M., and Busis, N. A. (1996). Myofibrillar myopathy with abnormal foci of desmin positivity. I. Light and electron microscopy analysis of 10 cases. *J. Neuropathol. Exp. Neurol.* 55:549–562.

Nakano, S., Shinde, A., Kawashima, S., Nakamura, S., Akiguchi, I., and Kimura, J. (2001). Inclusion body myositis. *Neurology* 56:87–93.

Nowak, K., Wattanasirichaigoon, D., Goebel, H. H., Wilce, M., Pelin, K., Donner, K., Jacob, R., Hübner, C., Oexle, K., Anderson, J., Verity, C. M., North, K. N., Iannaccone, S. T., Müller, C. R., Nürnberg, P., Muntoni, F., Sewry, C., Hughes, I.,

Sutphen, R., Lacson, A. G., Swoboda, K. J., Vigneron, J., Wallgren-Pettersson, C., Beggs, A. H., and Laing, N. B. (1999). Mutations in the skeletal muscle α–actin gene in patients with actin myopathy and nemaline myopathy. *Nat. Genet.* 23:208–212.

Oldfors, A., and Lindberg, C. (1999). Inclusion body myositis. *Curr. Opin. Neurol.* 12:527–533.

Oldfors, A., Darin, N., and Martinsson, T. (2002). Autosomal dominant myosin heavy chain IIa myopathy. In: Karpati, G. (ed.), *Structural and Molecular Basis of Skeletal Muscle Diseases*. ISN Neuropath Press, Basel, pp. 85–87.

Osborn, M., and Goebel, H. H. (1983). The cytoplasmic bodies in a congenital myopathy can be stained with antibodies to desmin, the muscle-specific intermediate filament protein. *Acta Neuropathol. (Berl.)* 62:149–152.

Parain, K., Herasse, M., Monnier, N., Lunardi, J., Marty, I., Romero, N. B., and Ferreiro, A. (2004). Proteins implicated in Ca^{2+} homeostasis: expression patterns in genetically characterized core myopathies (abstract G.P.8.01). *Neuromuscul. Disord.* 14:588.

Rappaport, L., Contard, F., Samuel, J. L., Delcayre, C., Marotte, F., Tomé, F., and Fardeau, M. (1988). Storage of phosphorylated desmin in a familial myopathy. *FEBS Lett.* 231:421–425.

Reichmann, H., Goebel, H. H., Schneider, C., and Toyka, K. V. (1997). Familial mixed congenital myopathy with rigid spine phenotype. *Muscle Nerve* 20:411–417.

Sabatelli, M., Bertini, E., Ricci, E., Salviati, G., Magi, S., Papacci, M., and Tonali, P. (1992). Peripheral neuropathy with giant axons and cardiomyopathy associated with desmin type intermediate filaments in skeletal muscle. *J. Neurol. Sci.* 109:1–10.

Sahgal, V., and Sahgal, S. (1977). A new congenital myopathy: a morphological, cytochemical and histochemical study. *Acta Neuropathol. (Berl.)* 37:225–230.

Schröder, R., Goudeau, B., Simon, M. C., Fischer, D., Eggermann, T., Clemen, C. S., Li, Z., Reimann, J., Xue, Z., Rudnik-Schöneborn, S., Zerres, K., van der Ven, P. F., Fürst, D. O., Kunz, W. S., and Vicart, P. (2003). On noxious desmin: functional effects of a novel heterozygous desmin insertion mutation on the extrasarcomeric desmin cytoskeleton and mitochondria. *Hum. Mol. Genet.* 12:657–669.

Schröder, J. M., Durling, H. J., and Laing, N. B. (2004). Actin myopathy with nemaline bodies, intranuclear rods, and a heterozygous mutation in *ACTA1* (Asp154Asn). *Acta Neuropathol. (Berl.)* 108:250–256.

Selcen, D., Krueger, B. R., and Engel, A. G. (2002). Familial cardioneuromyopathy with hyaline masses and nemaline rods: a novel phenotype. *Ann. Neurol.* 51:224–234.

Selcen, D., and Engel, A. G. (2003). Myofibrillar myopathy caused by novel dominant negative αB-crystallin mutations. *Ann. Neurol.* 54:804–810.

Selcen, D., and Engel, A. G. (2004). Mutations in myotilin cause myofibrillar myopathy. *Neurology* 62:1363–1371.

Selcen, D., Ohno, K., and Engel, A. G. (2004). Myofibrillar myopathy: clinical, morphological and genetic studies in 63 patients. *Brain* 127:439–451.

Sjöberg, G., Saavedra-Matiz, C. A., Rosen, D. R., Wijsman, E. M., Borg, K., Horowitz, S. H., and Sejersen, T. (1999). A missense mutation in the desmin rod domain is associated with autosomal distal myopathy, and exerts a dominant negative effect on filament formation. *Hum. Mol. Genet.* 8:2191–2198.

Sparrow, J. C., Nowak, K. J., Durling, H. J., Beggs, A. H., Wallgren-Pettersson, C., Romero, N. B., Nonaka, I., and Laing, N. B. (2003). Muscle disease caused by mutations in the skeletal muscle alpha-actin gene (*ACTA1*). *Neuromuscul. Disord.* 13:519–531.

Tajsharghi, H., Thornell, L.-E., Darin, N., Martinsson, T., Kyllerman, M., Wahlström, J., and Oldfors, A. (2002). Myosin heavy chain IIa gene mutation E706K is pathogenic and its expression increases with age. *Neurology* 58:780–786.

Tajsharghi, H., Thornell, L.-E., Lindberg, C., Lindvall, B., Henriksson, K. G., and Oldfors, A. (2003). Myosin storage myopathy associated with a heterozygous missense mutation in *MYH7*. *Ann. Neurol.* 54:494–500.

Vicart, P., Caron, A., Guicheney, P., Li, Z., Prévost, M.-C., Faure, A., Chateau, D., Chapon, F., Tomé, F., Dupret, J.-M., Paulin, D., and Fardeau, M. (1998). A missense mutation in the alpha-B crystallin chaperone gene causes a desmin-related myopathy. *Nat. Genet.* 20:92–95.

Viegas-Péquignot, E., Lin, Z. L., Dutrillaux, B., Apiou, F., and Paulin, D. (1989). Assignment of human desmin gene to band 2q35 by nonradioactive *in situ* hybridization. *Hum. Genet.* 83:33–36.

Vita, G., Migliorato, A., Baradello, A., Mazzeo, A., Rodolico, C., Falsaperla, R., and Messina, C. (1994). Expression of cytoskeletal proteins in central core disease. *J. Child Neurol.* 124:71–76.

Wallgren-Pettersson, C., Pelin, K., Nowak, K. J., Muntoni, F., Romero, N. B., Goebel, H. H., North, K. N., Beggs, A. H., Laing, N. B., and the International Consortium on Nemaline Myopathy (2004). Genotype-phenotype correlations in nemaline myopathy caused by mutations in the genes for nebulin and skeletal muscle α-actin. *Neuromuscul. Disord.* 14:461–470.

Yunis, E. J., and Samaha, F. J. (1971). Inclusion body myositis. *Lab. Invest.* 25:240–248.

19

Muscular Dystrophies and Protein Mutations

Mariz Vainzof and Mayana Zatz

Abstract

The neuromuscular disorders are a heterogeneous group of genetic diseases, causing a progressive loss of the motor ability. In the past decade, mutations in several genes have been identified, resulting in the deficiency or loss of function of different important proteins. Complementary biochemical and immunohistologic analyses have localized these proteins in several compartments of the muscle fiber: in the sarcolemmal muscle membrane (dystrophin, sarcoglycans, dysferlin, caveolin 3), extracellular matrix (α_2-laminin, collagen VI), in the sarcomere (telethonin, myotilin, titin), in the muscle cytosol (calpain 3, FRPR, TRIM32), and in the nucleus [emerin, lamin A/C, survival motor neuron gene (SMN) protein]. In the muscular dystrophies group, the allelic X-linked Duchenne and Becker forms are caused by mutations in the dystrophin gene. Among the Limb-girdle forms, seven autosomal dominant [limb-girdle muscular dystrophy (LGMD)1A to LGMD1G] and twelve autosomal recessive (LGMD2A-2L) are already known. The genes for the LGMD1 types were mapped respectively at 5q22-q34 (myotilin), 1q11–21 (lamin A/C), 3p25 (caveolin-3), 6q23, 5q31, 7q, and 4p21. Among the LGMD2 forms, the four clinically more severe types, the sarcoglycanopathies (LGMD2C-2E, SGpathies), mapped at 17q21, 4q12, 13q12, and 5q33, encode respectively for α-sarcoglycan (α-SG), β-SG, γ-SG, and δ-SG, which are glycoproteins of the sarcoglycan subcomplex of the dystrophin-glycoprotein complex (DGC). The other eight adult forms are LGMD2A at 15q (calpain 3), LGMD2B, at 2p31 (dysferlin), LGMD2G at 17q11–12 (telethonin), LGMD 2H at 9q31–33 (TRIM32), LGMD2I at 19q13.3 (Fukutin-related protein; -FKRP), and LGMD2J at 2q24 (titin), LGMD2K at 9q–4 (POMT1) and LGMD2L at 9q31 (Fukutin). Among the congenital forms, the congenital muscular dystrophy (CMD)1A with α_2-laminin deficiency (at 6q2) is the most common accounting for about 50% of the cases. Protein studies are of utmost importance for the elucidation of the pathophysiology of each genetic disorder involved. It also can be used for the differential diagnosis and to direct the search for gene mutations, mainly because, in addition to genetic heterogeneity, most of the known genes are very large and present a wide variability of pathogenic mutations. Genotype-phenotype correlation through the analysis of the effect of different mutations on protein expression and on phenotypic variability contributes to the understanding of gene function.

19.1. Introduction

Protein studies are of utmost importance for enhancing our understanding on genotype-phenotype correlations, as well as for diagnostic purposes, in particular in neuromuscular disorders where many genes are involved. The neuromuscular disorders are a heterogeneous group of genetic diseases, causing a progressive loss of the motor ability. In the past decade, mutations in several genes have been identified, resulting in the deficiency or loss of function of different important proteins. Biochemical and imunohistologic analyses have localized these proteins in several compartments of the muscle fiber: in the sarcolemmal muscle membrane (dystrophin, sarcoglycans, dysferlin, caveolin 3), extracellular matrix (α_2-laminin, collagen VI), in the sarcomere (telethonin, myotilin, titin), in the muscle cytosol (calpain 3, FRPR, TRIM32), and in the nucleus [emerin, lamin A/C, survival motor neuron gene (SMN) protein] (Table 19.1).

Table 19.1. Proteins involved in the different neuromuscular genetic diseases

Group of proteins	Proximal muscular dystrophy phenotypes	Mode of inheritance	Gene location	MIM
Plasma membrane				
Dystrophin	DMD/BMD	XL	Xp21	310200
α-Sarcoglycan	LGMD2D	AR	17q12	600119
β-Sarcoglycan	LGMD2E	AR	4q12	604286
γ-Sarcoglycan	LGMD2C	AR	13q12	253700
δ-Sarcoglycan	LGMD2F	AR	5q33-34	601287
Dysferlin	LGMD2B/Miyoshi myopathy	AR	2p13	253601
Caveolin 3	LGMD1C	AD	3p25	601253
Extracellular matrix				
Laminin α_2	Merosin-deficient CMD	AR	6q2	156225
Collagen VI	Ullrich's CMD, Bethlem myopathy	AR	21q22	254090
Cytosolic				
Calpain 3	LGMD2A	AR	15q15.1	253600
TRIM 32	LGMD2H	AR	9q31-34	254110
Myotubularin	Myotubular myopathy	XL	Xq28	310400
Involved in glycosylation:				
Fukutin	Fukuhama CMD/LGMD2L	AR	9q31-33	253800
FKRP	LGMD2I/CMD1C	AR	19q1	606596
POMFnT1	Muscle-eye-brain CMD	AR	1p32-34	253280
POMT1	Walker-Warburg CMD/LGMD2K	AR	9q34	236670
Sarcomeric proteins				
Titin	LGMD2J	AR	2q	
Myotilin	LGMD1A	AD	5q22-34	159000
Telethonin	LGMD2G	AR	17q11-12	601954
Actin	Nemaline myopathy	AD/AR	1q42	101800 256030
Tropomyosin 3	Nemaline myopathy	AD/AR	1q21-23	161800
Tropomyosin 2	Nemaline myopathy	AD	9p13	190990
Nebulin	Nemaline myopathy	AR	2q21-22	256030
Troponin T1	Nemaline myopathy	AR	19q13	190990
Nuclear proteins				
Emerin	Emery-Dreifuss MD	XL	Xq28	310300
Lamin A/C	LGMD1B	AD	1q11-21	159001
SMN	Spinal muscular atrophy	AR	5q11-13	253300

DMD, Duchenne muscular dystrophy; BMD, Becker MD; LGMD, limb-girdle MD; CMD, congenital MD; XL, X-linked inheritance; AD/AR, autosomal dominant/recessive inheritance. MIM: Reference number in McKusick VA, Mendelian Inheritance in Man. Catalogs of autosomal dominant, autosomal recessive, and X-linked phenotypes (see http://www.ncbi.nlm.nih.gov/Omim/).

Different approaches have been used to study proteins including quantification, localization by reaction with specific antibodies, and assays for specific biological or enzymatic activities. Protein studies by analyses of the status of different components of muscle fibers can be used for the differential diagnosis and elucidation of the pathophysilogy of these diseases. Here, we will focus on the most recent findings in proteins analysis and on our own studies in the Brazilian population.

At least 30 different forms of muscular dystrophy (MD) have been identified to date. Duchenne (DMD) and Becker (BMD) muscular dystrophies are allelic conditions caused by mutations in the dystrophin gene at Xp21 (Hoffman *et al.*, 1987; Koenig *et al.*, 1988) (see Table 19.1). The limb-girdle muscular dystrophies (LGMD) include a heterogeneous group of progressive disorders mainly affecting the pelvic and shoulder girdle musculature, ranging from severe forms with onset in the first decade of life and rapid progression, to milder forms of later onset and slower progression [for a review, see Bushby (1999); Zatz *et al.* (2000), (2003)]. Inheritance may be autosomal dominant (LGMD1) or recessive (LGMD2). During the past decade, 19 LGMD genes, 7 autosomal dominant (AD) and 12 autosomal recessive (AR), have been mapped. The AD forms are relatively rare and probably represent less than 5% of all LGMD (Bushby, 1999; Zatz *et al.*, 2000; 2003). The seven AD-LGMD forms are LGMD1A at 5q22, coding for the protein myotilin (Hauser *et al.*, 2000), LGMD1B at 1q11, coding for lamin A/C (Van der Kooi *et al.*, 1997), LGMD1C at 3p25, coding for caveolin-3 (McNally *et al.*, 1998; Minetti *et al.*, 1998), LGMD1D at 6q23 (Messina *et al.*, 1997), LGMD1E at 7q (Speer *et al.*, 1999), LGMD1F at 5q31 (Feit *et al.*, 1998), and LGMD1G at 4 p21, recently mapped in a Brazilian family (Starling *et al.*, 2004).

The protein products of the AR forms have been already identified. Four genes, localized at 17q21, 4q12, 13q12, and 5q33, respectively, code for α-sarcoglycan α-SG, β-SG, γ-SG, and δ-SG, which are glycoproteins of the sarcoglycan subcomplex of the dystrophin-glycoprotein complex (DGC) (Ervasti *et al.*, 1990; Yoshida & Ozawa, 1990). Mutations in these genes cause LGMD2C, 2D, 2E, and 2F, respectively, and constitute a distinct subgroup of LGMD (i.e., the sarcoglycanopathies). Among the clinically milder forms, LGMD2A, at 15q15.1, codes for calpain 3; LGMD2B, at 2p31, codes for dysferlin; and LGMD2G, at 17q11-12, codes for the sarcomeric telethonin. The Fukutin-related protein gene (FKRP), mapped at 19q13.3, was identified as the gene responsible for the LGMD2I form, as well as the severe form of congenital muscular dystrophy (MDC1C); the protein TRIM32 has been identified as the gene product of the LGMD2H form at 9q31-33. LGMD2J was described in the Finnish population as the result of autosomal recessive mutations in the titin gene [for reviews, see Bushby (1999); Zatz *et al.* (2000); Vainzof and Zatz (2003)]. LGMD2K is caused by mutations in the POMT1gene at 9q-34 and LGMD2L was associated to the fukutin gene at 9q31 very recently.

Protein studies can provide important information for the elucidation of the pathophysiology of each genetic disorder involved. It also can be used for the differential diagnosis and to direct the search for gene mutations, mainly because, in addition to genetic heterogeneity, most of the known genes are very large and presenting a wide variability of pathogenic mutations. Genotype-phenotype correlation through the analysis of the effect of different mutations on protein expression and on phenotypic variability contributes to the understanding of gene function (Anderson, 2001; Vainzof *et al.*, 2003b).

19.2. Sarcolemmal Proteins

19.2.1. Xp21 Muscular Dystrophies

Dystrophin, the site of the primary defect in DMD/BMD (Hoffman *et al.*, 1987; Koenig *et al.*, 1988), is a large 427-kDa rod-shaped subsarcolemmal protein. Its amino terminus domain binds to actin, and the cystein-rich and carboxyl terminus domains link dystrophin to a complex

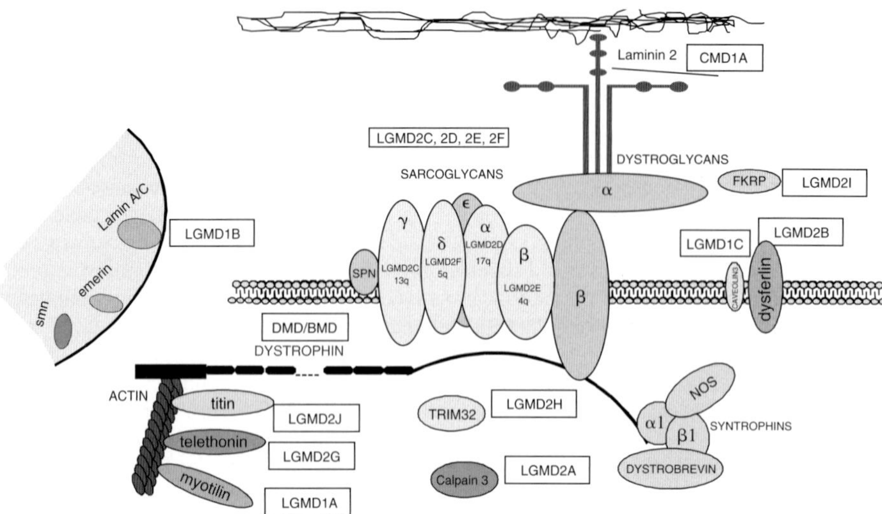

Figure 19.1. Schematic representation of proteins from the sarcolemma, the sarcomere, the cytosol, and the nucleus involved in the process of muscle degeneration in neuromuscular disorders. DMD, Duchenne muscular dystrophy; BMD, Becker MD; LGMD, limb-girdle MD; CMD, congenital MD.

of glycoproteins in the sarcolemma, the dystrophin-glycoprotein complex (DGC), which is vital for normal muscle function (Campbell and Kahl, 1989; Ervasti *et al.*, 1990; Yoshida and Ozawa, 1990). The DGC forms a bridge across the muscle membrane, between the inner cytoskeleton (dystrophin) and the basal lamina (merosin). It has been suggested that DGC stabilizes the sarcolemma and protects muscle fibers from long-term contraction-induced damage and necrosis. The DGC consists of dystroglycan (DG), sarcoglycan (SG), and syntrophin/dystrobrevin subcomplexes (Fig. 19.1). In addition to having a mechanical and structural function, the DGC has been also suggested to play a role in cellular communication (Hack *et al.*, 2000). Furthermore, it interacts with the sarcomeric network by binding dystrophin to F-actin [for a review, see Cohn and Cambell (2000)].

A complete analysis of the effect of dystrophin deficiency in the muscle is detailed in Chapter 20 of this volume.

Clinically, DMD patients show delayed onset of walking, a progressive weakness with difficulty in running, climbing stairs and jumping; frequent falls, and marked calf hypertrophy. The progressive muscle degeneration causes the loss of ambulation at about age 10 and death usually before the end of the second decade due to respiratory or cardiac failure. The milder BMD variant shows a less severe but more variable phenotype, ranging from a slightly less severe DMD-like condition to a very mild condition. Some patients may be able to walk throughout their lives. A patient is classified as DMD when wheelchair-dependent before the age of 13 years, and as BMD when able to walk beyond age 16 [for a review, see Bakker and Van Ommen (1998)].

The dystrophin gene is the largest human gene characterized to date, composed by 2.4 Mb and the major muscle transcript consists of 79 exons, spliced in a 14-kb mature RNA. The exons size varies between 30 and 300 bp and these are sparsely distributed among very large introns, which are responsible for the overall size of the gene. The coding region encompasses ~11,000 bp encoding a 3685-amino-acid protein, with a molecular weight of 427 kDa (Monaco *et al.*, 1986; Koenig *et al.*, 1987). Alternative splicing results in several different and/or shorter transcripts, and the diverse isoforms are expressed differentially in the following tissues: Dp427m (muscle), Dp427c (cortical), Dp427p (Purkinje cells), Dp260 (retina), Dp140 (central nervous system), Dp116 (perifpheral nerve), Dp71 (liver, brain, etc.), and Dp40 (early embryonic cells) [revision in den Dunnen (2001)].

The complete spectrum of mutations has not been fully defined, but the largest category of mutations in DMD/BMD patients are intragenic deletions, which facilitates molecular diagnosis. Approximately 60–65% of the mutations that cause DMD/BMD are large deletions, 5–10% are duplications, which can include one or more exons, while the remainder are point mutations or small rearrangements, which can be either nonsense or located at splice sites. All of these mutations lead to deficiency of the protein dystrophin in muscle (Hoffman *et al.*, 1987). In patients with deletions, the clinical differences observed between DMD and BMD are due to the effect of the mutations in the coding reading frame. In-frame deletions produce a smaller than normal, semi-functional protein, while frame-shifting deletions generally produce a few incorrect amino acids before a premature stop codon is generated. Frameshift mutations or stop codons tend to result in non-sense-mediated decay of mutant mRNA, and any grossly abnormal protein that might be produced is usually degraded rapidly (Frischmeyer and Dietz, 1999). Patients with intermediate severity can present both frame-shifting or in-frame mutations, the latter one being in functionally more important regions than the typical BMD deletions, while the frame-shifting mutations in these patients may show a higher level of somatic restoration of the reading frame via exon skipping.

The protein dystrophin can be detected in the sarcolemmal muscle membrane through imunohistochemical techniques, labeling sections from muscle biopsy with antibodies to different dystrophin domains. In normal muscle, a clear and uniform labeling at the periphery of all fibers can be observed, while in muscle from DMD patients the protein is absent or very reduced. A variable proportion (4–30%) of dystrophin-positive isolated or grouped fibers (called revertent fibers) are observed in most DMD patients, mainly with the N-terminal antibody. This small amount of dystrophin can be observed as faint bands by Western blotting but is not correlated with the clinical course (Nicholson *et al.*, 1990; Vainzof *et al.*, 1990; 1991). This observation should be taken into account before assessing the effect of any therapeutic trial with the replacement of dystrophin through gene or cell therapy. Intermediate DMD/BMD muscle demonstrates weak labeling on the majority of fibers, or variation in the intensity of the labeling. Usually, no dystrophin labeling is observed in DMD through Western blot methodology, although faint bands of several smaller sized products can be detectable associated with some particular mutations. In BMD patients, bands of abnormal size or abundance can be detected and are very useful for diagnostic purpose (Nicholson *et al.*, 1990; Vainzof *et al.*, 1992; 1995a). In the majority of the DMD/BMD cases, there is a correlation between the amount of dystrophin and the severity of the phenotype (Hoffman *et al.*, 1988; Nicholson *et al.*, 1990).

Approximately one third of DMD cases are caused by new mutations while the remaining two thirds are inherited through carrier mothers who are usually asymptomatic. Female relatives of DMD/BMD patients are thus at risk of carrying the mutations detected in the index case of the family. Approximately 60–80% of DMD/BMD carriers have elevated serum creatine kinase (CK) levels, and before the molecular era this test was the only one available for carriers identification. Currently, using molecular techniques it is possible to screen for the same mutation identified in the affected boy, the females from the families, allowing a more accurate genetic counseling. In the females, however, DNA analysis must be quantitative because of the presence of the mutation in a heterogygose state. The identification of the mutation in the index case also permits one to offer to the family an accurate prenatal diagnosis in a very early stage of pregnancy through DNA analysis in a chorionic villus biopsy done at the 10th to 11th week of gestation.

A protein homologous to dystrophin, utrophin, is encoded by a gene on chromosome 6 (Khurana *et al.*, 1990). The expression of utrophin is high in neuromuscular junction, but in the immature and in the dystrophic muscle, utrophin can be also detected in the sarcolemma (Helliwell *et al.*, 1992, Vainzof *et al.* 1995c). Although very high levels of utrophin have been found in the muscle from patients with severe phenotype of DMD (Vainzof *et al.*, 1995c), the overexpression of utrophin is considered as a potential possibility of therapy for DMD.

19.2.2. Sarcoglycanopathies

The sarcoglycanopathies are usually the most severe forms of LGMD. In Brazilian patients, they account for about 32% of classified LGMD-affected patients but their relative proportions vary in different populations (Zatz *et al.*, 2003). The four known components of the SG complex include α-SG, β-SG, γ-SG, and δ-SG. They are assembled in a complex that is inserted into the membrane. Mutations in one of the four SG proteins cause LGMD2C, 2D, 2F, and 2E, respectively. Many different mutations have already been identified in all the sarcoglycan genes, including missense, splicing, nonsense, small and large gene deletions (listed at http://www.dmd.nl).

Severe Duchenne-like presentations tend to be more common among these patients, with onset occurring early in childhood and confinement to a wheelchair before the age of 16. In the Brazilian population, the majority of the severely affected LGMD patients have a Sgpathy (Vainzof *et al.*, 1999). Nevertheless, milder courses have also been described in LGMD 2C, 2D, and 2E patients. An intriguing fact is that a few frequent alterations, like c.229C > T (R77C) in the α-SG (LGMD 2D) and c.521delT in the γ-SG genes, have been associated with both severe and mild forms. Therefore, the establishment of precise genotype-phenotype correlations has proved to be a challenge for most researchers working on sarcoglycanopathies (Vainzof *et al.*, 1999; Moreira *et al.*, 2003). Our group recently reported the spectrum of mutations in 35 Brazilian sarcoglycanopathy families, and the most common ones were α-SG: exon 3, C229C > T; β-SG: exon 3/4; γ-SG: exon 6, C521delT; and δ-SG: exon 7, C656delC, corresponding with 78%, 75%, 100%, and 80% of the disease alleles of the respective LGMD forms. Because obtaining a muscle biopsy in patients in advanced stages maybe very difficult, a multiplex DNA test for the concomitant screening for these five exons was established. It allows one to confirm the diagnosis in about 60% of patients with a clinical course compatible with a sarcoglycanopathy (Gouveia *et al.*, 2006).

In most muscle biopsies from patients with a sarcoglycanopathy, the primary loss or deficiency of any one of the four sarcoglycans, β-SG and δ-SG in particular, leads to a secondary deficiency of the whole subcomplex (Vainzof *et al.*, 1996; Bushby, 1999; Bonnemann, 1999; Hack *et al.*, 2000; Zatz *et al.*, 2000). However, exceptions may occur, such as the deficiency of γ-SG with a partial preservation of the other three SG in LGMD2C (Vainzof *et al.*, 1996) or the partial deficiency of only α-SG with the retention of the other three in LGMD2D (Bonnemann, 1999; Vainzof *et al.*, 2000b). The observation of a complete deficiency of one sarcoglycan with partial deficiency of the others may help to indicate which gene should be first screened for mutations.

Patients with primary sarcoglycan mutations and deficiency of the protein may have a secondary reduction in dystrophin, particularly patients with primary γ-SG deficiency. This suggests that γ-SG might interact more directly with dystrophin (Vainzof *et al.*, 1996). Therefore, a relative dystrophin deficiency may occur in non-Becker MD, and this should be taken into consideration for differential diagnosis.

19.2.3. Dysferlinopathies

Dysferlin is a ubiquitously expressed 230-kDa molecule localized in the periphery of muscle fibers, linked to the sarcolemmal membrane (Anderson *et al.*, 1999). Dysferlin can be detected also in blood, skin, and in chorionic villus biopsies (Vainzof *et al.*, 2001; Ho *et al.*, 2002).

Two distinct phenotypes are associated with mutations in this gene: Miyoshi myopathy, with a predominately distal muscle wasting (Liu *et al.*, 1998), and LGMD2B, with a proximal weakness (Bashir *et al.*, 1998). So far, due to the large size of the dysferlin gene (55 exons), only a few mutations have been identified, and no apparent hot spot for mutations has been found. Therefore, at this moment, muscle protein analysis is more effective than DNA analysis for diagnostic purposes of LGMD2B.

Protein analyses in LGMD2B have shown a total deficiency of dysferlin both by immunofluorescence and Western blotting. Although a partial deficiency has been reported in LGMD2B patients (Anderson *et al.*, 1999), dysferlin deficiency seems to be specific to LGMD2B in our patients and has not been seen as a secondary effect in other forms of MD (Vainzof *et al.*, 2001).

A normal localization and a normal molecular weight (MW) for dysferlin have been found in DMD and sarcoglycanopathies, suggesting no interaction between dysferlin and the DGC. However, a possible association between dysferlin and caveolin 3 has been described (Matsuda *et al.*, 2001). Mutations in the CAV-3 gene lead to a typical loss of caveolin immunolabeling, suggesting a dominant negative effect (Minetti *et al.*, 1998), and have been associated with the rare form of autosomal dominant (AD) muscular dystrophy LGMD1C, as well as rippling muscle disease.

19.3. Cytosolic Proteins

19.3.1. Calpainopathy

Calpain 3, a muscle-specific 94-kDa calcium-activated neutral protease 3, binds to titin. It is a cysteine protease that plays a role in the disassembly of sarcomeric proteins, but it may also have a regulatory role in the modulation of transcription factors. The loss of its function leads to activation of other proteases.

Mutations in the calpain 3 gene at 15q15 cause the calpainopathy LGMD2A (Richard *et al.*, 1995). LGMD2A is the most frequent form of LGMD in several populations, accounting for about 30% of the identified forms. LGMD2A patients present a wide range of distinct pathogenic mutations distributed along the entire length of the calpain 3 gene. However, some recurrent mutations are found more frequently in some populations such as the Basque, the Japanese, Brazilians, and Italians (Paula *et al.*, 2002; Zatz *et al.*, 2003; Fanin *et al.*, 2004). The study of calpain 3 protein in muscle at this time can only be carried out by Western blotting because the antibodies available do not react on sections. LGMD2A patients can show total, partial, or, more rarely, no calpain 3 deficiency at all, but no direct correlation has been observed between the amount of calpain and the severity of the phenotype. Very low levels or no expression of calpain 3 was observed in European and Brazilian patients with a clinical course varying from mild to severe (Anderson *et al.*, 1998; Vainzof *et al.*, 2000a). LGMD2A patients with missense mutations may present three faint or almost normal bands of 94-kDA calpain (Vainzof *et al.*, 2000a; Paula *et al.*, 2002), suggesting that some mutations may affect protein function without eliminating the protein from muscle. In fact, in a screening of 58 patients with primary LGMD2A, it has been recently shown that 80% had calpain 3 deficiency in the muscle while 20% had normal amount of the protein (Fanin *et al.*, 2004).

Histologic analysis of muscle biopsies from some asymptomatic and early stage patients with LGMD2A showed a consistent but unusual pattern with isolated fascicles of degenerating fibers in an almost normal muscle. These findings suggest that a peculiar pattern of focal degeneration occurs in calpainopathy, independent of the type of mutation or the amount of calpain 3 in the muscle (Vainzof *et al.*, 2003a). This has been very recently confirmed in calpain 3 knockout mice, which presented similar atrophic features, small foci of muscular necrosis, and abnormal sarcomere formation (Kramerova *et al.*, 2004).

Some unexpected results were also observed in calpainopathy, such as the confirmation of LGMD2A in a family with five affected sibs with a clinical diagnosis of Kugelberg-Welander (KW) type spinal muscular atrophy (Starling *et al.*, 2003). Calpain 3 analysis in the muscle showed a total deficiency of the protein and the screening of mutations in the calpain 3 gene disclosed the R769Q mutation (Amish mutation) in homozygosity. In addition, normal serum CK was also found in unrelated Brazilian patients with calpain mutations identified on both alleles. These results suggest that the spectrum of phenotypic variability in calpainopathy may be broader than suspected. Therefore, screening for mutations in the calpain 3 gene should be done in any atypical patients with a suspected diagnosis of KW with no detected deletion in the SMN gene, even if the serum CK is within normal levels (Starling *et al.*, 2003).

The analysis of calpain 3 in other forms of LGMD has shown a normal amount in sarcoglycanopathies (Anderson *et al.*, 1998; Vainzof *et al.*, 2000a). In addition, normal SG proteins were observed in calpain 3–deficient LGMD2A patients (Vainzof *et al.*, 1996), suggesting no direct relationship between calpain 3 and the SG complex. The observation of normal calpain bands in LGMD2G patients also suggest no correlation with telethonin (Vainzof *et al.*, 2002b).

However, an unexpected reduction of calpain 3 was observed in LGMD2B patients, suggesting a possible association between calpain 3 and dysferlin, which requires further study (Anderson *et al.*, 2000; Vainzof *et al.*, 2000a; Vainzof *et al.*, 2001). Subsequently, several studies have shown that a secondary reduction of calpain 3 may occur in other myopathies such as LGMD2I (Bushby and Beckmann, 2002) and LGMD2J (Haravuori *et al.*, 2001). Therefore, as deficiency of calpain 3 in the muscle can be secondary, a recent study based on protein analysis and molecular screening for mutation showed that the estimated *a priori* probability of having LGMD2A depends on the amount of calpain 3 in the muscle, varying among 84%, when the deficiency is total, to about 20%, when the amount of the protein is reduced in the muscle (Fanin *et al.*, 2004).

19.3.2. Fukutin-Related Protein

A Fukutin-related protein (FKRP) was recently identified, with sequence similarities to a family of proteins involved in the glycosylation of cell surface molecules (Brockington *et al.*, 2001). Mutations in the FKRP gene were identified in both LGMD-2I and congenital muscular dystrophy type 1C (CMD-1C) (Brockinton *et al.*, 2001a, 2001b), both previously mapped to an identical region on chromosome 19q13.3. CMD is characterized by onset of symptoms within the first few months of life, and in the 1C form there is inability to walk. The LGMD-2I form is characterized by a high variability in clinical course, with a spectrum of phenotypes ranging from a Duchenne-like disease course including cardiomyopathy to milder phenotypes with a slow progression (Brockinton *et al.*, 2001b).

Secondary protein abnormalities are common in this group of diseases, including a reduction of laminin α_2 labeling, mainly on immunoblots, and reduced Western blot labeling for laminin β_1. A variable reduction of α-DG expression was also observed in skeletal muscle biopsies from affected individuals, with a reduction in molecular weight observed by immunoblotting, which could indicate an association with a glycosylation defect. The analysis of 10 proteins in 13 molecularly classified Brazilian LGMD2I patients (Paula *et al.*, 2003) showed a significant reduction of α_2-laminin in almost all, with no direct correlation with the secondary reduction of α-DG or the type of mutation (Yamamoto *et al.*, 2003).

Therefore, until a proper antibody against FKRP is developed, muscle protein analysis showing a relative secondary deficiency of α-DG and laminin α_2 might suggest but does not allow a diagnosis of LGMD2I without molecular analysis.

19.3.3. Myotubularin

Myotubularin is the protein product of the X-linked MTM1 gene. Mutations in this gene cause the most severe neonatal (and lethal) form of myotubular myopathy (MTM) [for a review, see Vainzof (2001)]. Myotubular myopathy is a congenital muscle disorder characterized by severe hypotonia, generalized muscle weakness, and respiratory distress in affected males. The majority of the patients die during the first year of life.

Muscle histologic analysis shows the presence of small, rounded muscle fibers, with centrally located nuclei surrounded by a halo devoid of contractile elements, resembling fetal myotubes. It has been suggested that impaired muscle maturation underlies the pathology. About 192 different pathogenic mutations have been described in the MTM1 gene. The classic severe form is usually associated with a complete loss of protein function, due to truncating mutations or missense mutations in the phosphatase domain of the protein.

The myotubularin protein is cytoplasmic and associated with the plasma membrane. Through computer-based analysis of the predicted protein structure, myotubularin has been classified as a member of the protein tyrosine phosphatase (PTP) family. It has been suggested that myotubularin is required for muscle cell differentiation and that it exerts an important regulatory function during myogenesis.

19.4. Sarcomeric Proteins

The sarcomere is the unit of skeletal and cardiac muscle contraction. In the past few years, many studies have focused on the role of skeletal and cardiac muscle proteins (Gregorio *et al.*, 1999) and mutations in genes coding for several sarcomeric proteins such as telethonin (Moreira *et al.*, 2000) and myotilin (Hauser *et al.*, 2000) have been detected in different forms of muscular dystrophies. In addition, mutations in five different sarcomeric genes, actin, tropomyosin 3 and 2, nebulin, and troponin T1, have been identified as the cause of nemaline myopathy (NM), a congenital myopathy characterized by the presence of rod-like bodies inside muscle fibers [for a review, see Wallgreen-Pettersson and Laing (2003)]. Clinically, NM is associated with muscle weakness, respiratory and feeding difficulties at birth, an elongated face, and skeletal deformities. The autosomal recessive (AR) form appears to be the most common one and is caused by mutations in the nebulin gene. Nebulin is a large protein of 800 kDa, and its C-terminal integrates into the Z-disk while the N terminal projects into the I band. Nebulin may provide a molecular template to regulate the length of the actin filament. Protein studies on most patients with mutations in the nebulin gene has shown the presence of the protein, mainly with the C-terminal antibody (Gurgel-Giannetti *et al.*, 2001; Wallgren-Petterson and Laing., 2003). We have however identified a patient affected by nemaline myopathy, with a deficiency of only the SH3 domain of nebulin by Western blot analysis (Gurgel-Giannetti *et al.*, 2002).

More recently, homozygous mutations in the titin gene previously known to be responsible for AD tibial muscular dystrophy, a form prevalent in Finland, were found to cause a proximal limb-girdle muscular dystrophy, LGMD2J. A secondary reduction in calpain 3 was observed in these patients (Haravuori *et al.*, 2001).

The identification of new protein components of the sarcomere, such as ZASP, FATZ, and Ankrd2, using the two-hybrid or other protein interaction techniques, is continuously increasing. The role of most of these proteins is still unknown. However, so far no disease has been associated with any of them, suggesting that total deficiency is either very rare or incompatible with life, and therefore that they play an essential role in the constitution of the sarcomere.

19.4.1. Myotilinopathy

Myotilin is a novel 57-kDa cytoskeletal protein expressed in skeletal and cardiac muscle that colocalizes with α-actinin in the sarcomeric I-band. Its N-terminal sequence is unique, but the C-terminal half contains two Ig-like domains homologous to domains in titin. Titin acts as a molecular ruler for the assembly of the sarcomere by providing spatially defined binding sites for other sarcomeric proteins. Mutations in the myotilin gene cause the AD form of LGMD1A, but muscle myotilin protein is apparently normal in affected patients (Hauser *et al.*, 2000).

We have recently studied a 14-year-old girl with a clinical phenotype of LGMD and a deficiency of myotilin in the muscle. However, no mutation in the coding region of the myotilin gene was identified, suggesting that it may lie in noncoding regions or that the deficiency of myotilin may be secondary to some other primary (still unidentified) defect (Vainzof *et al.*, 2004).

Very recently, missense mutations were found in about 10% of patients with myofibrillar myopathy, all in the serine-rich exon 2 of MYOT, where the two previously identified LGMD1A mutations are located. As the patients showed peripheral neuropathy, cardiomyopathy, and distal weakness greater than proximal weakness, these phenotypes are now part of the spectrum of myotilinopathy (Selcen and Engel, 2004).

19.4.2. Telethoninopathy

Telethonin is a sarcomeric protein of 19 kDa present in the Z disk of the sarcomere of striated and cardiac muscle (Valle *et al.*, 1997). Mutations in the telethonin gene cause LGMD2G (Moreira *et al.*, 2000). Telethonin is one of the substrates of the serine kinase domain of titin. The specific function of telethonin and its interaction with other muscle proteins is still unknown.

Protein analysis in six Brazilian patients from four unrelated families showed that the first one was a compound heterozygote for two null mutations while the other three were homozygotes for the same frameshift mutation in the telethonin gene present in one of the alleles from the first family. Muscle biopsies from patients belonging to the four families showed a deficiency of the telethonin protein in muscle. We have recently screened 200 additional patients with clinical diagnosis of LGMD for mutation in the two exons of the telethonin gene. We identified three more families, homozygotes for the same mutations (Lima *et al.*, 2004), confirming thus that LGMD2G is relatively rare also in our population, accounting for about 4% of the classified AR-LGMD forms.

Although no other patient homozygous for telethonin mutations has been reported to date in other populations, telethonin mutations in one allele have been found among Italian patients (V. Nigro, personal communication). This is not surprising as the first identified Brazilian telethonin family is of Italian background.

Additional protein studies on LGMD2G patients have shown normal expression of dystrophin, sarcoglycans, dysferlin, calpain 3, and titin. However, a reduction in the amount of the recently identified sarcomeric protein myopaladin was observed in muscle fibers from the LGMD2G patients, suggesting an interaction between telethonin and myopaladin (Vainzof *et al.*, 2002a). Furthermore, immunofluorescence analysis for α-actinin-2 and myotilin showed a normal cross-striation pattern, suggesting that at least part of the Z-line of the sarcomere is preserved. Ultrastructural analysis confirmed the maintenance of the integrity of the sarcomeric architecture. Therefore, mutations in the telethonin gene do not seem to alter the sarcomere integrity (Vainzof *et al.*, 2002b).

The analysis of telethonin in muscle biopsies from patients with LGMD2A, LGMD2B, sarcoglycanopathies, and DMD showed normal localization, suggesting that the deficiency of calpain, dysferlin, sarcoglycans, and dystrophin does not seem to alter telethonin expression

(Vainzof *et al.*, 2002b). Telethonin was clearly present in the rods in muscle fibers from patients with NM, confirming its localization in the Z-line of the sarcomere.

19.5. Nuclear Proteins

Emerin is a membrane-spanning 34-kDa serine-rich protein that is a component of the nuclear lamina associated with the nuclear envelope. Mutations in the emerin gene cause the X-linked Emery Dreifuss dystrophy (Bione *et al.*, 1994). One postulated function of emerin is to stabilize the nuclear membrane and provide structural support in an environment in which the nucleus is subjected to mechanical stress.

Lamin A/C belongs to a family of nuclear laminar proteins and colocalizes with emerin. It is coded by a gene on chromosome 1q11-23. Abnormalities in the gene for lamin A are associated with an autosomal dominant illness that is clinically identical to emerin deficiency, the LGMD1B form (Bonne *et al.*, 1999).

The survival motor neuron gene (SMN) protein is the product responsible for the spinal muscular atrophy gene (SMA), an autosomal recessive neuromuscular disorder characterized by degeneration of motor neurons of the spinal cord. In humans, the SMN gene is found as two almost identical copies on chromosome 5, denoted SMN1 (telomeric) and SMN2 (centromeric), differing by just five nucleotides that undergo alternative splicing of exons 5 and 7 (Lefebvre *et al.*, 1995). SMN is both a cytoplasmic protein and a nuclear protein located in structures called "gems" (Young *et al.*, 2001). Studies characterizing the SMN protein are emerging.

19.6. Extracellular Matrix Proteins

Laminin 2 is a constituent of the basal lamina that links to dystroglycan and that provides structural support in the extracellular matrix. It is composed of three chains: alpha 2, beta 1, and gamma 1. Laminin α_2 deficiency due to mutations in the LAMA2 gene at 6q2 is the cause of the AR congenital muscular dystrophy 1A (CMD-1A) (Hillaire *et al.*, 1994). Laminin α_2 is totally deficient in muscle biopsies from patients with the severe congenital dystrophy phenotype. Partial deficiencies have been described in patients with heterogeneity in the clinical picture (Vainzof *et al.*, 1995b; Sewry *et al.*, 1997). A partial deficiency of only the 300-kDa more sensitive α_2-laminin antibody was also detected in patients with the classical LGMD clinical course (Naom *et al.*, 2000). In addition, secondary deficiency of α_2-laminin has been found in CMD1C and LGMD2I (Brockington *et al.*, 2001a, 2001b). Screening for mutations in the LAMA2 gene can help elucidate the primary or secondary etiology of these deficiencies.

As the LAMA2 gene is very large and heterogeneous for mutations, the analysis of the protein in muscle biopsy has been largely used for diagnosis. In addition to the basal lamina of muscle fibers, α_2-laminin is also detected in skin biopsies as well as in chorionic villi and therefore is a very useful marker and a powerful tool for prenatal diagnosis, as reported in several families with a previously affected CMD1A child (Muntoni *et al.*, 1995; Yamamoto *et al.*, 2004).

In addition to α_2-laminin deficiency, marked alterations in the glycosylation of α-dystroglycan have been very recently associated with some forms of severe congenital MD (Fukuyama MD, Walker-Warburg, muscle-eye-brain, and FKRP forms). Mutations in genes with glucosyltransferase activity have been identified as responsible for these diseases, suggesting that abnormal processing of α-dystroglycan may be central to the pathogenesis of a significant number of genetic conditions (Muntoni *et al.*, 2002). Recessive mutations in the COL6A2 gene cause

Ullrich CMD, a severe muscle weakness disease associated with multiple joint contractures and deficiency of collagen VI in tissues. On the other hand, dominant mutations in the same gene may cause Bethlem myopathy, a skeletal muscle myopathy with a large spectrum of clinical variability also associated with contractures. Therefore, AR or AD mutations in the same gene may cause clinically different diseases (Pepe *et al.*, 2002).

19.7. Protein Study for Differential Diagnosis

Testing for defective protein expression is a powerful tool for deciding where to start the search for gene mutations, particularly in the more prevalent forms of neuromuscular disorders (Vainzof *et al.*, 2003b; Zatz *et al.*, 2003). Male patients with suspected X-linked dystrophy are first tested for deletions in the dystrophin gene by DNA analysis in blood samples, because this is a less invasive test. The identification of a gene deletion will confirm the diagnosis of DMD/BMD in about 60% of cases. If no deletion is detected, a muscle biopsy is required for the analysis of muscle proteins in an attempt to elucidate the possible diagnosis. Dystrophin is the first protein to be tested, using N- and C-terminal epitope antibodies. A significant deficiency of dystrophin is suggestive of DMD or severe BMD. Complementary Western blot analysis will allow one to estimate the amount of protein present and a possible prognosis based on dystrophin quantity and the presence or absence of the different domains.

If an autosomal form is suspected, complementary studies for α-SG and γ-SG are conducted. If a deficiency of any of the SG is detected, additional studies for β-SG and δ-SG are done to confirm a possible sarcoglycanopathy. For the final diagnosis of a sarcoglycanopathy, mutation screening should start with α-SG because this is the cause of the most prevalent sarcoglycanopathy. If γ-SG is predominately absent, a γ-sarcoglycanopathy should be suspected first.

Another possibility for male patients who are dystrophin-positive but with early contractures and cardiomyopathy could be Emery-Dreyfuss MD. In such cases, immunofluorescence analysis for emerin should be done, and a negative nuclear labeling would confirm the diagnosis.

Additional immunofluorescence analyses for α_2-laminin should be done in more severely affected patients. Partial α_2-laminin deficiencies might be associated with a diagnosis of MDC1C or LGMD2I, but the deficiency needs to be confirmed by mutational analysis. Western blot analysis for α-dystroglycan glycosylation can also add some information about new genes related to alterations in this process, such as LGMD2I, and other forms of congenital MD.

In adult onset MD forms, multiplex Western blot analysis for dystrophin, calpain 3, and dysferlin, as well as telethonin analysis by IF has proved to be very useful for preliminary diagnosis of muscular dystrophy. A dystrophin deficiency is first suggestive of Xp21 MD dystrophy. However, small reductions can also be observed in sarcoglycanopathies and LGMD2I. Calpain 3 may occur as a secondary effect of other forms of MD, but dysferlin and telethonin deficiency have been shown so far to be the consequence of their respective primary gene defect. Therefore the absence of dysferlin or telethonin in a muscle biopsy strongly suggests a diagnosis of LGMD2B or LGMD2G, respectively.

When a deficiency of calpain 3 is detected and no mutations in the CAPN3 gene are found, the possibility of dysferlin, FKRP, and titin-associated conditions should be considered, mainly if the deficiency is partial.

If no protein or DNA alterations are found in patients with a clinical diagnosis of LGMD, the possibility of SMA should be considered due to the clinical overlap between these diseases. In addition, the possibility of a calpainopathy should be considered in atypical SMA patients with no SMN deletion.

For the remaining proteins described here (mainly the sarcomeric ones), for which no clear deficiency has been previously described associated with their respective primary defect, the available protein tests are not useful for diagnostic purposes. In this respect, the development of new methodologies for the identification of subtle protein alteration are of utmost importance, not only for diagnosis but also for understanding interactions among proteins in order to elucidate genotype-phenotype interaction in neuromuscular diseases.

Acknowledgments

The collaboration of the following persons is gratefully acknowledged: Dr. Maria Rita Passos-Bueno, Dr. Eloisa S. Moreira, Dr. Rita C.M. Pavanello, Dr. Ivo Pavanello, Dr. Acary S.B. Oliveira, Dr. Edmar Zanotelli, Dr. Helga Silva, Dr. Lydia U. Yamamoto, Dr. Flavia de Paula, Telma Gouveia, Luciana Luchessi, Viviane P. Muniz, Bruno Lima, Natassia Vieira, Alessandra Starling, Marta Cánovas, Antonia Cerqueira, and Constancia Urbani. We would also like to thank the following researchers, who kindly provided specific antibodies: L.V.B. Anderson, A. Beggs, J. Chamberlain, L.M. Kunkel, C. Bonnemann, E.E McNally, V. Nigro, Z. Labeit, O. Carpen, G. Faulkner, G. Valle, E. Hoffmann, R. Worton, E. Ozawa, K.Campbell, N. Thi Man, and G. Morris. Very special thanks are due to Dr. Zubrzycka-Gaarn (in memoriam), Dr. Peter Ray, and Dr. Ron Worton.

References

Anderson, L.V.B. (2001). Multiplex western blot analysis of the muscular dystrophy proteins. In: Bushby K.M.D. & Anderson L.V.B. (Editors), *Muscular Dystrophy: Methods and Protocols*. (Number 43 in the Methods in Molecular Medicine series.) Humana Press, Totowa, NJ, pp. 369–386.

Anderson, L.V., Davison, K., Moss, J.A., Richard, I., Fardeau, M., Tome, F.M., Hubner, C., Lasa, A., Colomer, J. Beckmann, J.S. (1998). Characterization of monoclonal antibodies to calpain 3 and protein expression in muscle from patients with limb-girdle muscular dystrophy type 2A. *Am. J. Pathol.* 153: 1169–1179.

Anderson, L.V., Davison, K., Moss, J.A., Young, C., Cullen, M.J., Walsh, J., Johnson, M.A., Bashir, R., Britton, S., Keers, S., Argov, Z., Mahjneh, I., Fougerousse, F., Beckmann, J.S., Bushby, K.M. (1999). Dysferlin is a plasma membrane protein and is expressed early in human development. *Hum. Mol. Genet.* 8: 855–861.

Anderson, L.V., Harrison, R.M., Pogue, R., Vafiadaki, E., Pollitt, C., Davison, K., Moss, J.A., Keers, S., Pyle, A., Shaw, P.J., Mahjneh, I., Argov, Z., Greenberg, C.R., Wrogemann, K., Bertorini, T., Goebel, H.H., Beckmann, J.S., Bashir, R., Bushby, K.M. (2000). Secondary reduction in calpain 3 expression in patients with limb girdle muscular dystrophy type 2B and Miyoshi myopathy (primary dysferlinopathies). *Neuromusc. Disord.* 10: 553–559.

Bakker, E. and Van Ommen, G.B. (1998). Duchenne and Becker muscular dystrophy (DMD and BMD). In: Emery A.E.H. (Editor), *Neuromuscular Disorders: Clinical and Molecular Genetics*. John Wiley & Sons, New York, pp. 59–85.

Bashir, R., Britton, S., Strachan, T., Keers, S., Vafiadaki, E., Lako, M., Richard, I., Marchand, S., Bourg, N., Argov, Z., Sadeh, M., Mahjneh, I., Marconi, G., Passos-Bueno, M.R., Moreira, E.S,, Zatz, M., Beckmann, J.S., Bushby, K. (1998). A gene related to the C. elegans spermatogenesis factor fer-1 is mutated in patients with limb-girdle muscular dystrophy type 2B (LGMD2B). *Nat. Genet.* 20: 37–42.

Bione, S., Maestrini, E., Rivella, S., Manzini, M., Regis, S., Romeo, G., Toniolo, D. (1994). Identification of a novel X-linked gene responsible for Emery-Dreifuss muscular dystrophy. *Nat. Genet.* 8: 323–327.

Bonne, G., Di Barletta, M.R., Varnous, S., Becane, H.M., Hammouda, E.H., Merlini, L., Muntoni, F., Greenberg, C.R., Gary, F., Urtizberea, J.A., Duboc, D., Fardeau, M., Toniolo, D., Schwartz, K. (1999). Mutations in the gene encoding lamin A/C cause autosomal dominant Emery-Dreifuss muscular dystrophy. *Nat. Genet.* 21: 285–288.

Bönnemann, C.G. (1999). Limb-girdle muscular dystrophies: an overview. *J. Child. Neurol.* 14: 31–33.

Brockington, M., Blake, D.J., Prandini, P., Brown, S.C., Torelli, S., Benson, M.A., Ponting, C.P., Estournet, B., Romero, N.B., Mercuri, E., Voit, T., Sewry, C.A., Guicheney, P., Muntoni, F. (2001a). Mutations in the fukutin-related protein gene (FKRP) cause a form of congenital muscular dystrophy with secondary laminin alpha2 deficiency and abnormal glycosylation of alpha-dystroglycan. *Am. J. Hum. Genet.* 69: 1198–1209.

Brockington, M., Yuva, Y., Prandini, P., Brown, S.C., Torelli, S., Benson, M.A., Herrmann, R., Anderson, L.V., Bashir, R., Burgunder, J.M., Fallet, S., Romero, N., Fardeau, M., Straub, V., Storey, G., Pollitt, C., Richard, I., Sewry, C.A., Bushby, K., Voit, T., Blake, D.J., Muntoni, F. (2001b). Mutations in the fukutin-related protein gene (FKRP) identify limb girdle muscular dystrophy 2I as a milder allelic variant of congenital muscular dystrophy MDC1C. *Hum. Mol. Genet.* 10: 2851–2859.

Bushby, K.M.D. (1999). The limb-girdle muscular dystrophiesmultiple genes, multiple mechanisms. *Hum. Mol. Genet.* 8: 1875–1882.

Bushby, K. and Beckmann, J. (2002). The 105th ENMC workshoppathogenesis in the non-sarcoglycan limb-girdle muscular dystrophies. *Neurom. Disord.* 43(1): 80–90

Campbell, K.P. and Kahl, S.D. (1989). Association of dystrophin and an integral membrane glycoprotein. *Nature* 338: 259–262.

Cohn, R.D. and Campbell, K.P. (2000). Molecular basis of muscular dystrophies. *Muscle Nerve* 23: 1456–1471.

Den Dunnen, J.T. (2001). Point Mutation detection in the dystrophin gene. In: Bushby K.M.D. & Anderson L.V.B. (Editors), *Muscular Dystrophy: Methods and Protocols*. (Number 43 in the Methods in Molecular Medicine series.) Humana Press, Totowa, NJ, pp. 85–109.

Ervasti, J.M., Ohlendieck, K., Kahl, S.D., Gaver, M.G., Campbell, K.P. (1990). Deficiency of a glycoprotein component of the dystrophin complex in dystrophic muscle. *Nature* 345: 315–319.

Fanin, M., Fulizio, L., Nascimeni, A.C., Spinazzi, M., Piluzo, G., Ventriglia, V.M., Ruzza, G., Siciliano, G., Trevisan, C.P., Politano, L., Nigro, V., Angelini, C. (2004). Molecular diagnosis in LGMD2A: Mutation analyisis or protein testing? *Hum. Mutat.* 24: 52–62.

Feit, H., Silbergleit, A., Schneider, L.B., Gutierrez, J.A., Fitoussi, R.P., Reyes, C., Rouleau, G.A., Brais, B., Jackson, C.E., Beckmann, J.S., Seboun, E. (1998). Vocal cord and pharyngeal weakness with autosomal dominant distal myopathy: clinical description and gene localization to 5q31. *Am. J. Hum. Genet.* 63: 1732–1742.

Frischmeyer, P.A., Diets, H.C. (1999). Nonsense-mediated mRNA decay in health and disease. *Hum. Mol. Genet.* 8: 1893–1900.

Gouveia, T.L.F., Paim, J.F.O., Zatz, M., Vainzof, M. (2006). Multiplex analysis for mutations in sarcoglycan genes. *Diag. Mol. Pathol.* 15(2): 95–100.

Gregório, C.C., Granzier, H., Sorimachi, H., Labeit, S. (1999). Muscle assembly: A titanic achievemen? *Curr. Opin. Cell Biol.* 11: 18–25.

Gurgel-Giannetti, J., Reed, U., Bang, M.L., Pelin, K., Donner, K., Marie, S., Carvalho, M., Fireman, M.A., Zanoteli, E., Oliveira, A.S., Zatz, M., Wallgren-Pettersson, C., Labeit, S., Vainzof, M. (2001). Nebulin expression in Nemaline myopathy. *Neuromusc. Disord.* 11:154–162.

Gurgel-Giannetti, J., Bang, M.L., Reed, U., Marie, S., Zatz, M., Labeit, S., Vainzof M. (2002). Lack of the C-terminal domain of nebulin in a patient with nemaline myopathy. *Muscle Nerve* 25(5): 747–52.

Hack, A.A., Groh, M.E., McNally, E.M. (2000). Sarcoglycans in muscular dystrophy. *Microsc. Res. Tech.* 48: 167–180.

Haravuori, H., Vihola, A., Straub, V., Auranen, M., Richard, I., Marchand, S., Voit, T., Labeit, S., Somer, H., Peltonen, L., Beckmann, J.S., Udd, B. (2001). Secondary calpain3 deficiency in 2q-linked muscular dystrophy: Titin is the candidate gene. *Neurology* 56: 869–877.

Hauser, M.A., Horrigan, S.K., Salmikangas, P., Viles, K.D., Tim, R.W., Torian, U.M., Anu, T. (2000). A mutation in the Myotilin gene causes limb-girdle muscular dystrophy 1A. *Hum. Mol. Genet.* 9: 2141–2147.

Helliwell, R.R., Nguyen, T.M., Morris, G.E., Davies, K.E. (1992). The dystrophin-related protein, utrophin, is expressed on the sarcolemma of regenerating human skeletal muscle fibres in dystrophies and inflammatory myopathies. *Neuromusc. Disord.* 2: 177–184.

Hillaire, D., Leclerc, A., Faure, S., Topaloglu, H., Chiannilkulchai, N., Guicheney, P., Grinas, L., Legos, P., Philpot, J., Evangelista, T. (1994). Localization of merosin-negative congenital muscular dystrophy to chromosome 6q2 by homozygosity mapping. *Hum. Mol. Genet.* 3: 1657–1661.

Ho, M., Gallardo, E., McKenna-Yasek, D., De Luna, N., Illa, I., Brown Jr, R.H. (2002). A novel, blood-based diagnostic assay for limb girdle muscular dystrophy 2B and Miyoshi myopathy. *Ann. Neurol.* 51: 129–133.

Hoffman, E.P., Brown, R.H., Kunkel, L.M. (1987). Dystrophin: The protein product of the Duchenne muscular dystrophy locus. *Cell* 51: 919–928.

Koenig, M., Hoffman, E.P., Bertelson, C.J., Monaco, A.P., Feener, C.A. and Kunkel, L.M. (1987). Complete cloning of the Duchenne muscular dystrophy (DMD) cDNA and preliminary genomic organization of the DMD gene in normal and affected individuals. *Cell* 50: 509–517.

Koenig, M., Monaco, A.P., Kunkel, L.M. (1988). The complete sequence of dystrophin predicts rod-shaped cytoskeletal protein. *Cell* 53: 219–228.

Kramerova, I., Kudryashova, E., Tidball, J.G., Spencer, M.J. (2004). Null mutation of calpin 3 (p94) in mice causes abnormal sarcomere formation *in vivo* and *in vitro*. *Hum. Mol. Genet.* 13: 1373–1388.

Khurana, T.S., Hoffman, E.P., Kunkel, L.M. (1990). Identification of a chromosome 6 encoded dystrophin related protein. *J. Biol. Chem.* 263: 16717–16720.

Lefebvre, S., Burglen, L., Reboullet, S., Clermont, O., Burlet, P., Viollet, L., Benichou, B., Cruaud, C., Millasseau, P., Zeviani, M. (1995). Identification and characterization of a spinal muscular atrophy-determining gene. *Cell* 80: 155–165.

Liu, J., Aoki, M., Illa, I., Wu, C., Fardeau, M., Angelini, C., Serrano, C., Urtizberea, J.A., Hentati, F., Hamida, M.B., Bohlega, S., Culper, E.J., Amato, A.A., Bossie, K., Oeltjen, J., Bejaoui, K., McKenna-Yasek, D., Hosler, B.A., Schurr, E., Arahata, K., de Jong, P.J., Brown, R.H. Jr. (1998). Dysferlin, a novel skeletal muscle gene, is mutated in Miyoshi myopathy and limb girdle muscular dystrophy. *Nat. Genet.* 20: 31–36.

Matsuda, C., Hayashi, Y.K., Ogawa, M., Aoki, M., Murayama, K., Nishino, I., Nonaka, I., Arahata, K., Brown, R.H. Jr. (2001). The sarcolemmal proteins dysferlin and caveolin-3 interact in skeletal muscle. *Hum. Mol. Genet.* 10: 1761–1766.

McNally, E.M., Moreira, E.S., Duggan, D.J., Bonnemann, C.G., Lisanti, M.P., Lidov, H.G., Vainzof, M., Passos-Bueno, M.R., Hoffman, E.P., Zatz, M., Kunkel, L.M. (1998). Caveolin-3 in muscular dystrophy. *Hum. Mol. Genet.* 7: 871–878.

Messina, D.I., Speer, M.C., Pericak-Vance, M.A., McNally, E.M. (1997). Linkage of familial dilated cardiomyopathy with conduction defect and muscular dystrophy to chromosome 6q23. *Am. J. Hum. Genet.* 61: 909–917.

Minetti, C., Sotgia, F., Bruno, C., Scartezzini, P., Broda, P., Bado, M., Masetti, E., Mazzocco, M., Egeo, A., Donati, M.A., Volonte, D., Galbiati, F., Cordone, G., Bricarelli, F.D., Lisanti, M.P., Zara, F. (1998). Mutations in the caveolin-3 gene cause autosomal dominant limb-girdle muscular dystrophy. *Nat. Genet.*18: 365–368.

Monaco, A.P., Neve, R.L., Colletti-Feener, C., Bertelson, C.J., Kurnit, D.M., Kunkel, L.M. (1986). Isolation of candidate cDNAs for portions of the Duchenne muscular dystrophy gene. *Nature* 323: 646–650.

Moreira, E.S., Wiltshire, T.J., Faulkner, G., Nilforoushan, A., Vainzof, M., Suzuki, O.T., Valle, G., Reeves, R., Zatz, M., Passos-Bueno, M.R., Jenne, D.E. (2000). Limb-girdle muscular dystrophy type 2G (LGMD2G) is caused by mutations in the gene encoding the sarcomeric protein telethonin. *Nat. Genet.* 24: 163–166.

Moreira, E.S., Vainzof, M., Suzuki, O.T., Pavanello, R.C.M., Zatz, M., Passos-Bueno, M.R. (2002). Genotype/phenotype correlations in 35 Brazilian families with sarcoglycanopathies, including the description of three novel mutations. *J. Med. Genet.* 40(2): E12.

Muntoni, F., Brockington, M., Blake, D.J., Torelli, S., Brow, S.C. (2002). Defective glycosylation in muscular dystrophy. *Lancet* 360(9343): 419–421.

Muntoni, F., Sewry, C., Wilson, L., Angelini, C., Trevisan, C.P., Brambati, B., Dubowitz, V. (1995). Prenatal diagnosis in congenital muscular dystrophy. *Lancet* 345: 591.

Naom, I., D'Alessandro, M., Sewry, C.A., Jardine, P., Ferlini, A., Moss, T., Dubowitz, V., Muntoni, F. (2000). Mutations in the laminin alpha2-chain gene in two children with early-onset muscular dystrophy. *Brain* 123: 31–41.

Nicholson, L.V., Johnson, M.A., Gardner-Medwin, G., Bhattacharya, S., Harris, J.B. (1990). Heterogeneity of dystrophin expression in patients with Duchenne and Becker muscular dystrophy. *Acta Neuropathol.* 80: 239–250.

Paula, F., Vainzof, M., Passos-Bueno, M.R., Pavanello, R.C.M., Matioli, S.R., Anderson, L.V.B., Nigro, V., Zatz, M. (2002). Clinical variability in calpainopathy: What makes the difference? *Eur. J. Hum. Genet.* 10(2): 825–832.

Paula, F., Vieira, N., Starling, A., Yamamoto, L.U., Lima, B., Pavanello, R.C.M., Vainzof, M., Nigro, V., Zatz, M. (2003). Asympromatic carriers for homozygous novel mutations in the FKRP gene: the other end of the spectrum. *Eur. J. Hum. Genet.* 11(12): 923–930.

Pepe, G., Bertini, E., Bonaldo, P., Bushby, K., Giusti, B., de Visser, M., Guicheney, P., Lattanzi, G., Merlini, L., Muntoni, F., Nishino, I., Nonaka, I., Yaou, R.B., Sabatelli, P., Sewry, C., Topaloglu, H., Van der Kooi, A. (2002). Bethlem myopathy (BETHLEM) and Ullrich scleroatonic muscular dystrophy: 100th ENMC international workshop. *Neuromusc. Disord.* 12(10): 984–993.

Richard, I., Broux, O., Allamand, V., Fougerousse, F., Chiannilkulchai, N., Bourg, N., Brenguier, L., Devaud, C., Pasturaud, P., Roudaut, C. (1995). Mutations in the proteolytic enzyme calpain 3 cause limb-girdle muscular dystrophy type 2A. *Cell* 81: 27–40.

Selcen, D., Engel, A.G. (2004). Mutations in myotilin cause myofibrillar myopathy. *Neurology* 62: 1248–1249.

Sewry, C.A., Naom, I., D'Alessandro, M., Sorokin, L., Bruno, S., Wilson, L.A., Dubowitz, V., Muntoni. F, (1997). Variable clinical phenotype in merosin-deficient congenital muscular dystrophy associated with differential immunolabelling of two fragments of the laminin alpha 2 chain. *Neuromusc. Disord.* 7: 169–175.

Speer, M.C., Vance, J.M., Grubber, J.M., Lennon Graham, F., Stajich, J.M., Viles, K.D., Rogala, A., McMichael, R., Chutkow, J., Goldsmith, C., Tim, R.W., Pericak-Vance, M.A. (1999). Identification of a new autosomal dominant limb-girdle muscular dystrophy locus on chromosome 7. *Am. J. Hum. Genet.* 64: 556–562.

Starling, A., Paula, F., Silva, H., Vainzof, M., Zatz, M. (2003). Calpainopathy: how broad is the spectrum of clinical variability? *J. Mol. Neurosci.* 21: 233–236.

Starling, A., Kok, R., Passos-Bueno, M.R., Vainzof, M., Zatz, M. (2004). A new form of autosomal dominant limb-girdle muscular dystrophy (LGMD1G) with progressive fingers and toes flexion limitation maps to chromosome 4p21. *Eur. J. Hum. Genet.* 12(12): 1033–1040.

Vainzof, M., Pavanello, R.C.M., Pavanello, I., Passos-Bueno, M.R., Rapaport, D., Zatz, M. (1990). Dystrophin immunostaining in muscles from patients with different types of muscular dystrophy: A Brazilian study. *J. Neurol. Sci.* 98: 221–233.

Vainzof, M., Zubrzycka-Gaarn, E.E., Rapaport, D., Passos-Bueno, M.R., Pavanello, I., Zatz, M. (1991). Immunofluorescence dystrophin study in Duchenne dystrophy through the concomitant use of two antibodies directed against the carboxy-terminal and the amino-terminal region of the protein. *J. Neurol. Sci.* 101: 141–147.

Vainzof, M., Passos-Bueno, M.R., Rapaport, D., Pavanello, R.C.M., Bulman, D.E., Zatz, M. (1992). Additional dystrophin fragment in Becker muscular dystrophy patients: Correlation with the pattern of DNA deletions. *Am. J. Med. Genet.* 44: 382–384.

Vainzof, M., Passos-Bueno, M.R., Pavanello, R.C.M., Zatz, M. (1995a). Is dystrophin always altered in Becker muscular dystrophy patients? *J. Neurol. Sci.* 131: 99–104.

Vainzof, M., Reed, U.C., Schwartzman, J.S., Pavanello, R.C.M., Passos-Bueno, M.R., Zatz, M. (1995b). Deficiency of merosin (Laminin M or α2) in congenital muscular dystrophy associated with cerebral white matter alterations. *Neuropediatrics* 26: 293–297.

Vainzof, M., Passos-Bueno, M.R., Nguyen thi, M, Zatz, M. (1995c). Absence of correlation between utrophin localization and quantity and the clinical severity in Duchenne dystrophy (DMD). *Am. J. Med. Genet.* 58: 305–309.

Vainzof, M., Passos-Bueno, M.R., Moreira, E.S., Pavanello, R.C.M., Marie, S.K., Andreson, L.V.B., Bonnemann, C.G., McNally, E.M., Nigro, V., Kunkel, L.M., Zatz, M. (1996). The sarcoglycan complex in the six autosomal recessive limb-girdle (AR-LGMD) muscular dystrophies. *Hum. Mol. Genet.* 5: 1963–1969.

Vainzof, M., Passos-Bueno, M.R., Pavanello, R.C.M., Marie, S.K., Zatz, M. (1999). Sarcoglycanophathy is responsible for 68% of severe autosomal recessive limb-girdle muscular dystrophy. *J. Neurol. Sci.* 164: 44–49.

Vainzof, M., Anderson, L.V.B., Moreira, E.S., Paula, F., Pavanello, R.C.M., Passos-Bueno, M.R., Zatz, M. (2000a). Calpain 3: Characterization of the primary defect in LGMD2A and analysis of its secondary effect in other LGMDs. *Neurology* 54: 436.

Vainzof, M., Moreira, E.S., Canovas, M., Suzuki, O.T., Pavanello, R.C.M., Costa, C.S., Passos-Bueno, M.R., Zatz, M. (2000b). Partial α-sarcoglycan deficiency associated with the retention of the SG complex in a LGMD2D family. *Muscle Nerve* 23: 984–988.

Vainzof, M. (2001). Congenital myopathies. In: Carakushansky G (Editor), *Doenças Genéticas em Pediatria*. Guanabara Koogan S/A, Rio de Janeiro, Brazil.

Vainzof, M., Pavanello, R.C.M., Anderson, L.V.B., Moreira, E.S., Passos-Bueno, M.R., Zatz, M. (2001). Dysferlin analysis in autosomal recessive limb-girdle muscular dystrophies (AR-LGMD). *J. Mol. Neurosci.* 17: 71–80.

Vainzof, M., Canovas, M., Labeit, S., Bang, M.L., Faulkner, G., Valle, G., Zatz, M.(2002a). Sarcomeric myopalladin study in limb-girdle muscular dystrophies. *J. Neurol. Sci.* 199: S33.

Vainzof, M., Moreira, E.S., Suzuki, O.T., Faulkner, G., Valle, G., Beggs, A.H., Carpen, O., Ribeiro, A.F., Zanoteli, E., Gurgel-Gianneti, J., Tsanaclis, A.M., Silva, H.C., Passos-Bueno, M.R., Zatz, M. (2002b). Telethonin protein expression in neuromuscular disorders. *BBA. Genetic Basis of Diseases* 1588: 33–40.

Vainzof, M., Zatz, M. (2003). Protein defects in Neuromuscular disorders. *Br. J. Med. Biol. Res.* 36(5): 543–555.

Vainzof, M., Paula, F., Tsanaclis, A.M., Zatz, M. (2003a). The effect of calpain 3 deficiency in the pattern of muscle degeneration in the earliest stages of LGMD2A. *J. Clin. Pathol.* 56: 624–626.

Vainzof, M., Passos-Bueno, M.R., Zatz, M. (2003b). Immunological methods for the analysis of protein expression in neurmuscular diseases. In: Potter, NT (Editor). *Methods in Molecular Medicine-Neurogenetics: Methods and Protocols*. Humana Press, Totowa, NJ, pp. 355–378.

Vainzof, M., Kossugue, P.M., Yamamoto, L.U., Gouveia, T.L.F., Santos, B.G.C., Lima, B.L., Gurgel-Giannetti, J., Moza, M., Carpen, O. (2004). Heterogeneity in myotilin expression in a patient with Limb Girdle Muscular Dystrophy. *Neuromusc. Disord.* 14: 606.

Valle, G., Faulkner, G., De Antoni, A., Pacchioni, B., Pallavicini, A., Pandolfo, D., Tiso, N., Toppo, S., Trevisan, S., Lanfranchi, G. (1997). Telethonin, a novel sarcomeric protein of heart and skeletal muscle. *FEBS Lett.* 415: 163–168.

Van der Kooi, A., Van Meegen, M., Ledderhof, T.M., McNally, E.M., de Visser, M., Bolhuis, P.A. (1997). Genetic localization of a newly recognized autosomal dominant limb-girdle muscular dystrophy with cadiac involvement (LGMD1B) to chromosome 1q11-21. *Am. J. Hum. Genet.* 60: 891–895.

Wallgren-Pettersson, C., Laing, N.G. (2003). Report of the 109rd ENMC International Workshop: 5th Workshop on Nemaline Myopathy. *Neuromusc. Disord.* 13: 501–507.

Yamamoto, L.u., Canovas, M., Pavanello, R.C.M., Lima, B.L., Paula, F., Vieira, N.F., Zatz, M., Vainzof, M. (2003). Muscle protein alterations in patients with mutations in the Fukutin Related Protein gene. *Am. J. Hum Genet.* 73(5): 556.

Yamamoto, L.U., Gollop, T.R., Naccache, N.F., Pavanello, R.C.M., Zanotelli, E., Zatz, M., Vainzof, M. (2004). Protein and DNA analysis for the pre natal diagnosis of α-2-laminin deficient congenital muscular dystrophy. *Diag. Mol. Pathol.* 13(3): 167–171.

Yoshida, M., Ozawa, E. (1990). Glycoprotein complex anchoring dystrophin to sarcolemma. *J. Biochem.* 108: 748–752.

Young, P.J., Le, T.T., Dunckley, M., Nguyen, T.M., Burghes, A.H., Morris, G.E. (2001). Nuclear gems and Cajal (coiled) bodies in fetal tissues: Nucleolar distribution of the spinal muscular atrophy protein, SMN. *Exp. Cell Res.* 265: 252–261.

Zatz, M., Vainzof, M., and Passos-Bueno, M.R. (2000). Limb-girdle muscular dystrophy: one gene with different phenotypes, one phenotye with different genes. *Curr. Opin. Neurol.* 13: 511–517.

Zatz, M., Starling, A.S., Paula, F., Vainzof, M. (2003). The ten autosomal recessive Limb-Girdle Muscular Dystrophies. *Neuromusc. Disord.* 13(7–8): 532–544.

20

The Functional Consequences of Dystrophin Deficiency in Skeletal Muscles

Jean-Marie Gillis

Abstract

Duchenne muscular dystrophy (DMD) is caused by mutation(s) affecting the dystrophin gene that codes for a cytoskeletal protein. Dystrophin interacts with cytoskeletal actin filaments and with a complex of transmembrane glycoproteins. This assembly disintegrates when dystrophin is absent. The purpose of this chapter is to review the structural and functional disorders observed in dystrophin-deficient muscles in order to understand their role and evaluate their importance in the pathologic processes that eventually lead to fiber necrosis. A preliminary brief presentation outlines the main facts of DMD and its genetics in humans. As an enormous wealth of data have been obtained from studies of the *mdx* mouse, a natural mutant lacking dystrophin, an extensive description and analysis of the dystrophic phenotype of this animal model of DMD is presented. The review covers the fields of histology, regeneration, mechanics, electrophysiology, intracellular calcium homeostasis, protease activation, protein over/down expression, inflammatory response and microcirculation in *mdx* muscles. An attempt is made to distinguish between initial pathologic processes, thought to be specific to the lack of dystrophin, from downstream events. The loss of the mechanical strength of the fiber plasma membrane resulting from the lack of dystrophin and of its associated protein appears to be the primary defect responsible for initiating the pathologic cascade of muscle dystrophy. Consequences of this situation for the development of compensatory therapies is discussed.

20.1. Introduction

In Chapter 19, Vainzof and Zatz have presented a comprehensive review of the various types of muscular dystrophies due to identified mutations affecting specific proteins. In most cases, however, the nature of the pathophysiologic mechanisms linking gene mutations to the disease phenotypes is utterly obscure. Duchenne muscular dystrophy (DMD), caused by mutations in the dystrophin gene, resulting in the absence of the protein in affected muscle fibers, is certainly the disease where the underlying pathophysiologic mechanism is the best documented. The topic of this chapter is to review the wealth of available information to clarify the mechanistic links between dystrophin deficiency and fiber necrosis.

The prime defect in DMD concerns only one protein, dystrophin. However, it is a highly interactive protein establishing links with the cytoskeletal actin, in the sub-sarcolemmal space, with a transmembrane complex of heteromeric glycoproteins, and with a constitutive form of NO synthase through other proteins (synthrophin, dystrobrevin), and the list of partners is possibly longer. The highly complex assembly of proteins brought together by dystrophin, in and near the fiber membrane, disorganizes in its absence and the constituents are dispersed and/or quickly degraded in the cytosol. Therefore, one has to keep in mind that the functional consequences of the absence of dystrophin are in fact resulting from the absence of dystrophin and of its numerous associated proteins.

In DMD patients, the clinical situation is dominated by the dramatic and progressing waste of skeletal muscles, severely impairing motricity, upright standing, and respiratory functions. Cardiac dysfunction occurs late in the progress of the disease, while smooth muscle disorders (e.g., gastroinstestinal problems) are comparatively mild and infrequent. Therefore, this review will be essentially focused on understanding the dystrophic processes in skeletal muscles.

The range of the experimental investigations on DMD patients is very limited for obvious ethical reasons. Fortunately, several animal models (mouse, dog, and cat) of DMD are available and extensively studied. Most of the results presented and discussed here come from studies of the spontaneous mutant *mdx* mouse where muscle fibers are devoid of dystrophin as are fibers of human patients. However, no one of the muscle dystrophies developed in these animal models is exactly similar to the human disease, for which the term *Duchenne muscular dystrophy* must be exclusively reserved. This calls for caution when extrapolating results and conclusions from research on animal models to the human situation. This is still more compelling for results obtained on simplified preparations, such as cell cultures.

When dystrophin was discovered as a result of the cloning of its gene, its function in normal muscle was completely unknown. Most of our current views of the role of dystrophin come from studies of the defects observed in its absence. Thus, at the end of this review, a tentative presentation of the physiologic roles of dystrophin will find its place. Correlatively, the wealth of information collected on dystrophin-deficient muscles provides the scientific background for rational developments of therapies.

20.2. Duchenne Muscular Dystrophy

20.2.1. Clinical Description

Early in the second half of the 19th century, Edward Meryon in England and Guillaume Duchenne in France reported clinical descriptions of boys suffering from progressive muscle dystrophy, with muscle wasting [see history in Emery and Emery (1995)]. Duchenne noticed that in young boys, muscle weakness of the legs was associated with apparent muscle hypertrophy of the calves, followed in older age by progressive fibrosis. In the series of papers he devoted to the disease, Duchenne coined the terms "pseudohypertrophic" and "myosclerotic" to emphasize these particularities (Duchenne, 1868). Later, William Gowers made a detailed description of the characteristic way affected boys used their arms to climb up their body when going from a lying position to a standing one (the "Gowers sign"; Fig. 20.1).

At birth, a Duchenne boy looks quite normal and his motor behavior develops as expected. The first symptoms are usually noticed by the parents when the boy is 2–3 years old. Difficulties in running, climbing stairs, and frequent falls become more and more concerning. In families with no previous history of muscle dystrophy, it is not infrequent that while having noticed these symptoms, parents play them down and delay a thorough medical examination so that the correct diagnosis is made late, with the dreadful possibility that another affected boy could be born in the meantime.

Around the age of 10, muscle weakness, joint stiffness, and osteoporotic fractures impose wheelchair dependence. After the loss of walking, spine deviations develop with serious impairment

Figure 20.1. The Gowers sign: the characteristic way a young DMD patient uses his arms to stand up. Original drawing by W.R. Gowers (1879) reproduced in Emery and Emery (1995), with permission of the *Royal Society of Medicine Press* (London).

of the respiratory function, often requiring surgical corrections. Toward the end of the second decade, the patient needs respiratory assistance, often involving tracheotomy, as diaphragm and other respiratory muscles are severely affected. Terminal respiratory failure, often aggravated by a late cardiomyopathy, is usual.

Diagnosis of a severe myopathy is based on the clinical symptoms and the detection of an elevation of enzymes of muscular origin, in the plasma: creatine phosphokinase (CPK), specially the MB isoform (Vainzof *et al.*, 1985), lactate dehydrogenase (LHD), pyruvate kinase (PK), and aldolase. Elevated levels of structural muscle proteins are also found (myosin B and troponin T and I). These findings are hallmarks of muscle wasting but are not specific of a particular form of dystrophy. In spite of the absence of clinical sign in early life, systematic screening of newborn boys has revealed elevated CPK level in *a posteriori* diagnosed Duchenne boys (Scheuerbrandt *et al.*, 1986). Muscle biopsy is essential for a DMD diagnosis. Histopathology reveals foci of muscle degeneration with necrotic fibres (Fig. 20.2) and the presence of macrophage that can falsely suggest inflammatory myositis. Some fibers show a wedge-shape and localized region of hypercontraction (the "delta lesion") that is believed to be the first, irreversible step of the necrotic process. Fast-glycolytic fibers (histologic type IIB) are the first to

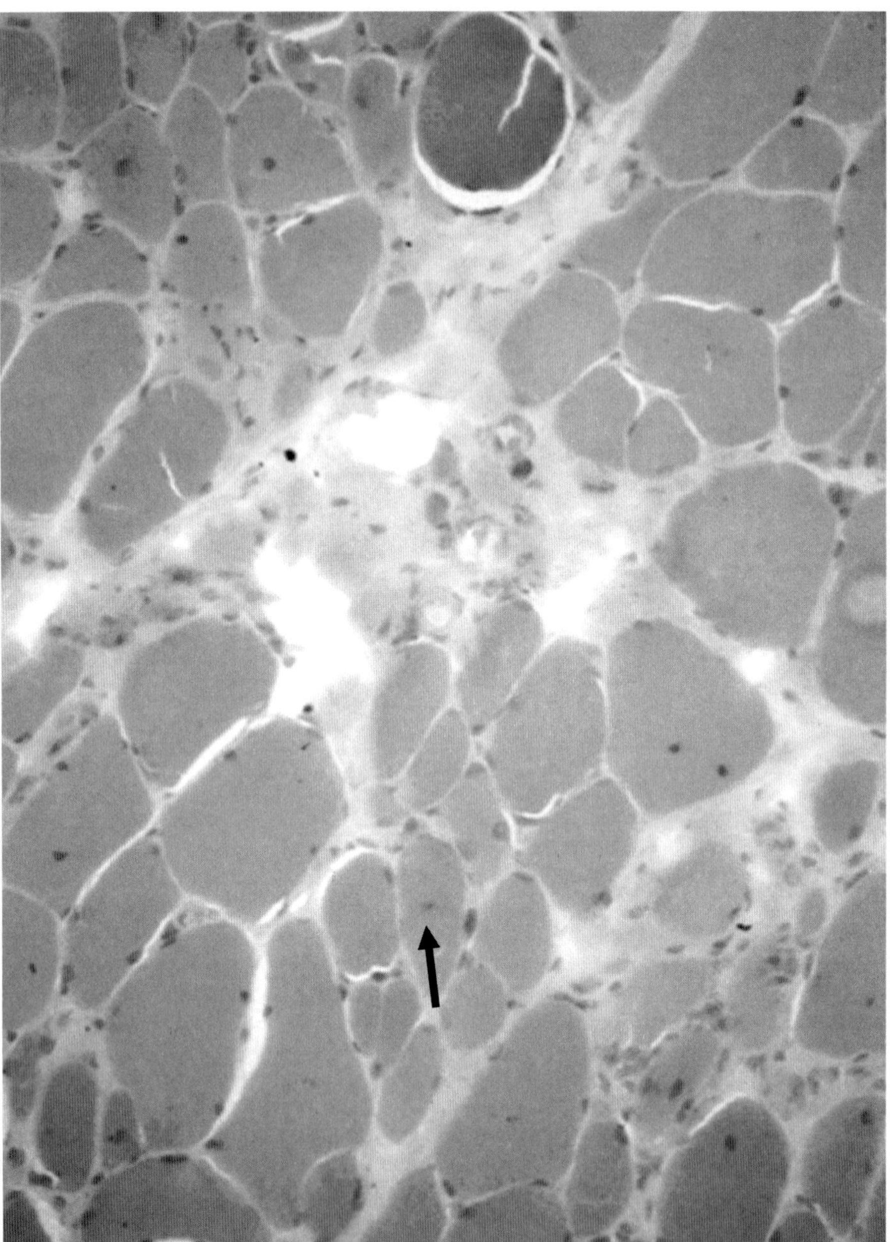

Figure 20.2. Transverse section through muscle fibers of a biopsy from the quadriceps muscle of a 7-year-old DMD patient. Hematoxylin-eosine staining. The picture is centered on a necrosis focus. Note the large differences in fiber diameters and the presence of centronucleated fibers (arrow). Image kindly provided by Dr. P.Van den Bergh (Louvain, Belgium).

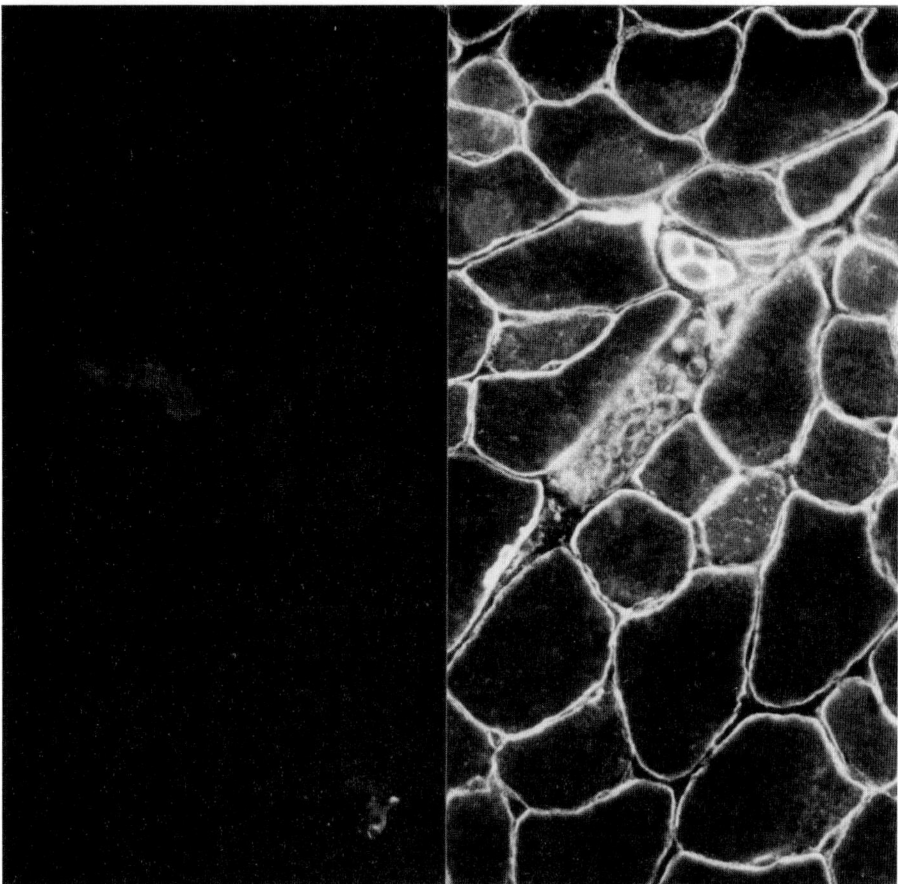

Figure 20.3. EDL muscle from mice. Right: Immunolabeling of dystrophin on the periphery of normal skeletal muscle fibers. Note also the positive reaction of the wall of small vessels (smooth muscles). Left: Absence of labeling in the EDL of a *mdx* mouse (see Section 20.2). Identical results are obtained on human muscles.

disappear from a muscle biopsy (Webster *et al.*, 1988). In young boys, biopsies also reveal sites of muscle regeneration with small-diameter, centronucleated myotubes and fibers, so that fiber diameters are very variable (Fig. 20.2). Later, these regenerating foci disappear, while fat deposits and fibrosis become prominent. It is the absence of immunologic detection of dystrophin (Fig. 20.3), both in muscle sections and muscle extracts (by Western blots), that provides the definite diagnosis of DMD.

A quantitative and objective assessment of muscle function is not easy to obtain in DMD boys, especially after the loss of walking. The group of Goubel has developed a chair for testing various properties of flexor muscles of the arm (Cornu *et al.*, 2001) like maximal voluntary isometric contraction (MVC) and resistance to fatigue at 75% MVC. Application of quick release during different levels of voluntary contraction allows calculating the series elastic stiffness, while application of sinusoidal length changes provides total joint stiffness. In a study involving 22 DMD boys (9–21 years old) and 15 healthy boys (9–15 years old) (Cornu *et al.*, 2001), it was observed that MVC was, on the average, reduced by seven times in DMD boys, while series elastic stiffness and total joint stiffness was markedly increased. The values of the latter properties were widely variable in DMD boys, reflecting the particular state of the disease in each patient. These studies are essential not only to monitor the progress of the disease in a given boy but also to get an objective assessment of the possible functional benefit of pharmacological (e.g., steroids) or nutritional (e.g., creatine) (Louis *et al.*, 2003) treatments of DMD.

Nuclear magnetic resonance (NMR) techniques offer noninvasive ways of studying DMD muscles *in situ*. By using ^{31}P NMR spectroscopy, the energetic status (PCr and ATP) of a group of muscles (e.g., calf muscles) can be quantified (Argov *et al.*, 2000) while glucose and lipid metabolism can be studied by ^{1}H-NMR (Boesch and Kreis, 2000). The latter can also be used to provide *in vivo* imaging of muscle mass and a quantitative visualization of degenerative foci, fat deposits, and fibrosis extension (Fig. 20.4). NMR is thus a very valuable tool for longitudinal studies.

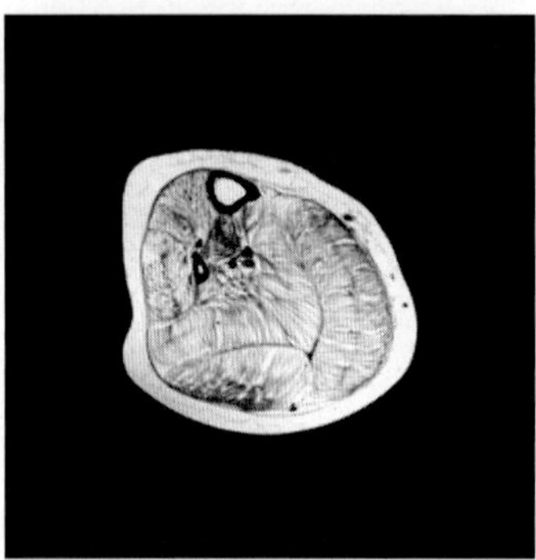

Duchenne disease, 13-year-old boy, T1 weighted image

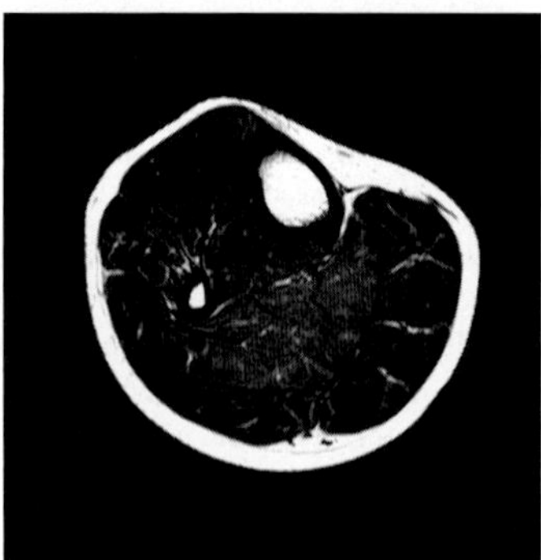

Normal volunteer, 18-year-old male, T1 weighted image

Figure 20.4. Top: ^{1}H-NMR image of a transverse section of the leg of a 13-year-old DMD patient, showing important fat infiltration of the muscle tissue. Bottom: The fat signal is pratically absent from the NMR image in a 19-year-old normal boy. (T1-weighted image: In this technique, both water and fat contribute to the image, but with a clearly higher signal intensity for fat.) Image kindly provided by F. Schick, J. Machann, and I. Mader (Tübingen and Freiburg, Germany).

20.2.2. Essentials of Human Genetics of DMD

DMD results from mutation(s) in the dystrophin gene located in the p21 region of the X chromosome (see Chapter 19) and is thus a sex-linked genetic disease. It affects essentially boys, with a frequency of 1/3500 male births. Extremely rare cases of DMD girls have been reported and resulted from chromosome translocations involving Xp21 and an autosome (Verellen-Dumoulin *et al.*, 1984). As patients die without progeny, the disease should spontaneously disappear. This is not the case and the stable frequency results from neomutations (either in the affected boy or in his mother). It can be *a priori* calculated that neomutations would account for one third of the cases (Nussbaum *et al.*, 2001).

In the cases of patients with a DMD history in the family, the mothers are thus heterozygous for the mutation and will transmit it to one half of their boys (who will be affected) and one half of their daughters (who will be carriers). About two thirds of the carrier women, while clinically unaffected, present elevated plasmatic CPK or PK levels .

When an affected boy has no record of DMD history in his family, it is very important to know if the neomutation occurred in him or in his mother. In the latter case, further boys could be affected and girls could be carriers. Analysis of DNA from mother and child will provide the clues. However, in about 10–15% of the mothers of such an "isolated" DMD case, DNA analysis (routinely performed on lymphocytes) could reveal a normal dystrophin gene, while the mother may carry a mutation in some of her germ-line cells (mosaicism) (van Essen *et al.*, 1992). In these cases, CPK levels are normal and noninformative. Prenatal diagnosis would then be the only way to know if a male foetus is affected.

20.3. The mdx Mouse and Other Animal Models of Duchenne Myopathy

The *mdx* mouse is a spontaneous mutant of the C57 black mouse, discovered by Bulfied *et al.* in 1984 (Bulfield *et al.*, 1984). Muscle fibers are lacking dystrophin (Fig. 20.3) as a consequence of a point mutation in exon 23, resulting in a premature stop codon (Sicinski *et al.*, 1989). In spite of the lack of dystrophin, the situation of *mdx* mice is much less severe than for human patients, a puzzling observation that has not received a satisfactory explanation so far. Clinical signs of muscle dystrophy are barely conspicuous when observing the spontaneous behavior of the animals but become more obvious when testing the muscle performance either *in vivo* or on isolated muscles (see Section 20.4.2). Muscles of the *mdx* mouse do not show the fat infiltration and progressive fibrosis so important in muscle of DMD patients in their second decade. The *mdx* mouse is fertile and its life span practically normal. Breeding and selection has provided *mdx* strains for laboratory investigations where males and females are homozygous for the mutation. A more detail description of the *mdx* phenotype will follow (see Section 20.4). The *mdx* mouse is the most studied animal model of DMD. It combines easy availability, breeding and maintenance; the size of the animal and of its individual muscles fits with many experimental protocols and apparatus for both *in vivo* and *in vitro* studies. Moreover, mouse is the favored mammal for transgenic interventions (knockout, -in, and overexpression).

The *dmx* mutant dog (Golden retriever) also lacks dystrophin but suffers from an extremely severe muscle dystrophy (Valentine *et al.*, 1992). Muscle wastage and weakness is dramatic, with progressive fibrosis. Death is premature and breeding very uncertain.

Two strains of dystrophic cats have been identified (Carpenter *et al.*, 1989; Kohn *et al.*, 1993). The dystrophic phenotype here is rather odd and dominated by hypertrophy affecting muscles of the tongue, the esophagus, and the diaphragm causing life-threatening obstructions.

These animal models and human patients share a common original defect: the lack of dystrophin so that the general term *dystrophinopathy* is adequate to encompass the various phenotypes. However, no one exactly mimics the human DMD. Arguably, this could be seen as the weak point of the animal studies. On the contrary, the various phenotypes can be considered as nature's experiments, as opportunities to compare genotype/phenotype relationships that will eventually provide unexpected clues for elucidating the mechanism of the dystrophic process, the compensatory reactions that tend to limit or delay it, and, hopefully, open the way to scientifically designed therapies.

20.4. The Dystrophic Phenotype of the mdx Mouse

Our current understanding of the role of dystrophin and of the consequences of its absence come essentially from studies of the *mdx* mouse. A throughout description of its phenotype will be presented here.

Dystrophinopathy follows a rather specific evolution in the *mdx* mouse. At birth, pups have normal muscles and plasma CPK is also within normal limits. This situation lasts until about the time of weaning (around days 18–20). Then a wave of intense muscle degeneration develops in a few days with a 10-fold increase of plasma CPK. Soon after, an active process of muscle regeneration is activated and, by day 45, numerous regenerated fibers can be observed. Adult mice have a nearly normal muscle mass and strength, though discrete signs of active dystrophy are still present. This sequence of events is best documented by histologic studies.

20.4.1. Histology

20.4.1.1. Overview

Degeneration usually occurs at discrete foci and is accompanied by an inflammatory reaction with microphage infiltration. In longitudinal sections of degenerating muscles, localized, wedge-shaped zones of hypercontraction are often observed ("delta lesions"); they are considered as the first irreversible step leading to fiber necrosis and suggest a membrane damage with massive penetration of extracellular Ca^{2+} ions.

Regenerating fibers are first detected as narrow myotubes with a chain of nuclei in central position, showing a strong basophilic reaction and a temporary expression of the embryonic/neonate isoforms of myosin. They progressively differentiate into young, still narrow, myofibers with a well organized transverse striation and eventually to apparently fully differentiated adult fibers of large diameter (~20 μm). Thus sections of regenerating muscles typically show a wide range of fiber diameter.

When regeneration takes place in a normal muscle (e.g., following a mechanical trauma), centronucleation is a transitory event, lasting at most a week. In fully regenerated fibers, nuclei have moved to the periphery, under the plasma membrane, a location typical of adult, healthy fibers. In the *mdx* mouse, regenerated fibers remain centronucleated for months, for reasons not elucidated so far. A typical leg muscle of a 3-month-old *mdx* mouse is made of 80% of centronucleated fibers, witnessing that a sequence of degeneration-regeneration took place in the past. However, the long persistence of centronucleation in regenerated fibers introduces a time lag and makes it a poor index to assess the dystrophic status at the time of the biopsy (this situation is specific to the mouse and is observed neither in other animal models nor in humans).

In spite of an apparently normal motor behavior, *mdx* mice suffer from continuing muscle dystrophy, albeit at a very low level, during most of their adult life. This is apparent in histology sections (discrete degeneration foci, large range of fiber diameters) and by the persistence of an elevated level of CPK (Coulton *et al.*, 1988) though the latter tends to normalize in old mice, > 12 months). The intensity of the dystrophic process seems very variable from mouse to mouse. It has been proposed that the low level of dystrophy resulted from the nearly perfect match of a very active regeneration versus an equally active degeneration, but the absence of detectable embryonic/neonate isoform of myosin (DiMario *et al.*, 1991) suggests that both processes are operating at basic rates in adult mouse. However, if regeneration is prevented by local γ-irradiation, the fiber content of limb muscles decays with a half-time of about 75 days [see discussion in Gillis (1999)].

In spite of the fact that every muscle of the animal is lacking dystrophin, the topography of structural damages is very variable. Diaphragm is the most affected muscle, with conspicuous fibrous development (Stedman *et al.*, 1991). This may be related to its continuous activity [breathing rate in mouse is 100/min; see Section 20.4.2.1 and Gillis (1996)]. Locomotion muscles are moderately affected, show little or no fibrosis, and tend to be slightly hypertrophied. Toe and finger muscles are still less affected (e.g., the flexor digitorum brevis shows only 25% of centronucleated fibers). Extraocular muscles (Karpati and Carpenter, 1986) and the striated part of the esophagus (Boland *et al.*, 1995) do not show any sign of dystrophy. Both muscles are made of fibers of very small diameter (10–13 μm), a structural feature that seems to make fibers much less susceptible to mechanical damage (a possible application of Laplace's law: the rupture pressure of a vessel is inversely proportional to its radius); smooth muscles are not affected in the *mdx* mouse (Boland, 1993), possibly for the same reason. Obviously, the mere lack of dystrophin does not automatically induce fiber necrosis.

20.4.1.2. The Regeneration Mechanism and Its Potential

Skeletal muscle regeneration is obtained by the activation of dormant mononucleated cells called "satellite cells." They are located along the muscle fibers, in sandwich between the fiber plasma membrane and the extracellular basal lamina, so that they can be easily taken for fiber nuclei. Satellite cell nuclei account for about 3% of the Hoechst-stainable nuclei associated with a given fiber. They can be distinguished by their specific immunolabeling with CD34 antibody (Nicole *et al.*, 2003). In response to fiber necrosis, a muscle-specific insulin-like growth factor is expressed and contributes to satellite cell activation.(Hill and Goldspink, 2003). Satellite cells become dedicated myoblasts that proliferate, fuse into elongated myotubes that eventually differentiate into genuine muscle fibers. This is accompanied by the sequential appearance and disappearance of embryonic and neonate myosin isoforms, eventually replaced by the adult isoforms of fast and slow myosin. The regenerated fibers carry their own new batch of satellite cells. It is now well documented that satellite cells form a heterogeneous population with subtypes oriented toward either differentiation (myotube formation) or self-renewal (new satellite cells). The various subtypes seem to be in a dynamic equilibrium, depending on the redox conditions (Chakravarthy *et al.*, 2001).

A local IM injection of snake venom or bupivacaine provides an experimental procedure to study *in situ* fiber regeneration after necrosis. The process can be also reproduced in a Petri dish from a muscle fragment [see, e.g., Bockhold *et al.* (1998)], at least up to the step of cross-striated and spontaneously beating myotubes, if cocultured with a nervous explant (Imbert *et al.*, 1995). There is no indication that the regeneration process would be different in humans.

It has been hypothesized that the eventual takeover of degeneration over regeneration is due to the exhaustion of the regenerating potential, as actively dividing satellite cells become progressively senescent (Webster and Blau, 1990), possibly as a result of telomere shortening. Indeed, in human patients (DMD and limb-girdle muscle dystrophy), it has been shown that the rate of telomere loss (187 bp/year) was 14 times greater than in healthy subjects (13 bp/year) (Decary *et al.*, 2000). However, in the *mdx* mouse, the situation seems different: for unknown reason(s), laboratory mice have a very long telomere and repeated mitosis do not shorten it significantly (Kipling, 1997).

20.4.2. Mechanical Properties of Dystrophin-Deficient Muscles

The following description will be restricted to muscles of the young adult mousse (3–5 months old), when dystrophy has reached a low, steady state situation.

20.4.2.1. *In Vivo* Studies

Many tests have been developed to study the mechanical performances of mice *in vivo*, but only those that have been applied to *mdx* mice will be reported here. (1) The "escape test" is a measure of a sudden burst of muscle activity of the mouse that tries to escape when its tail is gently pinched (Carlson and Makiejus, 1990) while the tail is attached to a force transducer. This produced the equivalent of a short, near maximal, isometric tetanus. Typically, the force developed (per unit weight of the animal) is reduced by 30–40% in the *mdx* mouse. (2) The "wire test" monitors acute muscle endurance: the mouse is suspended by its forelimbs to a long metallic wire. After each fall, the animal is resuspended and number of falls is recorded during 180 s. Wild-type mice have an average score of 9/10 that drops to 4/10 for *mdx* mice (Raymackers *et al.*, 2003). (3) Free runs in a mouse wheel show that *mdx* mouse reduce their voluntary activity by 30% (pups) and 60% (adults) (Carter *et al.*, 1995). (4) The "treadmill test" evaluates the midterm muscle performance and endurance by forcing the mouse to run forward on a horizontal floor running backwards; time is recorded until the animal feels exhausted and stops running. If the floor is inclined (10–25 degrees downhill), running includes so-called eccentric contractions to control the fall. *mdx* mice perform badly in this test: they show rapidly signs of exhaustion, contrary to normal mouse (Carter *et al.*, 1994). (5) The "rotarod test" evaluates the sensory-motor coordination together with muscle endurance: the mouse is placed on a rotating rod (3 cm diameter) and the time before falling is recorded. The test is made more exacting by increasing the rotation speed (e.g., from 5 rpm to 25 rpm). All these *in vivo* tests are easy to perform, repeatable, noninvasive, and are potentially discriminant, for example, to judge of the success of a given therapy. They can also be complemented by measurements of CPK levels and histologic analysis of selected muscles, removed after the test (Bansal *et al.*, 2003). However, these tests concern very complex muscle activities (as in the rotarod test) and their interpretation in terms of the causal limiting factor(s) is questionable. Moreover, they are much influenced by the genetic background, and various substrains of the *mdx* mice behave very differently in one or several of these tests.

In vivo tests are useful tools for investigating how exercise affects the dystrophic status and to document the concept of "activity-induced damages."

20.4.2.2. Studies of Isolated Muscles

The mechanical characteristics of dystrophin-lacking muscles are best studied on isolated muscle preparations. They include the "classical" fast and slow limb muscles: the extensor digitorum longus (EDL) and the soleus, respectively; isolated fibers from the toe muscle flexor digitorum brevis (FDB) and dissected strips of the diaphragm (the severity of dystrophy in the latter makes it the closest model to DMD muscles).

Isometric force in twitch and tetanus: Forces (mN) developed in these conditions are practically of normal amplitude in EDL and soleus *mdx* muscles. However, these muscles are systematically hypertrophied. Therefore when force is normalized to the muscle cross-sectional area (mN cm^{-2}), an ~20% deficit is observed. Thus, *mdx* muscles are pseudo-hypertrophic, a feature reported for the calves of young Duchenne boys. The origin of the normalized force deficit is unclear. Examination of muscle cross-section does not reveal necrosis and fibrous area of the size required to quantitatively explaining the deficit. It was reported that the total number of cross bridges per square millimetre and the elementary force per cross bridge were both lower in *mdx* muscle (diaphragm) (Coirault *et al.*, 2003). On the other hand, older investigations showed that the ratio of heat production over force during a tetanus was normal in *mdx* EDL (Kometani *et al.*, 1990) as was its force-velocity relationship (Maréchal and Beckers-Bleukx, 1988), suggesting that the energy coupling in actomyosin ATPase is not fundamentally affected by the lack of dystrophin. In EDL, tetanus relaxation is delayed by about 20%; this may reflect a slower uptake of calcium ions by the sarcoplasmic reticulum or/and a consequence of a progressive shift of the myosin heavy chain isoform toward slower ones. Recently, an investigation of the [Ca^{2+}] transients elicited by an action potential showed a reduction of amplitude in mdx fibers together with a slower return to the basal level (Woods *et al.*, 2004), suggesting that incomplete, but prolonged activation of actomyosin by Ca^{2+} is responsible for the reduction of both isometric force and relaxation speed.

20.4.2.3. Susceptibility to Eccentric Contractions

The most dramatic consequence of the lack of dystrophin is revealed during a series of "eccentric contractions," when a maximally tetanized muscle is forcibly stretched. At the time of the length change, a peak of force is generated that largely exceeds the isometric tetanic plateau. However, when the muscle is stimulated again, the force developed is reduced and the reduction increases with the number of previous eccentric contractions (Moens *et al.*, 1993). This experimental protocol is illustrated in Figure 20.5 where it is seen that after six eccentric contractions, the relative force drop reached 56%, at the seventh contraction.

Normal muscles are not insensitive to eccentric contractions, but in the same conditions as illustrated in Figure 20.5, the force drop is only 10–12%. Allen and his co-workers have extensively studied the mechanism(s) underlying the force drop seen after eccentric contractions in normal muscles (Balnave and Allen, 1995). They show that a reduced Ca^{2+} release (during tetanus) combined with a reduced Ca^{2+} sensitivity of the contractile machinery were the main factors behind the force drop, while localized structural damages were relatively rare. Severe sarcomere disorganization occurs only after very large stretches (+50% of optimal length) (Balnave *et al.*, 1997).

The high susceptibility of dystrophin-lacking muscles to high mechanical stress can be seen as the macroscopic consequence of the disorganization of the macromolecular assembly assuring the physical continuity between the fiber cytoskeleton and its extracellular matrix, through the plasma membrane. The intra-extracellular continuity made by the $\alpha 7\beta 1$ integrin is still present in *mdx* muscle, but, alone, it seems insufficient [see discussion in Rybakova *et al.* (2000)]. The ensuing membrane fragility makes it prone to localized lesions, in conditions of the high mechanical stress generated during eccentric contractions. These lesions can be directly visualized by the penetration of dyes (orange Procion, Evans blue) that are normally impermeant. In a wide variety of conditions, it was shown that the percentage of dye-penetrated fibers was linearly related to the force drop after a standard series of eccentric contractions (Fig. 20.6). However, the force drop (in %) is always greater than the percentage of dye-penetrated fibers. This suggests that, in *mdx* fibers, the force drop observed after eccentric contractions has a complex origin: it combines the effects of structural damages (with dye penetration) and of functional alterations (without dye penetration) present in normal muscles (Balnave and Allen, 1995). Most likely damaged fibers become first nonexcitable as the localized lesions impede the propagation

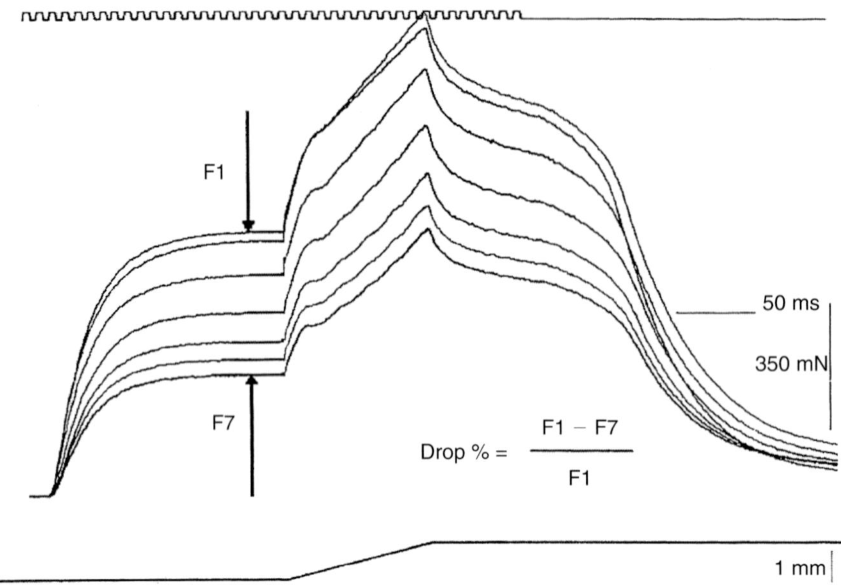

Figure 20.5. The effect of excentric contractions on the isometric force (middle traces) developed by the quadriceps muscle of a *mdx* mouse. The first part of the contraction is isometric; it is followed by a imposed lengthening of the activated muscle that produced a peak of force largely exceeding the maximal isometric force (phase of "eccentric contraction"). When stimulation ends, relaxation takes place. After a 2-minute rest interval, the same sequence of isometric-eccentric contraction is again elicited. Note the progressive drop of isometric force as contractions are repeated (here at the seventh contraction). Top trace: Stimuli. Bottom trace: length change. Reproduced from Deconinck *et al.* (1996), with permission of the *Proceedings of the National Academy of Sciences of the United States of America*.

of the action potential. Later, in some fibers, high fluxes of Ca^{2+} from the extracellular medium probably induces irreversible supercontraction, but in others, membrane resealing may occur. Indeed, in *mdx* mice submitted to eccentric contractions, part of the large force drop observed just after the exercise can be recovered during the next few hours, while fibers keep the fluorescent extracellular marker trapped. This strongly suggests that some "resealing" of the membrane defects can take place and it was very recently shown that muscle fibers are equipped with a dysferlin/Ca^{2+}-dependent mechanism able to reseal membrane lesions in a matter of a few minutes (Bansal *et al.*, 2003).

Eccentric contractions are not laboratory oddities. They occur in many integrated voluntary contractions to brake a member movement (e.g., the flexion of the arm can be slowed down by eccentric contractions of the extensor muscles). This occurs in climbing down the stairs (a controlled fall), in walking, in ballistic movements (to reach a target), even during the respiratory cycle: the elastic recoil of the thorax, during the first phase of expiration, is slowed down by eccentric contractions of the diaphragm (Remmers, 1976), a situation that might explain the severe dystrophy of this muscle. The particularly high susceptibility of dystrophin-lacking fibers to eccentric contraction is the pathophysiologic basis of the so-called "activity-induced damages"; moreover, it is the hallmark of dystrophinopathy and of some (not all) forms of sarcoglycan deficiency (Hack *et al.*, 1999; Hack *et al.*, 2000). It is not observed in other forms of even more severe myopathy like the *dy/dy* dystrophy (laminin-2 deficiency) (Head *et al.*, 2004), *Smn*

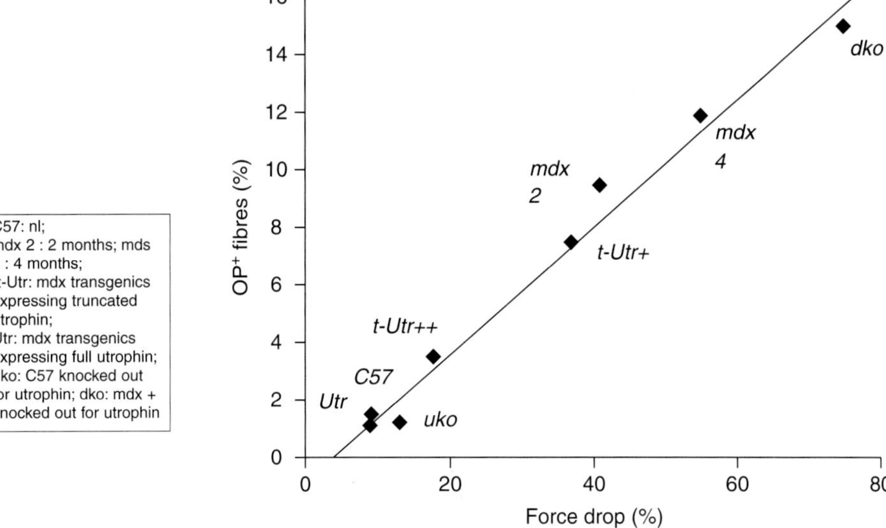

Figure 20.6. Linear correlation between the amplitude of the force drop after a standard series of eccentric contractions and the percentage of fibers penetrated by an extracellular dye (orange procion, OP^+ fibers). Points come from different lines of *mdx* mice, utrophin$^{-/-}$ *mdx* mice, and transgenic *mdx* mice overexpressing utrophin (see Section 20.4.8.2). Reproduced from Deconinck (1998).

gene defect myopathy (Nicole *et al.*, 2003), and calpain3$^{-/-}$ myopathy (Fougerousse *et al.*, 2003), to name the cases where the effect of eccentric contractions was studied.

20.4.3. Electrophysiologic Properties

In the absence of dystrophin and its associated proteins, the neuromuscular junction show subtle morphologic and functional alterations affecting the density and localization of ACh-receptors and the quantal release of ACh (Carlson and Roshek, 2001) However, clinically, the motor activity of the *mdx* mouse shows no signs of failure of the neuromuscular transmission. Propagation of the muscle action potential is normal in *mdx* muscles (Woods *et al.*, 2004). However, De Luca *et al.* (2001) reported that the threshold depolarization for mechanical activation was about 5 mV more negative for fibers of the *mdx* EDL. This result possibly reflects changes in the multivariate system which determines the moment-to-moment and point-to-point values of the intracellular concentrations of Ca^{2+} ions, resulting in more Ca^{2+} available for triggering contraction, but the exact parameter(s) involved remains uncertain.The same authors observed that *mdx* muscles recovered a normal value of the depolarization threshold when 60 mM taurine was added externally. Possibly, this addition restored a normal taurine content (reduced by 14% in *mdx* muscles) and favored a faster Ca^{2+} uptake by the sarcoplasmic reticulum (Huxtable and Bressler, 1973).

A transient increase in the membrane conductance for chloride ions (G_{Cl}) has been observed in *mdx* fibers, but this is limited to the period of regeneration that follows the first degenerative episode (from 2 months of age) and is no longer detected at 5 months. This observation might reflect more the action of regeneration-promoting factors than the lack of dystrophin (De Luca *et al.*, 1997).

20.4.4. Calcium and Dystrophinopathy

It has long been known that muscle biopsies from Duchenne patients have a twofold increase in their total calcium content (Bodensteiner and Engel, 1978); the same is found in muscles of the adult *mdx* mouse (Gailly *et al.*, 1993a). The exact location (extra- and/or intracellular) of this excess calcium is not established.

20.4.4.1. The Cytosolic [Ca^{2+}], at Rest

The concentration of free Ca^{2+} ions in the fiber cytosol can be measured using permeant or cell-injected Ca^{2+} indicators derived from the specific Ca chelator EGTA. The group of Fura-indicators display changes of fluorescence within the physiologic 20–500 nM range of Ca^{2+} concentrations. They are used either in the free acid form (to be injected) or in the permeant acetoxylmethylester (AM) form that has to be cleaved by cellular esterases to liberate the free form. In manually dissected single fibers from *mdx* FDB and injected with Fura-2, the cytosolic Ca^{2+} concentration is about twice the value observed in normal fibers (Hopf *et al.*, 1996; Turner *et al.*, 1988). It has been proposed that this elevation activates Ca-dependent proteases (e.g., calpains) leading to fiber necrosis. However, in fibers isolated after collagenase digestion of the extracellular matrix and loaded with the permeant Fura-2AM, the intracellular [Ca^{2+}] is well within normal values 20–80 nM. (De Backer *et al.*, 2002; Gailly *et al.*, 1993a; Head, 1993). Though the origin of this discrepancy is still a matter of controversy, it may arise from the experimental procedure for fiber isolation: a mechanical stress on fibers obtained by manual dissection is practically unavoidable. When the fiber membrane is submitted to such a stress by swelling in a hypotonic solution, $[Ca^{2+}]_i$ increases and the amplitude of the changes is larger in *mdx* fibers. Moreover, problems in obtaining the calibration parameters for the fluorescence signals could also be involved (Gailly *et al.*, 1993a).

20.4.4.2. Mechano-sensitive Calcium Channels

A particular class of Ca^{2+} channels has been described in the membrane of adult muscle fibers. They are voltage-independent (i.e., show a linear, ohmic I/V relationship), have a unitary conductance of ~8 pS, and respond to pressure changes by increasing activity (i.e., by increasing their open probability). Selectivity for Ca^{2+} is moderate and K^+, Na^+, Ba^{2+}, and Mn^{2+} are allowed through the channel. In *mdx* fibers, the basal activity of these channels is higher than in the normal one (Franco-Obregón and Lansman, 1994). This means that a higher influx of Ca^{2+} is taking place and possibly could contribute to increase $[Ca^{2+}]_i$, either in the bulk of the cytosol or locally (e.g., in the submembranous space). These two possibilities have been investigated experimentally.

The influx of Ca^{2+} can be estimated by following the extinction of the fluorescence signal of the intracellular Ca^{2+} indicator when Mn^{2+} ions are substituted for Ca^{2+} in the external medium. It has been shown that in various experimental conditions where the basal Ca^{2+} influx was increased by 10-fold (largely exceeding the difference between basal rates in *mdx* and normal fibers), the cytosol [Ca^{2+}] remained within normal values, both in normal and *mdx* fibers (De Backer *et al.*, 2002). Thus homeostasic mechanisms in *mdx* fibers are able to cope with an increased Ca^{2+} influx to prevent a rise of the cytosol [Ca^{2+}]. Recently, it was reported that undissociated fibers of the thin expiratory muscle trangularis sterni of *mdx* mouse show normal values of resting Ca^{2+} influx (Carlson *et al.*, 2003), in spite of severe dystrophy, suggesting that the collagenase dissociation could unmask differences between normal and dystrophin-lacking fibers that may not be detectable *in situ*.

The pathophysiologic role of these voltage-independent/mechano-sensitive Ca^{2+} channels is not clarified. It seems that dystrophin contributes to their fine tuning as their activity is increased in its absence, suggesting a role of dystrophin in the transduction of mechanical signal(s). In normal myotubes, a localized mechanical lesion of the membrane (~5 μm) induces a higher activity of these channels in the surrounding area (McCarter and Steinhardt, 2000), a phenomenon linked to a local increase of proteolysis as it is inhibited by antiproteases. It was then proposed that this higher channel activity contributes to initiate the dystrophic cascade. However, in transgenic *mdx* mice expressing moderate levels of utrophin (see Section 20.4.8.2), the activity of these Ca^{2+} channels is normalized while muscle dystrophy is still present (Squire *et al.*, 2002). An alternative view emerges from experiments showing that localized membrane lesions can be repaired by the movement of internal vesicles toward the lesion and their fusion with the plasma membrane (Bansal *et al.*, 2003). This mechanism is dysferlin-dependent and requires an influx of external Ca^{2+}. The latter could be provided by the higher activity of the Ca^{2+} channels, which could then be considered as partners of the resealing process.

20.4.4.3. The Submembranous [Ca^{2+}]

The submembranous [Ca^{2+}] was estimated indirectly, by its effect on Ca^{2+}-activated K^+ channels acting as local [Ca^{2+}] sensors. The activity of these channels in response to a depolarizing pulse is higher when the local [Ca^{2+}] is higher. It was observed that the threshold depolarizing pulse to obtain just detectable activity of these K^+ channels was lower in *mdx* fibers, suggesting a higher local [Ca^{2+}], estimated to be threefold higher than in normal fibers (Mallouk *et al.*, 2000). This was obtained in fibers clamped at 0 mV membrane potential. However, when fibers were maintained at negative membrane potentials similar to the physiologic ones (−60 to −80 mV), the activity of the K^+ channels in response to a standard depolarizing pulse was identical for *mdx* and normal fibers (Mallouk and Allard, 2002). The difference between the two experimental conditions (0 mV vs. –80 mV) is that in fully depolarized fibers, the Na^+/Ca^{2+} exchanger (Balnave and Allen, 1998) does no longer contribute to the extrusion of Ca^{2+} ions and may even be working in the reverse mode, allowing further Ca^{2+} influx. The suppression of the contribution of one of the partners of intracellular [Ca^{2+}] homeostasis might explain the observed increase in the submembranous space. This would not be the case in physiologic conditions. Quite recently, near membrane cytosolic [Ca^{2+}] has been investigated using the membrane-bound indicator FFP-18 (Han *et al.*, 2006). No difference was observed between normal and mdx fibers (fully polarized, collagenase isolated). Thus, the question of an elevated [Ca^{2+}] in the sub-membrane space of mdx fibers is not a settled issue.

20.4.4.4. Other Components of Ca^{2+} Homeostasis

The ability of the sarcoplasmic reticulum from *mdx* muscles to actively take up Ca^{2+} was investigated both on isolated vesicles (Kargacin, 1996) and *in situ* (Takagi *et al.*, 1992), with conflicting results. The former authors reported a reduction of the uptake rate, while the latter concluded that the sarcoplasmic/endoplasmic reticulum calcium ATPase (SERCA) pumping activity is normal *in situ*, in spite of an increased leakage that is compensated by an increased pumping rate.

Another partner of Ca^{2+} homeostasis in fast mouse muscle is the soluble Ca-Mg binding protein parvalbumin, with a 10^4 higher affinity for Ca^{2+} over Mg. It has been hypothesized that, in *mdx* muscle, parvalbumin contributes to compensate for the increased Ca^{2+} influx (Gailly *et al.*, 1993b) In both normal and *mdx* fibers, the absence of parvalbumin does not affect the resting $[Ca^{2+}]_i$ (Raymackers *et al.*, 2003), excluding a major role of parvalbumin on this parameter, an expected conclusion, as parvalbumin is essentially Mg-loaded in resting fibers. During muscle activity, when $[Ca^{2+}]_i$ reaches micromolar values, parvalbumin loses Mg^{2+} and binds Ca^{2+}, accelerating muscle relaxation (Gillis, 1985; Hou *et al.*, 1991). Parvalbumin expression is only significant in very small mammals (Heizmann *et al.*, 1982). By its high concentration within the muscle cytosol (mM), parvalbumin acts a strong Ca^{2+} buffer that is not available in muscles of larger animals. Could this situation explain why the *mdx* mouse only suffers from a mild dystrophy compared with DMD patients? A thorough study of mice deficient both in dystrophin and parvalbumin shows only a slight aggravation of the dystrophic status (Raymackers *et al.*, 2003), excluding that parvalbumin confers a major advantage.

20.4.4.5. [Ca^{2+}] Transient During Activity

In response to single electrical stimulation, transient peak heights were reported as identical in normal and *mdx* fibers (Turner *et al.*, 1988; Tutdibi *et al.*, 1999) and the kinetics of the transients, normal (Head, 1993) or slighly retarded (Turner *et al.*, 1988). This point was reinvestigated using a mixture of Ca^{2+} indicators covering a wide range of $[Ca^{2+}]_i$. The peaks of the [Ca^{2+}] transients were reduced in *mdx* fibers, suggesting an impairment of the efficacy of the excitation-contraction coupling. The the return to basal level pf [Ca^{2+}] was somewhat delayed in *mdx* fibers, so that the time-integral of the [Ca^{2+}] changes over 100 ms was roughly similar to normal fibers (Woods *et al.*, 2004) (see also Section 20.4.2). Thus, activation by itself does not produce a significant intracellular Ca^{2+} overload.

Finally, what is the role played by Ca^{2+} ions in dystrophinopathy? The answer is not straightforward and depends on the amplitude of the Ca^{2+} influx. On the one hand, dystrophin-lacking fibers, in spite of the fact that they are submitted to a moderately increased Ca^{2+} influx, possess robust homeostasic mechanisms able to cope with the situation. On the other hand, the fragility of the dystrophin-dystro-sarcoglycan depleted membrane makes it prone to physical lesions resulting from the high mechanical stress generated during contractions, especially eccentric contractions. In that situation, larger influxes of extracellular Ca^{2+} may overtake the capacity of the homeostasic mechanisms and uncontrolled activation of proteolysis initiates the necrosis cascades (Fig. 20.7). Indeed, an experimental demonstration that chronic exercise leads to

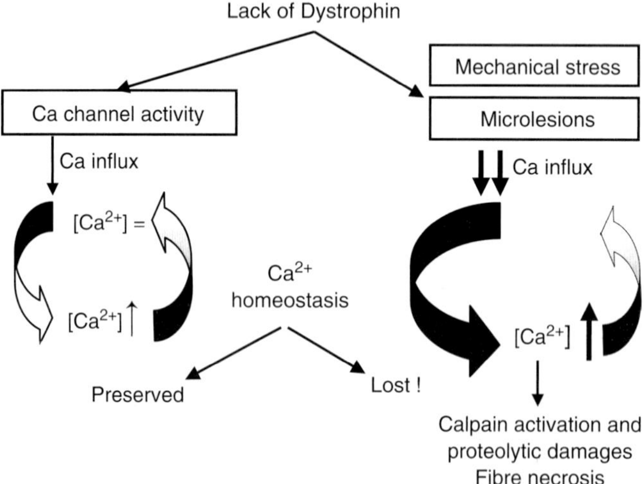

Figure 20.7. Lack of dystrophin and the problem of intracellular Ca^{2+} homeostasis. Left: Maintenance of a normal [Ca^{2+}], in spite of an increased Ca^{2+} influx. Right: Loss of Ca^{2+} homeostasis as a consequence of massive influx of extracellular Ca^{2+} caused by stress-induced lesions in fragilized, dystrophin-lacking membrane. Reproduced from Raymackers (2003).

alteration of calcium homeostasis has been published recently (Fraysse *et al.*, 2004). The role of activity-induced damages thus seems essential in engaging the dystrophic process.

20.4.5. Calpains and Proteolytic Activities

The observation that soleus muscles from *mdx* mice show a twofold larger efflux of tyrosine than normal muscle, which recovered a normal rate when external [Ca^{2+}] was reduced to 1/10, led to the hypothesis that dystrophinopathy was initiated by Ca-activation of intracellular proteolysis (Turner *et al.*, 1988). Calpains, a class of Ca^{2+}-dependent cysteine proteases, were the obvious candidates for that role. Skeletal muscles contain the ubiquitous m-calpain and μ-calpain (activated at mM and μM Ca^{2+} concentration, respectively) and a muscle specific form, calpain 3 [for an extended review, see Goll *et al.* (2003)]. Calpains can undergo a Ca^{2+}-activated autolysis, long thought to be the first obligatory step for activation, but this point is still debated (Cottin *et al.*, 2001). In muscles, specific targets for calpain proteolytic action are mostly cytoskeletal proteins (e.g., talin, vinculin, spectrin, filamin, desmin, and vimentin). The Z-line region, where calpains are located, seems particularly susceptible to their action. Activation of calpain activity is observed during the fusion of several myoblasts into a myotube (Dourdin *et al.*, 1997). Most likely calpains are involved in the reorganization of the cytoskeleton that this process entails.

Calpain activities have been essentially studied by *in vitro* assays (using the ability of muscle extracts to hydrolyze β-casein) and very little is known of their *in situ* activity and regulation. Two points are particularly puzzling: First, the level of Ca^{2+} concentration needed for activating even the μ-calpain is practically never present in a healthy living cell; second, a natural inhibitor of calpain, calpastatin, is also present in quantities usually exceeding that of calpains, so that the latter could be basically inhibited (Spenser and Tidball, 1992).

A tool for measuring calpain activity within living cells has been developed by attaching a chromophore to a permeant dipeptide that becomes fluorescent after cleavage (Rosser *et al.*, 1993). It was used on cultured myotubes, normal and *mdx* (Alderton and Steinhardt, 2000). Activity varied greatly from myotube to myotube and was not obviously different between normal and *mdx* preparation. It could be inhibited by leupeptin, a permeant, nonspecific protease inhibitor. However, in normal myotubes, activity progressively declined with aging of the culture, while *mdx* myotubes maintained a higher activity for longer times. Whatever the pathophysiologic interpretation of these results, they showed that a detectable calpain-like activity is present in quiescent myotubes, where the cytosol concentration of Ca^{2+} would be well below 200 nM, that is, much lower than the Ca^{2+} concentration needed to activate μ-calpain, on *in vitro* assays. Unfortunately, so far, no *in situ* measurements on adult fibers have been reported.

Are calpains involved in the cascade leading to dystrophy? As an early, initiating agent, the answer is uncertain as the question of a chronic elevation of cytosolic [Ca^{2+}] is debated; as a downstream event, when intracellular Ca^{2+} homeostasis is lost, activation of calpain is quite likely. Dystrophin-deficient mice resulting from the crossing of *mdx* mice with transgenic mice expressing very high level of the calpain inhibitor calpastatin show a marked reduction of various signs of dystrophy (Spencer and Mellgren, 2002), even if membrane fragility remains uncorrected. Aleviation of the dystrophic status of *mdx* mice has also been obtained with the synthetic calpain inhibitors BN 82270 (Burdi *et al.*, 2006).

20.4.6. Microcirculation

Muscle microcirculation is greatly increased during exercise through a NO-dependent vasodilatation (Lau *et al.*, 1998) that counteracts α-adrenergic influences. The fact that the membrane localization of NO-synthase is lost in the absence of dystrophin (Brenman *et al.*, 1996) suggests that the reactive vasodilatation could be impaired (Thomas *et al.*, 1998). This in turn would reduce oxygenation, CO_2 elimination, and normal pH recovery after activity. *In vivo*, this recovery was significantly slowed down in exercising *mdx* mice, as studied by ^{31}P-NMR spectroscopy (Dunn *et al.*, 1992), though pH recovery was normal after contractions of isolated *mdx* muscles with artificial perfusion (Decostre *et al.*, 2002). The difference illustrates the importance of the local conditions of extracellular washout. Moreover, NO increases glucose transport in muscle (Balon and Nadler, 1997) and acts as a fast reactive oxygen species scavenger (see Section 20.4.7), beneficial functions probably impaired in dystrophin-lacking fibers. In DMD patients, the important fibrosis reaction and fat deposits worsen the impairment of the muscle microcirculation.

20.4.7. Inflammatory Reactions

Signs of inflammation are commonly observed in muscle biopsies from DMD patients and *mdx* mice, around the foci of fiber necrosis, suggesting a nonspecific inflammatory response to fiber death. However, experimental evidence has accumulated suggesting that the numerous inflammatory cells, of both myeloid and lymphoid origin, could play an active role in promoting myopathy [for a review, see Spencer *et al.* (2001)]. Indeed, it is remarkable that the population of immune cells invading dystrophin-deficient muscle is utterly dissimilar from the population observed after mechanical injury in normal muscle. Moreover, the invasion by immune cells is an early event, clearly detectable from the onset of the first degenerative episode in *mdx* mice.

Myeloid cell population present in dystrophin-lacking muscles includes macrophages, neutrophils, eosinophils, and mast cells. They present a high potential of generation of reactive oxygen species (ROS) and, indeed, elevated concentrations of lipid peroxidation products were detected in the serum of *mdx* mice (Ragusa *et al.*, 1997). Moreover, the effects of ROS could be exaggerated in *mdx* muscle by the reduced production of NO, acting as a scavenger. If this production is normalized as in transgenic mice expressing high levels of NO synthase, many signs of dystrophy are reduced (macrophage invasion and activation, plasmatic CPK level) (Wehling *et al.*, 2001). Finally, an elevated concentration of mast cells can promote interstitial fibrosis (Granchelli *et al.*, 1995).

T-cells are represented by an elevated concentration of $CD8^+$ and $CD4^+$ cells expressing activation markers. The latter characteristics suggests an autoimmune component, and expression of MHC class I has been observed in DMD fibers (Appleyard *et al.*, 1985) (while normal muscle do not express MHC). Conversely, an early (at 6 days of age) depletion of $CD4^+$ and/or $CD8^+$ cells greatly reduced the histologic signs of the disease in muscles examined at the usual time of massive degeneration. The same was observed in *mdx* mice where the perforin-mediated killing has been genetically suppressed (Spencer *et al.*, 1997). One of the effects of T cells is to induce apoptosis through liberation of perforin. Various markers can be used to visualize cells undergoing apoptosis: *in situ* labeling of DNA fragmentation, expression of ubiquitin, and penetration of vital dyes (e.g., Evans blue, a late and nonspecific marker). Apoptosis was detected in *mdx* muscle prior the beginning of the first degenerative episode (Tidball, 1995); the highest occurrence of apoptotic nuclei coincided with the height of the degeneration wave and decreased afterwards, though remaining higher than in normal muscle where the event is extremely rare. Parvalbumin deficiency in *mdx* muscles is also associated with a higher frequency of apoptosis (Raymackers *et al.*, 2003). *mdx* mice housed in a cage with an exercise wheel show a much higher frequency of apoptosis after continuous running (Sandri *et al.*, 1997). After exercise, apoptotic nuclei are observed both in myo fibers and in interstitial cells, including satellite cells (Rossini *et al.*, 2000); the latter localization may indicate a reduction of the regeneration potential.

In summary, studies of the inflammatory status in *mdx* mice suggest that invasion and activation of inflammatory cells occur early in the life of the animal, most likely soon after the beginning of fiber necrosis, but that they are not the *primum movens* of the dystrophinopathy cascade. However, by the combined action of apoptosis induction, liberation of cytokines and reactive oxygen species, and promotion of fibrosis, the inflammatory response can greatly amplify the initial pathologic process resulting from the lack of dystrophin. The use of anti-inflammatory drugs as an adjuvant therapy in DMD finds here its justification. However, a differential study of the evolution of lymph node morphology and cell populations during the degeneration phase (3–5 weeks of age) and the subsequent regeneration period in *mdx* mice suggest that the immune system may mitigate myonecrosis by increased production of immunoglobulin and interferon-γ (Lagrota-Candido *et al.*, 2002).

20.4.8. Gene Up- and Down-Regulations

20.4.8.1. Overview

Differential analysis of normal and dystrophic muscles revealed up- or downregulations of several tens of genes. Detailed lists can be found in Bakay *et al.* (2002), Boer *et al.* (2002), and Rouger *et al.* (2002). To make a correct and useful interpretation of this information, one has to distinguish between gene up/down regulations that contribute to the pathologic phenotype, those that are mere consequences of it, and those that reveals activation of compensatory mechanisms that tend to limit the gravity of the disease. Two examples of the latter cases are (1) the regeneration of fibers from the activation of satellite cells (described in Section 20.4.1.2) and (2) the upregulation of the expression of utrophin.

20.4.8.2. Utrophin

As introduced in the previous chapter, utrophin is a protein presenting a fairly high degree of homology with dystrophin, especially in the NH_2- and COOH-terminals that are the most important functional regions for binding to actin and to β-dystroglycan, respectively. The main structural difference is a slightly shorter rod domain [for a review, see Blake *et al.* (2002)]. Utrophin is encoded on chromosome 6 (Khurana *et al.*, 1990) and is thus not affected in DMD. Utrophin is expressed in many organs and in young developing muscle fibers, but its expression is restricted to myotendinous and neuromuscular junctions where it is precisely colocalized with ACh-receptors in adult, differentiated fibers. Its physiologic role remains obscure as null-mutant mice for utrophin do not present notable pathology of the neuromuscular transmission, in spite of some structural alterations (Deconinck *et al.*, 1997b). Utrophin is upregulated in the absence of dystrophin, in the muscle of both *mdx* mice (Matsumara *et al.*, 1993) and of DMD patients (Mizumo *et al.*, 1993; Vainzof *et al.*, 1995) and is present along the whole length and periphery of the fibers. The structural homologies between dystrophin and utrophin led to the hypothesis that the latter could compensate, at least partially, for the lack of dystrophin and delay the progress of the diseases (Karpati, 1997). Indeed, double null-mutant mice, deficient in both dystrophin and utrophin, present a dramatically severe form of muscle dystrophy and an early death (Deconinck *et al.*, 1997c; Deconinck *et al.*, 1998) but that can be rescued by reexpression of utrophin (Rafael *et al.*, 1998).

In transgenic *mdx* mice overexpressing high amounts of utrophin, a complete correction of the dystrophic parameters is obtained. This includes a correct relocalization of the dystro-sarcoglycan complex in the membrane, a quasi disappearance of the centronucleated fibers, the normalization of the CPK plasma level, and a complete recovery of the muscle force and resistance to eccentric contractions (Deconinck *et al.*, 1997a; Tinsley *et al.*, 1998). Thus utrophin can act as a very efficient surrogate for dystrophin, able to correct structural and functional disorders of dystrophinopathy, an observation that opens therapeutic perspectives. This was challenged by authors reporting that DMD patients who presented significant (spontaneous) overexpression of utrophin in their fibers showed no obvious amelioration of their dystrophic status in comparison with other patients where utrophin expression was minimal (Vainzof *et al.*, 1995). It should be kept in mind that, if utrophin is to replace dystrophin, it must be present in amounts sufficient to reestablish the mechanical strength of the muscle membrane. To assess correctly the amounts of utrophin detected on Western blots, quantification should be made with known amounts of the protein (Rybakova *et al.*, 2002). In the transgenic *mdx* mouse line ("Fiona") where recovery was complete (Tinsley *et al.*, 1998), utrophin expression exceeded by 10-fold the spontaneous overexpression in the *mdx* mouse (it was first reported that expression in the Fiona line was ~3-fold over *mdx*, but it turned out to be 10-fold, after precise quantification). Below that level, recovery was partial or negligible (Deconinck *et al.*, 1997a; Squire *et al.*, 2002). Most likely, the overexpressions of utrophin observed in DMD patients did not attain the threshold amount to confer recovery of the mechanical resistance.

20.5. Conclusion

Dystrophin plays a major role in organizing a large heteromeric, transmembranous complex linking the cytoskeletal actin filaments to the extracellular matrix. Most likely, this complex provides mechanical resistance to the muscle membrane and ensures transmission of lateral components of the contractile force. In dystrophin-deficient fibers, the complex disorganizes and the muscle membrane is no longer able to sustain the high mechanical stress generated during contraction. Mechanical disruption of the membrane is, most likely, the first pathologic event triggering fiber necrosis. In turn, this generates downstream and amplifying consequences; for example, loss of intracellular Ca^{2+} homeostasis, activation of Ca-dependent proteolysis, inflammatory reactions, apoptosis, fibrosis, and vascular problems. For a while, muscle degeneration is compensated by active regeneration from a pool of dormant muscle-dedicated stem cells, but eventually degeneration takes over.

Besides its structural role, dystrophin seems to act as a mechanical sensor conveying information to mechanical-sensitive ionic channels. Whether the loss of this function contributes to the dystrophic process is still unresolved.

Discovery of the product of the DMD gene, dystrophin, in 1987 opened utterly new avenues in our understanding of muscle physiology and of Duchenne muscular dystrophy. The thorough analysis of the *mdx* mouse phenotype greatly contributed to our understanding of the role of dystrophin, which is primarily structural. Even if pharmacological therapies (e.g., anti-inflammatory steroids) (Kinali *et al.*, 2002) proved effective in moderating the dystrophic process, the definite cure of DMD must aim at restoring the mechanical strength of the muscle membrane. In the *mdx* mouse, this has been first achieved by the reexpression of dystrophin, either from external transgenes carried by viral vectors [see review by Wells and Wells (2002)] or by skipping the mutated exon 23 to restore the reading frame for the dystrophin gene translation (Goyenvalle *et al.*, 2004; Lu *et al.*, 2003). The latter is a most promising approach as 43% of DMD patients could potentially benefit from single exon skipping (Aartsma-Rus *et al.*, 2004). A second approach, also very successful in the *mdx* mouse, was to restore membrane resistance by overexpressions of either utrophin, the dystrophin homologue protein that reassembles the dystro-sarcoglycan complex (Tinsley *et al.*, 1998), or of the $\alpha7\beta1$ integrin that also links the cytoskeleton to the extracellular matrix, in parallel with the dystrophin complex (Burkin *et al.*, 2001). These two latter cases exemplify the importance—and the relative lack of specificity—of the structural/mechanical factor.

References

Aartsma-Rus, A., Janson, A.A., Kaman, W.E., Bremmer-Bout, M., van Ommen, G.J., den Dunnen, J.T. and van Deutekom, J.C. (2004). Antisense-induced multiexon skipping for Duchenne muscular dystrophy makes more sense. *Am. J. Hum. Genet.* 74:83–92.

Alderton, J.M. and Steinhardt, R.A. (2000). Calcium influx through calcium leak channels is responsible for the elevated levels of calcium-dependent proteolysis in dystrophic myotubes. *J. Biol. Chem.* 275:9452–9460.

Appleyard, S.T., Dunn, M.J., Dubowitz, V. and Rose, M.L. (1985). Increased expression of HLA ABC class I antigens by muscle fibres in Duchenne muscular dystrophy, inflammatory myopathy, and other neuromuscular disorders. *Lancet* 325:361–363.

Argov, Z., Lofberg, M. and Arnold, D.L. (2000). Insights into muscle diseases gained by phosphorus magnetic resonance spectroscopy. *Muscle Nerve* 23:1316–1334.

Bakay, M., Zhao, P., Chen, J. and Hoffman, E.P. (2002). A web-accessible complete transcriptome of normal human and DMD muscle. *Neuromuscul. Disord.* 12 Suppl 1:S125–141.

Balnave, C.D. and Allen, D.G. (1995). Intracellular calcium and force in single mouse muscle fibres following repeated contractions with stretch. *J. Physiol.* 488 (Pt 1):25–36.

Balnave, C.D. and Allen, D.G. (1998). Evidence for Na^+/Ca^{2+} exchange in intact single skeletal muscle fibers from the mouse. *Am. J. Physiol.* 274:C940–946.

Balnave, C.D., Davey, D.F. and Allen, D.G. (1997). Distribution of sarcomere length and intracellular calcium in mouse skeletal muscle following stretch-induced injury. *J. Physiol.* 502 (Pt 3):649–659.

Balon, T.W. and Nadler, J.L. (1997). Evidence that nitric oxide increases glucose transport in skeletal muscle. *J. Appl. Physiol.* 82:359–363.

Bansal, D., Miyake, K., Vogel, S.S., Groh, S., Chen, C.C., Williamson, R., McNeil, P.L. and Campbell, K.P. (2003). Defective membrane repair in dysferlin-deficient muscular dystrophy. *Nature* 423:168–172.

Blake, D.J., Weir, A., Newey, S.E. and Davies, K.E. (2002). Function and genetics of dystrophin and dystrophin-related proteins in muscle. *Physiol. Rev.* 82:291–329.

Bockhold, K.J., Rosenblatt, J.D. and Partridge, T.A. (1998). Aging normal and dystrophic mouse muscle: analysis of myogenicity in cultures of living single fibers. *Muscle Nerve* 21:173–183.

Bodensteiner, J.B. and Engel, A.G. (1978). Intracellular calcium accumulation in Duchenne dystrophy and other myopathies: a study of 567,000 muscle fibres in 114 biopsies. *Neurology* 28:439–446.

Boer, J.M., de Meijer, E.J., Mank, E.M., van Ommen, G.B. and den Dunnen, J.T. (2002). Expression profiling in stably regenerating skeletal muscle of dystrophin-deficient mdx mice. *Neuromuscul. Disord.* 12 Suppl 1:S118–124.

Boesch, C. and Kreis, R. (2000). Observation of intramyocellular lipids by 1H-magnetic resonance spectroscopy. *Ann. N. Y. Acad. Sci.* 904:25–31.

Boland, B., Himpens, B., Casteels, R. and Gillis, J.M. (1993). Lack of dystrophin but normal calcium homeostasis in the smooth muscle from dystrophic mdx mice. *J. Muscle Res. Cell Motil.* 14:133–139.

Boland, B., Himpens, B., Denef, J.F. and Gillis, J.M. (1995). Site-dependent pathological differences in smooth muscles and skeletal muscles of the adult mdx mouse. *Muscle Nerve* 18:649–657.

Brenman, J.E., Chao, D.S., Gee, S.H., McGee, A.W., Craven, S.E., Santillano, D.R., Wu, Z., Huang, F., Xia, H., Peters, M.F., Froehner, S.C. and Bredt, D.S. (1996). Interaction of nitric oxide synthase with the postsynaptic density protein PSD-95 and alpha1-syntrophin mediated by PDZ domains. *Cell* 84:757–767.

Bulfield, G., Siller , W.G., Wight, P.A.L. and Moore, K.J. (1984). X chromosome-linked muscular dystrophy (mdx) in the mouse. *Proc. Natl. Acad. Sci. USA* 81:1189–1192.

Burdi, R., Didonna, M.P., Pignol, B., Nico, B., Mangieri, D., Rolland, J.F., Camerino, C., Zallone, A., Ferro, P., Andreetta, F., Confalonieri, P. and De Luca, A. (2006). First evaluation of the potential effectiveness in muscular dystrophy of a novel chimeric compound, BN 82270, acting as calpain-inhibitor and anti-oxidant. *Neuromuscular Disorders* 16:237–248.

Burkin, D.J., Wallace, G.Q., Nicol, K.J., Kaufman, D.J. and Kaufman, S.J. (2001). Enhanced expression of the alpha 7 beta 1 integrin reduces muscular dystrophy and restores viability in dystrophic mice. *J. Cell Biol.* 152:1207–1218.

Carlson, C.G. and Makiejus, R.V. (1990). A noninvasive procedure to detect muscle weakness in the mdx mouse. *Muscle Nerve* 13:480–484.

Carlson, C.G. and Roshek, D.M. (2001). Adult dystrophic (mdx) endplates exhibit reduced quantal size and enhanced quantal variation. *Pflugers Arch.* 442:369–375.

Carlson, C.G., Gueorguiev, A., Roshek, D.M., Ashmore, R., Chu, J.S. and Anderson, J.E. (2003). Extrajunctional resting Ca2+ influx is not increased in a severely dystrophic expiratory muscle (triangularis sterni) of the mdx mouse. *Neurobiol. Dis.* 14:229–239.

Carpenter, J.L., Hoffman, E.P., Romanul, F.C., Kunkel, L.M., Rosales, R.K., Ma, N.S., Dasbach, J.J., Rae, J.F., Moore, F.M. and McAfee, M.B. (1989). Feline muscular dystrophy with dystrophin deficiency. *Am. J. Pathol.* 135:909–919.

Carter, G.T., Kikuchi, N., Abresch, R.T., Walsh, S.A., Horasek, S.J. and Fowler WM, J.R. (1994). Effects of exhaustive concentric and eccentric exercise on murine skeletal muscle. *Arch. Phys. Med. Rehabil.* 75:555–559.

Carter, G.T., Wineinger, M.A., Walsh, S.A., Horasek, S.J., Abresch, R.T. and Fowler WM, J.r. (1995). Effect of voluntary wheel-running exercise on muscles of the mdx mouse. *Neuromuscul. Disord.* 5:323–332.

Chakravarthy, M.V., Spangenburg, E.E. and Booth, F.W. (2001). Culture in low levels of oxygen enhances *in vitro* proliferation potential of satellite cells from old skeletal muscles. *Cell. Mol. Life Sci.* 58:1150–1158.

Coirault, C., Pignol, B., Cooper, R.N., Butler-Browne, G., Chabrier, P.E. and Lecarpentier, Y. (2003). Severe muscle dysfunction precedes collagen tissue proliferation in mdx mouse diaphragm. *J. Appl. Physiol.* 94:1744–1750.

Cornu, C., Goubel, F. and Fardeau, M. (2001). Muscle and joint elastic properties during elbow flexion in Duchenne muscular dystrophy. *J. Physiol.* 533:605–616.

Cottin, P., Thompson, V.F., Sathe, S.K., Szpacenko, A. and Goll, D.E. (2001). Autolysis of mu- and m-calpain from bovine skeletal muscle. *Biol. Chem.* 382:767–776.

Coulton, G.R., Morgan, J.E., Partridge, T.A. and Sloper, J.C. (1988). The mdx mouse skeletal muscle myopathy: I a histological, morphometric and biochemical investigation. *Neuropathol. Appl. Neurobiol.* 14:53–70.

De Backer, F., Vandebrouck, C., Gailly, P. and Gillis, J.M. (2002). Long-term study of Ca2+ homeostasis and of survival in collagenase-isolated muscle fibres from normal and mdx mice. *J. Physiol.* 542:855–865.

De Luca, A., Pierno, S. and Camerino, D.C. (1997). Electrical properties of diaphragm and EDL muscles during the life of dystrophic mice. *Am. J. Physiol.* 272:C333–340.

De Luca, A., Pierno, S., Liantonio, A., Cetrone, M., Camerino, C., Simonetti, S., Papadia, F. and Camerino, D.C. (2001). Alteration of excitation-contraction coupling mechanism in extensor digitorum longus muscle fibres of dystrophic mdx mouse and potential efficacy of taurine. *Br. J. Pharmacol.* 132:1047–1054.

Decary, S., Hamida, C.B., Mouly, V., Barbet, J.P., Hentati, F. and Butler-Browne, G.S. (2000). Shorter telomeres in dystrophic muscle consistent with extensive regeneration in young children. *Neuromuscul. Disord.* 10:113–120.

Deconinck, A.E., Potter, A.C., Tinsley, J.M., Wood, S.J., Vater, R., Young, C., Metzinger, L., Vincent, A., Slater, C.R. and Davies, K.E. (1997b). Postsynaptic abnormalities at the neuromuscular junctions of utrophin-deficient mice. *J. Cell Biol.* 136:883–894.

Deconinck, A.E., Rafael, J.A., Skinner, J.A., Brown, S.C., Potter, A.C., Metzinger, L., Watt, D.J., Dickson, J.G., Tinsley, J.M. and Davies, K.E. (1997c). Utrophin-dystrophin-deficient mice as a model for Duchenne muscular dystrophy. *Cell* 90:717–727.

Deconinck, N. (1998). *Functional evaluation of gene therapy in a mouse model of Duchenne muscular dystrophy*. PhD Thesis: Catholic University of Louvain. Bruxelles, Belgium.

Deconinck, N., Rafael, J.A., Beckers-Bleukx, G., Kahn, D., Deconinck, A.E., Davies, K.E. and Gillis, J.M. (1998). Consequences of the combined deficiency in dystrophin and utrophin on the mechanical properties and myosin composition of limb and respiratory muscles of the mouse. *Neuromuscul. Disord.* 8:362–370.

Deconinck, N., Ragot, T., Maréchal, G., Pérricaudet, M. and Gillis, J.M. (1996). Functional protection of dystrophic mouse (mdx) muscles after adenovirus-mediated transfer of a dystrophin minigene. *Proc. Natl. Acad. Sci. USA* 93:3570–3574.

Deconinck, N., Tinsley, J., De Backer, F., Fisher, R., Kahn, D., Phelps, S., Davies, K. and Gillis, J.M. (1997a). Expression of truncated utrophin leads to major functional improvements in dystrophin-deficient muscles of mice. *Nat. Med.* 3:1216–1221.

Decostre, V., Gailly, P., Debaix, H., Colson-Van Schoor, M., Cao, M.L. and Gillis, J.M. (2002). Intracellular pH regulation in isolated fast-twitch skeletal muscle from dystrophin-deficient mouse. *Neuromuscul. Disord.* 12:447–456.

DiMario, J.X., Uzman, A. and Strohman, R.C. (1991). Fiber regeneration is not persitent in dystrophic (MDX) mouse sleletal muscle. *Develop. Biol.* 148:314–321.

Dourdin, N., Brustis, J.J., Balcerzak, D., Elamrani, N., Poussard, S., Cottin, P. and Ducastaing, A. (1997). Myoblast fusion requires fibronectin degradation by exteriorized m-calpain. *Exp. Cell Res.* 235:385–394.

Duchenne, G.B.A. (1868). Recherches sur la paralysie musculaire pseudo-hypertrophique ou paralysie myo-sclérotique. *Archives générales de Médecine* 11:5–25, 179–209, 305–321, 421–443, 552–588.

Dunn, J.F., Tracey, I. and Radda, G.K. (1992). A 31P-NMR study of muscle exercice metabolism in mdx mice : evidence for abnormal pH regulation. *J. Neurol. Sci.* 113:108–113.

Emery, A.E.H. and Emery, M.L.H. (1995). *The history of a genetic diseases: Duchenne Muscular Dystrophy or Meryon's Disease*. The Royal Society of Medecine Press Ltd, London.

Fougerousse, F., Gonin, P., Durand, M., Richard, I. and Raymackers, J.M. (2003). Force impairment in calpain 3-deficient mice is not correlated with mechanical disruption. *Muscle Nerve* 27:616–623.

Franco-Obregón, A.J. and Lansman, J.B. (1994). Mechanosensitive ion channels in skeletal muscle from normal and dystrophic mice. *J. Physiol.* 481:299–309.

Fraysse, B., Liantonio, A., Cetrone, M., Burdi, R., Pierno, S., Frigeri, A., Pisoni, M., Camerino, C. and De Luca, A. (2004). The alteration of calcium homeostasis in adult dystrophic mdx muscle fibers is worsened by a chronic exercise *in vivo*. *Neurobiol. Dis.* 17:144–154.

Gailly, P., Boland, B., Himpens, B., Casteels, R. and Gillis, J.M. (1993a). Critical evaluation of cytosolic calcium determination in resting muscle fibres from normal and dystrophic (mdx) mice. *Cell Calcium* 14:473–483.

Gailly, P., Hermans, E., Octave, J.N. and Gillis, J.M. (1993b). Specific increase of genetic expression of parvalbumin in fast skeletal muscle of mdx mice. *FEBS Lett.* 326:272–274.

Gillis, J.M. (1985). Relaxation of vertebrate skeletal muscle. A synthesis of the biochemical and physiological approaches. *Biochim. Biophys. Acta* 811:97–145.

Gillis, J.M. (1996). The mdx mouse: why diaphragm? *Muscle Nerve* 19:1230.

Gillis, J.M. (1999). Understanding dystrophinopathies: an inventory of the structural and functional consequences of the absence of dystrophin in muscles of the mdx mouse [In Process Citation]. *J. Muscle Res. Cell Motil.* 20:605–625.

Goll, D.E., Thompson, V.F., Li, H., Wei, W. and Cong, J. (2003). The calpain system. *Physiol. Rev.* 83:731–801.

Goyenvalle, A., Vulin, A., Fougerousse, F., Leturcq, F., Kaplan, J.C., Garcia, L. and Danos, O. (2004). Rescue of dystrophic muscle through U7 snRNA-mediated exon skipping. *Science* 306:1796–1799.

Granchelli, J.A., Pollina, C. and Hudecki, M.S. (1995). Duchenne-like myopathy in double-mutant mdx mice expressing exaggerated mast cell activity. *J. Neurol. Sci.* 131:1–7.

Hack, A.A., Cordier, L., Shoturma, D.I., Lam, M.Y., Sweeney, H.L. and McNally, E.M. (1999). Muscle degeneration without mechanical injury in sarcoglycan deficiency. *Proc. Natl. Acad. Sci. USA* 96:10723–10728.

Hack, A.A., Lam, M.Y., Cordier, L., Shoturma, D.I., Ly, C.T., Hadhazy, M.A., Hadhazy, M.R., Sweeney, H.L. and McNally, E.M. (2000). Differential requirement for individual sarcoglycans and dystrophin in the assembly and function of the dystrophin-glycoprotein complex. *J. Cell Sci.* 113 (Pt 14):2535–2544.

Han, R., Grounds, M.D. and Bakker, A.J. (2006). Measurement of sub-membrane [Ca^{2+}] in adult myofibers and cytosolic [Ca^{2+}] in myotubes from normal and *mdx* mice using the Ca^{2+} indicator FFP-18. *Cell Calcium* 40:299–307.

Head, S.I. (1993). Membrane potential, resting calcium and calcium transients in isolated muscle fibres from normal and dystrophic mice. *J. Physiol.* 469:11–19.

Head, S.I., Bakker, A.J. and Liangas, G. (2004). EDL and soleus muscles of the C57BL6J/dy2j laminin-{alpha}2-deficient dystrophic mouse are not vulnerable to eccentric contractions. *Exp. Physiol.* 89:531–539.

Heizmann, C.W., Berchtold, M.W. and Rowlerson, A.M. (1982). Correlation of parvalbumin concentration with relaxation speed in mammalian muscles. *Proc. Natl. Acad. Sci. USA* 79:7243–7247.

Hill, M. and Goldspink, G. (2003). Expression and splicing of the insulin-like growth factor gene in rodent muscle is associated with muscle satellite (stem) cell activation following local tissue damage. *J. Physiol.* 549:409–418.

Hopf, F.W., Turner, P.R., Denetclaw, W.F., Reddy, P. and Steinhardt, R.A. (1996). A critical evaluation of resting intracellular free calcium regulation in dystrophic mdx muscle. *Am. J. Physiol.* 271:C1325–C1339.

Hou, T.T., Johnson, J.D. and Rall, J.A. (1991). Parvalbumin content and Ca2+ and Mg2+ dissociation rates correlated with changes in relaxation rate of frog muscle fibres. *J. Physiol. (London)* 441:285–304.

Huxtable, R.J. and Bressler, R. (1973). Effect of taurine on a muscle intracellular membrane. *Biochim. Biophys. Acta* 323:573–578.

Imbert, N., Cognard, C., Duport, G., Guillou, C. and Raymond, G. (1995). Abnormal calcium homeostasis in Duchenne muscular dystrophy myotubes contracting *in vitro*. *Cell Calcium* 18:177–186.

Kargacin, M.E., Kargacin, G.J. (1996). The sarcoplasmic reticulum calcium pump is functionally altered in dystrophic muscles. *Biochim. Biophys. Acta* 1290:4–8.

Karpati, G. and Carpenter, S. (1986). Small-caliber skeletal muscle fibers do not suffer deleterious consequences of dystrophic gene expression. *Am. J. Med. Genet.* 25:653–658.

Karpati, G. (1997). Utrophin muscles is on the action. *Nat. Med* 3:22–23.

Khurana, T.S., Hoffman, E.P. and Kunkel, L.M. (1990). Identification of a chromosome 6-encoded dystrophin-related protein. *J. Biol. Chem.* 265:16717–16720.

Kinali, M., Mercuri, E., Main, M., Muntoni, F. and Dubowitz, V. (2002). An effective, low-dosage, intermittent schedule of prednisolone in the long-term treatment of early cases of Duchenne dystrophy. *Neuromuscul. Disord.* 12 Suppl 1:S169–174.

Kipling, D. (1997). Telomere structure and telomerase expression during mouse development and tumorigenesis. *Eur. J. Cancer* 33:792–800.

Kohn, B., Guscetti, F., Waxenberger, M. and Augsburger, H. (1993). Muscular dystrophy in a cat. *Tierarztl. Prax.* 21:451–457.

Kometani, K., Tsugeno, H. and Yamada, K. (1990). Mechanical and energetic properties of dystrophic (mdx) mouse muscle. *Jpn. J. Physiol.* 40:541–549.

Lagrota-Candido, J., Vasconcellos, R., Cavalcanti, M., Bozza, M., Savino, W. and Quirico-Santos, T. (2002). Resolution of skeletal muscle inflammation in mdx dystrophic mouse is accompanied by increased immunoglobulin and interferon-gamma production. *Int. J. Exp. Pathol.* 83:121–132.

Lau, K.S., Grange, R.W., Chang, W.J., Kamm, K.E., Sarelius, I. and Stull, J.T. (1998). Skeletal muscle contractions stimulate cGMP formation and attenuate vascular smooth muscle myosin phosphorylation via nitric oxide. *FEBS Lett.* 431:71–74.

Louis, M., Lebacq, J., Poortmans, J.R., Belpaire-Dethiou, M.C., Devogelaer, J.P., Van Hecke, P., Goubel, F. and Francaux, M. (2003). Beneficial effects of creatine supplementation in dystrophic patients. *Muscle Nerve* 27:604–610.

Lu, Q.L., Mann, C.J., Lou, F., Bou-Gharios, G., Morris, G.E., Xue, S.A., Fletcher, S., Partridge, T.A. and Wilton, S.D. (2003). Functional amounts of dystrophin produced by skipping the mutated exon in the mdx dystrophic mouse. *Nat. Med.* 9:1009–1014.

Mallouk, N., Jacquemond, V. and Allard, B. (2000). Elevated subsarcolemmal Ca2+ in mdx mouse skeletal muscle fibers detected with Ca2+-activated K+ channels. *Proc. Natl. Acad. Sci. USA* 97:4950–4955.

Mallouk, N. and Allard, B. (2002). Ca(2+) influx and opening of Ca(2+)-activated K(+) channels in muscle fibers from control and mdx mice. *Biophys. J.* 82:3012–3021.

Maréchal, G. and Beckers-Bleukx, G. (1988). Force-velocity and myosin isozymes in muscles of dystrophic (mdx) mice. *Arch. Int. Physiol. Biochim.* 96:P16.

Matsumara, K., Ervasti, J., Ohlendieck, K., Kahl, S. and Campbell, K. (1993). Association of dystrophin-related protein with dystrophin associated proteins in mdx mouse muscle. *Nature* 360:588–591.

McCarter, G.C. and Steinhardt, R.A. (2000). Increased activity of calcium leak channels caused by proteolysis near sarcolemmal ruptures. *J. Membr. Biol.* 176:169–174.

Mizumo, Y., Nonaka, I., Hirai, S. and Ozawa, E. (1993). Reciprocal expression of dystrophin and utrophin in muscles of Duchenne muscular dystrophy patients, female DMD carriers and control subjects. *J. Neurol. Sci.* 119:43–52.

Moens, P., Baatsen, P.H.W.W. and Maréchal, G. (1993). Increased susceptibility of EDL muscles from mdx mice to damage induced by contractions with stretch. *J. Muscle Res. Cell Motil.* 14:446–451.

Nicole, S., Desforges, B., Millet, G., Lesbordes, J., Cifuentes-Diaz, C., Vertes, D., Cao, M.L., De Backer, F., Languille, L., Roblot, N., Joshi, V., Gillis, J.M. and Melki, J. (2003). Intact satellite cells lead to remarkable protection against Smn gene defect in differentiated skeletal muscle. *J. Cell Biol.* 161:571–582.

Nussbaum, R.L., McInnes, R.R. and Willard, H.F. (2001). *Thompson & Thompson Genetics in Medicine*. W.B.Saunders, Philadelphia. (6th ed.).

Rafael, J.A., Tinsley, J.M., Potter, A.C., Deconinck, A.E. and Davies, K.E. (1998). Skeletal muscle-specific expression of a utrophin transgene rescues utrophin-dystrophin deficient mice. *Nat. Genet.* 19:79–82.

Ragusa, R.J., Chow, C.K. and Porter, J.D. (1997). Oxidative stress as a potential pathogenic mechanism in an animal model of Duchenne muscular dystrophy. *Neuromuscul. Disord.* 7:379–386.

Raymackers, J.M. (2003). *Pathogenic mechanisms in animal models of Duchenne muscular dystrophy*. PhD Thesis: Catholic University of Louvain. Bruxelles, Belgium.

Raymackers, J.M., Debaix, H., Colson-VanSchoor, M., De Backer, F., Tajeddine, N., Schwaller, B., Gailly, P. and Gillis, J.M. (2003). Consequence of parvalbumin deficiency in the mdx mouse : histological, biochemical and mechanical phenotype of a new double mutant. *Neuromuscul. Disord.* 13:376–387.

Remmers, J.E. (1976). Analysis of ventilatory responses. *Chest* 70 Suppl: 134–137.

Rosser, B.G., Powers, S.P. and Gores, G.J. (1993). Calpain activity increases in hepatocytes following addition of ATP. Demonstration by a novel fluorescent approach. *J. Biol. Chem.* 268:23593–23600.

Rossini, K., Dona, A., Sandri, M., Destro, C., Dona, M. and Carraro, U. (2000). Time-course of exercise and apotosis in dystrophin-deficient muscle of mice. *Basic Appl. Myol.* 10:33–38.

Rouger, K., Le Cunff, M., Steenman, M., Potier, M.C., Gibelin, N., Dechesne, C.A. and Leger, J.J. (2002). Global/temporal gene expression in diaphragm and hindlimb muscles of dystrophin-deficient (mdx) mice. *Am. J. Physiol. Cell. Physiol.* 283:C773–784.

Rybakova, I.N., Patel, J.R. and Ervasti, J.M. (2000). The dystrophin complex forms a mechanically strong link between the sarcolemma and costameric actin. *J. Cell Biol.* 150:1209–1214.

Rybakova, I.N., Patel, J.R., Davies, K.E., Yurchenco, P.D. and Ervasti, J.M. (2002). Utrophin binds laterally along actin filaments and can couple costameric actin with sarcolemma when overexpressed in dystrophin-deficient muscle. *Mol. Biol. Cell* 13:1512–1521.

Sandri, M., Podhorska-Okolov, M., Geromel, V., Rizzi, C., Arslan, P., Franceschi, C. and Carraro, U. (1997). Exercise induces myonuclear ubiquitination and apoptosis in dystrophin-deficient muscle of mice. *J. Neuropathol. Exp. Neurol.* 56:45–57.

Scheuerbrandt, G., Lundin, A., Lovgren, T. and Mortier, W. (1986). Screening for Duchenne muscular dystrophy: an improved screening test for creatine kinase and its application in an infant screening program. *Muscle Nerve* 9:11–23.

Sicinski, P., Geng, Y., Ryder-Cook, A.S., Barnard, E.A., Darlison, M.G. and Barnard, B.J. (1989). The molecular basis of muscular dystrophy in the mdx mouse: a point mutation. *Science* 244:1578–1580.

Spenser, M.J. and Tidball, J.G. (1992). Calpain concentration is elevated althought net calcium-dependent proteolysis is suppressed in dystrophin-deficient muscle. *Exp. Cell Res.* 203:107–114.

Spencer, M.J., Walsh, C.M., Dorshkind, K.A., Rodriguez, E.M. and Tidball, J.G. (1997). Myonuclear apoptosis in dystrophic mdx muscle occurs by perforin-mediated cytotoxicity. *J. Clin. Invest.* 99:2745–2751.

Spencer, M.J., Montecino-Rodriguez, E., Dorshkind, K. and Tidball, J.G. (2001). Helper (CD4(+)) and cytotoxic (CD8(+)) T cells promote the pathology of dystrophin-deficient muscle. *Clin. Immunol.* 98:235–243.

Spencer, M.J. and Mellgren, R.L. (2002). Overexpression of a calpastatin transgene in mdx muscle reduces dystrophic pathology. *Hum. Mol. Genet.* 11:2645–2655.

Squire, S., Raymackers, J.M., Vandebrouck, C., Potter, A., Tinsley, J., Fisher, R., Gillis, J.M. and Davies, K.E. (2002). Prevention of pathology in mdx mice by expression of utrophin: analysis using an inducible transgenic expression system. *Hum. Mol. Genet.* 11:3333–3344.

Stedman, H.H., Sweeney, H.L., Shrager, J.B., Maguire, H.C., Panetteri, R.A., Petrof, B., Narusawa, M., Leferovich, J.M., Sladky, J.T. and Kelly, A.M. (1991). The mdx mouse diaphragm reproduces the degenerative changes of Duchenne muscular dystrophy. *Nature* 352:536–539.

Takagi, A., Kojima, S., Ida, M. and Araki, M. (1992). Increased leakage of calcium ion from the sarcoplasmic reticulum of the mdx mouse. *J. Neurol. Sci.* 110:160–164.

Thomas, G.D., Sander, M., Lau, K.S., Huang, P.L., Stull, J.T. and Victor, R.G. (1998). Impaired metabolic modulation of alpha-adrenergic vasoconstriction in dystrophin-deficient skeletal muscle. *Proc. Natl. Acad. Sci. USA* 95:15090–15095.

Tidball, J.G., Albrecht, D.E., Lokensgard, B.E. , Spencer, M.J. (1995). Apoptosis precedes necrosis of dystrophin-deficient muscle. *J. Cell Sci.* 108:2197–2204.

Tinsley, J., Deconinck, N., Fisher, R., Kahn, D., Phelps, S., Gillis , J.M. and Davies, K. (1998). Expression of full-length utrophin prevents muscular dystrophy in mdx mice. *Nat. Med.* 4:1441–1444.

Turner, P.R., Westwood, T., Regen, C.M. and Steinhardt, R.A. (1988). Increased protein degradation results from elevated free calcium levels found in muscle from mdx mice. *Nature* 355:735–738.

Tutdibi, O., Brinkmeier, H., Rudel, R. and Fohr, K.J. (1999). Increased calcium entry into dystrophin-deficient muscle fibres of MDX and ADR-MDX mice is reduced by ion channel blockers. *J. Physiol.* 515:859–868.

Vainzof, M., Passos-Bueno, M.R., Man, N. and Zatz, M. (1995). Absence of correlation between utrophin localization and quantity and the clinical severity in Duchenne/Becker dystrophies. *Am. J. Med. Genet.* 58:305–309.

Vainzof, M., Zatz, M. and Otto, P.A. (1985). Serum CK-MB activity in progressive muscular dystrophy: is it of nosologic value? *Am. J. Med. Genet.* 22:81–87.

Valentine, B.A., Winand, N.J. and Pradhan, D. (1992). Canine X-linked muscular dystrophy as an animal model of Duchenne muscular dystrophy. *Am. J. Med. Genet.* 42:352–356.

van Essen, A.J., Abbs, S., Baiget, M., Bakker, E., Boileau, C., van Broeckhoven, C., Bushby, K., Clarke, A., Claustres, M., Covone, A.E., *et al.* (1992). Parental origin and germline mosaicism of deletions and duplications of the dystrophin gene: a European study. *Hum. Genet.* 88:249–257.

Verellen-Dumoulin, C., Freund, M., De Meyer, R., Laterre, C., Frederic, J., Thompson, M.W., Markovic, V.D. and Worton, R.G. (1984). Expression of an X-linked muscular dystrophy in a female due to translocation involving Xp21 and non-random inactivation of the normal X chromosome. *Hum. Genet.* 67:115–119.

Webster, C., Silberstein, L., Hays, A.P. and Blau, H.M. (1988). Fast muscle fibers are preferentially affected in Duchenne muscular dystrophy. *Cell* 52:502–513.

Webster, C. and Blau, H.M. (1990). Accelerated age-related decline in replicative life-span of Duchenne muscular dystrophy myoblasts: implications for cell and gene therapy. *Somat. Cell Mol. Genet.* 16:557–565.

Wehling, M., Spencer, M.J. and Tidball, J.G. (2001). A nitric oxide synthase transgene ameliorates muscular dystrophy in mdx mice. *J. Cell Biol.* 155:123–131.

Wells, D.J. and Wells, K.E. (2002). Gene transfer studies in animals: what do they really tell us about the prospects for gene therapy in DMD? *Neuromuscul. Disord.* 12 Suppl 1:S11–22.

Woods, C.E., Novo, D., DiFranco, M. and Vergara, J.L. (2004). The action potential-evoked sarcoplasmic reticulum calcium release is impaired in mdx mouse muscle fibres. *J. Physiol.* 557:59–75.

21

Eye Lens Proteins and Cataracts

Roger John Willis Truscott

Abstract

At first sight, the lens of the eye would appear to be an ideal environment for amyloid fibril formation. The protein concentration is the highest of any tissue in the body, the proteins are very long-lived, present in slightly acidic conditions and subjected over time to extensive truncation and post-translational modification. In addition, it has been demonstrated that lens crystallins can readily be induced to form amyloid fibrils *in vitro*. The situation may be further exacerbated after the onset of age-related nuclear cataract, which is characterized by massive oxidation of cysteine and methionine residues, accompanied by protein unfolding. Despite this, there is as yet no evidence for amyloid fibril formation in either the aged or the cataract human lens. Paradoxically, the reason may have to do with the supramolecular ordered β-sheet array that crystallins adopt once they are packed into mature fiber cells. This extended matrix in normal lenses displays some of the classic features normally associated with amyloid, for example, staining with Congo red and thioflavine T.

21.1. Lens Crystallins

A number of proteins have been recruited by different species to act as structural proteins in the lens (Wistow and Piatigorsky, 1987) (Table 21.1). A high concentration is necessary in order to provide an adequate refractive index, so one requirement for such a protein is high solubility.

Because the presence of the cellular machinery, such as that associated with protein synthesis and degradation in normal cells, appears to be incompatible with transparency to visible light, all organelles have been dispensed with in the lens center in favor of an alternative strategy that relies upon the long-term stability of the proteins. Thus proteins once synthesized by the lens cells, and subsequently packaged into mature fiber cells, are then largely on their own. In many cases, the proteins chosen are oxido-reductases and these sometimes retain vestigial activity (Wistow *et al.*, 1987).

21.2. Human Lens Crystallins

The major structural proteins present in the human lens are the α, β, and γ crystallins. The β and γ crystallins are related in structure (Fig. 21.1) and seem to be derived from an ancestral microbial stress protein (Jaenicke and Slingsby, 2001). In the microbial crystallin homologues, protein stability is enhanced by high-affinity binding of Ca^{2+}. Ca^{2+} binding by lens proteins is very important for the integrity of the lens (Duncan and Jacob, 1984; Truscott *et al.*, 1990),

Table 21.1. Examples of enzymes and stress proteins recruited to act as structural proteins in the lenses of different animal species [updated from Piatigorsky and Wistow (1989)]

Crystallins	Species	Enzyme/stress protein
Ubiquitous		
α	Vertebrates	Small heat shock protein
β and γ	Vertebrates	Microbial stress protein
Taxon specific		
	Scallop	Aldehyde dehydrogenase
	Birds, reptiles	Argininosuccinate lyase
	Birds, crocodiles	Lactate dehydrogenase
	Frogs	NADPH reductases
	Rabbit	Hydroxyacyl CoA dehydrogenase
	Guinea-pig	Alcohol dehydrogenase
	Squid	Glutathione *S*-transferase
	Monkey	Betaine-homocysteine methyl transferase
		NADPH quinone oxidoreductase
	Camel	Lactate dehydrogenase
	Platypus	

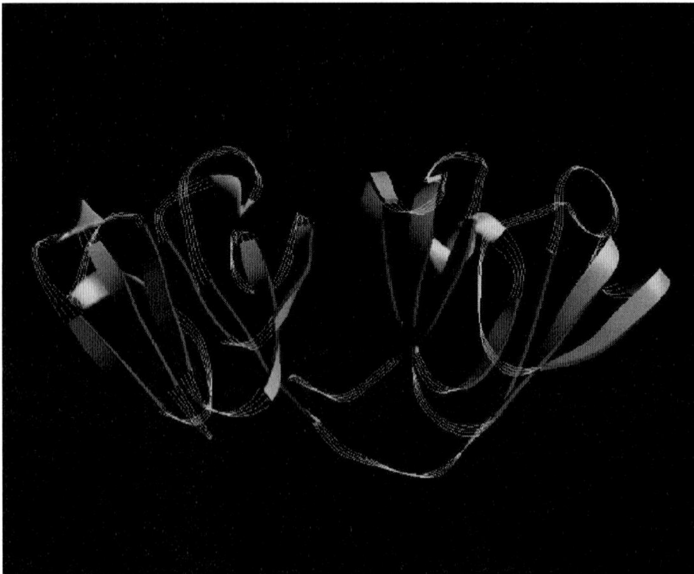

Figure 21.1. The structure of γB crystallin. This stable protein is synthesized prenatally and is a major component of the adult lens nucleus. The two homologous domains are linked by a six-residue peptide.

because the free Ca^{2+} concentration is normally maintained in the low micromolar range, whereas the total concentration of Ca^{2+} in the lens is millimolar. α Crystallin belongs to the family of small heat shock proteins (Horwitz, 2003). It acts as a chaperone in that it binds partially denatured proteins and, by forming high-molecular-weight aggregates, it maintains solubility. In the human lens there are two homologous α crystallin polypeptides: αA crystallin and αB crystallin. Both of these are phosphorylated to a minor extent (Spector *et al.*, 1985).

Three γ crystallins predominate: γC, γD, and γS crystallin (Lampi *et al.*, 1997), and, in the human lens, these are found together with five β crystallin polypeptides: βB1, βB2, βB3, βA1/A3, and βA4 (Lampi *et al.*, 1997). The γ crystallins are present as monomers (~20 kDa) whereas the β crystallins elute as two peaks by gel chromatography of lens extracts, corresponding with approximate masses of 60 kDa and 180 kDa. βB1 is found only in the high-molecular-weight fraction.

In addition to the crystallins, which comprise more than 90% of the proteins, the lens contains the other enzymes, receptors, and proteins typically associated with normal cells. Of special relevance to the lens is a high concentration of the integral membrane water channel, aquaporin 0 (Reizer *et al.*, 1993), connexins 43, 46, and 50, as well as a lens-specific cytoskeletal element (Quinlan *et al.*, 1996). Humans also manifest a lens-specific enzyme that glucosylates 3-hydroxykynurenine to produce the major primate lens UV filter, 3-hydroxykynurenine glucoside (van Heyningen, 1971a; Wood and Truscott, 1994).

21.3. Lens Growth and Differentiation

The human lens grows continuously throughout life. In adults, each lens adds approximately 1.5 mg per year of life (Harding, 1991). This process involves the differentiation of epithelial cells at the perimeter (equator) of the lens. Here the cuboidal epithelial cells begin to elongate, organelles are lost, and a great synthesis of structural proteins and membranes takes place leading ultimately to the formation of mature fiber cells that can stretch from the back to the front of the lens. Such cells may reach almost 1 cm in length and can be viewed simplistically as membranes filled with a high concentration of proteins. The center, or nucleus, of the lens therefore corresponds with that part of the lens that was synthesized before birth and the outer part (the cortex) to a region that is constantly growing by the addition of cells to a preexisting core. No cells are shed by the lens, and there is no evidence for significant protein synthesis or turnover in the lens nucleus.

The types of crystallins synthesized alter subtly after birth. γ Crystallins are predominately synthesized before birth. This class is replaced postnatally to some extent by γS crystallin that has a more open flexible structure (Cooper *et al.*, 1994b). A key feature in the design of a functional lens is protein packing, and short-range spatial order is important for transparency (Delaye and Tardieu, 1983). There are two aspects to this, which are related: dehydration of newly formed fiber cells and the interaction of individual crystallin polypeptides. Almost nothing is known about these two features.

In the outer, metabolically active region of the lens, the anterior epithelial and cortical fiber cells are more typical of other cells in the body. Here the protein concentration is less than 20% (Huizinga *et al.*, 1989). By contrast, the mature fiber cells of the nucleus contain 35% wet weight of protein. In man, this figure remains unchanged throughout our lifetime (Heys *et al.*, 2004), whereas in other species such as the rat, cow, and mouse, the concentration of protein continues to increase over time. A 2-year-old rat lens, for example, contains more than 65% protein! Species also vary greatly in protein content. Those that have evolved with the capacity to change the shape of the lens to assist focusing (e.g., birds and higher primates) generally have been found to have lower protein concentrations than those animals that do not (e.g., fish and rodents).

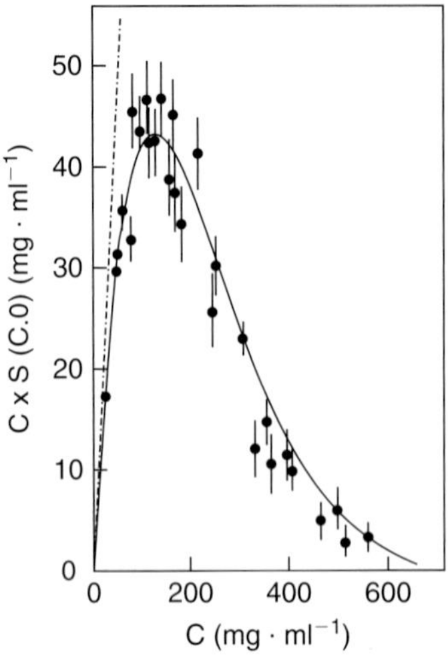

Figure 21.2. Light-scattering intensity as a function of protein concentration (C). Cytoplasmic extracts of calf lens cortices were examined by x-ray forward scattering (Delaye, 1983). The dashed line depicts the linear dependence of scattering expected from Rayleigh's equation if crystallins acted as independent scatterers. These data suggest a liquid or glass-like order of the crystallins in the lens that contributes to the transparency of the lens at physiologic protein concentrations.

As noted earlier, the short-range interactions of these highly concentrated protein solutions may be important for lens transparency. A simple experiment in which lens proteins were progressively diluted illustrates this (Fig. 21.2). Because three classes of proteins are present in the lens, one could assume that the packing and subunit interactions between individual crystallin polypeptides may also be important in lens function and transparency. Little is known about this (Cooper *et al.*, 1994a).

21.4. Factors that Promote Crystallin Stability Within the Lens

Proteins in the lens nucleus was synthesized prenatally. Because sight is the most important of our senses for survival, it is perhaps unsurprising that during the course of evolution, certain features were selected in order to enhance the stability of these proteins. In the case of humans, the crystallins must survive for several decades.

1. *Lens proteins*: Lens crystallins are inherently stable proteins (Jaenicke and Slingsby, 2001). A suite of proteases have been isolated from the lens and this at first may seem at odds with the overall function of a tissue that relies on the long-term integrity of its structural proteins. It is likely that many, if not all, of these proteases are relics of those that were required for the total digestion of organelles during terminal differentiation of the fiber cells. Native crystallins are not readily digested, although 10-kDa fragments of γ crystallin have been isolated from human lenses that correspond with the two "halves" of the molecule. In this case, cleavage has occurred at the peptide linking the two domains. Acetylation of the N-terminal residues of the polypeptides of the β and α crystallins presumably acts to protect these sites from exopeptidases. The acetylated N-terminal residues of the α crystallin subunits appear to be hidden in the interior of the high-molecular-weight α crystallin aggregate (600–800 kDa), the form in which it is extracted from lenses.

2. *Chaperones*: α Crystallin represents approximately one third of the total protein in the human lens. In 1992, Horwitz (Horwitz, 1992) discovered that α crystallin could prevent the heat-induced precipitation of either β or γ crystallins. An example of this simple experiment is

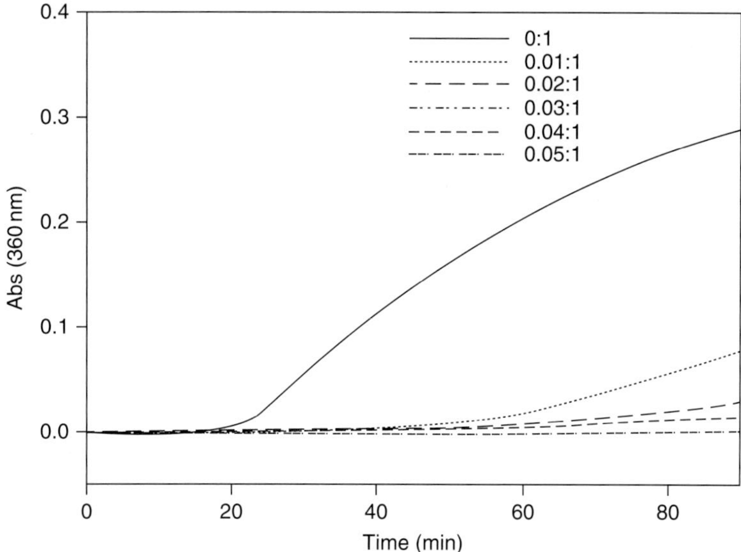

Figure 21.3. α Crystallin acting as a chaperone. A β-crystallin fraction isolated from bovine lenses by gel chromatography was incubated at 55°C in 50 mM phosphate buffer, pH 7.2, with different amounts of α crystallin. Protein precipitation was monitored using light scattering at 360 nm. The figures in the box are the ratios of the masses of α crystallin to β crystallin, respectively, used in each case.

presented in Fig. 21.3. It is very likely, but difficult to prove, that this is one of its functions in the lens. Purification of the high-molecular-weight α crystallin fraction from human lenses yields both subunits in the ratio of approximately 3:1 (αA:αB). Both the αA and αB crystallin polypeptides act as chaperones, however their efficacies depend on the target proteins, the particular assay conditions employed, and, to some extent, the modification (e.g., phosphorylation) of the α crystallin (Aquilina *et al.*, 2004). This apparent heterogeneity could have functional importance in the lens.

3. *Glutathione*: Oxidative damage may be one of the most important insults that the nuclear proteins need to be protected from. In support of this supposition, the lens has the highest glutathione concentration of any tissue in the body (Giblin, 2000), and this antioxidant is synthesized actively in the cortex. In response to an external oxidative stress, the pentose phosphate pathway can be upregulated to provide an increased flux of NADPH, the cofactor for glutathione reductase (Reddy and Giblin, 1984). As a result, primate lenses can withstand incubation in 1 mM H_2O_2. In part, this heightened resistance to external oxidative stress can be traced to levels of glutathione reductase that far exceed those of other animals (Rathbun *et al.*, 1986).

4. *Photooxidation and UV filters*: The main role of the lens is to focus and transmit light. Proteins are, however, not unprotected from damage associated with daily UV irradiation. Primates have developed a pathway for the production of small-molecular-weight UV filters based on metabolites of the amino acid tryptophan (van Heyningen, 1971b; Wood and Truscott, 1993). As a result, the lens macromolecules are exposed to little damaging radiation because the cornea transmits almost no light below 300 nm, and radiation in the 300–400 nm band is effectively removed by the UV filters.

5. *Oxygen*: In the lens center, there are no cellular organelles and there is very little metabolic activity. Hence there is no requirement for oxygen. One strategy to avoid oxidative damage, and therefore to prolong the stability of the proteins, would be to limit the amount of oxygen in the nucleus (Barbazetto *et al.*, 2004; McNulty *et al.*, 2004). Indeed the oxygen concentration in the center of the lens is less than 10 μM. Mitochondria that are present in a narrow zone just under the surface of the lens actively utilize oxygen and this appears to be the means by which nuclear oxygen is held at such a low level (McNulty *et al.*, 2004). Underscoring the importance of this element for the integrity of the lens proteins, exposure of patients (Palmquist *et al.*, 1984) or animals (Giblin *et al.*, 1995) to hyperbaric oxygen leads to the development of nuclear opacities.

6. *Temperature*: It is a little known fact that the temperature of the anterior portion of the eye is approximately 2°C lower than that of core body temperature (Weale, 1981). This is due to evaporative cooling from the cornea and to the distance from the blood supply. A slightly decreased temperature may act to enhance crystallin stability over a period of many years.

21.5. Aging

Cataract is chiefly a disease of the elderly, and it has been estimated that all people will suffer from cataract if they live long enough (Taylor, 2003). One only has to open one's eyes to see that many body features change with age, so it is perhaps surprising that there is little evidence for substantial age-related metabolic changes in the lens outer cortex, particularly with regard to the enzymes involved in glutathione regeneration (Rathbun and Bovis, 1986). Glutathione synthesis may, however, be impaired in older lenses by diminished cysteine transport (Rathbun and Murray, 1991).

The levels of free UV filters decrease linearly with age at 12% per decade (Bova *et al.*, 2001), and this may leave older lenses more prone to UV-induced photodamage. By contrast after middle age, the levels of UV filters that are covalently bound to the crystallins in the lens nucleus increase substantially (Hood *et al.*, 1999; Vazquez *et al.*, 2002) and because such modified proteins can absorb UVA, this renders them more susceptible to photooxidation by wavelengths of light that penetrate the cornea (Parker *et al.*, 2004).

The key feature with regard to the subsequent development of nuclear cataract would seem to be the onset of an internal barrier at middle age that effectively uncouples the lens nucleus from the cortex (Moffat *et al.*, 1999; Sweeney and Truscott, 1998). An increased residence time for small unstable molecules, such as UV filters, within the central zone of older lenses allows chemical reactions to take place that are almost undetectable in younger lenses. Such small reactive molecules may bind to nuclear proteins. In addition, because the flux of glutathione from the cortex is now limited, the nucleus becomes vulnerable to oxidative damage (Truscott, 2000). Oxidation is the hallmark of age-related nuclear (ARN) cataract.

21.6. Protein Changes with Age

Despite the many mechanisms in place to minimize damage to those lens proteins synthesized early in life, nuclear proteins isolated from the lenses of normal adults show evidence of enormous modification. This is most clearly illustrated by 2D gel electrophoresis (Garland *et al.*, 1996; Lampi *et al.*, 1998). A remarkable conclusion from these studies is that there appears to be no intact crystallins left in the nucleus of even a young adult. The post-translational modifications are numerous and include racemization (Masters *et al.*, 1977), phosphorylation (Ma *et al.*, 1998), deamidation (Takemoto and Boyle, 1998), truncation (Lampi *et al.*, 1998), methylation (Lapko *et al.*, 2003), and thiolation (Lou, 2000).

Glycation is a more controversial modification in the lens as some researchers have obtained data to suggest that it increases with age, however careful analysis has revealed that the levels of fructosyllysine (Dunn *et al.*, 1989) and total glycation (Patrick *et al.*, 1990) in human lens proteins are relatively constant with respect to age. On the other hand, the amount of carboxymethyl lysine (Dunn *et al.*, 1991) and a cyclic adduct thought to arise from attachment of methylglyoxal to arginine residues (Ahmed *et al.*, 2003) do increase almost linearly with age. The source of these modifying agents is not yet clear, but ascorbate or lipids (Odani *et al.*, 1998) are possible precursors. By contrast, oxidation of nuclear protein does not appear to be a feature of normal lens aging (Truscott, 2005).

It would not be surprising if such extensive changes led to alterations in the appearance and physical properties of the lens nucleus. Fortunately, these extensive molecular changes do not markedly alter transparency, although older lenses take on a yellow hue that appears to be the result of the progressive protein binding of UV filters (Hood *et al.*, 1999) and other products (Argirov *et al.*, 2004) that may possibly be derived from the breakdown of ascorbate.

Although transparency of the tissue is retained, these changes to macromolecules have optical and physical correlates. For example, there is an almost linear decline in accommodative amplitude starting in childhood [see Heys *et al.* (2004) for a graph of data from several sources], and this can be matched to a decrease in the ability of isolated human lenses to undergo deformation (Glasser and Campbell, 1998). Accommodation is the process by which ciliary muscle tension in the eye is released allowing the young lens to become more rounded and in this way to refract light to a greater extent (Glasser *et al.*, 2001). This ocular mechanism allows us to focus on nearby objects when we are young. After age 45–50, there is a total loss of focusing power (presbyopia) that may be linked primarily to changes in the properties of the lens, although this is not yet universally accepted.

It has only been recently that changes in the physical properties of the lens nucleus have been characterized. Dynamic mechanical analysis measurements have revealed an approximately 3 order of magnitude increase in the stiffness of the lens nucleus over our lifetime (Heys *et al.*, 2004). In teenage years, the high protein concentration in the nucleus of the lens appears to resemble that of a very viscous fluid. Because the lens nucleus must undergo major deformation in order for the lens to accommodate (Dubbelman *et al.*, 2003; Koretz *et al.*, 1997), this fluid-like state may facilitate such changes in shape. As a corollary, the marked increase in nuclear stiffness, which is most evident between the teens and age 60, may be the primary reason for presbyopia. This biochemical process is a boon for spectacle manufacturers. As noted earlier, the gross alteration in physical properties is not due to compaction of the human nucleus as has been documented in other species.

21.7. Protein Solubility

Extraction of human lenses has revealed that the amount of protein that is insoluble in aqueous buffer increases almost linearly with age (Coghlan and Augusteyn, 1977). A problem with such assays is that the lens must necessarily be destroyed in the process of homogenization, and it is therefore unclear to what extent the extraction method, and choice of buffer, influences the yield and therefore to what extent one can extrapolate these data to the intact lens. Recent NMR data however may corroborate the extraction results, because measurement of T2 in the nucleus of intact human lenses has revealed a linear decline with age (Moffat *et al.*, 2002). T2 is correlated with the concentration of soluble protein. The authors concluded that their NMR data could reflect a decrease in refractive index in the lens nucleus with age, however it is not known if "insoluble" protein can contribute to the refractive index of the lens.

It is probable that a time-dependent denaturation of crystallins is responsible for the steady increase in insoluble protein. If such a process were taking place, one may expect that the major lens protein and chaperone, α crystallin, would bind to the denaturing polypeptides. Evidence in support of this hypothesis has been obtained from analysis of both insoluble (Lund *et al.*, 1996) and soluble protein fractions. After age 40, no soluble α crystallin can be detected in the human lens nucleus (McFall-Ngai *et al.*, 1985; Roy and Spector, 1976). βB2 crystallin has also been suggested to assist the solubility of other crystallins, in particular other β crystallin polypeptides such as βB1 crystallin, with which it associates.

21.8. Age-Related Nuclear Cataract

Cataract is an opacity of the lens that interferes significantly with vision. It is the leading cause of world blindness and is currently treated by surgical removal of the opaque lens and its replacement by a plastic one. There are two main types of cataract: nuclear and cortical. Cortical cataract is a whitish opacity in the outer half of the lens and may be associated with a local loss of osmotic regulation.

Age-related nuclear (ARN) cataract is characterized by the progressive oxidation, coloration, and insolubilization of lens proteins that starts, and is most noticeable, in the lens center. The brown color of the lens nucleus, which appears to be specific to primates, has been used as a means for classifying ARN cataract lenses (Pirie, 1968). The insolubility is fundamentally different from that in normal lenses detailed in the previous section. In ARN cataract, a fraction appears that is insoluble even in 8 M urea, or 6 M guanidine hydrochloride (Truscott and Augusteyn, 1977a). Only by the addition of mercaptoethanol, or dithiothreitol, can these proteins be dissolved. Even then, when advanced ARN cataract lenses are extracted, a significant fraction of the lens protein remains insoluble in urea.

The proportion of protein that is insoluble in 8 M urea increases progressively with the worsening of the cataract until it represents approximately 50% of the total nuclear protein (Truscott and Augusteyn, 1977a). Enzymic digestion has demonstrated that α crystallin is a major component of this insoluble protein (Chen *et al.*, 1997). The requirement for addition of a reducing agent to achieve solubility has been interpreted as reflecting the formation of inter-chain disulfide bonds. This conclusion is consistent with the almost total loss of sulfhydryl groups in the proteins isolated from advanced ARN cataract lenses (Truscott and Augusteyn, 1977c). In agreement with this, crystallin polypeptides have been found disulfide-linked to membranes from ARN cataract, but not normal, lenses (Kodama and Takemoto, 1988). In the proteins from such cataract lenses, approximately half of the methionine residues have been converted to methionine sulfoxide (Garner and Spector, 1980; Truscott and Augusteyn, 1977b).

An overall change from a reducing environment in the center of a normal lens to one that is oxidative in ARN cataract is consistent with all the features that are known, including the pattern of hydroxylation of amino acid residues. Addition of oxygen atoms to the side chains of both aliphatic and aromatic amino acids appears to result from exposure to the hydroxyl radical (Fu *et al.*, 1998). Unsurprisingly, these large changes to the proteins are linked to unfolding (Harding, 1991).

It has been proposed that the key feature predisposing human lenses to subsequent ARN cataract is not a large decrease in the capacity of older normal, or even cataract, lenses to manufacture, or regenerate, antioxidants such as glutathione in the lens cortex. In agreement with this, the activities of the two enzymes involved in glutathione synthesis are unchanged in nuclear cataract (Rathbun *et al.*, 1993). Rather it is the breakdown in communication between the cortex and the nucleus (Truscott, 2000).

21.9. Amyloid or Amorphous Aggregate?

As noted in the previous section, the majority, perhaps all, of the proteins in the nuclei of advanced ARN cataract lenses have undergone extensive modification. Because these proteins were initially tightly folded, it is likely that the modifications result in partially folded conformations. These likely retain regions of the original β-sheet structure within their partially folded structure. As detailed by Uversky and Fink (Uversky and Fink, 2004), Rochet and Lansbury (Rochet and Lansbury, 2000), Dobson (2004), and others (e.g., Kelly, 1998), such partially folded conformations appear to be intermediates in the subsequent transformation of the proteins into either amyloid fibrils, soluble oligomers, or amorphous aggregates.

The formation of amyloid fibrils from human α-synuclein is thought to be a contributing factor in Parkinson's disease. This process is facilitated by molecular crowding and the oxidation of certain methionine residues also significantly affects fibrillation (Uversky *et al.*, 2002). Metals may also play a role in fibril formation (Yamin *et al.*, 2003). All of these features are also implicated in the formation of ARN cataract: the importance of redox-active metals being supported by the discovery of hydroxyl radical damage and other data (e.g., Garner *et al.*, 2000).

Is there any evidence for amyloid fibrillar structures in either older normal or ARN cataract lenses? Frederikse found that normal mouse lenses stained with Congo red and thioflavine T, beginning at the zone where cellular organelles are lost (Frederiske, 2000). This also corresponds with the region of the lens where crystallins are packed into mature fiber cells. This finding presumably reflects a stacked β-sheet supramolecular order of the cytosolic proteins rather than amyloid fibers themselves, as until now, no amyloid fibers have been isolated from lenses. If indeed fibrils are absent, it would appear that the highly modified protein present in cataract lenses is largely in the form of an unstructured aggregate.

21.10. Cataract and Alzheimer's Disease

The concentration of amyloid β peptide (Aβ) in human lenses is comparable with that in the cerebral cortex (Goldstein *et al.*, 2003), and oxidative stress appears to increase both the lenticular amount of this peptide and the amyloid precursor protein (Frederikse *et al.*, 1996). As noted by Harding, Alzheimer's disease (AD) and cataract share a number of features including occurrence late in life and the accumulation of intractable protein (Harding, 1998). Although αB crystallin was originally thought to be confined to the lens, it has been detected in a number of other tissues and has been found in elevated levels in the plaques that characterize AD. αB Crystallin can interact with Aβ to increase its neurotoxicity.

An unusual equatorial supranuclear cataract that colocalizes with Aβ immunoreactivity in the cell cytoplasm may be specific for AD patients (Goldstein *et al.*, 2003). This enhanced immunoreactivity was not associated with amyloid fibrils. Despite this finding, there is evidence that amyloid can form in the lens. For example, a murine γB mutant (Sandilands *et al.*, 2002) displays a cataract reported to be composed of amyloid-like inclusions. The purified γB crystallin itself is unstable and forms fibrils under conditions where the wild-type protein remains soluble (Sandilands *et al.*, 2002).

21.11. Conclusion

ARN cataract can be considered as a conformational disorder, albeit of an unusual type. The proteins present in advanced nuclear cataract lenses are clearly very different from those that were synthesized originally, and it is almost certain that the raft of extensive covalent modifications undergone by these proteins is responsible for the opacification that is the clinical manifestation of cataract. There is, however, a caveat. It is also clear that essentially all of the proteins in the center of normal aged human lenses have also been extensively modified and are likely to be denatured to some extent. In this sense, aging of the normal lens may also be considered to be a conformational disorder, and the progressive loss of accommodation with age, resulting ultimately in presbyopia, is perhaps its recognizable symptom. A presbyopic lens extracted from a normal donor over age 50 is still transparent, despite being inflexible.

It is still not clear exactly what changes are responsible at the molecular level for converting an old clear lens into an opaque cataractous lens. Very likely oxidation is involved, as lens proteins in old normal lenses show little, if any, evidence of oxidation, and the extent of oxidation of both cysteine and methionine residues increase progressively as the cataract worsens. On the other hand, it should be noted that oxidation itself may be insufficient to induce opacification of the lens, as the cysteine residues in lens proteins of other species such the rodents (Kuck *et al.*, 1982) are almost totally oxidized in older animals, and yet the lenses remain transparent. Oxidation by H_2O_2, hydroxyl radical, and other reactive oxygen species, coupled with novel post-translational modifications consequent upon the conversion of the nuclear region into an oxidative environment is a likely scenario. The maintenance of a reducing environment in the lens nucleus thus becomes crucial for the prevention of cataract. If a reducing milieu can be retained, then the enormous modifications, and presumably conformational changes, that take place to the proteins in a normal lens can apparently be tolerated without a loss of transparency. If this is indeed true, the reason for this resilience could be traced to the high concentration of crystallins and their supramolecular order within the lens.

Acknowledgments

Michael Friedrich is thanked for providing the data in Figure 21.3. Roger Truscott is a National Health and Medical Research Council Senior Research fellow. This work was supported by grants from the NIH (no. R01EY013570) and NHMRC (no. 307615).

References

Ahmed, N., Thornalley, P. J., Dawczynski, J., Franke, S., Strobel, J., Stein, G., and Haik, G. M. (2003). Methylglyoxal-derived hydroimidazolone advanced glycation end-products of human lens proteins. *Invest Ophthalmol Vis Sci* 44: 5287–5292.

Aquilina, J. A., Benesch, J. L., Ding, L. L., Yaron, O., Horwitz, J., and Robinson, C. V. (2004). Phosphorylation of alpha B-crystallin alters chaperone function through loss of dimeric substructure. *J Biol Chem* 279: 28675–28680.

Argirov, O. K., Lin, B., and Ortwerth, B. J. (2004). 2-ammonio-6-(3-oxidopyridinium-1-yl)hexanoate (OP-lysine) is a newly identified advanced glycation end product in cataractous and aged human lenses. *J Biol Chem* 279: 6487–6495.

Barbazetto, I. A., Liang, J., Chang, S. H., Zheng, L., Spector, A., and Dillon, J. P. (2004). Oxygen tension in the rabbit lens and vitreous before and after vitrectomy. *Exp Eye Res* 78: 917–924.

Bova, L. M., Sweeney, M. H., Jamie, J. F., and Truscott, R. J. (2001). Major changes in human ocular UV protection with age. *Invest Ophthalmol Vis Sci* 42: 200–205.

Chen, Y. C., Reid, G. E., Simpson, R. J., and Truscott, R. J. (1997). Molecular evidence for the involvement of alpha crystallin in the colouration/crosslinking of crystallins in age-related nuclear cataract. *Exp Eye Res* 65: 835–840.

Coghlan, S. D., and Augusteyn, R. C. (1977). Changes in the distribution of proteins in the aging human lens. *Exp Eye Res* 25: 603–611.

Cooper, P. G., Aquilina, J. A., Truscott, R. J., and Carver, J. A. (1994a). Supramolecular order within the lens: 1H NMR spectroscopic evidence for specific crystallin-crystallin interactions. *Exp Eye Res* 59: 607–616.

Cooper, P. G., Carver, J. A., Aquilina, J. A., Ralston, G. B., and Truscott, R. J. (1994b). A 1H NMR spectroscopic comparison of gamma S- and gamma B-crystallins. *Exp Eye Res* 59: 211–220.

Delaye, M., and Tardieu, A. (1983). Short-range order of crystallin proteins accounts for eye lens transparency. *Nature* 302: 415–417.

Dobson, C. M. (2004). Experimental investigation of protein folding and misfolding. *Methods* 34: 4–14.

Dubbelman, M., Van der Heijde, G. L., Weeber, H. A., and Vrensen, G. F. (2003). Changes in the internal structure of the human crystalline lens with age and accommodation. *Vision Res* 43: 2363–2375.

Duncan, G., and Jacob, T. J. (1984). Calcium and the physiology of cataract. *Ciba Found Symp* 106: 132–152.

Dunn, J. A., McCance, D. R., Thorpe, S. R., Lyons, T. J., and Baynes, J. W. (1991). Age-dependent accumulation of N epsilon-(carboxymethyl)lysine and N epsilon-(carboxymethyl)hydroxylysine in human skin collagen. *Biochemistry* 30: 1205–1210.

Dunn, J. A., Patrick, J. S., Thorpe, S. R., and Baynes, J. W. (1989). Oxidation of glycated proteins: age-dependent accumulation of N epsilon-(carboxymethyl)lysine in lens proteins. *Biochemistry* 28: 9464–9468.

Frederikse, P. H., Garland, D., Zigler Jr. J. S., and Piatigorsky, J. (1996). Oxidative stress increases production of beta-amyloid precursor protein and beta-amyloid in mammalian lenses, and beta-amyloid has toxic effects on lens epithelial cells. *J Biol Chem* 271: 10169–10174.

Frederiske, P. H. (2000). Amyloid-like protein structure in mammalian ocular lenses. *Curr Eye Res* 20: 462–468.

Fu, S., Dean, R., Southan, M., and Truscott, R. (1998). The hydroxyl radical in lens nuclear cataractogenesis. *J Biol Chem* 273: 28603–28609.

Garland, D. L., Duglas-Tabor, Y., Jimenez-Asensio, J., Datiles, M. B., and Magno, B. (1996). Nucleus of the human lens: demonstration of a highly characteristic protein pattern by two-dimensional electrophoresis and a new method of lens dissection. *Exp Eye Res* 62: 285–291.

Garner, B., Davies, M. J., and Truscott, R. J. (2000). Formation of hydroxyl radicals in the human lens is related to the severity of nuclear cataract. *Exp Eye Res* 70: 81–88.

Garner, M. H., and Spector, A. (1980). Selective oxidation of cysteine and methionine in normal and senile cataractous lenses. *Proc Natl Acad Sci USA* 77: 1274–1277.

Giblin, F. J., Padgaonkar, V. A., Leverenz, V. R., Lin, L. R., Lou, M. F., Unakar, N. J., Dang, L., Dickerson, J. E., Jr., and Reddy, V. N. (1995). Nuclear light scattering, disulfide formation and membrane damage in lenses of older guinea pigs treated with hyperbaric oxygen. *Exp Eye Res* 60: 219–235.

Giblin, F. J. (2000). Glutathione: a vital lens antioxidant. *J Ocul Pharmacol Ther* 16: 121–135.

Glasser, A., and Campbell, M. C. (1998). Presbyopia and the optical changes in the human crystalline lens with age. *Vision Res* 38: 209–229.

Glasser, A., Croft, M. A., and Kaufman, P. L. (2001). Aging of the human crystalline lens and presbyopia. *Int Ophthalmol Clin* 41: 1–15.

Goldstein, L. E., Muffat, J. A., Cherny, R. A., Moir, R. D., Ericsson, M. H., Huang, X., Mavros, C., Coccia, J. A., Faget, K. Y., Fitch, K. A., *et al.* (2003). Cytosolic beta-amyloid deposition and supranuclear cataracts in lenses from people with Alzheimer's disease. *Lancet* 361: 1258–1265.

Harding, J. J. (1991). *Cataract Biochemistry, Epidemiology and Pharmacology*, 1st ed (London, Chapman and Hall).

Harding, J. J. (1998). Cataract, Alzheimer's disease, and other conformational diseases. *Curr Opin Ophthalmol* 9: 10–13.

Heys, K. R., Cram, S. L., and Truscott, R. J. W. (2004). Massive increase in the stiffness of the human lens nucleus with age: the basis for presbyopia? *Mol Vision* 10: 956–963.

Hood, B. D., Garner, B., and Truscott, R. J. (1999). Human lens coloration and aging. Evidence for crystallin modification by the major ultraviolet filter, 3-hydroxy-kynurenine O-beta-D-glucoside. *J Biol Chem* 274: 32547–32550.

Horwitz, J. (1992). Alpha-crystallin can function as a molecular chaperone. *Proc Natl Acad Sci USA* 89: 10449–10453.

Horwitz, J. (2003). Alpha-crystallin. *Exp Eye Res* 76: 145–153.

Huizinga, A., Bot, A. C. C., de Mull, F. F. M., Vrensen, G. F. J. M., and Greve, J. (1989). Local variation of absolute water content of human and rabbit eye lenses measured by Raman microspectroscopy. *Exp Eye Res* 48: 487–496.

Jaenicke, R., and Slingsby, C. (2001). Lens crystallins and their microbial homologs: structure, stability, and function. *Crit Rev Bioch Mol Biol* 36: 435–499.

Kelly, J. W. (1998). The alternative conformations of amyloidogenic proteins and their multistep assembly pathways. *Curr Opin Struct Biol* 8: 101–106.

Kodama, T., and Takemoto, L. (1988). Characterization of disulfide-linked crystallins associated with human cataractous lens membranes. *Invest Ophthalmol Vis Sci* 29: 145–149.

Koretz, J. F., Cook, C. A., and Kaufman, P. L. (1997). Accommodation and presbyopia in the human eye. Changes in the anterior segment and crystalline lens with focus. *Invest Ophthalmol Vis Sci* 38: 569–578.

Kuck, J. F., Yu, N. T., and Askren, C. C. (1982). Total sulfhydryl by raman spectroscopy in the intact lens of several species: variations in the nucleus and along the optical axis during aging. *Exp Eye Res* 34: 23–37.

Lampi, K. J., Ma, Z., Shih, M., Shearer, T. R., Smith, J. B., Smith, D. L., and David, L. L. (1997). Sequence analysis of betaA3, betaB3, and betaA4 crystallins completes the identification of the major proteins in young human lens. *J Biol Chem* 272: 2268–2275.

Lampi, K. J., Ma, Z., Hanson, S. R., Azuma, M., Shih, M., Shearer, T. R., Smith, D. L., Smith, J. B., and David, L. L. (1998). Age-related changes in human lens crystallins identified by two-dimensional electrophoresis and mass spectrometry. *Exp Eye Res* 67: 31–43.

Lapko, V. N., Smith, D. L., and Smith, J. B. (2003). Methylation and carbamylation of human gamma-crystallins. *Protein Sci* 12: 1762–1774.

Lou, M. F. (2000). Thiol regulation in the lens. *J Ocul Pharmacol Ther* 16: 137–148.

Lund, A. L., Smith, J. B., and Smith, D. L. (1996). Modifications of the water-insoluble human lens alpha-crystallins. *Exp Eye Res* 63: 661–672.

Ma, Z., Hanson, S. R., Lampi, K. J., David, L. L., Smith, D. L., and Smith, J. B. (1998). Age-related changes in human lens crystallins identified by HPLC and mass spectrometry. *Exp Eye Res* 67: 21–30.

Masters, P. M., Bada, J. L., and Zigler, J. J. (1977). Aspartic acid racemisation in the human lens during ageing and in cataract formation. *Nature* 268: 71–73.

McFall-Ngai, M. J., Ding, L. L., Takemoto, L. J., and Horwitz, J. (1985). Spatial and temporal mapping of the age-related changes in human lens crystallins. *Exp Eye Res* 41: 745–758.

McNulty, R., Wang, H., Mathias, R., Ortwerth, B. J., Truscott, R. J. W., and Bassnett, S. (2004). Regulation of tissue oxygen levels in the mammalian lens. *J Physiol* 559: 883–898.

Moffat, B. A., Landman, K. A., Truscott, R. J., Sweeney, M. H., and Pope, J. M. (1999). Age-related changes in the kinetics of water transport in normal human lenses. *Exp Eye Res* 69: 663–669.

Moffat, B. A., Atchison, D. A., and Pope, J. M. (2002). Explanation of the Lens Paradox. *Optom Vis Sci* 79: 148–150.

Odani, H., Shinzato, T., Usami, J., Matsumoto, Y., Brinkmann Frye, E., Baynes, J. W., and Maeda, K. (1998). Imidazolium crosslinks derived from reaction of lysine with glyoxal and methylglyoxal are increased in serum proteins of uremic patients: evidence for increased oxidative stress in uremia. *FEBS Lett* 427: 381–385.

Palmquist, B. M., Philipson, B., and Barr, P. O. (1984). Nuclear cataract and myopia during hyperbaric oxygen therapy. *Br J Ophthalmol* 68: 113–117.

Parker, N. R., Jamie, J. F., Davies, M. J., and Truscott, R. J. W. (2004). Protein-bound kynurenine is a photosensitiser of oxidative damage. *Free Radic Biol Med* 37: 1479–1489.

Patrick, J. S., Thorpe, S. R., and Baynes, J. W. (1990). Nonenzymatic glycosylation of protein does not increase with age in normal human lenses. *J Gerontol* 45: B18–23.

Pirie, A. (1968). Color and solubility of the proteins of human cataracts. *Invest Ophthalmol* 7: 634–650.

Quinlan, R. A., Carter, J. M., Sandilands, A., and Prescott, A. R. (1996). The beaded filament of the eye lens—an unexpected key to intermediate filament structure and function. *Trends Cell Biol* 6: 123–126.

Rathbun, W. B., and Bovis, M. G. (1986). Activity of glutathione peroxidase and glutathione reductase in the human lens related to age. *Curr Eye Res* 5: 381–385.

Rathbun, W. B., Bovis, M. G., and Holleschau, A. M. (1986). Species survey of glutathione peroxidase and glutathione reductase: search for an animal model of the human lens. *Ophthalmic Res* 18: 282–287.

Rathbun, W. B., and Murray, D. L. (1991). Age-related cysteine uptake as rate-limiting in glutathione synthesis and glutathione half-life in the cultured human lens. *Exp Eye Res* 53: 205–212.

Rathbun, W. B., Schmidt, A. J., and Holleschau, A. M. (1993). Activity loss of glutathione synthesis enzymes associated with human subcapsular cataract. *Invest Ophthalmol Vis Sci* 34: 2049–2054.

Reddy, V. N., and Giblin, F. J. (1984). Metabolism and function of glutathione in the lens. *Ciba Found Symp* 106: 65–87.

Reizer, J., Reizer, A., and Saier, M. H. J. (1993). The MIP family of integral membrane channel proteins: sequence comparisons, evolutionary relationships, reconstructed pathway of evolution, and proposed functional differentiation of the two repeated halves of the proteins. *Crit Rev Biochem Mol Biol* 28: 235–257.

Rochet, J.-C., and Lansbury, P. T. J. (2000). Amyloid fibrillogenesis: themes and variations. *Curr Opin Struct Biol* 10: 60–68.

Roy, D., and Spector, A. (1976). Absence of low-molecular-weight alpha crystallin in nuclear region of old human lenses. *Proc Natl Acad Sci USA* 73: 3484–3487.

Sandilands, A., Hutcheson, A. M., Long, H. A., Prescott, A. R., Vrensen, G., Loster, J., Klopp, N., Lutz, R. B., Graw, J., Masaki, S., *et al.* (2002). Altered aggregation properties of mutant gamma-crystallins cause inherited cataract. *EMBO J* 21: 6005–6014.

Spector, A., Chiesa, R., Sredy, J., and Garner, W. (1985). cAMP-dependent phosphorylation of bovine lens alpha-crystallin. *Proc Natl Acad Sci USA* 82: 4712–4716.

Sweeney, M. H., and Truscott, R. J. (1998). An impediment to glutathione diffusion in older normal human lenses: a possible precondition for nuclear cataract. *Exp Eye Res* 67: 587–595.

Takemoto, L., and Boyle, D. (1998). Deamidation of specific glutamine residues from alpha-A crystallin during aging of the human lens. *Biochemistry* 37: 13681–13685.

Taylor, H. (2003). Eye care for the future: the Weisenfeld lecture. *Invest Ophthalmol Vis Sci* 44: 1413–1418.

Truscott, R. J., Marcantonio, J. M., Tomlinson, J., and Duncan, G. (1990). Calcium-induced opacification and proteolysis in the intact rat lens. *Invest Ophthalmol Vis Sci* 31: 2405–2411.

Truscott, R. J. (2000). Age-related nuclear cataract: a lens transport problem. *Ophthalmic Res* 32: 185–194.

Truscott, R. J. W., and Augusteyn, R. C. (1977a). Changes in human lens proteins during nuclear cataract formation. *Exp Eye Res* 24: 159–170.

Truscott, R. J. W., and Augusteyn, R. C. (1977b). Oxidative changes in human lens proteins during senile nuclear cataract formation. *Biochim Biophys Acta* 492: 43–52.

Truscott, R. J. W., and Augusteyn, R. C. (1977c). The state of sulphydryl groups in normal and cataractous human lenses. *Exp Eye Res* 25: 139–148.

Truscott, R. J. W. (2005). Age-related nuclear cataract: oxidation is the key. *Exp Eye Res* 80: 709–725.

Uversky, V. N., and Fink, A. L. (2004). Conformational constraints for amyloid fibrillation: the importance of being unfolded. *Biochem Biophys Acta* 1698: 131–153.

Uversky, V. N., Yamin, G., Souillac, P. O., Goers, J., Glaser, C. B., and Fink, A. L. (2002). Methionine oxidation inhibits fibrillation of human alpha-synuclein *in vitro*. *FEBS Lett* 517: 239–244.

van Heyningen, R. (1971a). Fluorescent derivatives of 3-hydroxy-L-kynurenine in the lens of man, the baboon and the grey squirrel. *Biochem J* 123: 30–31.

van Heyningen, R. (1971b). Fluorescent glucoside in the human lens. *Nature* 230: 393–394.

Vazquez, S., Aquilina, J. A., Jamie, J. F., Sheil, M. M., and Truscott, R. J. (2002). Novel protein modification by kynurenine in human lenses. *J Biol Chem* 277: 4867–4873.

Weale, R. A. (1981). Human ocular aging and ambient temperature. *Br J Ophthalmol* 65: 869–870.

Wistow, G., and Piatigorsky, J. (1987). Recruitment of enzymes as lens structural proteins. *Science* 236: 1554–1556.

Wistow, G. J., Mulders, J. W., and de Jong, W. W. (1987). The enzyme lactate dehydrogenase as a structural protein in avian and crocodilian lenses. *Nature* 326: 622–624.

Wood, A. M., and Truscott, R. J. W. (1993). UV filters in human lenses: tryptophan catabolism. *Exp Eye Res* 56: 317–325.

Wood, A. M., and Truscott, R. J. W. (1994). Ultraviolet filter compounds in human lenses: 3-hydroxykynurenine glucoside formation. *Vision Res* 34: 1369–1374.

Yamin, G., Glaser, C. B., Uversky, V. N., and Fink, A. L. (2003). Certain metals trigger fibrillation of methionine-oxidized alpha-synuclein. *J Biol Chem* 278: 27630–27635.

V

Altered Protein Structure and Changes in Cellular/Nuclear Function

22

Glutamine/Asparagine-Rich Regions in Proteins and Polyglutamine Diseases

Hitoshi Okazawa

Abstract

Glutamine/asparagine-rich sequences are physiologically used for prion-based propagation of yeast phenotype mediated by Ure2 and Sup35, which accommodate yeast cells to changes of the nutritional condition. In addition, the repeat sequences are critically involved in pathogenesis of polyglutamine diseases. In this chapter, I review general outlines of the physiologic and pathologic roles of glutamine-rich and asparagine-rich sequences in proteins.

22.1. Physiologic Roles of Glutamine/Asparagine-Rich Regions in Proteins

The amino acids (AAs) are generally grouped into four categories, acidic, basic, uncharged polar, and nonpolar AAs, according to properties of the side chains. Asparagine, glutamine, serine, threonine, and tyrosine possess an uncharged polar side chain. Amino acids of this group are relatively hydrophilic, although they are not charged at neutral pH. Therefore, uncharged polar AAs are usually located on the outside of proteins (Alberts *et al.*, 1994), which could be a protein-protein interaction module. Only the length of backbone carbon number of the side chain is different between asparagine and glutamine (Fig. 22.1), thus biochemical features of these two amino acids are highly homologous. Glutamine/asparagine-rich sequences attract attention because they are linked to two major groups of neurodegeneration: prion diseases and polyglutamine diseases.

Glutamine/asparagine-rich region is essential to cause aggregation of yeast prion proteins such as Ure2 and Sup35 and is sufficient for prion-based propagation of yeast phenotype mediated by these proteins (Wickner, 1994; Ter-Avanesyan *et al.*, 1994; Maisison *et al.*, 1995, 1997). The aggregation property of glutamine/asparagine-rich region is tightly linked to epigenetic propagation of specific phenotypes [*URE3*] and [PSI^+] of yeast [for review see Uptain and Lindquist (2002)], as described in previous chapters. In addition to [*URE3*] and [PSI^+], other yeast proteins such as Rnq1 also seem to utilize asparagine/glutamine-rich sequence for prion-based genetic transmission (Uptain and Lindquist, 2002).

Gutamine/asparagines also compose repeat sequences in some proteins. Polyglutamine repeat sequence is especially important because its expansion leads to hereditary neurodegeneration. CAG repeats that are translated into polyglutamine tract sequences are unstable during replication, thus repeat expansion could occur *de novo* during mitosis and meiosis. Once abnormal expansion is obtained, it is transmitted to subsequent generations. Polyglutamine-containing peptides are cleaved from full-length proteins, change the conformation, and aggregate. Up to

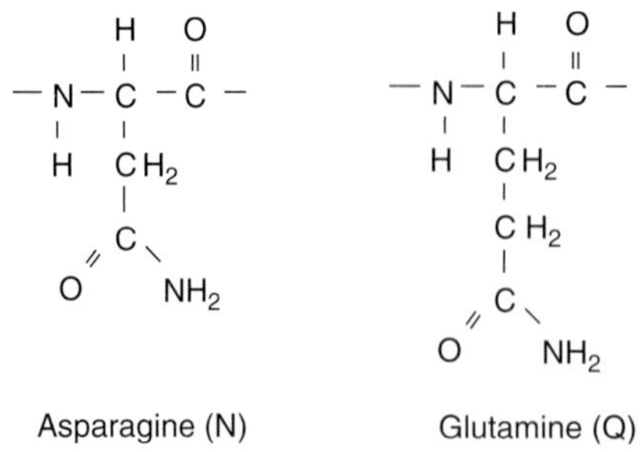

Asparagine (N) Glutamine (Q)

Figure 22.1. Chemical structures of asparagine and glutamine.

now, the expansion is known to cause nine neurodegenerative disorders: Huntington's disease (HD), spinocerebellar atrophy-1 (SCA1), SCA2, SCA3, SCA6, SCA7, SCA17, dentato-rubro-pallido-luysian atrophy (DRPLA), and Kennedy's disease (bulbospinal muscular atrophy).

Meanwhile, polyglutamine repeat sequences seem to have some physiologic functions that are not directly linked to aggregation. Computer search reveals a number of proteins possessing glutamine tracts. When we searched a nonredundant human database with 40 repeats of glutamine by using NCBI BLAST, 6748 genes were retrieved as hits. It is noteworthy that most of the genes included in the list are nuclear proteins involved in transcription and RNA modification (Table 22.1). They are forkhead transcription factor, TFIID, TATA box binding factor, mediator of RNA polymerase subunit 12 homologue/trap230, SWI/SNF matrix-associated factor, androgen receptor, nuclear receptor coactivator 3, transcription factor Brn-2, and so on (Table 22.1). However, polyglutamine repeats seem to be evolutionally new, because, for example, TATA binding protein possesses the polyglutamine tract only in mammals and the repeat number increases rapidly from mouse to human. These findings collectively suggest that the polyglutamine repeat sequence might play some roles in the protein function especially in higher animals.

We have searched binding proteins to the polyglutamine tract sequence by using yeast two-hybrid screening. As the result, we have cloned multiple genes containing polar amino acid–rich regions instead of a specific gene (Imafuku *et al.*, 1998). This means that the polyglutamine tract could act as a protein-interacting motif whereas the binding cannot be so specific by itself. Among the proteins interacting with the polyglutamine tract, we found a novel nuclear protein polyglutamine-tract binding protein-1 (PQBP-1), which turned out to be involved in transcription and RNA splicing (Waragai *et al.*, 1999; Okazawa *et al.*, 2002) and to be a causative gene of mental retardation (Kalscheuer *et al.*, 2003; Lenski *et al.*, 2003). In the case of a POU transcription factor the initials are taken from the four proteins first seen to have POU domain: ***P***it-1 (also called GHF1), a pituitary-specific factor that activates the genes encoding growth hormone, prolactin, and other pituitary proteins; ***O***ct1, a ubiquitous protein that recognizes a certain eight-base-pair sequence called the octa box, and ***O***ct2, the B-cell-specific protein that recognizes the octa box and activates immunoglobulin genes; and ***U***NC-86, a nematode gene product involved in determining neuronal cell fates, Oct-2, the polyglutamine tract sequence was shown to be a transcriptional activation domain acting synergistically with a proline-rich sequence within the molecule (Tanaka *et al.*, 1994). The result also seems to suggest that polyglutamine tract sequence by itself is not highly selective for the interaction partner but it could act as an assistant interaction module.

22.2. Conformational Changes of Yeast Prion Proteins

Prion protein-like confomational change of yeast proteins was first proposed by Reed Wickner in 1994 (Wickner, 1994). Genetic determinant of [*URE3*], a phenotype where cells take

Table 22.1. Representative proteins possessing polyglutamine tract sequences

Forkhead box P2	3e-16
TATA box binding protein	3e-16
Transcription factor IID	9e-16
Mediator of RNA polymerase II, subunit 12 homologue (= TRAP230/ARC240/OPA containing protein)	5e-14
Ataxin-1	6e-14
Nuclear transcription factor/not	2e-13
MECT/MAML2 fusion protein	2e-13
Myeloid/lymphoid or mixeded-line	7e-13
ALR	7e-13
Androgen receptor	3e-12
Nuclear receptor coactivator 6	1e-11
SWI/SNF-like protein 2	1e-11
Transcriptional activator hSNF2a	1e-11
TNRC11	1e-11
Transcriptional coactivator AIB3	1e-11
CAGH16	2e-11
Nuclear coactivator 3	2e-11
P400 SWI/SNF-related protein	5e-11
E1A binding protein p400	5e-11
MN1 protein (meningioma 1)	1e-10
THAP domain containing protein	3e-10
BMP-2 inducible kinase isoform	5e-10
TPA inducible protein	7e-10
Small-conductance calcium-activated potassium channel SK3	1e-9
Mastermind-like 3	2e-9

Top 25 proteins searched by BLAST in nonredundant human protein database was with Q_{40} sequence.

up ureido-succinic acid (USA), had not been known for about 25 years. The trait of transmission was clearly non-Mendelian (Lacroute, 1971), whereas it was only known that a certain cytoplasmic factor was involved in transmission (Aigle and Loacroute, 1975). The finding first demonstrated that prion-like conformational change of Ure2 protein underlies the epigenetic transmission of [*URE3*] phenotype. Normal yeast cells do not take up USA when the medium is rich in nitrogen sources, because Ure2 protein represses transcriptional regulator Gln3 that activates expression of nitrogen catabolism genes. When Urep2 is transformed into prion and becomes nonfunctional, [*URE3*], mutants take up USA even in the nitrogen-source-rich medium [for review see Tuite and Cox (2003)].

Another yeast prion protein, Sup35p, is the yeast counterpart of the animal translational termination factor eRF3. Sup35p is involved in [PSI^+], which was originally identified as a cytoplasmic suppressor of super-suppressors by Brian Cox in 1965 (Cox 1965). Sup35p is soluble in the [psi^+] state whose conformation is similar to mammalian PrP^c, but it becomes insoluble in the [PSI^+] state. Sup35p in [PSI^+] yeast cells forms insoluble fibrillary amyloid aggregates just like mammalian PrP^{Sc}, resulting in depletion of the translational termination factors. The depletion allows ribosomes to read through stop codons (nonsense suppression), and this mechanism is used to overcome certain nutritional deficiencies of the medium. For example, in the absence of adenine, ade1 cells carrying a nonsense mutation in an essential gene for adenine synthesis cannot survive with the [ps^-] state of Sup35p. However, in the [PSI^+] state, the nonsense mutation is suppressed as described above and yeast cells can survive as a revertant.

In mammalian prion proteins, propagation of prion forms is triggered by interaction of normal PrP^c with mutant PrP^{Sc}, though it is not yet clear whether oligomer of PrP^{Sc} acts as a seed or monomer of PrP^{Sc} acts as a template for conformational change (Tuite and Cox, 2003). The propagation manner is considered to be similar in human and yeast prion proteins. Transformation to prion form seems to occur spontaneously at 10^{-6} to 10^{-7} frequencies during mis- and re-folding of yeast proteins (Lund and Cox, 1981; Liu and Lindquist, 1999). The seed is transmitted to daughter cells as propagons. Native Ure2p or Sup35 assemble with the preformed seed to form amyloid fibrils reacting with Congo red and thioflavine T that are resistant to proteolysis (Taylor *et al.*, 1999; Thaul *et al.*, 1999). Because Congo red specifically reacts with the cross β-structure (Khurana *et al.*, 2001), it is generally believed that α-helical structure of Ure2p is converted into β-sheet structure. The conformational change or sequestration into amyloid fibrils results in the loss of function of native proteins, which leads to the phenotypic change of yeast. However, it is of note that native structure might be retained in the full-length Ure2p assembled into fibrils judging from the data with atomic force microscopy and FTIR spectroscopy (Bousset *et al.*, 2002).

Conformational change and propagation of yeast prion proteins are modified by chaperone heat shock proteins (Hsp). Two groups independently showed that Hsp104 plays a critical role in propagation of [PSI^+] state (Patino *et al.*, 1996; Paushkin *et al.*, 1996). In the absence of Hsp104, prion form of Sup35p is not maintained. In addition, Hsp104 destabilizes Sup35p aggregates and seems to make propagon. Hsp70 family proteins and Hsp40 affect polymerization process independently or by collaborating with Hsp104 [for review see Tuite and Cox (2003)].

22.3. Conformational Change of Polyglutamine Disease Proteins

Conformational change of polyglutamine disease proteins is not for epigenetic transmission of phenotypes but is involved in the pathology of neurodegeneration. Expansion of the polyglutamine repeat sequence occurs during meiosis. The expansion of polyglutamine tract lowers the energy threshold to change conformation of disease protein. *De novo* expansion in SCA7 could be 12 CAG/transmission (Lebre and Brice, 2003). Variation in repeat numbers seems to occur among different tissues although it may not be able to explain the selective neuronal death in different regions of the brain (Watase *et al.*, 2003).

Although various conformations are proposed as aggregating forms of polyglutamine disease proteins, the most prevailing model is "polar zipper" proposed by Max Perutz and collaborators (Perutz *et al.*, 1994). Synthetic polyglutamine peptides easily change the conformation, become insoluble in water, and form aggregates. Perutz and collaborators rendered deca-L-gluatmine peptide soluble by attaching two aspartate residues to N-terminal and two lysine residues to C-terminal. They observed that the peptide forms worm-like structure of 70–120 Å thick and several hundred Å long. X-ray diffraction showed a pattern suggesting cross-β-structure. Atomic model showed that the β-strands tend to be linked between main chain and side chain by hydrogen bonds (Perutz *et al.*, 1994). From these observations, they proposed anti-parallel β-sheets (polar zipper) [for review see Perutz (1996)]. Further analyses using computer have suggested additional models including parallel, hairpin, and compact randam coil structures made of β-sheets (see review Ross *et al.*, 2003). Perutz and collaborators improved their polar zipper to cylindrical β-sheets made of 20 polyglutamines per turn (Perutz *et al.*, 2002). Hydrogen bonds help short cylinders to interact with each other and develop amyloid nanotubes.

22.4. Cellular Responses and Dysfunctions in Polyglutamine Diseases

Because detailed cellular dysfunctions induced by mutant polyglutamine disease proteins will be described in the following chapters, I would like to overview the general concepts of cellular responses to mutant polyglutamine disease proteins and of the cellular dysfunction induced by those proteins.

22.4.1. Cellular Responses to Abnormal Polyglutamine Proteins

Disease proteins carrying an abnormally expanded polyglutamine tract are translated in endoplasmic reticulum (ER). The mutant proteins form abnormal conformations and become the target of chaperone proteins, which try to unfold and refold the polyglutamine disease proteins. If the effort is unsuccessful, chaperone proteins induce ER stress response. A part of disease proteins in abnormal structures are transported to the cytoplasm or further to the nucleus and cleaved by proteinases. Ubiquitination occurs in these subcellular compartments, and ubiquitinated proteins are recognized as the target of proteasome degradation. However, they are sometimes resistant to degradation and are accumulated at centrosome where factors for protein degradation are concentrated (aggresome). In this case, it is known that aggregated polyglutamine proteins disturb the proteasomal function. A part of proteins after cleavage moves to the nucleus and forms relatively firm nuclear inclusions. It is not yet clear at which step the polyglutamine peptide is most toxic for neurons. However, recent notions have suggested strongly that monomer or oligomer in abnormal conformations but not final aggregates is toxic. Recent findings on the metabolism of polyglutamine proteins will be reviewed in the following (Fig. 22.2).

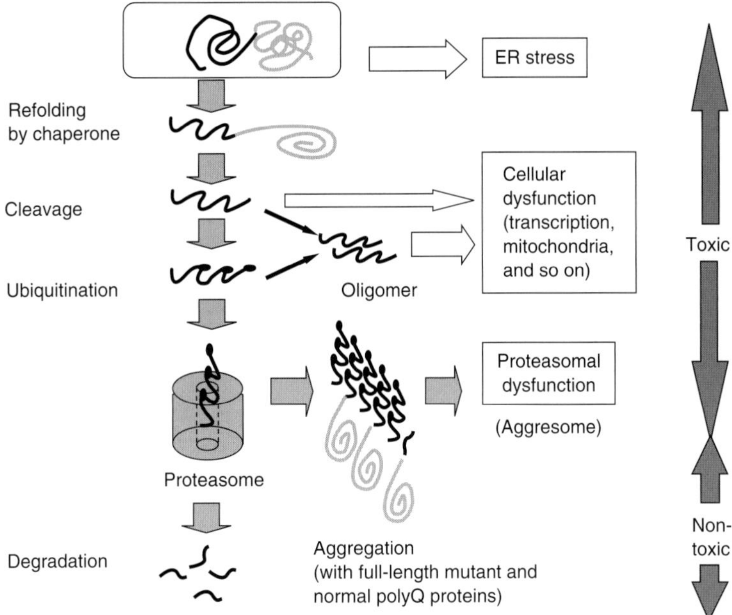

Figure 22.2. Basic metabolism of mutant polyglutamine disease proteins.

22.4.1.1. Processing of Polyglutamine Proteins

It is known that huntingtin, ataxin-3, ataxin-6, ataxin-7, atrophin/DRPLA product, and androgen receptor are cleaved into a peptide containing the polyglutamine sequence [reviewed by Gusella and MacDonald (2000); Zoghbi and Orr (2000); Taylor *et al.* (2002); Ross *et al.* (2002); Bates (2003); Okazawa (2003)]. It is also likely that other polyglutamine disease gene products are also processed. The proteolytic enzymes reported up to now include caspase-3 for androgen receptor (LaFevre-Bernt and Ellerby 2003), caspase-1 for ataxin-3 (Berke 2004), calpain family (calpain-1, 5, 7, 10) and caspase-3 for huntingtin (Kim *et al.*, 2001; Gafni *et al.*, 2004). The cleaved polyglutamine peptides containing polyglutamine tract were detected in patient brain samples or in model systems. It is generally believed that shorter polyglutamine peptides are more toxic and more prone to aggregate. Accordingly, some groups reported that inhibition of the cleavage protects animal models from progress of pathology (Gafni *et al.*, 2004).

It was not clear whether protein cleavage is similarly performed in the cytoplasm and in the nucleus. However, (Lunkes *et al.*, 2002) reported recently that nuclear aggregation contained a specifically short peptide of about 50 kDa cleaved by aspartic endopeptidases while cytoplasmic inclusion body contains longer peptides (Lunkes *et al.*, 2002). The result suggests either that only short peptides could move to the nucleus or that nucleus-specific endopeptidases cleave huntingtin in the nucleus.

22.4.1.2. Ubiquitination of Polyglutamine Protein

Polyglutamine proteins are highly ubiquitinated in neurons. This central dogma is derived from immunohistochemical observation of cytoplasmic and nuclear inclusion bodies and from Western blot analysis of intracellular mutant proteins. Meanwhile, it is not so clear which types of cleaved peptides (or noncleaved native protein) are ubiquitinated. It is also not known whether the abnormal proteins/peptides are ubiquitinated in the nucleus or in the cytoplasm of neurons. Ubiquitin ligase for each polyglutamine protein seems to be different according to the published results. So far, several ubiquitin ligases are known to interact with polyglutamine disease proteins or peptides. The first example is E6-AP/Ube3a, which promotes ubiquitination of ataxin-1 (Cummings *et al.*, 1999). When Ube3 E3 ubiquitin ligase mutant mice and SCA1 transgenic mice were mated, nuclear inclusion formation was remarkably inhibited and, surprisingly at that time, the pathology of Purkinje cells became much worse. This means that E6-AP/Ube3 is critically involved in ubiquitination of ataxin-1. In addition, from the findings emerged the recently prevailing theory that preaggregates are more toxic for cells. There are some other proteins involved in ubiquitination. A1Up (ataxin-1 interacting ubiquitin-like protein) could be involved in ubiquitination of ataxin-1 (Davidson *et al.*, 2000) because it contains an ubiquitin-like region (Ubl), four heat shock chaperonin-binding motifs, and an ubiquitin-associated domain (UBA). The Ubl domain interatcts with S5a subunit of 19S proteasome and the UBA domain binds polyubiquitin chains. These results suggest that A1Up regulates degradation of ataxin-1 (Riley *et al.*, 2004).

22.4.1.3. Proteasome-mediated Degradation of Polyglutamine Proteins

It had been believed that ubiquitinated polyglutamine proteins or peptides are transported to and degraded by proteasome (Cummings, 1998). However, Venkatraman *et al.* recently showed that small polyglutamine peptides containing expanded repeats are hardly digested by eukaryotic proteasome (Venkatraman *et al.*, 2004). Although their experiments were based on *in vitro* study, if it is true, we should better assume that ubiquitinated peptides mainly accumulate rather than being cleared from cells. The result might be relevant to the finding that accumulated polglutamine proteins disturb the function of proteasome (Waelter *et al.* 2001). Peptides that could not be digested by proteasome accumulate around the proteasome complex. The accumulation domain corresponds with centrosome where breakdown of microtubles are actively conducted. This region is therefore called also aggresome (Waelter *et al.*, 2001). The result again stimulates the discussion whether ubiquitination protects or damages neurons. If ubiquitination promotes protein degradation or segregation from interacting partners, ubiquitination is supportive for cells. Meanwhile, if not, ubiquitination might be potentially toxic for neurons.

22.4.1.4. Nuclear Transport of Polyglutamine Proteins

In some polyglutamine diseases, normal form of the disease protein is located in the nucleus. For example, Kennedy's disease is caused by repeat expansion of a nuclear hormone receptor, androgen receptor. SCA17 is caused by mutation of TATA binding protein, an essential nuclear factor binding to TATA box and initiating transcription. Furthermore, most of the mutant polyglutamine disease proteins are definitely transported to the nucleus because they are actually aggregated in the nucleus.

However, our knowledge about their nuclear transport is clearly insufficient. In the case that normal forms are nuclear proteins, the mutant proteins may be transported into the nucleus by ordinary nuclear transport system through nuclear pore complex. However, there seem to be some additional factors that accelerate the nuclear import because mutant proteins are accumulated in the nucleus even if the normal forms are located predominately in the cytoplasm. Furthermore, nuclear export of mutant polyglutamine proteins is not yet reported.

22.4.1.5. Aggregation of Polyglutamine Proteins

Short peptides containing polyglutamine tract sequence easily change the conformation as described above, bind mutually, and form aggregates in the cytoplasm or in the nucleus when the repeat length exceeds a disease-specific threshold. The aggregates are called *inclusion body*, which is one of the most important pathologic hallmarks of polyglutamine diseases. Threshold of the repeat number for the onset of human patients is different among polyglutamine diseases. In the case of huntingtin, *in vitro* aggregation revealed that the threshold locates between 32 and 37. On the other hand, in the case of SCA6, the threshold locates between 18 and 20. Although the reason is not clarified, the difference probably comes from the properties of surrounding sequences. Aggregation is not only dependent on the length of repeats but also on time and concentration of mutant proteins (Scherzinger *et al.*, 1999). In addition full-length mutant polyglutamine proteins and normal polyglutamine proteins are co-aggregated (Dyer and McMurray, 2001; Busch *et al.*, 2003).

22.4.1.6. Aggregation Sites in the Cell

Mutant polyglutamine proteins are aggregated in the cytoplasm and also in the nucleus. The focus of aggregation in the cytoplasm merges on centrosome (Johnston *et al.*, 1998), the organizing center of microtubles for cell division where protein complex formation and protein degradation are actively conducted. Numerous chaperone proteins including Hsp70 and Hsp27 as well as proteasome proteins are located in the centrosome. Mutant proteins are probably transported to centrosome for degradation. However, they are rather resistant to degradation by proteasome, and overflowed proteins accumulate in the centrosome. Centrosome embedded by protein aggregation is sometimes termed *aggresome*.

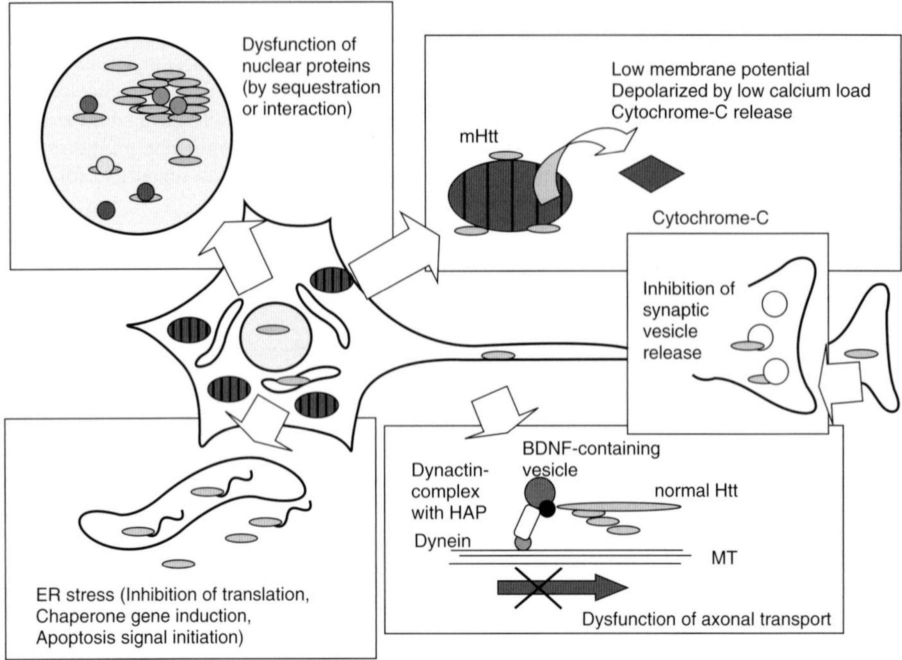

Figure 22.3. Aggregation sites of mutant polyglutamine disease proteins in neurons and cellular dysfunctions induced by mutant proteins.

In the nucleus, aggregates are not localized to visible nuclear domains. Instead, they accumulate in nuclear domains that can be detected by specific marker proteins. For example, ataxin-1 is known to accumulate in promyelocytic leukemia (PML) body (Skinner *et al.*, 1997) and also nuclear body of PQBP-1 (Okazawa *et al.*, 2002). Functional relevance of accumulation of mutant proteins in specific nuclear domains is not completely understood so far.

22.4.2. Cellular Dysfunction by Abnormal Polyglutamine Proteins

Mutant polyglutamine disease proteins are known to disturb multiple cellular functions (Fig. 22.3). ER stress is linked to apoptosis in non-neuronal cells. Proteasomal dysfunction will also trigger the apoptotic cascade if the capacity to refold or degrade abnormal proteins becomes insufficient. It is also reported that mutant proteins associate with mitochondrial membrane.

22.4.2.1. Transcriptional Disturbance

Transcriptional dysfunction is considered to be a key element of the polyglutamine disease pathology. It is because most polyglutamine disease proteins accumulate in the nucleus and because they interact with many nuclear proteins including transcription factors, transcription cofactors (coactivators and corepressors), and splicing factors [for review see Okazawa (2003); Sugars and Rubinsztein (2003)]. It is not definitely understood whether functional disturbance of nuclear factors is induced by interaction with soluble polyglutamine proteins or by sequestration into insoluble aggregates. However, recent findings have shown that nuclear proteins shuttle between inclusion body and nuclear matrix (Kim *et al.*, 2002; Stenoien *et al.*, 2002). In addition, inclusion formation may not be toxic for neurons. These results seem to suggest that interaction with soluble form is essential for the transcription pathology. Although a few numbers of transcription-related factors have been highlighted, it is not known whether neuronal dysfunction is mediated by a specific transcription factor or by multiple transcription factors. We observed that general transcription is repressed by polyglutamine mutant proteins, supporting the latter hypothesis (Hoshino *et al.*, 2004).

22.4.2.2. ER Stress

Mutant polyglutamine proteins induce endoplasmic reticulum (ER) stress that protects cells from protein misfolding. It ceases translation, induces chaperone gene upregulation, and triggers signaling of apoptosis. The first pathway is mediated by PERK (protein kinase regulated by RNA (PKR)-like ER kinase), the second is mediated by IRE1 (inositol-requiring protein) and ATF6 (activating transcription factor 6), and the third is mediated by IRE-JNK (c-Jun N-terminus kinase) cascade or CHOP (C/EBP-homologous protein). These responses basically stop *de novo* synthesis of toxic proteins and select the pathway to suicide when refolding of toxic proteins fails. In other words, when synthesis of unfolded or misfolded proteins overflows the threshold of refolding, cell death signaling is triggered in ER.

22.4.2.3. Proteasome Dysfunction

Proteasomal dysfunction is also induced by aggregation of mutant proteins. As described in the former section, mutant polyglutamine proteins gather to centrosome where a number of proteasomal proteins are focused. The translocation by itself is rational, whereas the elongated polyglutamine peptides are hardly cleaved by proteasome (Venkatraman *et al.*, 2004) and therefore the mutant peptide accumulates in the functional area of proteasome. When large nonsoluble inclusion bodies are formed, the region is called *aggresome* (Johnston *et al.*, 1998). At this stage, proteasomal function is disturbed by the aggregation.

22.4.2.4. Mitochondrial Stress

Recent findings from many groups suggest that polyglutamine proteins induce mitochondrial stress. In addition to the results that polyglutamine proteins trigger cytochrome C release from mitochondria (Jana *et al.*, 2001), it is reported that mutant huntingtin proteins accumulate at mitochondrial membrane and they subsequently induce depolarization of mitochondrial membrane potential (Panov *et al.*, 2002). Choo *et al.* reconfirmed that mutant huntingtin protein associates with outer mitochondrial membrane and lowers the Ca^{2+} threshold to open mitochondrial permeability transition pore (Choo *et al.*, 2004).

22.4.2.5. Disturbance of Axonal Transport

Recent notions from htt-interacting molecules such as HAP1 (huntingtin-associated protein 1) and HIP1 (huntingtin-interacting protein 1) suggested involvement of huntingtin in axonal transport. HAP, which was cloned by two-hybrid screening, is predominately expressed in the brain and in neurons. It is known to associate with p150^{Glued}, a subunit of dynactin complex (Engelender *et al.*, 1997; Gutekunst *et al.*, 1998). Dynactin complex associates with dynein, an anterograde kinetomotor protein moving on microtubles, and transports vesicles. Recently, it was shown that brain-derived neurotrophic factor (BDNF)-containing vesicles are carried by the complex including HAP1 (Gauthier *et al.*, 2004). Furthermore, htt protein also associates with the complex. In the case of mutation, the association between the transport complex and microtubles is reduced, and therefore the transport of BDNF is decreased. The reduction of transport and resultant release of BDGF might be relevant to the pathology.

22.4.2.6. Disturbance of Synaptic Vesicle Release

It is reported that presynaptic button is a place where mutant huntingtin proteins are prone to aggregate (Li *et al.*, 2000). The Li group tested the synaptic dysfunction hypothesis and found that the glutamate transmitter release is disturbed in HD models (Li *et al.*, 2003).

22.4.2.7. Depletion of Normal Polyglutamine Proteins

Mutant polyglutamine disease proteins sequester their interaction partners into inclusion bodies, and the sequestration is suspected to deplete physiologic proteins from functional fields. However, as discussed in the previous section, anchoring strength to inclusion body is different among interaction partners. Normal polyglutamine disease proteins seem to be most strongly adhered to inclusions, thus the depletion could be relevant to the pathology when normal protein is essential for neuronal survival [for review see Cattaneo *et al.* (2001)]. For example, normal huntingtin that is essential for neuronal function is tightly bound to inclusion body (Busch *et al.*, 2003).

22.4.2.8. Neuronal Death or Dysfunction?

It is still controversial whether neuronal dysfunction or neuronal death causes the symptoms of polyglutamine disease patients. Recent results from transgenic mouse models have shown that symptomatic onset precedes neuronal death in the brain. Transgenic mice show motor dysfunction before the decrease of neurons is observed, suggesting strongly that neuronal dysfunction causes symptoms [for review see Okazawa (2003); Sugars and Rubinsztein (2003)]. In this sense, it is of note that mental retardation, which does not accompany neuronal death in youth, and polyglutamine diseases share common pathologic molecules such as cyclic-AMP-responsive element (CREB)-binding protein (CBP) and PQBP-1. Both proteins are sequestered into polyglutamine inclusion bodies in neurons (Kazantsev *et al.*, 1999; McCampbell *et al.*, 2000; Steffan *et al.*, 2000; Okazawa *et al.*, 2002; Busch *et al.*, 2003) to cause dysfunction of both molecules in transcription. At the same time, mutations of CBP and PQBP-1 are the causes of Rubinstein-Taybi syndrome (Murata *et al.*, 2001; Kalkhoven *et al.*, 2003) and Renpenning syndrome (Kalscheuer *et al.*, 2004; Lenski *et al.*, 2004), respectively. Therefore, disturbance of neuronal transcription may underlie the symptoms of neurodegenerative disorders before neuronal cell death occurs. Meanwhile, we do not know which kind of changes are induced in neurons after a long term of transcriptional dysfunction.

22.5. Conclusion

In this chapter, general outlines of the physiologic and pathologic roles of glutamine-rich and asparagine-rich regions have been reviewed. The conformation-dependent aggregation of disease proteins is common, but they induce overlapped but partially specific cellular responses and dysfunctions in different polyglutamine diseases. Detailed pathomechanisms of representative polyglutamine diseases will be reviewed in the following chapters. I have not discussed roles of glutamine-rich and asparagines-rich regions in human prion diseases that are described in the previous chapters.

References

Alberts, B., Bray, D., Lewis, J., Raff, M., Roberts, K., and Watson, J. D. (1994). Molecular Biology of the Cell. 3rd edition, Garland, New York and London.

Aigle, M., and Lacroute, F. (1975). Genetic aspects of [URE3], a non-Mendelian, cytoplasmically-inherited mutation in yeast. *Mol Gen Genet.* 136:327–335.

Arrasate, M., Mitra, S., Schweitzer, E. S., Segal, M. R., and Finkbeiner, S. (2004). Inclusion body formation reduces levels of mutant huntingtin and the risk of neuronal death. *Nature.* 431:805–810.

Bates, G. (2003). Huntingtin aggregation and toxicity in Huntington's disease. *Lancet.* 361:1642–1644.

Berke, S. J., Schmied, F. A., Brunt, E. R., Ellerby, L. M., and Paulson, H. L. (2004). Caspase-mediated proteolysis of the polyglutamine disease protein ataxin-3. *J Neurochem.* 89:908–918.

Bousset, L., Thompson, N. H., Radford, S. E., and Melki, R. (2002). The yeast prio Ure2P retains its native α-helical conformation upon assembly into protein fibrils in vitro. *EMBO J.* 21:2903–2911.

Busch, A., Engemann, S., Lurz, R., Okazawa, H., Lehrach, H., and Wanker, E. E. (2003). Mutant huntingtin promotes the fibrillogenesis of wild-type huntingtin: A potential mechanism for loss of huntingtin function in Huntington's disease. *J Biol Chem.* 278:41452–41461.

Cattaneo, E., Rigamonti, D., Goffredo, D., Zuccato, C., Squitieri, F., and Sipione, S. (2001). Loss of normal huntingtin function: New developments in Huntington's disease research. *Trends Neurosci.* 24:182–188.

Cox, B. S. (1965). [PSI$^+$], a cytoplasmic suppressor of super-suppressor in yeast. *Heredity.* 20:505–521.

Choo, Y. S., Johnson, G. V., MacDonald, M., Detloff, P. J., and Lesort, M. (2004). Mutant huntingtin directly increases susceptibility of mitochondria to the calcium-induced permeability transition and cytochrome c release. *Hum Mol Genet.* 13:1407–1420.

Cummings, C. J., Mancini, M. A., Antalffy, B., DeFranco, D. B., Orr, H. T., and Zoghbi, H. Y. (1998). Chaperone suppression of aggregation and altered subcellular proteasome localization imply protein misfolding in SCA1. *Nat Genet.* 19:148–154.

Cummings, C. J., Reinstein, E., Sun, Y., Antalffy, B., Jiang, Y., Ciechanover, A., Orr, H. T., Beaudet, A. L., and Zoghbi, H. Y. (1999). Mutation of the E6-AP ubiquitin ligase reduces nuclear inclusion frequency while accelerating polyglutamine-induced pathology in SCA1 mice. *Neuron.* 24:879–892.

Davidson, J. D., Riley, B., Burright, E. N., Duvick, L. A., Zoghbi, H. Y., and Orr, H. T. (2000). Identification and characterization of an ataxin-1-interacting protein: A1Up, a ubiquitin-like nuclear protein. *Hum Mol Genet.* 9:2305–2312.

Dyer, R. B., and McMurray, C. T., (2001). Mutant protein in Huntington disease is resistant to proteolysis in affected brain. *Nat Genet.* 29:270–278.

Engelender, S., Sharp, A. H., Colomer, V., Tokito, M. K., Lanahan, A., Worley, P., Holzbaur, E. L., and Ross, C. A. (1997). Huntingtin-associated protein 1 (HAP1) interacts with the p150Glued subunit of dynactin. *Hum Mol Genet.* 6:2205–2212.

Gafni, J., Hermel, E., Young, J. E., Wellington, C. L., Hayden, M. R., and Ellerby, L. M. (2004). Inhibition of calpain cleavage of huntingtin reduces toxicity: Accumulation of calpain/caspase fragments in the nucleus. *J Biol Chem.* 279: 20211–20220.

Gauthier, L. R., Charrin, B. C., Borrell-Pages, M., Dompierre, J. P., Rangone, H., Cordelieres, F. P., De Mey, J., MacDonald, M. E., Lessmann, V., Humbert, S., and Saudou, F. (2004). Huntingtin controls neurotrophic support and survival of neurons by enhancing BDNF vesicular transport along microtubules. *Cell.* 118:127–138.

Gusella, J. F., and MacDonald, M. E. (2000). Molecular genetics: Unmasking polyglutamine triggers in neurodegenerative disease. *Nat Rev Neurosci.* 1:109–115.

Gutekunst, C. A., Li, S. H., Yi, H., Ferrante, R. J., Li, X. J., and Hersch, S. M. (1998). The cellular and subcellular localization of huntingtin-associated protein 1 (HAP1): Comparison with huntingtin in rat and human. *J Neurosci.* 18:7674–7686.

Hoshino, M., Tagawa, K., Okuda, T., and Okazawa, H. (2004). General transcriptional repression by polyglutamine disease proteins is not directly linked to the presence of inclusion bodies. *Biochem Biophys Res Commun.* 313:110–116.

Imafuku, I., Waragai, M., Takeuchi, S., Kanazawa, I., Kawabata, M., Mouradian, M.M., and Okazawa, H. (1998) Polar amino acid-rich sequences bind to polyglutamine tracts. *Biochem Biophys Res Commun* 253: 16–20

Jana, N. R., Zemskov, E. A., Wang, G. h., and Nukina, N. (2001). Altered proteasomal function due to the expression of polyglutamine-expanded truncated N-terminal huntingtin induces apoptosis by caspase activation through mitochondrial cytochrome c release. *Hum Mol Genet.* 10:1049–1059.

Johnston, J. A., Ward, C. L., and Kopito, R. R. (1998). Aggresomes: A cellular response to misfolded proteins. *J Cell Biol.* 143:1883–98.

Kalkhoven, E., Roelfsema, J. H., Teunissen, H., den Boer, A., Ariyurek, Y., Zantema, A., Breuning, M. H., Hennekam, R. C., and Peters, D. J. (2003). Loss of CBP acetyltransferase activity by PHD finger mutations in Rubinstein-Taybi syndrome. *Hum Mol Genet.* 12:441–450.

Kalscheuer, V. M., Freude, K., Musante, L., Jensen, L. R., Yntema, H. G., Gecz, J., Sefiani, A., Hoffmann, K., Moser, B., Haas, S., Gurok, U., Haesler, S., Aranda, B., Nshedjan, A., Tzschach, A., Hartmann, N., Roloff, TC., Shoichet, S., Hagens, O., Tao, J., Van Bokhoven, H., Turner, G., Chelly, J., Moraine, C., Fryns, J. P., Nuber, U., Hoeltzenbein, M., Scharff, C., Scherthan, H., Lenzner, S., Hamel, B. C., Schweiger, S., and Ropers, H. H. (2003). Mutations in the polyglutamine binding protein 1 gene cause X-linked mental retardation. *Nat Genet.* 35:313–315.

Kazantsev, A., Preisinger, E., Dranovsky, A., Goldgaber, D., and Housman, D. (1999). Insoluble detergent-resistant aggregates form between pathological and nonpathological lengths of polyglutamine in mammalian cells. *Proc Natl Acad Sci USA* 96:11404–11409.

Kim, Y. J., Yi, Y., Sapp, E., Wang, Y., Cuiffo, B., Kegel, K. B., Qin, Z. H., Aronin, N., and DiFiglia, M. (2001). Caspase 3-cleaved N-terminal fragments of wild-type and mutant huntingtin are present in normal and Huntington's disease brains, associate with membranes, and undergo calpain-dependent proteolysis. *Proc Natl Acad Sci USA* 98:12784–12789.

Kim, S., Nollen, E. A. A., Kitagawa K., Bindokas, V. P., and Morimoto, R. I. (2002). Polyglutamine protein aggregates are dynamic. *Nat Cell Biol.* 4:826–831.

Khurana, R., Uversky, V. N., Nielsen, L., and Fink, A. L. (2001). Is Congo Red an amyloid-specific dye?. *J Biol Chem.* 276:22715–22721.

Lacroute, F. (1971). Non-Mendelian mutation allowing ureidosuccinic acid uptake in yeast. *J Bacteriol.* 106:519–522.

LaFevre-Bernt, M. A., and Ellerby, L. M. (2003). Kennedy's disease. Phosphorylation of the polyglutamine-expanded form of androgen receptor regulates its cleavage by caspase-3 and enhances cell death. *J Biol Chem.* 278:34918–34924.

Lebre, A. S., and Brice, A. (2003). Spinocerebellar ataxia 7 (SCA7). *Cytogenet Genome Res.* 100:154–163.

Lenski, C., Abidi, F., Meindl, A., Gibson, A., Platzer, M., Frank Kooy R., Lubs, H. A., Stevenson, R. E., Ramser, J., and Schwartz, C. E. (2004). Novel truncating mutations in the polyglutamine tract binding protein 1 gene (PQBP1) cause Renpenning syndrome and X-linked mental retardation in another family with microcephaly. *Am J Hum Genet.* 74:777–780.

Li, H., Li, S. H., Johnston, H., Shelbourne, P. F., and Li, X.J. (2000). Amino-terminal fragments of mutant huntingtin show selective accumulation in striatal neurons and synaptic toxicity. *Nat Genet.* 25:385–389.

Li, H., Wyman, T., Yu, Z. X., Li, S. H., and Li, X. J. (2003). Abnormal association of mutant huntingtin with synaptic vesicles inhibits glutamate release. *Hum Mol Genet.* 12:2021–2030.

Lunkes, A., Lindberg, K. S., Ben-Haim, L., Weber, C., Devys, D., Landwehrmeyer, G. B., Mandel, J. L., and Trottier, Y. (2002). Proteases acting on mutant huntingtin generate cleaved products that differentially build up cytoplasmic and nuclear inclusions. *Mol. Cell* 10:259–169.

McCampbell, A., Taylor, J. P., Taye, A. A., Robitschek, J., Li, M., Walcott, J., Merry, D., Chai, Y., Paulson, H., Sobue, G., and Fischbeck, K. H. (2000). CREB-binding protein sequestration by expanded polyglutamine. *Hum Mol Genet.* 9:2197–2202.

Murata, T., Kurokawa, R., Krones, A., Tatsumi, K., Ishii, M., Taki, T., Masuno, M., Ohashi, H., Yanagisawa, M., Rosenfeld, M. G., Glass, C. K., and Hayashi, Y. (2001). Defect of histone acetyltransferase activity of the nuclear transcriptional coactivator CBP in Rubinstein-Taybi syndrome. *Hum Mol Genet.* 10:1071–1076.

Okazawa, H. (2003). Polyglutamine diseases: a transcription disorder? *Cell Mol Life Sci.* 60:1427–1439.

Okazawa, H., Rich, T., Chang, A., Lin, X., Waragai, M., Kajikawa, M., Enokido, Y., Komuro, A., Kato, S., Shibata, M., Hatanaka, H., Mouradian, M. M., Sudol, M., and Kanazawa, I. (2002). Interaction between mutant ataxin-1 and PQBP-1 affects transcription and cell death. *Neuron.* 34:701–713.

Panov, A. V., Gutekunst, C. A., Leavitt, B. R., Hayden, M. R., Burke, J. R., Strittmatter, W. J., and Greenamyre, J. T. (2002). Early mitochondrial calcium defects in Huntington's disease are a direct effect of polyglutamines. *Nat Neurosci.* 5:731–736.

Perutz, M. F., Johnston, T., Suzuki, M., and Finch, J. T. (1994). Glutamine repeats as polar zippers: their possible role in inherited neurodegenerative diseases. *Proc Natl Acad Sci USA* 91:5355–5358.

Perutz, M. F. (1996). Glutamine repeats and inherited neurodegenerative diseases: molecular aspects. *Curr Opin Struct Biol.* 6:848–858.

Perutz, M. F., Finch, J. T., Berriman, J., and Lesk, A. (2002). Amyloid fibers are water-filled nanotubes. *Proc Natl Acad Sci USA* 99:5591–5595.

Ptino, M. M., Liu, J. J., Glover, J. R., and Lindquist, S. (1996). Support for the prion hypothesis for inheritance of phenotypic trait in yeast. *Science.* 273: 622–626.

Paushkin, S. V., Kushnirov, V. V., Smirnov, V. N., and Ter-Avanesyan, M. D. (1996). Propagation of the yeast prion-like [psi+] determinant is mediated by oligomerization of the SUP35-encoded polypeptide chain release factor. *J. EMBO* 15:3127–3134.

Riley, B. E., Xu, Y., Zoghbi, H. Y., and Orr, H.T. (2004). The effects of the polyglutamine repeat protein ataxin-1 on the UbL-UBA protein A1Up. *J Biol Chem.* 279:42290–42301.

Ross, C. A. (2002). Polyglutamine pathogenesis: Emergence of unifying mechanisms for Huntington's disease and related disorders. *Neuron.* 35:819–822.

Ross, C. A., Poirier, M. A., Wanker, E. E., and Amzel, M. (2003). Polyglutamine fibrilogenesis: The pathway unfold. *Proc Natl Acad Sci USA* 100:1–3.

Scherzinger, E., Sittler, A., Schweiger, K., Heiser, V., Lurz, R., Hasenbank, R., Bates, G. P., Lehrach, H., and Wanker, E. E. (1999). Self-assembly of polyglutamine-containing huntingtin fragments into amyloid-like fibrils: Implications for Huntington's disease pathology. *Proc Natl Acad Sci USA* 96:4604–4609.

Skinner, P. J., Koshy, B. T., Cummings, C. J., Klement, I. A., Helin, K., Servadio, A., Zoghbi, H. Y., and Orr, H. T. (1997). Ataxin-1 with an expanded glutamine tract alters nuclear matrix-associated structures. *Nature*. 389:971–974.

Steffan, J. S., Kazantsev, A., Spasic-Boskovic, O., Greenwald, M., Zhu, Y. Z., Gohler, H., Wanker, E. E., Bates, G. P., Housman, D. E., and Thompson, L. M. (2000). The Huntington's disease protein interacts with p53 and CREB-binding protein and represses transcription. *Proc Natl Acad Sci USA* 97:6763–6768.

Stenoien, D. L., Mielke, M., and Mancini, M. A. (2002). Intranuclear ataxin1 inclusions contain both fast- and slow-exchanging components. *Nat Cell Biol*. 4:806–810.

Sugars, K. L., and Rubinsztein, D. C. (2003). Transcriptional abnormalities in Huntington disease. *Trends Genet*. 19:233–238.

Taylor, K. L., Cheng, N., Williams, R. W., Steven, A. C., and Wichner, R. B. (1999). Prion domain initiation of amyloid formation in vitro from nativ Ure2p. *Science*. 283:1339–1343.

Taylor, J. P., Hardy, J., and Fischbeck, K. H. (2002). Toxic proteins in neurodegenerative disease. *Science*. 296:1991–1995.

Ter-Avanesyan, M. D., Dagkesamanskaya, A. R., Kushnirov, V. V., and Smirnov, V. N. (1994). The SUP35 omnipotent suppressor gene is involved in the maintenance of the non-Mendelian determinat [PSI^+] in the yeast Saccharomyces cerevisiae. *Genetics*. 137:671–676.

Thual, C., Komar, A. A., Bousset, L., Fernandez-Bellot, E., Cullin, C., and Melki, R. (1999). Structural characterization of Saccharomyces cerevisiae prion-like protein Ure2p. *J Biol Chem*. 274:13666–13674.

Tuite, M. F., and Cox, B. S. (2003). Propagation of yeast prions. *Nat Rev Mol Cell Biol*. 4:878–889.

Uptain, S. M., and Lindquist, S. (2002). Prions as protein-based genetic elements. *Annu Rev Microbiol*. 56:703–741.

Venkatraman, P., Wetzel, R., Tanaka, M., Nukina, N., and Goldberg, A. L. (2004). Eukaryotic proteasomes cannot digest polyglutamine sequences and release them during degradation of polyglutamine-containing proteins. *Mol Cell*. 14:95–104.

Waelter, S., Scherzinger, E., Hasenbank, R., Nordhoff, E., Lurz, R., Goehler, H., Gauss, C., Sathasivam, K., Bates, G. P., Lehrach, H., and Wanker, E. E. (2001). The huntingtin interacting protein HIP1 is a clathrin and alpha-adaptin-binding protein involved in receptor-mediated endocytosis. *Hum Mol Genet*. 10:1807–1817.

Watase, K., Venken, K. J., Sun, Y., Orr, H. T., and Zoghbi, H. Y. (2003). Regional differences of somatic CAG repeat instability do not account for selective neuronal vulnerability in a knock-in mouse model of SCA1. *Hum Mol Genet*. 12:2789–2795.

Wickner, R. B. (1994). [URE3] as an altered Ure2 peortein: Evidence for a prion analog in Saccharomyces cerevisiae. *Science*. 264:566–569.

Zoghbi, H. Y., and Orr, H. T. (2000). Glutamine repeats and neurodegeneration. *Annu Rev Neurosci*. 23:217–247.

23

Mechanistic Insights into the Polyglutamine Ataxias

Victor M. Miller and Henry L. Paulson

Abstract

Several hypotheses have been advanced in order to explain the molecular mechanisms by which expanded polyglutamine (polyQ) triggers neurodegeneration. In this chapter, we discuss the experimental evidence supporting the leading hypotheses in the field. In particular, we focus on the mechanisms by which abnormal protein folding, oligomerization, and aggregation may impair neuronal function and survival in the dominant spinocerebellar ataxias (SCAs) caused by polyQ expansion.

23.1. Introduction

The hereditary ataxias comprise a class of progressive, neurodegenerative disorders that share the major clinical symptom of progressive motor incoordination affecting gait and limb function (Schols *et al.*, 2004; Taroni and DiDonato, 2004). At least 35 distinct genetic loci exist for inherited ataxia, including both dominant and recessive forms of disease (Taroni and DiDonato, 2004). The dominantly inherited forms of progressive ataxia are, by convention, termed *spinocerebellar ataxias* (SCAs) because the affected brain regions typically extend beyond the cerebellum to the brain stem, spinal cord, and, in some cases, basal ganglia. At least nine SCAs result from the expansion of unstable DNA repeats within otherwise unrelated disease genes. In six of these (SCAs 1, 2, 3, 6, 7, and 17), the mutation is an expansion of a CAG triplet repeat in the protein coding region of the gene (Taroni and DiDonato, 2004). This repeat encodes an abnormally long, uninterrupted polyglutamine (polyQ) tract in the respective disease proteins—hence the designation, polyQ diseases. Expansion of this homopolymeric amino acid motif leads to abnormal accumulation and aggregation of the disease protein in neurons with associated neuronal dysfunction and eventual neuronal cell loss (Zoghbi and Orr, 2000).

23.2. Clinical and Molecular Genetic Features of polyQ Ataxias

The six dominant SCAs mentioned above comprise most of the nine known polyQ diseases, a group of progressive neurodegenerative disorders that also includes Huntington's disease (HD), spinal bulbar muscular atrophy (SBMA), and dentatorubral-pallidoluyisian atrophy (DRPLA). Clinically, HD and SBMA differ from the SCAs while DRPLA shows some overlap with SCAs (ataxia being part of the syndrome in DRPLA). HD is a progressive neuropsychiatric disorder characterized by mood, cognitive and motor abnormalities (primarily chorea) but *not* ataxia, and SBMA is a form of adult onset motor neuron degeneration in which disease is strictly limited to motor neurons. These marked clinical differences despite a shared mutational mechanism (expanded polyQ) speak to the importance of disease protein context in pathogenesis (Taylor *et al.*, 2002; Zoghbi and Orr, 2000).

Collectively, the polyQ ataxias are a major cause of neurogenetic diseases. Some of the individual polyQ ataxias also represent the most common forms of dominantly inherited ataxia in certain ethnic populations; for example, SCA3 in Portugal and SCA1 in Northern Russia. In the United States, SCAs 1, 2, 3, and 6 together constitute more than half of all families with dominant ataxia, and all four are polyQ ataxias (Schols *et al.*, 2004). Although the most common clinical presentation of polyQ SCA is progressive ataxia, a wide and varying range of additional symptoms or signs can occur. Depending on the disease, these can include ophthalmoplegia, abnormal saccades, retinal degeneration, pyramidal and extrapyramidal features, dementia, amyotrophy, peripheral neuropathy, and parkinsonism (Taroni and DiDonato, 2004; Zoghbi and Orr, 2000).

Like the diverse clinical phenotypes, the pattern and extent of neurodegeneration can vary greatly, sometimes even within the same affected family. This largely reflects difference in the size of the expansion in the affected individual. All polyQ ataxias show, to some degree, degeneration in the cerebellum (most commonly Purkinje cell loss) and brain stem (except SCA6). Some SCAs are characterized by prominent degeneration elsewhere, such as the retina in SCA7 and cerebral cortex in SCA17, not shared by the other members of the group (Schols *et al.*, 2004; Taroni and DiDonato, 2004). These differences highlight fundamental quandaries in understanding the pathogenesis of these disorders. Despite widespread expression of the disease proteins, why are only neurons affected? And what mediates the selective vulnerability of specific neuronal subtypes in each disease when they share a common genetic lesion? The answers lie in the particular molecular structure and function of the various disease proteins upon which a novel, toxic gain-of-function is imparted by the elongated polyQ tract.

At the molecular level, the polyQ ataxias represent important models for studying both the basis of trinucleotide DNA repeat instability and the biology of protein misfolding and aggregation. The accumulation and aggregation of mutant proteins in the polyQ ataxias parallels molecular events in other classes of inherited neurologic disease including hereditary forms of Parkinson's disease, Alzheimer's disease (AD), frontotemporal dementia, and amyotrophic lateral sclerosis (Taylor *et al.*, 2002). Indeed, in most age-related neurodegenerative diseases, abnormal protein accumulation occurs in affected brain regions. For example, Alzheimer's disease brains show aberrant accumulation of tau and β-amyloid, and Parkinson's disease brains contain Lewy bodies rich in α-synuclein (Taylor *et al.*, 2002). Though these aggregates differ in structure and composition from the intranuclear, perinuclear, and neuritic inclusions found in polyQ SCAs, fundamental biophysical insights gained from the study of misfolded polyQ proteins have shed light on the properties of other neurodegenerative disease proteins, and vice versa (Kayed *et al.*, 2003). Investigation of both the adaptive and pathologic cellular responses to mutant polyQ proteins may also highlight pathways that trigger neurodegeneration in other neurodegenerative diseases.

In many respects, then, polyQ ataxias are test cases for understanding a broader set of human neurologic diseases. In the remaining sections of this chapter, we discuss the leading hypotheses put forth to explain the molecular pathogenesis of polyQ SCAs. We review recent experimental evidence for and against each hypothesis and pay particular attention to ways in which abnormal protein folding, oligomerization, and aggregation may impair neuronal function and survival.

23.3. Toxic Aggregates—or not?

The toxic nature of polyQ aggregates remains controversial. At the very minimum, aggregates detectable in human disease brain represent biological markers of a complex set of pathophysiologic, cellular disturbances. These physiologic disturbances are the subject of subsequent sections. Here we discuss some of the most recent evidence implicating polyQ aggregates themselves as bad actors, innocent bystanders, or even protective structures.

A unifying feature of polyQ diseases is the presence in neurons of microscopically visible proteinaceous inclusions containing the mutant protein. These inclusions are seen in select regions of disease brain, varying among the various polyQ diseases (Zoghbi and Orr, 2000). Their discovery almost a decade ago immediately suggested that protein misfolding was central to polyQ pathogenesis (DiFiglia *et al.*, 1997). Consistent with this view, purified forms of recombinant, mutant polyQ proteins show an intrinsic tendency to adopt amyloid-like conformations and form aggregates in the absence of any cellular components or other proteins (Chen *et al.*, 2001; Chen *et al.*, 2002). The repeat threshold required to initiate aggregation *in vitro* closely matches the threshold length that triggers disease in humans (~35–40). Still longer repeats are even more aggregate-prone *in vitro* and cause earlier onset disease, more widespread pathology, and earlier death in affected humans and mouse models (Chen *et al.*, 2001; Chen *et al.*, 2002). Cell and animal models ranging from bacteria to worms, flies, fish, and mice have been employed to study polyQ disease, and nearly all of these develop macroscopic aggregates as part of the phenotype. These complementary lines of evidence point to a fundamental biophysical property of expanded polyQ proteins that is somehow related to their ability to cause disease.

Cell-based and animal models of polyQ disease typically rely on the expression of soluble, monomeric polyQ proteins to initiate the phenotype. This makes it difficult to study aggregate toxicity with precision because native monomer, misfolded monomer, oligomeric or partially aggregated intermediates, and mature aggregates may all coexist in the same cell. To overcome this limitation and directly address aggregate toxicity, Yang *et al.* delivered preformed aggregates of Q20 or Q42 polypeptides to cultured cells and assessed cell death (Yang *et al.*, 2002). An attached fluorescein group allowed the polyQ peptide to be visualized by microscopy, and a nuclear localization signal (NLS) was added to force nuclear entry. Peptides containing 20 or 42 repeats could be induced to form fibrillar aggregates. Purified aggregates (Q20 or Q42) added to the culture medium were readily taken up by cells, and dramatic cell death was observed when aggregates entered the nucleus. Interestingly, polyQ aggregates proved to be toxic to cells whether they were derived from pathogenic or normal length repeats, whereas monomeric forms of the same two polyQ peptides were nontoxic. Nuclear localization was critical for toxicity, and aggregates of a non-polyQ protein were not at all toxic even when delivered to the nucleus (Yang *et al.*, 2002). This cell-based study strengthens the case for direct toxicity of aggregates. It is important to recognize, however, that aggregates of the sort used in this study are biochemically defined, insoluble, pure polyQ complexes that may differ fundamentally from the inclusions that form in cells; the latter are heterogenous collections of diverse proteins whose accumulation is likely driven, in part, by cellular processes. Thus, this study does not answer whether the type of inclusions seen in disease are toxic, inert, or protective structures.

While inclusions are frequently found in polyQ disease brain, their distribution correlates rather poorly with regions of degeneration both in affected humans and transgenic mouse models. In a knock-in mouse model of SCA7, for example, degeneration did not correlate with macroscopic aggregate formation (Yoo *et al.*, 2003). To directly assess the role of nuclear aggregates in SCA1, Klement *et al.* generated transgenic mice expressing ataxin-1-77Q in which the self-association region essential for ataxin-1 dimerization had been deleted (Klement *et al.*, 1998). Despite having no detectable ataxin-1 inclusions, these mice developed ataxia and Purkinje cell pathology similar to mice expressing full-length ataxin-1–82Q. In contrast, mice expressing mutant ataxin-1 engineered to remain in the cytoplasm (due to distruption of its nuclear localization signal) developed no behavioral or neuropathologic phenotype (Klement *et al.*, 1998). Thus, nuclear localization of ataxin-1 is necessary to cause disease in transgenic mice while nuclear aggregation of ataxin-1 is not.

In a recent carefully controlled, cell-based study, Finkbeiner and colleagues followed individual primary neurons over time to determine the temporal and causal relationship of polyQ aggregate formation to cell death (Arrasate *et al.*, 2004). Neurons were transfected with an expression vector encoding a fluorescently tagged polyQ-containing fragment of the HD disease protein, huntingtin, and then repeatedly observed by a robotically controlled miscroscope (Arrasate *et al.*, 2004). The most interesting result of the study? Neurons in which visible aggregates formed were less likely to die than neurons in which the mutant fragment remained diffusely present in the cell. This result argues that large, visibly detectable aggregates may play a protective role. An interesting, unanswered question is whether the diffusely present protein that proved to be toxic in this experiment existed in a monomeric, oligomeric, or prefibrillar form.

Taken together, these studies demonstrate that visible aggregates are not required to initiate disease in neurons. That said, large aggregates of mutant protein are clearly not a feature of normal neurons. In the end, probably the best that can be said is that the relationship between aggregates and disease is complex. However, visible aggregates themselves are neither necessary nor sufficient for polyQ proteins to exert toxicity. Their conspicuous and pervasive association with disease ensures that aggregates and the process by which they arise will remain a subject of study. It is clear that the field must, and in many respects already has, become more sophisticated than merely counting aggregate burden when assessing disease phenotypes.

The study of polyQ aggregates, the disease proteins that form them, and the cells in which they occur has generated the many additional hypotheses about pathogenic mechanisms discussed further in the following sections. In general, these hypotheses are not mutually exclusive with a mechanistic model in which protein misfolding plays a pivotal role.

23.4. PolyQ Ataxias as Channelopathies

Neurons expend a great deal of cellular energy segregating ions across the cell membrane in order to initiate and propagate action potentials and to regulate the release of signaling molecules. A small body of literature suggests that structural features of oligomeric polyQ proteins could allow them to form aberrant ion channels that compromise neuronal physiology. If correct, then the polyQ SCAs would join the so-called channelopathies. Channel formation has been reported for several proteins that are known to adopt amyloid-like conformations (Kayed *et al.*, 2003; Kayed *et al.*, 2004). Channel formation by polyQ proteins was reported by two groups: Hirakura *et al.* and Monoi *et al.* They described the formation of stable, membrane-spanning channels that allowed for unregulated passage of ions (Hirakura *et al.*, 2000; Monoi *et al.*, 2000). Importantly, the threshold for ion channel formation by recombinant polyQ protein closely paralleled the repeat length known to cause disease in humans: a polyQ protein with 40 repeats could form long-lived ion channels but one with 29 repeats could not. A structural model for polyQ channels was proposed in which Q37 (the approximate threshold repeat length in several polyQ diseases) was just sufficient to span the lipid bilayer and form a passage for ions (Hirakura *et al.*, 2000; Monoi *et al.*, 2000).

Further evidence consistent with channel formation by polyQ proteins was reported by Panov *et al.* (Panov *et al.*, 2002). This group showed decreased mitochondrial membrane potential

Table 23.1. Molecular Features of the PolyQ SCAs

Disease	Gene / locus	Mutant $(CAG)_n$	Gene product	Inclusion pathology	Protein function
SCA1	SCA1 / 6p23	39–82	Ataxin-1	NI	Unknown Possible RNA binding motifs Phosphorylation dependent Toxicity in mice
SCA2	SCA2 / 12p24	36–63	Ataxin-2	CI	Unknown Interacts with a polyA binding protein Possible role in RNA metabolism
SCA3	SCA3 / 14q32.1	53–84	Ataxin-3	NI	Deubiquitinating enzyme Ubiquitin binding protein
SCA6	CACNA1A / 19p13	21–33	P/Q-type calcium channel a1A subunit	CI-perinuclear	Voltage gated calcium channel Plasma membrane ion channel
SCA7	SCA7 / 3p14	37–306	Ataxin-7	NI	Unknown Associated with histone Acetyltransferase complex
SCA17	TBP / 6q27	43–63	TATA binding protein	NI	Basal transcription factor
DRPLA	DRPLA / 12p13.31	49–84	Atrophin-1	NI	Unknown Fly homologue is a transcriptional co-repressor

SCA, spinocerebellar ataxia; DRPLA, dentatorubral-pallidoluyisian atrophy; TBP, TATA binding protein; NI, nuclear inclusions; CI, cytoplasmic inclusions.

and depolarization at reduced Ca^{2+} levels in HD lymphoblasts compared with mitochondria from control patients. Mitochondria isolated from brains of transgenic mice expressing a pathogenic polyQ tract exhibited similar membrane dysfunction. Electron microscopy revealed mutant polyQ protein localized to the mitochondrial membrane. Similar defects could be reproduced by adding recombinant, expanded polyQ protein to mitochondria. These results suggested that expanded polyQ tracts are capable of depolarizing mitochondria, leaving the cell vulnerable to metabolic stress and apoptotic cascades. The presence of polyQ proteins at the mitochondrial membrane hints that this might occur via a channel-forming mechanism (Panov *et al.*, 2002).

The one unquestioned channelopathy among the polyQ ataxias is SCA6. SCA6 results from an expansion in the *CACNA1A* gene that encodes a voltage-gated calcium channel subunit (Zhuchenko *et al.*, 1997). In this case, the mechanism presumably differs from that discussed above: expanded polyQ tract would alter function of an existing ion channel rather than form a novel channel-like structure that inserted into the membrane. Uncertainty exists regarding the mechanism of toxicity in SCA6. Does SCA6 result from a more general toxic gain-of-function due to expanded polyQ or from a polyQ-induced loss of normal channel function? A couple of facts provide some clues. The much smaller range of CAG repeat expansions in SCA6 differs from the typical disease range observed in other polyQ SCAs (see Table 23.1). Knockout of the *CACNA1A* gene in mice results in an ataxic phenotype and, in tissue culture, expression of the channel subunit with expanded polyQ results in a shift in the voltage dependence of channel inactivation (Fletcher *et al.*, 1996; Matsuyama *et al.*, 1999). These lines of evidence support a role for direct channel dysfunction in SCA6. But, as in the other polyQ SCAs, abnormal accumulations of the disease protein have been found in SCA6 patient brains suggesting that a toxic oligomeric species generated by the polyQ expansion may contribute to pathogenesis in this disease. The SCA6 "aggregates" differ from those described in most other polyQ disease in that they have not been reported in the nucleus and there is no biochemical evidence that they consist of insoluble complexes (Ishikawa *et al.*, 1999).

PolyQ-mediated formation of aberrant ion channels could explain the mechanism by which mutant polyQ protein oligomerization injures neurons. The fact that at least one polyQ disease (SCA6) is caused by an expansion in an ion channel squares with the reasonable supposition that neurons are especially vulnerable to dysfunction of ion channels. The evidence is strong that recombinant amyloidogenic proteins, including polyQ, can form a pore that allows the passage of ions. What is lacking, however, is compelling evidence that these channel structures actually form when the polyQ tract occurs in the context of a full-length disease protein (or even a cleaved fragment of the disease protein). Physical constraints dictated by the surrounding protein context may preclude, or severely limit, channel formation by polyQ. Nevertheless, even a low level of channel formation that compromises the membrane potential of critical organelles, like mitochondria, might sensitize neurons to secondary insults.

23.5. Mitochondrial Dysfunction

Neurons depend on a constant and abundant supply of ATP. As the ATP-producing organelle, the mitochondrion is essential to neuronal function and survival. In addition to its role in energy metabolism, the mitochondrion also participates in programmed cell death by integrating pro-apoptotic and anti-apoptotic signals. As a by-product of energy production, mitochondria generate the majority of reactive oxygen species (ROS) in the cell. Oxidative damage by free radicals has been implicated in many late onset, neurodegenerative conditions as well as aging more generally. Mitochondria are thus pivotal to both the life and death of neurons. Even relatively modest impairment of mitochondria could trigger neuronal dysfunction and death in the polyQ diseases.

Early evidence of energy impairment in polyQ disease came from studies of cerebral glucose metabolism using positron emission tomography scans in HD patients. Decreased glucose metabolism was detected in regions heavily affected in HD (Alavi *et al.*, 1986). Decreased respiratory chain complex activities have been detected in HD brain, platelets, fibroblasts, and muscle (Browne *et al.*, 1997; Gu *et al.*, 1996). Similar defects have also been seen in the R6/2 mouse model of HD (Tabrizi *et al.*, 2000). The glycolytic enzyme GAPDH has been reported to bind huntingtin and is downregulated in the brains of some HD mouse models, though this is not likely to be a primary element in pathogenesis (Burke *et al.*, 1996). Studies using mitochondrial toxins have produced patterns of degeneration mimicking features of polyQ disease. For example, 3-nitropropionic acid (3-NP), an irreversible inhibitor of succinate dehydrogenase, produces lesions resembling the neuropathologic features of HD when administered to rodents (Beal *et al.*, 1993).

Therapeutic trials have been undertaken based on the assumption that energy impairment plays an important role in polyQ neurodegeneration. The electron transport molecule, ubiquinone (or coenzyme Q10), showed promise by slowing disease progression in HD mouse models (Ferrante *et al.*, 2002), but failed to show a statistically significant benefit in a well controlled human trial of HD subjects (Huntington Study Group, 2001). Oral administration of creatine has also been attempted on the theory that it should increase phosphocreatine levels and provide a buffer against cellular energy depletion. Creatine was protective in HD mouse models but showed no benefit in an SCA1 mouse model, raising the possibilities that energy impairment plays a disease-specific role in pathogenesis or that particular mouse models are more sensitive to mitochondrial energy defects (Ferrante *et al.*, 2000; Kaemmerer *et al.*, 2001). To date, there is no definitive evidence that efforts to alleviate deficits in mitochondrial energetics will alter the disease phenotype in humans (Verbessem *et al.*, 2003). Energy depletion in neurons can lead to metabolic derangements, generation of free radicals, and crippling of critical ion channels, any of which could cause malfunction and death of brain cells. For these reasons, mitochondrial energy production remains a viable avenue of investigation in the pathogenesis of the polyQ SCAs.

23.6. Transcriptional Dysregulation

A significant body of work underlies the hypothesis that expanded polyQ triggers disease by dysregulating gene expression at the transcriptional level (Hughes and Olson, 2001). Central to this hypothesis is the observation that expanded polyglutamine engages in aberrant protein-protein interactions, including some with important transcription factors. Toxicity could result from soluble interactions that interfere with transcription factor activity or from depletion of one or more transcription factors via their sequestration into polyQ aggregates. Either route might result in inappropriate or reduced activity at specific promoters or in perturbations of chromatin modification by histone acetyltransferases and other enzymes. Several lines of evidence support the involvement of polyQ-containing proteins in transcription. Many transcription factors contain polyQ or glutamine-rich domains, and polyQ tracts themselves can serve as transcriptional activators (Gerber *et al.*, 1994). Indeed, polyQ expansion in one transcription factor, TATA binding protein (TBP), causes SCA17, and the SCA7 disease protein ataxin-7 appears to be part of a histone acetyltransferase-containing complex (Helmlinger *et al.*, 2004; McMahon *et al.*, 2005; Nakamura *et al.*, 2001; Scheel *et al.*, 2003).

PolyQ proteins can interact directly with transcription factors, and nuclear entry of the polyQ protein seems to be important for toxicity in several polyQ diseases (Klement *et al.*, 1998). Reported interactions include TATA-binding protein (TBP) (in its nonexpanded form), the

Drosophila eyes-absent protein (EYA), CREB-binding protein (CBP), p53, nuclear receptor co-repressor (N-CoR), mSin3A and TAFII130, and repressor-element-1 transcription factor–neuron-restrictive silencer factor (REST–NRSF) (Perez *et al.*, 1998; Boutell *et al.*, 1999; Shimohata *et al.*, 2000; Steffan *et al.*, 2000; Steffan *et al.*, 2001). In some cases, the transcription factor is mislocalized or sequestered in inclusions. For instance, TBP localizes to nuclear inclusions in human SCA3 disease brain and TAFII130 localizes to inclusions in DRPLA and HD (Perez *et al.*, 1998; Shimohata *et al.*, 2000). CBP is found in the inclusions of several polyQ SCAs including SCA1 and SCA3. In live cell models, however, dramatic differences in the mobility of CBP was found between inclusions formed by ataxin-1 and ataxin-3 suggesting that immunodetection of a transcription factor in inclusions in fixed tissue constitutes merely circumstantial evidence for inactivation by sequestration in aggregates (Chai *et al.*, 2002; Stenoien *et al.*, 2002). For some transcription factors, interactions with polyQ proteins are known to inhibit function. For instance, mutant huntingtin represses TAFII130 promoters and mutant ataxin-3 can repress CBP-dependent transcription in cell models (Chai *et al.*, 2001; Shimohata *et al.*, 2000).

Microarray analyses of brain tissue from a panel of polyQ mouse models consistently demonstrate transcriptional alterations. There is some overlap among the disease models regarding the functional categories of dysregulated genes, but the overlap is less robust on a gene by gene basis (Sugars and Rubinsztein, 2003). It is also unclear how well the observed changes in mRNA levels correlate with corresponding protein levels. Moreover, these experiments pose a challenge in discerning which gene expression changes are directly related to polyQ pathogenesis and which are compensatory responses by sick neurons. In favor of the transcriptional dysregulation model, overexpression of CBP was able to suppress polyglutamine toxicity in a fly model of disease while loss of CBP function in mice resulted in a phenotype that partly mirrored that of mice overexpressing mutant polyQ proteins (Taylor *et al.*, 2003).

A second but related explanation for transcriptional dysregulation centers on the effects of mutant polyQ proteins on histone acetylation. Transcription factors known to interact with polyQ proteins, CBP and p300/CBP-associated factor (P/CAF), possess histone acetyltransferase activity. PolyQ proteins bind their acetyltransferase domains and inhibit activity *in vitro*. In cells expressing polyQ proteins, alterations in histone acetylation are often observed. Interestingly, administration of histone deacetylase (HDAC) inhibitors rescued polyQ toxicity in cells and flies (Steffan *et al.*, 2000; Steffan *et al.*, 2001). Although alterations in histone acetylation were not observed at baseline, administration of the HDAC inhibitor SAHA to the R6/2 mouse model of polyQ toxicity resulted in modest slowing of the phenotype (Hockly *et al.*, 2003). HDAC inhibitors are currently being explored in a human trial of HD.

In summary, transcriptional dysregulation remains a viable hypothesis to explain polyQ disease pathogenesis. Differences in the protein context provided by the various polyQ disease proteins could lead to overlapping, yet nonidentical, changes that might account for disease-specific features. Moreover, subtle transcriptional changes could contribute to pathogenesis during the presumably prolonged period of neuronal dysfunction before overt neuronal loss. Identified transcriptional changes (and affected transcription factors) also may provide defined targets for developing small molecules to alter transcription as possible therapy.

23.7. Failures in Protein Quality Control

Cells must ensure that nascent proteins fold correctly and deal with the refolding of proteins damaged by physiologic stress or mutations. Heat-shock protein chaperones (HSPs) facilitate folding and maintain proteins in a soluble conformation. If the native conformation of a protein cannot be achieved, futile refolding efforts by molecular chaperones either continue or the protein is targeted for degradation (Hartl and Hayer-Hartl, 2002). For many damaged or misfolded proteins, the principal route for protein destruction is the ubiquitin-proteasome pathway (UPP) (Berke and Paulson, 2003). Abnormally folded proteins tend to aggregate. If the concentration of misfolded proteins exceeds cellular folding and degradative capacity, these proteins can form insoluble, intracellular aggregates reminiscent of those seen in the polyQ SCAs. Together with the UPP, molecular chaperones (principally the Hsp70 and Hsp40 classes) carry out the major protein quality control (QC) functions in the cell (Berke and Paulson, 2003; Hartl and Hayer-Hartl, 2002). Several lines of evidence are consistent with a disease mechanism in which the burden of misfolded polyQ proteins overwhelms this QC capacity in neurons.

Hsp chaperones localize to polyQ aggregates in patient tissues and cellular and animal models (Paulson *et al.*, 1997; Cummings *et al.*, 1998). This suggests, but certainly does not prove, that the protein aggregates result from protein misfolding. The sequestering of chaperones into aggregates may decrease the soluble pool of functioning chaperones, thereby lowering the overall protein folding capacity of the cell. This in turn could result in an environment that favors further misfolding and aggregation over refolding and degradation. Overexpression of Hsp70 and Hsp40 chaperones reduces aggregation and cell death in mammalian cells and rescues neurodegeneration in *Drosophila* models of polyQ SCAs (Cummings *et al.*, 1998; Warrick *et al.*, 1999). In flies, chaperone overexpression alters the biochemical properties of aggregates, rendering them detergent soluble, though visible inclusions remain (Chan *et al.*, 2000). These findings support the hypothesis that polyQ proteins do in fact compromise the folding capacity of cells, resulting in toxic oligomeric species. Genetic screens in *C. elegans* and yeast also point to a role for chaperones in buffering the toxicity of polyQ proteins, although genes in many other pathways were also found to be involved (Nollen *et al.*, 2004; Willingham *et al.*, 2003). Unfortunately, transgenic overexpression of Hsp70 chaperones has yielded only marginal benefit in polyQ mouse models, suggesting that reduced chaperone activity does not fully explain the pathology seen in polyQ SCAs (Cummings *et al.*, 2001; Hansson *et al.*, 2003; Hay *et al.*, 2004).

Irretrievably misfolded proteins typically become ubiquitinated and delivered to the proteasome for degradation. There is evidence that UPP function declines with age, paralleling the typically later onset of symptoms in the polyQ SCAs (Goto *et al.*, 2001). Because inclusions in polyQ disorders are ubiquitinated and sequester proteasome components, this suggested to researchers that a failure of the UPP might underlie disease. Indeed, eukaryotic proteasomes are incapable, *in vitro*, of fully digesting peptides containing polyQ repeats (Venkatraman *et al.*, 2004). Moreover, in cell-based proteasome reporter assays, expression of pathogenic polyQ proteins caused impairment of the UPP (Bence *et al.*, 2001). These results suggest that attempts by the cell to digest pathogenic polyQ tracts might cause the release of aggregation-prone fragments or that polyQ proteins themselves might stall the proteasome, rendering it unavailable for normal QC functions. To test this theory directly, Zoghbi and colleagues recently crossed a transgenic mouse harboring a proteasome reporter gene to an SCA7 knock-in model. In a well-controlled set of experiments, no proteasome inhibition was found in the degenerating retina of SCA7 mice (Bowman *et al.*, 2005). Why the contradictory results for proteasome reporter assays in cell-based versus mouse studies? A likely explanation is that in transiently transfected cells, a massive aggregate burden accumulates rapidly, whereas in the *in vivo* model, neuropathologic features and protein accumulation develop much more slowly. At this point, despite solid *in vitro* and cell-based data, there is no compelling *in vivo* evidence that mutant polyQ proteins inhibit the proteasome in affected tissues.

We have recently discovered that the C-terminal Hsp70-interacting protein (CHIP) may participate at a crucial junction in the handling of misfolded polyQ. As both a co-chaperone and ubiquitin ligase, CHIP links the two major arms of protein quality control, molecular chaperones and the ubiquitin-proteasome system. We found that CHIP overexpression rescues mutant, polyQ-induced phenotypes in several *in vitro* and non-mammalian animal models. In a transgenic, polyQ, mouse model, CHIP haploinsufficiency markedly accelerated the phenotype (Miller *et al.*, 2005). This demonstrates that the critical components of neuronal QC that modulate polyQ disease are yet to be fully investigated, particularly with respect to the interplay between molecular chaperones and the UPP.

Protein QC is particularly critical for postmitotic neurons, which in humans function for decades in the face of high metabolic demands, environmental insults, and age-related physiologic changes. As outlined above, considerable experimental evidence indicates that chaperones and the UPP are involved in the response to mutant polyQ proteins. At the same time, the most recent data in animal models suggest that the simple explanation that mutant polyQ protein chokes off, or overwhelms, neuronal QC cannot completely explain pathogenesis. Instead, more subtle impairment of QC may be one of several insults that the neuron faces when chronically exposed to expanded polyQ.

23.8. Impaired Axonal Transport

Histopathologic analyses of some polyQ disease brains show widespread neuritic inclusions, raising the possibility that perturbation of axonal transport contributes to pathogenesis (DiFiglia *et al.*, 1997). Impairment of the axonal transport system in other neurologic disease is already well established. Mutations in the kinesin gene *KIF1B* cause Charcot-Marie-Tooth disease type 2A, which is characterized by dysfunction of peripheral neurons, possibly due to reduced kinesin-dependent transport of synaptic vesicle precursors (Zhao *et al.*, 2001). Mutation of the neuronal kinesin heavy-chain gene, KIF5A, causes hereditary spastic paraplegia (Reid *et al.*, 2002). A transgenic mouse model of amyotrophic lateral sclerosis shows abnormalities in microtubule-based transport and degeneration of motor neurons (Williamson and Cleveland, 1999). In the polyQ SCAs, impairment of axonal transport could occur through several mechanisms. Aggregated disease proteins could physically block transport in narrow axons. Mutant polyQ proteins could interact aberrantly with transport pathway proteins and thus titrate them away from their normal transport functions. Finally, some of the disease proteins may normally function in axonal transport, and this function could be directly impaired by polyQ expansion.

In HD mouse models and human patient brains, dystrophic neurites in the striatum are consistently observed. These neurites exhibit characteristic features of blocked axons such as prominent swellings with vesicle and organelle accumulation associated with aggregates (DiFiglia *et al.*, 1997). In SCA6, axonal accumulations have been observed and are reported to contain neurofilaments and other materials indicative of a failure in transport (Yang *et al.*, 2000). These findings are reminiscent of the phenotype caused by motor protein mutants in *Drosophila*. Recently, expression of mutant polyQ proteins in *Drosophila*, including ataxin-3 fragments, was found to cause similar axonal blockages. Mutant polyQ also reduced levels of key motor proteins (Gunawardena *et al.*, 2003). Though the molecular mechanism was less well defined, experiments in giant squid axons and mammalian tissue culture cells showed inhibition of anterograde and retrograde transport by truncated versions of huntingtin or the androgen receptor (Szebenyi *et al.*, 2003). These observations suggest that aggregated disease proteins physically block transport and titrate motors proteins away from their normal functions.

In the case of the HD protein huntingtin, *Drosophila* studies suggest that the normal protein interacts with the transport machinery to aid in vesicular transport (Szebenyi *et al.*, 2003). Precisely how huntingtin associates with the axonal transport machinery is unclear, but its interaction with huntingtin associated protein 1 (HAP1), a protein that interacts with the p150 subunit of dynactin, may provide a link. Mutations in the *Drosophila* HAP1-like protein, Milton, causes axonal transport defects implicating the huntingtin-HAP1 complex in this process (Szebenyi *et al.*, 2003). The SCA3 protein, ataxin-3, interacts with histone deacetylase 6 and dynein during transport of misfolded proteins, thus linking ataxin-3 to cytoskeletal transport processes in non-neuronal cultured cells (Burnett and Pittman, 2005). A role for normal ataxin-3 in neuronal cargo transport, however, has not been established. Though details are still lacking, it seems plausible that some polyQ disease proteins—notably huntingtin—directly play a role in cytoskeletal transport of vesicles, organelles, or other protein cargo, and that polyQ expansion impairs these functions (Gunawardena *et al.*, 2003).

Neurons rely on axonal transport to appropriately partition organelles, protein complexes, signaling molecules, and neuroprotective and repair molecules. Because effective axonal transport is so vital to neurons, the hypothesis that defects caused by mutant polyQ proteins could account for selective neuronal vulnerability in these diseases is intriguing (Gunawardena and Goldstein, 2005). The prominence of neuritic inclusions and dystrophic neurites in model systems and some disease brains strongly suggests that polyQ proteins cause pathologic changes in neurites. Indeed, recent studies suggest that impaired axonal transport plays a more general role in neurodegenerative diseases such as Alzheimer's disease (Stokin *et al.*, 2005). Clearly, however, impaired transport does not account for the pathology seen in all polyQ diseases. In some cases, such as the nuclear localized ataxin-1, the disease protein is never present in neurites. This makes it unlikely that transport is directly affected by the disease protein (whether soluble or aggregated) or that it normally functions in that process. Further study is required to build on the already promising evidence that impaired axonal transport plays a major role in certain polyQ diseases and perhaps an indirect, accessory role in others.

23.9. Conclusion

Despite sharing a common mutation and several core clinical features, the polyQ diseases are a diverse and complex group of disorders. The polyQ ataxias differ widely in their clinical features, and the mutation occurs in entirely different protein contexts in each case with no sequence similarity outside the polyQ region (Zoghbi and Orr, 2000). The threshold repeat length and the range of pathogenic repeats also differs somewhat across the diseases; witness the upper limit for SCA6 of 33 versus >300 for SCA7 (Table 23.1). All show accumulation of the disease protein in neurons, yet these inclusions differ in their molecular and biochemical properties and subcellular localization; for example, exclusively nuclear inclusions in SCA1 versus cytoplasmic inclusions in SCA6. The normal functions of the disease proteins range from fairly well characterized transcription factors to more poorly understood ubiquitin binding proteins to proteins implicated in RNA homeostasis (Table 23.1).

Given this diversity, perhaps it is not surprising that no clear and convincing winner has emerged among the contenders for the title of most relevant pathogenic mechanism. While there are many differences between the individual polyQ SCAs, they do share an obvious, common mutation associated with abnormal protein conformation and aggregation. It is intellectually appealing to suppose that, in all polyQ diseases, common mechanisms are at work to seal the fate of afflicted neurons. But it is more likely that the detailed biology of each disease protein will dictate which, if any, of the mechanisms we discussed predominates in a given disease. Indeed it may be overly simplistic to assume that the failure of any one cellular process is solely to blame for the devastating phenotype. For instance, subtle or partial failure in protein quality control might create a favorable environment for aggregation. The aggregated protein could then interact with and slowly deplete transcription factors altering expression of essential genes. As the neuron struggles to cope with this loss of homeostasis, the aggregates grow and clog the axon.

Ultimately, the neuron's physiology becomes so impaired that mitochondria depolarize and trigger the neuron's death. This projected scenario, while perhaps not correct in its details, underscores the point that a multiplicity of insults may conspire to injure neurons. Ultimately, the proof will come in the therapeutic pudding. Can compounds be found that block one or more of these pathways and lead to treatment? The various hypotheses described here are fostering targeted therapies based on experimental observations. Which approaches reverse or slow disease in animal models and affected humans will bring us powerful empirical evidence in favor of the right pathogenic mechanisms.

Acknowledgments

Work on polyQ disease in our laboratory was supported by the National Institutes of Health (NIH NS38712 and NS47535) and grants from the Ataxia MJD Research Project and the National Ataxia Foundation.

References

Alavi, A., Dann, R., Chawluk, J., Alavi, J., Kushner, M., and Reivich, M. (1986). Positron emission tomography imaging of regional cerebral glucose metabolism. *Semin Nucl Med* 16: 2–34.

Arrasate, M., Mitra, S., Schweitzer, E. S., Segal, M. R., and Finkbeiner, S. (2004). Inclusion body formation reduces levels of mutant huntingtin and the risk of neuronal death. *Nature* 431: 805–810.

Beal, M. F., Brouillet, E., Jenkins, B. G., Ferrante, R. J., Kowall, N. W., Miller, J. M., Storey, E., Srivastava, R., Rosen, B. R., and Hyman, B. T. (1993). Neurochemical and histologic characterization of striatal excitotoxic lesions produced by the mitochondrial toxin 3-nitropropionic acid. *J Neurosci* 13: 4181–4192.

Bence, N. F., Sampat, R. M., and Kopito, R. R. (2001). Impairment of the ubiquitin-proteasome system by protein aggregation. *Science* 292: 1552–1555.

Berke, S. J., and Paulson, H. L. (2003). Protein aggregation and the ubiquitin proteasome pathway: gaining the UPPer hand on neurodegeneration. *Curr Opin Genet Dev* 13: 253–261.

Boutell, J. M., Thomas, P., Neal, J. W., Weston, V. J., Duce, J., Harper, P. S., and Jones, A. L. (1999). Aberrant interactions of transcriptional repressor proteins with the Huntington's disease gene product, huntingtin. *Hum Mol Genet* 8: 1647–1655.

Bowman, A. B., Yoo, S. Y., Dantuma, N. P., and Zoghbi, H. Y. (2005). Neuronal dysfunction in a polyglutamine disease model occurs in the absence of ubiquitin-proteasome system impairment and inversely correlates with the degree of nuclear inclusion formation. *Hum Mol Genet* 14: 679–691.

Browne, S. E., Bowling, A. C., MacGarvey, U., Baik, M. J., Berger, S. C., Muqit, M. M., Bird, E. D., and Beal, M. F. (1997). Oxidative damage and metabolic dysfunction in Huntington's disease: selective vulnerability of the basal ganglia. *Ann Neurol* 41: 646–653.

Burke, J. R., Enghild, J. J., Martin, M. E., Jou, Y. S., Myers, R. M., Roses, A. D., Vance, J. M., and Strittmatter, W. J. (1996). Huntingtin and DRPLA proteins selectively interact with the enzyme GAPDH. *Nat Med* 2: 347–350.

Burnett, B. G., and Pittman, R. N. (2005). The polyglutamine neurodegenerative protein ataxin 3 regulates aggresome formation. *Proc Natl Acad Sci USA* 102: 4330–4335.

Chai, Y., Wu, L., Griffin, J. D., and Paulson, H. L. (2001). The role of protein composition in specifying nuclear inclusion formation in polyglutamine disease. *J Biol Chem* 276: 44889–44897.

Chai, Y., Shao, J., Miller, V. M., Williams, A., and Paulson, H. L. (2002). Live-cell imaging reveals divergent intracellular dynamics of polyglutamine disease proteins and supports a sequestration model of pathogenesis. *Proc Natl Acad Sci USA* 99: 9310–9315.

Chan, H. Y., Warrick, J. M., Gray-Board, G. L., Paulson, H. L., and Bonini, N. M. (2000). Mechanisms of chaperone suppression of polyglutamine disease: selectivity, synergy and modulation of protein solubility in Drosophila. *Hum Mol Genet* 9: 2811–2820.

Chen, S., Berthelier, V., Yang, W., and Wetzel, R. (2001). Polyglutamine aggregation behavior *in vitro* supports a recruitment mechanism of cytotoxicity. *J Mol Biol* 311: 173–182.

Chen, S., Ferrone, F. A., and Wetzel, R. (2002). Huntington's disease age-of-onset linked to polyglutamine aggregation nucleation. *Proc Natl Acad Sci USA* 99: 11884–11889.

Cummings, C. J., Mancini, M. A., Antalffy, B., DeFranco, D. B., Orr, H. T., and Zoghbi, H. Y. (1998). Chaperone suppression of aggregation and altered subcellular proteasome localization imply protein misfolding in SCA1. *Nat Genet* 19: 148–154.

Cummings, C. J., Sun, Y., Opal, P., Antalffy, B., Mestril, R., Orr, H. T., Dillmann, W. H., and Zoghbi, H. Y. (2001). Overexpression of inducible HSP70 chaperone suppresses neuropathology and improves motor function in SCA1 mice. *Hum Mol Genet* 10: 1511–1518.

DiFiglia, M., Sapp, E., Chase, K. O., Davies, S. W., Bates, G. P., Vonsattel, J. P., and Aronin, N. (1997). Aggregation of huntingtin in neuronal intranuclear inclusions and dystrophic neurites in brain. *Science* 277: 1990–1993.

Ferrante, R. J., Andreassen, O. A., Jenkins, B. G., Dedeoglu, A., Kuemmerle, S., Kubilus, J. K., Kaddurah-Daouk, R., Hersch, S. M., and Beal, M. F. (2000). Neuroprotective effects of creatine in a transgenic mouse model of Huntington's disease. *J Neurosci* 20: 4389–4397.

Ferrante, R. J., Andreassen, O. A., Dedeoglu, A., Ferrante, K. L., Jenkins, B. G., Hersch, S. M., and Beal, M. F. (2002). Therapeutic effects of coenzyme Q10 and remacemide in transgenic mouse models of Huntington's disease. *J Neurosci* 22: 1592–1599.

Fletcher, C. F., Lutz, C. M., O'Sullivan, T. N., Shaughnessy, J. D., Jr., Hawkes, R., Frankel, W. N., Copeland, N. G., and Jenkins, N. A. (1996). Absence epilepsy in tottering mutant mice is associated with calcium channel defects. *Cell* 87: 607–617.

Gerber, H. P., Seipel, K., Georgiev, O., Hofferer, M., Hug, M., Rusconi, S., and Schaffner, W. (1994). Transcriptional activation modulated by homopolymeric glutamine and proline stretches. *Science* 263: 808–811.

Goto, S., Takahashi, R., Kumiyama, A. A., Radak, Z., Hayashi, T., Takenouchi, M., and Abe, R. (2001). Implications of protein degradation in aging. *Ann N Y Acad Sci* 928: 54–64.

Gu, M., Gash, M. T., Mann, V. M., Javoy-Agid, F., Cooper, J. M., and Schapira, A. H. (1996). Mitochondrial defect in Huntington's disease caudate nucleus. *Ann Neurol* 39: 385–389.

Gunawardena, S., and Goldstein, L. S. (2005). Polyglutamine diseases and transport problems: deadly traffic jams on neuronal highways. *Arch Neurol* 62: 46–51.

Gunawardena, S., Her, L. S., Brusch, R. G., Laymon, R. A., Niesman, I. R., Gordesky-Gold, B., Sintasath, L., Bonini, N. M., and Goldstein, L. S. (2003). Disruption of axonal transport by loss of huntingtin or expression of pathogenic polyQ proteins in Drosophila. *Neuron* 40: 25–40.

Hansson, O., Nylandsted, J., Castilho, R. F., Leist, M., Jaattela, M., and Brundin, P. (2003). Overexpression of heat shock protein 70 in R6/2 Huntington's disease mice has only modest effects on disease progression. *Brain Res* 970: 47–57.

Hartl, F. U., and Hayer-Hartl, M. (2002). Molecular chaperones in the cytosol: from nascent chain to folded protein. *Science* 295: 1852–1858.

Hay, D. G., Sathasivam, K., Tobaben, S., Stahl, B., Marber, M., Mestril, R., Mahal, A., Smith, D. L., Woodman, B., and Bates, G. P. (2004). Progressive decrease in chaperone protein levels in a mouse model of Huntington's disease and induction of stress proteins as a therapeutic approach. *Hum Mol Genet* 13: 1389–1405.

Helmlinger, D., Hardy, S., Sasorith, S., Klein, F., Robert, F., Weber, C., Miguet, L., Potier, N., Van-Dorsselaer, A., Wurtz, J. M., *et al.* (2004). Ataxin-7 is a subunit of GCN5 histone acetyltransferase-containing complexes. *Hum Mol Genet* 13: 1257–1265.

Hirakura, Y., Azimov, R., Azimova, R., and Kagan, B. L. (2000). Polyglutamine-induced ion channels: a possible mechanism for the neurotoxicity of Huntington and other CAG repeat diseases. *J Neurosci Res* 60: 490–494.

Hockly, E., Richon, V. M., Woodman, B., Smith, D. L., Zhou, X., Rosa, E., Sathasivam, K., Ghazi-Noori, S., Mahal, A., Lowden, P. A., *et al.* (2003). Suberoylanilide hydroxamic acid, a histone deacetylase inhibitor, ameliorates motor deficits in a mouse model of Huntington's disease. *Proc Natl Acad Sci USA* 100: 2041–2046.

Hughes, R. E., and Olson, J. M. (2001). Therapeutic opportunities in polyglutamine disease. *Nat Med* 7: 419–423.

Huntington Study Group. (2001). A randomized, placebo-controlled trial of coenzyme Q10 and remacemide in Huntington's disease. *Neurology* 57: 397–404.

Ishikawa, K., Fujigasaki, H., Saegusa, H., Ohwada, K., Fujita, T., Iwamoto, H., Komatsuzaki, Y., Toru, S., Toriyama, H., Watanabe, M., *et al.* (1999). Abundant expression and cytoplasmic aggregations of [alpha]1A voltage-dependent calcium channel protein associated with neurodegeneration in spinocerebellar ataxia type 6. *Hum Mol Genet* 8: 1185–1193.

Kaemmerer, W. F., Rodrigues, C. M., Steer, C. J., and Low, W. C. (2001). Creatine-supplemented diet extends Purkinje cell survival in spinocerebellar ataxia type 1 transgenic mice but does not prevent the ataxic phenotype. *Neuroscience* 103, 713–724.

Kayed, R., Head, E., Thompson, J. L., McIntire, T. M., Milton, S. C., Cotman, C. W., and Glabe, C. G. (2003). Common structure of soluble amyloid oligomers implies common mechanism of pathogenesis. *Science* 300: 486–489.

Kayed, R., Sokolov, Y., Edmonds, B., McIntire, T. M., Milton, S. C., Hall, J. E., and Glabe, C. G. (2004). Permeabilization of lipid bilayers is a common conformation-dependent activity of soluble amyloid oligomers in protein misfolding diseases. *J Biol Chem* 279: 46363–46366.

Klement, I. A., Skinner, P. J., Kaytor, M. D., Yi, H., Hersch, S. M., Clark, H. B., Zoghbi, H. Y., and Orr, H. T. (1998). Ataxin-1 nuclear localization and aggregation: role in polyglutamine-induced disease in SCA1 transgenic mice. *Cell* 95: 41–53.

Matsuyama, Z., Wakamori, M., Mori, Y., Kawakami, H., Nakamura, S., and Imoto, K. (1999). Direct alteration of the P/Q-type Ca2+ channel property by polyglutamine expansion in spinocerebellar ataxia 6. *J Neurosci* 19: RC14.

McMahon, S. J., Pray-Grant, M. G., Schieltz, D., Yates, J. R., 3rd, and Grant, P. A. (2005). Polyglutamine-expanded spinocerebellar ataxia-7 protein disrupts normal SAGA and SLIK histone acetyltransferase activity. *Proc Natl Acad Sci USA* 102: 8478–8482.

Miller, V. M., Nelson, R. F., Gouvion, C. M., Williams, A., Rodriguez-Lebron, E., Harper, S. Q., Davidson, B. L., Rebagliati, M. R., and Paulson, H. L. (2005). CHIP suppresses polyglutamine aggregation and toxicity *in vitro* and *in vivo*. *J Neurosci.* 25(40): 9125–9161.

Monoi, H., Futaki, S., Kugimiya, S., Minakata, H., and Yoshihara, K. (2000). Poly-L-glutamine forms cation channels: relevance to the pathogenesis of the polyglutamine diseases. *J Biophys* 78: 2892–2899.

Nakamura, K., Jeong, S. Y., Uchihara, T., Anno, M., Nagashima, K., Nagashima, T., Ikeda, S., Tsuji, S., and Kanazawa, I. (2001). SCA17, a novel autosomal dominant cerebellar ataxia caused by an expanded polyglutamine in TATA-binding protein. *Hum Mol Genet* 10: 1441–1448.

Nollen, E. A., Garcia, S. M., van Haaften, G., Kim, S., Chavez, A., Morimoto, R. I., and Plasterk, R. H. (2004). Genome-wide RNA interference screen identifies previously undescribed regulators of polyglutamine aggregation. *Proc Natl Acad Sci USA* 101: 6403–6408.

Panov, A. V., Gutekunst, C. A., Leavitt, B. R., Hayden, M. R., Burke, J. R., Strittmatter, W. J., and Greenamyre, J. T. (2002). Early mitochondrial calcium defects in Huntington's disease are a direct effect of polyglutamines. *Nat Neurosci* 5: 731–736.

Paulson, H. L., Perez, M. K., Trottier, Y., Trojanowski, J. Q., Subramony, S. H., Das, S. S., Vig, P., Mandel, J. L., Fischbeck, K. H., and Pittman, R. N. (1997). Intranuclear inclusions of expanded polyglutamine protein in spinocerebellar ataxia type 3. *Neuron* 19: 333–344.

Perez, M. K., Paulson, H. L., Pendse, S. J., Saionz, S. J., Bonini, N. M., and Pittman, R. N. (1998). Recruitment and the role of nuclear localization in polyglutamine-mediated aggregation. *J Cell Biol* 143: 1457–1470.

Reid, E., Kloos, M., Ashley-Koch, A., Hughes, L., Bevan, S., Svenson, I. K., Graham, F. L., Gaskell, P. C., Dearlove, A., Pericak-Vance, M. A., *et al.* (2002). A kinesin heavy chain (KIF5A) mutation in hereditary spastic paraplegia (SPG10). *Am J Hum Genet* 71: 1189–1194.

Scheel, H., Tomiuk, S., and Hofmann, K. (2003). Elucidation of ataxin-3 and ataxin-7 function by integrative bioinformatics. *Hum Mol Genet* 12: 2845–2852.

Schols, L., Bauer, P., Schmidt, T., Schulte, T., and Riess, O. (2004). Autosomal dominant cerebellar ataxias: clinical features, genetics, and pathogenesis. *Lancet Neurol* 3: 291–304.

Shimohata, T., Nakajima, T., Yamada, M., Uchida, C., Onodera, O., Naruse, S., Kimura, T., Koide, R., Nozaki, K., Sano, Y., *et al.* (2000). Expanded polyglutamine stretches interact with TAFII130, interfering with CREB-dependent transcription. *Nat Genet* 26: 29–36.

Steffan, J. S., Kazantsev, A., Spasic-Boskovic, O., Greenwald, M., Zhu, Y. Z., Gohler, H., Wanker, E. E., Bates, G. P., Housman, D. E., and Thompson, L. M. (2000). The Huntington's disease protein interacts with p53 and CREB-binding protein and represses transcription. *Proc Natl Acad Sci USA* 97: 6763–6768.

Steffan, J. S., Bodai, L., Pallos, J., Poelman, M., McCampbell, A., Apostol, B. L., Kazantsev, A., Schmidt, E., Zhu, Y. Z., Greenwald, M., *et al.* (2001). Histone deacetylase inhibitors arrest polyglutamine-dependent neurodegeneration in Drosophila. *Nature* 413: 739–743.

Stenoien, D. L., Mielke, M., and Mancini, M. A. (2002). Intranuclear ataxin1 inclusions contain both fast- and slow-exchanging components. *Nat Cell Biol* 4: 806–810.

Stokin, G. B., Lillo, C., Falzone, T. L., Brusch, R. G., Rockenstein, E., Mount, S. L., Raman, R., Davies, P., Masliah, E., Williams, D. S., and Goldstein, L. S. (2005). Axonopathy and transport deficits early in the pathogenesis of Alzheimer's disease. *Science* 307: 1282–1288.

Sugars, K. L., and Rubinsztein, D. C. (2003). Transcriptional abnormalities in Huntington disease. *Trends Genet* 19: 233–238.

Szebenyi, G., Morfini, G. A., Babcock, A., Gould, M., Selkoe, K., Stenoien, D. L., Young, M., Faber, P. W., MacDonald, M. E., McPhaul, M. J., and Brady, S. T. (2003). Neuropathogenic forms of huntingtin and androgen receptor inhibit fast axonal transport. *Neuron* 40: 41–52.

Tabrizi, S. J., Workman, J., Hart, P. E., Mangiarini, L., Mahal, A., Bates, G., Cooper, J. M., and Schapira, A. H. (2000). Mitochondrial dysfunction and free radical damage in the Huntington R6/2 transgenic mouse. *Ann Neurol* 47: 80–86.

Taroni, F., and DiDonato, S. (2004). Pathways to motor incoordination: the inherited ataxias. *Nat Rev Neurosci* 5: 641–655.

Taylor, J. P., Hardy, J., and Fischbeck, K. H. (2002). Toxic proteins in neurodegenerative disease. *Science* 296: 1991–1995.

Taylor, J. P., Taye, A. A., Campbell, C., Kazemi-Esfarjani, P., Fischbeck, K. H., and Min, K. T. (2003). Aberrant histone acetylation, altered transcription, and retinal degeneration in a Drosophila model of polyglutamine disease are rescued by CREB-binding protein. *Genes Dev* 17: 1463–1468.

Venkatraman, P., Wetzel, R., Tanaka, M., Nukina, N., and Goldberg, A. L. (2004). Eukaryotic proteasomes cannot digest polyglutamine sequences and release them during degradation of polyglutamine-containing proteins. *Mol Cell* 14: 95–104.

Verbessem, P., Lemiere, J., Eijnde, B. O., Swinnen, S., Vanhees, L., Van Leemputte, M., Hespel, P., and Dom, R. (2003). Creatine supplementation in Huntington's disease: a placebo-controlled pilot trial. *Neurology* 61: 925–930.

Warrick, J. M., Chan, H. Y., Gray-Board, G. L., Chai, Y., Paulson, H. L., and Bonini, N. M. (1999). Suppression of polyglutamine-mediated neurodegeneration in Drosophila by the molecular chaperone HSP70. *Nat Genet* 23: 425–428.

Williamson, T. L., and Cleveland, D. W. (1999). Slowing of axonal transport is a very early event in the toxicity of ALS-linked SOD1 mutants to motor neurons. *Nat Neurosci* 2: 50–56.

Willingham, S., Outeiro, T. F., DeVit, M. J., Lindquist, S. L., and Muchowski, P. J. (2003). Yeast genes that enhance the toxicity of a mutant huntingtin fragment or alpha-synuclein. *Science* 302: 1769–1772.

Yang, Q., Hashizume, Y., Yoshida, M., Wang, Y., Goto, Y., Mitsuma, N., Ishikawa, K., and Mizusawa, H. (2000). Morphological Purkinje cell changes in spinocerebellar ataxia type 6. *Acta Neuropathol (Berl)* 100: 371–376.

Yang, W., Dunlap, J. R., Andrews, R. B., and Wetzel, R. (2002). Aggregated polyglutamine peptides delivered to nuclei are toxic to mammalian cells. *Hum Mol Genet* 11: 2905–2917.

Yoo, S. Y., Pennesi, M. E., Weeber, E. J., Xu, B., Atkinson, R., Chen, S., Armstrong, D. L., Wu, S. M., Sweatt, J. D., and Zoghbi, H. Y. (2003). SCA7 knockin mice model human SCA7 and reveal gradual accumulation of mutant ataxin-7 in neurons and abnormalities in short-term plasticity. *Neuron* 37: 383–401.

Zhao, C., Takita, J., Tanaka, Y., Setou, M., Nakagawa, T., Takeda, S., Yang, H. W., Terada, S., Nakata, T., Takei, Y., *et al.* (2001). Charcot-Marie-Tooth disease type 2A caused by mutation in a microtubule motor KIF1Bbeta. *Cell* 105: 587–597.

Zhuchenko, O., Bailey, J., Bonnen, P., Ashizawa, T., Stockton, D. W., Amos, C., Dobyns, W. B., Subramony, S. H., Zoghbi, H. Y., and Lee, C. C. (1997). Autosomal dominant cerebellar ataxia (SCA6) associated with small polyglutamine expansions in the alpha 1A-voltage-dependent calcium channel. *Nat Genet* 15: 62–69.

Zoghbi, H. Y., and Orr, H. T. (2000). Glutamine repeats and neurodegeneration. *Annu Rev Neurosci* 23: 217–247.

24

Molecular Pathogenesis of the Polyglutamine Disease: Spinal and Bulbar Muscular Atrophy

Erica S. Chevalier-Larsen and Diane E. Merry

Abstract

Spinal and bulbar muscular atrophy (SBMA) is an adult onset neurodegenerative disease that affects men beginning in the fourth to fifth decade of life. The disease results from the expansion of a CAG trinucleotide repeat in the first exon of the androgen receptor gene. Symptoms of SBMA include weakness and atrophy of the proximal limb and bulbar muscles as well as partial androgen insensitivity. Cellular markers of disease include the loss of spinal motor neurons and the formation of nuclear aggregates of mutant androgen receptor protein (NII). The mechanism for pathogenesis in SBMA is unknown, and there is currently no treatment for the disease. New transgenic mouse, fly, and cell models of SBMA have provided important insights into the role of testosterone binding as a critical initiator of disease. An understanding of androgen receptor (AR) metabolism upon ligand binding is likely to provide new insights into the upstream events in SBMA pathogenesis. Moreover, the creation of novel models of disease is providing insights into abnormal neuronal responses to expanded AR accumulation. Here we discuss these ideas in the context of both basic AR biology and novel therapeutic development for SBMA.

24.1. Polyglutamine Expansion Diseases

Polyglutamine expansion is responsible for a number of dominantly inherited neurodegenerative diseases including Huntington's disease (HD), spinocerebellar ataxia type 1, 2, 3, 6, 7, and 17, dentatorubropallidoluysian atrophy (DRPLA), and spinal and bulbar muscular atrophy (SBMA) (Cummings and Zoghbi, 2000; Zoghbi and Orr, 2000; Nakamura *et al.*, 2001). In these diseases, a repeated CAG motif, which encodes a polyglutamine tract, exceeds a threshold length and results in pathology. Age of onset and severity of symptoms are correlated with the length of the polyglutamine expansion; longer polyglutamine tracts result in more severe symptoms and earlier disease onset (Doyu *et al.*, 1992; La Spada *et al.*, 1992). In addition, expanded CAG tracts are unstable and exhibit a greater tendency to expand and contract during male meiosis (La Spada *et al.*, 1992). Thus, polyglutamine expansion diseases exhibit genetic anticipation, with successive generations often showing increasing CAG tract length and, consequently, earlier onset and greater severity of symptoms (Igarashi *et al.*, 1992). The extent of anticipation is limited in SBMA, however, because onset of the disease depends on the presence of testosterone (Katsuno *et al.*, 2002; Takeyama *et al.*, 2002; Chevalier-Larsen *et al.*, 2004).

The appearance of aggregates of the mutant protein is characteristic of polyglutamine expansion diseases (Davies *et al.*, 1997; DiFiglia *et al.*, 1997; Paulson *et al.*, 1997) and suggests that the expanded polyglutamine tract interferes with normal protein processing. Despite a unifying mechanism of mutation, the symptoms of polyglutamine expansion diseases differ depending on the gene containing the CAG expansion (Ross, 2002). Although the genes containing a CAG expansion mutation are ubiquitously expressed, specific, although somewhat overlapping, populations of neurons are affected in each disease. The reason for this selective toxicity remains unclear, although protein context, specifically pertaining to the normal metabolism and interactions of the mutant protein, may play a role in this selective neuronal specificity.

24.2. Spinal and Bulbar Muscular Atrophy

Spinal and bulbar muscular atrophy, currently the only X-linked polyglutamine expansion disease, results from expansion of a CAG tract, within the first exon of the androgen receptor (AR) gene, to a number greater than 40 (La Spada *et al.*, 1991). The frequency of SBMA is estimated at 1 in 40,000; however, it is likely that the disease is underdiagnosed. Men affected with SBMA typically begin to exhibit weakness and atrophy in the proximal limb muscles of the hip and shoulder girdles, as well as fasciculations of the tongue and bulbar muscles of the jaw, between 30 and 50 years of age (Kennedy *et al.*, 1968). Muscle cramping can precede these symptoms by several years. Patients with SBMA do not exhibit the spasticity and hyperreflexia commonly associated with upper motor neuron involvement. Subclinical impairment of sensory function is common (Sobue *et al.*, 1989). Death often occurs from respiratory infection secondary to aspiration of food and drink.

Patients with SBMA often manifest signs of androgen insensitivity, including gynecomastia, testicular atrophy, and infertility (Arbizu *et al.*, 1983). However, loss of androgen function alone is not responsible for the neuropathy seen in SBMA, as evidenced by the finding that individuals with complete androgen insensitivity syndrome (CAIS) do not manifest neurologic symptoms (Pinsky *et al.*, 1992; Quigley *et al.*, 1992). Disease instead results from the polyglutamine expansion mutation via the acquisition of a novel, toxic property (Merry *et al.*, 1998; Stenoien *et al.*, 1999; Diamond *et al.*, 2000). Despite this dominant toxic property, women rarely show symptoms of the disease, even in the homozygous state (Schmidt *et al.*, 2002), although some mild cases have been reported (Ferlini *et al.*, 1995; Mariotti *et al.*, 2000).

Pathologic signs of SBMA include loss of motor neurons in the brain stem and anterior horn of the spinal cord, and the presence of neuronal intranuclear inclusions (NII), nuclear aggregates of mutant androgen receptor (AR) protein (Sobue *et al.*, 1989; Li *et al.*, 1998a). In SBMA patients, NII contain only amino-terminal portions of the androgen receptor, suggesting aberrant cleavage or degradation of the AR, resulting in an insoluble amino-terminal protein fragment (Li *et al.*, 1998b). Extranuclear accumulations of expanded AR protein have recently been described (Adachi *et al.*, 2005); whether these aggregates are similarly comprised solely of amino-terminal AR regions remains to be determined. While the presence of extranuclear aggregates is more strongly correlated with CAG repeat length than are NII, their pathologic role remains unclear.

24.3. Androgen Receptor

The disease-causing polyglutamine expansion in SBMA is located within the amino-terminal region of the AR. The 919-amino acid AR protein is a member of the steroid/thyroid nuclear receptor superfamily. In the unbound state, the AR exists within an aporeceptor complex composed of the heat shock proteins Hsc70, Hsp40, Hsp90, p23, and immunophilins (Pratt and Welsh, 1994; Caplan *et al.*, 1995; Fang *et al.*, 1996; Fliss *et al.*, 1999). Both testosterone and dihydrotestosterone (DHT) bind to AR with high affinity via the carboxyl-terminal ligand-binding domain, causing a conformational change and dissociation from the aporeceptor complex (Hadley, 2000). Once ligand is bound, this conformational switch exposes the nuclear localization signal adjacent to the DNA-binding domain and directs the translocation of the AR to the nucleus. Dimerization of the AR in an antiparallel orientation occurs via interactions between the activation function 1 (AF1) and activation function 2 (AF2) domains, as well as by zinc finger interactions (Wong *et al.*, 1993). Within the nucleus, zinc fingers in the DNA binding domain mediate the binding of the AR dimer to DNA; binding serves to initiate and recruit the ordered assembly of components of the transcription machinery (Shang *et al.*, 2002; Huang *et al.*, 2003). The primary AR domain responsible for these events is the AF2 region within the ligand-binding domain; the conformational movement of alpha helix 12 upon hormone binding provides a binding face for the AF1 domain (and other motifs in the amino terminal region of the AR) and subsequently for members of the p160/SRC family of coactivators (He *et al.*, 2000; He *et al.*, 2004). p160 family protein binding is essential for the further recruitment of coactivators with histone modifying functions (e.g., CBP) and chromatin-remodeling complexes. This ordered assembly of transcriptional activators, coactivators, and chromatin remodeling complexes results in the recruitment of the basal transcription machinery (TFII protein complexes) and RNA polymerase II to the region of the transcriptional start site, leading to transcription initiation (Tsai and O'Malley, 1994; Ma *et al.*, 1999; Shang *et al.*, 2002; Huang *et al.*, 2003).

The role of the AR in sexual development and reproduction is well characterized, but its role in neurons is more elusive. The AR is known to be highly expressed in motor neurons (Sar and Stumpf, 1977; Sheridan, 1984; Sarrieau *et al.*, 1990) and has been shown to play a role in trophic support (Jones, 1988; Yu, 1989; Kujawa *et al.*, 1991; Varela *et al.*, 2000), as well as neuronal plasticity (Kurz *et al.*, 1986; Kujawa *et al.*, 1993; Matsumoto *et al.*, 1995; Matsumoto, 2001; Yang *et al.*, 2004), neuronal regeneration (Fargo and Sengelaub, 2004a, 2004b), and synapse elimination (Jordan, 1995). Androgens, specifically, DHT, modulate neuritic outgrowth, in part through regulation of neuritin gene expression (Marron *et al.*, 2005).

AR binds testosterone and DHT with greater affinity than other steroids. Of these two physiologic ligands of the AR, DHT may be as much as 10 times more potent than testosterone (Deslypere *et al.*, 1992; Grino *et al.*, 1990), although the effects of the different binding affinities on specific physiologic pathways may vary depending on the presence of enzymes that metabolize testosterone (e.g., aromatase, 3α-hydroxysteroid dehydrogenase, 5α-reductase).

24.4. Ligand Dependence

The relatively asymptomatic status of female carriers of SBMA has been somewhat of an enigma, given the fact that polyglutamine expansion confers a dominant toxic property to the mutant AR. Some asymptomatic female carriers of SBMA may exhibit skewed inactivation in blood lymphocytes of the X chromosome bearing expanded AR, suggesting that lyonization plays a role in protecting some women from developing advanced disease (Ishihara *et al.*, 2001). However, the report of homozygous females without symptoms (Schmidt *et al.*, 2002) indicates that X-inactivation is not the sole factor protecting females from disease. An early mouse model of SBMA used a truncated form of AR that lacked the carboxyl-terminal ligand-binding and DNA-binding domains (Abel *et al.*, 2001). These mice developed a generalized neuronopathy but failed to reproduce key aspects of the disease, including gender specificity and lower motor neuron-selective toxicity (Abel *et al.*, 2001), leading to the conclusion that the context of the full-length AR protein was important to these specific aspects of disease. In the past few years, an overwhelming amount of data from a variety of cell, fly, and mouse models of SBMA, utilizing the full-length AR, have revealed the importance of ligand binding in SBMA pathogenesis.

When treated with androgens, cells expressing full-length expanded AR showed an increase in AR aggregation (Grierson *et al.*, 2001; Darrington *et al.*, 2002; Walcott and Merry, 2002). The effects of androgen blockade differed depending on the model. Darrington *et al.* reported that treatment with the androgen antagonists flutamide and cyproterone acetate prevented or reduced aggregation, respectively (Darrington *et al.*, 2002). In contrast, flutamide failed to impact aggregation in an inducible PC12 cell model (Walcott and Merry, 2002). Results similar to those described by Walcott and Merry were observed in a fly model transgenic for full-length expanded AR: consumption of testosterone and flutamide both caused neurodegeneration and lead to inclusion formation (Takeyama *et al.*, 2002).

The development of transgenic mouse models of SBMA has promoted the clarification of a role for hormone in disease onset and progression. Several transgenic mouse models of SBMA have been developed using the full-length human expanded AR (Katsuno *et al.*, 2002; McManamny *et al.*, 2002; Chevalier-Larsen *et al.*, 2004; Sopher *et al.*, 2004). All of these models revealed that males exhibit greater severity and earlier onset of symptoms than do their female counterparts despite an autosomal pattern of transgene inheritance. These data argued strongly for the involvement of a factor, other than X-inactivation, in the gender specificity of SBMA. Following the characterization of these mouse models, numerous hormone manipulation experiments have confirmed this central tenet of SBMA pathogenesis, previously suggested from findings of cell and fly models of SBMA: androgens play a critical role in SBMA pathogenesis. Administration of exogenous testosterone to female animals exacerbated disease progression (Katsuno *et al.*, 2002). Furthermore, both surgical and chemical castration in males prevented the onset of symptoms (Katsuno *et al.*, 2002; Katsuno *et al.*, 2003) or lead to partial recovery of motor function (Chevalier-Larsen *et al.*, 2004). Consistent with the findings from cell and fly models (Takeyama *et al.*, 2002; Walcott and Merry, 2002), flutamide had no effect in diseased males (Katsuno *et al.*, 2003). These data confirmed the causative role of ligand in disease and revealed the potential for recovery upon its removal. In addition, both Katsuno *et al.* and Chevalier-Larsen *et al.* demonstrated that hormone removal resulted in a reduction in soluble and *insoluble* AR, suggesting that clearance of the aggregated protein is possible once hormone is removed (Katsuno *et al.*, 2002; Katsuno *et al.*, 2003; Chevalier-Larsen *et al.*, 2004).

The finding that flutamide failed to impact disease while methods of androgen removal ameliorated phenotype initially appears contradictory. However, these data actually reveal another key step in disease progression. While treatment with leuprorelin (Katsuno *et al.*, 2003) or castration (Katsuno *et al.*, 2002; Chevalier-Larsen *et al.*, 2004) removes testosterone, and cyproterone acetate (Darrington *et al.*, 2002) binds the AR but prevents its nuclear translocation, flutamide binds to the AR allowing and, at certain dosages, even facilitating its translocation to the nucleus. Thus, the data from these models suggests that ligand exerts its pathogenic effects by promoting disassociation from the aporeceptor complex and/or nuclear translocation of AR. While it is clear that X-inactivation plays a part in protecting females with mutated AR (Ishihara *et al.*, 2001), it is now equally apparent that disease onset and progression in males is dependent on ligand binding to AR. Whether testosterone or dihydrotestosterone has a more potent effect on pathogenesis remains to been seen. The role of estrogen as a potentially neuroprotective agent in SBMA also needs clarification.

24.5. Pathogenic Metabolism of the AR

The finding of a requirement for hormone binding in SBMA suggests some details of a model of pathogenesis. Upon hormone binding, the normal AR dissociates from the aporeceptor complex and translocates to the nucleus, where it binds DNA as a dimer to regulate the transcription of target genes. The finding that antagonists that block AR transcriptional function but promote its nuclear translocation fail to ameliorate disease in fly and mouse models of SBMA (Takeyama *et al.*, 2002; Katsuno *et al.*, 2003) and aggregation in a cell model of SBMA (Walcott and Merry, 2002) suggests that nuclear location but not AR transcriptional function is required for disease. Moreover, retention of the mutant AR in the cytoplasm through the use of a nuclear export signal prevented disease (Takeyama *et al.*, 2002), further implicating the nucleus as the site of abnormal AR metabolism. It should be mentioned, however, that results from one study suggest a role for the cytoplasm in SBMA pathogenesis. Thomas *et al.* (2004) showed that retention of *normal* AR in the cytoplasm, through the mutation of acetylation sites, caused its misfolding and cytoplasmic aggregation in a manner similar to some cell models of expanded AR (Thomas *et al.*, 2004). Whether this reveals a role for the cytoplasm in the pathogenic mechanism of SBMA awaits further testing.

The basis of the likely requirement for nuclear localization of the mutant AR is unknown. Studies of a truncated, expanded AR (Bailey *et al.*, 2001) revealed a reduction in proteasomal turnover of the mutant, expanded AR that was alleviated with increased expression of the molecular chaperones Hsp70 and Hsp40. Indeed, the presence of NII in cells and animal models expressing full-length AR further suggests a nuclear site of the refractory degradation of the mutant AR. The importance of the nucleus may be related to the inability of nuclear proteasomes to eliminate the toxic protein; in addition, whether the expanded AR can be efficiently exported from the nucleus is also unknown. Indeed, deletion of the nuclear localization signal not only eliminated nuclear inclusions but also resulted in a decrease in steady-state AR levels (Holder and Merry, unpublished observations). Lastly, treatment of full-length AR-expressing cells with a compound (celastrol) that induces the expression of Hsp70 and Hsp40 resulted in a decrease in nuclear aggregation and steady-state AR protein (McBride *et al.*, 2004; Katsuno *et al.*, 2005; Waza *et al.*, 2005), further suggesting a role of degradative processes in the expanded AR altered metabolism.

While this discussion has focused on the altered degradation of the AR in evaluating the role of the nucleus in mutant AR metabolism, additional studies will be needed to determine if subcellular localization specificity may also involve the cellular stress response (Cowan *et al.*, 2003) or ER stress (Thomas *et al.*, 2005).

24.6. Transcriptional (Dys)Regulation

The signs of partial androgen insensitivity in SBMA patients suggest that a partial loss of AR function may contribute to disease pathogenesis. Studies of cell culture models of SBMA have demonstrated an inverse relationship between polyglutamine tract size and transcriptional activity of the AR (Mhatre *et al.*, 1993; Nakajima *et al.*, 1996; Grierson *et al.*, 2001). Another cell model demonstrated that expanded polyglutamine altered AR-dependent gene expression (Lieberman *et al.*, 2002). The majority of the identified AR-activated genes were downregulated by AR polyglutamine expansion, indicating a partial loss of transcriptional function, a state that could be attributed to changes in post-transcriptional modification of AR, alterations in AR degradation, or both (Lieberman *et al.*, 2002). In addition, the production of a truncated form of expanded AR lacking the ligand binding domain while retaining the DNA binding region raised the possibility of transcriptional activation in the absence of hormonal regulation (Butler *et al.*,

1998). Taken together, these data indicate that polyglutamine expansion of the AR results in transcriptional dysregulation. Indeed, downregulation of tubulin, a target of AR transcription, has been suggested as possible contributor to neurodegeneration (Butler *et al.*, 2001).

NII in SBMA have been shown to sequester several proteins other than AR, including heat shock proteins, protease components, and transcriptional regulators (Li *et al.*, 1998a; McCampbell *et al.*, 2000). cAMP response element binding protein (CREB)-binding protein (CBP), a histone acetyltransferase, is found in NII of cell and animal models, as well as autopsy material from patients with SBMA (McCampbell *et al.*, 2000). Indeed, CBP sequestration is found in models of other polyglutamine diseases, including SCA3 (McCampbell *et al.*, 2000) and Huntington's disease (Nucifora *et al.*, 2001). Reduction of soluble CBP by sequestration (McCampbell *et al.*, 2000; Nucifora *et al.*, 2001) or increased turnover (Jiang *et al.*, 2003) is coincident with a state of general hypoacetylation, a condition that is recovered in cells upon increased expression of CBP (McCampbell *et al.*, 2000; Nucifora *et al.*, 2001) or treatment with histone deacetylase (HDAC) inhibitors (McCampbell *et al.*, 2001). Treatment with HDAC inhibitors ameliorated symptoms and increased acetylation in a transgenic mouse model of SBMA (Minamiyama *et al.*, 2004), in a fly polyglutamine model (Steffan *et al.*, 2001), and in mouse models of Huntington's disease (Hockly *et al.*, 2003). Alterations in CBP expression may therefore represent a unified mechanism of pathogenesis among polyglutamine diseases. While CBP mediates histone acetylation in the regulatory regions important for transcription of many genes, one particular target of CBP transcriptional regulation, vascular endothelial growth factor (VEGF), is of particular interest. Expression of VEGF, a neurotrophic factor previously shown to play an important role in motor neuron survival (Oosthuyse *et al.*, 2001; Van Den Bosch *et al.*, 2004), is reduced prior to manifestation of symptoms in a mouse model of SBMA (Sopher *et al.*, 2004). Further, overexpression of CBP or treatment with VEGF reduced cell death in motor neuron culture model of SBMA (Sopher *et al.*, 2004). While VEGF is unlikely to be the sole target of CBP-dependent transcription that plays a role in polyglutamine pathogenesis, it represents one candidate for therapeutic targeting. The identification of other genes transcriptionally dysregulated in SBMA (Lieberman *et al.*, 2002) may lead to additional strategies for disease amelioration, not only for SBMA but for other polyglutamine diseases as well.

24.7. Proteolysis

NII from patient tissue contain only amino-terminal portions of the androgen receptor (Li *et al.*, 1998a), suggesting that either certain AR epitopes are masked by aggregation or the AR is aberrantly cleaved, resulting in an insoluble amino-terminal protein fragment. Precedent for the latter comes from studies of polyglutamine-expanded huntingtin protein; dissolution of NII from a cell model of HD produced a fragment of huntingtin (Lunkes *et al.*, 2002). Soluble fragments of other polyglutamine proteins have been observed in mouse models of SCA7 (Garden *et al.*, 2002), HD (Zhou *et al.*, 2003), and DRPLA (Schilling *et al.*, 1999); whether such soluble species represent amyloidogenic fragments that participate in aggregation is unclear.

Data from several SBMA cell models points to the formation of an amino-terminal AR protein fragment derived from full-length expanded AR. Cell models expressing full-length expanded AR yielded a ~75-kDa protein fragment (Abdullah *et al.*, 1998; Butler *et al.*, 2001), the production of which was associated with increased cell death (Abdullah *et al.*, 1998). The finding in an inducible model of SBMA (Walcott and Merry, 2002) of testosterone-dependent NII containing only amino-terminal epitopes suggests that this cell model recapitulated the altered AR metabolism found in patients (Li *et al.*, 1998a; Li *et al.*, 1998b). Indeed, purification and dissolution of NII from these cells reveals an approximately 50-kDa amino-terminal fragment of the

AR (Rosemiller and Merry, unpublished observations). Further proteolysis of an already truncated expanded AR, resulting in a ~15-kDa AR fragment, was reported by Merry *et al.* (Merry *et al.*, 1998). Whether these fragments represent pathogenic species is unclear.

The AR has been shown to be a substrate for the pro-apoptotic caspase-3 (Kobayashi *et al.*, 1998; Ellerby *et al.*, 1999), suggesting that cleavage of the expanded AR protein and its resulting fragment may be the result of caspase activity associated with cell death. Inhibition of caspase-3-mediated AR cleavage, either by mutation of the cleavage site at Asp146 (Ellerby *et al.*, 1999) or the mitogen-activated protein (MAP) kinase phosphorylation site at Ser514 (LaFevre-Bernt and Ellerby, 2003), blocked aggregate formation and reduced cell death in a cell model of disease. These data support the idea that the fragment formed from caspase cleavage of expanded AR is toxic, as preventing its formation ameliorated cell death. Further investigation is needed to determine which species of expanded AR is harmful.

Components of the ubiquitin proteasome system (UPS), such as the 26S proteasome (Poletti *et al.*, 1998; Stenoien *et al.*, 1999; Bailey *et al.*, 2002), and ubiquitin (Li *et al.*, 1998a) itself are sequestered into NII, implicating deficits in proteasomal degradation of expanded polyglutamine in the pathogenesis of SBMA and other polyglutamine diseases (Chai *et al.*, 1999; Wyttenbach *et al.*, 2000). Indeed, recent data indicating that proteins containing expanded polyglutamine tracts are generally refractory to (Venkatraman *et al.*, 2004) or inefficiently degraded by (Holmberg *et al.*, 2004) proteasomal processing strongly support this idea.

Aside from, or perhaps because of, its abnormal proteolysis, substantial data from cell models suggest that expanded polyglutamines may cause alterations to the normal function of the UPS (Bence *et al.*, 2001; Jana *et al.*, 2001; Venkatraman *et al.*, 2004; Bennett *et al.*, 2005). The co-localization of UPS proteins, as well as heat shock proteins, with NII may represent a cellular effort to refold or degrade insoluble AR. This effort may exhaust the resources of the UPS, depleting proteasomal function and preventing degradation of AR or other proteins. In neuronal cell models of Huntington's disease, the presence of expanded polyglutamine protein has been shown to reduce proteasomal function in general (Bence *et al.*, 2001; Jana *et al.*, 2001; Bennett *et al.*, 2005), as well as inhibit the proteolysis of a specific proteasomal protein substrate, p53 (Jana *et al.*, 2001). Furthermore, expanded polyglutamine expression may interfere with proteasome function under conditions of cellular stress (Ding *et al.*, 2002). Inhibition of the UPS increased insoluble AR in cell and fly models of SBMA (Bailey *et al.*, 2002; Chan *et al.*, 2002) and increased neurodegeneration (Chan *et al.*, 2002). However, at least one *in vivo* model of polyglutamine disease, SCA7, was recently shown to develop in the absence of proteasomal dysfunction (Bowman *et al.*, 2005). Thus, while a role for the UPS in the abnormal metabolism of expanded polyglutamines in general and of expanded AR in particular is well substantiated, whether a pathogenic role exists in SBMA for a proteasomal inability to degrade non-polyglutamine substrates remains to be seen.

Expanded polyglutamine tracts have been found to serve as better substrates for transglutaminase (Kahlem *et al.*, 1996). Moreover, expanded AR was shown to differ from normal AR in the combination of intra- and intermolecular cross-links in which it participates (Mandrusiak *et al.*, 2003). The formation of isopeptide bonds through transglutaminase activity suggests a mechanism by which proteins bearing an expanded polyglutamine tract might interfere with proteasome function; isopeptide bonds would be indigestible by the proteasome and may lead to disruption of normal proteolytic function. *In vitro*, expanded polyglutamine protein was insoluble in the presence of transglutaminase activity (Karpuj *et al.*, 1999) and inhibition of transglutaminase activity rescued proteasome dysfunction in a cell model of SBMA (Mandrusiak *et al.*, 2003). Moreover, transglutaminase inhibition has proved effective as a therapy in a mouse model of Huntington's disease (Karpuj *et al.*, 2002). An increase in transglutaminase-dependent isopeptide bonds were observed in neuronal tissues in a mouse model of SBMA (Mandrusiak *et al.*, 2003), suggesting that transglutaminase activity may indeed represent a pathogenic mechanism *in vivo*.

24.8. Axonal Transport and Cytoskeletal Alterations

Relatively high levels of AR are expressed in motor neurons of the brain stem and spinal cord (Sar and Stumpf, 1977; Sheridan, 1984; Sarrieau *et al.*, 1990). However, the AR is found at appreciable levels in other brain regions as well (Puy *et al.*, 1995; Shah *et al.*, 2004), without producing related neurologic symptoms. The reasons for this selective motor neuron susceptibility are unclear. Motor neurons are highly dependent on axonal transport over large distances; this dependence offers a possible molecular mechanism for selective motor neuron sensitivity. Abnormal distribution of the anterograde motor protein, dynein, and one of its cargoes, mitochondria, has been observed in dystrophic neurites containing aggregated expanded AR in a testosterone-treated motor neuron cell model of SBMA (Piccioni *et al.*, 2002). Neurites appeared swollen around AR aggregates; kinesin and mitochondria were localized to the aggregates, suggesting defective anterograde movement of these cargoes, perhaps as a result of physical axonal blockage. Other studies have also demonstrated that anterograde *and* retrograde transport are inhibited by expression of expanded AR (Szebenyi *et al.*, 2003). In contrast with the findings of Piccioni *et al.* (2002), Szebenyi *et al.* (2003) saw no evidence of AR aggregation nor altered distribution of transport proteins, suggesting that physical blockade of the transport machinery may not be necessary for aberrations in axonal transport. It is interesting to note that alterations in cellular transport have also been observed with the expression of expanded huntingtin protein in cell and fly models (Gunawardena *et al.*, 2003; Szebenyi *et al.*, 2003), suggesting yet another unified mechanism for cellular dysfunction in polyglutamine diseases. It is also notable that in the human population a mutation in p150, a component of the dynein/dynactin complex, produces a slowly progressing form of human lower motor neuron disease with bulbar manifestations phenotypically similar in many respects to SBMA (Puls *et al.*, 2003). Moreover, the genetic disruption of the dynein/dynactin complex, by overexpression of the transport motor component dynamitin (LaMonte *et al.*, 2002) or mutation of dynein heavy chain (Hafezparast *et al.*, 2003), is sufficient to cause slowly progressive motor neuron disease in mice. Whether these mutations lead to disease through the altered transport of specific cargoes or through alternative disruptions in cell physiology will require further investigation.

Recent studies have revealed a reduction in unphosphorylated NF-H in the cell bodies of motor neurons of the spinal cord in symptomatic transgenic SBMA males; recovery of unphosphorylated NF-H immunoreactivity coincides with partial recovery of motor function and a reduction in NII load (Chevalier-Larsen *et al.*, 2004). The linkage of this pathologic marker with disease symptoms suggests a pathway that is affected in the disease process. Preliminary evidence suggests that altered localization of NF-H phosphorylation may be responsible for these alterations. While phosphorylated NF-H (pNF-H) is found in filamentous structures in both cortex and spinal cord in wild-type male mice, there is a reduction in the filamentous distribution of pNF-H in SBMA male mice and the presence of diffuse pNF-H in the nuclei of these neurons (Chevalier-Larsen *et al.*, unpublished results). It is possible that changes in NF-H regulation could have substantial downstream repercussions. Changes in NF-H expression have been shown to affect expression of other neurofilament subunits (Tu *et al.*, 1995), while changes in NF-H phosphorylation have been shown to modulate its proteolysis (Pant, 1988; Schumacher *et al.*, 1999; Huh *et al.*, 2002), its association with microtubules (Saito *et al.*, 1995), and its axonal transport (Jung *et al.*, 2000). Such disruption could affect cytoskeletal integrity, axonal caliber, and/or axonal transport. Further experiments will serve to elucidate the role of altered NF-H in SBMA.

24.9. Prospective Therapies

Recent animals models of SBMA have demonstrated the potential for recovery of motor function after removal or inhibition of androgens (Katsuno *et al.*, 2002; Katsuno *et al.*, 2003;

Chevalier-Larsen *et al.*, 2004). These and other models have also indicated that neurologic symptoms preceded neuronal loss, indicating that neuronal dysfunction is likely to be responsible for pathogenesis (Katsuno *et al.*, 2002; Chevalier-Larsen *et al.*, 2004). The most effective treatments will be administered when motor neurons are still present; the timing and rate of neuronal death in SBMA patients, however, is unknown. In animal models, the point at which intervention results in recovery varies widely, depending on the model and its rate of disease progression. Intervention via androgen removal was only effective in early symptomatic stages in a rapidly progressing model (Katsuno *et al.*, 2002; Katsuno *et al.*, 2003), while a slowly progressive model showed partial recovery even in aged, late symptomatic animals (Chevalier-Larsen *et al.*, 2004). More research is needed to determine when treatment should be administered in human patients to achieve the greatest effect. The availability of several models with variable courses of disease progression may prove useful in determining when intervention should occur.

Although no treatment for SBMA is currently available, several potential therapies are suggested by the studies discussed here. Given the critical role of ligand in disease progression and the general lack of efficacy of androgen receptor antagonists in cell and animal models, drugs that eliminate or reduce hormone levels are a logical choice for therapeutic intervention. The efficacy of leuprorelin has already been demonstrated in a SBMA mouse model (Katsuno *et al.*, 2003), and clinical trials with this drug are ongoing in Japan. Reducing circulating levels of the more potent AR ligand, DHT, by inhibiting 5-alpha reductase may also prove effective. Finasteride, a type II 5-alpha reductase inhibitor, or dutasteride, which inhibits both type I and type II, are U.S. FDA approved for use in androgenic alopecia and benign prostatic hyperplasia, respectively. Either may be an effective treatment for SBMA. Alternatively, reduction of dihydrotestosterone might also be achieved through modulation of 3-alpha hydroxysteroid dehydrogenase (3-alpha HSD) activity through the use of selective serotonin reuptake inhibitors (SSRIs). In the presence of the SSRI, fluoxetine, for instance, the efficiency of 3alpha-HSD-dependent conversion of dihydrotestosterone to androstanediol is increased substantially (Griffon and Mellon, 1999).

Modulation of androgens is not the only treatment shown to be effective in animal models of SBMA. Administration of sodium butyrate, an HDAC inhibitor, ameliorates symptoms and increases DNA acetylation within a narrow range of doses (higher doses exhibited toxicity) (Minamiyama *et al.*, 2004). While RNA transcription at multiple genetic loci may be enhanced by HDAC inhibition, upregulation of one or a small number of specific proteins may be all that is necessary to reverse neuronal dysfunction. For example, increasing levels of VEGF might promote neuronal health and function. Increasing levels of VEGF in a mouse model of ALS via lentiviral delivery increased motor function and survival (Azzouz *et al.*, 2004; Storkebaum *et al.*, 2005). It is unlikely, however, that VEGF represents the only relevant functional pathway in motor neurons affected in SBMA. It may be that increasing levels of the upstream regulator of VEGF, CBP, would serve to enhance expression of the many genes regulated by this coactivator protein.

The finding that increasing expression of molecular chaperones through the overexpression of Hsp70 and Hsp40 (Bailey *et al.*, 2002) or treatment with compounds that activate the heat shock response (Hafezparast *et al.*, 2003; Kieran *et al.*, 2004; McBride *et al.*, 2004) suggests that increasing molecular chaperone expression via either of these two routes would likely prove therapeutically beneficial. Indeed, expression of Hsp70 in a transgenic mouse ameliorated disease in one SBMA mouse model (Adachi *et al.*, 2003). It may also be beneficial to alter components of the UPS, in order to facilitate effective clearing of mutant AR protein; such an approach will require a more detailed understanding of the molecular and biochemical basis of inefficient polyglutamine degradation.

Over the past few years, research in SBMA and related polyglutamine diseases has illuminated several cellular mechanisms involved in disease pathogenesis. Moreover, models are in place to further elucidate the molecular and cellular basis of polyglutamine toxicity. Any number of these pathways could be exploited for therapeutic intervention. The development of several animal models that recapitulate key aspects of disease provides an excellent medium for testing potential treatments. Further, the differences between these models may prove advantageous in screening therapies: quickly progressing models will allow rapid assessment of efficacy while more slowly progressing models will provide information on long-term effects of treatment and help to refine an appropriate time course for intervention.

References

Abdullah, A. A. R., Trifiro, M. A., Panet-Raymond, V., Alvarado, C., de Tourreil, S., Frankel, D., Schipper, H. M., and Pinsky, L. (1998). Spinobulbar muscular atrophy: polyglutamine-expanded androgen receptor is proteolytically resistant *in vitro* and processed abnormally in transfected cells. *Hum Mol Genet* 7:379–384.

Abel, A., Walcott, J., Woods, J., Duda, J., and Merry, D. E. (2001). Expression of expanded repeat androgen receptor produces neurologic disease in transgenic mice. *Hum Mol Genet* 10:107–116.

Adachi, H., Katsuno, M., Minamiyama, M., Sang, C., Pagoulatos, G., Angelidis, C., Kusakabe, M., Yoshiki, A., Kobayashi, Y., Doyu, M., and Sobue, G. (2003). Heat shock protein 70 chaperone overexpression ameliorates phenotypes of the spinal and bulbar muscular atrophy transgenic mouse model by reducing nuclear-localized mutant androgen receptor protein. *J Neurosci* 23:2203–2211.

Adachi, H., Katsuno, M., Minamiyama, M., Waza, M., Sang, C., Nakagomi, Y., Kobayashi, Y., Tanaka, F., Doyu, M., Inukai, A., Yoshida, M., Hashizume, Y., and Sobue, G. (2005). Widespread nuclear and cytoplasmic accumulation of mutant androgen receptor in SBMA patients. *Brain* 128:659–670.

Arbizu, T., Santamaria, J., Gomez, J. M., Quilez, A., and Serra, J. P. (1983). A family with adult spinal and bulbar muscular atrophy, X-linked inheritance and associated testicular failure. *J Neurol Sci* 59:371–382.

Azzouz, M., Ralph, G. S., Storkebaum, E., Walmsley, L. E., Mitrophanous, K. A., Kingsman, S. M., Carmeliet, P., and Mazarakis, N. D. (2004). VEGF delivery with retrogradely transported lentivector prolongs survival in a mouse ALS model. *Nature* 429:413–417.

Bailey, C. K., Andriola, I. F., Kampinga, H. H., and Merry, D. E. (2002). Molecular chaperones enhance the degradation of expanded polyglutamine repeat androgen receptor in a cellular model of spinal and bulbar muscular atrophy. *Hum Mol Genet* 11:515–523.

Bence, N. F., Sampat, R. M., and Kopito, R. R. (2001). Impairment of the ubiquitin-proteasome system by protein aggregation. *Science* 292:1552–1555.

Bennett, E. J., Bence, N. F., Jayakumar, R., and Kopito, R. R. (2005). Global impairment of the ubiquitin-proteasome system by nuclear or cytoplasmic protein aggregates precedes inclusion body formation. *Mol Cell* 17:351–365.

Bowman, A. B., Yoo, S.-Y., Dantuma, N. P., and Zoghbi, H. Y. (2005). Neuronal dysfunction in a polyglutamine disease model occurs in the absence of ubiquitin-proteasome system impairment and inversely correlates with the degree of nuclear inclusion formation. *Hum Mol Genet* 14:679–691.

Butler, R., Leigh, P. N., and Gallo, J. M. (2001). Androgen-induced up-regulation of tubulin isoforms in neuroblastoma cells. *J Neurochem* 78:854–861.

Butler, R., Leigh, P. N., McPhaul, M. J., and Gallo, J. M. (1998). Truncated forms of the androgen receptor are associated with polyQ expansion in X-linked spinal and bulbar muscular atrophy. *Hum Mol Genet* 7:121–127.

Caplan, A. J., Langley, E., Wilson, E. M., and Vidal, J. (1995). Hormone-dependent transactivation by the human androgen receptor is regulated by a dnaJ protein. *J Biol Chem* 270:5251–5257.

Chai, Y., Koppenhafer, S. L., Shoesmith, S. J., Perez, M. K., and Paulson, H. L. (1999). Evidence for proteasome involvement in polyglutamine disease: localization to nuclear inclusions in SCA3/MJD and suppression of polyglutamine aggregation in vitro. *Hum Mol Genet* 8:673–682.

Chan, H. Y., Warrick, J. M., Andriola, I., Merry, D., and Bonini, N. M. (2002). Genetic modulation of polyglutamine toxicity by protein conjugation pathways in Drosophila. *Hum Mol Genet* 11:2895–2904.

Chevalier-Larsen, E. S., O'Brien, C. J., Wang, H., Jenkins, S. C., Holder, L., Lieberman, A. P., and Merry, D. E. (2004). Castration restores function and neurofilament alterations of aged symptomatic males in a transgenic mouse model of spinal and bulbar muscular atrophy. *J Neurosci* 24:4778–4786.

Cowan, K. J., Diamond, M. I., and Welch, W. J. (2003). Polyglutamine protein aggregation and toxicity and linked to the cellular stress response. *Hum Mol Genet* 12:1377–1391.

Cummings, C., and Zoghbi, H. (2000). Fourteen and counting: unraveling trinucleotide repeat diseases. *Hum Mol Genet* 9:909–916.

Darrington, R. S., Butler, R., Leigh, P. N., McPhaul, M. J., and Gallo, J. M. (2002). Ligand-dependent aggregation of polyglutamine-expanded androgen receptor in neuronal cells. *Neuroreport* 13:2117–2120.

Davies, S. W., Turmaine, M., Cozens, B. A., DiFiglia, M., Sharp, A. H., Ross, C. A., Scherzinger, E., Wanker, E. E., Mangiarini, L., and Bates, G. P. (1997). Formation of neuronal intranuclear inclusions underlies the neurological dysfunction in mice transgenic for the HD mutation. *Cell* 90:537–548.

Diamond, M. I., Robinson, M. R., and Yamamoto, K. R. (2000). Regulation of expanded polyglutamine protein aggregation and nuclear localization by the glucocorticoid receptor. *Proc Natl Acad Sci USA* 97:657–661.

DiFiglia, M., Sapp, E., Chase, K. O., Davies, S. W., Bates, G. P., Vonsattel, J. P., and Aronin, N. (1997). Aggregation of huntingtin in neuronal intranuclear inclusions and dystrophic neurites in brain. *Science* 277:1990–1993.

Ding, Q., Lewis, J. J., Strum, K. M., Dimayuga, E., Bruce-Keller, A. J., Dunn, J. C., and Keller, J. N. (2002). Polyglutamine expansion, protein aggregation, proteasome activity, and neural survival. *J Biol Chem* 277:13935–13942.

Doyu, M., Sobue, G., Mukai, E., Kachi, T., Ysauda, T., Mitsuma, T., and Takahashi, A. (1992). Severity of X-linked recessive bulbosopinal neuronopathy correlates with size of the tandem CAG repeat in androgen receptor gene. *Ann Neurol* 32:707–710.

Ellerby, L. M., Hackam, A. S., Propp, S. S., Ellerby, H. M., Rabizadeh, S., Cashman, N. R., Trifiro, M. A., Pinsky, L., Wellington, C. L., Salvesen, G. S., Hayden, M. R., and Bredesen, D. E. (1999). Kennedy's disease: caspase cleavage of the androgen receptor is a crucial event in cytotoxicity. *J Neurochem* 72:185–195.

Fang, Y., Fliss, A. E., Robins, D. M., and Caplan, A. J. (1996). Hsp90 regulates androgen receptor hormone binding affinity *in vivo*. *J Biol Chem* 271:28697–28702.

Fargo, K. N., and Sengelaub, D. R. (2004a). Testosterone manipulation protects motoneurons from dendritic atrophy after contralateral motoneuron depletion. *J Comp Neurol* 469:96–106.

Fargo, K. N., and Sengelaub, D. R. (2004b). Exogenous testosterone prevents motoneuron atrophy induced by contralateral motoneuron depletion. *J Neurobiol* 60:348–359.

Ferlini, A., Patrosso, M. C., Guidetti, D., Merlini, L., Uncini, A., Ragno, M., Plasmati, R., Fini, S., Repetto, M., Vezzoni, P., and Forabosco, A. (1995). Androgen receptor gene (CAG)n repeat analysis in the differential diagnosis between Kennedy Disease and other motoneuron disorders. *Am J Med Genet* 55:105–111.

Fliss, A. E., Rao, J., Melville, M. W., Cheetham, M. E., and Caplan, A. J. (1999). Domain requirements of DnaJ-like (Hsp40) molecular chaperones in the activation of a steroid hormone receptor. *J Biol Chem* 274:34045–34052.

Garden, G. A., Libby, R. T., Fu, Y. H., Kinoshita, Y., Huang, J., Possin, D. E., Smith, A. C., Martinez, R. A., Fine, G. C., Grote, S. K., Ware, C. B., Einum, D. D., Morrison, R. S., Ptacek, L. J., Sopher, B. L., and La Spada, A. R. (2002). Polyglutamine-expanded ataxin-7 promotes non-cell-autonomous purkinje cell degeneration and displays proteolytic cleavage in ataxic transgenic mice. *J Neurosci* 22:4897–4905.

Grierson, A. J., Shaw, C. E., and Miller, C. C. (2001). Androgen induced cell death in SHSY5Y neuroblastoma cells expressing wild-type and spinal bulbar muscular atrophy mutant androgen receptors. *Biochim Biophys Acta* 1536:13–20.

Gunawardena, S., Her, L.-S., Brusch, R. G., Laymon, R. A., Niesman, I. R., Gordesky-Gold, B., Sintasath, L., Bonini, N. M., and Goldstein, L. S. B. (2003). Disruption of axonal transport by loss of huntingtin or expression of pathogenic polyQ proteins in Drosophila. *Neuron* 40:25–40.

Hafezparast, M., Klocke, R., Ruhrberg, C., Marquardt, A., Ahmad-Annuar, A., Bowen, S., Lalli, G., Witherden, A. S., Hummerich, H., Nicholson, S., Morgan, P. J., Oozageer, R., Priestley, J. V., Averill, S., King, V. R., Ball, S., Peters, J., Toda, T., Yamamoto, A., Hiraoka, Y., Augustin, M., Korthaus, D., Wattler, S., Wabnitz, P., Dickneite, C., Lampel, S., Boehme, F., Peraus, G., Popp, A., Rudelius, M., Schlegel, J., Fuchs, H., Hrabe de Angelis, M., Schiavo, G., Shima, D. T., Russ, A. P., Stumm, G., Martin, J. E., and Fisher, E. M. C. (2003). Mutations in dynein link motor neuron degeneration to defects in retrograde transport. *Science* 300:808–812.

He, B., Kemppainen, J., and Wilson, E. M. (2000). FXXLF and WXXLF sequences mediate the NH2-terminal interaction with the ligand binding domain of the androgen receptor. *J Biol Chem* 275:22986–22994.

He, B., Gampe, R. T., Jr., Kole, A. J., Hnat, A. T., Stanley, T. B., An, G., Stewart, E. L., Kalman, R. I., Minges, J. T., and Wilson, E. M. (2004). Structural basis for androgen receptor interdomain and coactivator interactions suggests a transition in nuclear receptor activation function dominance. *Mol Cell* 16:425–438.

Hockly, E., Richon, V. M., Woodman, B., Smith, D. L., Zhou, X., Rosa, E., Sathasivam, K., Ghazi-Noori, S., Mahal, A., Lowden, P. A., Steffan, J. S., Marsh, J. L., Thompson, L. M., Lewis, C. M., Marks, P. A., and Bates, G. P. (2003). Suberoylanilide hydroxamic acid, a histone deacetylase inhibitor, ameliorates motor deficits in a mouse model of Huntington's disease. *Proc Natl Acad Sci USA* 100:2041–2046.

Holmberg, C. I., Staniszewski, K. E., Mensah, K. N., Matouschek, A., and Morimoto, R. I. (2004). Inefficient degradation of truncated polyglutamine proteins by the proteasome. *EMBO J* 23:4307–4318.

Huang, Z. Q., Li, J., Sachs, L. M., Cole, P. A., and Wong, J. (2003). A role of cofactor-cofactor and cofactor-histone interactions in targeting p300, SWI/SNF and mediator for transcription. *EMBO J* 22:2146–2155.

Huh, J. W., Laurer, H. L., Raghupathi, R., Helfaer, M. A., and Saatman, K. E. (2002). Rapid loss and partial recovery of neurofilament immunostaining following focal brain injury in mice. *Exp Neurol* 175:198–208.

Igarashi, S., Tanno, Y., Onodera, O., Yamazaki, M., Sato, S., Ishikawa, A., Mayatani, N., Nagashima, M., Ishikawa, Y., Sahashi, K., Ibi, T., Miyatake, T., and Tsuji, S. (1992). Strong correlation between the number of CAG repeats in androgen receptor genes and the clinical onset of features of spinal and bulbar muscular atrophy. *Neurology* 42:2300–2302.

Ishihara, H., Kanda, F., Nishio, H., Sumino, K., and Chihara, K. (2001). Clinical features and skewed X-chromosome inactivation in female carriers of X-linked recessive spinal and bulbar muscular atrophy. *J Neurol* 248:856–860.

Jana, N. R., Zemskov, E. A., Wang, G.-h., and Nukina, N. (2001). Altered proteasomal function due to the expression of polyglutamine-expanded truncated N-terminal huntingtin induces apoptosis by caspase activation through mitochondrial cytochrome c release. *Hum Mol Genet* 10:1049–1059.

Jiang, H., Nucifora, F. C., Jr., Ross, C. A., and DeFranco, D. B. (2003). Cell death triggered by polyglutamine-expanded huntingtin in a neuronal cell line is associated with degradation of CREB-binding protein. *Hum Mol Genet* 12:1–12.

Jones, K. J. (1988). Steroid hormones and neurotrophism: relationship to nerve injury. *Metab Brain Dis* 3:1–18.

Jordan CL, W. S., and Arnold AP (1995). Androgenic, not estrogenic, steroids alter neuromuscular synapse elimination in the rat levator ani. *Brain Rex Dev Brain Res* 84:215–224.

Jung, C., Yabe, J. T., and Shea, T. B. (2000). C-terminal phosphorylation of the high molecular weight neurofilament subunit correlates with decreased neurofilament axonal transport velocity. *Brain Res* 856:12–19.

Kahlem, P., Terre, C., Green, H., and Djian, P. (1996). Peptides containing glutamine repeats as substrates for transglutaminase-catalyzed cross-linking: relevance to diseases of the nervous system. *Proc Natl Acad Sci USA* 93:14580–14585.

Karpuj, M. V., Garren, H., Slunt, H., Price, D. L., Gusella, J., Becher, M. W., and Steinman, L. (1999). Transglutaminase aggregates huntingtin into nonamyloidogenic polymers, and its enzymatic activity increases in Huntington's disease brain nuclei. *Proc Natl Acad Sci USA* 96:7388–7393.

Karpuj, M. V., Becher, M. W., Springer, J. E., Chabas, D., Youssef, S., Pedotti, R., Mitchell, D., and Steinman, L. (2002). Prolonged survival and decreased abnormal movements in transgenic model of Huntington disease, with administration of the transglutaminase inhibitor cystamine. *Nat Med* 8:143–149.

Katsuno, M., Adachi, H., Kume, A., Li, M., Nakagomi, Y., Niwa, H., Sang, C., Kobayashi, Y., Doyu, M., and Sobue, G. (2002). Testosterone reduction prevents phenotypic expression in a transgenic mouse model of spinal and bulbar muscular atrophy. *Neuron* 35:843–854.

Katsuno, M., Adachi, H., Doyu, M., Minamiyama, M., Sang, C., Kobayashi, Y., Inukai, A., and Sobue, G. (2003). Leuprorelin rescues polyglutamine-dependent phenotypes in a transgenic mouse model of spinal and bulbar muscular atrophy. *Nat Med* 9:768–773.

Katsuno, M., Sang, C., Adachi, H., Minamiyama, M., Waza, M., Tanaka, F., Doyu, M., and Sobue, G. (2005). Pharmacological induction of heat-shock proteins alleviates polyglutamine-mediated motor neuron disease. *Proc Natl Acad Sci USA* 102:16801–16806.

Kennedy, W. R., Alter, M., and Sung, J. H. (1968). Progressive proximal spinal and bulbar muscular atrophy of late onset: a sex-linked recessive trait. *Neurology* 18:671–680.

Kieran, D., Kalmar, B., Dick, J. R. T., Riddoch-Contreras, J., Burnstock, G., and Greensmith, L. (2004). Treatment with arimoclomol, a coinducer of heat shock proteins, delays disease progression in ALS mice. *Nat Med* 10:402–405.

Kobayashi, Y., Miwa, S., Merry, D. E., Kume, A., Mei, L., Doyu, M., and Sobue, G. (1998). Caspase-3 cleaves the expanded androgen receptor protein of spinal and bulbar muscular atrophy in a polyglutamine repeat length-dependent manner. *BBRC* 252:145–150.

Kujawa, K. A., Emeric, E., and Jones, K. J. (1991). Testosterone differentially regulates the regenerative properties of injured hamster facial motor neurons. *J Neurosci* 11:3898–3908.

Kujawa, K. A., Jacob, J. M., and Jones, K. J. (1993). Testosterone regulation of the regenerative properties of injured rat sciatic motor neurons. *J Neurosci Res* 35:268–273.

Kurz, E. M., Sengelaub, D. R., and Arnold, A. P. (1986). Androgens regulate the dendritic length of mammalian motoneurons in adulthood. *Science* 232:395–398.

La Spada, A. R., Wilson, E. M., Lubahn, D. B., Harding, A. E., and Fischbeck, K. H. (1991). Androgen receptor gene mutations in X-linked spinal and bulbar muscular atrophy. *Nature* 353:77–79.

La Spada, A. R., Roling, D., Harding, A. E., Warner, C. L., Speigel, R., Hausmanowa-Petrusewicz, I., Yee, W.-C., and Fischbeck, K. H. (1992). Meiotic stability and genotype-phenotype correlation of the expanded trinucleotide repeat in X-linked spinal and bulbar muscular atrophy. *Nat Genet* 2:301–304.

LaFevre-Bernt, M. A., and Ellerby, L. M. (2003). Kennedy's disease. Phosphorylation of the polyglutamine-expanded form of androgen receptor regulates its cleavage by caspase-3 and enhances cell death. *J Biol Chem* 278:34918–34924.

LaMonte, B. H., Wallace, K. E., Holloway, B. A., Shelly, S. S., Ascano, J., Tokito, M., Van Winkle, T., Howland, D. S., and Holzbaur, E. L. (2002). Disruption of dynein/dynactin inhibits axonal transport in motor neurons causing late-onset progressive degeneration. *Neuron* 34:715–727.

Li, M., Miwa, S., Kobayashi, Y., Merry, D. E., Yamamoto, M., Tanaka, F., Doyu, M., Hashizume, Y., Fischbeck, K. H., and Sobue, G. (1998a). Nuclear inclusions of the androgen receptor protein in spinal and bulbar muscular atrophy. *Ann Neurol* 44:249–254.

Li, M., Nakagomi, Y., Kobayashi, Y., Merry, D. E., Tanaka, F., Doyu, M., Mitsuma, T., Hashizume, Y., Fischbeck, K. H., and Sobue, G. (1998b). Nonneural nuclear inclusions of androgen receptor protein in spinal and bulbar muscular atrophy. *Am J Pathol* 153:695–701.

Lieberman, A. P., Harmison, G., Strand, A. D., Olson, J. M., and Fischbeck, K. H. (2002). Altered transcriptional regulation in cells expressing the expanded polyglutamine androgen receptor. *Hum Mol Genet* 11:1967–1976.

Lunkes, A., Lindenberg, K. S., Ben-Haiern, L., Weber, C., Devys, D., Landwehrmeyer, G. B., Mandel, J.-L., and Trottier, Y. (2002). Proteases acting on mutant huntingtin generate cleaved products that differentially build up cytoplasmic and nuclear inclusions. *Mol Cell* 10:259–269.

Ma, H., Hong, H., Huang, S. M., Irvine, R. A., Webb, P., Kushner, P. J., Coetzee, G. A., and Stallcup, M. R. (1999). Multiple signal input and output domains of the 160-kilodalton nuclear receptor coactivator proteins. *Mol Cell Biol* 19:6164–6173.

Mandrusiak, L. M., Beitel, L. K., Wang, X., Scanlon, T. C., Chevalier-Larsen, E., Merry, D. E., and Trifiro, M. A. (2003). Transglutaminase potentiates ligand-dependent proteasome dysfunction induced by polyglutamine-expanded androgen receptor. *Hum Mol Genet* 12:1497–1506.

Mariotti, C., Castellotti, B., Pareyson, D., Testa, D., Eoli, M., Antozzi, C., Silani, V., Marconi, R., Tezzon, F., Siciliano, G., Marchini, C., Gellera, C., and Donato, S. D. (2000). Phenotypic manifestations associated with CAG-repeat expansion in the androgen receptor gene in male patients and heterozygous females: a clinical and molecular study of 30 families. *Neuromuscul Disord* 10:391–397.

Marron, T. U., Guerini, V., Rusmini, P., Sau, D., Brevini, T. A., Martini, L., and Poletti, A. (2005). Androgen-induced neurite outgrowth is mediated by neuritin in motor neurones. *J Neurochem* 92:10–20.

Matsumoto, A., Arai, Y., Urano, A., and Hyodo, S. (1995). Molecular basis of neuronal plasticity to gonadal steroids. *Functional Neurol* 10:59–76.

Matsumoto, A. (2001). Androgen stimulates neuronal plasticity in the perineal motoneurons of aged male rats. *J Comp Neurol* 430:389–395.

McBride, J. D., Gilbert, S. A., and Merry, D. E. (2004). The effects of celastrol on nuclear inclusion load and heat shock protein expression in a cell model of spinal and bulbar muscular atrophy. Program No. 226.18 2004 Abstract Niewer/Itinerary Planner. Washington, DC: Society for Neuroscience.

McCampbell, A., Taylor, J. P., Taye, A. A., Robitschek, J., Li, M., Walcott, J., Merry, D., Chai, Y., Paulson, H., Sobue, G., and Fischbeck, K. H. (2000). CREB-binding protein sequestration by expanded polyglutamine. *Hum Mol Genet* 9:2197–2202.

McCampbell, A., Taye, A. A., Whitty, L., Penney, E., Steffan, J. S., and Fischbeck, K. H. (2001). Histone deacetylase inhibitors reduce polyglutamine toxicity. *Proc Natl Acad Sci USA* 98:15179–15184.

McManamny, P., Chy, H. S., Finkelstein, D. I., Craythorn, R. G., Crack, P. J., Kola, I., Cheema, S. S., Horne, M. K., Wreford, N. G., O'Bryan, M. K., De Kretser, D. M., and Morrison, J. R. (2002). A mouse model of spinal and bulbar muscular atrophy. *Hum Mol Genet* 11:2103–2111.

Merry, D. E., Kobayashi, Y., Bailey, C. K., Taye, A. A., and Fischbeck, K. H. (1998). Cleavage, aggregation and toxicity of the expanded androgen receptor in spinal and bulbar muscular atrophy. *Hum Mol Genet* 7:693–701.

Mhatre, A. N., Trifiro, M. A., Kaufman, M., Kazemi-Esfarjani, P., Figlewicz, D., Rouleau, G., and Pinsky, L. (1993). Reduced transcriptional regulatory competence of the androgen receptor in X-linked spinal and bulbar muscular atrophy. *Nat Genet* 5:184–188.

Minamiyama, M., Katsuno, M., Adachi, H., Waza, M., Sang, C., Kobayashi, Y., Tanaka, F., Doyu, M., Inukai, A., and Sobue, G. (2004). Sodium butyrate ameliorates phenotypic expression in a transgenic mouse model of spinal and bulbar muscular atrophy. *Hum Mol Genet* 13:1183–1192.

Nakajima, H., Kimura, F., Nakagawa, T., Furutama, D., Shinoda, K., Shimizu, A., and Ohsawa, N. (1996). Transcriptional activation by the androgen receptor in X-linked spinal and bulbar muscular atrophy. *J Neurol Sci 1*42:12–16.

Nakamura, K., Jeong, S. Y., Uchihara, T., Anno, M., Nagashima, K., Nagashima, T., Ikeda, S., Tsuji, S., and Kanazawa, I. (2001). SCA17, a novel autosomal dominant cerebellar ataxia caused by an expanded polyglutamine in TATA-binding protein. *Hum Mol Genet* 10:1441–1448.

Nucifora, J. F. C., Sasaki, M., Peters, M. F., Huang, H., Cooper, J. K., Yamada, M., Takahashi, H., Tsuji, S., Troncoso, J., Dawson, V. L., Dawson, T. M., and Ross, C. A. (2001). Interference by huntingtin and atrophin-1 with CBP-mediated transcription leading to cellular toxicity. *Science* 291:2423–2428.

Oosthuyse, B., Moons, L., Storkebaum, E., Beck, H., Nuyens, D., Brusselmans, K., Van Dorpe, J., Hedllings, P., Gorselink, M., and Heymans, S. (2001). Deletion of the hypoxia-response element in the vascular endothelial growth factor prmoter causes motor neuron degeneration. *Nat Genet* 28:131–138.

Pant, H. C. (1988). Dephosphorylation of neurofilament proteins enhances their susceptibility to degradation by calpain. *J Biochem* 256:665–668.

Paulson, H. L., Perez, M. K., Trottier, Y., Trojanowski, J. Q., Subramony, S. H., Das, S. S., Vig, P., Mandel, J.-L., Fischbeck, K. H., and Pittman, R. N. (1997). Intranuclear inclusions of expanded polyglutamine protein in spinocerebellar ataxia type 3. *Neuron* 19:1–20.

Piccioni, F., Pinton, P., Simeoni, S., Pozzi, P., Fascio, U., Vismara, G., Martini, L., Rizzuto, R., and Poletti, A. (2002). Androgen receptor with elongated polyglutamine tract forms aggregates that alter axonal trafficking and mitochondrial distribution in motor neuronal processes. *J FASEB* 16:1418–1420.

Pinsky, L., Trifiro, M., Kaufman, M., Beitel, L. K., Mhatre, A., Kazemi-Esfarjani, P., Sabbaghian, N., Lumbroso, R., Alvarado, C., Vasiliou, M., and Gottlieb, B. (1992). Androgen resistance due to mutation of the androgen receptor. *Clin Invest Med* 15:456–472.

Poletti, A., Coscarella, A., Negri-Cesi, P., Colciago, A., Celotti, F., and Martini, L. (1998). 5 alpha-reductase isozymes in the central nervous system. *Steroids* 63:246–251.

Pratt, W. B., and Welsh, M. J. (1994). Chaperone functions of the heat shock proteins associated with steroid receptors. *Cell Biol* 5:83–93.

Puls, I., Jonnakuty, C., LaMonte, B. H., Holzbaur, E. L., Tokito, M., Mann, E., Floeter, M. K., Bidus, K., Drayna, D., Oh, S. J., Brown, R. H., Jr., Ludlow, C. L., and Fischbeck, K. H. (2003). Mutant dynactin in motor neuron disease. *Nat Genet* 33:455–456.

Puy, L., MacLusky, N. J., Becker, L., Karsan, N., Trachtenberg, J., and Brown, T. J. (1995). Immunocytochemical detection of androgen receptor in human temporal cortex characterization and application of polyclonal androgen receptor antibodies in frozen and paraffin-embedded tissues. *J Steroid Biochem Mol Biol* 55:197–209.

Quigley, C. A., Friedman, K. J., Johnson, A., Lafreniere, R. G., Silverman, L. M., Lubahn, D. B., Brown, T. R., Wilson, E. M., Willard, H. F., and French, F. S. (1992). Complete deletion of the androgen receptor gene: definition of the null phenotype of the androgen insensitivity syndrome and determination of carrier status. *J Clin Endocr Metab* 74:927–933.

Ross, C. A. (2002). Polyglutamine pathogenesis: emergence of unifying mechanisms for Huntington's disease and related disorders. *Neuron* 35:819–822.

Saito, T., Shima, H., Osawa, Y., Nagao, M., Hemmings, B. A., Kishimoto, T., and Hisanaga, S. (1995). Neurofilament-associated protein phosphatase 2A: its possible role in preserving neurofilaments in filamentous states. *Biochemistry* 34:7376–7384.

Sar, M., and Stumpf, W. E. (1977). Androgen concentration in motor neruons of crainial nerves and spinal cord. *Science* 19:77–79.

Sarrieau, A., Mitchell, J. B., Lal, S., Olivier, A., Quirion, R., and Meaney, M. J. (1990). Androgen binding sites in human temporal cortex. *Neuroendocrinology* 51:713–716.

Schilling, G., Wood, J. D., Duan, K., Slunt, H. H., Gonzales, V., Yamada, M., Cooper, J. K., Margolis, R. L., Jenkins, N. A., Copeland, N. G., Takahashi, H., Tsuji, S., Price, D. L., Borchelt, D. R., and Ross, C. A. (1999). Nuclear accumulation of truncated atrophin-1 fragments in a transgenic mouse model of DRPLA. *Neuron* 24:275–286.

Schmidt, B. J., Greenberg, C. R., Allingham-Hawkins, D. J., and Spriggs, E. L. (2002). Expression of X-linked bulbospinal muscular atrophy (Kennedy disease) in two homozygous women. *Neurology* 59:770–772.

Schumacher, P. A., Eubanks, J. H., and Fehlings, M. G. (1999). Increased calpain I-mediated proteolysis, and preferential loss of dephosphorylated NF200, following traumatic spinal cord injury. *Neuroscience* 91:733–744.

Shah, N. M., Pisapia, D. J., Maniatis, S., Mendelsohn, M. M., Nemes, A., and Axel, R. (2004). Visualizing sexual dimorphism in the brain. *Neuron* 43:313–319.

Shang, Y., Myers, M., and Brown, M. (2002). Formation of the androgen receptor transcription complex. *Mol Cell* 9:601–610.

Sheridan, P. J. (1984). Autoradiographic localization of steroid receptors in the brain. *Clin Neuropharmacol* 7:281–295.

Sobue, G., Hashizume, Y., Mukai, E., Hirayama, M., Mitsuma, T., and Takahashi, A. (1989). X-linked recessive bulbospinal neronopathy: a clinicopathological study. *Brain* 112:209–232.

Sopher, B. L., Thomas, P. S., Jr., LaFevre-Bernt, M. A., Holm, I. E., Wilke, S. A., Ware, C. B., Jin, L. W., Libby, R. T., Ellerby, L. M., and La Spada, A. R. (2004). Androgen receptor YAC transgenic mice recapitulate SBMA motor neuronopathy and implicate VEGF164 in the motor neuron degeneration. *Neuron* 41:687–699.

Steffan, J. S., Bodai, L., Pallos, J., Poelman, M., McCampbell, A., Apostol, B. L., Kazantsev, A., Schmidt, E., Zhu, Y.-Z., Greenwald, M., Kurokawa, R., Housman, D. E., Jackson, G. R., Marsh, J. L., and Thompson, L. M. (2001). Histone deacetylase inhibitors arrest polyglutamine-dependent neurodegeneration in Drosophila. *Nature* 413:739–743.

Stenoien, D. L., Cummings, C. J., Adams, H. P., Mancini, M. G., Patel, K., DeMartino, G. N., Marcelli, M., Weigel, N. L., and Mancini, M. A. (1999). Polyglutamine-expanded androgen receptors form aggregates that sequester heat shock proteins, proteasome components and SRC-1, and are suppressed by the HDJ-2 chaperone. *Hum Mol Genet* 8:731–741.

Storkebaum, E., Lambrechts, D., Dewerchin, M., Moreno-Murciano, M. P., Appelmans, S., Hoh, H., Van Damme, P., Rutten, B., Man, W. Y., De Mol, M., Wyns, S., Manka, D., Vermeulen, K., Van Den Bosch, L., Mertens, N., Schmitz, C., Robberecht, W., Conway, E. M., Collen, D., Moons, L., and Carmeliet, P. (2005). Treatment of motoneuron degeneration by intracerebroventricular delivery of VEGFin a rat model of ALS. *Nat Neurosci* 8:85–92.

Szebenyi, G., Morfini, G. A., Babcock, A., Gould, M., Selkoe, K., Stenoien, D. L., Young, M., Faber, P. W., MacDonald, M. E., McPhaul, M. J., and Brady, S. T. (2003). Neuropathogenic forms of huntingtin and androgen receptor inhibit fast axonal transport. *Neuron* 40:41–52.

Takeyama, K., Ito, S., Yamamoto, A., Tanimoto, H., Furutani, T., Kanuka, H., Miura, M., Tabata, T., and Kato, S. (2002). Androgen-dependent neurodegeneration by polyglutamine-expanded human androgen receptor in Drosophila. *Neuron* 35:855–864.

Thomas, M., Dadgar, N., Aphale, A., Harrell, J. M., Kunkel, R., Pratt, W. B., and Lieberman, A. P. (2004). Androgen receptor acetylation site mutations cuase trafficking defects, misfolding, and aggregation similar to expanded glutamine tracts. *J Biol Chem* 279:8389–8395.

Thomas, M., Yu, Z. X., Dadgar, N., Varambally, S., Yu, J., Chinnaiyan, A. M., and Lieberman, A. P. (2005). The unfolded protein response modulates toxicity of the expanded glutamine androgen receptor. *J Biol Chem* 280:21264–21271.

Tsai, M.-J., and O'Malley, B. W. (1994). Molecular mechanisms of action of steroid/thryoid receptor superfamily members. *Annu Rev Biochem* 63:451–486.

Tu, P. H., Elder, G., Lazzarini, R. A., Nelson, D., Trojanowski, J. Q., and Lee, V. M. (1995). Overexpression of the human NFM subunit in transgenic mice modifies the level of endogenous NFL and the phosphorylation state of NFH subunits. *J Cell Biol* 129:1629–1640.

Van Den Bosch, L., Sorkebaum, E., Vleminckx, V., Moons, L., Vanopdenbosch, L., Scheveneels, W., Carmeliet, P., and Robberecht, W. (2004). Effects of vascular endothelial growth factor (VEGF) on motor neuron degeneration. *Neurobiol Dis* 17:21–28.

Varela, C. R., Bengston, L., Xu, J., MacLennan, A. J., and Forger, N. G. (2000). Additive effects of ciliary neurotrophic factor and testosterone on motoneuron survival; differential effects on motoneuron size and muscle morphology. *Exp Neurol* 165:384–393.

Venkatraman, P., Wetzel, R., Tanaka, M., Nukina, N., and Goldberg, A. L. (2004). Eukaryotic proteasomes cannot digest polyglutamine sequences and release them during degradation of polyglutamine-containing proteins. *Mol Cell* 14:95–104.

Walcott, J. L., and Merry, D. E. (2002). Ligand promotes intranuclear inclusions in a novel cell model of spinal and bulbar muscular atrophy. *J Biol Chem* 277:50855–50859.

Waza, M., Adachi, H., Katsuno, M., Minamiyama, M., Sang, C., Tanaka, F., Inukai, A., Doyu, M., and Sobue, G. (2005). 17-AAG, an Hsp90 inhibitor, ameliorates polyglutamine-mediated motor neuron degeneration. *Nat Med* 11:1088–1095.

Wong, C. I., Zhou, Z. X., Sar, M., and Wilson, E. M. (1993). Steroid requirement for androgen receptor dimerization and DNA binding. Modulation by intramolecular interactions between the NH2-terminal and steroid-binding domains. *J Biol Chem* 268:19004–19012.

Wyttenbach, A., Carmichael, J., Swartz, J., Furlong, R. A., Narain, Y., Rankin, J., and Rubinsztein, D. C. (2000). Effects of heat shock, heat shock protein 40 (HDJ-2), and proteasome inhibition on protein aggregation in cellular models of Huntington's disease. *Proc Natl Acad Sci USA* 97:2898–2903.

Yang, L. Y., Verhovshek, T., and Sengelaub, D. R. (2004). Brain-derived neurotrophic factor and androgen interact in the maintenance of dendritic morphology in a sexually dimorphic rat spinal nucleus. *Endocrinology* 145:161–168.

Yu, W. A. (1989). Administration of testoterone attenuates neronal loss following axotomy in the brainstem motor nuclei of female rats. *J Neurosci* 9:3908–3914.

Zhou, H., Cao, F., Wang, Z., Yu, Z. X., Nguyen, H. P., Evans, J., Li, S. H., and Li, X. J. (2003). Huntingtin forms toxic NH2-terminal fragment complexes that are promoted by the age-dependent decrease in proteasome activity. *J Cell Biol* 163:109–118.

Zoghbi, H. Y., and Orr, H. T. (2000). Glutamine repeats and neurodegeneration. *Annu Rev Neurosci* 23:217–247.

VI

Post-Translational Modification and Protein Conformational Diseases

25

Protein Glycation and Cataract: A Conformational Disease

John J. Harding

Abstract

Increased glycation is associated with aging and complications of diabetes. So it is not surprising that glycation has been a subject of intensive study. Most emphasis is on long-lived proteins, mostly structural, because they are exposed to the sugars for a longer time than enzymes and other nonstructural proteins in most tissues. However, glycation is not specific and is not restricted to structural proteins. Besides leading to the functional impairment of modified proteins, glycation was shown to produce significant structural alterations, resulting in modified proteins with properties similar to those of "molten-globule" intermediates of protein folding and unfolding pathways. Evidence supporting the role of nonenzymic post-translational modification of lens proteins in cataract is overviewed.

25.1. Glycation

Glycation is the process by which sugars react with proteins. The reaction(s) proceed without the need for an enzyme and are not specific. They are the inevitable consequence of having proteins surrounded by sugar molecules in cells. Any protein will in time react with any sugar. This is not to say that all proteins and all sugars react equally rapidly, because they do not, and glucose may have achieved its central role in metabolism because it is the least reactive sugar in these uncontrolled pathways.

The functions of proteins can be impaired by glycation, and in conditions such as diabetes, where sugar concentrations are increased, the extent of glycation and the degree of damage increase in parallel. Glycation may be the main cause of the slowly developing complications of diabetes. Sugar concentrations and glycation also increase in aging.

25.1.1. The Glycation Reactions

The chemical reactions called *glycation* start as the simple formation of a Schiff base from a sugar in its open chain form and a protein amino group and proceed to a complex set of reactions to form colored, fluorescent, and cross-linking species. The early glycation reactions are shown in Figure 25.1. The first step leads to a Schiff base (glycosylamine), which then undergoes the Amadori rearrangement to form a ketoamine. After the Amadori product, the reactions become more varied and complicated leading to components that are termed *advanced glycation end-products* (AGEs). The structures of some AGEs are shown in Figure 25.2. The emphasis of

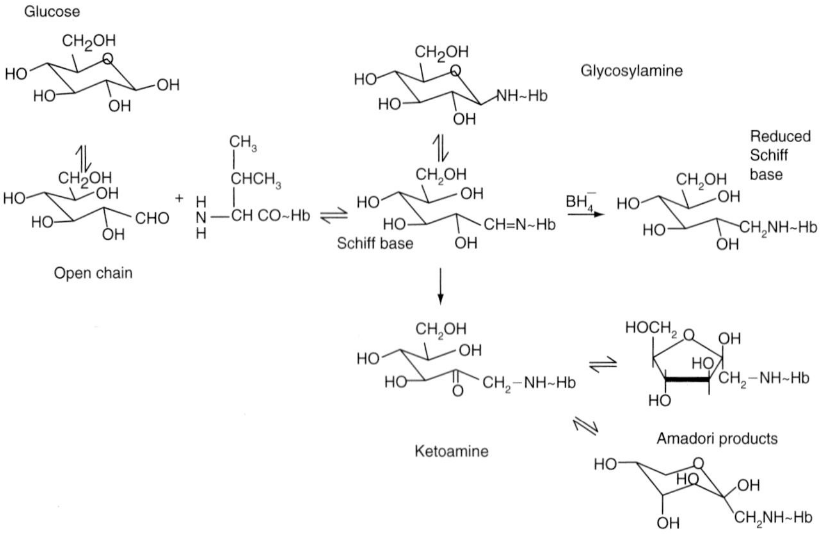

Figure 25.1. Early glycation reactions: the formation of the Schiff base and Amadori compound. Taken from Harding (1985).

research in recent years has been on AGEs (Thorpe and Baynes, 2003), but early glycation products can have a damaging effect and may be more important in damaging nonstructural proteins such as enzymes.

Glycation is not catalyzed by enzymes and consequently is not specific: almost any protein will react with almost any sugar. Experimental studies have utilized a wide variety of proteins and many sugars from glucose, the least reactive, and other hexoses, to sugar phosphates, ribose (Paul *et al.*, 1998), methylglyoxal (Ahmed *et al.*, 2003), glyoxal, ascorbate and its oxidation product dehydroascorbate (Argirov *et al.*, 2003), and other sugars. The reactivity of individual sugars depends largely on the proportion that exists in the open-chain form. Identification of new AGEs continues (Argirov *et al.*, 2003, 2004), and some major AGEs may be derived from lipid (Januszewski *et al.*, 2003). It is generally assumed that only a single sugar can attach to each amino group but there is some evidence for diglycation (Blakytny *et al.*, 1997).

25.1.2. The Effects of Glycation

Increased glycation is associated with aging and complications of diabetes so it is not surprising that glycation is damaging to proteins. Most emphasis was on long-lived proteins, mostly structural, because they are exposed to the sugars for a longer time than enzymes and other nonstructural proteins in most tissues. Very quickly it was apparent that the extent of glycation was approximately doubled in diabetes for a variety of proteins including collagen, myelin basic protein, lens crystallins, lens capsule, low-density lipoprotein (LDL), as well as hemoglobin (Harding, 1985). Similar increases were found in aging. Glycation caused yellowing, aggregation, and cross-linking associated with the increased presence of AGEs (Raza and Harding, 1991; Biemel *et al.*, 2002). These were late changes and it was possible under milder conditions to see

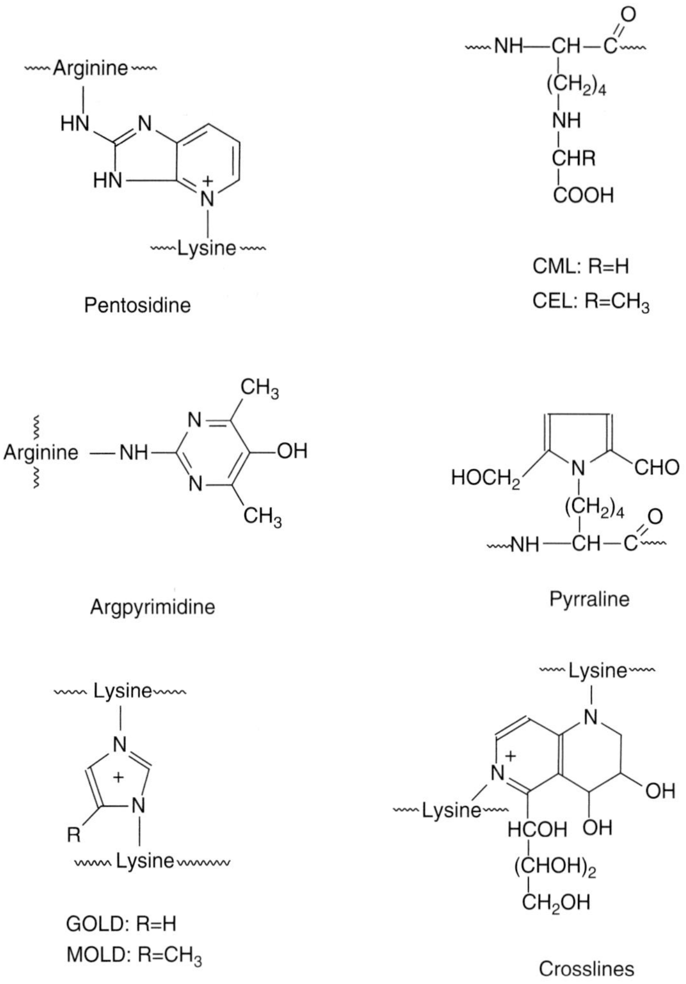

Figure 25.2. Structures of some AGEs: pentosidine, argpyrimidine, carboxymethyllysine (CML), carboxyethyllysine (CEL), pyrraline, glyoxal-lysine dimer (GOLD), methylglyoxal-lysine dimer (MOLD), and crosslines. Taken from Jakus and Rietbrock (2004).

conformational change to the proteins (Beswick and Harding, 1987). Glycation causes apoptosis of isolated pancreatic β-cells (Kaneto *et al.*, 1996).

Not only did general glycation increase with diabetes but also the extent of glycation, and especially of AGE formation, increased with the severity of diabetic complications (Monnier *et al.*, 1986, 1992; Singh *et al.*, 2001). Furthermore, glycation increases in the lens with age and cataract (see Section 25.4), and in other diseases such as Alzheimer's disease as discussed in the next section.

25.2. Cataract: A Conformational Disease

The lens is a transparent tissue in the eye between vitreous and aqueous humors and responsible for focusing incoming light on to the retina (Fig. 25.3). Cataract is opacity of the ocular lens sufficient to impair vision. It is the major cause of visual impairment and blindness worldwide. An advanced cataract can appear completely opaque, and the idea was put forward that the protein-rich tissue had achieved the state of a boiled egg with its protein entirely denatured. Later it was shown more formally that conformational change occurs at an earlier stage during the cataractogenic process (Harding, 1972). Thiol groups buried in the native proteins become more reactive and the protein becomes more susceptible to proteolysis. This was one of the first demonstrations of conformational change in disease, but now cataract has been joined by a variety of diseases where altered protein conformation and aggregation play a central role. Many of these diseases are discussed in this volume.

The next question for cataract researchers was to decide how unfolded protein came to accumulate in the lens, which contains very old proteins as a result of the peculiar way in which it grows. The lens is almost completely cellular with all cells contained within a capsule and never escaping (Fig. 25.4). Lining the anterior capsule is a single layer of epithelial cells, which divide

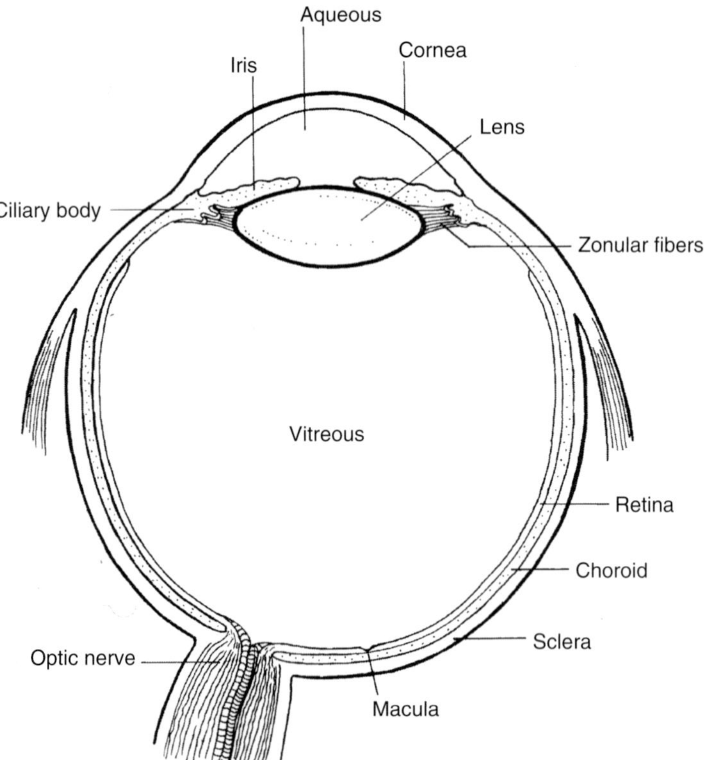

Figure 25.3. A cross section through the eye. Drawn by John Cronin.

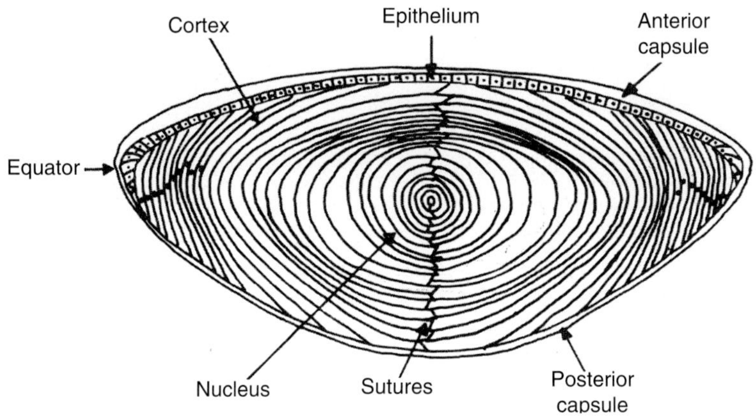

Figure 25.4. A cross section through the lens. Drawn by John Cronin.

and move to the equator where they elongate to form the long thin fiber cells that fill most of the adult lens. During elongation they lose their organelles, including ribosomes, and thus are unable to synthesize protein. As a consequence, the mature fiber cells contain some of the oldest proteins in the body, therefore it is unlikely that the unfolded protein in cataract, especially the common nuclear cataract, is derived from new protein. More likely the problem is caused by native protein becoming unfolded.

Over the years many ideas for causes of unfolding of lens proteins in cataract have been suggested (Table 25.1). Most of these are simple nonenzymic post-translational modifications of the proteins. All of these have been identified in human cataract, either only in cataract and not in clear lenses or at later stages of cataract (Harding, 1991). Glycation heads the list and has been studied perhaps more than others but all could play a similar role and could be responsible for the additive effects seen in this multifactorial disease. An interesting feature for protein chemists is that most of these changes involve anodic shifts, a decrease in positive charge or increase in negative charge, which could be relevant in additive effects. Anodic shifts were reported for many proteins in aging and disease (Harding, 1985). Diabetes is the most consistent risk factor for cataract adding more plausibility to the idea that glycation is an important cause of human cataract (Ederer *et al.*, 1981; Harding *et al.*, 1993). Glycated hemoglobin correlates with lens opacity in human studies (Leske *et al.*, 1999).

Parallels have been drawn between changes in cataract and those in Alzheimer's disease (Harding, 1993, 1997), and increased glycation of amyloid plaques and neurofibrillary tangles was found in Alzheimer's disease (Smith *et al.*, 1994; Vitek *et al.*, 1994; Yan *et al.*, 1994). This is consistent with the possibility that diabetes may be a risk factor for Alzheimer's disease (Ott *et al.*, 1999; Arvanitakis *et al.*, 2004). Increased glycation was also noted in other conformational diseases, Parkinson's and Pick's diseases, and in the lesions of age-related macular degeneration (Castellani *et al.*, 1996; Kimura *et al.*, 1996; Ishibashi *et al.*, 1998; Muench *et al.*, 2000). *In vitro*, it has been shown that glycation of albumin and lens crystallins can lead to formation of amyloid-like structures (Bouma *et al.*, 2003; O'Brien and Harding, unpublished results). Glycation causes partial unfolding of proteins providing the flexibility essential for amyloid formation (Uversky and Fink, 2004). β-Amyloid protein was found in lenses from Alzheimer patients and associated with cataract (Goldstein *et al.*, 2003).

Table 25.1. Proposed causes of unfolding of proteins in cataract

Glycation
Carbamylation
Steroid adduct formation
Truncation
Deamidation
Racemization
Mixed disulfide formation
Protein-protein disulfide formation
Methionine oxidation

25.3. Lens Proteins: Crystallins

The lens, as a largely cellular structure, has all the essential proteins found in any tissue but in particular it has a set of structural proteins called *crystallins*, whose predominance means that simple separations of mammalian lens supernatants by size-exclusion chromatography provide reasonably pure preparations of the three major crystallin groups (Fig. 25.5). γ-Crystallin is a compact monomeric protein of 20 kDa, whereas β- and α-crystallins are aggregates of up to 800 kDa. After being assigned a purely structural role for the best part of a century, α-crystallin was identified as a molecular chaperone (Horwitz, 1992). It protects proteins against thermal and reduction-induced aggregation, and it protects enzymes against inactivation induced by glycation, and other post-translational modifications (Blakytny and Harding, 1992a; Ganea and Harding, 1995; Hook and Harding, 1998). It assisted the renaturation of glyceraldehyde 3-phosphate dehydrogenase from guanidine (Ganea and Harding, 2000). β- and γ-crystallins are part of the same superfamily and can be called the βγ-crystallins.

α-, β-, and γ-crystallins have predominately β-sheet structure although in non-mammalian species there are crystallins related to enzymes, many with helical structure. The structures of some βγ-crystallins are known to high resolution (Wistow *et al.*, 1983; Bax *et al.*, 1990; Najmudin *et al.*, 1993), and the symmetry and surface charge network may contribute to their stability.

25.4. Glycation of Crystallins

The first demonstration of glycation of lens proteins was probably by van Heyningen (1969) who noted that radiolabeled glyceraldehyde became bound to the proteins as they became yellow-brown. Glycation, formerly called nonenzymic glycosylation, of hemoglobin had just been identified and that of collagen was soon to follow (Harding, 1985). Glycation of both α- and ε-amino groups was demonstrated, and van Heyningen had also noted a loss of thiols. Now it is clear that reaction between sugars and proteins is inevitable wherever they come together. Glycation of lens proteins is readily accomplished *in vitro* with a variety of sugars including glucose and glucose 6-phosphate (Stevens *et al.*, 1978), xylose, galactose, and fructose (Monnier *et al.*, 1981), glucosamine (Ajiboye and Harding, 1989), ribose (Liang and Rossi, 1990), and ascorbate (Bensch *et al.*, 1985). Glyoxal and methylglyoxal take part in similar reactions (Argirova and Breipohl, 2002). Glycation by fructose has been of particular interest because greatly increased levels are found in diabetic lens and it reacts faster than glucose. Ribose, methylglyoxal, and ascorbate are even more reactive.

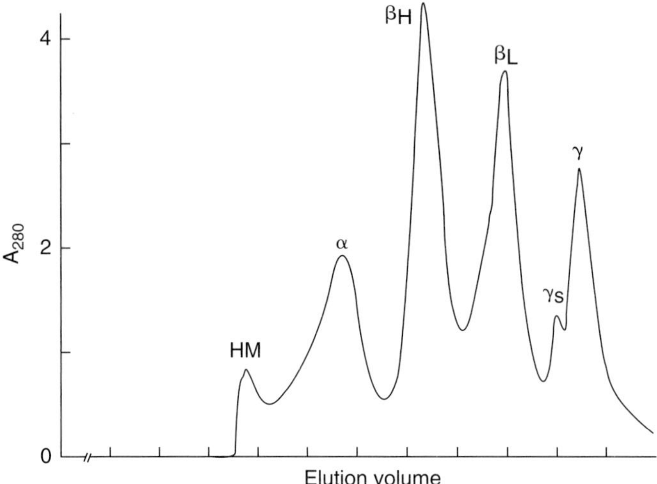

Figure 25.5. Size-exclusion chromatography of bovine lens proteins. After a small peak of high-molecular-weight protein, α-crystallin is followed by βH-, βL-, and γ-crystallins (Beswick and Harding, unpublished results).

Glycation *in vitro* causes yellowing (van Heyningen, 1969; Raza and Harding, 1991), increased fluorescence (Raza and Harding, 1991), aggregation (Stevens *et al.*, 1978), cross-linking (Argirova and Breipohl, 2002), and an increased susceptibility to thiol oxidation (Stevens *et al.*, 1978) which could be interpreted as conformational change. Conformational change was shown later by tryptophan fluorescence, circular dichroism, and by an increased susceptibility to tryptic digestion (Beswick and Harding, 1987; Liang and Rossi, 1990; Kumar *et al.*, 2004a). All these changes were of great interest because they correspond with changes already noted in cataract. Protein carbonyls, often taken as an indicator of oxidative change, increase with glycation (Argirova and Breipohl, 2002). In addition to the general effects of yellowing, cross-linking and aggregation glycation has more specific effects on lens proteins, for example glycation by the more reactive sugars diminishes the chaperone function of α-crystallin (Cherian and Abraham, 1995; van Boekel *et al.*, 1996; Plater *et al.*, 1997). More recently, Kumar *et al.* (2004b) reported that α-crystallin modified by methylglyoxal or glyoxal appeared to have enhanced chaperone activity in traditional aggregation assays, but decreased function in nonaggregation assays, so they warned against sole use of the popular aggregation assays.

In vivo glycation has been found in normal lens and increases in the diabetic rat (Stevens *et al.*, 1978), in normal aging (Chiou *et al.*, 1981; Swamy and Abraham, 1987), and in cataract (Monnier *et al.*, 1979; Pande *et al.*, 1979; Ansari *et al.*, 1980; Kasai *et al.*, 1983; Perry *et al.*, 1987; Oimomi *et al.*, 1988). Up to 20% of lens polypeptides may contain glucose-lysine adducts in cataracts from human diabetics. Glycated α-crystallin from cataracts from human diabetics had loss of tertiary structure but retained native secondary structure (Liang and Chylack, 1987) just as found for the protein glycated *in vitro* (Beswick and Harding, 1987). Lens fluorescence, largely attributable to glycation, was associated with levels of glycated hemoglobin over 14 years and also with diabetic complications (Kessel *et al.*, 2004).

All crystallin groups are susceptible to glycation. It is likely that all amino groups on crystallins would react with sugar given time but among the more reactive groups is the α-amino group of γB-crystallin (Pennington and Harding, 1994). Limited glycation of α-crystallin is restricted to the flexible C-terminal extensions (Blakytny *et al.*, 1997), but under extreme conditions seven lysines of αA-crystallin and ten lysines of αB-crystallin were glycated by glucose (Ortwerth *et al.*, 1992). Similarly, M glucose and fructose glycate most of the lysine residues of βB2-crystallin (Zhao *et al.*, 1996). It is perhaps relevant that the α- and β-crystallins have blocked α-amino groups as do some enzymes susceptible to glycation (e.g., superoxide dismutase: see next section) thus being protected from glycation at this reactive site.

In most tissues, the proteins turn over rapidly and there is little time for the accumulation of late glycation products, but this is not true of the long-lived proteins of the lens. Yellowing and fluorescence increase with age apparently as the result of AGE formation (Cheng *et al.*, 2002). Carboxymethyllysine (Fig. 25.2) was identified in lens and increases in human lens with age (Ahmed *et al.*, 1986; Dunn *et al.*, 1989), as do other AGEs (Araki *et al.*, 1992; Ahmed *et al.*, 2003). Another AGE, crossline, increases with age in rat lens protein, and even more is found in protein from diabetic rat lens (Obayashi *et al.*, 1996). Methylglyoxal hydroimidazolone 1 (MG-H1), 2-ammonio-6-(3-oxidopyridinium-1-yl)hexanoate (OP-lysine), argpyrimidine, and other AGEs also increase with cataract and are major glycation products in human lens (Wilker *et al.*, 2001; Ahmed *et al.*, 2003; Argirov *et al.*, 2004). AGE cross-link compounds increase in brown cataracts (Biemel *et al.*, 2002).

Glycation is not specific and is not restricted to structural proteins. In the next section, results on glycation of enzymes are discussed.

25.5. Glycation of Enzymes

Although long-lived structural proteins were the original focus of glycation studies, the lack of specificity means that any protein, including enzymes, could be affected. Indeed many structural proteins, like the α- and β-crystallins and collagen, have blocked α-amino groups and thus protection of the most reactive groups for glycation. It can be said that mammalian crystallins have been selected as protected proteins that can survive in cells devoid of protein turnover, but enzymes also exist in these cells and may be more susceptible to glycation and similar modifications. Enzyme activity is generally lower in the lens nucleus than in the cortex, but what is remarkable is that there is any enzyme activity at all in human lens nucleus where there is no protein turnover.

The activity of many enzymes is decreased in cataract (Harding, 1991). Glycation of Cu-Zn superoxide dismutase, sorbitol dehydrogenase, alcohol dehydrogenase, aspartate aminotransferase, and Ca-ATPase *in vivo* has been reported in several tissues, and glycation increases in diabetes, leading to inactivation of the enzymes (Arai *et al.*, 1987a, 1987b; Shilton and Walton, 1991; Oda *et al.*, 1994; Hoshi *et al.*, 1996; Okada *et al.*, 1997; Bidasee *et al.*, 2004). Radiolabeled glucose was incorporated into superoxide dismutase as the activity fell *in vitro*, and the reactive lysine residues were identified. Erythrocyte activity of superoxide dismutase and catalase are considerably decreased in human diabetes, both types I and II (Abou-Seif and Youssef, 2004). Many enzymes can be inactivated *in vitro* by many different sugars and related carbonyl compounds (Table 25.2). Where different sugars have been used, the relative rates of inactivation were usually in order of the proportion of the open chain form of the sugars. In general, inactivation of enzymes could be achieved at lower concentrations of sugars than were required to see conformational change or cross-linking of structural proteins. The glycation reaction sites have been identified for a number of enzymes (Watkins *et al.*, 1985; Arai, 1987a, 1987b; Adachi *et al.*, 1992; Fujita *et al.*, 1998).

Several enzymes could be inactivated by protein that had been pre-glycated with methylglyoxal or ascorbate presumably because dicarbonyls or similar reactive compounds were released from the glycated protein (Morgan *et al.*, 2002; O'Brien and Harding, unpublished results).

In addition to crystallins and enzymes, other lens proteins are glycated, for example, calmodulin and a channel protein (Evcimen and Nebioglu, 1996; Prabhakaram *et al*., 1996). The glycation decreased the binding of calmodulin to aquaporin and could thus affect calcium-mediated intracellular processes leading to opacification (Swamy-Mruthinti, 2001).

The various pathways leading from elevated levels of glucose in the lens to cataract are illustrated in Figure 25.6. High glucose levels lead to increased levels of other more reactive sugars, and all glycate the crystallins, enzymes, and other proteins causing conformational changes and loss of function. This causes an accumulation of unfolded protein and thus of aggregates and to loss of glutathione and oxidation. Glycation of channel proteins and NaK-ATPase leads to ionic imbalance, water influx, and vacuoles that scatter light. All paths lead to cataract. Similar changes may well occur in other tissues.

Table 25.2. Some enzymes inactivated by sugars and related carbonyl compounds

Enzyme	Sugar	Reference
Alanine aminotransferase	fruct, ribose and glyceraldehyde	V
Alkaline phosphatase	glc	G
δ-Aminolevulinic dehydratase	glc and 7 others	R
Aspartate aminotransferase	fruct, ribose, MG, and glyceraldehyde	A, T, U
Catalase	glc, fruct, G6P, ribose	J
Cu-Zn superoxide dismutase	glc	B
Esterase	glc, G6P, fruct, ribose	M
Fructose 1,6 P2 aldolase	MG	C
Fumarase	fruct	O
β-Galactosidase	glc	G
Glucose 6-phosphate dehydrogenase	glc, gal, fruct	G, H
Glutathione peroxidase	3-Deoxyglucosone	X
Glutathione reductase	glc, fruct, G6P and MG	D, P
Glyceraldehyde 3-phosphate dehydrogenase	MG, F6P, G6P, fruct	C, L
Lactate dehydrogenase	MG, ascorbate	P, S
Malate dehydrogenase	glc, G6P, fruct	I
NaK-ATPase	G6P, fruct, MG	E
Phosphoglyceromutase	MG	C
Phosphoglycerate kinase	MG	C
Phosphofructokinase	MG	C
Pyruvate kinase	fruct	N
Ribonuclease	glc	F
Rubisco	ascorbate	W
Sorbitol dehydrogenase	glc, fruct	Q
Superoxide dismutase	glc, G6P, fruct	J, K

Sugars: glc = glucose; G6P = glucose-6-phosphate; F6P = fructose 6-phosphate; fruct = fructose; MG = methylglyoxal.

References: A: Drsata *et al.* (2002); B: Arai *et al.* (1987a, b); C: Leoncini *et al.* (1980); D: Blakytny and Harding (1992a); E: Garner and Spector (1986); F: Watkins *et al.* (1985); G: Ganea (1988); H: Ganea and Harding (1995); I: Heath *et al.* (1996); J: Yan and Harding (1997); K: Arai *et al.* (1987a,b); L: Hook and Harding (1997); M: Yan and Harding (1999); N: Ganea and Harding (1999); O: Hook and Harding (1998); P: Morgan *et al.* (2002); Q: Hoshi *et al.* (1996); R: Caballero *et al.* (1998); S: Argirova and Breipohl (2002); T: Okada *et al.* (1997); U: Seidler and Kowalewski (2003); V: Beranek *et al.* (2001); W: Yamauchi *et al.* (2002); X: Niwa and Tsukushi (2001).

25.6. Primary Targets of Glycation

After the emphasis of glycation studies on structural proteins had progressed to enzymes and other proteins with clearly defined functions, it was clearly not sufficient for each researcher to show that his or her favorite protein could be glycated and that this reaction impaired its function in some way. This would be true of almost every protein. The question arises as to which proteins in any tissue are the most vulnerable to glycation. Which proteins constitute the weak links when sugars are the main danger?

In the studies of crystallins described in Section 25.4, the concentrations of sugars required to produce perceptible change were usually in excess of 100 mM, whereas the inactivation of enzymes was achieved in the same sort of time period by concentrations of 5 or 10 mM. This would indicate that enzymes are likely to be the weak links, but demonstrating this is not easy.

After limited glycation of a total lens supernatant, most of the sugar was bound to proteins that themselves became bound to an enzyme affinity column, suggesting that enzymes were the primary targets (Yan and Harding, 2003). Further separation of the bound protein on gels and amino acid sequence analysis only demonstrated γIIIa and γIIIb-crystallin. Although these crystallins may be important, it is likely that enzymes were among the primary targets but each enzyme would be a minor component of the protein subjected to sequencing. More studies of this kind, perhaps using other techniques, are required to establish the primary targets in lens and other tissues.

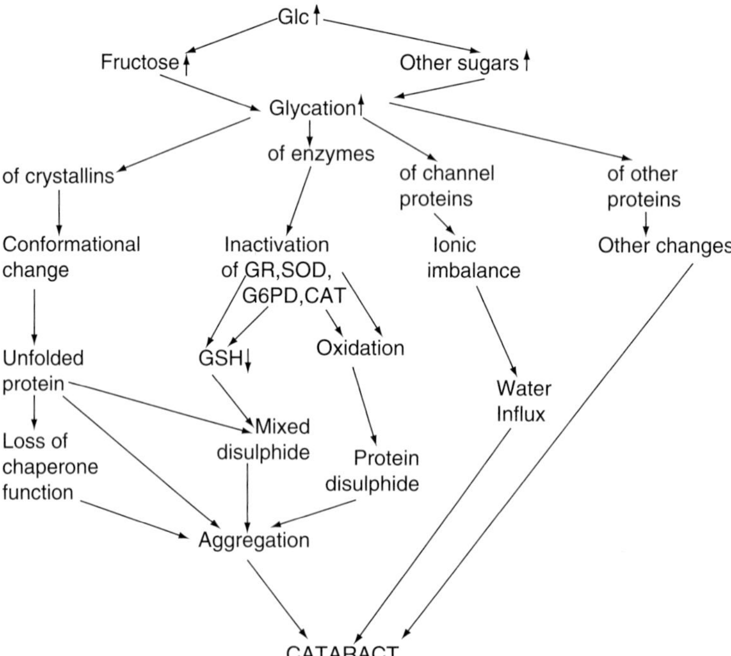

Figure 25.6. Scheme summarizing many changes seen in cataract and linking them by putative causal pathways. The key role for glycation and conformational change is apparent. CAT, catalase; G6PD, glucose 6-phosphate dehydrogenase; GR, glutathione reductase; GSH, reduced glutathione; SOD, superoxide dismutase.

25.7. Future Prospects

Future research in this area may include further attempts to identify AGE structures but more especially to identify quantitatively important structures given that many of those most discussed, such as pentosidine, are present in tiny amounts and cannot be seen during careful mass spectrometric examination of cataract lenses.

Further studies of cataract as a conformational disease should establish its relationship to other conformational diseases. To what extent do they share common etiologic pathways? What can lens researchers learn from those studying other conformational diseases and vice versa? The glycation of lens proteins has been thoroughly investigated, and it might be anticipated that in the future, emphasis will turn to the specialized proteins such as enzymes and channel proteins. For the lens, and for other tissues, it will be important to determine which proteins are most susceptible to glycation; not only which are most readily glycated but also which are most easily damaged as a result of glycation?

Finally, protection against glycation will become a very fertile field of research, not dealt with in this review, but there are already published reviews (Monnier, 2003; Thornalley, 2003; Kyselova *et al.*, 2004) and another will appear shortly (Harding and Ganea, 2006).

The obvious way to prevent excessive glycation in diabetes is to control blood sugar levels more tightly, but in addition there are several types of agent to prevent glycation, including aspirin and ibuprofen. Aspirin, ibuprofen, and paracetamol (acetaminophen) protected against cataract in diabetic rats (Swamy and Abraham, 1989; Blakytny and Harding, 1992b) and may protect against human cataract (Cotlier and Sharma, 1981; van Heyningen and Harding, 1986; Harding and van Heyningen, 1988; Harding *et al.*, 1989; Mohan *et al.*, 1989; Christen *et al.*, 1998). There is also evidence for beneficial effects of aspirin-like drugs in ischemic heart disease, stroke, Alzheimer's disease, colorectal cancer, and prostate cancer, adding strength to the view that these diseases may have overlapping etiologies and that it may be possible to find common therapies (Harding, 2001). It may be that glycation is part of the common etiology.

Compounds that react with the sugars and remove them from reaction with protein have been proposed as antiglycation agents. These include amino acids (Sulochana *et al.*, 1998) and peptides such as carnosine (Hipkiss *et al.*, 1995; Argirova and Argirov, 2003). Aminoguanidine was introduced as a compound to react with the Amadori product and block further reactions (Bucala *et al.*, 1991; 1994; Monnier, 2003; Thornalley, 2003; Ganea, 2004). It decreased cataract development in moderately diabetic rats (Swamy-Mruthinti *et al.*, 1996). Other compounds are being studied and some have reached clinical trial. Yet other compounds may reverse glycation or prevent its consequences.

References

Abou-Seif, M. A. and Youssef, A. (2004). Evaluation of some biochemical changes in diabetic patients *Clin. Chim. Acta* 346: 161–170.

Adachi, T., Ohta, H., Hayashi, K., Hirano, K. and Marklund, S. (1992). The site of nonenzymic glycation of human extracellular-superoxide dismutase *in vitro*. *Free Radic. Biol. Med.* 13: 205–210.

Ahmed, M. U., Thorpe, S. R. and Baynes, J. W. (1986). Identification of N^{ε}-carboxymethyllysine as a degradation product of fructoselysine in glycated protein. *J. Biol. Chem.* 261: 4889–4894.

Ahmed, N., Thornalley, P. J., Dawczynski, J., Franke, S., Strobel, J. Stein, G. and Haik, G. M. (2003). Methylglyoxal-derived hydroimidazolone advanced glycation end-products of human lens proteins. *Invest. Ophthalmol. Vis. Sci.* 44: 5287–5292.

Ajiboye, R. and Harding, J. J. (1989). The non-enzymic glycosylation of bovine lens proteins by glucosamine and its inhibition by aspirin, ibuprofen and glutathione. *Exp. Eye Res.* 49: 31–41.

Ansari, N. H., Awasthi, Y. G., and Srivastava, S. K. (1980). Role of glycosylation in disulphide formation and cataractogenesis. *Exp. Eye Res.* 31: 9–19.

Arai, K., Iizuka, S., Tada, Y., Oikawa, K. and Taniguchi, N. (1987a). Increase in the glucosylated form of erythrocyte Cu-Zn superoxide dismutase in diabetes and close association of the non-enzymic glucosylation with the enzyme activity. *Biochim. Biophys. Acta* 924: 292–296.

Arai, K., Maguchi, S., Fujii, S., Ishibashi, H., Oikawa, K. and Taniguchi, N. (1987b). Glycation and inactivation of human Cu-Zn superoxide dismutase: identification of the *in vitro* glycated sites. *J. Biol. Chem.* 262: 16969–16972.

Araki, N., Ueno, N., Chakrabarti, B., Morino, Y. and Horiuchi, S. (1992). Immunochemical evidence for the presence of advanced glycation endproducts in human lens proteins and its positive correlation with ageing. *J. Biol. Chem.* 267: 10211–10214.

Argirov, O.K., Lin, B., Oleson, P. and Ortwerth, B. J. (2003). Isolation and characterization of a new advanced glycation endproduct of dehydroascorbic acid and lysine. *Biochim. Biophys. Acta* 1620: 235–244.

Argirov, O.K., Lin, B. and Ortwerth, B. J. (2004). 2-Ammonio-6-(3-oxidopyridinium-1-yl)hexanoate (OP-lysine) is a newly identified advanced glycation endproduct in cataractous and aged human lenses. *J. Biol. Chem.* 279: 6487–6495.

Argirova, M. and Argirov, O. (2003). Inhibition of ascorbic acid-induced modifications in lens proteins by peptides. *J. Pep. Res.* 9: 170–176.

Argirova, M. and Breipohl, W. (2002). Comparison between modifications of lens proteins resulted from glycation with methylglyoxal, glyoxal, ascorbate and fructose. *J. Biochem. Mol. Toxicol.* 16: 140–145.

Arvanitakis, Z., Wilson, R. S., Bienias, J. L., Evans, D. A.and Bennett, D. A. (2004). Diabetes mellitus and risk of Alzheimer disease and decline in cognitive function. *Arch. Neurol.* 61: 661–666.

Bax, B, Lapatto, R., Nalini, V., Driessen, H., Lindley, P. F., Mahadevan, D., Blundell, T. L. and Slingsby, C. (1990). X-ray analysis of B2-crystallin and evolution of oligomeric lens proteins. *Nature* 347: 776–780.

Bensch, K. G., Fleming, J. E. and Lohmann, W. (1985). The role of ascorbic acid in senile cataract. *Proc. Natl. Acad. Sci. USA* 82: 7193–7196.

Beranek, M., Drsata, J. and Palicka, V. (2001). Inhibitory effect of glycation on catalytic activity of alanine aminotransferase. *Mol. Cell. Biochem.* 218: 35–39.

Beswick, H. T. and Harding, J. J. (1987). Conformational changes induced in lens α- and γ-crystallins by modification by glucose 6-phosphate. Implications for cataract. *J. Biochem.* 246: 761–769.

Bidasee, K. R., Zhang, Y., Shao, C. H., *et al.* (2004). Diabetes increases formation of advanced glycation endproducts on sarco(endo)plasmic reticulum Ca^{2+}-ATPase. *Diabetes* 53: 463–467.

Biemel, K. M., Friedl, D. A. and Lederer, M. O. (2002). Identification and quantification of major Maillard cross-links in human serum albumin and lens protein. *J. Biol. Chem.* 277: 24907–24915.

Blakytny, R. and Harding, J. J. (1992a). Glycation (non-enzymic glycosylation) inactivates glutathione reductase. *J. Biochem.* 288: 303–307.

Blakytny, R. and Harding, J. J. (1992b). Prevention of cataract in diabetic rats by aspirin, paracetamol (acetaminophen) and ibuprofen. *Exp. Eye Res.* 54: 509–518.

Blakytny, R., Carver, J. A., Harding, J. J., Kilby, G. W. and Sheil, M. M. (1997). A spectroscopic study of bovine α-crystallin: investigation of flexibility of the C-terminal extension, chaperone activity, and evidence for diglycation. *Biochim. Biophys. Acta* 1343: 299–315.

Bouma, B., Kroon-Batenburg, L. M. J., Wu, Y.-P., Brunges, B., Posthuma, G., Kranenburg, O., de Groot, P., Voest, E. E. and Gebbink, M.F.B.J. (2003). Glycation induces formation of amyloid cross-beta structure in albumin. *J. Biol. Chem.* 278: 41810–41819.

Bucala, R., Vlassara, H. and Cerami, A. (1991). Advanced glycation endproducts. In: *Post-translational modification of proteins*. (Eds. Harding, J. J. and Crabbe, M.J.C.) CRC Press, Boca, Raton. pp. 53–79.

Bucala, R., Vlassara, H. and Cerami, A. (1994). Advanced glycation endproducts: role in diabetic and non-diabetic vascular disease. *Drug Dev. Res.* 32: 77–89.

Caballero, F. A., Gerez, E. N., Polo, C. F., Vazquez, E. S. and Batlle, A. (1998). Reducing sugars trigger -aminolevulinic dehydratase inactivation: evidence of *in vitro* aspirin prevention. *Gen Pharmacol* 31: 441–445.

Castellani, R., Smith, M. A., Richey, P.L. and Perry, G. (1996). Glycoxidation and oxidative stress in Parkinson disease and diffuse Lewy body disease. *Brain Res.* 737: 195–200.

Cheng, R., Lin, B. and Ortwerth, B. J. (2002). Rate of formation of AGEs during ascorbate glycation and during aging in human lens tissue. *Biochim. Biophys. Acta* 1587: 65–74.

Cherian, M. and Abraham, E. C. (1995). Decreased molecular chaperone property of α-crystallins due to post-translational modifications. *Biochem. Biophys. Res. Commun.* 208, 675–679.

Chiou, S.-H., Chylack, L. T., Tung, W. H. and Bunn, H. F. (1981). Nonenzymic glycosylation of bovine lens crystallins. Effect of aging. *J. Biol. Chem.* 256: 5176–5180.

Christen, W.G., Manson, J. E., Glynn, R.J. *et al.* (1998). Low-dose aspirin and risk of cataract and subtypes in a randomised trial of US physicians. *Ophthalmic Epidemiol.* 5: 133–142.

Cotlier, E. and Sharma, Y. G. (1981). Aspirin and senile cataracts in rheumatoid arthritis. *Lancet* I: 607.

Drsata, J., Beranek, M., Pali, V. and Caron, C. (2002) Inhibition of aspartate aminotransferase by glycation *in vitro* under various conditions. *J. Enz. Inhib. Med. Chem.* 17: 31–36.

Dunn, J. A., Patrick, J. S., Thorpe, S. R. and Baynes, J. W. (1989). Oxidation of glycated protein: age-related accumulation of N^{ε}-(carboxymethyl)lysine in lens proteins. *Biochemistry* 28: 9464–9468.

Ederer, F., Hiller, R. and Taylor, H. R. (1981). Senile lens changes and diabetes in two population studies. *Am. J. Ophthalmol.* 91: 381–395.

Evcimen, N.D. and Nebioglu, S. (1996). Calmodulin glycation in diabetic rat lenses. *Can. J. Physiol. Pharmacol.* 74: 1287–1293.

Fujita, T., Suzuki, K., Tada, T., Yoshihara, Y., Hamaoka, R., Uchida, K., Matuo, Y., Sasaki, T., Hanafusa, T. and Taniguchi, N. (1998). Human erythrocyte bisphosphoglycerate mutase: Inactivation by glycation *in vivo* and *in vitro*. *J. Biochem.* 124: 1237–1244.

Ganea, E. (1988). Non-enzymic glycosylation of beta-galactosidase, alkalaine phosphatase, and glucose 6-phosphate dehydrogenase. *Rev. Roum. Biochim.* 25: 101–106.

Ganea, E. (2004). Prevention of insulin glycation by metabolic and synthetic inhibitors. In: *Cellular dysfunction in atherosclerosis and diabetes*. (Simionescu, M, Sima, A. and Popov, D. Eds.) Romanian Academy Publishing House. pp. 205–214.

Ganea, E. and Harding, J. J. (1995). Molecular chaperones protect against glycation-induced inactivation of glucose 6-phosphate dehydrogenase. *Eur. J. Biochem.* 231: 181–185.

Ganea, E. and Harding, J. J. (1999). Glycation-inactivated pyruvate kinase can be protected by alpha-crystallin acting as a molecular chaperone in an enzyme-chaperone complex. *Proc. Rom. Acad. Series B.* 1: 39–44.

Ganea, E. and Harding, J. J. (2000). α-Crystallin assists the renaturation of glyceraldehyde-3-phosphate dehydrogenase. *J. Biochem.* 345: 467–472.

Garner, M. H. and Spector, A. (1986). ATP hydrolysis kinetics by Na,K-ATPase in cataract. *Exp. Eye Res.* 42: 339–348.

Goldstein, L. E., Muffat, J. A., Cherny, R. A., Moir, R. D., Ericsson, M. H., Huang, X., Mavros, C., Coccia, J. A., Faget, K. Y., Fitch, K. A., Masters, C. L., Tanzi, R. E., Chylack, L. T. and Bush, A. I. (2003). Cytosolic β-amyloid deposition and supranuclear cataracts in lenses from people with Alzheimer's disease. *Lancet* 361: 1258–1265.

Harding, J. J. (1972). Conformational changes in human lens proteins in cataract. *J. Biochem.* 129: 97–100.

Harding, J. J. (1985). Nonenzymatic post-translational modification of proteins *in vivo*. *Advan. Protein Chem.* 37: 247–334.

Harding, J. J. and van Heyningen, R. (1986). Do aspirin-like drugs protect against cataract? *Lancet* 1: 1111–1113.

Harding, J. J. and van Heyningen, R. (1988). Drugs, including alcohol, that act as risk factors for cataract, and possible protection against cataract by aspirin-like analgesics and cyclopenthiazide. *Br. J. Ophthalmol.* 72: 809–814.

Harding, J. J., Egerton, M. and Harding, R. S. (1989). Protection against cataract by aspirin, paracetamol and ibuprofen. *Acta. Ophthalmol.* 67: 518–524.

Harding, J. J. (1991). *Cataract: Biochemistry, Epidemiology and Pharmacology*. Chapman and Hall, London.

Harding, J. J. (1993). Alzheimer disease (AD), modified proteins, and aspirin. *Alzheimer Dis. Assoc Disorders* 7: 55 –58.

Harding, J. J., Egerton, M., van Heyningen, R. and Harding, R. S. (1993). Diabetes, glaucoma, sex and cataract: analysis of combined data from two case-control studies. *Br. J. Ophthalmol.* 77: 2–6.

Harding, J. J. (1997). Alzheimer disease and cataract: common threads. *Alzheimer Dis. Assoc. Disorders* 11: 123.

Harding, J. J. (2001). Can drugs or micronutrients prevent cataract? *Drugs Ageing* 18: 473–486.

Harding, J. J. and Ganea, E. (2006). Protection against glycation and similar post-translational modifications of proteins. *Biochem. Biophys. Acta.* 1764: 1436–1446.

Heath, M. M., Rixon, K. C. and Harding, J. J. (1996). Glycation-induced inactivation of malate dehydrogenase protection by aspirin and a lens molecular chaperone, alpha-crystallin. *Biochim. Biophys. Acta* 13115: 176–184.

Hipkiss, A. R., Michaelis, J. and Syrris, P. (1995). Non-enzymatic glycosylation of the dipeptide L-carnosine, a potential anti-protein-cross-linking agent. *FEBS Lett.* 371: 81–85.

Hook, D. W. A. and Harding, J. J. (1997). Inactivation of glyceraldehyde 3-phosphate dehydrogenase by sugars, prednisolone-21-hemisuccinate, cyanate and other small molecules. *Biochim. Biophys. Acta* 1362: 232–242.

Hook, D. W. A. and Harding, J. J. (1998). Protection of enzymes by α-crystallin acting as a molecular chaperone. *Int. J. Biol. Macromol.* 22: 295–306.

Horwitz, J. (1992). α-Crystallin can function as a molecular chaperone. *Proc. Natl. Acad. Sci. U.S.A.* 89: 10449–53.

Hoshi, A., Takahashi, M., Fujii, J., Myint, T., Kaneto, H., Suzuki, K., Yamasaki, Y., Kamada, T. and Taniguchi, N. (1996). Glycation and inactivation of sorbitol dehydrogenase in normal and diabetic rats. *J. Biochem.* 318: 119–123.

Ishibashi, T., Murata, T., Hangai, M., Nagai, R., Horiuchi, S., Lopez, P. F., Hinton, D. R. and Ryan, S.J. (1998). Advanced glycation endproducts in age-related macular degeneration. *Arch. Ophthalmol.* 116: 1629–1632.

Jakus, V. and Rietbrock, N. (2004). Advanced glycation end-products and the progress of diabetic vascular complications. *Physiol. Res.* 53: 131–142.

Januszewski, A.S., Alderson, A. L., Metz, T. O., Thorpe, S. R. and Baynes, J. W. (2003). Role of lipids in chemical modification of proteins and development of complications in diabetes. *Biochem. Soc. Trans.* 31: 1414–1416.

Kaneto, H., Fujii, J., Myint, T., *et al.* (1996) Reducing sugars trigger oxidative modification and apoptosis in pancreatic β-cells by provoking oxidative stress through the glycation reaction. *J. Biochem.* 320: 855–863.

Kasai, K., Nakamura, T., Kase, N., *et al.* (1983). Increased glycosylation of proteins from cataractous lenses in diabetes. *Diabetologia* 25: 36–38.

Kessel, L., Sander, B., Dalgaard, P. and Larsen, M. (2004). Lens fluorescence and metabolic control in type I diabetic patients: a 14 year follow up study. *Br. J. Ophthalmol.* 88: 1169–1172.

Kimura, T., Ikeda, K., Takamatsu, J., *et al.* (1996). Identification of advanced glycation endproducts of the Maillard reaction in Pick's disease. *Neurosci. Lett.* 219: 95–98.

Kumar, M. S., Mrudula, T., Mitra, N. and Reddy, G. B. (2004a). Enhanced degradation and decreased stability of eye lens α-crystallin upon methylglyoxal modification. *Exp. Eye Res.* 79: 577–583.

Kumar, M. S., Reddy, P. Y., Surolia, I. and Reddy, G.B. (2004b). Effect of dicarbonyl browning on α-crystallin chaperone-like activity: physiological significance and caveats of *in vitro* aggregation assays. *J. Biochem.* 379: 273–282.

Kyselova, Z., Stefek, M. and Bauer, V. (2004). Pharmacological prevention of diabetic cataract. *J. Diabetes Complications* 18: 129–140.

Leske, M. C., Wu, S., Hennis, A., *et al.* (1999). Diabetes, hypertension, and central obesity as cataract risk factors in a black population. *Ophthalmology* 106: 35–41.

Leoncini, G., Maresca, M. and Bonsignore, A. (1980). The effect of methylglyoxal on the glycolytic enzymes. *FEBS Lett.* 117: 17–18.

Liang, J. N. and Chylack, L. T. (1987). Spectroscopic study on the effects of nonenzymis glycosylation on human α-crystallin. *Invest. Ophthalmol. Vis. Sci.* 28: 790–794.

Liang, J.N. and Rossi, M. T. (1990). *In vitro* non-enzymatic glycation and formation of browning products in the bovine lens α-crystallin. *Exp. Eye Res.* 50: 367–371.

Mohan, M., Sperduto, R. D., Angra, S. K., *et al.* (1989). India-US case-control study of age-related cataracts. *Arch. Ophthalmol.* 107: 670–676.

Monnier, V. M., Stevens, V. J. and Cerami, A. (1979). Non-enzymatic glycosylation, sulphydryl oxidation, and aggregation of lens proteins in experimental sugar cataracts. *J. Exp. Med.* 150: 1098–1107.

Monnier, V. M., Stevens, V. J. and Cerami, A. (1981). Maillard reactions involving proteins and carbohydrates *in vivo*: relevance to diabetes mellitus and ageing. *Prog. Fd. Nutr. Sci.* 5: 315–327.

Monnier, V. M., Vishwanath, V., Frank, K. E., Elmets, C. A., Dauchot, P. and Kohn, R. R. (1986). Relation between complications of type I diabetes mellitus and collagen-linked fluorescence. *N. Engl. J. Med.* 314: 403–408.

Monnier, V. M., Sell, D. R., Miyata, S., Nagaraj, R. H., Odetti, P. and Lapolla, A. (1992). Advanced Maillard reaction products as markers for tissue damage in diabetes and uraemia: relevance to diabetic nephropathy. *Acta Diabetol.* 29: 130–135.

Monnier, V. M. (2003). Intervention against the Maillard reaction *in vivo*. *Arch. Biochem. Biophys.* 419: 1–15.

Muench, G., Luth, H. J., Wong, A., Arendt, T., Hirsch, E., Ravid, R. and Riederer, P. (2000). Crosslinking of α-synuclein by advanced glycation endproducts-an early pathophysiological step in Lewy body formation? *J. Chem. Neuroanat.* 20: 253–257.

Morgan, P. E., Dean, R. T. and Davies, M. J. (2002). Inactivation of cellular enzymes by carbonyls and protein-bound glycation/glycoxidation products *Arch. Biochem. Biophys.* 403: 259–269.

Najmudin, S., Nalini, V., Driessen, H. P. C., Slingsby, C. Blundell, T. L. Moss, D. S., and Lindley, P. F. (1993). Structure of the bovine eye lens protein γB(γII) crystallin at 1.47 A. *Acta Cryst.* D49: 223–233.

Niwa, T. and Tsukushi, S. (2001). 3 Deoxyglucosone and AGEs in uremic complications: inactivation of glutathione peroxidase by 3-deoxyglucosone. *Kidney Int.* 59, Suppl. 78: S37-S41.

Obayashi, H., Nakano, K., Shigeta, H., *et al.* (1996). Formation of crossline as a fluorescent advanced glycation endproduct *in vitro* and *in vivo*. *Biochem. Biophys. Res Commun.* 226: 37–41.

Oda, A., Bannai, C., Yamaoka, T., Katori, T. Matsushima, T., and Yamashita, K. (1994). Inactivation of Cu, Zn-superoxide dismutase by *in vitro* glycosylation and in erythrocytes of diabetic patients. *Horm. Metab. Res.* 26: 1–4.

Oimomi, M., Maeda, Y., Hata, F., *et al.* (1988). Glycation of cataractous lens in non-diabetic senile subjects and in diabetic patients. *Exp. Eye Res.* 46: 415–420.

Okada, M., Murakami, Y. and Miyamoto, E. (1997). Glycation and inactivation of aspartate aminotransferase in diabetic rat tissues. *J. Nutr. Sci. Vitaminol.* 43: 463–469.

Ortwerth, B. J., Slight, S. H., Prabhakaram, M., Sun, Y. and Smith, J. B. (1992). Site-specific glycation of lens crystallins by ascorbic acid. *Biochim. Biophys. Acta* 1117: 207–215.

Ott, A., Stolk, R. P., van Harskamp, F., Pols, H. A. P., Hofman, A. and Breteler, M. M. B. (1999). Diabetes mellitus and the risk of dementia. *Neurology* 53: 1937–1942.

Pande, A., Garner, W. H. and Spector, A. (1979). Glucosylation of human lens protein and cataractogenesis. *Biochem. Biophys. Res. Commun.* 89: 1260–1266.

Paul, R.G., Avery, N. C., Slatter, D.L., Sims, T. J. and Bailey, A. J. (1998). Isolation and characterization of advanced glycation end products derived from the *in vitro* reaction of ribose and collagen. *J. Biochem.* 330: 1241–1248.

Pennington, J. and Harding, J. J. (1994). Identification of the site of glycation of γ-II-crystallin by (^{14}C)-fructose. *Biochim. Biophys. Acta* 1226: 163–167.

Perry, R. E., Swamy, M. S. and Abraham, E.C. (1987). Progressive changes in lens crystallin glycation and high-molecular-weight aggregate formation leading to cataract development in streptozotocin-diabeteic rats. *Exp. Eye Res.* 44: 269–282.

Plater, M. L., Goode, D. and Crabbe, M. J. C. (1997). Ibuprofen protects α-crystallin against post-translational modification by preventing protein crosslinking. *Ophthalmic Res.* 29: 421–428.

Prabhakaram, M., Katz, M. L. and Ortwerth, B. J. (1996). Glycation mediated crosslinking between alpha-crystallin and MP26 in intact lens membranes. *Mech. Ageing Dev.* 91: 65–78.

Raza, K. and Harding, J. J. (1991). Non-enzymic modification of lens proteins by glucose and fructose: effects of ibuprofen. *Exp. Eye Res.* 52: 205 -212.

Seidler, N. W. and Kowalewski, C. (2003). Methylglyoxal-induced glycation affects protein topography. *Arch. Biochem. Biophys.* 410: 149–154.

Shilton, B. H. and Walton D. J. (1991). Sites of glycation of human and horse liver alcohol dehydrogenase *in vivo*. *J. Biol. Chem.* 266: 5587–5592.

Singh, R., Barden, A., Mori, T. and Beilin, L. (2001). Advanced glycation endproducts: a review. *Diabetologia* 44: 129–146.

Smith, M. A., Taneda, S., Richey, P. L., Miyata, S., Yan, S-D., Stern, D., Sayre, L. M., Monnier, V. M. and Perry, G. (1994). Advanced Maillard reaction end products are associated with Alzheimer disease pathology. *Proc. Natl. Acad. Sci. USA* 91: 5710–5714.

Stevens, V. J., Rouzer, C. A., Monnier, V. M. and Cerami, A. (1978). Diabetic cataract formation: potential role of glycosylation of lens crystallins. *Proc. Natl. Acad. Sci. USA* 75: 2918–2922.

Sulochana, K. N., Punitham, R. and Ramakrishnan, S. (1998). Beneficial effect of lysine and amino acids on cataractogenesis in experimental diabetes through possible anti-glycation of lens proteins. *Exp. Eye Res.* 67: 597–601.

Swamy, M. S. and Abraham, E. C. (1987). Lens protein composition, glycation and high molecular weight aggregation in aging rats. *Invest. Ophthalmol. Vis. Sci.* 28: 1693–1701.

Swamy, M. S. and Abraham, E. C. (1989). Inhibition of lens crystallin glycation and high molecular weight aggregate formation by aspirin *in vitro* and *in vivo*. *Invest. Ophthalmol. Vis. Sci.* 30: 1120–1126.

Swamy-Mruthinti, S. (2001). Glycation decreases calmodulin binding to lens transmembrane protein MIP. *Biochim. Biophys. Acta* 1536: 64–72.

Swamy-Mruthinti, S., Green, K. and Abraham, E. C. (1996). Inhibition of cataracts in moderately diabetic rats by aminoguanidine. *Exp Eye Res.* 62: 505–512.

Thornalley, P. J. (2003). Use of aminoguanidine (Pimagedine) to prevent the formation of advanced glycation endproducts. *Arch. Biochem. Biophys.* 419: 31–40.

Thorpe, S. R. and Baynes, J. W. (2003). Maillard reaction products in tissue proteins: new products and new perspectives. *Amino Acids* 25: 275–281.

Uversky, V. N. and Fink, A. L. (2004). Conformational restraints for amyloid fibrillation: the importance of being unfolded. *Biochim. Biophys. Acta.* 1698: 131–153.

Van Boekel, M. A. M., Hoogakker, S. E., Harding, J. J. and de Jong, W. W. (1996). The influence of some post-translational modifications on the chaperone-like activity of α-crystallin. *Ophthalmol. Res.* 28 (Suppl 1): 32–38.

Van Heyningen, R. (1969). The metabolism of D-glyceraldehyde by the lens. *J. Biochem.* 112: 211–220.

Van Heyningen, R. and Harding, J. J. (1986). Do aspirin-like analgesics protect against cataract? *Lancet* I: 1111–1113.

Vitek, M. P., Bhattacharya, K., Glendening, J. M., Stopa, E., Vlassara, H., Bucala, R., Manogue, K. and Cerami, A. (1994). Advanced glycation end products contribute to amyloidosis in Alzheimer disease. *Proc. Natl. Acad. Sci. U.S.A.* 91: 4766–4770.

Watkins, N. G., Thorpe, S. R. and Baynes, J. W. (1985). Glycation of amino groups in protein. *J. Biol. Chem.* 260: 10629–10636.

Wilker, S. C., Chellan, P., Arnold, B. M. and Nagaraj, R. H. (2001). Chromatographic quantification of argpyrimidine, a methylglyoxal-derived product in tissue proteins: comparison with pentosidine. *Anal. BioChem.* 290: 353–358.

Wistow, G., Turnell, B., Summers, L., Slingsby, C., Moss, D., Miller, L., Lindley, P. F. and Blundell, T. L. (1983). X-ray analysis of the eye lens protein γ-crystallin at 1.9 A resolution. *J. Mol. Biol.* 170: 175–202.

Yamauchi, Y., Ejiri, Y. and Tanaka, K. (2002). Glycation by ascorbic acid causes loss of activity of ribulose-1,5-biphosphate carboxylase/oxygenase and its increased susceptibility to proteases. *Plant Cell Physiol.* 43: 1334–1341.

Yan, H. and Harding, J. J. (1997). Glycation-induced inactivation and loss of antigenicity of catalase and superoxide dismutase. *J. Biochem.* 328: 599–605.

Yan, H. and Harding, J. J. (1999). Inactivation and loss of antigenicity of esterase by sugars and a steroid. *Biochim. Biophys. Acta.* 1454: 183 –190.

Yan, H. and Harding, J. J. (2003). γ-Crystallin is the primary target of glycation in the bovine lens incubated under physiological conditions. *J. Biochem.* 374: 677–685.

Yan, S.-D., Chen, X., Schmidt, A.-M., Brett, J., Godman, G., Zou, Y.-S., Scott, C. W., Caputo, C., Frappier, T., Smith, M. A., Perry, G., Yen, S.-H. and Stern, D. (1994). Glycated tau protein in Alzheimer disease: a mechanism for induction of oxidant stress. *Proc. Natl. Acad. Sci. USA* 91: 7787–7791.

Zhao, H-R., Smith, J. B., Jiang, X-Y, and Abraham, E. C. (1996). Sites of glycation of βB2-crystallin by glucose and fructose. *Biochem. Biophys. Res. Commun.* 229: 128–133.

26

Defective Glycosylation and Muscular Dystrophies

Paul T. Martin

Abstract

A number of forms of congenital muscular dystrophy (CMD) have now been identified that involve defects in the glycosylation of dystroglycan with O-mannosyl-linked glycans. Unlike many neurologic disorders, altered glycosylation in CMDs does not cause aberrant protein aggregation or loss of expression. Instead, defects in glycosylation inhibit the binding of extracellular matrix proteins such as laminin to dystroglycan on neuronal and muscle cell membranes, thereby causing cellular pathology. Overexpression of one gene implicated in CMD, *LARGE*, can increase dystroglycan glycosylation and restore its function in cells taken from CMD patients. Thus, stimulating protein glycosylation may also be an avenue to treatment of these disorders.

26.1. Introduction

Glycosylation is a powerful class of post-translational modifications that can modulate both the chemical and biological properties of proteins. Most biologists view protein glycosylation within the rather narrow context of its role in protein folding and secretion; modification of many proteins with N-linked carbohydrate structures is essential for their proper folding in the endoplasmic reticulum and migration through the Golgi apparatus to the plasma membrane. Indeed, defects in pathways such as N-linked glycosylation cause complex human diseases that display multiorgan pathology due to multiple protein defects (Freeze, 2001). Increasingly, however, protein glycosylation is being shown to play direct roles in regulating essential protein functions such as adhesion, signaling, or enzyme activity. For example, O-linked N-acetylglucosamine (GlcNAc) has now been shown be present on a very large numbers of proteins at sites of serine or threonine phosphorylation (e.g., RNA polymerase II) where it can regulate protein signaling functions such as transcription (Vosseller *et al.*, 2002), O-Linked fucose structures are essential for the proper signaling of Notch and related proteins (Haltiwanger, 2002), polysialic acid is essential for some functions of neural cell adhesion molecule (Eckhardt *et al.*, 2000), and heparan sulfate is essential for fibroblast growth factor signaling (Esko and Selleck, 2002). Additionally, a large number of proteins (including galectins, siglecs, and selectins) bind glycoproteins via their carbohydrate structures and act as lectins (Varki, 1999). Thus, it should not be surprising that aberrant glycosylation would affect proteins in ways that would lead to pathology. The kinds of pathology involved can range from alteration of a protein's biophysical properties (such as aberrant folding or aggregation) to more specific deficits in protein function. There is perhaps no better example in the latter category in the past half decade than the function of the

O-mannosyl pathway on dystroglycan and its role in congenital muscular dystrophy. This story demonstrates both the high degree of specificity that defects in glycosylation can give rise to as well as the remarkable way in which a glycosylation pathway and its functions can be defined by human genetics.

26.2. Dystroglycan and the Dystrophin-Associated Glycoprotein Complex

The glycosylation-dependent congenital muscular dystrophies (CMDs) are a class of neuromuscular disorders that all are defined by alterations in the glycosylation of alpha dystroglycan. Dystroglycan protein is synthesized as a single polypeptide that is then post-translationally cleaved into an alpha and a beta chain (Ibraghimov-Beskrovnaya *et al.*, 1992). Alpha dystroglycan is a densely glycosylated peripheral membrane protein that binds to components of the extracellular matrix including laminin, agrin, and perlecan, as well as to membrane proteins such as neurexins and to infectious agents such as viruses and bacteria [for review, see Martin (2003)] (Fig. 26.1). The carbohydrate chains comprising the O-linked structures in the middle third of alpha dystroglycan are necessary for the binding of extracellular ligands (Ervasti and Campbell, 1993; Michele *et al.*, 2002). Alpha dystroglycan also binds tightly to beta dystroglycan, a transmembrane glycoprotein (Ervasti and Campbell, 1991). The intracellular domain of beta dystroglycan, in turn, binds dystrophin or its autosomal homologue utrophin. Dystrophin (and utrophin) bind to filamentous actin. The sarcoglycans are additional transmembrane components of this protein complex, which in aggregate is referred to as the dystrophin-associated glycoprotein complex. Members of the dystrophin-associated glycoprotein complex, including dystroglycan, also bind a myriad of potential signaling components, including dystrobrevins, syntrophins, neuronal nitric oxide synthase, G proteins, MAP kinases, adaptor proteins (Grb2), and PI3-Akt (phosphatidylinositol-3-v-Akt thymoma viral oncogene homologue/protein kinase B) kinases. Thus, dystroglycan may be important not only in regulating the stability of the muscle membrane but also in matrix signaling.

Loss of expression of most proteins within the dystrophin-glycoprotein complex causes muscular dystrophy [for review, see Blake *et al.* (2002)]. Muscular dystrophy is a term that collectively refers to a neuromuscular disorders associated with the wasting of skeletal muscles. Other tissues, cardiac muscle in particular, can also be severely affected. Loss of dystrophin protein expression causes Duchenne muscular dystrophy, a severe X-linked myopathy. Becker muscular dystrophy, which has a milder disease progression, results from partial loss of dystrophin expression or function. Loss of sarcoglycans causes forms of limb-girdle muscular dystrophy, and loss of laminin alpha 2 expression results in congenital muscular dystrophies MCD1A or MCD1B. Unlike the muscular dystrophies involving defective glycosylation of dystroglycan, the forms of muscular dystrophy involving dystrophin, sarcoglycans, and laminins are due to loss of protein expression and as such are not diseases of glycosylation. Glycosylation, however, may still impact the underlying molecular and cellular pathology in these diseases (Nguyen *et al.*, 2002).

The proper glycosylation of dystroglycan protein with O-linked carbohydrates is essential for its function in binding to the extracellular matrix (Ervasti and Campbell, 1993), which is in turn central to its role in maintaining membrane stability. These O-linked carbohydrates structures are a mix of O-linked mannose chains of the type NeuAcα2,3Galβ1,4GlcNAcβ1,2Manα-O (NeuAc, N-acetylneuraminic acid; Gal, galactose; Man, mannose) and O-linked N-acetylgalactosamine (GalNAc) structures of the type Galβ1,3GalNAcα-O (Chiba *et al.*, 1997; Sasaki *et al.*, 1998; Smalheiser *et al.*, 1998). While carbohydrates that are O-linked via GalNAc are common on many

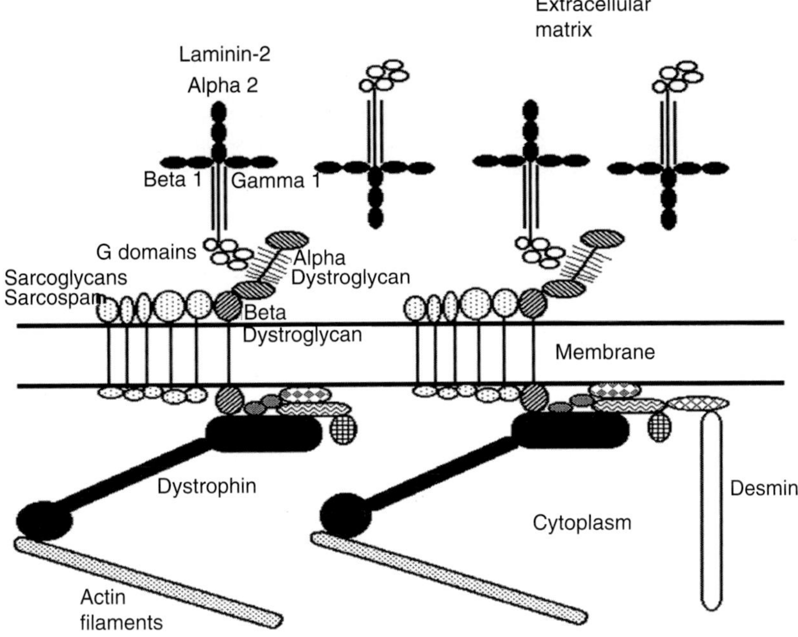

Figure 26.1. Dystroglycan glycosylation and its place in the dystrophin-glycoprotein complex. Laminin, which is present in the extracellular matrix surrounding the myofiber membrane, binds to α dystroglycan. This interaction requires the O-mannosyl-linked carbohydrates present in the middle third of the α dystroglycan molecule. α dystroglycan binds to β dystroglycan, which spans the muscle membrane. The intracellular domain of β dystroglycan binds to dystrophin, which links the complex both to actin and to number of signaling components. The sarcoglycans are additional transmembrane components within the dystrophin-associated glycoprotein complex.

mammalian glycoproteins, there are only several glycoproteins reported to have O-linked mannose [for review, see Martin (2003)]. There are roughly 50 sites for O-linked glycosylation in the middle third of the dystroglycan molecule (Ibraghimov-Beskrovnaya *et al.*, 1992). Thus, this region is intensely glycosylated with such structures, much as is seen with other mucin-like proteins, and this explains dystroglycan is about 50% carbohydrate by molecular weight. The expression of these two types of O-linked carbohydrate chains has been found on all three forms of dystroglycan protein where carbohydrate sequencing has been reported (Chiba *et al.*, 1997; Sasaki *et al.*, 1998; Smalheiser *et al.*, 1998). Additional carbohydrate structures that may exist on these O-linked chains are α1,3 linked fucose (Smalheiser *et al.*, 1998), terminal β1,4 linked GalNAc (Xia *et al.*, 2002), which is normally confined to expression at the neuromuscular junction in skeletal muscle (Martin *et al.*, 1999), and HNK-1 (SO4-3GlcNAcβ1,3Galβ1,4GlcNAc) (SO4, sulfate; GlcA, glucuronic acid) (Chiba *et al.*, 1997).

Ervasti and Campbell (1993) showed that removal of the O-linked chains on alpha dystroglycan inhibits the ability of laminin to bind to the protein, effectively divorcing the extracellular matrix from the muscle membrane. Removal of N-linked chains, by contrast, had no effect. This early biochemical demonstration of the importance of the O-linked carbohydrate chains has been confirmed in a dramatic way by the discovery that defects in genes that modify this aspect of dystroglycan glycosylation also inhibit the binding of laminin (Michele *et al.*, 2002) and cause muscular dystrophy (Kobayashi *et al.*, 1998; Brockington *et al.*, 2001a; Brockington *et al.*, 2001b; Yoshida *et al.*, 2001; Beltran-Valero de Bernabe *et al.*, 2002; Longman *et al.*, 2003). Unlike many types of glycosylation that affect protein folding or stability, this defect does not inhibit the expression of dystroglycan protein at the skeletal muscle membrane (Michele *et al.*, 2002). Rather, the hypoglycosylated protein that is expressed at the membrane is not functional in binding to the extracellular matrix (Michele *et al.*, 2002). Thus, unlike most of the other disorders described in this book, which involve abnormal aggregation or folding of proteins, the glycosylation-dependent CMDs are defects in protein conformation.

26.3. Glycosylation-Dependent Congenital Muscular Dystrophy

Thus far, five genes have been shown to cause forms of human CMD where dystroglycan glycosylation is affected. These include protein O-mannosyltransferase I (POMT1)–Walker-Warburg syndrome (WWS) (Beltran-Valero de Bernabe *et al.*, 2002), protein O-mannosyl β1, 2 N-acetylglucosaminyltransferase I (POMGnT1)–muscle-eye-brain (MEB) disease (Yoshida *et al.*, 2001), fukutin-Fukuyama congenital muscular dystrophy (FCMD) (Kobayashi *et al.*, 1998), Fukutin-related protein (FKRP)–congenital muscular dystrophy 1C (MDC1C) (Brockington *et al.*, 2001a), limb-girdle muscular dystrophy 2I (LGMD2I) (Brockington *et al.*, 2001b), and LARGE–congenital muscular dystrophy 1D (MDC1D) (Longman *et al.*, 2003) (Fig. 26.2). All of these are autosomal recessive disorders.

POMT1 (protein O-mannosyltransferase I) was identified in WWS by a candidate gene approach based on its homology with yeast O-mannosyltransferases (Beltran-Valero de Bernabe *et al.*, 2002). The enzymes encoded by these genes transfer mannose in an O-linkage to serine or threonine residues of proteins. In the original report, nonsense, frameshift, and missense mutations were identified in POMT1 in 6 of 30 unrelated WWS patients. Thus, POMT1 mutations only account for a fraction of WWS cases, and this means that defects in other genes must also cause this disease. POMT1 encodes a 3.1-kb mRNA that is predicted to produce an integral membrane protein of the endoplasmic reticulum. POMT1 transcripts are ubiquitously expressed,

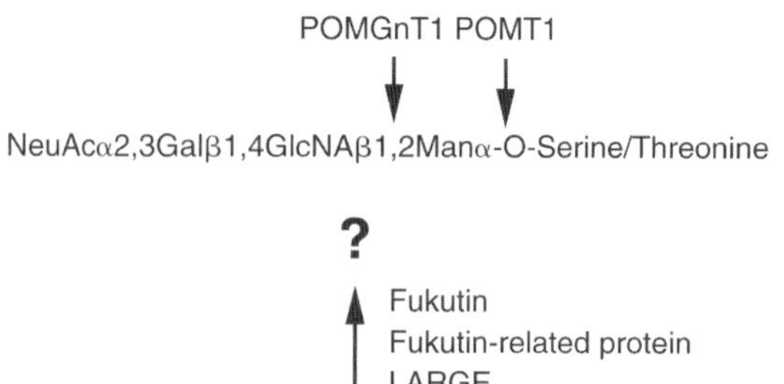

Figure 26.2. Genes affecting dystroglycan glycosylation associated with congenital muscular dystrophy. The O-mannosyl linked carbohydrates on α dystroglycan in skeletal muscle are composed of a linear chain of four carbohydrates. There are five genes known to cause congenital muscular dystrophy that alter the glycosylation of dystroglycan on its O-linked chains. POMT1 (defective in Walker-Warburg syndrome) synthesizes the O-mannosyl linkage and POMGnT1 (defective in muscle-eye-brain disease) synthesizes the β1,2GlcNAc linkage. Three other genes also likely affect this pathway: fukutin (defective in Fukuyama congenital muscular dystrophy), fukutin-related protein (FKRP; defective in congenital muscular dystrophy 1C and limb-girdle muscular dystrophy 2I), and LARGE (defective in congenital muscular dystrophy 1D). The function of these three genes is not currently known.

with highest expression in testes and fetal brain, and the POMT1 gene has been localized to human chromosome 9q34.1.

Base substitutions, deletions, and frameshift mutations in the POMGnT1 (protein O-mannosyl β1,2 N-acetylglucosaminyl transferase I) gene have been shown to cause muscle-eye-brain (MEB) disease (Yoshida *et al.*, 2001). POMGnT1 gene defects associated with disease are all loss of function changes (Manya *et al.*, 2003; Zhang *et al.*, 2003). POMGnT1 encodes a 3-kb mRNA that is ubiquitously expressed, as well as a 3.4-kb transcript that is expressed in spinal cord, lymph node, and trachea. POMGnT1 is predicted to be a Golgi-associated type II transmembrane protein, and the gene has been mapped to human chromosome 1p33-34.

Fukutin defects give rise to FCMD, a disease found almost exclusively in Japan (Kobayashi *et al.*, 1998). Patients of Japanese descent usually carry a retrotransposonal insertion in the 3′ untranslated region of the fukutin gene, though other point mutations have also been described (Kondo-Iida *et al.*, 1999). Fukutin encodes a 4-kb mRNA most highly expressed in fetal brain, but it is present in many adult tissues, including skeletal muscle, and the gene has been mapped to human chromosome 9q31.

Fukutin-related protein (FKRP) mutations cause either a severe form of congenital muscular dystrophy (MDC1C) (Brockington *et al.*, 2001a) or limb-girdle muscular dystrophy 2I (LGMD2I) (Brockington *et al.*, 2001b), which is more variable in its clinical presentation. FKRP encodes a 4-kb transcript that is highly expressed in brain, muscle, placenta, and heart, and is weakly expressed in numerous other tissues. The name fukutin-related protein is unfortunate, as it implies that this protein is highly related to fukutin, which it is not. In addition, the functional relationship between these proteins is not known. FKRP has been mapped to human chromosome 19q13.3.

A single compound mutation/deletion in the LARGE gene has been reported in a child with severe mental retardation, white matter changes, and congenital muscular dystrophy, termed MDC1D (Longman *et al.*, 2003). LARGE encodes a 4.3-kb cDNA that is predicted to be a transmembrane tandem glycosyltransferase (i.e., it has two potential glycosyltransferase domains). A deletion in LARGE is also the genetic defect in the myodystrophy (myd) mouse (Grewal *et al.*, 2001), a mouse with muscular dystrophy and brain pathology. The LARGE gene has been mapped to human chromosome 22q12.3-q13.1.

There are two findings that are common to all of the congenital muscular dystrophies caused by defects in these five genes; First, all are associated with an apparent underglycosylation of alpha dystroglycan in skeletal muscle. Second, all have muscular dystrophy as a main clinical finding. There are features of these disorders that are also quite distinct. WWS, MEB, and FCMD patients have a myriad of brain and eye abnormalities in addition to muscular dystrophy. Brain findings include coarse gyri with an abnormally nodular surface ("cobblestone cortex"), cerebellar cysts, pontocerebellar hypoplasia, absent or reduced corpus collosum, hydrocephalus, and hypomyelination of the white matter. The defective migration of cortical neurons that gives rise to the aberrant patterning of the cortex is most likely secondary to defects in the pial glia-limitans membrane. Eye abnormalities include congenital glaucoma, retinal dysplasia or detachment, microopthalmia, myopia, atrophy of the optic nerve, buphthalmos, and anterior chamber defects. These findings are distinct from the more classical muscular dystrophies like Duchenne, which have few neurologic findings. The muscular dystrophies caused by fukutin-related protein also appear to be distinct in that MDC1C and LGMD2I typically do not present with this spectrum of brain and eye findings (Brockington *et al.*, 2001a, 2001b). A recent report by Muntoni and colleagues, however, demonstrates severe CNS defects in several patients with FKRP mutations, and these defects mimic those observed in WWS and MEB (Beltran-Valero de Bernabe *et al.*, 2004). Thus, the lack of brain and eye pathology in most MDC1C (and LGMD2I) cases may merely be a reflection of the fact that those patients maintain partial FKRP function. Until more patients with LARGE mutations are reported, it is difficult to know the extent to which this report places LARGE in the spectrum of findings found in WWS, MEB, FCMD, and MDC1C. However, the myodystrophy (myd) mouse contains a deletion in LARGE that is most likely a loss of LARGE function (Grewal *et al.*, 2001). These mice have defects in cortical migration that are similar, though not nearly as severe, as WWS, as well as muscular dystrophy.

Almost all experiments to date support the contention that defects in O-mannosylation are entirely due to loss of dystroglycan function. This is consistent with the biochemical evidence suggesting that O-mannosylation is restricted to a small number of proteins in mammals (see Martin, 2003). Loss of dystroglycan is lethal in the mouse at an early embryonic stage of development due to loss of Reichert's membrane (Williamson *et al.*, 1997). Recent reports demonstrate that loss of fukutin (Takeda *et al.*, 2003) and POMT1 (Willer *et al.*, 2004) also result in embryonic lethality in the same time period, with POMT1 having defects in Reichert's membrane. Reichert's membrane is a rodent-specific structure. As such, the embryonic lethality in these animals is probably not relevant to human disease. Nevertheless, the similar phenotype of fukutin, POMT1, and dystroglycan in mice suggests that fukutin and POMT1 are essential to dystroglycan function. Chimeric mice lacking fukutin in some cells also mimicked the spectrum of findings in FCMD, including disorganization of the laminar pattern of the cortex, intrahemispheric fusion, hippocampal and cerebellar dysgenesis, retinal detachment, and severe muscular dystrophy (Takeda *et al.*, 2003). In addition, tissue-specific loss of dystroglycan protein in mouse brain (Moore *et al.*, 2002), peripheral nerve (Saito *et al.*, 2003), and skeletal muscle (Cohn *et al.*, 2002) using the Cre-Lox system at later developmental stages also gives rise to most of the findings in WWS, MEB, and FCMD. Collectively, these results strongly suggest that the clinical findings in these disorders are caused by the direct impact of the glycosylation defect on dystroglycan function. Consistent with this hypothesis, Campbell and colleagues have shown that hypoglycosylated dystroglycan taken from skeletal muscle of FCMD, MEB, and WWS patients, as well as from the myd mouse, does not bind laminin or other matrix components with significant affinity (Michele *et al.*, 2002).

26.4. Function of CMD Genes

Because LARGE, fukutin, FKRP, POMT1, and POMGnT1 all affect dystroglycan glycosylation and cause human disease, there is intense interest in understanding how these genes function to affect the glycosylation of the dystroglycan protein. It is clear from recent studies that POMT1 and POMGnT1 are glycosyltransferases that can respectively synthesize the first two O-linked carbohydrates of the O-mannosyl carbohydrate chains on alpha dystroglycan (Fig. 26.2). Based on homology with yeast, two O-mannosyltransferase genes have been identified in humans, POMT1 and POMT2 (see Willer *et al.*, 2003). Recently, POMT1 was shown to modify a recombinant form of alpha dystroglycan with O-linked mannose (Manya *et al.*, 2004). POMT2 co-expression was necessary for the induction of O-mannosyltransferase activity, suggesting that these two proteins work in an oligomeric complex. This is consistent with a recent RNAi study in flies that also showed that phenotype of POMT2 knockdown mutant flies had the same phenotype (twisted abdomen) as classical POMT1 mutant flies (Ichimiya *et al.*, 2004). This data suggests that POMT2 mutations should cause WWS (van Reeuwikcl *et al.*, 2001). In the original WWS report, POMT1 mutations were shown to cause 20% of WWS cases, but no mutations in POMT2 were identified in any of the remaining patients (Beltran-Valero de Bernabe *et al.*, 2002). Thus, it remains to be seen if POMT2 defects cause CMD or if POMT2 is an essential gene in humans.

POMGnT1 was shown by Schachter and colleagues (Zhang *et al.*, 2002) and by Endo and colleagues (Yoshida *et al.*, 2001) to be a uridine diphosphate (UDP)-GlcNAc: β1,

2Manα-O-N-acetylglucosaminyl transferase. This enzyme is highly specific for O-linked mannose and does not transfer GlcNAc onto the mannose linkages typically found on N-linked oligosaccharides. Thus, POMGnT1, like POMT1, is relatively specific for dystroglycan, as dystroglycan is one of only several mammalian proteins shown to contain O-linked mannose. Reductions in POMGnT1 activity occur in MEB (Zhang *et al.*, 2003; Manya *et al.*, 2003). Thus, there is no doubt that this gene functions to affect the glycosylation of alpha dystroglycan and that this activity is lost in MEB. The subsequent two linkages that decorate the O-linked mannose chains on dystroglycan, NeuAcα2,3Galβ1,4, would be made by sialyltransferases and galactosyltransferases, respectively. Many of the genes that encode these two classes of enzymes have been cloned and studied for many years, and none of these genes are implicated in CMDs. Because fukutin, FKRP, and LARGE do not have any homology to known α2,3 sialyltransferase or β1,4 galactosyltransferase genes, they are not likely to possess such activities.

While loss of fukutin and FKRP expression clearly alter the extent of dystroglycan glycosylation, the primary sequence of these genes is not homologous to other known glycosyltransferases. Thus, their role in the O-mannose pathway is very unclear. Fukutin and FKRP are expressed either in the endoplasmic reticulum or Golgi apparatus. There is disagreement in the literature as to which protein is in which compartment (Matsumoto *et al.*, 2004; Esapa *et al.*, 2002). Regardless, their expression in these organelles is consistent with their ability to interact with and potentially modulate POMT1, POMGnT1, LARGE, or other glycosyltransferases. How they accomplish this is an important aspect of the glycosylation puzzle that remains to be solved. One obvious possibility is that FKRP or fukutin are required for POMT1, POMT2, POMGnT1, or LARGE activity. FKRP and/or fukutin could alter the processivity of these enzymes or binding of these enzymes to substrates such as UDP-GlcNAc or GDP-Man. Either of these scenarios could lower the density of carbohydrates present on alpha dystroglycan. Loss of fukutin or FKRP could also cause the appropriate capping of the four carbohydrate O-mannose chains with carbohydrates that could not be further modified. Another possibility is that loss of fukutin or FKRP function would alter the migration time of dystroglycan through the secretory pathway. This could lower the extent of glycosylation by reducing the time in which POMT1 or related enzymes could act on dystroglycan.

LARGE has a domain structure that suggests that it has two glycosyltransferase domains, though what these domains actually do is still unknown. Clearly, LARGE alters both dystroglycan glycosylation and dystroglycan function; Campbell and colleagues have shown that overexpression of LARGE in primary cells taken from patients with MEB, WWS, or FCMD can increase the glycosylation of alpha dystroglycan and rescue laminin binding (Barresi *et al.*, 2004). If the O-mannosyl carbohydrate chains are absent (or very reduced) on alpha dystroglycan taken from WWS muscle, LARGE should not be able to extend those chains further if it is specific to O-mannosyl glycans. LARGE, however, is clearly able to increase dystroglycan glycosylation in WWS muscle cells to a level that is equivalent to or even greater than that seen in wild-type muscle. The high degree of alpha dystroglycan glycosylation in these experiments suggests the intriguing possibility that LARGE may not act on the O-mannose pathway at all. For example, LARGE could modify the O-linked GalNAc chains on alpha dystroglycan or modify N-linked carbohydrates. Alternatively, LARGE could replace POMT1 function in activating POMT2 or related glycosyltransferase activities, thereby overcoming the O-mannose defect in WWS. While there is no evidence that dystroglycan contains glycosaminoglycans, this would be another possible explanation for the robust glycosylation in these carbohydrate-deficient forms of the protein.

Indeed, the tandem glycosyltransferase domain structure of LARGE is reminiscent of enzymes involved in disaccharide synthesis of glycosaminoglycans (Esko and Selleck, 2002). While alpha dystroglycan has been reported not to have such structures, it is possible that the high induction of glycosylation resulting from LARGE overexpression could be due to synthesis of a polydisaccharide repeat such as poly-N-acetyllactosamine. Mouse laminin-1 from the Engelbreth-Holm-Swarm (EHS) tumor, for example, has such chains (see Martin, 2003). The synthesis of disaccharide repeats by LARGE would allow it to glycosylate dystroglycan to a very high degree, much as can occur with glycosoaminoglycan synthesis. Indeed, overexpression of LARGE in WWS muscle cells increases dystroglycan glycosylation beyond wild-type levels (Barresi *et al.*, 2004), which would be consistent with this mode of glycosylation.

26.5. Sufficiency of Carbohydrates for Ligand Binding

Another important issue yet to be resolved is whether the O-linked carbohydrates that are missing on dystroglycan in CMDs are directly responsible for binding to matrix proteins such as laminin or if other aspects of the dystroglycan polypeptide are also required. Laminins are large trimeric proteins composed of an alpha, beta, and gamma chain (see Fig. 26.1). Most evidence now suggests that it is the five globular G domains at the C-terminus of the alpha chain that are sufficient for laminin binding to alpha dystroglycan (Shimizu *et al.*, 1999; Talts *et al.*, 1999; Talts *et al.*, 2000). Likewise, G domain fragments on agrin and perlecan bind to dystroglycan in the absence of other protein domains (Gesemann *et al.*, 1996; Talts *et al.*, 1999). The primary laminin alpha chain expressed along myofibers is laminin alpha 2. The G domains of laminin alpha 2 are sufficient to bind to alpha dystroglycan in the 10–50 nM range, while recombinant agrin G domains bind dystroglycan at 2–5 nM (Gesemann *et al.*, 1998; Talts *et al.*, 1999). These same G domains also can bind heparin and sulfatides with high affinity (Gesemann *et al.*, 1996; Talts *et al.*, 1999). Thus, the G domains of laminin bind to carbohydrates and carbohydrate modifications. In addition, some evidence suggests that sialic acid inhibits laminin binding to dystroglycan (Chiba *et al.*, 1996), which would argue for direct carbohydrate binding. Agrin G domains also can bind neoglycoconjugates such as those found on O-linked mannose on alpha dystroglycan (Xia and Martin, 2002). Thus, there is a lot of data, albeit highly indirect, to support to notion that the loss of O-linked glycosylation on dystroglycan affects laminin binding by inhibiting its interaction directly with O-linked carbohydrates.

Because the O-linked structures on alpha dystroglycan are likely to occur on multiple serines and theronines, these carbohydrate structures may also be required for the proper conformation of the alpha dystroglycan protein. As such, they may also be essential to induce conformational changes in alpha dystroglycan that allow laminin interactions with the dystroglycan polypeptide. Indeed, Campbell and colleagues provided evidence suggesting that the N-terminal domain of dystroglycan, which is separate from the region containing O-linked carbohydrates, is required for laminin binding (Kanagawa *et al.*, 2004). This study suggests that the model for laminin binding to dystroglycan is more complex than a simple binding to the O-linked mannose carbohydrate chains. The evidence in this regard comes from an analysis of laminin binding to recombinant fragments of dystroglycan containing deletions of its various protein domains. This data is interpreted as showing that O-linked carbohydrates alone are insufficient for laminin binding in the absence of the N-terminal fragment, which presumably does not contain such structures. It is also possible, however, that the N-terminus of dystroglycan is required for proper glycosylation of the mucin domain on these recombinant protein fragments. This would still allow the direct laminin-carbohydrate model to hold. Indeed, cleavage of the N-terminus on an already glycosylated full-length alpha dystroglycan protein by furin was recently shown not to inhibit laminin binding (Singh *et al.*, 2004), which again would be consistent with this model.

26.6. Conclusion

There is a wealth of biochemical and genetic data to show that genes involved in the O-mannose glycosylation pathway cause forms of congenital muscular dystrophy. Because these carbohydrates are required for the binding of ligands such as laminin to dystroglycan, these gene defects likely act to cause disease by severing the binding of extracellular matrix proteins in the muscle basal lamina to dystroglycan in the muscle membrane. Both brain and eye defects in these disorders are also likely to come from defective binding of proteins to dystroglycan. Because the underglycosylated dystroglycan protein is still expressed at the plasma membrane in these disorders, the glycosylation-dependent CMDs likely represent disorders of protein conformation, as opposed to disorders of protein folding, aggregation, or expression. In the coming years, there will be new genes identified in this pathway that cause CMDs, and much work remains on understanding the function of the already known genes and their relationship to one another. Because defects in glycosylation pathways are often highly pleiotropic due to the common nature of this type of post-translational modification, the relatively specific nature of the O-mannose pathway to dystroglycan makes it one of the best systems in which to understand not only the contribution of glycosylation to disease but the mechanisms involved in the fundamentally important process of protein glycosylation. Because this defect does not affect protein expression in the muscle cell membrane, it also is an excellent model to test functional roles for carbohydrates in processes such as ligand binding. The demonstration that manipulations of protein glycosylation can also have a therapeutic benefit in models of these disorders makes the understanding of these processes all the more important.

26.7. Note Added in Proof

Mutations in POMT2 have now been shown to cause Walker Warburg syndrome and associated glycosylation defects in alpha dystroglycan.

Acknowledgments

This work was supported by grants from the NIH (AR050202) and the Muscular Dystrophy Association to P.T.M.

References

Barresi, R., Michele, D.E., Kanagawa, M., Harper, H.A., Dovico, S.A., Satz, J.S., Moore, S.A., Zhang, W., Schachter, H., Dumanski, J.P., Cohn, R.D., Nishino, I., and Campbell, K.P. (2004). LARGE can functionally bypass alpha dystroglycan glycosylation defects in distinct congenital muscular dystrophies. *Nat. Med.* 10:696–703.

Beltran-Valero de Bernabe, D., Currier, S., Steinbrecher, A., Celli, J., Beusekom, E., van er Zwaag, B., Kayersili, H., Merlini, L., Chitayat, D., Dobyns, W.B., *et al.* (2002). Mutations in the O-mannosyltransferase gene POMT1 give rise to the severe neuronal migration disorder Walker-Warburg syndrome. *Am. J. Hum. Genet.* 71:1033–1043.

Beltran-Valero de Bernabe, D., Voit, T., Longman, C., Steinbrecher, A., Straub, V., Yuva, Y., Hermann,, R., Sperner, J., Korenke, C, Diesen, C. *et al.* (2004). Mutations in the FKRP gene can cause muscle eye brain disease and Walker Warburg syndrome. *J. Med. Gen.* 41:e61.

Blake, D.J., Weir, A., Newey, S.E. and Davies, K.E. (2002). Function and genetics of dystrophin and dystrophin-associated proteins in muscle. *Physiol. Rev.* 82:291–329.

Brockington, M., Blake, D.J., Prandini, P., Brown, S.C., Torelli, S., Benson, M.A., Ponting, C.P., Estournet, B., Romero, N.B., Mercuri, E. *et al.* (2001a). Mutations in the fukutin-related protein gene (FKRP) cause a form of congenital muscular dystrophy with secondary laminin alpha 2 deficiency and abnormal glycosylation of alpha dystroglycan. *Am. J. Hum. Genet.* 69:1198–1209.

Brockington, M., Yuva, Y., Prandini, P., Torelli, S., Benson, M.A., Hermann, R., Anderson, L.V., Bashir, R., Burgunder, J. M., Fallet, S., *et al.* (2001b) Mutations in the fukutin-related protein gene (FKRP) identify limb girdle muscular dystrophy 2I as a milder allelic variant of congenital muscular dystrophy MDC1C. *Hum. Mol. Genet.* 10:2851–2859.

Chiba, A., Matsumura, K., Yamada, H., Inazu, T., Shimizu, T., Kusonoki, S., Kanazawa, I., Kobata, A. and Endo, T. (1997). Structures of sialylated O-linked oligosaccharides of bovine peripheral nerve alpha dystroglycan. *J. Biol. Chem.* 272:2156–2162.

Cohn, R.D., Henry, M.D., Michele, D.E., Barresi, R., Saito, F., Moore, S.A., Flanagan, J.D., Skwarchuk, M.W., Robbins, M.E., Mendell, J.R., Williamson, R.D. and Campbell, K.P. (2002). Disruption of dag1 in differentiated skeletal muscle reveals a role for dystroglycan in muscle regeneration. *Cell* 110:639–648.

Eckhardt, M., Bukalo, O.,Chazal, G., Wang, L., Gordis, C., Schachner, M., Gerardy-Schahn, R., Cremer, H., and Dityatev, A. (2000). Mice deficient in the polysialyltransferase ST8SiaIV/PST-1 allow discrimination of the roles of neural cell adhesion molecule protein and polysialic acid in neural development and synaptic plasticity. *J. Neurosci.* 20:5234–5244.

Ervasti, J.M. and Campbell, K.P. (1991). Membrane organization of the dystrophin-glycoprotein complex. *Cell* 66:1121–1131.

Ervasti, J.M. and Campbell, K.P. (1993). A role for the dystrophin-glycoprotein complex as a transmembrane linker between actin and laminin. *J. Cell Biol.* 122:809–823.

Esapa, CT., Benson, M.A., Schroder, J.E., Martin-Rendon, E., Brockington, M., Brown, S.C., Muntoni, F., Kroger, S. and Blake, D.J. (2002). Functional requirements for fukutin-related protein in the Golgi apparatus. *Hum. Mol. Genet.* 11:3319–3331.

Esko, J.D. and Selleck, S.B. (2002). Order out of chaos:assembly of ligand binding sites in heparan sulfate. *Annu. Rev. Biochem.* 71:435–471.

Freeze, H.H. (2001). Update and perspectives on congenital disorders of glycosylation. *Glycobiology* 11:129R–143R.

Gesemann, M., Cavalli, V., Denzer, A.J., Brancaccio, A., Schumacher, B., and Ruegg, M. A. (1996). Alternative splicing of agrin alters its binding to heparin, dystroglycan, and the putative agrin receptor. *Neuron* 16:755–767.

Gesemann, M., Brancaccio, A., Schumacher, B., and Ruegg, M.A. (1998). Agrin is a high-affinity binding protein of dystroglycan in non-muscle tissue. *J. Biol. Chem.* 273:600–605.

Grewal, P.K., Holzfeind, P.J., Bittner, R.J. and Hewitt, J.E. (2001). Mutant glycosyltransferase and altered glycosylation of alpha dystroglycan in the myodystrophy mouse. *Nat. Genet.* 28:151–154.

Haltiwanger, R.S. (2002). Regulation of signal transduction pathways in development by glycosylation. *Curr. Opin. Struct. Biol.* 12:593–598.

Ibraghimov-Beskrovnaya, O., Ervasti, J.M., Leveille, C.J., Slaughter, C.A., Sernett, S.W. and Campbell, K.P. (1992). Primary structure of dystrophin-associated glycoproteins linking dystrophin to the extracellular matrix. *Nature* 355:696–702.

Ichimiya, T., Manya, H., Ohmae, Y., Yoshida, H., Takahashi, K., Ueda, R., Endo, T. and Nishihara, S. (2004). The twisted abdomen phenotype of Drosophila POMT1 and POMT2 mutants coincides with their heterophilic protein O-mannosyltransferase activity. *J. Biol. Chem.* 279:42638–42647.

Kanagawa, M., Saito, F., Kunz, S., Yoshida-Moriguchi, T., Barresi, R., Kobayashi, Y.M., Muschler, J., Dumanski, J.P., Michele, D.E., Oldstone, M.B.A. and Campbell, K.P. (2004). Molecular recognition by LARGE is essential for expression of functional dystroglycan. *Cell* 117:953–964.

Kobayashi, K., Nakahori, Y., Miyake, M., Matsumura, K., Kondo-Iida, E., Nomura, Y., Segawa, M., Yoshioka, M., Saito, K, Osawa, M. *et al.* (1998). An ancient retrotransposonal insertion causes Fukuyama-type congenital muscular dystrophy. *Nature* 394:388–392.

Kondo-Iida, E., Kobayashi, K., Watanabe, M., Sasaki, J., Kumagai, T., Koide, H., Saito, K., Osawa, M., Nakamura, Y. and Toda, T. (1999). Novel mutations and genotype-phenotype relationships in 107 families with Fukuyama-type congenital muscular dystrophy (FCMD). *Hum. Mol. Genet.* 8:2303–2309.

Longman, C., Brockington, M., Torelli, S., Jimenez-Mallebrera, C., Kennedy, C., Khalil, N., Feng, L., Saran, R.K., Voit, T., Merlini, L., Sewry, C.A., Brown, S.C. and Muntoni, F. (2003). Mutations in the human LARGE gene cause MDC1D, a novel form of congenital muscular dystrophy with severe mental retardation and abnormal glycosylation of alpha-dystroglycan. *Hum. Mol. Genet.* 12: 2853–2861.

Manya, H., Sakai, K., Kobayashi, K., Taniguchi, K., Kawakita, M., Toda, T. and Endo T. (2003). Loss-of-function of an N-acetylglucosaminyltransferase, POMGnT1, in muscle-eye-brain disease. *Biochem. Biophys. Res. Commun.* 306:93–97.

Manya, H., Chiba, A., Yoshida, A., Wang, X., Chiba, Y., Jaigami, Y., Margolis, R. and Endo, T. (2004). Demonstration of mammalian protein O-mannosyltransferase activity: coexpression of POMT1 and POMT2 required for enzymatic activity. *Proc. Natl. Acad. Sci. USA* 101:500–505.

Martin, P.T. (2003) Dystroglycan glycosylation and its role in matrix binding in skeletal muscle. *Glycobiology* 13:55R–66R.

Martin, P.T., Scott, L.J.C., Porter, B.E. and Sanes, J.R. (1999) .Distrinct structures and functions of related pre-and postsynaptic carbohydrates at the mammalian neuromuscular junction. *Mol. Cell. Neurosci.* 13:105–118.

Matsumoto, H., Noguchi, S., Sugie, K., Ogawa, M., Murayama, K., Hayashi, Y.K. and Nishino, I. (2004). Subcellular localization of fukutin and fukutin-related protein in muscle cells. *J. Biochem.* 135:709–712.

Michele, D.E., Barresi, R., Kanagawa, M., Saito, F., Cohn, R.D., Satz, J.S., Dollar, J., Nishino, I., Kelley, R.I., Somer, H. *et al.* (2002). Post-translational disruption of dystroglycan-ligand interactions in congenital muscular dystrophies. *Nature* 418:417–422.

Moore, S.A., Saito, F., Chen, J., Michele, D.E., Henry, M.D., Messing, A., Cohn, R.D., Ross-Barta, S.E., Westra, S., Williamson, R.A., Hoshi, T. and Campbell, K.P. (2002). Deletion of brain dystroglycan recapitulates aspects of congenital muscular dystrophy. *Nature* 418:422–425.

Nguyen, H.H., Jayasinha, V., Xia, B., Hoyte, K. and Martin, P.T. (2002). Overexpression of the cytotoxic T cell GalNAc transferase in skeletal muscle inhibits muscular dystrophy in the mdx mouse. *Proc. Natl. Acad. Sci. USA* 99:5616–5621.

Saito, F., Moore, S.A., Barresi, R., Henry, M.D., Messing, A., Ross-Barta, S.E., Cohn, R.D., Williamson, R.A., Sluka, K.A., Sherman, D.L. *et al.* (2003). Unique role of dystroglycan in peripheral nerve myelination, nodal structure, and sodium channel stabilization. *Neuron* 38:747–758.

Sasaki, T., Yamada, H., Matsumura, K., Shimizu, T., Kobata, A. and Endo, T. (1998) .Detection of O-mannosyl glycans in rabbit skeletal muscle alpha-dystroglycan. *Biochem. Biophys. Acta* 1425:599–606.

Shimizu, H., Hosokawa, H., Ninomiya, H., Miner, J.H., and Masaki, T. (1999). Adhesion of cultured bovine aortic endothelial cells to laminin-1 mediated by dystroglycan. *J. Biol. Chem.* 274:11995–12000.

Singh, J., Itahana, Y., Knight-Krajewski, S., Kanagawa, M., Campbell, K.P., Bissel, M.J. and Muschler, J. (2004). Proteolytic enzymes and altered glycosylation modulate dystroglycan function in carcinoma cells. *Cancer Res.* 64:6152–6159.

Smalheiser, N.R., Haslam, S.M., Sutton-Smith, M., Morris, H.R. and Dell, A. (1998). Structural analysis of sequences O-linked to mannose reveals a novel Lewis x structure in cranin (dystroglycan) purified from sheep brain. *J. Biol. Chem.* 273:23689–23703.

Takeda, S., Kondo, M., Sasaki, J., Kurahashi, H., Kano, H., Arai, K., Misaki, K., Fukui, T., Kobayashi, K., Tachikawa, M., Imamura, M., Nakamura, Y., Shimizu, T., Murakami, T., Sunada, Y., Fujikado, T., Matsumura., K., Terashima, T. and Toda, T. (2003). Fukutin is required for maintenance of muscle integrity, cortical histiogenesis and normal eye development. *Hum. Mol. Genet.* 12:1449–1459.

Talts, J.F., Andac, Z., Gohring, W., Brancaccio, A., and Timpl, R. (1999). Binding of the G domains of laminin alpha 1 and alpha 2 chains and perlecan to heparin, sulfatides, alpha dystroglycan, and several extracellular matrix proteins. *J. EMBO* 18:863–870.

Talts, J.F., Sasaki, T., Miosge, N., Gohring, W., Mann, K., Mayne, R., and Timpl, R. (2000). Structural and functional analysis of the recombinant G domain of the laminin alpha 4 chain and its proteolytic processing in tissues. *J. Biol. Chem.* 275:35192–35199.

van Reeuwijck, Janssen, M., van den Elzen, C., Beltran-Valero de Bernabe, D., Sabatelli, P., Merlini, L., Boon, M., Scheffer, H., Brockington, M., Muntoni, F., Huynen, M. A., Verrips, A., Walsh, C. A., Barth, P. G., Brunner, H. G., and van Bokhoven, H. (2002) POMT2 mutations cause alpha dystroglycan hypoglycosylation and Walker-Warburg syndrome. *J. Med. Genet.* 42: 907–912.

Varki, A. (1999). Discovery and classification of animal lectins. In Varki, A., Cummings, R., Esko, J., Freeze, H., Hart, G., and Marth, J. (eds.) *Essentials of Glycobiology*, Cold Spring Harbor Laboratory Press, New York, pp. 333–343.

Vosseller, K., Sakabe, K., and Hart, G.W. (2002). Diverse regulation of protein function by O-GlcNAc: a nuclear and cytoplasmic carbohydrate post-translational modification. *Curr. Opin. Chem. Biol.* 6:851–857.

Willer, T., Camern Valero, M., Tanner, W., Cruces, J. and Strahl, S. (2003). O-mannosyl glycans: from yeast to novel associations with human disease.*Curr. Opin. Struct. Biol.* 13:621–630.

Willer, T., Prados, B., Falcon-Perez, J.M., Renner-Muller, I., Przemeck, G.K.H., Lommel, M., Coloma, A., Camen Valero, M., Hrabe de Angelis, M., Tanner, W., Wolf, E., Strahl, S. and Cruces, J. (2004). Targeted disruption of the Walker Warburg syndrome gene Pomt1 in mouse results in embryonic lethality. *Proc. Natl. Acad. Sci. USA* 101:14126–14131.

Williamson, R.A., Henry, M.D., Daniels, K.J., Hrstka, R.F., Lee, J.C., Sunada, Y, Ibraghimov-Beskrovnaya, O. and Campbell, K.P. (1997). Dystroglycan is essential for early embryonic development:disruption of Reichert's membrane in Dag-1 null mice. *Hum. Mol. Genet.* 6:831–841.

Xia, B. and Martin, P.T. (2002). Modulation of agrin binding and activity by the CT and related carbohydrate antigens. *Mol. Cell. Neurosci.* 19:539–551.

Xia, B., Hoyte, K., Kammesheidt, A., Deerinck, T., Ellisman, M. and Martin, P.T. (2002). Overexpression of the CT GalNAc transferase in skeletal muscle alters myofiber growth, neuromuscular structure, and laminin expression. *Dev. Biol.* 242:58–73.

Yoshida, A., Kobayashi, K., Manya, H., Taniguchi, K., Kano, H, Mizuno, M., Inazu, T., Mitsuhashi, M., Takahashi, S., Takeuchi, M. *et al.* (2001). Muscular dystrophy and neuronal migration disorder caused by mutations in a glycosyltransferase, POMGnT1. *Dev. Cell* 1:717–724.

Zhang, W., Betel, D., and Schachter, H. (2002). Cloning and expression of a novel UDP-GlcNAc:alpha-D-mannoside beta1,2-N-acetylglucosaminyltransferase homologous to UDP-GlcNAc:alpha-3-D-mannoside beta 1,2-N-acetylglucosaminyltransferase I. *J. Biochem.* 361:153–162.

Zhang, W., Vajsar, J., Cao, P., Breningstall, G., Diesen, C., Dobyns, W., Hermann, R., Leheshoki, A.E., Steinbrecher, A., Talim, B., Toda, T. Topaloglu, H., Voit, T. and Schachter, H. (2003). Enzymatic diagnostic test for muscle-eye-brain type congenital muscular dystrophy using commercially available reagents. *Clin. Biochem.* 36:339–344.

Index

Printed in the United States of America